Springer Series in Chemical Physics

Volume 119

The purpose of this series is to provide comprehensive up-to-date monographs in both well established disciplines and emerging research areas within the broad fields of chemical physics and physical chemistry. The books deal with both fundamental science and applications, and may have either a theoretical or an experimental emphasis. They are aimed primarily at researchers and graduate students in chemical physics and related fields.

More information about this series at http://www.springer.com/series/676

Kaoru Yamanouchi · Sergey Tunik · Vladimir Makarov
Editors

Progress in Photon Science

Recent Advances

Springer

Editors
Kaoru Yamanouchi
Department of Chemistry
The University of Tokyo
Tokyo, Japan

Sergey Tunik
Institute of Chemistry
Saint Petersburg State University
Saint Petersburg, Russia

Vladimir Makarov
Physics Faculty
Lomonosov Moscow State University
Moscow, Russia

ISSN 0172-6218
Springer Series in Chemical Physics
ISBN 978-3-030-05973-6 ISBN 978-3-030-05974-3 (eBook)
https://doi.org/10.1007/978-3-030-05974-3

Library of Congress Control Number: 2018964252

This Springer imprint is published by the registered company Springer Nature Switzerland AG
The registered company address is: Gewerbestrasse 11, 6330 Cham, Switzerland

Preface

In October, 2014, School of Science, the University of Tokyo (UTokyo), jointly with Department of Civil Engineering, School of Engineering, established the Students and Researchers Exchange Program in Sciences (STEPS) with Lomonosov Moscow State University (MSU) and Saint Petersburg State University (SPbU) with financial support from the Re-Inventing Japan Project, which was renamed later and is now called Inter-University Exchange Project, of the Ministry of Education, Culture, Sports, Science and Technology. Since then, we have been working to promote the STEPS program not only by exchanging students and researchers but also by holding scientific symposia and gatherings.

With support from the STEPS program, we held the 1st STEPS Symposium on Photon Science on March 21 and 22, 2015 in the auditorium of the Chemistry Main Building on the Hongo Campus of the University of Tokyo, and the contents of the 28 lectures given at the symposium were compiled as a form of a book entitled "Progress in Photon Science—Basics and Applications," which was published in 2017 as the 115th volume of Springer Series in Chemical Physics. In order to keep the momentum we gained through the fruitful discussion we had during the 1st symposium, followed by a number of international research cooperation among the attendees, we decided to keep holding STEPS Symposium on Photon Science in cooperation of the research groups involved in the STEPS program.

The second STEPS Symposium on Photon Science was held on March 14–16, 2016, in Hotel New Peterhof in Saint Petersburg, Russia, and invited talks were given by the following 33 researchers: Andrey Fedyanin, Kirill Grigoriev, Maria G. Khrenova, Olga Kosareva, Konstantin Kouzakov, Kuzechkin Nikolay, Yuri V. Popov, Andrey Stolyarov from MSU; Alex Andreev, Larisa Gulina, Alexander Konev, Alina Manshina, Igor Mashek, Alexander Pastor, Yuriy Petrov, Sergey Pulkin, Yulia Shakirova, Nikolay Timofeev, Olesya Tomashenko, Oleg Vyvenko from SPbU; Toshiaki Ando, Shinichi Fukahori, Takashi Hiroi, Reika Kanya, Kuniaki Konishi, Erik Lötstedt, Koji Nakabayashi, Katsunori Nakai, Tamas Szidarovzky, Makoto Takamoto, Kana Yamada, Kaoru Yamanouchi, and Junji Yumoto from UTokyo.

The third symposium of this symposium series was held on March 11 and 12, 2018, in Hotel Korston Moscow in Moscow, Russia, and invited talks were given by the following 36 researchers: Alexei Balakin, Andrey Fedotov, Kirill Grigoriev, Alexei Grum-Grzhimailo, Elena Gryzlova, Olga Kosareva, Konstantin Kouzakov, Vladimir Makarov, Vyacheslav Morozov, Yuri V. Popov, Andrei Savel'ev, Svyatoslav Shlenov, Andrey Stolyarov, Aleksandr Ushakov from MSU; Elena Grachova, Ivan Ignatiev, Alexander Konev, Alina Manshina, Alexey Povolotskiy, Sergey Pulkin, Mikhail Ryazantsev, Yulia Shakirova, Sergey Tunik, Yuriy Tverjanovich, Oleg Vyvenko from SPbU; Takanobu Amano, Takashi Hiroi, Reika Kanya, Kuniaki Konishi, Erik Lötstedt, Norikatsu Mio, Koji Nakabayashi, Jun Okabayashi, Takeaki Ozawa, Tatsuya Tsukuda, and Kaoru Yamanouchi from UTokyo.

All of the lectures in these two symposia were well-prepared and were very exciting and informative. We therefore thought it appropriate to record the topics at the forefront of photon science addressed in the lectures in the form of a second volume of "Progress in Photon Science," which was later entitled as "Progress in Photon Science—Recent Advances," and extended an invitation to the researchers who gave a presentation at the 3rd symposium to write a review style chapter for this second volume. Thanks to their enthusiastic responses, we were able to compile as many as 25 peer-reviewed chapters. Each chapter of this book begins with an introductory part, in which a clear and concise overview of the topic and its significance is given and moves onto a description of the authors' most recent research results.

The chapters of this volume cover a diverse range of the research field of photon science, and the topics may be grouped into five categories: light and light propagation (Chaps. 1–5), biological applications (Chaps. 6–9), atoms and molecules (Chaps. 10–15), plasma and relativistic phenomena (Chaps. 16–19), and nanostructures and functional material (Chaps. 20–25). We hope this volume will convey the excitement of a variety of activities in photon science to the readers and stimulate interdisciplinary interactions among researchers, thus paving the way to explorations of new frontiers in photon science. We also hope this book will be regarded as a valued record of the STEPS program, facilitating its further promotion.

We would like to take this opportunity to thank all of the authors who kindly contributed to "Progress in Photon Science—Recent Advances," by describing their most recent work at the frontiers of photon science. We also thank the reviewers who have read the submitted manuscripts carefully.

One of the co-editors (KY) thanks Ms. Kaoru Kikuchi, a secretary at Center for Ultrafast Intense Laser Science, School of Science, the University of Tokyo, and Ms. Mihoshi Abe, a secretary to my research group, for their help with the preparation of this book.

Finally, we would like to thank Ms. Valeria Shatilova and Ms. Takako Jono at the STEPS office of School of Science, the University of Tokyo for helping us prepare for the 2nd and 3rd STEPS Symposium on Photon Science, and Dr. Claus

Ascheron, former Physics Editor at Springer-Verlag at Heidelberg, who retired this February, for his kind support for our second volume of "Progress in Photon Science" in the Springer Series in Chemical Physics.

The 2nd STEPS Symposium on Photon Science in Saint Petersburg (2017)

The 3rd STEPS Symposium on Photon Science in Moscow (2018)

Tokyo, Japan — Kaoru Yamanouchi
Saint Petersburg, Russia — Sergey Tunik
Moscow, Russia — Vladimir Makarov

Contents

Contributors

Takanobu Amano Department of Earth and Planetary Science, University of Tokyo, Tokyo, Japan

A. A. Andreev Saint Petersburg State University, St. Petersburg, Russia; ELI-ALPS, Szeged, Hungary; MBI, Berlin, Germany

Anatoliy Averyanov Institute of Chemistry, St. Petersburg State University, St. Petersburg, Russia

Mikhail Balabas Department of General Physics 1, Saint-Petersburg State University, St. Petersburg, Russia

A. V. Balakin Faculty of Physics & International Laser Center, Lomonosov Moscow State University, Moscow, Russia; Institute on Laser and Information Technologies, Branch of the Federal Scientific Research Centre "Crystallography and Photonics" of Russian Academy of Sciences, Shatura, Moscow, Russia

Mikhail Balmakov Institute of Chemistry, St. Petersburg State University, St. Petersburg, Russia

Vitaly M. Boitsov Saint-Petersburg Academic University – Nanotechnology Research and Education Centre RAS, Saint Petersburg, Russia

Anton Bondarenko Institute of Physics, St. Petersburg State University, St. Petersburg, Petrodvoretz, Russia

Eugene Borisov Research Centre "Optical and Laser Methods of Researches of Matter", Saint-Petersburg State University, St. Petersburg, Russia

Oleg B. Chakchir Saint-Petersburg Academic University – Nanotechnology Research and Education Centre RAS, Saint Petersburg, Russia

Pavel S. Chelushkin Institute of Chemistry, St. Petersburg State University, Saint Petersburg, Russia

Szymon Chorazy Faculty of Chemistry, Jagiellonian University, Kraków, Poland

O. Chuluunbaatar Joint Institute for Nuclear Research, Dubna, Moscow, Russia; Institute of Mathematics, National University of Mongolia, UlaanBaatar, Mongolia

P. Decleva Dipartimento di Scienze Chimiche e Farmaceutiche, Universita' di Trieste, Trieste, Italy

M. S. Dzhidzhoev Faculty of Physics & International Laser Center, Lomonosov Moscow State University, Moscow, Russia

Marina Evdokimova Saint-Petersburg State University, Saint-Petersburg, Russia

A. Galstyan Institute of Condensed Matter and Nanosciences, Université Catholique de Louvain, Louvain-la-Neuve, Belgium

V. M. Gordienko Faculty of Physics & International Laser Center, Lomonosov Moscow State University, Moscow, Russia

K. S. Grigoriev Physics Faculty, Moscow State University, Moscow, Russia

Kirill S. Grigoriev Faculty of Physics, International Laser Center, Lomonosov Moscow State University, Moscow, Russia

Alexei N. Grum-Grzhimailo Skobeltsyn Institute of Nuclear Physics, Lomonosov Moscow State University, Leninskie Gory, Moscow, Russia

Elena V. Gryzlova Skobeltsyn Institute of Nuclear Physics, Lomonosov Moscow State University, Leninskie Gory, Moscow, Russia

Masahiro Hoshino Department of Earth and Planetary Science, University of Tokyo, Tokyo, Japan

Mikhail Ilyshin St. Petersburg State Institute of Technology, St. Petersburg, Russia

Denis Ivanov Department of General Physics 1, Saint-Petersburg State University, St. Petersburg, Russia

I. E. Ivanov Faculty of Physics & International Laser Center, Lomonosov Moscow State University, Moscow, Russia

Tatiana Ivanova Department of General Physics 1, Saint-Petersburg State University, St. Petersburg, Russia

Masanori Iwamoto Department of Earth and Planetary Science, University of Tokyo, Tokyo, Japan

Alexey Kalinichev Research Centre "Optical and Laser Methods of Researches of Matter", Saint-Petersburg State University, St. Petersburg, Russia

V. P. Kandidov Faculty of Physics and International Laser Center, M.V. Lomonosov Moscow State University, Moscow, Russia

Tsuyoshi Kato Department of Chemistry, School of Science, The University of Tokyo, Tokyo, Japan

Ilya Kolesnikov Saint-Petersburg State University, Saint-Petersburg, Russia

Alexander Konev Saint-Petersburg State University, Saint-Petersburg, Russia

Alexander S. Konev Institute of Chemistry, St. Petersburg State University, St. Petersburg, Russian Federation

Irina Korshok Department of General Physics 1, Saint-Petersburg State University, St. Petersburg, Russia

Olga G. Kosareva Faculty of Physics and International Laser Center, Lomonosov Moscow State University, Moscow, Russia; Lebedev Physical Institute of the Russian Academy of Sciences, Moscow, Russia Institute of Modern Optics, Nankai University, Tianjin, China

Konstantin A. Kouzakov Faculty of Physics, Lomonosov Moscow State University, Moscow, Russia

N. A. Kuzechkin Institute on Laser and Information Technologies, Branch of the Federal Scientific Research Centre "Crystallography and Photonics" of Russian Academy of Sciences, Shatura, Moscow, Russia; Faculty of Physics & International Laser Center, Lomonosov Moscow State University, Moscow, Russia

Zs. Lecz ELI-ALPS, Szeged, Hungary

Erik Lötstedt Department of Chemistry, School of Science, The University of Tokyo, Tokyo, Japan

V. A. Makarov Physics Faculty, Moscow State University, Moscow, Russia; International Laser Center, Moscow State University, Moscow, Russia

Vladimir A. Makarov Faculty of Physics, Moscow State University, Moscow, Russia; International Laser Center, Moscow State University, Moscow, Russia

Alina Manshina Institute of Chemistry, St. Petersburg State University, Petrodvorets, St. Petersburg, Russia

Yosuke Matsumoto Department of Physics, Chiba University, Inage-ku, Chiba, Japan

Nikita G. Mikheev Faculty of Physics, M.V. Lomonosov Moscow State University, Moscow, Russia

Norikatsu Mio Institute for Photon Science and Technology, University of Tokyo, Bunkyo-ku, Tokyo, Japan

S. K. Mishra ELI-ALPS, Szeged, Hungary Physical Research Laboratory (PRL), Ahmedabad, India

Vyacheslav B. Morozov Faculty of Physics, International Laser Center, M.V. Lomonosov Moscow State University, Moscow, Russia

F. Mota-Furtado Department of Mathematics, Royal Holloway, University of London, Egham, Surrey, UK

Koji Nakabayashi Department of Chemistry, School of Science, The University of Tokyo, Bunkyo-ku, Tokyo, Japan

Dmitrii M. Nikolaev Saint-Petersburg Academic University – Nanotechnology Research and Education Centre RAS, Saint Petersburg, Russia

Shin-ichi Ohkoshi Department of Chemistry, School of Science, The University of Tokyo, Bunkyo-ku, Tokyo, Japan

Jun Okabayashi Research Center for Spectrochemistry, The University of Tokyo, Tokyo, Japan

Andrei N. Olenin Faculty of Physics, International Laser Center, M.V. Lomonosov Moscow State University, Moscow, Russia

Takeaki Ozawa Department of Chemistry, School of Science, The University of Tokyo, Tokyo, Japan

P. F. O'Mahony Department of Mathematics, Royal Holloway, University of London, Egham, Surrey, UK

Vladislav V. Pankratov Faculty of Physics and International Laser Center, Lomonosov Moscow State University, Moscow, Russia; Lebedev Physical Institute of the Russian Academy of Sciences, Moscow, Russia

Maksim Panov Institute of Chemistry, St. Petersburg State University, St. Petersburg, Russia

Maxim S. Panov Institute of Chemistry, Saint Petersburg State University, Saint Petersburg, Russia

Nikolai A. Panov International Laser Center, Lomonosov Moscow State University, Moscow, Russia

Nikolay A. Panov Faculty of Physics and International Laser Center, Lomonosov Moscow State University, Moscow, Russia; Lebedev Physical Institute of the Russian Academy of Sciences, Moscow, Russia

I. A. Perezhogin Technological Institute for Superhard and Novel Carbon Materials, Troitsk, Russia; International Laser Center, Moscow State University, Moscow, Russia

B. Piraux Institute of Condensed Matter and Nanosciences, Université Catholique de Louvain, Louvain-la-Neuve, Belgium

Yu. V. Popov Skobeltsyn Institute of Nuclear Physics, Lomonosov Moscow State University, Moscow, Russia; Joint Institute for Nuclear Research, Dubna, Moscow, Russia

N. N. Potravkin International Laser Center, Moscow State University, Moscow, Russia

Anastasia Povolotckaia Saint-Petersburg State University, Saint-Petersburg, Russia

Alexey Povolotskiy Saint-Petersburg State University, Saint-Petersburg, Russia

Nickolai Pulkin Department of Laser Technologies and Systems, ITMO University, St. Petersburg, Russia

Sergey Pulkin Department of General Physics 1, Saint-Petersburg State University, St. Petersburg, Russia

Mikhail N. Ryazantsev Institute of Chemistry, Saint Petersburg State University, Saint Petersburg, Russia; Saint-Petersburg Scientific Center of the Russian Academy of Sciences, Saint Petersburg, Russia

Andrei B. Savel'ev Faculty of Physics and International Laser Center, Lomonosov Moscow State University, Moscow, Russia

Svetlana Savel'eva Department of General Physics 1, Saint-Petersburg State University, St. Petersburg, Russia

Daniil E. Shipilo Faculty of Physics and International Laser Center, Lomonosov Moscow State University, Moscow, Russia; Lebedev Physical Institute of the Russian Academy of Sciences, Moscow, Russia

Georgy M. Shishkov Faculty of Physics, Lomonosov Moscow State University, Moscow, Russia

A. P. Shkurinov Faculty of Physics & International Laser Center, Lomonosov Moscow State University, Moscow, Russia; Institute on Laser and Information Technologies, Branch of the Federal Scientific Research Centre "Crystallography and Photonics" of Russian Academy of Sciences, Shatura, Moscow, Russia; The National University of Science and Technology MISiS, Moscow, Russia

S. A. Shlenov Faculty of Physics and International Laser Center, M.V. Lomonosov Moscow State University, Moscow, Russia

Vladislav Shoev Department of General Physics 1, Saint-Petersburg State University, St. Petersburg, Russia

Andrey A. Shtyrov Saint-Petersburg Academic University – Nanotechnology Research and Education Centre RAS, Saint Petersburg, Russia

P. M. Solyankin Institute on Laser and Information Technologies, Branch of the Federal Scientific Research Centre “Crystallography and Photonics” of Russian Academy of Sciences, Shatura, Moscow, Russia; Faculty of Physics & International Laser Center, Lomonosov Moscow State University, Moscow, Russia

Sergey P. Tunik Institute of Chemistry, St. Petersburg State University, Saint Petersburg, Russia

Vladimir G. Tunkin Faculty of Physics, M.V. Lomonosov Moscow State University, Moscow, Russia

Andrey Tverjanovich Institute of Chemistry, St. Petersburg State University, St. Petersburg, Russia

Yuriy Tverjanovich Institute of Chemistry, St. Petersburg State University, St. Petersburg, Russia

Svetlana Uvarova Department of General Physics 1, Saint-Petersburg State University, St. Petersburg, Russia

E. V. Vasilyev Faculty of Physics and International Laser Center, M.V. Lomonosov Moscow State University, Moscow, Russia

Dmitrii Venediktov Department of General Physics 1, Saint-Petersburg State University, St. Petersburg, Russia

Vladimir Venediktov Department of General Physics 1, Saint-Petersburg State University, St. Petersburg, Russia; Laser Measurement and Navigation Systems Department, Electrotechnical University “LETI”, St. Petersburg, Russia

Sergey Yu. Vyazmin Saint-Petersburg Academic University – Nanotechnology Research and Education Centre RAS, Saint Petersburg, Russia

Oleg Vyvenko Institute of Physics, St. Petersburg State University, St. Petersburg, Petrodvoretz, Russia

Dmitrii V. Yakovlev Faculty of Physics, International Laser Center, M.V. Lomonosov Moscow State University, Moscow, Russia

Igor P. Yakovlev St. Petersburg State Chemical Pharmaceutical Academy, Saint Petersburg, Russia

Kaoru Yamanouchi Department of Chemistry, School of Science, The University of Tokyo, Tokyo, Japan

Hideaki Yoshimura Department of Chemistry, School of Science, The University of Tokyo, Tokyo, Japan

I. A. Zhvaniya Faculty of Physics & International Laser Center, Lomonosov Moscow State University, Moscow, Russia

Part I
Light and Light Propagation

Chapter 1
Polarization Singularities Nucleation in the Self-focusing of an Elliptically Polarized Laser Beam in Kerr Medium and Isotropic Phase of Nematic Liquid Crystal

Vladimir A. Makarov, Kirill S. Grigoriev, Nikolai A. Panov, Olga G. Kosareva and Georgy M. Shishkov

Abstract The possibility of C-points formation is shown in the process of the self-focusing of an originally homogeneously elliptically polarized Gaussian beam in an isotropic nonlinear medium and in isotropic phase of nematic liquid crystal, the temperature of which is close to the temperature of nematic-isotropic phase transition. In the case of axial symmetry of incident beam's intensity profile the generation of C-lines, which have the shape of circumference, is possible in separate planes, which are perpendicular to the axis of the beam. If the axial symmetry of the incident beam's intensity profile is broken C-lines become three-dimensional curves and there is a propagation coordinate domain, in which two pairs of C-points exist. The total topological charge of these four C-points equals zero.

V. A. Makarov (✉) · K. S. Grigoriev · O. G. Kosareva · G. M. Shishkov
Faculty of Physics, Lomonosov Moscow State University, Leninskie Gory 1, Moscow 119991, Russia
e-mail: vamakarov@phys.msu.ru

K. S. Grigoriev
e-mail: ksgrigoriev@ilc.edu.ru

O. G. Kosareva
e-mail: kosareva@physics.msu.ru

G. M. Shishkov
e-mail: gm.shishkov@physics.msu.ru

V. A. Makarov · K. S. Grigoriev · N. A. Panov
International Laser Center, Lomonosov Moscow State University, Leninskie Gory 1, 119991 Moscow, Russia
e-mail: napanov@ilc.edu.ru

K. Yamanouchi et al. (eds.), *Progress in Photon Science*, Springer Series in Chemical Physics 119, https://doi.org/10.1007/978-3-030-05974-3_1

1.1 Introduction

The phenomenon of the self-focusing (SF) of electromagnetic waves was predicted by Askaryan in 1962 in general form, more than fifty years ago. In his paper devoted to this effect, he wrote [1] "The action of the light beam on the medium can become so strong, that the properties of the medium within the beam differ from those outside the beam. It causes waveguide propagation of light beam and cancels the divergence of the beam due to diffraction. This phenomenon can be named as the self-focusing of the electromagnetic beam". SF is described by the system of nonlinear parabolic equations for the slowly varying amplitudes $E_{\pm} = (E_x \pm i E_y)/\sqrt{2}$ of the circularly polarized components of the electric field. The first efforts to analyze the spatial dynamics of the intensity and polarization distributions of the propagating laser light were taken in [2–6]. The system of equations for the dimensionless waist sizes $f_{\pm}$ of the circularly polarized partial beams was solved numerically in [5, 6] and the monotonic and non-monotonic regimes of $f_{\pm}$ behavior were found. The oscillation and other regimes of polarization ellipse's ellipticity degree dependence on the propagation coordinate were derived in [7]. The self-focusing of an elliptically polarized light beam with the Gaussian profile in an isotropic medium with spatial nonlocality of nonlinear optical response is studied in [8, 9]. The examination of the system of equations for $E_{\pm}$ using the method of moments showed that the distribution of the intensity of an originally homogeneously elliptically polarized beam loses its Gaussian shape in the process of light propagation in nonlinear medium and its polarization distribution becomes non-uniform in the plane of the beam cross-section [8]. The obtained dependencies of the light intensity, the ellipticity degree of the polarization ellipse and the angle of orientation of the major axis of the polarization ellipse on the spatial coordinates were found to be in a reasonable accordance with numerical simulations data near the axis of the beam [9, 10].

More than forty years ago SF of laser pulses with duration of few tens picoseconds in isotropic phase of nematic liquid crystal (NLC) at temperature close to the temperature of nematic-isotropic phase transition (NIPT) was of a special interest due to peculiarities of the light propagation dynamics. Among them one can point out significant growth of nonlinear optical response (a few orders of magnitude higher compared to ordinary liquids) and the dramatic increase of its relaxation time when approaching the temperature of NIPT, the existence of low limit of focused beam waist size regardless of initial power of laser radiation, its unusual robustness—the beam did not split into separated filaments [11, 12], transformation of elliptical polarization of laser beam into linear during its propagation [13]. Theoretical research of elliptically polarized beam SF in isotropic phase of NLC at temperature close to NIPT demonstrated [14] that many of the above peculiarities are caused by strong nonlocality of nonlinear optical response in this specific temperature domain. The nonlocality itself is caused by inter-molecular correlations near NIPT and medium heating. The theory of optical response of liquid crystal was developed and comprehensively described in [15, 16]. Flexibility of liquid crystals makes them one of the most prominent media for changing and designing of phase and polarization profiles of propagating laser radiation.

The complicated spatial evolution of the intensity and polarization during light SF lets one to suppose the formation of the points of polarization singularity (C-points) in which the elliptical polarization of the propagating radiation degenerates into a circular one [17, 18]. C-points are differentiated by their topological charge and type. The topological charge is the variation of orientation angle of the major axis of the polarization ellipse calculated along a small closed loop surrounding the C-point and normalized on 2π (customarily, this quantity can be either 1/2 or $-1/2$) [17]. Each topological type of C-point corresponds to the qualitatively different distribution of the polarization ellipses near the point of singularity. In literature three topological types are known: lemon (topological charge 1/2), monstar (topological charge 1/2) and star (topological charge $-1/2$) [17]. In three dimensions, the loci of these points are the lines, which are called C-lines.

Liquid crystal optical elements are now widely used in experimental singular optics. The researches of last ten years have shown that optical vortices may appear when light propagates in NLC with topological defects—disclinations that spontaneously appear in films under degenerate planar anchoring [19], generated by external fields [20, 21], or even light-driven quasi-static electric field [22]. Recent paper [23] reports an effective way to generate optical vortices in wide spectrum range using Bragg reflection of the light from the layer of cholesteric liquid crystal with topological defect. Finally, optical vortices and polarization singularities can naturally arise as a result of optical anisotropy in LC without topological defects [24, 25].

Light singularities are of special interest in nonlinear optics. Structurally stable singularities formation was experimentally observed in several nonlinear optical processes in crystals: photorefraction [26, 27], Pockels effect [28–30], Faraday effect [31]. A number of theoretical researches discover the complicated dynamics of C-points generated on the surface and in the bulk of isotropic media with spatial dispersion of quadratic nonlinearity by the fundamental beams with polarization singularities in sum-frequency generation and second harmonic generation processes [32, 33]. Optical singularities behavior in filamentation and supercontinuum generation processes were studied in [34–36]. Until recently, the possibility of polarization singularities formation in Gaussian beam, self-focusing in isotropic medium with cubic nonlinearity, was not researched, except for [10]. However, the specific range of the nonlinear medium's parameters considered in this paper did not allow polarization singularities to appear.

In the present work we study the possibility of C-points formation in the process of SF of an initially uniformly elliptically polarized Gaussian beam in an isotropic medium with cubic nonlinearity (without frequency and spatial dispersion) and in isotropic phase of NLC near the temperature of NIPT.

1.2 Basic Equations of the Self-focusing of Elliptically Polarized Laser Beam in an Isotropic Medium and in an Isotropic Phase of NLC

The propagation along z axis of an elliptically polarized light beam in an isotropic medium with cubic nonlinearity is described by the following system of equations [8, 9] for the slowly varying complex amplitude $E_{\pm}(x, y, z)$ of circularly polarized wave:

$$\frac{\partial E_{\pm}}{\partial z} + \frac{i}{2k}\Delta_{\perp} E_{\pm} = -i\sigma(|E_{\pm}|^2 + 2|E_{\mp}|^2)E_{\pm}. \tag{1.1}$$

Here $\sigma = 2\pi\omega^2\chi^{(3)}_{xxyy}/kc^2$ is proportional to the component of the local nonlinear susceptibility tensor $\hat{\chi}^{(3)}(\omega;\omega,\omega,-\omega)$, k—is the wave number, the transversal Laplacian $\Delta_{\perp} = \partial^2/\partial r^2 + r^{-1}\partial/\partial r$, where $r = (x^2+y^2)^{1/2}$.

In isotropic phase of NLC the propagation of elliptically polarized light beam is described by two parabolic equations for $E_{\pm}(x, y, z)$:

$$\frac{\partial E_{\pm}}{\partial z} + \frac{i}{2k}\Delta_{\perp} E_{\pm} = -\frac{2\pi i k}{3n^2}\Delta\chi(QE_{\pm} + q_{\pm}E_{\mp}). \tag{1.2}$$

This system generalizes the equation used in [14] for describing the linearly polarized light. Here n is a linear refractive index of NLC in the isotropic phase, $Q = Q_{xx} + Q_{yy}$, $q_{\pm} = Q_{xx} - Q_{yy} \pm 2iQ_{xy}$, where Q_{ij} is a traceless three-dimensional symmetrical real tensor of NLC ordering parameter [37], which determines the anisotropic term $\Delta\chi$ of medium linear dielectric susceptibility tensor $\hat{\chi}$, induced by electromagnetic field, $\chi_{ij} = (\mathrm{Tr}\{\hat{\chi}\}\delta_{ij} + 2\Delta\chi Q_{ij})/3$. The equations for Q_{ij} in stationary case are obtained by varying the expression for free energy of NLC in isotropic phase close to NIPT in the presence of electromagnetic field [11, 12, 15, 16]. Following [37], we represent the density F of free energy as a Taylor series, including main terms up to the second power of Q_{ij} tensor invariants (and its spatial derivatives):

$$F = F_0 + \frac{1}{2}[a\Delta T Q_{ij}^2 + L_1(\partial_i Q_{jk})^2] - \frac{1}{3}\Delta\chi Q_{ij}E_i^* E_j. \tag{1.3}$$

In (1.3), F_0 is free energy density of isotropic phase of NLC, L_1 and a are parameters that do not depend on temperature $T = \mathrm{const}$, $\Delta T = T - T^*$, where T^* is slightly below the actual NIPT temperature [37, 38]. Using the properties of the tensor of NLC ordering parameter $\mathrm{Tr}\{\hat{Q}\} = 0$, $Q_{ij} = Q_{ji}$ and varying (1.3), we obtain the equations for Q and $q_{\pm}$:

$$\begin{aligned}(a\Delta T - L_1\Delta_{\perp})Q &= \frac{\Delta\chi}{18}(|E_+|^2 + |E_-|^2),\\ (a\Delta T - L_1\Delta_{\perp})q_{\pm} &= \frac{\Delta\chi}{3}E_{\pm}E_{\mp}^*.\end{aligned} \tag{1.4}$$

Let an initially homogenously polarized beam with plane wavefront and Gaussian-like intensity profile fall normally on flat medium surface ($z = 0$). Its circularly polarized components are then given by the expressions:

$$E_{\pm}(x, y, z = 0) = I_0^{1/2}(1 \pm M_0)^{1/2}\exp[-(x^2/b^2 + y^2)/r_0^2]. \quad (1.5)$$

Here r_0 is beam waist size, I_0 is maximum of normalized intensity $I(x, y, z) = (|E_+|^2 + |E_-|^2)/2$ in plane $z = 0$. In (1.5), constant $M_0 = M(x, y, z = 0)$ where ellipticity degree of polarization ellipse $M(x, y, z) = (|E_+|^2 - |E_-|^2)/2I_0$. Varying M_0 it is possible to obtain any polarization state of incident radiation: from right hand circularly polarized ($M_0 = -1$) through linearly polarized ($M_0 = 0$) to left-hand circularly polarized ($M_0 = 1$) light. Parameter M_0 is actually the third normalized Stokes parameter. The real constant $b \geq 1$ characterizes stretching of the beam intensity profile along x-axis. It is readily shown that problems (1.1), (1.5) and (1.2), (1.4), (1.5) are symmetrical with respect to simultaneous substitutions: M_0 to $-M_0$, $q_{\pm}$ to $q_{\mp}$, and $E_{\pm}$ to $E_{\mp}$. For this reason, we will further assume that $M_0 \geq 0$.

1.3 Polarization Singularities in the SF in an Isotropic Kerr Medium

It is natural to expect the C-lines formation with left-hand circular polarization when $|E_-(x, y, z)|^2 \ll |E_+(x, y, z)|^2$. The equation $r_m(z_m)$, determining the series of curves, can be easily obtained [39] from limiting condition $E_-(x, y, z) = 0$ and the system (1.1) in the following form:

$$r_m(z_m) = -\left(\frac{\partial^2 E_-}{\partial r^2}\right)^{-1}_{\substack{r=r_m,\\ z=z_m}} \left(\frac{\partial E_-}{\partial r}\right)_{\substack{r=r_m,\\ z=z_m}}, \quad (1.6)$$

where $m = 1, 2, 3, \ldots$. In the case of axial symmetry ($b = 1$) separate C-line with number "m" has the shape of a circumference $r_m(z_m)$ with radius r_m that fully lies in the plane $z = z_m$. The physical reasons of the C-line formation at distances z_m from the nonlinear medium border are discussed in detail in [39].

We illustrate the C-lines formation by the numerical simulations, when axially symmetric uniformly elliptically polarized incident beam has the normalized intensity $I_0 \sigma L_d = 1.7$ ($L_d = kr_0^2$) and ellipticity degree of polarization ellipse $M_0 = 0.85$. The first C-line $r_1(z_1)$ appears at the distance z_1 and lies in the interval $1.46L_d < z_1 < 1.47L_d$. It is interesting that in this range of propagation coordinate the distribution of the transversal component of the electromagnetic energy flow (its general form can be found, for example, in [35]) changes significantly with increasing z. For $z < z_1$ the beam cross-section is divided into four zones (Fig. 1.1a): "gray" center (I), "white" internal ring (II), "gray" external ring (III) and "white" peripheral zone (IV). The direction of transversal flow of energy is indicated by

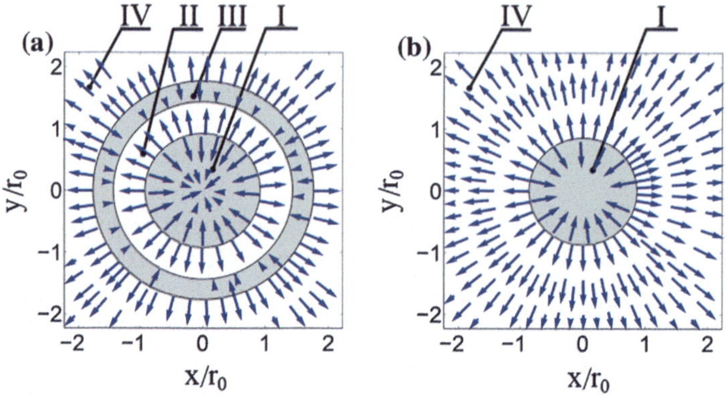

Fig. 1.1 Transversal distribution of electromagnetic energy flow in the plane $z/L_d = 1.46$ (**a**) and $z/L_d = 1.47$ (**b**)

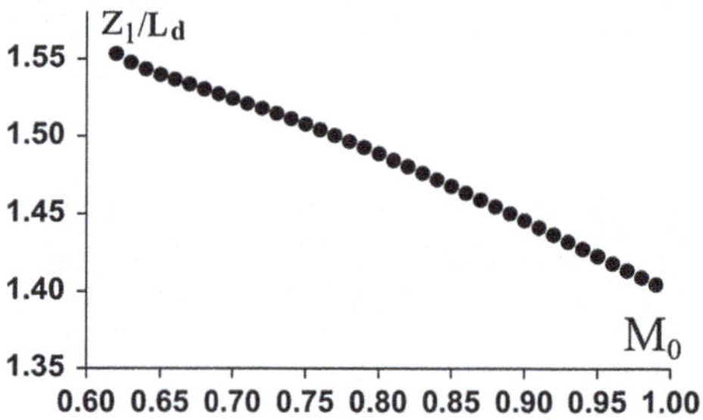

Fig. 1.2 Dependency of distance z_1/L_d between the medium's border and the plane, containing C-line, on the initial ellipticity degree M_0. The normalized intensity $I_0\sigma L_d = 1.7$

arrows. As the z-coordinate increases the distribution of the transversal component of the electromagnetic energy flow changes significantly (Fig. 1.1b): the zone II is expanding, consuming zone III. As a result for $z > z_1$ only the central zone (I), in which the light energy flows toward the beam's axis, and the peripheral zone (IV), where the energy flows away, are presented in laser beam cross-section.

The appearance of C-line $r_1(z_1)$ takes place during SF of light with $|M_0|$ close to 1. In [8] the SF was considered for the almost linearly polarized beams, and this is the reason why the C-lines did not appear. Figure 1.2 illustrates typical dependency $z_1(M_0)$ when the normalize intensity is set. The closer the state of polarization is to the circular one, the closer to the medium border the C-line arises.

If the axial symmetry of an incident beam is broken ($b > 1$), C-line transform from the circumference that fully lies in the beam cross-sections into a three dimensional curve. It intersects a continuous set of the transversal cross-sections of the propagating beam, leading to the formation of C-points. Figure 1.3 shows the dependencies $z_n(b)$ (squares) and $z_a(b)$ (circles), where two pairs of C-points with opposite topological charges are nucleated and annihilated correspondingly. The length of the segment $[z_n(b),\ z_a(b)]$ is increasing as the parameter b grows.

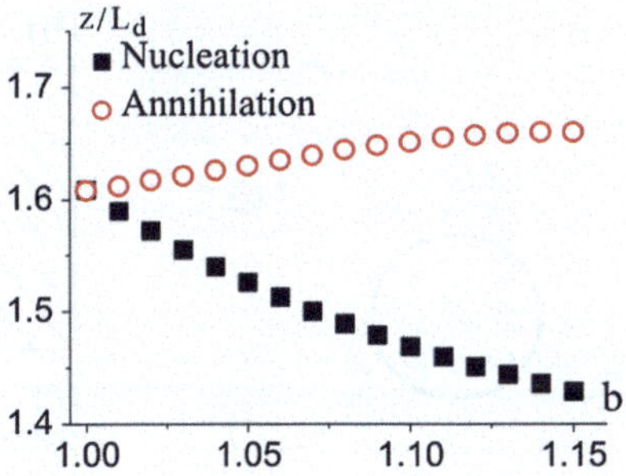

Fig. 1.3 Dependency of nucleation coordinate z_n/L_d (squares) and annihilation coordinate z_a/L_d (circles) of C-points on the axes ratio of the incident beam elliptical intensity profile in the case of $M_0 = 0.85$ and $I_0\sigma L_d = 1.7$

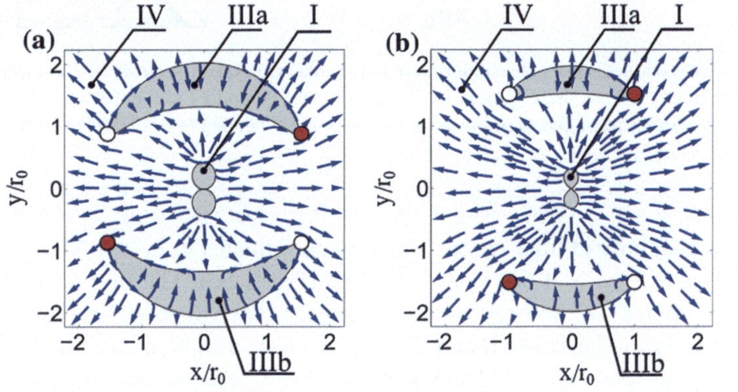

Fig. 1.4 Transversal distributions of energy flow in the cross-section of the propagating beam with the parameters $b = 1.03$, $M_0 = 0.85$, $I_0\sigma L_d = 1.7$ at distance $z/L_d = 1.57$ (**a**) and $z/L_d = 1.60$ (**b**). Filled and empty circles mark the C-points with topological charge $1/2$ and $-1/2$ correspondingly

The generated C-points are marked by filled (charge 1/2) and empty (charge $-1/2$) circles in Fig. 1.4, where the distribution of transverse electromagnetic energy flow is shown for two values of the propagation coordinate from the segment $[z_n(b),\ z_a(b)]$. C-points located near the sharp ends of the crescents and in the vicinity of the singularities the energy flow is vortex-like. As z-coordinate increases, zones IIIa and IIIb in Fig. 1.4a are getting smaller (Fig. 1.4b). When the zones IIIa and IIIb vanish, the C-points collide and annihilate each other.

Figure 1.5a shows the distribution of polarization ellipses for the same incident beam parameters for which Fig. 1.4a is represented. In Fig. 1.5a each ellipse has the same eccentricity and orientation as the polarization ellipse in the point in the laser beam cross-section, corresponding to the ellipse center and the black curve marks the e^{-1} intensity level. The distributions of the polarization lines, which are tangent in every point to the major axis of the corresponding polarization ellipse, are

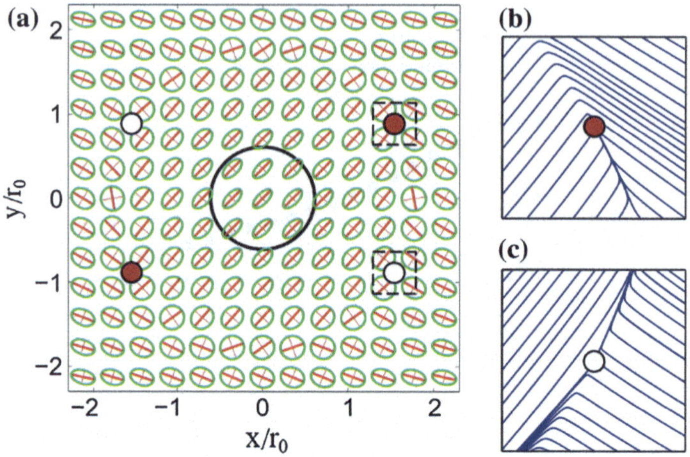

Fig. 1.5 Polarization ellipses distribution in the transversal cross section of the beam (**a**) and polarization lines close to the C-points in dashed squares (**b** and **c**) at the distance $z/L_d = 1.57$. Filled (empty) circles mark the C-points with topological charge 1/2 (−1/2). The beam's parameters are the same as in Fig. 1.4a

shown in the vicinity of the C-points with the coordinates $x/r_0 = 1.54$, $y/r_0 = 0.89$ (Fig. 1.5b) and $x/r_0 = 1.54$, $y/r_0 = -0.89$ (Fig. 1.5c). The first C-point has the topological charge 1/2 and represents a monstar type (three polarization lines are meeting at the singularity point and they lie inside a straight angle). The second one has the topological charge −1/2 and is of a star type, because three polarization lines are meeting at the point, but they do not lie inside a straight angle.

1.4 Polarization Singularities in the SF in an Isotropic Phase of NLC Near the Temperature of INPT

In the hypothetical limit case $L_1 = 0$ (which is actually impossible for any liquid crystal), (1.2), (1.4), can be simplified to those describing the evolution of $E_{\pm}$ during SF in the isotropic medium with Kerr-like nonlinearity without spatial dispersion [8]. Even in this particular case, circularly polarized components $E_{\pm}$ can propagate in medium very differently (qualitatively different regimes of $E_{\pm}$ evolution are thoroughly described in [8]). Various criteria of SF collapse of propagating radiation lead to various expressions for the threshold intensity. The appearance of C-points is possible in certain cases [39] if the polarization of the incident beam is close to circular. The phenomena described in [39] are caused by the aberrations of the propagating beam in light-induced nonlinear Gaussian-like lens, which are studied in general in [40] and particularly in [41].

At sufficiently small L_1, (1.4) have approximate solutions

$$Q \approx \frac{\Delta\chi}{18a\Delta T}\left(1+\frac{L_1}{a\Delta T}\Delta_\perp\right)\left(|E_+|^2+|E_-|^2\right),$$
$$q_\pm \approx \frac{\Delta\chi}{3a\Delta T}\left(1+\frac{L_1}{a\Delta T}\Delta_\perp\right)\left(E_\pm E_\mp^*\right), \tag{1.7}$$

from which it is evident that intensity growth in the beam center and the collapse of the beam, caused by SF, lead to corresponding growth of absolute values of transversal derivatives of the electromagnetic field near the beam axis. Being negative, the summands with transversal derivatives slow down the growth of absolute values of Q and $q_\pm$, which is followed by their decrease, so the beam starts defocusing. This process can happen several times during beam propagation. As a result, a multi-focal structure is formed inside NLC [14]. If the nonlocality of NLC is strong enough, its defocusing impact is noticeable even at early stages of beam propagation and there is no sharp increase of beam intensity, which is typical to self-focusing. In this paper, we consider such values of dimensionless power $P = \pi I_0(k\omega\Delta\chi)^2/(54n^2a\Delta T)$, $\eta = L_1/(a\Delta T r_0^2)$ and M_0, that provide the multi-focal structure formation in the bulk of NLC. We stress out that all three values are determined both by NLC properties and incident beam parameters, so one can easily control the self-focusing regime in a real experiment.

First of all, we consider axially symmetrical incident beams ($b = 1$ in (1.5)). Owing to the symmetry rules, lines of purely circular polarization (C-lines) can only be circles, the planes of which are perpendicular to the beam axis. Centers of the circles lie on the beam axis as well. In numerical simulations carried out in $\{r, z\}$ space, we looked for special points (r_i, z_i) $(i = 1, 2, 3, \ldots)$ in which E_+ (E_-) equals zero and $\text{Arg}\{E_+\}$ $(\text{Arg}\{E_-\})$ is not determined. These points correspond to C-lines described above, as the polarization ellipse there degenerates into a circle.

When $M_0 \sim 1$ the following inequality is valid $|E_+(r, 0)| \gg |E_-(r, 0)|$ and the power of the right-hand circularly polarized component is much greater than that of the left-hand one at NLC border. Thus, the propagation dynamics of E_+ and E_- are significantly different. The mean period of multi-focal structure (sequence of maxima and minima) of dependency $|E_-(0, z)|$ is much shorter than that of dependency $|E_+(0, z)|$. The SF of partial left-hand polarized beam E_- is much weaker: its peak intensity variation is not so big compared to the right-hand circularly polarized component E_+. Due to the $E_\pm$ cross-interaction, there are short-period minor perturbations in $|E_+(0, z)|$ dependency as well that are caused by $|E_-(0, z)|$ fast oscillations.

Our numerical simulations show that ring-shaped C-lines with left-hand (right-hand) polarization appear close to the cross-sections, in which the peak intensity of left-hand (right-hand) circularly polarized component attains local maxima. The left column in Fig. 1.6 shows typical dependencies $\arg\{E_+(r, z)\}$ (a), $|E_\pm(0, z)|$ (b) and $\arg\{E_-(r, z)\}$ (c) in case $M_0 = 0.85$. Two-dimensional gray-scale diagrams show the phase distributions of circularly polarized partial beams in r, z space. Circles mark the points where the phase is not determined. Each point is connected by a thin

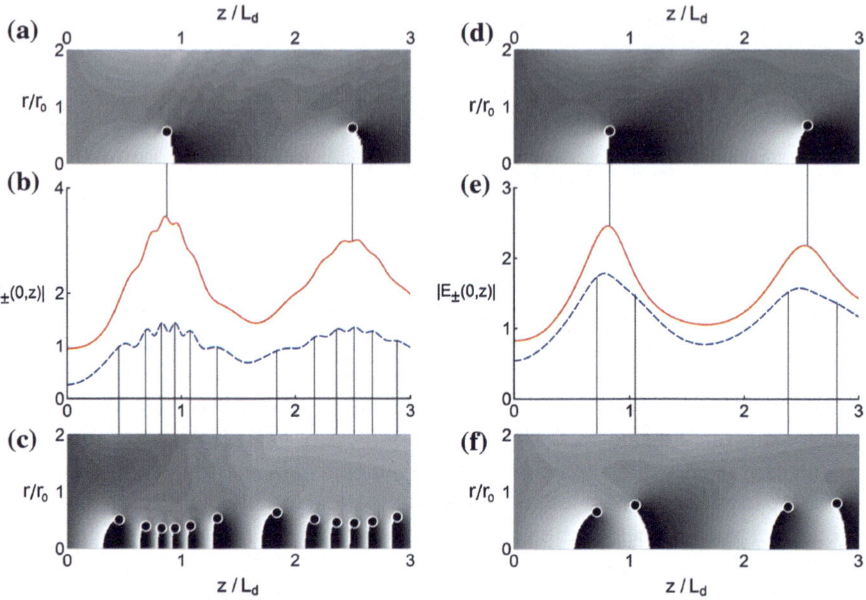

Fig. 1.6 The dependencies of the field parameters of the beam in the bulk of NLC at $P = 2$, $\eta = 0.009$, and $M_0 = 0.85$ (**a–c**) and $\eta = 0.07$ and $M_0 = 0.4$ (**d–f**). Subfigures (**a**, **d**) show the evolution of phase radial distribution of the right-hand circularly polarized component and (**c**, **f**) show the same of the left-hand polarized one. The black color corresponds to the zero phase, and the white color corresponds to phase 2π. Subfigures (**b**, **e**) show the peak intensities of the right-hand (solid line) and the left-hand (dashed line) circularly polarized components of the beam. Thin vertical lines designate cross-sections, containing C-lines. Their intersections with plane containing z-axis are marked by circles in (**a**, **c**, **d**, **f**)

vertical line with the corresponding dependency of right-hand circularly polarized (solid line) or left-hand circularly polarized (dashed line) partial beam peak intensity on propagation coordinate z.

It can be derived from (1.2), (1.4) that in the extreme case of the right-hand circularly polarized incident radiation ($M_0 = 1$) its polarization state remains constant during propagation. However, even in this case the circumferences on which the intensity of the beam is zero (phase singularities) appear in the cross-sections that are close to multi-focal structure maxima. Short-period perturbations of $|E_{+}(0, z)|$ dependency are absent in this particular case.

When the polarization state is not close neither to circular nor to linear one ($0 < M_0 < 1$), the periods of multi-focal structure formed in NLC can be close enough (Fig. 1.6e). In this case, $|E_{+}(r, 0)| > |E_{-}(r, 0)|$ and the left-hand polarization singularities (zeros of E_{+}) are formed more rarely—only in the cross-sections that are close to $|E_{+}(0, z)|$ maxima, which is characteristic to the case $M_0 \sim 1$. Right-hand polarization singularities (zeros of E_{-}) are formed more often (Fig. 1.6f), but their z-coordinates are shifted away from positions of peak intensity maxima of E_{-}

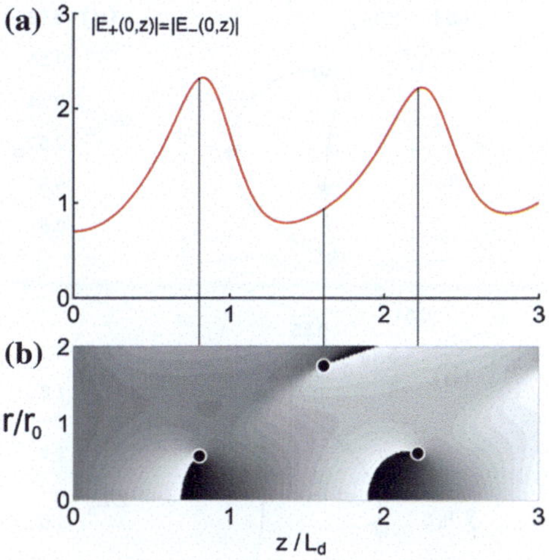

Fig. 1.7 Peak amplitude (**a**) and phase distribution (**b**) of the initially linearly polarized beam in NLC ($P = 1.7, \eta = 0.06$). The black color corresponds to the zero phase, and the white color corresponds to phase 2π. Thin vertical lines designate cross-sections, containing lines of phase singularities. Their intersections with the plane containing z-axis are marked by circles

component. As M_0 increases and initial polarization of the beam is getting close to circular, the phase distributions $\arg\{E_\pm(r, z)\}$ are getting more and more different (see right and left columns in Fig. 1.6). The distance between left-handed C-lines is getting shorter and eventually the singularities occupy the cross-sections close to peak intensity maxima of E_- component (Fig. 1.6c).

As the polarization state of the incident beam is getting close to linear, the phase and peak intensity dependencies in right column of Fig. 1.6 become more and more identical. Generated C-lines remain circles centered at the beam axis and they lie close to the corresponding peak intensity maxima. C-lines of opposite polarization handedness are getting closer to each other. However, a distinctively new family of C-lines appear in the beam when $M_0 \to 0$. They have the same geometry as the "old" ones, but they are located close to local minima of $|E_\pm(0, z)|$ functions. The radii of these "new" C-lines are few times greater than those of the "old" ones. In the limit case of linearly polarized incident beam ($M_0 = 0$), the equality $|E_+(r, z)| = |E_-(r, z)|$ is valid in any point inside NLC. Thus, the complex amplitudes of circularly polarized components of the beam can attain zero values only simultaneously. In this case, phase singularities (points of zero intensity of light field) are formed. When M_0 slightly deviates from zero, the phase distribution of more powerful right-hand polarized component $\arg\{E_+(r, z)\}$ changes rapidly and the radii of left-hand polarized C-lines, located close to minima of multi-focal structure, dramatically increase. As a result, only these left-hand polarized C-lines, which lie close to maxima of $|E_+(0, z)| \approx |E_-(0, z)|$, are left near the axis of the beam. The radii of right-hand polarized C-lines, on which $E_- = 0$, change significantly slower, when M_0 increased from zero. Figure 1.7 shows typical dependencies of peak amplitudes $|E_\pm(0, z)|$ (a) and phase distribution $\arg\{E_\pm(r, z)\}$ (b) at $P = 1.7$, $\eta = 0.06$.

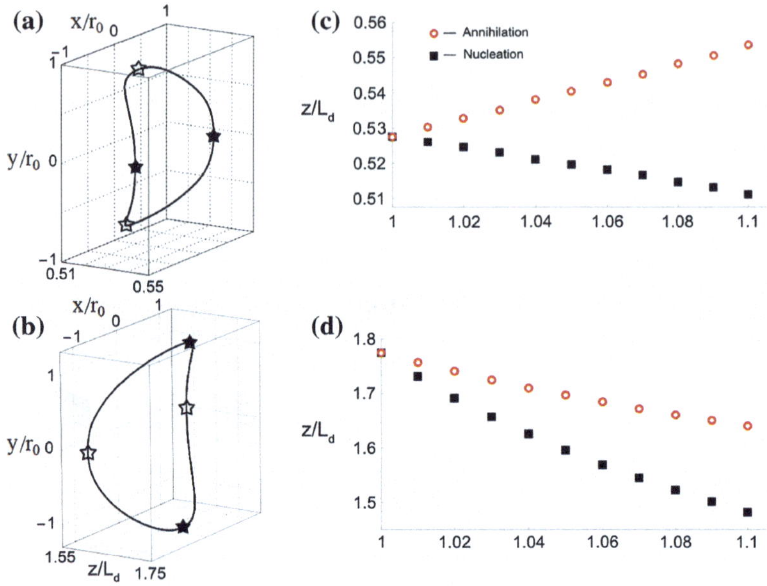

Fig. 1.8 Three-dimensional right-hand (**a**) and left-hand (**b**) C-lines at $b = 1.05$ and **c**, **d** dependencies of corresponding nucleation (squares) and annihilation (circles) z/L_d-coordinates on beam stretching parameter b. Other beam and medium parameters are $P = 2$, $\eta = 0.05$, $b = 1.05$, $M_0 = 0.85$. In (**a**, **b**), white stars show the points of singularities nucleation and black stars show the points of their annihilation

C-lines of both polarization handednesses emerging during SF of a light beam in the bulk of NLC are stable with respect to small disturbance of beam at medium border [42]. For example, stretching of beam transversal intensity profile along x-axis transforms initially planar ring-shaped C-lines into fully three-dimensional saddle-like curves. The examples of two of these C-lines with right-hand and left-hand circular polarization are shown in Fig. 1.8a, b. Each of the curves occupies a certain domain of transversal cross-sections $z_{ni} - z_{ai}$ and form points of polarization singularities (C-points) at intersections. In cross-sections with coordinates z_{ni}, pairwise nucleations of four C-points of opposite topological charges take place and in cross-sections with coordinates z_{ai} these C-points annihilate in pairs. The sum topological charge of nucleated C-points is zero. As the stretching parameter b in (1.5) increases and transversal intensity profile of the beam becomes more elliptical, C-lines get more deformed and domains of z-coordinate in which C-points exist are expanding (Fig. 1.8b, d). Since these C-lines are caused by wavefront aberrations of $E_\pm$ components, they look like the nodal lines that are discussed in [40, 41]. The expansion of the domain $z_{ni} - z_{ai}$ was previously shown for the C-lines that are studied in [39].

Other kind of light singularities is three-dimensional surfaces on which the polarization is linear (L-surfaces). The intersections of these surfaces with the cross-sections of the beam (L-lines) are separating the regions with right-hand and left-hand polarized light. In major part of space, the beam, which we consider in this paper, is right-hand polarized and does not contain L-lines. However, they inevitably appear in the vicinity of left-hand polarized C-lines, separating them from the rest of the beam. The thorough analysis of L-lines dynamics is more complicated compared to C-lines and should be the subject of a future investigation.

1.5 Conclusions

The possibility of generation of polarization singularities during SF of an initially uniformly elliptically polarized Gaussian beam in medium with cubic nonlinearity is shown by numerical simulations. If the incident beam's intensity profile has no axial symmetry the C-lines generation is possible in separate planes, perpendicular to the beam's axis. These lines have the shape of circumferences and the closer the initial beam polarization state is to the circular, the closer to the medium's border the first C-line is. If the incident beam's intensity profile has no axial symmetry, then there is a segment of propagation coordinate, in which two pairs of C-points exist and their total topological charge is zero. The growth of beam asymmetry leads to the expansion of this segment.

When initially homogenously polarized axially symmetric Gaussian light beam self-focuses in an isotropic phase of NLC at temperature close to the temperature of nematic-isotropic transition, C-lines with right-hand and left-hand circular polarization are formed in its cross-sections that are close to local extrema of peak intensities of corresponding circularly polarized components of the propagating field. These lines are circles centered at beam axis and they appear in wide range of NLC parameters and at any polarization ellipse initial ellipticity degree (except for $M_0 = 0$). They are also stable to small perturbations of incident beam transversal profile. If one stretches the intensity profile of the incident beam that way, it becomes elliptical, then C-lines become fully three dimensional curves. When intersecting with transversal cross-sections of the beam, they form four C-points, two of which having positive topological charge and two other having negative one.

Funding The authors acknowledge financial support from the Russian Foundation for Basic Research (Grant No. 16-02-00154).

References

1. G.A. Askar'yan, Effects of the gradient of a strong electromagnetic beam on electrons and atoms. JETP **15**, 1088 (1962)
2. S.A. Akhmanov, A.P. Sukhorukov, R.V. Khokhlov, Self-focusing and diffraction of light in a nonlinear medium. Sov. Phys. Uspekhi **10**, 609 (1968)
3. A.P. Sukhorukov, Thermal self-action of intense light waves. Sov. Phys. Uspekhi **13**, 410 (1970)
4. A.A. Chaban, Self-focusing of light in the Kerr effect. JETP Lett. **5**, 48 (1967)
5. W.G. Wagner, H.A. Haus, J.H. Marburger, Large-scale self-trapping of optical beams in the paraxial ray approximation. Phys. Rev. **175**, 256 (1968)
6. V. Nayyar, A. Kumar, Nonlinear dynamics of an elliptically polarized beam with elliptical irradiance distribution. Opt. Commun. **73**, 501 (1989)
7. S. Vlasov, V. Gaponov, I. Eremina, L. Piskunova, Self-focusing of wave beams with elliptical polarization. Radiophys. Quantum Electron. **21**, 358 (1978)
8. A.A. Golubkov, V.A. Makarov, Amplitude and polarization effects in self-focusing of laser radiation in media with spatial dispersion of nonlinearity. Radiophys. Quantum Electron. **31**, 737 (1988)
9. A.A. Golubkov, V.A. Makarov, I.A. Perezhogin, Formation of elliptically polarized ring-shaped electric field structures on the self-focusing of light in an isotropic medium featuring a spatially disperse nonlinearity. Mosc. Univ. Phys. Bull. **64**(1), 54 (2009)
10. V.A. Makarov, A.A. Golubkov, I.A. Perezhogin, S.S. Savvina, Polarization transformation during beam focusing in chiral liquid. Proc. SPIE **5333**, 30 (2004)
11. G.K.L. Wong, Y.R. Shen, Transient self-focusing in a NLC in the isotropic phase. Phys. Rev. Lett. **32**, 527 (1974)
12. E.G. Hanson, Y.R. Shen, G.K.L. Wong, Experimental study of self-focusing in a liquid crystalline medium. App. Phys. **14**, 65 (1977)
13. N.N. Zhukov, O.P. Zaskal'ko, A.S. Zalot'ko, V.F. Kitaeva, Transformation of elliptical polarization of light wave into linear polarization in the isotropic phase of a nematic liquid crystal. JETP Lett. **52**, 606 (1990)
14. S.M. Arakelyan, G.A. Vardanyan, V.A. Vysloukh, G.A. Lyakhov, V.A. Makarov, Yu.S. Chilingarian, Effect of spatial dispersion of nonlinearity on self-focusing of laser radiation in liquid crystals. Theory and numerical experiments. Radiophys. Quantum Electron. **22**, 36 (1979)
15. P.G. de Gennes,The physics of liquid crystals, in *The International Series of Monographs on Physics* (Clarendon Press, UK, 1974)
16. L.M. Blinov, *Structure and Properties of Liquid Crystals* (Springer, Germany, 2011)
17. M.R. Dennis, K. O'Holleran, M.J. Padgett, Singular optics: optical vortices and polarization singularities. Prog. Opt. **53**, 293 (2009)
18. J.F. Nye, M.V. Berry, Dislocations in wave trains. Proc. R. Soc. A **336**, 1605 (1974)
19. C. Loussert, U. Delabre, E. Brasselet, Manipulating the orbital angular momentum of light at the micron scale with nematic disclinations in a liquid crystal film. Phys. Rev. Lett. **111**, 037802 (2013)
20. E. Brasselet, C. Loussert, Electrically controlled topological defects in liquid crystals as tunable spin-orbit encoders for photons. Opt. Lett. **36**, 719 (2011)
21. E. Brasselet, Tunable optical vortex arrays from a single nematic topological defect. Phys. Rev. Lett. **108**, 087801 (2012)
22. R. Barboza, U. Bortolozzo, G. Assanto, E. Vidal-Henriquez, M.G. Clerc, S. Residor1, Harnessing optical vortex lattices in nematic liquid crystals. Phys. Rev. Lett. **111**, 093902 (2013)
23. J. Kobashi, H. Yoshida, M. Ozak1, Polychromatic optical vortex generation from patterned cholesteric liquid crystals. Phys. Rev. Lett. **116**, 253903 (2013)
24. A.D. Kiselev, R.G. Vovk, R.I. Egorov, V.G. Chigrinov, Polarization-resolved angular patterns of nematic liquid crystal cells: topological events driven by incident light polarization. Phys. Rev. A **78**, 033815 (2008)

25. A.D. Kiselev, V.G. Chigrinov, Optics of short-pitch deformed-helix ferroelectric liquid crystals: symmetries, exceptional points, and polarization-resolved angular patterns. Phys. Rev. E **90**, 042504 (2014)
26. A.V. Ilyenkov, A.I. Khiznyak, L.V. Kreminskaya, M.S. Soskin, M.V. Vasnetsov, Birth and evolution of wave-front dislocations in a laser beam passed through a photorefractive LiNbO3: Fe crystal. Appl. Phys. B **62**, 465 (1996)
27. A.V. Ilyenkov, L.V. Kreminskaya, M.S. Soskin, M.V. Vasnetsov, Birth, evolution and annihilation of phase singularities in the propagation of a laser beam passed through a self-focusing strontium barium niobate crystal. J. Nonlinear Opt. Phys. Mater. **6**, 169 (1997)
28. X. Lu, Z. Wu, W. Zhang, L. Chen, Polarization singularities and orbital angular momentum sidebands from rotational symmetry broken by the Pockels effect. Sci. Rep. **4**, 4865 (2014)
29. N.K. Viswanathan, V. Kumar, C.T. Samlan, Electro-optically tunable topological transformation, in *12th International Conference on Fiber Optics and Photonics, OSA,* p. T4C.4 (2014)
30. I. Skab, Y. Vasylkiv, I. Smaga, R. Vlokh, Spin-to-orbital momentum conversion via electro-optic Pockels effect in crystals. Phys. Rev. A **84**, 043815 (2011)
31. X. Lu, L. Chen, Vortex generation and inhomogeneous Faraday rotation of a nonparaxial gaussian beam in isotropic magneto-optic crystals. Opt. Lett. **39**, 3728 (2014)
32. K.S. Grigoriev, V.A. Makarov, I.A. Perezhogin, Polarization singularities in a sum-frequency light beam generated by a bichromatic singular beam in the bulk of an isotropic nonlinear chiral medium. Phys. Rev. A **92**, 023814 (2015)
33. K.S. Grigoriev, V.A. Makarov, I.A. Perezhogin, Formation of the lines of circular polarization in a second harmonic beam generated from the surface of an isotropic medium with nonlocal nonlinear response in the case of normal incidence. J. Opt. **18**, 014004 (2016)
34. S. Shier, P. Polynkin, J. Moloney, Self-focusing of femtosecond diffraction resistant vortex beams in water. Opt. Lett. **36**, 3834 (2011)
35. D.N. Neshev, A. Dreischuh, G. Maleshkov, M. Samoc, Y.S. Kivshar, Supercontinuum generation with optical vortices. Opt. Express **18**, 18368 (2010)
36. V.P. Kandidov, I.S. Golubtsov, O.G. Kosareva, Supercontinuum sources in a high-power femtosecond laserpulse propagating in liquids and gases. Quantum Electron. **34**, 348 (2004)
37. P.G. de Gennes, Phenomenology of short-range-order effects in the isotropic phase of nematic materials. Phys. Lett. A **30**, 454 (1969)
38. B. Van Roie, J. Leys, K. Denolf, C. Glorieux, G. Pitsi, J. Thoen, Weakly first-order character of the nematic-isotropic phase transition in liquid crystals. Phys. Rev. E **72**, 041702 (2005)
39. N.A. Panov, V.A. Makarov, K.S. Grigoriev, M.S., Yatskevitch, O.G. Kosareva, Generation of polarization singularities in the self-focusing of an elliptically polarized laser beam in an isotropic Kerr medium. Phys. D **332**, 73 (2016)
40. L.V. Kreminskaya, M.S. Soskin, A.I. Khizhnyak, The Gaussian lenses give birth to optical vortices in laser beams. Opt. Commun. **145**, 377, (1998)
41. S. Subota, V. Reshetnyak, M.S. Soskin, Phase singularity birth owing to Gaussian beam self-action in nematic liquid crystal. Mol. Cryst. Liq. Cryst. **375**, 481 (2002)
42. V.A. Makarov, K.S. Grigoriev, G.M. Shishkov, Polarization singularities in self-focusing of an elliptically polarized laser beam in an isotropic phase of nematic liquid crystal close to the temperature of phase transition. Mol. Cryst. Liq. Cryst. **650**, 23 (2017)

Chapter 2
Generation and Transformation of Light Beams with Polarization Singularities in Three-Wave Mixing Processes in Isotropic Nonlocal Medium

K. S. Grigoriev, I. A. Perezhogin, N. N. Potravkin and V. A. Makarov

Abstract The emerging and evolution of polarization singularities in various nonlinear optical processes in isotropic media was studied both analytically and numerically. The interacting light beams with non-uniform polarization were considered in parabolic approximation and the longitudinal components of their fields were taken into account. The generation of the nonlinear signal from the bulk and surface of the medium was shown to be caused by its chirality and nonlocal quadratic response.

2.1 Introduction

In the end of the 20th century a fundamental book "Natural Focusing and Fine Structure of Light" saw the light of the day [1]. The book summarized the research of the previous three decades that introduced a new kind of optics, devoted to the fine

---

K. S. Grigoriev (✉) · V. A. Makarov
Physics Faculty, Moscow State University, Leninskie Gory 1, Moscow, Russia
e-mail: ksgrigoriev@ilc.edu.ru

V. A. Makarov
e-mail: vamakarov@phys.msu.ru

I. A. Perezhogin
Technological Institute for Superhard and Novel Carbon Materials, Centralnaya 7a, Troitsk, Russia
e-mail: i.a.perezz@gmail.com

I. A. Perezhogin · N. N. Potravkin · V. A. Makarov
International Laser Center, Moscow State University, Leninskie Gory 1, Moscow, Russia
e-mail: i.a.perezz@gmail.com

N. N. Potravkin
e-mail: potrav@mail.ru

V. A. Makarov
e-mail: vamakarov@phys.msu.ru

K. Yamanouchi et al. (eds.), *Progress in Photon Science*, Springer Series in Chemical Physics 119, https://doi.org/10.1007/978-3-030-05974-3_2

intrinsic structure of the light fields. In particular, it was shown that a monochromatic homogeneously polarized light field can be pierced with curved lines, along which the intensity of the field is exactly zero and the phase of light is undetermined [2]. Analogous structures are found in inhomogeneously polarized light as well, and these are the lines, along which the polarization of light is purely circular (C-lines) [3]. The intersections of the singularity lines with a given plane are isolated points of undetermined phase (optical vortices) and purely circular polarization (C-points) respectively. Optical singularities, both in three and two dimensions, have a relation with topological defects of the light field and, generally, are structurally stable, i.e. they are not destroyed when the light field is slightly perturbed. Unusual distribution of the light in the vicinity of the singularities allows one to classify them by their topological features. There is a physical quantity q corresponding to each type of singularities that is undefined only in the singular points themselves. The normalized variation of the quantity, calculated along a small closed contour Γ surrounding a singularity, defines its topological charge:

$$Z_q = \frac{1}{2\pi} \oint\limits_{\Gamma} \mathrm{d}q. \tag{2.1}$$

For the optical vortices in homogeneously polarized light fields the quantity q is the phase of the field. In continuous fields the variations of the phase along the contour Γ are multiples of 2π and the topological charge of a phase singularity is an integer number. When the field is inhomogeneously polarized, the quantity q represents the angle between the major axis of the polarization ellipse and a fixed direction, perpendicular to the axis of the light beam. This angle is undetermined for C-points and, since the ellipse is identical to its 180° rotation, the topological charge of a C-point is either half-integer or integer number. Singularities with the least possible absolute value of topological charge (± 1 for optical vortices and $\pm 1/2$ for C-points) are structurally stable, while the singularities with greater charges tend to split into the elementary ones. Topological charge remains constant along the singularity line in three-dimensional space, unless two singularities get close to each other and are pairwise annihilated. The reverse process when two singularities are nucleated in a regular field is also possible. In each act of nucleation and annihilation the interacting singularities have opposite topological charges [1, 4].

Since the first theoretical research [2, 3] and until present day the optical singularities are of high interest in many branches of physics. The development of various experimental techniques allows one to create light fields with singularities [5–11]. Along with them, the singularities detection methods evolve and become more and more precise [12–15]. The behaviour of optical singularities was studied in isotropic random fields [16–18], fibers [19, 20], liquid crystals [21, 22] and even in the skylight [23]. The research in nonlinear optics involving light singularities was mainly focused on the interaction of scalar phase vortices in matter, for example, harmonics generation [24–26], parametric interaction [27, 28], nonlinear refraction [29, 30], filamentation [31, 32] and many other. The propagation of non-uniformly polarized light in nonlinear medium is described by a larger system of equations for orthogonally polarized components of the light and to find analytic solutions even of the most elementary problems is a challenging task. Nevertheless, the efforts are justified, as the complex vectorial structure of inhomogeneously polarized light can provide necessary conditions for the nonlinear transformation of light even in isotropic media, as it was shown in [33–41]. The chapter outlines these papers and presents a comparative description of the generation and evolution scenarios of the polarization singularities in three-wave mixing processes in the bulk and on the surface of nonlinear isotropic medium.

2.2 Bulk Sum-Frequency Generation

The sum frequency generation in the bulk of isotropic liquid was first reported in [42]. This three-wave mixing process is not forbidden since the frequencies of the fundamental beams are different and the tensor of local quadratic susceptibility is non-zero. The vector field of nonlinear polarization of the medium was shown to be

$$\tilde{\mathbf{P}}_3 = \chi^{(2)}[\tilde{\mathbf{E}}_1 \times \tilde{\mathbf{E}}_2], \tag{2.2}$$

where tildes denote the fast-oscillating fields (not their complex amplitudes) and $\chi^{(2)}$ is the only non-zero component of the susceptibility tensor. The sum-frequency generation goes effectively only in chiral liquids, the molecules of which are different from their mirror images, since $\chi^{(2)} = 0$ for the racemic solutions [42]. If $\tilde{\mathbf{E}}_1$ and $\tilde{\mathbf{E}}_2$ are simple plane waves, propagating collinearly along the same axis Oz, then the vectors $\tilde{\mathbf{E}}_1$ and $\tilde{\mathbf{E}}_2$ are orthogonal to the axis and their vector product is parallel to it. Hence, the vector field $\tilde{\mathbf{P}}_3$ in (2.2) is purely longitudinal and cannot generate a free signal wave in the bulk of the medium [43]. However, real laser beams are not entirely plane waves and even in the collinear geometry the sum-frequency generation can still take place due to the interaction of slightly non-collinear spatial Fourier harmonics of the beams. This interaction can be analyzed in the parabolic approximation, in which the fundamental beams are given by their slowly varying complex amplitudes $\mathbf{E}_1$ and $\mathbf{E}_2$. The evolution equations for these amplitudes are:

$$\frac{\partial \mathbf{E}_m}{\partial z} - \frac{\mathrm{i}}{2k_m}\left(\frac{\partial^2}{\partial x^2} + \frac{\partial^2}{\partial y^2}\right)\mathbf{E}_m = 0, \tag{2.3}$$

where $m = 1, 2$. In these equations we neglect the pump depletion and assume that each fundamental beam propagates linearly. The fast-oscillating electric field of the beams is

$$\tilde{\mathbf{E}}_m = \mathrm{Re}\left\{\mathbf{E}_m \exp(\mathrm{i}\omega_m t - \mathrm{i}k_m z)\right\}, \tag{2.4}$$

where ω_m are fundamental frequencies and k_m are corresponding fundamental wavenumbers. Using Maxwell equation div $\tilde{\mathbf{E}}_m = 0$ one can show that the fundamental beams are not entirely transverse and find their small longitudinal components in the following way:

$$E_{mz} \approx \frac{\mathrm{i}}{k_m} \operatorname{div} \mathbf{E}_{m\perp}, \tag{2.5}$$

where $\mathbf{E}_{m\perp}$ are projections of the amplitudes $\mathbf{E}_m$ on the Oxy plane. The ratio $|E_{mz}|/|\mathbf{E}_{m\perp}|$ has the magnitude of $\mu_m = \lambda_m/w_m$, where λ_m is the beam wavelength and w_m is its characteristic transverse size. The next summand in the approximation (2.5) is proportional to μ_m^3 and should not be taken into account in the parabolic approximation, used in (2.3).

Analytic expressions for the complex amplitude of the sum-frequency beam are found in three steps. First, one specifies the transverse profiles of the fundamental beams and calculates their longitudinal components. Second, one calculates the nonlinear polarization vector field $\tilde{\mathbf{P}}_3$ using the material equation (2.2) and extracts its complex amplitude from the equality $\tilde{\mathbf{P}}_3 = \mathrm{Re}\,\{\mathbf{P}_3 \exp(\mathrm{i}\omega_3 t - \mathrm{i}k_\Sigma z)\}$, where $\omega_3 = \omega_1 + \omega_2$ and $k_\Sigma = k_1 + k_2$. Third, one selects the solenoidal part of the $\tilde{\mathbf{P}}_3$ to get rid of its longitudinal spatial Fourier harmonics

$$\mathbf{P}_3^{(s)} = \mathbf{P}_{3\perp} + \frac{\mathrm{i}}{k_3} \operatorname{grad}_\perp P_{3z} \tag{2.6}$$

and uses it in the right part of the evolution equation for the complex amplitude of the signal beam

$$\frac{\partial \mathbf{E}_3}{\partial z} - \frac{\mathrm{i}}{2k_3}\left(\frac{\partial^2}{\partial x^2} + \frac{\partial^2}{\partial y^2}\right)\mathbf{E}_3 = \frac{2\pi \mathrm{i} k_3}{\varepsilon_3}\mathbf{P}_3^{(s)} \exp(\mathrm{i}\Delta k z), \tag{2.7}$$

where k_3 is the wavenumber at the sum frequency, $\Delta k = k_\Sigma - k_3$ is the phase mismatch and ε_3 is the linear dielectric permittivity at the sum frequency. Assuming that the border of the medium lies at $z = 0$ and $\mathbf{E}_3(z = 0) = 0$ one integrates the (2.7) using the Green's function.

2.2.1 Two Gaussian Fundamental Beams

The effective generation of the sum-frequency beam can be realized even with two Gaussian fundamental beams with homogeneous polarization. Their slowly varying envelopes have the following dependency on the transverse and propagation coordinates:

$$E_{m\pm}(x, y, z) = \sqrt{\frac{1 \pm M_m}{2}} \frac{E_{0m}}{\beta_m(z)} \exp\left(\pm \mathrm{i}\Psi_m - \frac{x^2 + y^2}{w_m^2 \beta_m(z)}\right). \tag{2.8}$$

The complex amplitudes are given by their components in the basis of circularly polarized waves:

$$\mathbf{E}_m = E_{m+}\frac{\mathbf{e}_x - \mathrm{i}\mathbf{e}_y}{\sqrt{2}} + E_{m-}\frac{\mathbf{e}_x + \mathrm{i}\mathbf{e}_y}{\sqrt{2}}, \tag{2.9}$$

where $E_{m\pm} = (E_{mx} \pm \mathrm{i}E_{my})/\sqrt{2}$ are complex amplitudes of right- and left-hand circularly polarized components of the beam. Parameters M_m in (2.8) are called ellipticity degrees and determine the polarization state of the beams: from $M_m = 1$ (right-hand circular) through $M_m = 0$ (linear) to $M_m = -1$ (left-hand circular). We assume that each beam has its own ellipticity degree, its own angle Ψ_m between the major axis of the ellipse and the Ox axis, its own amplitude E_{0m} and its own waist size w_m. Since the medium is isotropic, we can assume that $\Psi_1 = 0$ and $\Psi_2 = \Psi$ without loss of generality. For the sake of simplicity we assume that the waists have the same position l_0 in the bulk of the nonlinear medium. Then, the functions $\beta_m(z)$, governing the propagation of the fundamental beams, have the following form:

$$\beta_m(z) = 1 + \mathrm{i}\frac{z - l_0}{l_m}. \tag{2.10}$$

The diffraction lengths of the beams are $l_m = k_m w_m^2/2$.

To find the solenoidal part of nonlinear polarization field of the medium at the sum frequency we apply the method of separation in (2.6) to the material equation (2.2). The amplitudes of the transverse electric fields, which are substituted in the material equation, are given by (2.8) and their longitudinal component is obtained as in (2.5). As the result, we come to the following equation:

$$\begin{aligned} P_{3\pm}^{(s)} &= \pm P_A(x, y, z)\Big[k_\Sigma\sqrt{(1 \pm M_1)(1 \pm M_2)}(x \mp \mathrm{i}y)\mathrm{e}^{\pm \mathrm{i}\Psi} \\ &+ \left(k_1\sqrt{(1 \pm M_1)(1 \mp M_2)}\mathrm{e}^{\mp \mathrm{i}\Psi} + k_2\sqrt{(1 \mp M_1)(1 \pm M_2)}\mathrm{e}^{\pm \mathrm{i}\Psi}\right)(x \pm \mathrm{i}y)\Big]. \end{aligned} \tag{2.11}$$

The nonlinear polarization field is composed of transverse modes of first order and its intensity at the beam axis is zero. Expressions (2.11) are proportional to the same envelope function $P_A(x, y, z)$:

$$P_A(x, y, z) = \frac{1}{2}\frac{\chi^{(2)}E_{01}E_{02}}{k_\Sigma\beta_1(z)\beta_2(z)}\exp\left(-\frac{x^2 + y^2}{w_3^2\beta_3(z)}\right)d_{12}(z). \tag{2.12}$$

Here the effective waist size $w_3 = (w_1^{-2} + w_2^{-2})^{-1/2}$ and the effective propagation function $\beta_3(z) = (w_1^{-2} + w_2^{-2})[1/(w_1^2\beta_1(z)) + 1/(w_2^2\beta_2(z))]^{-1}$. It is worth mentioning, that the function

$$d_{12}(z) = \frac{1}{l_1\beta_1(z)} - \frac{1}{l_2\beta_2(z)} \tag{2.13}$$

is zero and no sum-frequency signal is generated if the diffraction lengths l_1 and l_2 are equal.

Complex amplitudes of the sum frequency beam, obtained by the integration of equation (2.7) have a form that is very similar to (2.11):

$$E_{3\pm} = \pm J(x, y, z)\Big[k_\Sigma\sqrt{(1 \pm M_1)(1 \pm M_2)}\mathrm{e}^{\pm \mathrm{i}\Psi}(x \mp \mathrm{i}y) + \Big(k_1\sqrt{(1 \pm M_1)(1 \mp M_2)}\mathrm{e}^{\mp \mathrm{i}\Psi} + k_2\sqrt{(1 \mp M_1)(1 \pm M_2)}\mathrm{e}^{\pm \mathrm{i}\Psi}\Big)(x \pm \mathrm{i}y)\Big], \tag{2.14}$$

where $J(x, y, z)$ is the following integral:

$$J(x, y, z) = \int_0^z \frac{\mathrm{d}z'}{B(z, z')} \frac{\mathrm{i}\pi k_3 \chi^{(2)} E_{01} E_{02}}{k_\Sigma \beta_1(z')\beta_2(z')} \exp\left(\mathrm{i}\Delta k z' - \frac{x^2 + y^2}{w_3^2 \beta_3(z') B(z, z')}\right). \tag{2.15}$$

This function is axially symmetric and has Gaussian-like transverse profile. The function $B(z, z')$ in the integral (2.15) is

$$B(z, z') = 1 + \frac{2\mathrm{i}(z - z')}{k_3 w_3^2 \beta_3(z')}. \tag{2.16}$$

Polarization distribution of the signal beam is determined by the expression in the square brackets in (2.14). All transverse polarization patterns of the sum frequency beam are similar to each other in its each cross-section. The orientation and ellipticity degree of polarization ellipses in the signal beam remains the same along every straight line passing through the axis of the beam. Thus, the polarization singularities may emerge in the signal beam along these lines in the case if one of two circularly polarized components $P_{3\pm}^{(s)}$ is zero along the line. Consider the general case when both fundamental beams are elliptically polarized. There are two values of the angle Ψ between the major axes of polarization ellipses of the first and second beam at which the signal beam contains a line of left-hand circular polarization if the ellipticity degrees of the beams satisfy two following inequalities:

$$\left| k_1\sqrt{\frac{1 - M_2}{1 + M_2}} - k_2\sqrt{\frac{1 - M_1}{1 + M_1}} \right| \le k_\Sigma, \tag{2.17}$$

$$k_1\sqrt{\frac{1 - M_2}{1 + M_2}} + k_2\sqrt{\frac{1 - M_1}{1 + M_1}} \ge k_\Sigma. \tag{2.18}$$

The violation of (2.17) leads to the violation of (2.18), and vice versa. If both beams are right-hand polarized ($M_1 > 0$ and $M_2 > 0$) then the inequality (2.18) is invalid and the signal beam is always right-hand polarized, regardless on the value of Ψ. The explicit solutions of inequalities (2.17)–(2.18) are rather cumbersome and hard to analyze. Figure 2.1 illustrates some examples of transverse cross-sections of the

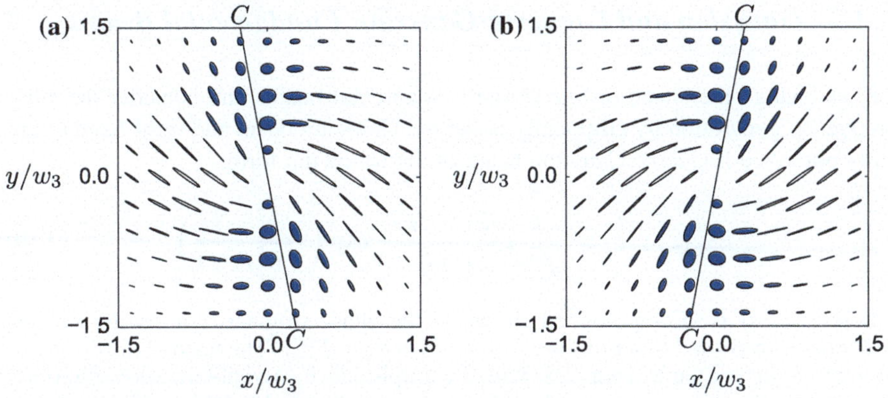

Fig. 2.1 Transverse polarization distributions in the sum frequency beam generated in the bulk of the isotropic medium by two elliptically polarized Gaussian beams with ellipticity degrees $M_1 = -0.5$, $M_2 = 0.2$. The angle between the axes of the beams is **a** $\Psi = 65.3°$, **b** $\Psi = 114.7°$. The wavenumbers ratio of the beams is $k_2/k_1 = 2$. Left-hand C-points lie along the line marked by $C - C$ symbols

sum frequency beam with line of left-hand polarization singularities in them. The ellipses in this figure represent the polarization state parameters of the beam in the centers of the ellipses: the intensity of the field, the eccentricity and the orientation of the ellipse. Filled blue ellipses are left-hand and opened ellipses are right-hand. Transverse coordinates are normalized on the effective waist size w_3. The C-points lie along a straight line in each cross-section of the signal beam and form a plane surface in three dimensions, in the points of which the polarization of light is purely circular The appearance of this non-generic locus of C-points in the cross-section of the beam is caused by the high symmetry of the medium and incident beam. The generated C-points are not structurally stable and disappear when any ellipticity degree M_i or orientation angle Ψ_i is slightly changed.

As it was mentioned earlier, one obtains these patterns by the variation of Ψ for certain values of M_1 and M_2. However, if one of the beam is right-hand circularly polarized, the variation of the angle Ψ does not change the polarization pattern of the signal beam. Instead, there exist a single value M^* of the ellipticity degree of the other beam at which the sum frequency beam contains left-hand polarization singularities. If $M_1 = 1$ then $M_2^* = (k_1^2 - k_\Sigma^2)/(k_1^2 + k_\Sigma^2)$ and if $M_2 = 1$ then $M_1^* = (k_2^2 - k_\Sigma^2)/(k_2^2 + k_\Sigma^2)$. If at least one of the fundamental beams is left-hand circularly polarized then the generation of left-hand polarization singularities in the signal beam is not possible. The analogous inequalities providing the possibility of right-hand C-points generation in the signal beam can be obtained from (2.17) to (2.18) by reverting all the "+" and "−" signs inside the radicals.

2.2.2 Gaussian and Laguerre-Gaussian Fundamental Beams

Let us change the shape of one of the fundamental beams and consider the sum-frequency generation by elliptically polarized Gaussian beam and right-hand circularly polarized Laguerre-Gaussian beam of the following form:

$$E_{2+}(x, y, z) = \frac{x + i\xi y}{\sqrt{2}w_2\beta_2(z)} \frac{E_l}{\beta_2(z)} \exp\left(-\frac{x^2 + y^2}{w_2^2\beta_2(z)}\right), \tag{2.19}$$

where $\xi = \pm 1$ is the topological charge of the phase singularity in the beam. The other parameters and functions are defined as in Sect. 2.2.1. The form of the Gaussian beam remains the same as in (2.8) with the amplitude E_g and polarization ellipticity degree M_0. The solenoidal part of nonlinear polarization field of the medium is found through the same steps as in Sect. 2.2.1 and its complex amplitude has the following form:

$$P_{3+}^{(s)}(x, y, z) = P_B(x, y, z) \left\{ \frac{k_\Sigma}{k_2}\sqrt{1 + M_0}(1 + \xi) + \sqrt{1 - M_0}(1 - \xi) \right.$$
$$\left. + (x + i\xi y)\left[\sqrt{1 - M_0}k_2(x + iy) + \sqrt{1 + M_0}k_\Sigma(x - iy)\right] d_{12}(z) \right\}, \tag{2.20}$$

$$P_{3-}^{(s)}(x, y, z) = -P_B(x, y, z)\sqrt{1 - M_0}\left\{ \frac{k_1}{k_2}(1 + \xi) + k_1(x + i\xi y)(x - iy)d_{12}(z) \right\}. \tag{2.21}$$

The nonlinear polarization field is composed of transverse modes of zero and second order. It is worth mentioning, that the function $d_{12}(z)$, defined here exactly as in (2.13), is the amplitude of higher transverse modes and vanishes if the diffraction lengths l_1 and l_2 are equal. Both expressions (2.20)–(2.21) are proportional to the same envelope function $P_B(x, y, z)$:

$$P_B(x, y, z) = \frac{1}{2\sqrt{2}} \frac{\chi^{(2)} E_g E_l}{k_\Sigma w_2 \beta_1(z)\beta_2^2(z)} \exp\left(-\frac{x^2 + y^2}{w_3^2\beta_3(z)}\right), \tag{2.22}$$

where w_3 and $\beta_3(z)$ are defined as in Sect. 2.2.1.

Analytic expressions for the complex amplitude of the signal beam are found by integrating (2.7) in which the right part is given by (2.20)–(2.22). The solution of the equation can be expressed in quadratures:

$$E_{3+}(x, y, z) = \sqrt{1 + M_0}(1 + \xi)\left[\frac{k_\Sigma}{k_2}J_0(x, y, z) + i\frac{k_\Sigma}{k_3}J_1(x, y, z)\right]$$
$$+ \sqrt{1 - M_0}(1 - \xi)\left[J_0(x, y, z) + i\frac{k_2}{k_3}J_1(x, y, z)\right]$$
$$+ (x + i\xi y)\left[\sqrt{1 - M_0}k_2(x + iy) + \sqrt{1 + M_0}k_\Sigma(x - iy)\right] J_2(x, y, z). \tag{2.23}$$

$$E_{3-}(x,y,z)=\sqrt{1-M_0}\left\{(1+2\xi)\left[\frac{k_1}{k_2}J_0(x,y,z)+\mathrm{i}\frac{k_1}{k_3}J_1(x,y,z)\right]\right.$$
$$\left.+k_1(x+\mathrm{i}\xi y)(x-\mathrm{i}y)J_2(x,y,z)\right\}. \tag{2.24}$$

The functions $J_{0,1,2}(x,y,z)$ are given by the following integrals:

$$J_0(x,y,z)=A\int_0^z\frac{\mathrm{d}z'}{B(z,z')}\frac{1}{\beta_1(z')\beta_2^2(z')}\exp\left(\mathrm{i}\Delta kz'-\frac{x^2+y^2}{w_3^2\beta_3(z')B(z,z')}\right), \tag{2.25}$$

$$J_1(x,y,z)=A\int_0^z\frac{\mathrm{d}z'}{B^2(z,z')}\frac{(z-z')d_{12}(z')}{\beta_1(z')\beta_2^2(z')}\exp\left(\mathrm{i}\Delta kz'-\frac{x^2+y^2}{w_3^2\beta_3(z')B(z,z')}\right), \tag{2.26}$$

$$J_2(x,y,z)=A\int_0^z\frac{\mathrm{d}z'}{B^3(z,z')}\frac{d_{12}(z')}{\beta_1(z')\beta_2^2(z')}\exp\left(\mathrm{i}\Delta kz'-\frac{x^2+y^2}{w_3^2\beta_3(z')B(z,z')}\right), \tag{2.27}$$

where

$$A=\frac{\mathrm{i}\sqrt{2}\pi k_3\chi^{(2)}E_gE_l}{2\varepsilon_3k_\Sigma w_2} \tag{2.28}$$

and $B(z,z')$ is defined in (2.16). Functions (2.25)–(2.27) are axially symmetric and have Gaussian-like asymptotic in transverse coordinates x and y. The signal beam, being a superposition of these functions, has only central symmetry: $E_{3\pm}(x,y,z)=E_{3\pm}(-x,-y,z)$. Finding the exact positions of the polarization singularities (solutions of $E_{3\pm}(x,y,z)=0$) at arbitrary z is not possible and linear interpolation methods are used to visualize the structure of C-lines in the sum-frequency beam.

The main properties of C-lines in the signal beam are inherited from the generating beam of nonlinear polarization $\mathbf{P}_3^{(s)}$ in the right-hand part of (2.7). Since the expressions for its circularly polarized components have much simpler form compared to the signal beam, one can readily find the number and positions of the singularities of the field $\mathbf{P}_3^{(s)}$ (the solutions of $P_{3\pm}^{(\mathrm{s})}(x,y,z)=0$). These singularities are called the *generator* points or G-points. There are several types of $\mathbf{P}_3^{(s)}$ polarization structures with different values of total topological charges of G-points. One can switch between these types by changing the ellipticity degree M_0 of Gaussian beam at the frequency ω_1 or the topological charge of the Laguerre-Gaussian beam at the frequency ω_2.

When topological charge $\xi=-1$, right-hand generators ($P_{3-}^{(\mathrm{s})}(x,y,z)=0$) lie only on the beam axis $x=y=0$. The polarization pattern of the signal beam has the same structure: only one right-hand C-line lies on its axis. Its topological charge is 1 and this is a polarization singularity of charge higher than elementary. Correct analysis of its evolution is beyond the scope of present research. In the case when $\xi = 1$ right-hand G-points lie only in the waist of the beam $z = l_0$ and they form a circle centered at the beam axis with the radius $r_C = [2(l_1 - l_2)/(kl_1l_2)]^{1/2}$ if l_1 is greater than l_2. The G-line of that kind cannot generate a C-line in a signal beam that would intersect its different cross-sections and form isolated C-points. Further on, we will focus only on the left-hand generators and singular points ($P_{3+}^{(s)}(x, y, z) = 0$ and $E_{3+}(x, y, z) = 0$).

Whatever the topological charge of the Laguerre-Gaussian beam is, there exists a critical value of the ellipticity degree M_0^*, that separates two regimes of C-points generation in considered nonlinear process. The exact value of M_0^* depends only on the fundamental wavenumbers ratio:

$$M_0^* = \frac{k_2^2 - k_\Sigma^2}{k_2^2 + k_\Sigma^2}. \tag{2.29}$$

If the product $\xi(M_0 - M_0^*)$ is less than zero there is a pair of left-hand G-points in each cross-section of the beam. They lie symmetrically with respect to the the Oz axis and their sum topological charge is ξ. These G-points form two isolated G-lines in the bulk of the medium. In the other case when $\xi(M_0 - M_0^*) > 0$ there are either four or zero G-points. One pair of them is positively charged and the other one is negatively charged, so their total topological charge is zero. Figure 2.2 illustrates two configurations of right-hand G-points and G-lines with zero and non-zero total topological charge. The singularities are marked with filled (charge $+1/2$) and empty (charge $-1/2$) circles, the stars in Fig. 2.2c designate the points in space in which two singularities are nucleated (white stars) or annihilated (black stars).

Numerical analysis of the expressions (2.23)–(2.24) show that in the case of small phase mismatch Δk in (2.7) the C-lines structure in the signal beam is close to the structure of G-lines in the field $\mathbf{P}_3^{(s)}$. However, as the phase mismatch increases, the C-lines are deformed and become more and more helical. The handedness of the helix is determined by the topological charge of C-point, its polarization handedness and the sign of the phase mismatch. This effect is demonstrated in Fig. 2.3, in which two C-lines configurations at different values of the phase mismatch are presented. The curvature of the C-lines in Fig. 2.3a is so strong, that they intersect some of the transverse cross-sections of the signal beam more than once, thus creating additional C-points. Total topological charge of the C-points remains equal to that of its generators in the field $\mathbf{P}_3^{(s)}$. In the other case when total topological charge of the G-points is zero, C-lines also form a closed loop in space, similar to the one shown in Fig. 2.2. The increase of the phase mismatch Δk leads to the deformation of the loop, the emerging of new loops and the connectivity changes in the old ones (Fig. 2.4).

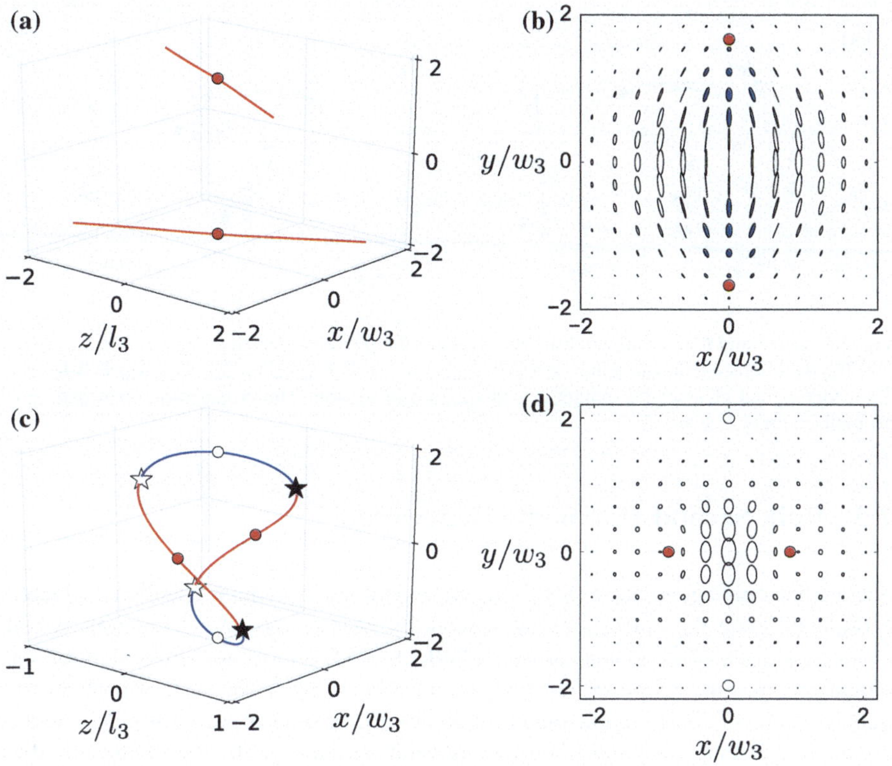

Fig. 2.2 Left-hand G-lines (**a**, **c**) and transverse distributions of the medium polarization field $\mathbf{P}_3^{(s)}$ in the waist of the fundamental beams ($z = 0$) (**b**, **d**) with non-zero (**a**, **b**) and zero (**c**, **d**) sum topological charges of G-points. The charge of the Laguerre-Gaussian fundamental vortex is $\xi = 1$, the other parameters are $w_2/w_1 = 2$, $k_2/k_1 = 1.4$, $M_0 = -0.7$ (**a**, **b**) and $w_2/w_1 = 0.3$, $k_2/k_1 = 2$, $M_0 = 0$ (**c**, **d**). G-lines with positive (negative) topological charge are colored in red (blue). The charges of the lines are also marked by filled (charge 1/2) and empty (charge −1/2) circles. White stars mark the points in space in which two singularity lines with opposite topological charges are born and the black stars mark the points of their pairwise annihilation

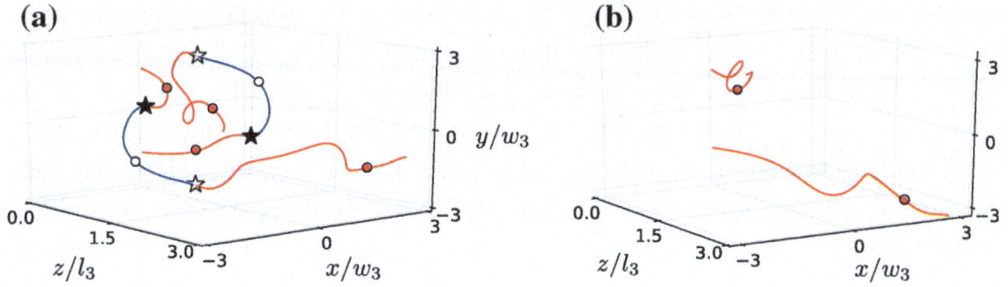

Fig. 2.3 Left-hand C-lines at two different values of the phase mismatch **a** $l_3\Delta k = -5$, **b** $l_3\Delta k = 5$ (**b**). The parameters of the fundamental beams are $w_2/w_1 = 2$, $k_2/k_1 = 1.4$, $\xi = 1$, $M_0 = -0.7$, $l_0 = 0$. Total topological charge of C-points is 1 in each cross-section. The propagation coordinate z is normalized on $l_3 = k_3 w_3^2/2$

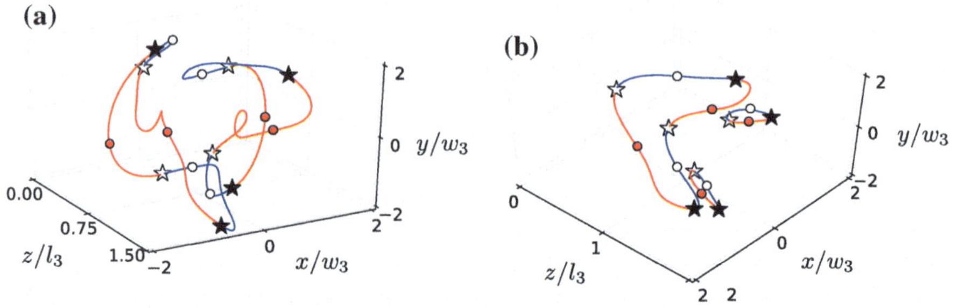

Fig. 2.4 Left-hand C-lines at two different values of the phase mismatch **a** $l_3 \Delta k = -8$, **b** $l_3 \Delta k = -4$. The parameters of the fundamental beams are $w_2/w_1 = 0.3, k_2/k_1 = 2, \xi = 1, M_0 = 0, l_0 = 1$. Total topological charge of C-points is 0 in each cross-section. The propagation coordinate z is normalized on $l_3 = k_3 w_3^2/2$

2.3 Bulk Second-Harmonic Generation

Second harmonic generation (SHG) in centrosymmetric media is usually considered forbidden, since the corresponding susceptibility tensor is zero. However, recent researches discussed few mechanisms of effective SHG even in such kind of media. Among them, one can point out the symmetry breaking in the surface layer of the medium [44, 45], SHG in plasma produced by focused laser pulse [46], degenerate six-photon mixing [47] and SHG in a twisted medium [48]. One of the possible mechanisms of SHG in isotropic liquid is the nonlocality of its quadratic response. The polarization vector $\tilde{\mathbf{P}}_3$ of the nonlocal medium is dependent not only on the fundamental electric field $\tilde{\mathbf{E}}_1$, but also on its gradient [49, 50]:

$$\tilde{\mathbf{P}}_3 = \gamma_1^{(2)} \operatorname{grad}(\tilde{\mathbf{E}}_1 \cdot \tilde{\mathbf{E}}_1) + \gamma_2^{(2)} (\tilde{\mathbf{E}}_1 \cdot \operatorname{grad}) \tilde{\mathbf{E}}_1, \tag{2.30}$$

where $\gamma_1^{(2)}$ and $\gamma_2^{(2)}$ are non-zero components of the tensor of nonlocal quadratic susceptibility of the medium. This tensor does not vanish even in racemic liquids. To calculate the slowly varying envelope of the second harmonic beam one has to do the same steps as in Sect. 2.2, including the calculation of longitudinal components of the electric field $\tilde{\mathbf{E}}_1$ and the separation of the solenoidal part of $\tilde{\mathbf{P}}_3$. The only thing that is changed is the material equation (2.2). Notably, the first summand in (2.30) is a purely potential field and it is cut out during the separation of the solenoidal part, so the second harmonic signal from the bulk of the medium is proportional only to the component $\gamma_2^{(2)}$.

2.3.1 Poincaré Beam

Second harmonic generation in the bulk of the medium has a few peculiarities, which the process of sum-frequency generation does not. First, the signal beam always consists of lower transversal modes, compared to the fundamental one. This is analogous to the vanishing of the sum-frequency signal in Sect. 2.2.1 and the disappearance of its higher transversal modes in Sect. 2.2.2 when the diffraction length of the fundamental beams are the same. Second, there is no signal beam generated in the medium if the fundamental beam is homogeneously polarized in its cross-section. Hence, to achieve effective second harmonic generation one has to use the heterogeneously polarized multi-mode fundamental beam. The simplest example of such kind of beam is Poincaré beam, which is a superposition of Gaussian and Laguerre-Gaussian transversal modes with opposite polarization handedness [51]:

$$E_-(x, y, z) = \frac{E_g}{\beta(z)} \exp\left(-\frac{x^2 + y^2}{w^2\beta(z)}\right), \tag{2.31}$$

$$E_+(x, y, z) = \frac{E_l}{\beta(z)} \frac{p(x + \mathrm{i}y) + q(x - \mathrm{i}y)}{\sqrt{2}w\beta(z)} \exp\left(-\frac{x^2 + y^2}{w^2\beta(z)}\right). \tag{2.32}$$

Here E_g, E_l are the amplitudes of the beam and p and q are two complex constants such that $|p|^2 + |q|^2 = 1$. The propagation function $\beta(z) = 1 + \mathrm{i}(z - l_0)/l_d$ is defined in the same way as in Sect. 2.2. For the simplicity of the forthcoming calculations we assume that all the modes have the same waist sizes w and their propagation behavior is characterized by the same diffraction length $l_d = kw^2/2$, where k is the wavenumber at the fundamental frequency. The constants p, q, can be rewritten as following:

$$p = \cos\theta/2, \qquad q = \mathrm{e}^{\mathrm{i}\eta} \sin\theta/2. \tag{2.33}$$

Here $\theta \in [0; \pi]$ and it determines the amplitude ratio of two Laguerre-Gaussian modes in the incident beam. The parameter η depends on the difference of the phases of these two modes as well as on the phase shift of the Gaussian mode. Since the Poincaré beam is circularly polarized only on its axis, a single left-hand polarization singularity lies in each cross-section of the beam. The topological charge of the C-point is determined by θ as $C = 0.5\,\mathrm{sgn}\{\cos\theta\}$ and does not depend on η. Figure 2.5 illustrates transverse distribution of the polarization ellipses in the waist of the Poincaré beam at different values of θ and η.

The expressions for the solenoidal part of medium nonlinear polarization and the signal beam at doubled frequency are quite simple and similar to each other. Therefore, we only present the final solution of (2.7) for the considered problem:

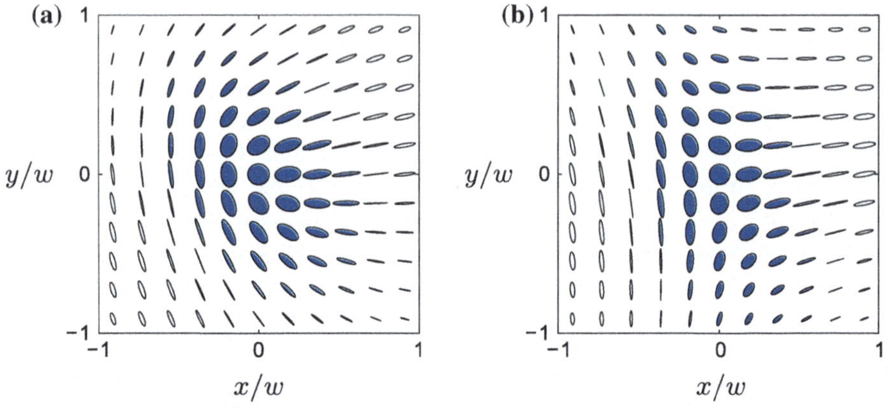

Fig. 2.5 Transverse polarization distributions in the waist of Poincaré beam. A left-hand polarization singularity lies at the axis of the beam ($x = y = 0$). The parameters of the beam are **a** $\theta = \pi/5$, $\eta = \pi/3$ (C-point charge 1/2) and **b** $\theta = 2\pi/3$, $\eta = \pi/6$ (C-point charge −1/2). The ratio of amplitudes of the modes $E_l/E_g = 2.5$

$$\mathbf{E}_3(x, y, z) = \gamma_2^{(2)} E_g E_l (q\mathbf{e}_+ - p\mathbf{e}_-) \times \int_0^z \mathrm{d}z' \frac{\mathrm{i}\pi\sqrt{2}k_3}{\varepsilon_3 w \beta^2(z') B_{\mathrm{SH}}(z, z')} \exp\left(\mathrm{i}\Delta k z' - \frac{2(x^2 + y^2)}{w^2 \beta(z') B_{\mathrm{SH}}(z, z')}\right). \tag{2.34}$$

Here k_3 and ε_3 are wavenumber and dielectric permittivity at the double frequency respectively, $\Delta k = 2k - k_3$ and

$$B_{\mathrm{SH}}(z, z') = 1 + \frac{4\mathrm{i}(z - z')}{k_3 w^2 \beta(z')}. \tag{2.35}$$

The signal beam is homogeneously polarized and has Gaussian-like transverse profile. Its polarization state is determined by the coefficients p and q of Laguerre-Gaussian part of the fundamental beam and the polarization handedness of its central area. The ellipticity degree of the polarization ellipses in the signal beam is

$$M_3 = q^2 - p^2. \tag{2.36}$$

If the topological charge of the C-point in the Poincaré beam is positive then the handedness of the polarization ellipses of the second harmonic beam is the same as in the center of the fundamental beam (Fig. 2.6a). Otherwise, the handedness of the signal polarization ellipse is opposite to that of the central area of the fundamental beam (Fig. 2.6b).

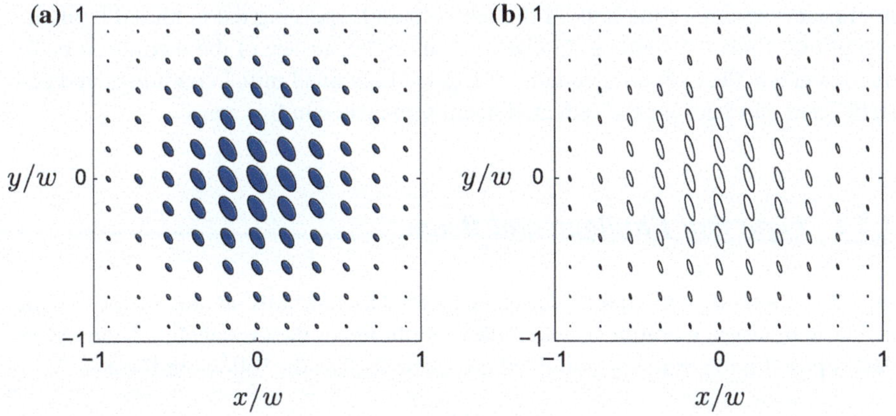

Fig. 2.6 Transverse polarization distributions of the second harmonic beam generated by Poincaré fundamental beam in the bulk of isotropic medium. The parameters of the fundamental beam are **a** $\theta = \pi/5$, $\eta = \pi/3$ and **b** $\theta = 2\pi/3$, $\eta = \pi/6$ (C-point charge $-1/2$). The ratio of amplitudes of the modes is $E_l/E_g = 2.5$

2.4 Surface Second-Harmonic Generation

The problem of surface three-wave mixing poses several issues which makes it a more complicated and interesting problem than nonlinear mixing in the bulk. One of the major tasks is to consider the responses of the bulk and thin surface layer of the medium separately [52, 53]. The transition layer, in which the symmetry of the medium is broken, is present in any kind of medium, but it should especially be taken into account when dealing with the media with spatially nonlocal nonlinear response. To do so, we use a method proposed in [54], which takes the impact of the surface layer into account by modification of the boundary conditions for the electromagnetic field at the combined frequency. The method states that there appears a discontinuity in tangent and normal components of the electromagnetic field that are proportional to the formally introduced surface current $\tilde{\mathbf{i}}$ of bound charges. This current oscillates at the combined frequency and for the processes three-wave mixing its phenomenological form is the following:

$$\tilde{\mathbf{i}}_s = \hat{\kappa}^{(2)} : \tilde{\mathbf{E}}\tilde{\mathbf{E}}. \tag{2.37}$$

Here $\hat{\kappa}^{(2)}$ is the material tensor of the surface layer and $\tilde{\mathbf{E}}$ is the electric field in the vacuum near the boundary of the medium. Like the surface of an isotropic medium this tensor has the external symmetry group ∞ and it also has the permutation symmetry of the two last indices due to the degeneration of the fundamental frequencies.

To find the distribution of the signal field at double frequency one has to calculate the values of the fundamental field near the surface of the medium on its both sides, using classical Fresnel rules. Then one uses the material equation (2.30) to find the nonlinear response of the bulk of the medium and, applying the modified boundary conditions, finds the signal field distribution on the surface of the medium. Finally, one integrates the evolution equation (2.7) with obtained initial conditions and zero right-hand part because the reflected beam propagates in free space.

2.4.1 Gaussian Fundamental Beam

Let the elliptically polarized Gaussian beam fall normally at the surface of the isotropic nonlocal medium. The complex amplitude of the propagating beam, being decomposed into circularly polarized components, has the following form:

$$E_{\pm}(x, y, z) = \sqrt{\frac{1 \pm M_0}{2}} \frac{E_0}{\beta(z)} \exp\left(-\frac{x^2 + y^2}{w^2 \beta(z)}\right). \tag{2.38}$$

The Oz axis of the beam is perpendicular to the surface and directed to the bulk of the medium and propagation function $\beta(z)$ is defined as in Sect. 2.3.1. For the sake of simplicity we assume that the waist of the beam is located directly on the surface of the medium.

It is convenient to express the complex amplitude of the reflected beam at the double frequency in cylindrical coordinate system linked with the axes Ox_1, Oy_1, Oz_1 of the reflected beam. Having $x_1 = -x = r_1 \cos\varphi_1$ and $y_1 = y = r_1 \sin\varphi_1$ one gets:

$$\begin{aligned} E_{3\pm}(r_1, \varphi_1, z_1) &= A(r_1, z_1)\sqrt{1 \mp M_0} \\ &\times \left\{\sqrt{1 \pm M_0}(D_{2\pm} - D_0)\mathrm{e}^{\pm i\varphi_1} + \sqrt{1 \mp M_0} D_{2\pm}\mathrm{e}^{\mp i\varphi_1}\right\} \end{aligned} \tag{2.39}$$

with coefficients $D_0 = n_3\kappa^{(2)}_{zxx} + i\omega\gamma^{(2)}_1/n_3$, $D_{2\pm} = n_1(\kappa^{(2)}_{yyz} \pm i\kappa^{(2)}_{xyz})$. Here n_1, n_3 are real refraction indices of the medium at the fundamental and doubled frequencies respectively. The envelope of the reflected beam is described by

$$A(r_1, z_1) = \frac{16\pi i\sqrt{2}E_0^2 r_1}{\beta^2(z_1)\omega w^2(1 + n_1)^2(1 + n_3)} \exp\left(-\frac{2r_1^2}{w^2\beta^2(z_1)}\right). \tag{2.40}$$

Each circularly polarized component of the signal beam in (2.39) is a superposition of two Laguerre-Gaussian modes of the first order with topological charges 1 and -1. The polarization pattern of the beam is the same in all its cross-sections and the parameters of the polarization ellipse do not depend on the polar radius r_1. Generally, the intensities of the circularly polarized components $E_{3\pm}$ are non-zero in every point of the beam cross-section except for the center ($x_1 = y_1 = 0$). However, if the absolute values of two coefficients of $\mathrm{e}^{\pm i\varphi_1}$ in (2.39) are equal for at least one of the components, then there is a straight line $\varphi_1 = \varphi_s$ in the cross-section of the beam on which the amplitude of this component reaches zero. If the other component does

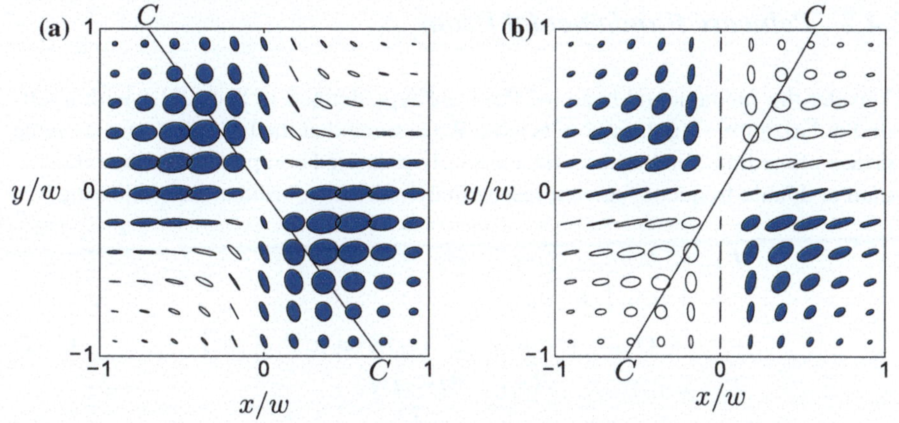

Fig. 2.7 Transverse polarization distributions in the signal beam at **a** $M_0 = 0.45$, **b** $M_0 = -0.02$. The parameters of the nonlinear medium are $n_1 = 1.33$, $n_2 = 1.35$, $\kappa^{(2)}_{yyz}/\kappa^{(2)}_{zxx} = 1.5$, $\kappa^{(2)}_{xyz}/\kappa^{(2)}_{zxx} = 0.3$, $\gamma_2^{(2)}/\kappa^{(2)}_{zxx} = 0.2$. Polarization singularities lie along the lines marked by $C - C$ symbols

not vanish on the same line then the polarization ellipses degenerate into circles on the line. Just like in Sect. 2.2.1, the generated locus of *C*-points is not structurally stable.

The ellipticity degree of the incident beam is the only parameter that can be easily varied and affects the polarization state of the beam at the double frequency. For almost any set of parameters of the nonlinear medium there are two values of ellipticity degree M_0:

$$M_0^{s\pm} = \pm \frac{|D_{2\pm}|^2 - |D_{2\pm} - D_0|^2}{|D_{2\pm}|^2 + |D_{2\pm} - D_0|^2} \tag{2.41}$$

and when $M_0 = M_0^{s+}$ (or $M_0 = M_0^{s-}$) there is a line of left-hand (or right-hand) circular polarization in the cross-sections of the signal beam. Figure 2.7 illustrates typical transverse distributions of polarization ellipses at the waist of the signal beam at these two specific values of ellipticity degree and the same parameters of the medium. Notice, that the absolute values of the ellipticity degree, given by (2.41) are not greater than unity, i.e. they always have physical sense. However, in the case when $M_0^{s\pm}$ reach $+1$ or -1, the signal beam is a homogeneously circularly polarized beam, which is not the case of polarization singularity. For example, this situation always occurs if the response of the transition layer of the medium is absent. In this case the boundary conditions for the second harmonic field become classical and $D_{2\pm}$ in (2.39) are zero.

2.4.2 Poincaré Fundamental Beam

Consider the fundamental beam of Poincaré type defined in Sect. 2.3.1 with a left-hand polarization singularity on its axis. We assume that the beam is focused on the surface of medium. In this case, the circularly polarized components of the reflected beam at double frequency are expressed in cylindrical coordinates as following:

$$\begin{aligned} E_{3+} &= A(\rho_1, z_1)\left\{\rho_1 \mathrm{e}^{-\mathrm{i}\varphi_1} D_{2+}\right. \\ &+ 0.5h\left[D_{1+}\beta(z_1)p\right. \\ &\left.\left.+ (D_0 - D_{2+})(2\rho_1^2\beta^{-1}(z_1)K(\varphi_1)\mathrm{e}^{\mathrm{i}\varphi_1} - p)\right]\right\}, \end{aligned} \tag{2.42}$$

$$\begin{aligned} E_{3-} &= hA(\rho_1, z_1)\left\{h\rho_1^2\beta^{-1}(z_1)D_{2-}\right. \\ &\times\left[\rho_1^2\beta^{-1}(z_1)K^2(\varphi_1^2)\mathrm{e}^{\mathrm{i}\varphi_1} - L(\varphi_1, z_1)\right] \\ &+ 0.5\left[-D_{1-}\beta(z_1)q+\right. \\ &\left.\left.+ (D_0 - D_{2-})(2\rho_1^2\beta^{-1}(z_1)K(\varphi_1)\mathrm{e}^{-\mathrm{i}\varphi_1} - q)\right]\right\}, \end{aligned} \tag{2.43}$$

where

$$A(\rho_1, z_1) = \frac{32\mathrm{i}\sqrt{2}\pi E_g^2}{\beta^2(z_1)\omega w(1+n_1)^2(1+n_3)} \exp\left(-\frac{2\rho_1^2}{\beta^2(z_1)}\right), \tag{2.44}$$

$$K(\varphi_1) = p\mathrm{e}^{-\mathrm{i}\varphi_1} + q\mathrm{e}^{\mathrm{i}\varphi_1}, \tag{2.45}$$

$$L(\varphi_1, z_1) = p[p\mathrm{e}^{-\mathrm{i}\varphi_1} + q(2 - \beta(z_1))\mathrm{e}^{\mathrm{i}\varphi_1}]. \tag{2.46}$$

In the above expressions $h = \sqrt{2}E_l/E_g$, $\rho = r_1/w$, $D_0 = n_3\kappa_{zxx}^{(2)} + i\omega\gamma_1^{(2)}/n_3$, $D_{1\pm} = n_1(\kappa_{yyz}^{(2)} \pm \mathrm{i}\kappa_{xyz}^{(2)}) + \mathrm{i}\omega\gamma_2^{(2)}/(n_1+n_3)$, $D_{2\pm} = n_1(\kappa_{yyz}^{(2)} \pm \mathrm{i}\kappa_{xyz}^{(2)})$. If $h = 0$ then the fundamental beam contains single circularly polarized Gaussian mode and (2.42)–(2.43) match with analogous (2.39) in Sect. 2.4.1. To find the polarization singularities positions one has to solve equations $E_{3\pm} = 0$, which is a complicated problem in general case. Thus, we further present the analytic solutions in some important particular cases and after that we outline some general features of the signal beam. The index "1", referring to the coordinate system of the reflected beam will be omitted for brevity.

First of all, let the surface response is negligibly small for some reasons (the tensor $\hat{\kappa}^{(2)} = 0$), in this case then the boundary conditions for the second harmonic field become classical. The solutions (ρ_+, φ_+) and (ρ_-, φ_-) of the equations $E_{3\pm} = 0$ can be expressed in the following implicit form:

$$p + q\mathrm{e}^{2\mathrm{i}\varphi_+} = \frac{[1 - \gamma_r\beta(z)]\beta(z)\cos\theta/2}{2\rho_+^2}, \tag{2.47}$$

$$p\mathrm{e}^{-2\mathrm{i}\varphi_-} + q = \frac{[1 + \gamma_r\beta(z)]\beta(z)\mathrm{e}^{\mathrm{i}\eta}\sin\theta/2}{2\rho_-^2}, \tag{2.48}$$

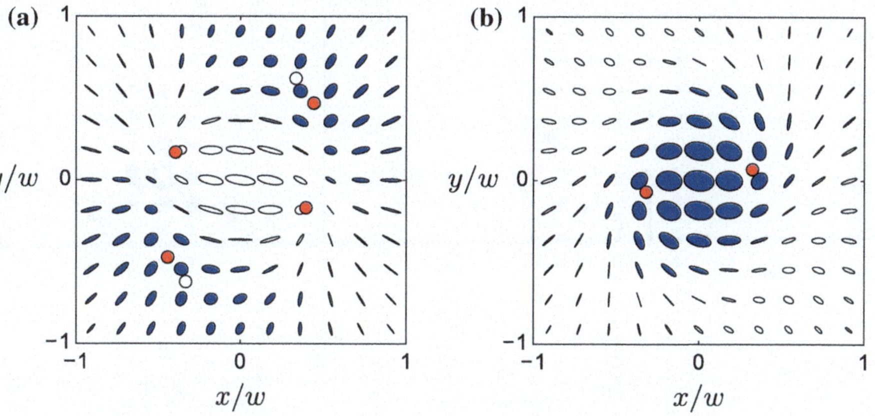

Fig. 2.8 Transverse polarization distributions in the signal beam at distance $0.55l_d$ above the surface of the nonlinear medium with purely bulk response at **a** $\theta = \pi/5$, **b** $\theta = 4\pi/5$ and $h = 1$, $\eta = 0$. The parameters of the nonlinear medium are $n_1 = 1.33$, $n_2 = 1.35$, $\hat{\kappa}^{(2)} = 0$, $\gamma_2^{(2)}/\gamma_1^{(2)} = 0.5$

where $\gamma_r = (1 + n_3/n_1)\gamma_2^{(2)}/\gamma_1^{(2)}$. The number of solutions of equations (2.47)–(2.48) is mainly dependent on the ratio of the absolute values of p and q in their left-hand parts, i.e. on the value of the angle θ in the definition (2.33). If $\theta < \pi/2$ (positive charge of the C-point in the fundamental beam) than there always are two values of polar angle φ_{1-} and $\varphi_{2-} = \varphi_{1-} + \pi$, for which one can find one real value of dimensionless polar radius ρ_-, so that (ρ_-, φ_{1-}) and (ρ_-, φ_{2-}) are solutions of (2.48). These solutions correspond to a pair of right-hand C-points in each cross section of the signal beam. They lie symmetrically with respect to the beam axis and their topological charges are $1/2$. On the other hand, at the same values of parameter θ the number of solutions of (2.47) is either 4 or 0, depending on the propagation coordinate z. This means that in certain transverse planes z of the signal beam there are two pairs of left-hand C-points. The points of each pair are located symmetrically with respect to the beam axis and the singularities of the one pair have positive topological charge $(1/2)$, while the singularities of the other pair have negative charge $(-1/2)$.

If the C-point of the incident beam has negative topological charge $(\theta > \pi/2)$ then the beam at doubled frequency has the structure analogous to that one of the case $\theta < \pi/2$, but the handedness and topological charges of the polarization singularities are swapped to the opposite. There is always a pair of left-hand C-points with topological charges $-1/2$ and possibly two more pairs of right-hand C-points with pairwise opposite topological charges. Figure 2.8 illustrates the transverse polarization distributions in the second harmonic beam for some of the described cases.

As it was mentioned before, the explicit expressions of the positions of all C-points in the reflected beam at doubled frequency at arbitrary values of θ, η and h are very cumbersome by themselves. However, using (2.42) and (2.43), one can readily find the sums of the topological charges of right- and left-handed singularities. It

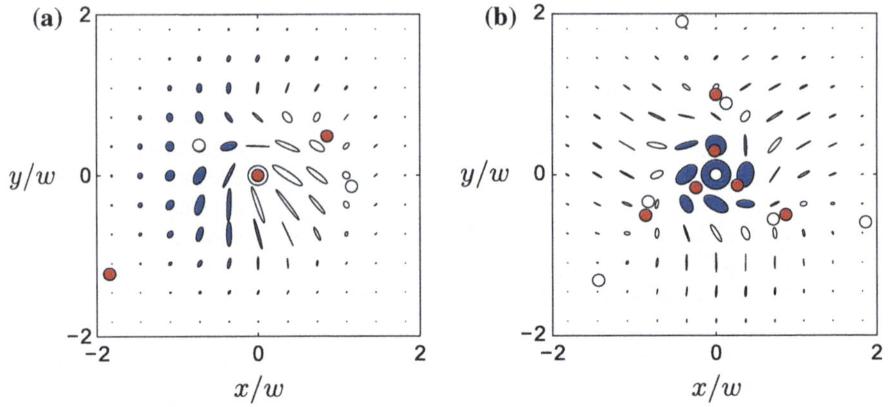

Fig. 2.9 Transverse polarization distributions in the signal beam at distance **a** $0.55l_d$ and **b** $0.75l_d$ above the surface of the nonlinear medium. The incident beam parameters are **a** $\theta = 0$, **b** $\theta = \pi$ and $h = 1.6$, $\eta = 0$. The parameters of the nonlinear medium are **a** $\kappa_{yyz}^{(2)}/\kappa_{zxx}^{(2)} = 0.75$, **b** $\kappa_{yyz}^{(2)}/\kappa_{zxx}^{(2)} = 0.15$ and $n_1 = 1.33$, $n_2 = 1.35$, $\kappa_{xyz}^{(2)}/\kappa_{zxx}^{(2)} = 0.3$, $\omega\gamma_1^{(2)}/\kappa_{zxx} = 0.5$, $\omega\gamma_2^{(2)}/\kappa_{zxx} = 0.25$

is related with the variation of the phases of the circularly polarized components $\arg\{E_\pm\}$ calculated along the circumference of sufficiently large radius in the beam's cross-section. Assuming $\rho \gg 1$ one can obtain the following:

$$\arg\{E_+\} \approx \arg\{A(\rho, z)\beta^{-1}(z)(D_0 - D_{2+})K(\varphi)e^{i\varphi}\}, \tag{2.49}$$

$$\arg\{E_-\} \approx \arg\{A(\rho, z)\beta^{-2}(z)D_{2-}K^2(\varphi)e^{i\varphi}\}. \tag{2.50}$$

In each transverse cross-section $z = \text{const}$ the variation of the phases (2.49) and (2.50) are the same. Following the calculations, analogous to those made while analyzing (2.47) and (2.48), one can find that the sum topological charge of right-hand (or left-hand) C-points of the signal beam is $1/2$ (or 0) when $\theta < \pi/2$ and $-3/2$ (or 1) when $\theta > \pi/2$. These sums do not depend on the parameter η, the ratio of Gaussian and Laguerre-Gaussian parts h and the parameters of the nonlinear medium. The sum charges of singularities in the signal beam are only dependent on the charge of the singularity in the incident beam. The values of sum charges for two cases of media

Table 2.1 The sum charges of the right-hand (the first number) and left-hand (the second number) C-points in the signal beam for the different types of the nonlinear medium's response and two possible charges of the singularity in the incident beam

Top. charge of the incident C-point	Sum top. charge (bulk response)	Sum top. charge (combined response)
1/2	1 and 0	1/2 and 0
−1/2	0 and 1	−3/2 and 1

with purely bulk and combined nonlinear response are outlined in Table 2.1. The typical transverse polarization distributions and C-lines structures of the reflected second harmonic beam are presented in Fig. 2.9 for the cases $\theta = 0$ and $\theta = \pi$.

2.5 Conclusions

Despite the high symmetry of isotropic medium there are still several mechanisms of sum-frequency and second harmonic generation in its bulk and from its surface. The necessary condition of the sum-frequency generation is the chirality of the medium, while second harmonic can be generated in the media with spatial nonlocality of its nonlinear response. In collinear geometry of input light beams interaction these processes are also provided by the vectorial structure and spatial finiteness of the beams. In some processes, like bulk sum-frequency generation and surface second harmonic generation, the nonlinear signal may contain polarization singularities even if the input beams are elliptically polarized Gaussian modes. In the case, when the fundamental radiation already has single optical singularity, the number of singularities in the nonlinear signal increases. The exact amount and topological properties of the generated singularities can be controlled by the polarization state of the input beams. In the process of bulk second harmonic generation the situation is opposite. In order to produce a nonlinear signal, one has to use a beam with polarization singularity, not the beams with uniform polarization distribution. The second harmonic beam has more simple structure compared to the input beam and is homogeneously polarized. The shape and handedness of its polarization ellipse can be controlled by the topological features of the input singularity.

The authors acknowledge financial support from the Russian Foundation for Basic Research (Grant No. 16-02-00154).

References

1. J.F. Nye, *Natural Focusing and Fine Structure of Light: Caustics and Wave Dislocations* (Institute of Physics Publishing, Bristol and Philadelphia, 1999)
2. J.F. Nye, M.V. Berry, Proc. R. Soc. A **336**(1605), 165 (1974). https://doi.org/10.1098/rspa.1974.0012
3. J.F. Nye, Proc. R. Soc. A **389**(1797), 279 (1983). https://doi.org/10.1098/rspa.1983.0109
4. M.R. Dennis, K. O'Holleran, M.J. Padgett, (Elsevier, 2009), pp. 293–363. https://doi.org/10.1016/S0079-6638(08)00205-9. http://www.sciencedirect.com/science/article/pii/S0079663808002059
5. L. Allen, M. Beijersbergen, R. Spreeuw, J. Woerdman, Phys. Rev. A **45**, 8185 (1992). https://doi.org/10.1103/PhysRevA.45.8185. http://link.aps.org/doi/10.1103/PhysRevA.45.8185
6. J. Strohaber, T.D. Scarborough, C.J.G.J. Uiterwaal, Appl. Opt. **46**(36), 8583 (2007). https://doi.org/10.1364/AO.46.008583. http://ao.osa.org/abstract.cfm?URI=ao-46-36-8583
7. E. Brasselet, N. Murazawa, H. Misawa, S. Juodkazis, Phys. Rev. Lett. **103**, 103903 (2009). https://doi.org/10.1103/PhysRevLett. 103.103903. http://link.aps.org/doi/10.1103/PhysRevLett.103.103903

8. F. Cardano, E. Karimi, L. Marrucci, C. de Lisio, E. Santamato, Opt. Express **21**(7), 8815 (2013). https://doi.org/10.1364/OE.21.008815. http://www.opticsexpress.org/abstract.cfm?URI=oe-21-7-8815
9. B. Yang, E. Brasselet, J. Opt **15**(4), 044021 (2013). http://stacks.iop.org/2040-8986/15/i=4/a=044021
10. B. Hao, J. Leger, Appl. Opt. **46**(33), 8211 (2007). https://doi.org/10.1364/AO.46.008211. http://ao.osa.org/abstract.cfm?URI=ao-46-33-8211
11. R. Won, Nat. Photon. **8**(1), 8 (2014). http://dx.doi.org/10.1038/nphoton.2013.358
12. O. Angelsky, I. Mokhun, A. Mokhun, M. Soskin, Phys. Rev. E **65**, 036602 (2002). https://doi.org/10.1103/PhysRevE.65.036602. http://link.aps.org/doi/10.1103/PhysRevE.65.036602
13. M. Burresi, R. Engelen, A. Opheij, D. van Oosten, D. Mori, T. Baba, L. Kuipers, Phys. Rev. Lett. **102**, 033902 (2009). https://doi.org/10.1103/PhysRevLett.102.033902. http://link.aps.org/doi/10.1103/PhysRevLett.102.033902
14. E. Brasselet, S. Juodkazis, Phys. Rev. A **82**, 063832 (2010). https://doi.org/10.1103/PhysRevA.82.063832. http://link.aps.org/doi/10.1103/PhysRevA.82.063832
15. K.E. Ballantine, J.F. Donegan, P.R. Eastham, Sci. Adv. **2**(4) (2016). https://doi.org/10.1126/sciadv.1501748. http://advances.sciencemag.org/content/2/4/e1501748
16. M.V. Berry, M.R. Dennis, Proc. R. Soc. A **457**(2005), 141 (2001). https://doi.org/10.1098/rspa.2000.0660. http://rspa.royalsocietypublishing.org/content/457/2005/141.abstract
17. M. Dennis, Opt. Commun. **213**(4–6), 201 (2002). https://doi.org/10.1016/S0030-4018(02)02088-6. http://www.sciencedirect.com/science/article/pii/S0030401802020886
18. F. Flossmann, K. O'Holleran, M.R. Dennis, M.J. Padgett, Phys. Rev. Lett. **100**, 203902 (2008). https://doi.org/10.1103/PhysRevLett.100.203902. http://link.aps.org/doi/10.1103/PhysRevLett.100.203902
19. A. Volyar, T. Fadeeva, Tech. Phys. Lett. **28**(2), 102 (2002). https://doi.org/10.1134/1.1458503. http://dx.doi.org/10.1134/1.1458503
20. C.N. Alexeyev, B.P. Lapin, M.A. Yavorsky, Phys. Rev. A **78**, 013813 (2008). https://doi.org/10.1103/PhysRevA.78.013813. http://link.aps.org/doi/10.1103/PhysRevA.78.013813
21. A.D. Kiselev, J. Phys. Cond. Matt. **19**(24), 246102 (2007). http://stacks.iop.org/0953-8984/19/i=24/a=246102
22. N.F. Smyth, W. Xia, J. Phys. B **45**(16), 165403 (2012). http://stacks.iop.org/0953-4075/45/i=16/a=165403
23. M.V. Berry, M.R. Dennis, R.L. Lee Jr., New J. Phys. **6**(1), 162 (2004). http://stacks.iop.org/1367-2630/6/i=1/a=162
24. Y. Toda, S. Honda, R. Morita, Opt. Express **18**(17), 17796 (2010). https://doi.org/10.1364/OE.18.017796. http://www.opticsexpress.org/abstract.cfm?URI=oe-18-17-17796
25. A. Stabinis, S. Orlov, V. Jarutis, Opt. Commun. **197**(4–6), 419 (2001). https://doi.org/10.1016/S0030-4018(01)01388-8. http://www.sciencedirect.com/science/article/pii/S0030401801013888
26. M. Zrch, C. Kern, P. Hansinger, A. Dreischuh, C. Spielmann, Nat. Phys. **8**, 743 (2012). https://doi.org/10.1038/nphys2397
27. A.P. Sukhorukov, A.A. Kalinovich, G. Molina-Terriza, L. Torner, Phys. Rev. E **66**, 036608 (2002). https://doi.org/10.1103/PhysRevE.66.036608. http://link.aps.org/doi/10.1103/PhysRevE.66.036608
28. F. Lenzini, S. Residori, F.T. Arecchi, U. Bortolozzo, Phys. Rev. A **84**, 061801 (2011). https://doi.org/10.1103/PhysRevA.84.061801. http://link.aps.org/doi/10.1103/PhysRevA.84.061801
29. A.V. Ilyenkov, A.I. Khiznyak, L.V. Kreminskaya, M.S. Soskin, M.V. Vasnetsov, Appl. Phys. B **62**(5), 465 (1996). https://doi.org/10.1007/BF01081045. http://dx.doi.org/10.1007/BF01081045
30. A. Dreischuh, D.N. Neshev, V.Z. Kolev, S. Saltiel, M. Samoc, W. Krolikowski, Y.S. Kivshar, Opt. Express **16**(8), 5406 (2008). https://doi.org/10.1364/OE.16.005406. http://www.opticsexpress.org/abstract.cfm?URI=oe-16-8-5406
31. G. Maleshkov, D.N. Neshev, E. Petrova, A. Dreischuh, J. Opt. **13**(6), 064015 (2011). http://stacks.iop.org/2040-8986/13/i=6/a=064015

32. V. Jukna, C. Milián, C. Xie, T. Itina, J. Dudley, F. Courvoisier, A. Couairon, Opt. Express **22**(21), 25410 (2014). https://doi.org/10.1364/OE.22.025410. http://www.opticsexpress.org/abstract.cfm?URI=oe-22-21-25410
33. S.N. Volkov, N.I. Koroteev, V.A. Makarov, J. Exp. Theor. Phys. **86**(4), 687 (1998). https://doi.org/10.1134/1.558527. http://dx.doi.org/10.1134/1.558527
34. S.N. Volkov, N.I. Koroteev, V.A. Makarov, Quant. Electron. **25**(12), 1183 (1995). http://stacks.iop.org/1063-7818/25/i=12/a=A14
35. V.A. Makarov, I.A. Perezhogin, N.N. Potravkin, Quant. Electron. **41**(2), 149 (2011). http://stacks.iop.org/1063-7818/41/i=2/a=A12
36. V. Makarov, I. Perezhogin, Opt. Commun. **281**(14), 3906 (2008). https://doi.org/10.1016/j.optcom.2008.03.064. http://www.sciencedirect.com/science/article/pii/S0030401808003234
37. V.A. Makarov, I.A. Perezhogin, Quant. Electron. **39**(7), 627 (2009). http://stacks.iop.org/1063-7818/39/i=7/a=A06
38. K.S. Grigoriev, V.A. Makarov, I.A. Perezhogin, N.N. Potravkin, Quant. Electron. **41**(11), 993 (2011). http://stacks.iop.org/1063-7818/41/i=11/a=A09
39. K.S. Grigoriev, V.A. Makarov, I.A. Perezhogin, Phys. Rev. A **92**, 023814 (2015). https://doi.org/10.1103/PhysRevA.92.023814. http://link.aps.org/doi/10.1103/PhysRevA.92.023814
40. K.S. Grigoriev, V.A. Makarov, I.A. Perezhogin, J. Opt. **18**(1), 014004 (2016). http://stacks.iop.org/2040-8986/18/i=1/a=014004
41. K.S. Grigoriev, N.Y. Kuznetsov, E.B. Cherepetskaya, V.A. Makarov, Opt. Express **25**(6), 6253 (2017). https://doi.org/10.1364/OE.25.006253. http://www.opticsexpress.org/abstract.cfm?URI=oe-25-6-6253
42. J.A. Giordmaine, Phys. Rev. **138**, A1599 (1965). https://doi.org/10.1103/PhysRev.138.A1599. http://link.aps.org/doi/10.1103/PhysRev.138.A1599
43. Y.R. Shen, *The Principles of Nonlinear Optics*. Wiley series in pure and applied optics (Wiley, 1984). http://books.google.ru/books?id=qYIpAQAAMAAJ
44. J.E. Sipe, D.J. Moss, H.M. van Driel, Phys. Rev. B **35**, 1129 (1987). https://doi.org/10.1103/PhysRevB.35.1129. http://link.aps.org/doi/10.1103/PhysRevB.35.1129
45. Y. Shen, Nature **337**, 519 (1989)
46. M. Beresna, P.G. Kazansky, Y. Svirko, M. Barkauskas, R. Danielius, App. Phys. Lett. **95**(12), 121502 (2009). https://doi.org/10.1063/1.3232235. http://aip.scitation.org/doi/abs/10.1063/1.3232235
47. M.M. Coles, M.D. Williams, D.L. Andrews, Opt. Express **21**(10), 12783 (2013). https://doi.org/10.1364/OE.21.012783. http://www.opticsexpress.org/abstract.cfm?URI=oe-21-10-12783
48. A. Bahabad, A. Arie, Opt. Express **15**(26), 17619 (2007). https://doi.org/10.1364/OE.15.017619. http://www.opticsexpress.org/abstract.cfm?URI=oe-15-26-17619
49. S.A. Akhmanov, G.A. Lyakhov, V.A. Makarov, V.I. Zharikov, Optica Acta **29**(10), 1359 (1982). https://doi.org/10.1080/713820767. http://dx.doi.org/10.1080/713820767
50. A.A. Golubkov, V.A. Makarov, J. Mod. Opt. **37**(9), 1531 (1990). https://doi.org/10.1080/09500349014551711. http://dx.doi.org/10.1080/09500349014551711
51. E.J. Galvez, B.L. Rojec, V. Kumar, N.K. Viswanathan, Phys. Rev. A **89**, 031801 (2014). https://doi.org/10.1103/PhysRevA.89.031801. http://link.aps.org/doi/10.1103/PhysRevA.89.031801
52. Y.R. Shen, Appl. Phys. B **68**(3), 295 (1999). https://doi.org/10.1007/s003400050622. http://dx.doi.org/10.1007/s003400050622
53. P. Guyot-Sionnest, W. Chen, Y.R. Shen, Phys. Rev. B **33**, 8254 (1986). https://doi.org/10.1103/PhysRevB.33.8254. http://link.aps.org/doi/10.1103/PhysRevB.33.8254
54. A.A. Golubkov, V.A. Makarov, Usp.-Phys. **38**(3), 325 (1995). http://stacks.iop.org/1063-7869/38/i=3/a=A06

Chapter 3
Spatio-Temporal and Spectral Transformation of Femtosecond Pulsed Beams with Phase Dislocation Propagating Under Conditions of Self-action in Transparent Solid-State Dielectrics

S. A. Shlenov, E. V. Vasilyev and V. P. Kandidov

Abstract Self-action of a femtosecond optical vortex in fused silica at a wavelength of 1900 nm is analyzed by means of numerical simulations. The formation of a multi-focus ring structure is demonstrated. We show that self-focusing in a ring of relatively large radius occurs without plasma generation. The frequency spectrum of the pulse is broadening mainly into the Stokes band.

3.1 Introduction

Self-focusing of femtosecond pulses may result in formation of extended filaments with high fluence [1]. The parameters of such structures are quite stable. In particular, stable peak intensity values of the order of magnitude 5×10^{13} W/cm^2 can be observed in the femtosecond pulse at distances significantly greater than the length of the beam waist [2, 3].

Femtosecond filamentation has been widely studied for Gaussian and other beams with a smooth phase. In such beams, a single filament is usually formed on the axis. In media with anomalous group velocity dispersion self-focusing in space is accompanied by self-compression in time, which leads to the formation of localized in space and time "light bullets" [4, 5]. In fused silica samples, such conditions arise for IR radiation at wavelengths greater than 1.3 μm [6].

S. A. Shlenov (✉) · E. V. Vasilyev · V. P. Kandidov
Faculty of Physics and International Laser Center, M.V. Lomonosov Moscow State University, Leninskie Gori 1, 119991 Moscow, Russia
e-mail: shlenov@physics.msu.ru

E. V. Vasilyev
e-mail: vasilev.evgeniy@physics.msu.ru

V. P. Kandidov
e-mail: kandidov@physics.msu.ru

K. Yamanouchi et al. (eds.), *Progress in Photon Science*, Springer Series in Chemical Physics 119, https://doi.org/10.1007/978-3-030-05974-3_3

On the other hand, there are beams with phase singularity, in which the spiral phase dislocation prevents the appearance of the light field on the beam axis, retaining its ring structure. These donut shaped laser beams, which are often called spiral laser beams or optical vortices, can be promising for such applications as tubular refractive index micromodifications, electrons accelerations, etc. [7]. Self-focusing of spiral laser beams in a medium with cubic nonlinearity was considered theoretically in [7, 8], where the dependence of critical power on the optical vortex topological charge m was obtained. The presence of phase dislocation significantly increases the value of the critical power. For example, the critical power for an optical vortex with topological charge $m = 1$ is four times higher than the critical power of a Gaussian beam. As the topological charge increases, the critical power becomes even higher.

The experimental observation of self-focusing of vortex beams, maintaining ring structure, was done in a water cell for 100 fs pulses at a wavelength of 800 nm [8, 9]. In studies of the self-action of femtosecond vortex beams, considerable attention is paid to the emergence of multiple filamentation. The fact is that the vortex beam is quite stable with respect to its radius, but in a nonlinear medium with cubic nonlinearity it undergoes azimuthal modulational instability. As far as we know, the first observation of the set of filaments in vortex beams was performed in [10]. The annular beam with phase singularity on the axis during self-focusing in sodium vapor formed a high-intensity ring, which then disintegrated into separate hot spots due to azimuthal instability. In [9] on the basis of experimental and theoretical studies, a formula was proposed for estimating the number of filaments arising after the collapse of a singular beam, depending on the excess of the peak power over the critical one and the value of the topological charge. The possibility of increasing the distance to the start of multiple filamentation of high-power femtosecond radiation in vortex beams was considered in [11]. It is shown numerically that in relatively wide beams propagating in atmosphere it is possible to transfer high fluence annular structure at a distance of hundreds of meters before it decays into a set of filaments. Improvement of longitudinal stability and reproducibility of filaments in vortex beams with energy of 2–30 mJ formed by special phase plates and propagating in atmosphere was demonstrated in [12]. Experimental studies of the behavior of optical vortices during filamentation in gases and numerical analysis in a stationary approximation without taking into account plasma generation showed the possibility of obtaining stable multiple filamentation regimes in beams with a lattice of optical vortices [13].

A detailed picture of the spatiotemporal dynamics of a femtosecond pulse in a vortex beam in fused silica at a wavelength of 800 nm was obtained in our paper [14]. It is shown that a tubular structure with a radius of 3–4 μm can be formed in the nonlinear focus, and the length of this structure significantly exceeds the length of the beam waist under linear propagation.

A number of papers are devoted to the study of frequency spectrum transformation of femtosecond pulses in vortex beams. Experimental spectra of supercontinuum during filamentation of beams with phase singularity in a variety of media (BK7, silica, water, CaF_2) were obtained in [15]. It was registered that in the process of supercontinuum generation, the profile of the annular beam practically did not change, and the arising of a small number of filaments resulted in characteristic interference fringes in the supercontinuum radiation.

Transformation of optical vortices into newly emerging spectral components due to phase modulation and four-wave mixing was demonstrated in [16]. The influence of inertia of plasma formation on the propagation stability of optical vortices was studied numerically in [17] under filamentation of beams with phase singularity in fused silica. It is shown that it leads to a quasi-soliton regime of the beam propagation, which is stable to perturbations destroying symmetry, and the multi-focus behavior of optical vortices was revealed.

It should be noted that currently experimental and numerical studies of femtosecond filamentation in vortex beams are mainly performed at wavelengths corresponding to the normal group velocity dispersion in a medium considered. On the other hand, tunable parametric amplifiers allow obtaining high-intensity femtosecond pulses in the mid-IR range, where group velocity dispersion of many transparent solid-state dielectrics is anomalous and where there is a self-compression of pulses. Self-action of pulses with spiral phase dislocation qualitatively changes the manifestation of self-focusing in a nonlinear medium, provoking the formation of tubular filaments.

In this paper, we investigate numerically the spatiotemporal dynamics and spectral characteristics of femtosecond optical vortices under self-action in fused silica glass at a wavelength of 1900 nm. In the first part of the paper, the mathematical model based on the approximation of slowly varying wave [18] is discussed in detail. This allows describing self-steepening of the optical pulse and corresponding transformation of its frequency spectrum. No less important for the correct description of the spatiotemporal dynamics of the pulse is taking into account the delayed response of the Kerr medium, for the description of which a damped oscillator model can be used [19]. The features of the self-action of optical vortices are discussed in the second part of the paper. In particular, the possibility of nonlinear self-focusing into a narrow ring structure without the appearance of plasma is shown.

3.2 Mathematical Model for Numerical Simulations of Optical Vortex Self-action

3.2.1 Nonlinear Wave Equation

The numerical simulation of optical vortex propagation is based on nonlinear wave equation, which could be obtained starting with Maxwell's system of equations.

Considering medium is dielectric ($\rho_f(\mathbf{r}, t) = 0$) and nonmagnetic ($\mathbf{B}(\mathbf{r}, t) = \mu_0 \mathbf{H}(\mathbf{r}, t)$) we get:

$$\Delta\mathbf{E}(\mathbf{r},t) - \nabla(\nabla\cdot\mathbf{E}(\mathbf{r},t)) = \frac{1}{\varepsilon_0 c^2}\frac{\partial \mathbf{j}(\mathbf{r},t)}{\partial t} + \frac{1}{c^2}\frac{\partial^2\mathbf{E}(\mathbf{r},t)}{\partial t^2} + \frac{1}{\varepsilon_0 c^2}\frac{\partial^2\mathbf{P}(\mathbf{r},t)}{\partial t^2}, \tag{3.1}$$

where $\mathbf{E}$ is electric field, $\mathbf{j}$ is current density and $\mathbf{P}$ is polarization, $c = 1/\sqrt{\varepsilon_0\mu_0}$. We examine each of (3.1) members one by one using several further approximations.

Assuming paraxiality, that is transverse wave number is much smaller than longitudinal one, $k_\perp \ll k_z$, and linear polarization of the electric field $\mathbf{E}$, we can use method of slowly varying amplitude $\mathbf{E}$ and current density $\mathbf{j}$:

$$\mathbf{E}(\mathbf{r},t) = \frac{1}{2}\mathbf{e}A(\mathbf{r},t)\exp\{i(\omega_0 t - k_0 z)\} + \text{c.c.} \tag{3.2}$$

$$\mathbf{j}(\mathbf{r},t) = \frac{1}{2}\mathbf{e}J(\mathbf{r},t)\exp\{i(\omega_0 t - k_0 z)\} + \text{c.c.} \tag{3.3}$$

where $A(\mathbf{r},t)$ and $J(\mathbf{r},t)$ are complex slowly varying amplitudes, $\mathbf{e}$ is a unit vector.

Using dipole approximation ($\chi^{(i)}(\mathbf{r},t) = \chi^{(i)}(t)$) we can write ($\otimes$ means convolution) polarization as:

$$\begin{aligned}\mathbf{P}(\mathbf{r},t) &= \mathbf{P}^{(1)}(\mathbf{r},t) + \mathbf{P}^{(2)}(\mathbf{r},t) + \mathbf{P}^{(3)}(\mathbf{r},t) + \cdots = \varepsilon_0\chi^{(1)}(t)\otimes\mathbf{E}(\mathbf{r},t) + \\ &+ \varepsilon_0\chi^{(2)}(t)\otimes\mathbf{E}(\mathbf{r},t)\mathbf{E}(\mathbf{r},t) + \varepsilon_0\chi^{(3)}(t)\otimes\mathbf{E}(\mathbf{r},t)\mathbf{E}(\mathbf{r},t)\mathbf{E}(\mathbf{r},t) + \cdots\end{aligned} \tag{3.4}$$

where $\chi^{(i)}(t)$—permittivity tensor of i-th order. We consider isotropic medium ($\mathbf{P}^{(2m)}(\mathbf{r},t) = 0,\ m \in \mathbb{N}$). Each $(n+2)$-th term of the polarization series (3.4) refers to n-th one as $\sim|I(\mathbf{r},t)|/|I_\text{a}|$, where $I_\text{a} \sim 5\times 10^{16}$ W/cm^2 is atomic intensity. Peak intensity inside the filament is about 5×10^{13} W/cm^2, so the ratio is of the order of $\sim 10^{-3}$. We can neglect nonlinearities of higher orders and write:

$$\mathbf{P}(\mathbf{r},t) = \mathbf{P}^{(1)}(\mathbf{r},t) + \mathbf{P}^{(3)}(\mathbf{r},t). \tag{3.5}$$

Therefore we will limit our consideration only to linear and cubic polarizations. Linear polarization can be written as

$$\begin{aligned}\mathbf{P}^{(1)}(\mathbf{r},t) &= \varepsilon_0\chi^{(1)}(t)\otimes\mathbf{E}(\mathbf{r},t) = \varepsilon_0\int_{-\infty}^{+\infty}\chi^{(1)}(\tau)\mathbf{E}(\mathbf{r},t-\tau)\mathrm{d}\tau = \\ &= \frac{1}{2}\mathbf{e}\exp\{i(\omega_0 t - k_0 z)\}\varepsilon_0\int_{-\infty}^{+\infty}\chi^{(1)}(\tau)A(\mathbf{r},t-\tau)\exp\{-i\omega_0\tau\}\mathrm{d}\tau.\end{aligned} \tag{3.6}$$

Using inverse Fourier transform of slowly varying electric field amplitude

$$A(\mathbf{r},t)=\frac{1}{2\pi}\int\limits_{-\infty}^{+\infty}\tilde{A}(\mathbf{r},\Omega)\exp\{i\Omega t\}\mathrm{d}\Omega, \tag{3.7}$$

forward Fourier transform of electric susceptibility

$$\tilde{\chi}^{(1)}(\omega)=\int\limits_{-\infty}^{+\infty}\chi^{(1)}(t)\exp\{-i\omega t\}\mathrm{d}t \tag{3.8}$$

and relation between susceptibility and wave vector $\tilde{\chi}^{(1)}(\omega)=k^2(\omega)c^2/\omega^2-1$, we obtain

$$\mathbf{P}^{(1)}(\mathbf{r},t)=\frac{1}{2}\mathbf{e}\exp\{i(\omega_0 t-k_0 z)\}\varepsilon_0 c^2\frac{1}{2\pi}\int\limits_{-\infty}^{+\infty}\frac{k^2(\omega_0+\Omega)}{\omega_0^2+2\omega_0\Omega+\Omega^2}\tilde{A}(\mathbf{r},\Omega)\times \\ \times\exp\{i\Omega t\}\mathrm{d}\Omega-\varepsilon_0\mathbf{E}(\mathbf{r},t), \tag{3.9}$$

where $k(\omega)=\omega n(\omega)/c$, $n(\omega)$—index of refraction. The linear part of the last term in (3.1) takes the form:

$$\frac{1}{\varepsilon_0 c^2}\frac{\partial^2\mathbf{P}^{(1)}(\mathbf{r},t)}{\partial t^2}=-\frac{1}{2}\mathbf{e}\exp\{i(\omega_0 t-k_0 z)\}\times \\ \times\frac{1}{2\pi}\int\limits_{-\infty}^{+\infty}k^2(\omega_0+\Omega)\tilde{A}(\mathbf{r},\Omega)\exp\{i\Omega t\}\mathrm{d}\Omega-\frac{1}{c^2}\frac{\partial^2\mathbf{E}(\mathbf{r},t)}{\partial t^2}. \tag{3.10}$$

The expression for cubic polarization consists of two terms related to the first and third harmonics:

$$\mathbf{P}^{(3)}(\mathbf{r},t)=\varepsilon_0\chi^{(3)}(t)\otimes\mathbf{E}(\mathbf{r},t)\mathbf{E}(\mathbf{r},t)\mathbf{E}(\mathbf{r},t)= \\ =\frac{1}{8}\mathbf{e}\exp\{i(3\omega_0 t-3k_0 z)\}\varepsilon_0\iiint\chi^{(3)}(\tau_1,\tau_2,\tau_3)A(\mathbf{r},t-\tau_1)A(\mathbf{r},t-\tau_2)A(\mathbf{r},t-\tau_3)\times \\ \times\exp\{-i\omega_0(\tau_1+\tau_2+\tau_3)\}\mathrm{d}\tau_1\mathrm{d}\tau_2\mathrm{d}\tau_3+ \\ +\frac{3}{8}\mathbf{e}\exp\{i(\omega_0 t-k_0 z)\}\varepsilon_0\iiint\chi^{(3)}(\tau_1,\tau_2,\tau_3)A(\mathbf{r},t-\tau_1)A(\mathbf{r},t-\tau_2)A^*(\mathbf{r},t-\tau_3)\times \\ \times\exp\{-i\omega_0(\tau_1+\tau_2-\tau_3)\}\mathrm{d}\tau_1\mathrm{d}\tau_2\mathrm{d}\tau_3. \tag{3.11}$$

Due to strong violation of wave synchronism, $\Delta k = 3k(\omega_0) - k(3\omega_0) \neq 0$, we neglect the third harmonic generation. Fourier transform of 3-th order electric susceptibility (similarly to (3.8)) and well-known relation between intensity and electric field amplitude $I(\mathbf{r}, t) = cn_0\varepsilon_0|A(\mathbf{r}, t)|^2/2$ yield

$$\mathbf{P}^{(3)}(\mathbf{r}, t) = \frac{3}{4}\mathbf{e}\exp\{i(\omega_0 t - k_0 z)\}\varepsilon_0\frac{\tilde{\chi}^{(3)}}{cn_0\epsilon_0}I(\mathbf{r}, t)A(\mathbf{r}, t). \tag{3.12}$$

The second time derivative of nonlinear polarization in (3.1) takes the form

$$\frac{1}{\varepsilon_0 c^2}\frac{\partial^2\mathbf{P}^{(3)}(\mathbf{r}, t)}{\partial t^2} = -\frac{1}{2}\mathbf{e}\exp\{i(\omega_0 t - k_0 z)\}\frac{2k_0^2}{n_0}\hat{T}^2\Delta n_k(\mathbf{r}, t)A(\mathbf{r}, t), \tag{3.13}$$

where

$$\hat{T} = 1 - \frac{i}{\omega_0}\frac{\partial}{\partial t} \tag{3.14}$$

is the operator of wave-nonstationarity [18], which provides more accurate description of short pulse propagation compared to commonly used approximation of this operator by unit.

In case of instant response the nonlinear (Kerr) addition to the refractive index $\Delta n_k(\mathbf{r}, t) = n_2 I(\mathbf{r}, t)$, where $n_2 = 3\tilde{\chi}^{(3)}/4cn_0^2\epsilon_0$.

Delayed (Raman) response can be taken into account by special convolution with oscillating kernel $H(\tau)$ [19]:

$$\Delta n_k(\mathbf{r}, t) = (1 - g)n_2 I(\mathbf{r}, t) + gn_2\int_0^{+\infty} H(\tau)I(\mathbf{r}, t - \tau)\mathrm{d}\tau, \tag{3.15}$$

where

$$H(t) = \Theta(t)\frac{1 + \Omega_\mathrm{R}^2\tau_k^2}{\Omega_\mathrm{R}\tau_k^2}\sin(\Omega_\mathrm{R} t)\exp\{-t/\tau_k\} \tag{3.16}$$

and g—weighting factor, $H(t)$—convolution kernel, $\Theta(t)$—Heaviside step function, Ω_R—rotating frequency of molecules, τ_k—characteristic response time.

Conduction current density $\mathbf{j}$ depends on plasma appearing in filamentation regime. Using simple Drude's model we can write motion equation for electron gas including elastic electron-ion interactions with collision frequency ν_ei:

$$\frac{\mathrm{d}\mathbf{j}(\mathbf{r}, t)}{\mathrm{d}t} = \frac{e^2}{m_\mathrm{e}}\mathbf{E}(\mathbf{r}, t)N_\mathrm{e}(\mathbf{r}, t) - \nu_\mathrm{ei}\mathbf{j}_\mathrm{e}(\mathbf{r}, t), \tag{3.17}$$

where N_e is free electron concentration; e and m_e—charge and mass of electron. Substituting (3.3) into (3.17), we obtain:

$$J(\mathbf{r},t)\omega_0\left(i+\frac{\nu_{ei}}{\omega_0}+\frac{1}{\omega_0}\frac{\partial}{\partial t}\right)=\frac{e^2}{m_e}N_e(\mathbf{r},t)A(\mathbf{r},t). \tag{3.18}$$

Note that $i+\nu_{ei}/\omega_0+\omega_0^{-1}\partial/\partial t=i\hat{T}+\nu_{ei}/\omega_0$. Factor ν_{ei}/ω_0 is small, because for dielectrics $\nu_{ei}\sim 10^{14}\ \mathrm{s}^{-1}$ and central frequency for near infrared radiation is about $\omega_0\sim 5\times 10^{14}\ \mathrm{s}^{-1}$. After some algebra we obtain expressions for conduction current density:

$$\mathbf{j}(\mathbf{r},t)=\frac{1}{2}\mathbf{e}\frac{e^2}{m_e}\frac{1}{i\omega_0+\frac{\partial}{\partial t}}\left(1+i\frac{\nu_{ei}}{\omega_0}\hat{T}^{-1}\right)N_e(\mathbf{r},t)A(\mathbf{r},t)\exp\{i(\omega_0 t-k_0 z)\} \tag{3.19}$$

and its derivative:

$$\frac{1}{\varepsilon_0 c^2}\frac{\partial \mathbf{j}(\mathbf{r},t)}{\partial t}=\frac{1}{2}\mathbf{e}\exp\{i(\omega_0 t-k_0 z)\}\times \\ \times\left(-\frac{2k_0^2}{n_0}\Delta n_{pl}(\mathbf{r},t)-i\hat{T}^{-1}\sigma(\mathbf{r},t)\right)A(\mathbf{r},t), \tag{3.20}$$

where we denoted $\omega_{pl}^2(\mathbf{r},t)=e^2N_e(\mathbf{r},t)/m_e\varepsilon_0$ as plasma frequency, $\Delta n_{pl}(\mathbf{r},t)=-\omega_{pl}^2(\mathbf{r},t)/2n_0\omega_0^2$ as nonlinear plasma addition to refractive index and $\sigma(\mathbf{r},t)=-\omega_{pl}^2(\mathbf{r},t)/c^2\times\nu_{ei}/\omega_0$ as bremsstrahlung cross-section.

Substituting the expressions (3.10), (3.13) and (3.20) to (3.1), assuming $\nabla\cdot\mathbf{E}(\mathbf{r},t)=0$ and expanding the Laplace operator to transverse and longitudinal parts $\Delta=\Delta_\perp+\partial^2/\partial z^2$ we obtain nonlinear wave equation, which can be written in retarded time $t'=t-k_1 z$ and $z'=z$ coordinates as

$$\frac{\partial^2 A(\mathbf{r},t')}{\partial z'^2}-2ik_0\left(1-\frac{ik_1}{k_0}\frac{\partial}{\partial t'}\right)\frac{\partial A(\mathbf{r},t')}{\partial z'}+\Delta_\perp A(\mathbf{r},t')= \\ =-\frac{2k_0^2}{n_0}\Delta n_{pl}(\mathbf{r},t)A(\mathbf{r},t')-i\hat{T}^{-1}\sigma(\mathbf{r},t)A(\mathbf{r},t')- \\ -\frac{1}{2\pi}\int_{-\infty}^{+\infty}\left(k^2(\omega_0+\Omega)-(k_0+k_1\Omega)^2\right)\tilde{A}(\mathbf{r},\Omega)\exp\{i\Omega t'\}\mathrm{d}\Omega- \\ -\frac{2k_0^2}{n_0}\hat{T}^2\Delta n_k(\mathbf{r},t')A(\mathbf{r},t'), \tag{3.21}$$

where $k_1=\mathrm{d}k/\mathrm{d}\omega|_{\omega=\omega_0}$. We neglect the second derivative of the field A on z' due to slowly varying amplitude approximation, assume the factor k_1/k_0 in brackets approximately equals to $1/\omega_0$ and redesignate t', z' back to t, z.

Equation (3.21) lacks dissipation terms. We can account for the energy loss due to plasma generation by adding the following equation:

$$-\frac{\partial I(\mathbf{r},t)}{\partial z}=\frac{\partial N_e(\mathbf{r},t)}{\partial t}K\hbar\omega_0, \tag{3.22}$$

where $K=\langle U_i/\hbar\omega_0+1\rangle$—number of photons on frequency ω_0 needed to put electron out from the ionization potential U_i.

Passing from intensity I to field amplitude A after some algebra we obtain:

$$2ik_0\frac{\partial A(\mathbf{r},t)}{\partial z}=-ik_0\frac{\frac{\partial N_e(\mathbf{r},t)}{\partial t}K\hbar\omega_0}{I(\mathbf{r},t)}A(\mathbf{r},t). \tag{3.23}$$

On the right-hand side of the (3.23) we can distinguish the nonlinear absorption coefficient as $\alpha(\mathbf{r},t)=(\partial N_e(\mathbf{r},t)/\partial t)K\hbar\omega_0/I(\mathbf{r},t)$. Adding linear extinction coefficient δ we finally obtain the nonlinear wave equation for simulation of optical vortex self-action:

$$\begin{aligned}2ik_0\frac{\partial A(\mathbf{r},t)}{\partial z}&=\hat{T}^{-1}\Delta_\perp A(\mathbf{r},t)+\\&+\hat{T}^{-1}\frac{1}{2\pi}\int_{-\infty}^{+\infty}\left(k^2(\omega_0+\Omega)-(k_0+k_1\Omega)^2\right)\tilde{A}(\mathbf{r},\Omega)\exp\{i\Omega t\}\mathrm{d}\Omega+\\&+\frac{2k_0^2}{n_0}\hat{T}\Delta n_k A(\mathbf{r},t)+\frac{2k_0^2}{n_0}\hat{T}^{-1}\Delta n_{\mathrm{pl}}A(\mathbf{r},t)+i\hat{T}^{-2}\sigma A(\mathbf{r},t)-\\&-ik_0(\alpha+\delta)A(\mathbf{r},t).\end{aligned} \tag{3.24}$$

Obtained wave equation contains several operators of wave nonstationarity $\hat{T}$ as in [18], but the important feature of this equation is the presence of terms responsible for medium ionization, depending on concentration of generated electrons. To calculate electron concentration N_e we need cinetic equation for plasma electrons. The increment of electron concentration is proportional to difference between neutrals concentration N_0 and current electron concentration N_e with ionization rate $W(I)$:

$$\frac{\partial N_e(\mathbf{r},t)}{\partial t}=W(I)(N_0-N_e(\mathbf{r},t)). \tag{3.25}$$

Electrons generation may be enhanced by the avalanche with nonlinear coefficient ν_i [20]:

$$\nu_i(I)=\frac{1}{U_i}\frac{e^2\nu_{ei}}{m_e(\omega_0^2+\nu_{ei}^2)}\frac{I}{cn_0\varepsilon_0}. \tag{3.26}$$

Recombination of plasma electrons is taken into account as a negative term which is proportional to current electrons concentration with some constant coefficient β.

Finally we obtain the cinetic equation for plasma electrons:

$$\frac{\partial N_e(\mathbf{r},t)}{\partial t}=W(I)(N_0-N_e(\mathbf{r},t))+\nu_i(I)N_e(\mathbf{r},t)-\beta N_e(\mathbf{r},t). \tag{3.27}$$

It should be noticed that ionization rate $W(I)$ is calculated according to the Keldysh model [21].

3.2.2 Problem Statement and Initial Conditions

Numerical simulations of optical vortex self-action are connected with experimental setup shown in Fig. 3.1.

Femtosecond laser system generates Gaussian beam, which travels through special vortex lens, providing annular beam with topological charge m at the focal plane, where the input face of the silica glass sample is at $z = 0$. Femtosecond pulse with vortex beam propagates further in the bulk fused silica experiencing self-action.

General case of the problem is computationally complex. If we consider pulse propagation before it breaks up due to azimuthal instability, we can save computing resources using axial symmetry of the beam and introducing one spatial coordinate $r = \sqrt{x^2 + y^2}$ instead of two cartesian coordinates x and y. It should be noticed that described approach forbids initialization of asymmetric vortex phase. But we can make a substitution $A(\mathbf{r}, z, t) = A'(r, z, t)\exp\{im\varphi\}$ in (3.24) and obtain a nonlinear wave equation for $A'(r, z, t)$. The only difference from (3.24) will be the new expression for the transverse laplacian:

$$\Delta_\perp = \frac{\partial^2}{\partial r^2} + \frac{1}{r}\frac{\partial}{\partial r} - \frac{m^2}{r^2}. \tag{3.28}$$

We suppose that at the focal plane of vortex lens there is an annular beam with topological charge m in a bandwidth-limited Gaussian pulse:

$$A_V(r, t) = A_0\left(\frac{r}{r_0}\right)^2 \exp\left\{-\frac{r^2}{2r_0^2}\right\}\exp\left\{-\frac{t^2}{2t_0^2}\right\}\exp\left\{im\varphi\right\}, \tag{3.29}$$

where $r_0 = 120\,\mu\text{m}$, $t_0 = 60$ fs, $\varphi = \arctan(y/x)$. Intensity and phase profiles of this vortex beam are shown in Fig. 3.2.

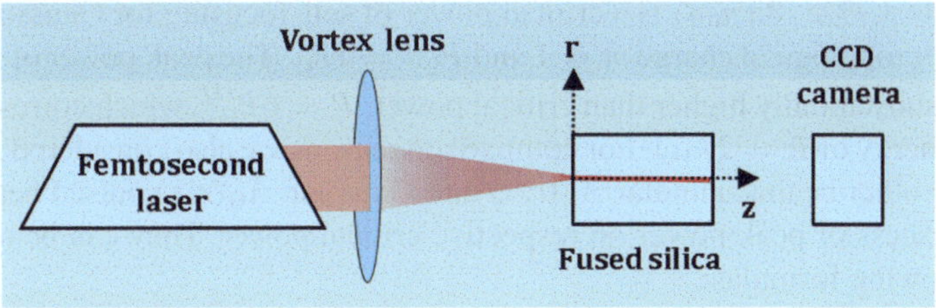

Fig. 3.1 Schematic experimental setup for numerical simulations of optical vortex self-action

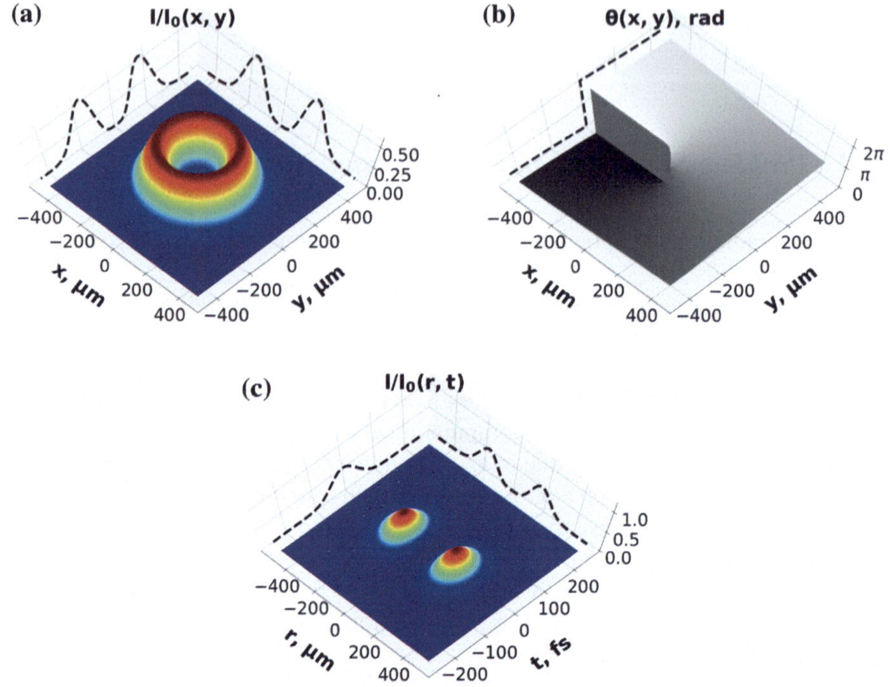

Fig. 3.2 Intensity (**a**) and phase (**b**) spatial distribution and spatiotemporal intensity distribution (**c**) for optical vortex with topological charge $m = 1$ at $z = 0$. Line profiles are shown by dashed lines

Self-action of optical vortex in fused silica is studied in the region of anomalous group velocity dispersion at the wavelength of 1900 nm, where $k_2 = -80$ fs^2/mm. The critical power of self-focusing for vortex beam with topological charge m can be calculated by formula [9]:

$$P_{\mathrm{V}}^{(m)} = \frac{2^{2m+1}\Gamma(m+1)\Gamma(m+2)}{2\Gamma(2m+1)} P_{\mathrm{G}}, \tag{3.30}$$

where $P_{\mathrm{G}} = 3.77\lambda^2/8\pi n_0 n_2$ is a critical power of self-focusing for Gaussian beam. We consider topological charge $m = 1$ and $P_{\mathrm{V}}^{(1)} = 4P_{\mathrm{G}}$. The peak power of the pulse was taken substantially higher than critical power, $P = 6P_{\mathrm{V}}^{(1)}$, which corresponds to the pulse energy of $E = 27$ μJ. For comparison reasons we also considered propagation of two other beams: annular $A_{\mathrm{R}}(r, t)$ and Gaussian $A_{\mathrm{G}}(r, t)$ pulsed beams with the same excess of peak power on respective critical power. They can be described by the following formulas:

$$A_{\mathrm{R}}(r, t) = A_0 \left(\frac{r}{r_0}\right)^2 \exp\left\{-\frac{r^2}{2r_0^2}\right\} \exp\left\{-\frac{t^2}{2t_0^2}\right\}, \tag{3.31}$$

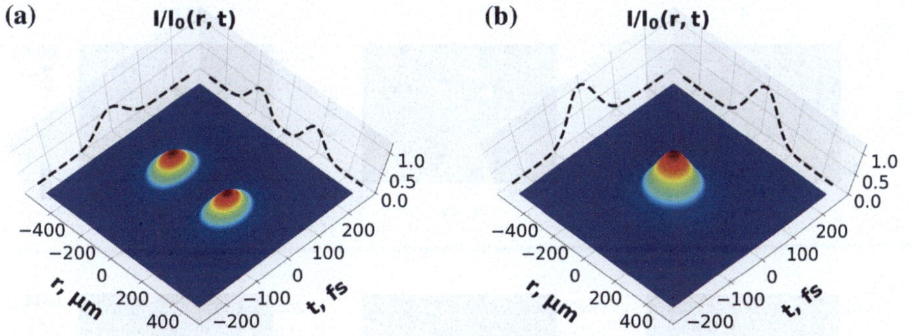

Fig. 3.3 Spatiotemporal profiles for annular (**a**) and Gaussian (**b**) beams at $z = 0$

$$A_{\mathrm{G}}(r, t) = A_0 \exp\left\{-\frac{r^2}{2r_0^2}\right\} \exp\left\{-\frac{t^2}{2t_0^2}\right\}. \quad (3.32)$$

Note that $A_{\mathrm{R}} = A_{\mathrm{G}}(r/r_0)^2$ and $A_{\mathrm{V}} = A_{\mathrm{R}} \exp\{im\varphi\}$. Spatiotemporal intensity distributions of annular and Gaussian beams are shown in Fig. 3.3.

3.3 Spatiotemporal Dynamics and Spectral Broadening of Optical Vortex in Fused Silica at 1900 nm

3.3.1 Evolution of Intensity Distribution in Vortex, Annular and Gaussian Beams

Figure 3.4 shows intensity distributions for different beams on the logarithmic scale with initial intensity $I_0 = 2.81 \times 10^{11}$ W/cm^2 at distances corresponding to local maxima along z axis.

In the initial stage of vortex beam propagation, the self-action yields narrowing ring of practically the same radius. This is clearly seen in Fig. 3.4a at $z = 1.41$ cm, when the first local maximum of intensity 1.5×10^{13} W/cm^2 is reached, its localization in time domain being shifted to the pulse tail. Two mechanisms are responsible for the pulse delay. Both of them are connected with Kerr effect. These are delayed nonlinear response and operator of the wave-nonstationarity. The further propagation of the pulse is accompanied by a decrease in the radius of the ring structure containing the focusing part of the pulse energy. The next intensity maximum of almost the same value is reached at $z = 4.68$ cm (Fig. 3.5). The global maximum is reached in a few millimeters. It exceeds 5×10^{13} W/cm^2 and the radius of the ring structure is reduced to approximately 10 μm. On the one hand, phase dislocation prevents energy localization on the optical axis. On the other hand, the achieved intensity values are sufficient for plasma generation, which defocuses pulse tail. Together these factors lead to the cessation of the emergence of high-intensity ring structures in the

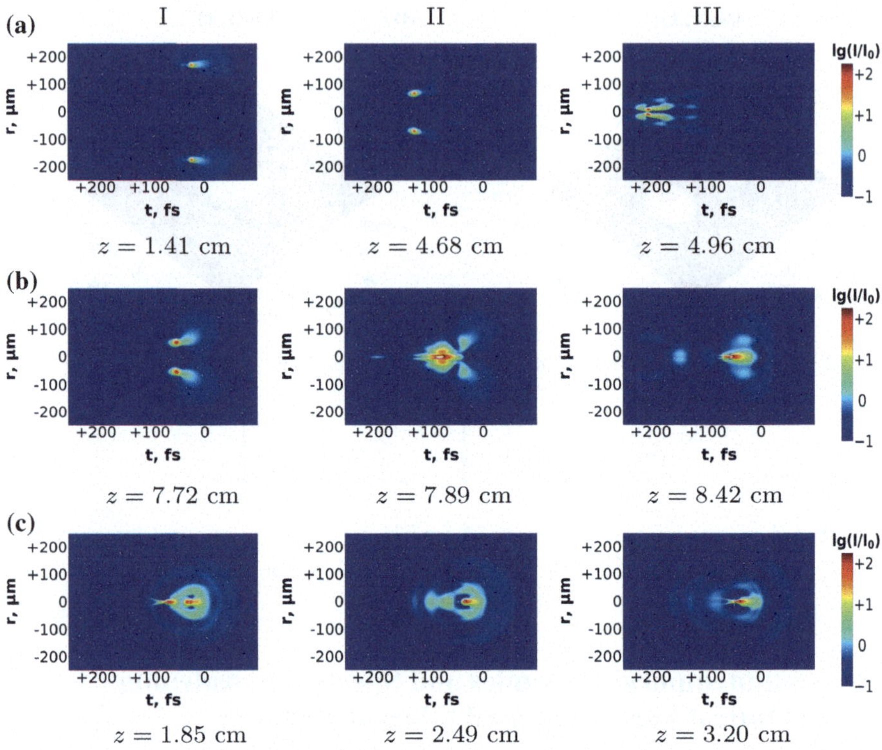

Fig. 3.4 Spatiotemporal intensity distributions for vortex (**a**), annular (**b**) and Gaussian (**c**) beams

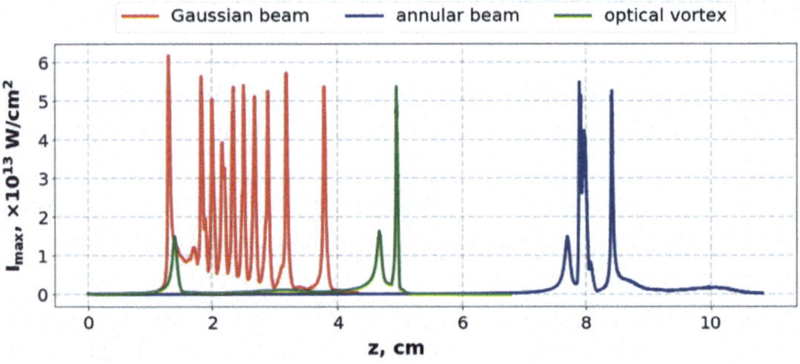

Fig. 3.5 Maximum intensity dependence on coordinate z along propagation distance for vortex, annular and Gaussian beams

beam profile and widening of the pulse (Fig. 3.4a III). Note that a noticeable plasma concentration (Fig. 3.6) is achieved only in the last narrowest ring with maximum intensity. Defocusing of the two previous high-intensity rings occurs without plasma generation.

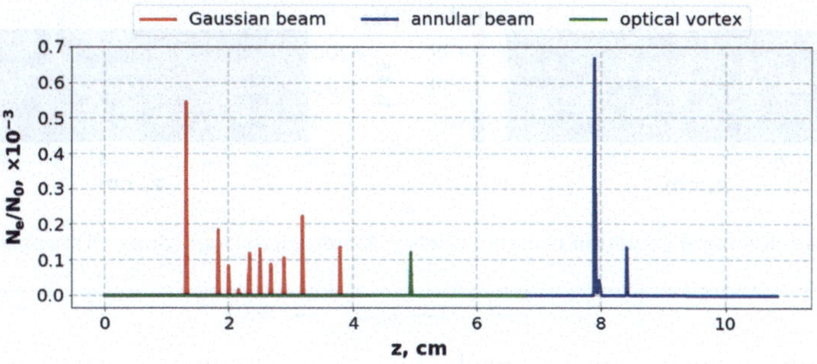

Fig. 3.6 Peak plasma concentration dependence on coordinate z along propagation distance for vortex, annular and Gaussian beams

The spatiotemporal dynamics of annular beam in the initial stage of self-action is qualitatively similar. Again we can see a narrowing ring in the cross section due to self-focusing (Fig. 3.4b I). However, because the critical power for the annular beam is significantly less than for the optical vortex (it is about 1.2 times higher than the critical power for the Gaussian beam), the absolute value of its peak power is also less. This leads to the fact that the maximum intensity in the narrowing ring is achieved at the significantly greater distance $z = 7.72$ cm. The absence of phase singularity ultimately leads to the appearance of the maximum intensity on the beam axis (Fig. 3.4b II). After that, the self-focusing of the annular beam becomes similar to the self-focusing of the Gaussian beam (Fig. 3.4c).

Intensity increases up to 5×10^{13} W/cm^2, plasma with concentration $0.6 \times 10^{-3} N_0$ appears. Peak plasma concentration is 6 times higher than in the vortex beam, intensity maximum being on the level of optical vortex, but it remains on longer time interval. Parts of this high-intensive structure self-focus, deviding it to two light bullets, which start to compete for pulse energy. Light bullets' life cycle is connected with movement towards pulse tail, so the last bullet finally extinguishes the first one, after which we can see approximately typical for gauss pulse evolution. Specified bullet dies at the tail of the pulse, background energy forms the new one at the front and the process repeats until there is enough power (Fig. 3.4b III).

Gaussian beam propagation starts with self-focusing at the optical axis at time slices slightly shifted to pulse tail due to delayed response of Kerr nonlinearity and influence of operator $\hat{T}$ (Fig. 3.4c I). The intensity reaches values 6×10^{13} W/cm^2, plasma electrons appear at the trail of the pulse (Figs. 3.5 and 3.6). Ring structure in the pulse cross-section is formed due to interference of radiation moving towards and backwards the optical axis. Nonlinear focus drifts backwards on time coordinate while there is enough power and then disappears (Fig. 3.4c II). Such relatively stable in space and time domain structures are usually cited as a "light bullets". Background energy gives the birth to new light bullet on the pulse front and the process repeats several times (Figs. 3.4c III and 3.7a). Anomalous group velocity dispersion keeps these spatiotemporal bullets from splitting into sub-pulses. The number of light bul-

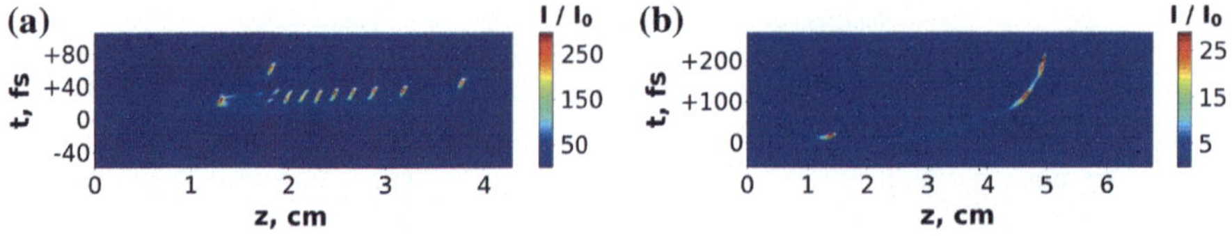

Fig. 3.7 Spatiotemporal evolution of light bullets in Gaussian (**a**) and vortex (**b**) beams

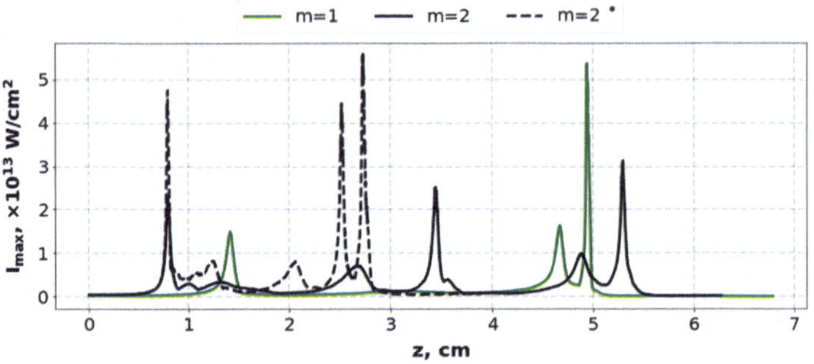

Fig. 3.8 Maximum intensity dependence on coordinate z along propagation distance for vortex with topological charge $m = 1$ (green line), $m = 2$ (black line) and $m = 2$ in the model without operator $\hat{T}$ in instant Kerr effect (dashed line)

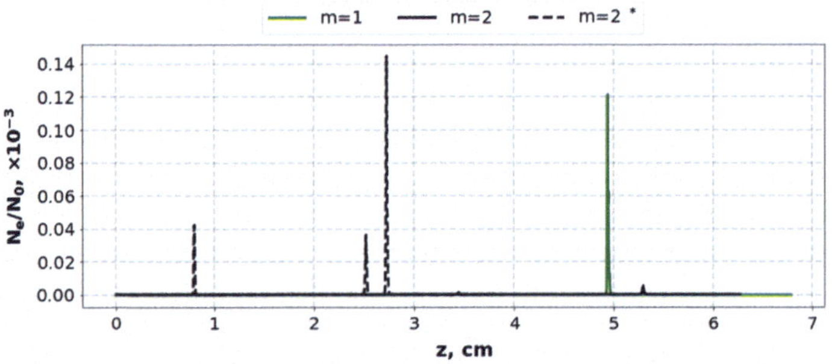

Fig. 3.9 Peak plasma concentration dependence on coordinate z along propagation distance for vortex with topological charge $m = 1$ (green line), $m = 2$ (black line) and with $m = 2$ in the model without operator $\hat{T}$ in instant Kerr effect (dashed line)

lets in the vortex beam is significantly less than in Gaussian one. They have an annular structure and quickly shift to the tail of the pulse (Fig. 3.7b).

The propagation of an optical vortex with a higher topological charge ($m = 2$) has the peculiarity that the maximum intensity along the propagation direction z is approximately twice less than that of a vortex with a charge $m = 1$. This is clearly seen in Fig. 3.8, which shows the maximum intensity for beams with different topological charges. This is enough for the maximum plasma concentration in a beam with $m = 2$ to be 30 times less than in a beam with $m = 1$ (Fig. 3.9). This case can be considered as an ionization free mode of self-focusing of the optical vortex. The peak intensity and concentration of the plasma can be "returned" to the "ionization values", if we cross out the operator of the wave-nonstationarity $\hat{T}$ from the mathematical model in the term describing Kerr nonlinearity (Figs. 3.8 and 3.9, dashed curve). This demonstrates the importance of this operator for the correct description of the self-action of a femtosecond optical vortex.

3.3.2 Fluence of Optical Vortex and Gaussian Beams

The spatial distribution of light energy is characterized by fluence

$$F(r, z) = \int_{-\infty}^{+\infty} I(r, z, t)\mathrm{d}t \tag{3.33}$$

Linear divergence of optical vortex beam is the same as Gaussian beam [22]. The diffraction length in fused silica glass for both beams is 6.8 cm. At smaller distances, the divergence of the optical vortex is weak (Fig. 3.10).

Fluence distributions for vortex and Gaussian beams in nonlinear medium are shown in Fig. 3.11.

Initial self-focusing of optical vortex taking place on $z = 1.4$ cm is characterized by ring narrowing. Its radius doesn't practically change, fluence inside the ring being increased. We can see the first nonlinear focus as a hot point in Fig. 3.11a at $z = 1.4$ cm. The intensity in this non-linear focus does not exceed $I = 1.5 \times 10^{13}$ W/cm^2, despite the fact that the plasma defocusing lens in this place is not formed. After passing the first nonlinear focus, most of the pulse energy flows to the beam axis. However, the pace of this flowing is different. Thus, two rings are formed in the cross section. The inner ring reaches the minimum radius in the vicinity of $z = 2$ cm.

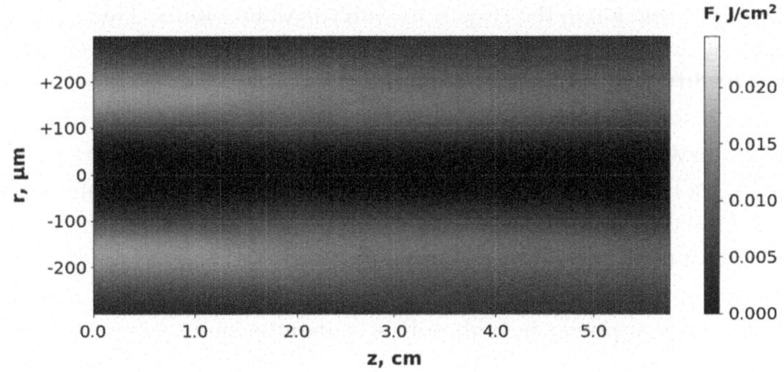

Fig. 3.10 Fluence distribution for optical vortex in case of linear propagation

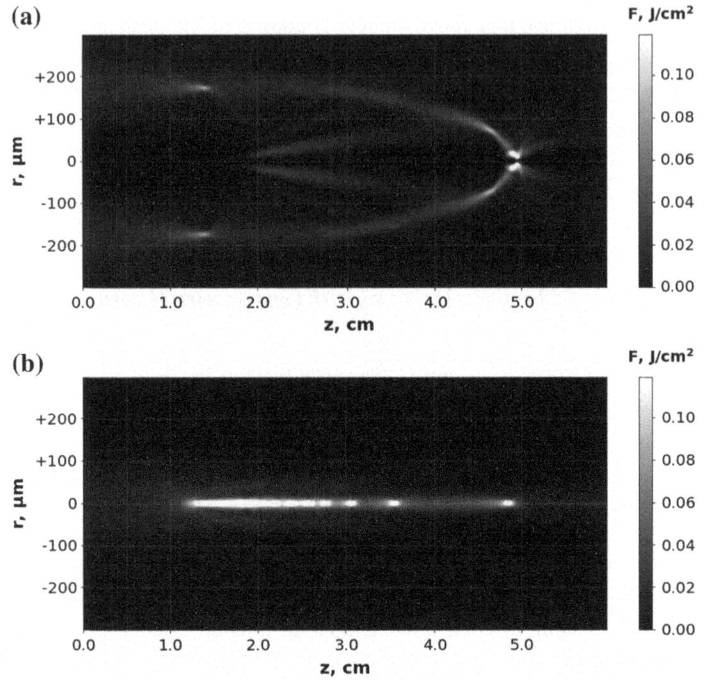

Fig. 3.11 Fluence distribution for optical vortex (**a**) and Gaussian beam (**b**)

However, the energy in it is not enough to form pronounced hot spots. Then the radius of this inner ring begins to grow, and fluence inside it decreases. The outer fluence ring at the same distances continues to shrink at a gradually increasing rate. Ultimately, this ring is narrowed to the smallest radius up to 10 μm, providing the nonlinear focus with the highest fluence $F = 0.31$ J/cm^2 at approximately $z = 5$ cm. The peak intensity in this location sufficient for plasma generation, which defocuses the tail of the pulse. Note that at a short distance before this point there is another local maximum of fluence in the ring of the intermediate radius. Thus, in the process of self-action of the optical vortex, a multi-focus structure is formed consisting of several rings of different radius arising at different distances z.

This multi-focus structure is significantly different from the self-focusing of a Gaussian beam in which a collapse occurs on the beam axis (Fig. 3.11b). In a Gaussian beam, moving foci and refocusing form a sequence of almost continuous hot spots with fluence $F = 0.48$ J/cm^2, in each of them a noticeable plasma concentration is reached. Note that the distance between the first and the last hot spots in both beams turns out to be the same. Recall that despite the significant energy difference, the excess over the critical power in both pulses is also the same.

3.3.3 Evolution of Frequency Spectrum and Energy Transformation in Optical Vortex

Spectral dynamics as a function of the propagation distance z for vortex beam is shown in Fig. 3.12 in logarithmic scale. It is clearly seen as due to self-phase modulation the spectrum begins to gradually expand mainly into the Stokes region. Sharp broadening of the spectrum occurs when the beam reaches a nonlinear focus at $z = 1.4$ cm. At this point, an annular light bullet is formed, which implies strong pulse self-steepening, and a noticeable broadening of the spectrum in the anti-Stokes region is observed. With further propagation, this process is repeated when other nonlinear foci are reached in the vicinity $z = 5$ cm.

To describe pulse spectrum dynamics quantitatively, we devided spectrum with full energy E_0 to three bands: central ($\lambda = 1900 \pm 145$ nm with energy E_c), anti-Stokes ($\lambda < 1755$ nm with energy E_a) and Stokes ($\lambda > 2045$ nm with energy E_s). Thus, the width of the central band is 5 times greater than the width of the input pulse spectrum ($1/e$ level). So, initially pulse energy E_0 is located entirely in the central band (Fig. 3.13).

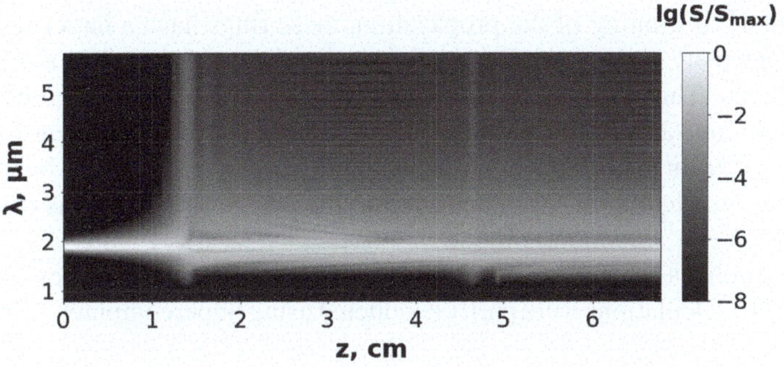

Fig. 3.12 Evolution of frequency spectrum for optical vortex in logarithmic scale

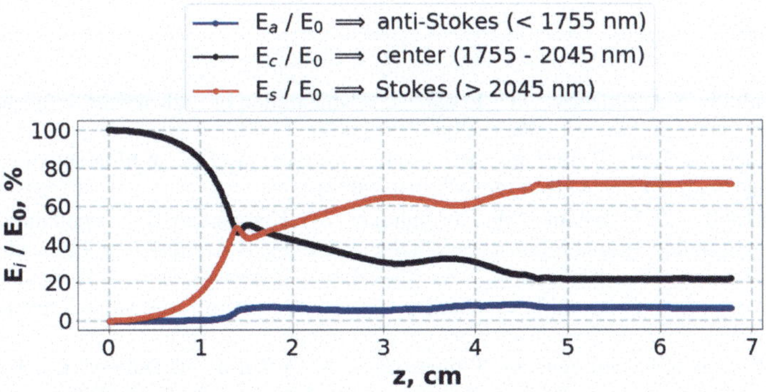

Fig. 3.13 Energy transformation from central spectral band (1745 – 2045 nm) towards anti-Stokes (<1745 nm) and Stokes (>2045 nm) regions

First centimeters of propagation distance are characterized by energy transformation mostly from central to Stokes region. After the first nonlinear focus at $z = 1.4$ cm most of the pulse energy is in the Stokes region, 5–10% of the total energy being transformed into the anti-Stokes region. In the end more than 70% of the energy is converted into Stokes region and less than 10% into anti-Stokes region.

3.4 Conclusions

In this paper, we considered the self-action in fused silica glass at a wavelength of 1900 nm of a femtosecond optical vortex—annular beam with a phase singularity under conditions of preserving the axial symmetry of its intensity profile. In computer simulations, we used the model including slowly varying wave approximation for propagation equation with the operator of wave-nonstationarity $\hat{T}$ and delayed response of Kerr nonlinearity. This turns out to be critical for correct describing the multi-focus structure of an optical vortex in a nonlinear medium.

It is shown that for a six-fold excess of the pulse peak power over the critical one, rings with high fluence are formed sequentially in nonlinear foci along the optical axis. At the beginning of the propagation, these rings have a larger radius. The maximum intensity in these nonlinear foci is several times lower than in filament of Gaussian beam. This is not sufficient to produce appreciable plasma concentration that results in ionisation free propagation of the pulse through nonlinear focus. The frequency spectrum of the pulse is broadening mainly into the Stokes band. A relatively strong broadening of the spectrum into the anti-Stokes region is observed in nonlinear foci.

This research was supported by the Russian Foundation for Basic Research, grant 18-02-00624. Calculations were partly conducted using Supercomputer "Lomonosov" in MSU.

References

1. V.P. Kandidov, S.A. Shlenov, O.G. Kosareva, Filamentation of high-power femtosecond laser radiation. Quantum Electron. **39**(3), 205–228 (2009)
2. A. Couairon, A. Lotti, Panagiotopoulos et al., Ultrashort laser pulse filamentation with Airy and Bessel beams. Proc. SPIE **8770**, 87701E (2013)
3. S.V. Chekalin, A.E. Dokukina, A.E. Dormidonov, V.O. Kompanets, E.O. Smetanina, V.P. Kandidov, Light bullets from a femtosecond filament. J. Phys. B At. Mol. Opt. Phys. **48**(9) (2015)
4. E.O. Smetanina, V.O. Kompanets, S.V. Chekalin, A.E. Dormidonov, V.P. Kandidov, Anti-Stokes wing of femtosecond laser filament supercontinuum in fused silica. Opt. Lett. **38**(1), 16–18 (2013)
5. I. Grazuleviciute, R. Suminas, G. Tamosauskas, A. Couairon, A. Dubetis, Carrier-envelope phase-stable spatiotemporal light bullets. Opt. Lett. **40**(16), 3719–3722 (2015)
6. I.H. Malitson, Interspecimen comparison of the refractive index of fused silica. J. Opt. Soc. Am. **55**(10), 1205–1209 (1965)

7. P. Sprangle, E. Esarev, J. Krall, Laser driven electron acceleration in vacuum, gases, and plasmas. Phys. Plasma **3**(5), 2183 (1996)
8. V.I. Kruglov, Yu.A. Logvin, V.M. Volkov, The theory of spiral laser beams in nonlinear media. J. Modern Opt. **39**(11), 2277–2291 (1992)
9. L.T. Vuong, T.D. Grow, A. Ishaaya, A.L. Gaeta, G.W.t Hooft, E.R. Eliel, G. Fibich, Collapse of optical vortices. Phys. Rev. Lett. **96**(13), 133901 (2006)
10. M.S. Bigelow, P. Zerom, R.W. Boyd, Breakup of ring beams carrying orbital angular momentum in sodium vapor. Phys. Rev. Lett. **92**(8), 083902–4 (2004)
11. A. Vincotte, L. Berge, Femtosecond optical vortices in air. Phys. Rev. Lett. **95**, 193901 (2005)
12. M. Fisher, C. Siders, E. Johnson, O. Andrusyak, C. Brown, M. Richardson, Control of filamentation for enhancing remote detection with laser induced breakdown spectroscopy. Proc. SPIE **6219**, 621907 (2006)
13. P. Hansinger, A. Dreischuh, G.G. Paulus, Vortices in ultrashort laser pulses. Appl. Phys. B **104**, 561 (2011)
14. E.V. Vasil'ev, S.A. Shlenov, Filamentation of an annular laser beam with a vortex phase dislocation in fused silica. Quantum Electron. **46**(11), 1002 (2016)
15. D.N. Neshev, A. Dreischuh, G. Maleshkov, M. Samoc, YuS Kivshar, Supercontinuum generation with optical vortices. Opt. Express **18**(17), 18368–18373 (2010)
16. P. Hansinger, G. Maleshkov, L. Garanovich, D.V. Skryabin, D.N. Neshev, White light generated by femtosecond optical vortex beams. JOSA B **33**(4), 681–690 (2016)
17. O. Khasanov, T. Smirnova, O. Fedotova, G. Rusetsky, O. Romanov, High-intensive femtosecond singular pulses in Kerr dielectrics. Appl. Opt. **51**(10), 198–207 (2012)
18. T. Brabec, F. Krausz, Nonlinear optical pulse propagation in the single-cycle regime. Phys. Rev. Lett. **78**, 3282 (1997)
19. K.J. Blow, D. Wood, Theoretical description of transient stimulated Raman scattering in optical fibers. IEEE J. Quantum Electron **25**(12), 2665–2673 (1989)
20. Yu.P. Raizer, *Gas Discharge Physics* (Springer-Verlag, Berlin, Heidelberg, 1991), p. 526
21. L.V. Keldysh, Ionization in the field of a strong electromagnetic wave. JETP **20**(5), 1307–1314 (1965)
22. S.G. Reddy, C. Permangatt, S. Prabhakar, A. Anwar, J. Banerji, R.P. Singh, Divergence of optical vortex beams. Appl. Opt. **54**, 6690 (2015)

Chapter 4
Picosecond Pulsed High-Peak-Power Lasers

Nikita G. Mikheev, Vyacheslav B. Morozov, Andrei N. Olenin, Vladimir G. Tunkin and Dmitrii V. Yakovlev

Abstract Recent trends in the use and development of advanced schemes of high peak-power picosecond lasers are reviewed. Pulsed (pulsed repetitive) high-peak-power picosecond lasers of millijoule and multi-millijoule single pulse level operating at reasonably high repetition rates are required in a number of scientific and technological applications. The developed approach utilizes active-passive mode-locked and negative feedback controlled oscillator that provides generation of stable, closed to transform limited pulses with pulse duration of 25 ps (with Nd:YAG) and 16 ps (with Nd:YLF). Oscillator—regenerative amplifier scheme based on the common diode-end-pumped laser crystal generates pulses up to 1.2 mJ with Nd:YAG and up to 2 mJ with Nd:YLF crystals. Two-pass Nd:YAG diode-end-pumped amplifier provides output radiation of 4 mJ single pulse energy at 300 Hz repetition rate, that was converted in the second harmonic with more than 60% efficiency. Numerical modeling allows adequate description of the pulse formation process. Using 300 μm thickness Fabry-Perot etalons with different reflection coatings inside oscillator provided generation of pulses with increased up to 120, 180 and 400 ps durations. Aberrative character of thermal lens and mode structure at end-pump geometry were analyzed using decomposition on embedded beams. It was supposed that res-

N. G. Mikheev · V. G. Tunkin
Faculty of Physics, M.V. Lomonosov Moscow State University, 119991 Moscow, Russia
e-mail: ng.mikheev@physics.msu.ru

V. G. Tunkin
e-mail: vladimir_tunkin@mail.ru

V. B. Morozov (✉) · A. N. Olenin · D. V. Yakovlev
Faculty of Physics, International Laser Center,
M.V. Lomonosov Moscow State University, 119991 Moscow, Russia
e-mail: morozov@phys.msu.ru

A. N. Olenin
e-mail: andrei_olenin@phys.msu.ru

D. V. Yakovlev
e-mail: dmyak@physics.msu.ru

K. Yamanouchi et al. (eds.), *Progress in Photon Science*, Springer Series in Chemical Physics 119, https://doi.org/10.1007/978-3-030-05974-3_4

onator stability range might be enhanced owing to adaptive action of the aberration lens. Optimized pulse diode-end-pumped double-pass amplifier schemes utilizing Nd:YLF, Nd:YAG and Nd:YVO_4 crystals are discussed.

4.1 Introduction

Powerful picosecond lasers are in growing use recent years in a wide range of research and practical applications. High-peak-power lasers were effectively used for studying fast dynamics of physico-chemical processes [1, 2], for nonlinear high-resolution time-domain spectroscopy [3, 4]. Modern high repetition rate and high average power picosecond lasers provide new level of micromachining technologies, based on high ablation rates, excellent accuracy and minimal thermal impact on the material at such processes as micro-drilling and surface structuring in precision manufacturing and engineering [5–8], microelectronics [9], semiconductor [10] and photovoltaic [11, 12] industries.

Picosecond lasers of the highest average power usually utilize standard MOPA (master oscillator and power amplifier) schemes starting from a cw oscillator, generating low energy pulses (of sub-μJ/nJ/pJ level) with high repetition frequency, typically 50–100 MHz, based on Nd-doped bulk crystals [13–17], Yb-doped fibers [18–20] or a gain-switched diode laser [21, 22]. Then the power can be boosted up to multi-W and kW level applying advanced diode pumped amplifier stages based on bulk crystal [5, 14, 23–25], fiber [21, 22, 24, 26–28], Innoslab [29, 30], thin disk [32–35] modules. Using a pulse picker scheme based on a Pockels cell and a polarizer switching selected pulses one can reduce repetition rate of the initial sequence down to desired frequency while maintaining at further amplification high average power, and accordingly, substantially increasing single pulse energy. This can be important e.g. for the efficiency of an ablation processes [5, 8]. For power amplification at several orders of magnitude, amplifier stages can utilize regenerative or multipass amplifier schemes. At high enough output repetition frequencies, the average laser power may be independent on the frequency.

Due to the limitations associated with the manifestation of nonlinear self-modulation effects and damage threshold in rather thin cores of optical fibers, high average power fiber amplifiers are usually seeded with radiation on the initial high repetition rates and the single pulse energy is within a few microjoules [26–28]. Induced thermal lens and birefringence due to thermomechanical stresses at high pump average power limit average output power of continuously pumped laser rod amplifiers while single pulse energies are usually well below the mJ level [5, 14]. Innoslab picosecond amplifiers represent advanced scheme for effective multi-pass amplification that can provide multi-Watt average output power combined with reasonably high single pulse energy. The latter, however, is limited due to narrow aperture of laser crystals and accordingly output beam diameter. The highest single pulse energy value, equal to only 0.8 mJ, was obtained using Nd:YVO_4 Innoslab [31]. Brightest sub-picosecond high thin-disc Yb:YAG amplifiers are capable of producing output of multi-Watt average power and multi-millijoule single pulse energy [34, 35]. It should be noted that due to high saturation fluence near 5 J/cm^2, operation conditions of disk Yb:YAG ps/fs amplifiers are far enough from the saturation regime. Thus, thin disk application is reasonable at either multi-kHz repetition rates or continuous wave regimes.

Some scientific and technological applications such as time-resolved laser spectroscopy [36], precise satellite laser ranging [37, 38], lunar laser ranging [39, 40], some special ablation issues [41], electron sources in photo-injectors [42, 43] and pulsed microwave sources [44], require picosecond lasers with single pulse energy of mJ and multi-mJ level operating at reasonably low repetition rates, typically within 1 kHz or somewhat higher. Despite the high peak power close to GW, average power amounts only to a few watts. Then, the schemes based on obtaining of millijoule level pulses using a thin disk amplifier or directly from thin disk oscillator [33] seem to be unreasonably complex and excessive. While diode pumped bulk crystal amplifier schemes remain acceptable [45]. Usual approach involves using seeding laser of low energy and subsequent effective amplification.

Seeding picoseconds radiation can be produced taking into account specific application conditions and requirements for wavelength, repetition rate, energy, pulse duration, spectral width etc. One can choose the most appropriate approach from the following:

(i) Mode-locked diode pumped oscillator on a Nd:YAG, Nd:YVO_4 or Nd:YLF crystal [13–17, 45].
(ii) Yb-doped mode-locked fiber laser [46] operating on the proper wavelength for further gaining in Nd:YAG, Nd:YVO_4 or Nd:YLF amplifiers.
(iii) Diode-pumped microchip laser [45, 47] operating in sub-ns or in sub-100-ps pulse regime.
(iv) Gain-switched diode laser [21, 22].
(v) Mode-locked semiconductor laser [48, 49].

Preparation of the seeding picosecond pulses train with the required repetition rate can be realized using two different operation regimes: continuous wave or pulsed (pulsed repetitive) (see Fig. 4.1).

In the continuous wave approach, seeding pulses are split off the self-reproducible pulse sequence of the mode-locked cw master oscillator by means of a high contrast pulse picker [5, 9, 14]. It should be noted that the establishment of stable cw operation regime may require a certain time after starting lasing, typically within millisecond time scale [45]. In this case, external triggering may be supported only with the optical pulse jitter value respective to arbitrary synchro-pulse comparable to the round trip time whereas the latter one should be long enough, at least of several nanoseconds, what is necessary for reliable pulse picker operation usually governed by high voltage pulses.

In the pulsed repetitive approach, unlike the cw one, each laser pulse starts from the noise level without any phase connection to the previous pulse. At pulsed pump, optical pulse formation is assumed to occur at the final stage of the pumping process

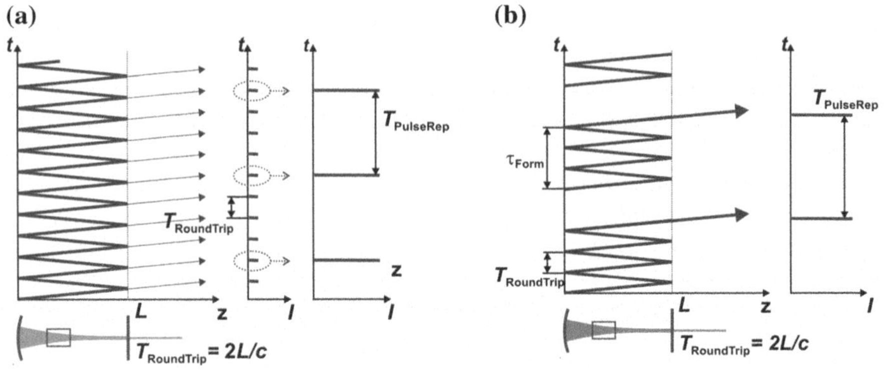

Fig. 4.1 Continuous wave (**a**) and pulsed (pulsed repetitive) (**b**) operation regimes

and can be quite short and unstable. Negative feedback schemes allow radical increase of output pulse-to-pulse stability [50].

There are a lot of approaches using active, passive mode-locking and combined schemes of cw and pulsed repetitive oscillators on Nd-doped crystals. Active mode-locking may be realized by means of acousto-optical [45, 51] or electro-optical [45, 52] modulators. Passive mode-locking in bulk laser crystal oscillator schemes now is mostly done by means of semiconductor saturable adsorber mirrors (SESAM) [53]. Active mode-locking maintains higher peak power inside oscillator while passive mode-locking usually provides simpler oscillator scheme and pulse shortening close to transform limited value [45, 54]. Standard schemes of passive and/or active mode-locking in pulsed repetitive lasers with bulk active crystals and long cavities allows generating near transform limited pulses with pulse width from several picoseconds to several tens picoseconds. Hybrid active-passive schemes can be also effectively used [55, 56]. Active mode-locking principally allows controlling, to some extent, generated picosecond pulse time position respective to the phase of periodic mode-locking voltage. The latter, in the pulsed repetitive regime, can be synthesized during each pulse formation period. Then the optical jitter respective to an external triggering signal can be principally reduced. In the combined active-passive mode-locked and negative feedback controlled scheme, optical jitter with respect to the external synchro-pulse was as low as 40 ps [57]. It should be also noted that this scheme allows stabilizing optical pulse formation process within ~5 μs after the generation uprising.

Diode-end-pumped passively Q-switched short cavity length microchip lasers based on composite laser crystals like Nd:YAG/Nd:Cr^{4+} give simple, compact and low-cost alternative to mode-locked laser systems for obtaining pulses of sub-nano and picoseconds pulse duration [58]. Short oscillator length provides effective pulse shortening at multi-kW peak power up to 290 ps [59], 218 ps [60]. Microchip lasers oscillate in a single longitudinal mode ensuring transform-limited pulses and provide high beam quality, linearly polarized output beams and outstanding pulse-to-pulse energy stability. Timing jitter value reflects time structure of the pump pulses and can be quite high, up to approximately 0.01 of pump pulse duration [56] that can be equal to several microseconds. Another type of microchip laser uses semiconductor saturable absorber mirrors (SESAM) and may produce pulses as short as 50 ps at 1.064 nm [61]. Shortest pulses of 37 ps were obtained in [62]. SESAM parameters can be specially designed according to the required properties and the absorption coefficient can be high enough, while SESAM has essentially low damage threshold in comparison with the composite crystals.

Advanced gain-switched laser diodes driven by radio-frequency pulses give potentially attractive universal solution of simple and compact picosecond seeding arrangement [63] adaptable to further amplifying stages.

Shortest way to rich millijoule single pulse energy level starting from the seed pulses is to use the regenerative amplification. It is well established scheme for generation of stable, high-peak-power ultra-short laser pulses since 1980s [64]. Active mode-locked cw flash-lamp pumped Nd:YLF laser provided 2.5 mJ, 37 ps pulses at 500 Hz repetition rate. The same scheme with Nd:YAG gain medium produced 80 ps pulses with 1 mJ energy [65]. 85 ps duration pulses of 1.1 mJ energy at 1.064 nm at up to 2 kHz repetition rates were produced by Nd:YAG regenerative amplifier with acousto-optic injection [66]. Pulses of 100 ps duration and of 60 mJ energy were achieved using Nd:YAG regenerative amplifier at 10 Hz repetition rate [67]. New generation of powerful Nd-doped laser systems was based on diode pumping, providing lower thermal loading of laser crystals, and passive mode-locking by means of SESAM, the schemes ensured reliable operation at higher power and repetition rates. Longitudinally diode-pumped Nd:YAG system worked at repetition rates up to 5 kHz, maximum single pulse energy of 1.1 mJ at 2 kHz was accompanied by the pulse lengthening from seeded 12 ps to output 25 ps, it was attributed to the gain narrowing [68]. Single pulse output energy of 0.35 mJ at a minimal pulse repetition rate of 5 kHz was achieved with a diode end-pumped Nd:YVO_4 regenerative amplifier, the output pulse duration was 19 ps, if seeded by a 7 ps oscillator, repetition rate can be extended up to 80 kHz [69]. In [70], 10 ps pulses of 0.5 mJ energy at 20 kHz were produced in Nd:YLF regenerative amplifier pumped with 42 W total power fiber coupled laser diodes. Single pulse energy up to 1.7 mJ was obtained with Nd:YVO_4 laser system with an adjustable pulse duration between 217 ps and 1 ns [71]. Pulses of more than 1 mJ energy and of 14.2 ps pulse width were produced with Nd:YVO_4 regenerative amplifier at 10 kHz repetition rate [16].

The use of laser diodes or laser diode arrays for pumping became fully justified at standard repetition rates higher than 100 Hz. They provide much higher efficiency, excellent stability and reduced thermal load on the active crystal [45, 58]. In comparison with the diode-side-pumped geometry, end-pumping produces pumped zone mainly in the central part of laser crystal [45, 72]. These gives better matching of laser mode with the pump profile and, accordingly, better pump efficiency. But however, end-pumping may be principally accompanied by essentially inhomogeneous intensity distribution in radial direction, aberrational component of thermal lens even at low pump powers, increased aberrations with the pump power, that leads to laser beam quality degradation and falling down of the pump power efficiency. Thus, appropriate analysis of diode pumping configurations looks quite important.

To summarize the introduction, we note that pulsed (pulsed repetitive) high-peak-power picosecond lasers of millijoule and multy-millijoule single pulse energy level operating at reasonably high repetition rates are required in a number of scientific and technological applications. The development of such systems based on neodymium-doped bulk laser crystals and advanced mode-locking schemes remains quite reasonable owing to characteristics appropriate for effective picosecond pulse amplification.

In the present paper, we consider schematic solutions and modeling maintaining development of such energy-effective, compact, robust and easy integrated pulsed diode-pumped picosecond lasers.

4.2 Experiment. Pulsed Repetitive Oscillator, Regenerative and Two-Pass Amplifiers

Simplified scheme of pulsed repetitive laser with operation control by means of active and passive mode-locking and negative feedback through induced radiation losses is illustrated by Fig. 4.2 and uses approach described in [55]. Stable transform limited pulses are formed in the master oscillator and then gained in regenerative amplifier. Oscillator cavity of 1.5 m length is formed by fully reflective mirrors M1, M3 and semiconductor saturable absorber mirror M2.

Regenerative amplifier cavity of similar length is based on mirrors M1, M3, M4 and polarizers P1 and P2. Both the oscillator and the regenerative amplifier cavities utilize one common active laser crystal. At convex M3 mirror with focal length about 1 m the resonator stability range with Nd:YAG continues up to ~100 Hz [73]. The present scheme may use replaceable crystals Nd:YAG and Nd:YLF of 10 mm length and 5 mm diameter, longitudinally pumped by means of fiber coupled qcw 70 W peak-power array of diode lasers [74] with wavelength centered on 808 nm (for Nd:YAG) or 806 nm (for Nd:YLF).

On the stage of pulse formation in the oscillator cavity, active mode-locking sinusoidal voltage is applied to one crystal of electro-optical modulator EOM on RTP crystals made under double-crystal thermo-compensated scheme. Sine wave frequency should correspond to the round trip time ~10 ns in the oscillator cavity. To reduce sine wave generator frequency, it is reasonable to fit it to the double round trip. Active mode-locking only can provide generation with single pulse duration of 300–500 ps. Additional pulse shortening is achieved by means of passive mode-locking which is realized with the saturable absorber mirror SAM (BATOP GmbH). To provide stable mode-locking at pulsed pump regime, SAM absorbance should be high enough. We used SAM with absorbance up to 13% and modulation depth up to 8%. All the elements: SAM, oscillator mirrors, RTP modulator are suited for operation on the wavelengths of 1064 and 1053 nm. Small part of oscillator radiation

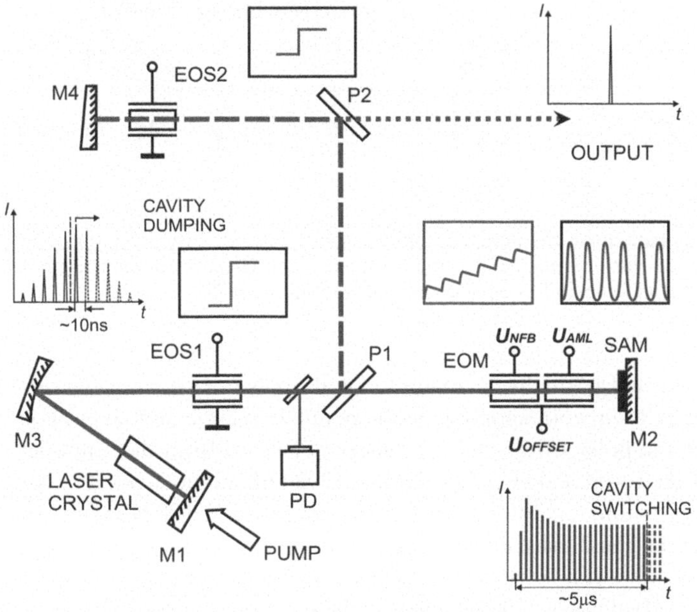

Fig. 4.2 Simplified optical scheme of picoseconds laser. M1–M4: laser mirrors; P1, P2: prism and thin-film polarizers; SAM: semiconductor saturable absorber mirror; EOM: two-crystal thermo-compensated electro-optical modulator; EOS1, EOS2: electro-optical switchers; PD: photodiode; U_{AML}: active-mode locking voltage; U_{OFFSET}: bias voltage; U_{NFB}: negative feedback voltage; lower right waveform: pulse development inside the oscillator; upper left waveform: gaining pulse in the regenerative amplifier after the cavity switching; upper right waveform: output pulse after the cavity dumping

is taken off by the beam-splitter S to the photodiode PD and the signal from the negative feedback loop is applied in proper polarity to the second crystal section of the EOM. Negative feedback results in Q-factor correction which is opposite to the oscillator radiation variations. This results in extending pulse formation time up to several microseconds. At the same time, this allows to support energy of pulse circulating in the oscillator on the level corresponding to the most effective shortening in the SAM. In order to improve negative feedback efficiency, feedback voltage is combined with additional adjustable offset voltage applied to the EOM. The resulting transmission function gets a complex shape and has one main maximum during the round trip. Maximum pulse shortening up to 25 ps (Nd:YAG) or 15 ps (Nd:YLF) is provided with SAM of 8% modulation depth.

Pulse formation in the oscillator takes up to 5 μs and utilizes only small part of the stored population inversion. After that, prepared picosecond pulse of approximately 1 μJ energy is directed from the oscillator to high-Q cavity of the regenerative amplifier. The cavity switching is fulfilled by means of electro-optical switcher EOS1 on DKDP crystal and polarizer P1. Remaining population inversion goes to the regenerative amplification stage. Single round trip unsaturated amplification factor is

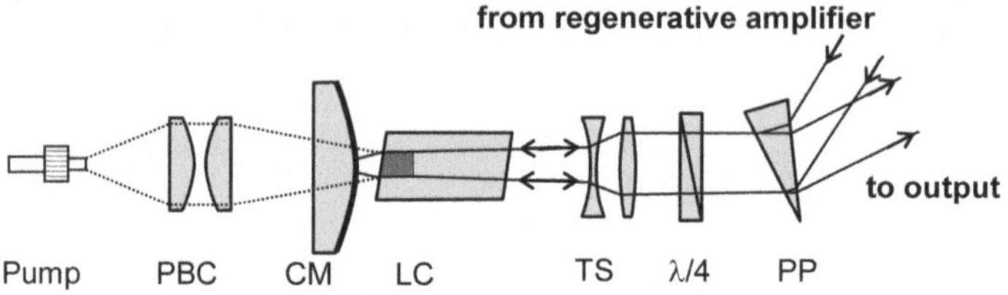

Fig. 4.3 Two-pass diode end-pumped amplifier optical scheme: PBC is a pump beam condenser, CM is a convex mirror with AR at pump wavelength and HR at laser wavelength, LC is a laser crystal, TS is a telescope, PP is a polarizing prism from calcite

equal to 2–2.5 and gradually decreases to 1 as the pulse energy rises. The amplification up to the maximum corresponding to the population depletion takes about 10 round trips. Then the pulse of up to 2 mJ energy is ejected from the amplifier cavity by means of electro-optical switcher EOS2 on DKDP and polarizer P2. Output single pulse energy within repetition rates stability range below 100 Hz is up to 1.2 mJ with Nd:YAG and up to 2 mJ with Nd:YLF crystals.

Stability range can be shifted to higher repetition rates if to compensate thermal lens in laser crystal, e.g., by means of replacing plane mirror M1, next to Nd:YAG laser crystal, by convex one. Installing M1 with curvature radius −0.8 m extended stable operation range up to approximately 400 Hz. However, repetition rate increase also resulted in higher aberrational losses of radiation (see Sect. 4.5) and in corresponding output energy falling down. At 300 Hz, maximal single pulse energy was ~0.5 mJ.

Further increase of single pulse energy can be realized either by using the same optical scheme with more powerful pump and larger pump area or by means of additional multipass amplifier stage. We arranged two-pass amplifier according to the scheme shown in Fig. 4.3. Nd:YAG crystal of 10 mm length and 5 mm diameter is longitudinally pumped through the convex mirror by fiber coupled qcw 100 W peak-power array of diode lasers [74] with wavelength centered at 808 nm. Telescope is used for reducing diameter of the input laser beam to ~0.5 mm and simultaneously to collimate the output beam. Reflecting convex mirror curvature radius is chosen −400 mm to compensate strong thermal lens in the amplifier crystal and to prevent focusing amplified output beam on the optical surfaces. At 300 Hz repetition rate, measured single pulse energy was 3.5 mJ with the beam quality $M^2 = 1.15$ (Fig. 4.4). Conversion of output beam with diameter of 1.2 mm into second harmonic yielded 2.4 mJ at 532 nm what corresponded to the efficiency of more than 65%.

4.3 Numerical Modeling. Pulse Formation in the Oscillator

Models describing the evolution of time pulse profile in lasers with active and passive mode-locking were proposed in [75]. Equations can be solved analytically in the cases of stationary generation with purely active or purely passive mode-locking. Under

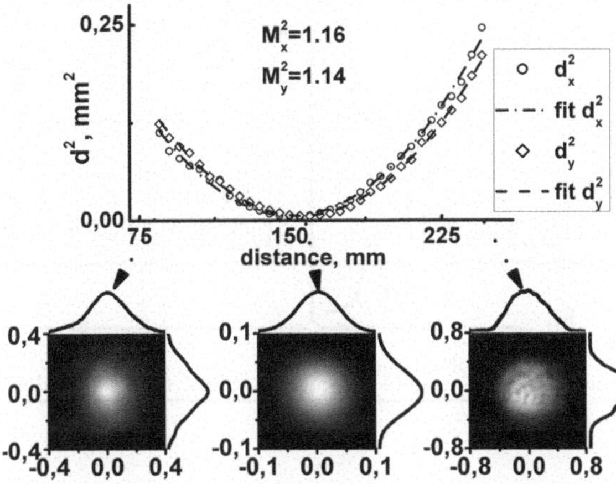

Fig. 4.4 Beam profiles after lens with focal distance of 15 cm at the output of Nd:YAG picosecond laser and two-pass diode end-pumped amplifier. Laser generates pulses with energy of 3.5 mJ at pulse repetition rate of 300 Hz

the joint action of two mechanisms, only numerical solution is possible. Simplified one-dimensional model was designed [76] for numerical calculation of generation process in advanced pulse-periodic high-peak-power picosecond lasers.

The model describes pulse formation governed with both active and passive mode-locking and negative feedback with adjustable modulation depth. Laser radiation is characterized by complex amplitude $\tilde{A}(z, t)$ of electric field depending on time t and axial coordinate z. Birefringent effect in the electro-optical modulator is described by phase shift $\Delta\varphi(z, t)$ between o and e field components. Elements of the optical system are considered as independent blocks numerated with index l taking values *EOM* (electro-optical modulator), *G* (gain medium), *P* (polarizer), *SAM* (saturable absorber mirror), *BS* (beam splitter). There are two kinds of elements: active and passive. Effect of the active ones depends on incoming radiation or electric signals, while the passive elements action remains fixed. Pulse propagation between two consecutive elements is considered as a transfer of the current pulse profile $\tilde{A}(z, t)$ with speed of light. Modification of radiation pulse when passing the next optical element can be generally expressed by:

$$\tilde{A}_n(z_l^{out}, t)/\tilde{A}_n(z_l^{in}, t) = K_l\Big[t, \Delta\varphi(z, t), \tilde{A}_n(z, t), U_{\mathrm{EOM}}(t), \ldots\Big] \tag{3.1}$$

$$\Delta\varphi_n(z_l^{out}, t) - \Delta\varphi_n(z_l^{in}, t) = \Delta\Phi_l\Big[t, \Delta\varphi(z, t), \tilde{A}_n(z, t), U_{\mathrm{EOM}}(t), \ldots\Big] \tag{3.2}$$

where z_l is coordinate of element l, n corresponds to a roundtrip number, indexes *in* and *out* denote values *before* and *after* modification respectively, K_l denotes complex amplitude modification factor and $\Delta\Phi_l$ describes phase shift between o and e components, attributed to element with index l, $U_{\mathrm{EOM}}(t)$ is voltage applied to EOM. Figure 4.5 illustrates the model structure corresponding to the experimental scheme.

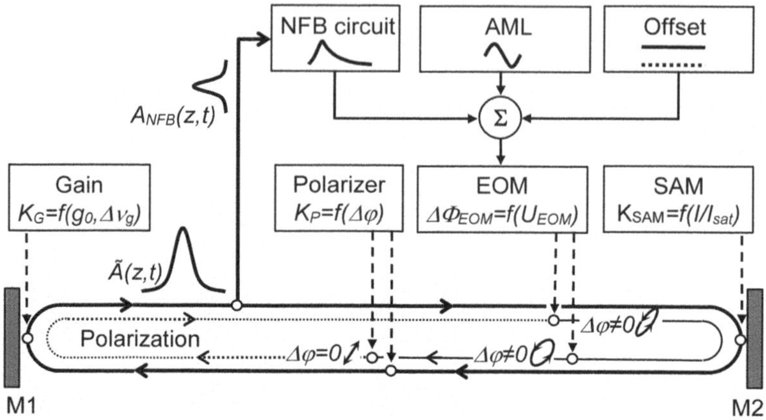

Fig. 4.5 Scheme of radiation transformation in the numerical model, corresponding to the real optical scheme. Outer contour illustrates transformation of complex amplitude and the inner contour shows phase transformations. Solid lines show area, where field amplitude $\tilde{A}(z, t)$, phase shift $\Delta\varphi(z, t)$ between *o* and *e* components in the EOM and electric voltage are calculated. Dotted line corresponds to region of constant phase. Elements of resonator affecting the radiation are considered as a dot sources and their locations are indicated with open circles. Vertical arrows indicate action of resonator elements on radiation. M1, M2 are resonator mirrors, g_0 is a gain in line center, $\Delta\nu_g$ is the gain bandwidth, I is a radiation intensity on SAM, I_{SAT} is the SAM saturation intensity

The details of the modeling approach can be found in the previous paper [76]. Typical pulse energy evolution governed by active and passive mode-locking, negative feed-back and bias voltage is shown in Fig. 4.6a. Enlarged fragment of the uprising beginning part between trip numbers from 50 to 90 is given in inset (b). On the first stage there is pulse formation by active mode–locking and very fast energy growth from noise level (c). Then, negative feedback stabilizes pulse energy at a certain level, pulse energy remains almost the same within round trip number region from 100 to 500, and the developed radiation pulse is gradually compressed by SAM and its intensity correspondingly grows. Pulse shortening with the round trip number is illustrated by insets (d) and (e).

Comparison of the calculated and experimental dependences, pulse width measurements data and additional verification of the model by means of "switching off" active mode-locking and negative feedback and comparing to the results of the pulse width estimations on the basis of [76] confirm understanding adequacy of the pulse formation processes and the described model validity.

4.4 Pulse Width Controlling with Fabry-Perot Etalon

Minimum pulse duration obtainable with mode-locking is determined by the inverse value of the gain spectrum width. Actually, generated pulses are somewhat longer due to several factors such as gain narrowing, phase self-modulation etc. which can

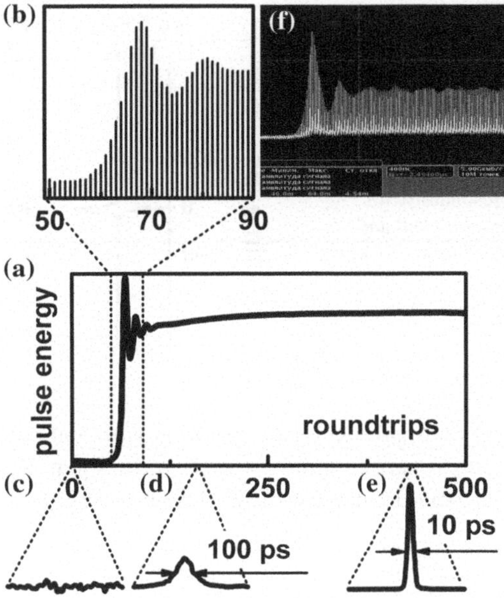

Fig. 4.6 Pulse evolution under active and passive mode-locking and negative feedback action. Pulse energy evolution within the generation process (**a**), pulse train snapshot during transformation from noise to steady level (**b**), pulse profile at different stages: noise at early generation stage (**c**), long pulse in the beginning of the stabilization region (**d**), short pulse formed at the end of the pulse formation (**e**) and experimental waveform of laser output (**f**)

be taken into account when making estimations for cw regime [75, 77]. In some practical cases, controllable variations in the pulse width can be required. To vary picosecond pulse width, one can use intracavity Fabry-Perot etalon. Pulse shortening was observed in the case of tuning etalon transparency minimum to the center of the gain spectrum owing to effective increase of the gain width [78]. On the contrary, overlaying intracavity etalon transmission peak with the gain spectrum results in spectral narrowing and corresponding pulse length extending. Pulse width tuning in the range 8–80 ps was made using uncoated etalons of different thicknesses [79].

In the present work, we realized generating picosecond pulses with variable durations in range of 16–400 ps by use of a thin Fabry-Perot etalons with different reflection coatings. The etalons were made from fused quartz plates with a diameter of 10 mm and 300 μm thickness. The thickness was chosen so that the optical round trip path inside the plate was at least several times shorter than the original pulse width. Partially reflected coatings were applied on both sides of the plates. Reflection coefficient *R* was made equal to 72, 79 and 89%. The etalons were installed into protective metal frames and then placed into oscillator shown in Fig. 4.2 close to the EOS1. Fine tuning of the etalon transmission line was carried out by its slight inclination. Figure 4.7 demonstrates results of the output pulse time profile measurements in the case of Nd:YLF laser crystals. The measurements were made using 2 ps time resolution streak camera. The leftmost picture shows scan of 16 ps length pulses obtained without etalon. Using etalons with the reflection of 72, 79 and 89%, we obtained pulse durations correspondingly 120, 180 and 400 ps.

The results obtained with Nd:YAG are similar to that with Nd:YLF. It should be noted that we used the same etalons in both cases but the fine tuning was realized

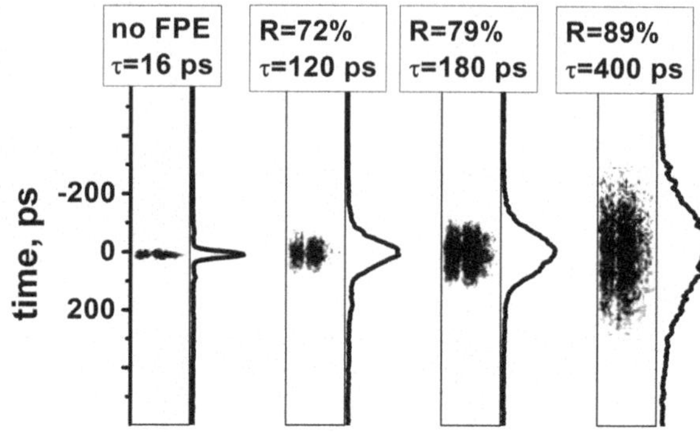

Fig. 4.7 Time profiles of output pulses recorded with streak-camera of 2 ps resolution and corresponding chronograms at different Fabry-Perot etalons coatings reflection values. Inside oscillator resonator with Nd:YLF laser crystal and active-passive mode-locking

at different inclinations. Laser operation with and without the etalons has the same pulse-to-pulse energy stability with the mean square value within 1%. While the time duration dispersion increases from 1.5 up to 3%.

4.5 Thermal Lens at the End-Pump Geometry

Using end-pump geometry usually provides better pump efficiency in comparison with the side-pumping but, at the same time, is accompanied by aberrational heat–induced lens due to inhomogeneous pump distribution over laser beam cross-section. Heat generation depends on average pump power and accordingly on pump pulse repetition rate, when pulsed (qcw) pump with fixed peak power is used. If pumped volume length is much shorter than resonator length then aberrational lens may be treated as a phase screen with nonparabolic radial phase shift dependence in an axially symmetrical resonator. Parabolic contribution can be governed with standard spherical lenses or mirrors. While non-parabolic part of phase distortion is quite small, one can consider it as additional diffraction losses for fundamental Gaussian mode [73]. At this condition, range of obtainable pulse repetition rates can be estimated under the assumption that the resonator is stable and generation of fundamental Gaussian mode is possible. With heat generation increase at growing up pump pulse repetition rate, this approach will not reflect the real picture. Laguerre-Gaussian beams which are self-reproducing on each roundtrip of resonator with spherical mirrors, do not satisfy mode stability criteria in aberrational resonator since wave front of each beam is spherical in any cross-section. Generally speaking, presence of an aberrational element in resonator leads to non-Gaussian fundamental and higher modes uprising. Each mode is uniquely defined by its specific field distribution in any cross-section. Such modes representation however is not illustrative and is difficult to analyze. One can assume more convenient and visual interpretation. Non-Gaussian mode field structure can be represented by decomposition into Laguerre-Gaussian beams with arbitrary initial Gaussian beam width. To make such decomposition more

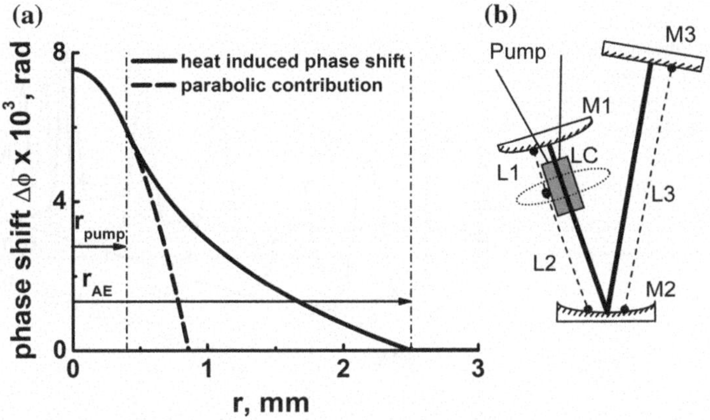

Fig. 4.8 **a** Simulation of heat induced phase shift in axially symmetrical Nd:YAG laser crystal at diode-end-pumping. Parabolic part inside pumped region is replaced by logarithmic dependence in the rest of crystal volume. Phase shift corresponds to 1 Hz and is scaled by means of multiplying on pump pulse repetition rate. **b** Resonator equivalent optical scheme: M1 is the convex mirror, $R1 = 1.5$ m; M2 is the concave mirror; $R2 = 2$ m; M3 is the plain mirror; LC is the laser crystal with heat induced lens. Distances are $L1 = 0.015$ m, $L2 = 0.7$ m, $L3 = 0.75$ m

reasonable one can choose beam width as equal to width of embedded Gaussian beam, which is in $\sqrt{M^2}$ times smaller than mode width in any cross-section [80]. In this case, the resonator mode field in some cross-section is represented by the complex decomposition coefficients. Squared amplitude coefficients represent intensities of the Laguerre-Gaussian components. In the axially symmetric case, the lowest component is Gaussian and the higher ones have different number of concentric rings. It should be noted that the mode decomposition remains unchanged when passing through nonaberrative elements or resonator sections and is modified at transition through the aberrative elements.

In our simulations, we use model of heat induced phase shift [73] that suggests homogeneous heat generation inside cylindrical pumped region in laser crystal with radial direction of heat flows. Phase profile that corresponds to diode-end-pumping process of Nd:YAG laser crystal with heating average power of 4 mW inside cylindrical region of radius 0.4 mm is shown in Fig. 4.8a. This heating power due to quantum defect corresponds to 1 Hz pump repetition rate and can be scaled by means of multiplying on pump pulse repetition rate. Resonator equivalent optical scheme that is used for picosecond pulses generation at pulsed diode end-pumping is shown in Fig. 4.8b. Real resonator optical scheme may contain several plain folding mirrors for compactness. Convex mirror M1 near laser crystal provides compensation of parabolic part of the heat induced lens and allows laser to oscillate in a certain range of the pump pulse repetition rates corresponding to the stability range. Choosing concave mirror M2 one can adjust the mode size inside the laser crystal that is necessary for good mode overlapping with the pump region.

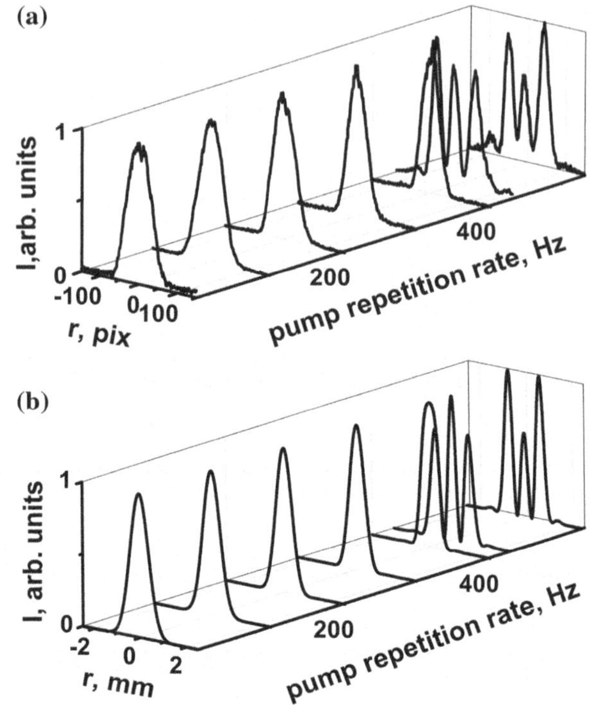

Fig. 4.9 Experimentally registered (**a**) and simulated with taken into account heat induced aberrational lens in laser crystal (**b**) beam profiles *at Nd:YAG laser output* depending on pump repetition rate

In the experiment, we varied pump pulse repetition rate and recorded beam profiles behind a plain folding mirror that is located in the middle of distance between LC and M2 and is not shown in Fig. 4.8b. Beam profiles are close to Gaussian at pulse repetition rate up to 400 Hz (average pump power 5 W) (see Fig. 4.9).

Then abrupt profile transformation occurs inside very narrow range of repetition rates. Generated mode gets a ring structure around the central part. To explain this effect one can analyze the resonator mode structure at heat induced aberrations. Let's sort modes in order of decrease of roundtrip modification factor which is expressed as:

$$G_n = |\gamma_n|^2 \frac{\int I_n(r) g(r) r \mathrm{d}r \int g(r) r \mathrm{d}r}{\int I_n(r) r \mathrm{d}r \int g^2(r) r \mathrm{d}r} \tag{5.1}$$

where n is a mode index, γ_n is the mode n eigenvalue, $I_n(r)$ is the mode intensity profile *at the laser crystal location*, $g(r)$ is the gain profile. Fraction in (5.1) is the pump and the mode overlap integral [45]. That is, at each pump pulse repetition rate (or pump average power) we classify the modes on the basis of their best amplification conditions. Generated mode corresponds to highest value of G_n. Figure 4.10 illustrates calculated intensity profiles for two modes with highest G_n at different pulse repetition rates: (a, b) well below, (c) slightly lower and (d) higher the frequency

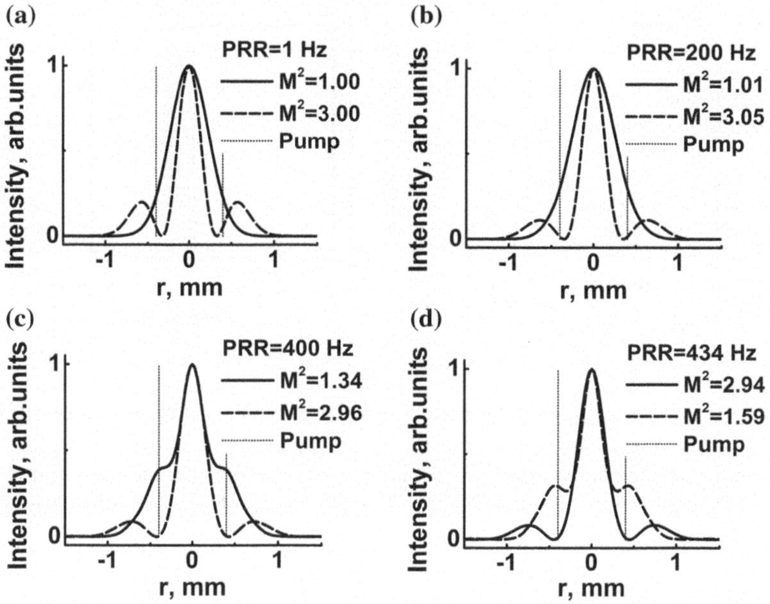

Fig. 4.10 Intensity profiles of two resonator modes *inside laser crystal* that have highest roundtrip modification factors G_n at different pump pulse repetition rates (PRR). Respective beam quality factors and the pump beam width are indicated for reference

of the mode change. Corresponding M^2 factor values are also indicated. Solid line profiles in Fig. 4.10 (*inside laser crystal*) and calculated profiles in Fig. 4.9b (*on the resonator output*) at proper pump pulse repetition rates correspond to the same generated modes. One can see that at repetition rate of 434 Hz generation of mode with higher M^2 factor containing more pronounced ring structure is maintained.

According to the beginning part of the current Section, we decompose the calculated mode with the highest G_n value into Laguerre-Gaussian beams with the embedded Gaussian beam width for each pump pulse repetition rate. Corresponding dependences for weights coefficients of four lowest Laguerre-Gaussian components are shown in Fig. 4.11.

Figure 4.11 illustrates that the aberrational mode corresponding to the best amplification conditions (that is, to the generated beam) is composed mainly of Gaussian component at the pulse repetition rate below 400 Hz. At higher pulse repetition rate, it radically changes its structure to several Laguerre-Gaussian component composition, and the Gaussian contribution drops sharply. Therefore, the sharp mode profile changing with the increase of pump pulse repetition rate takes place due to falling down roundtrip gain for quasi-Gaussian mode and its growing up for ring mode at fixed geometry of the pumped volume. This situation differs cardinally from the case of resonator with spherical mirrors when modes are Laguerre-Gaussian beams and

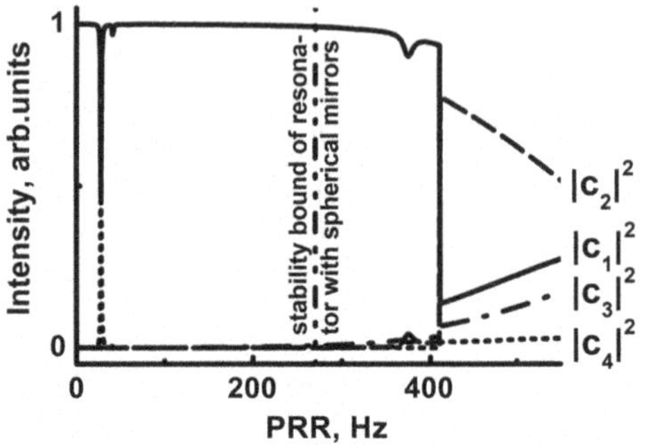

Fig. 4.11 Normalized intensities $|c_1|^2$, $|c_2|^2$, $|c_3|^2$, $|c_4|^2$ of four lowest Laguerre-Gaussian components as functions of pump pulse repetition rate (PRR) for the mode decomposition in the case of resonator with aberrational heat induced lens. $|c_1|^2$ corresponds to the Gaussian component

they keep transvers profiles proportions in every cross-section at any pump pulse repetition rate.

One more notice can be done for resonator with aberrational element that is characterized by phase distortion curve shown on Fig. 4.8a when parabolic shape inside homogeneously pumped region turns to slower logarithmic drop on the wings. In agreement with [81], any beam phase distortion has effective spherical contribution that depends on phase derivative averaged over beam intensity distribution. Considering heat-induced aberration corresponds to an effective lens with the optical power for the aberrational mode decreasing from the crystal axis to the periphery. That is, the greater the optical power of the lens at the center, the more radiation goes into the peripheral region of the mode, for which the lens turns out to be weaker. In comparison with the ideal case of spherical heat-induced lens in resonator when its optical power depends only on pump average power and does not vary in radial direction and resonator loses stability at certain pump pulse repetition rate, the aberrational element has in a manner "adaptive" property for passing beam. It means that the beam width increases relatively slow to the case of spherical lens when the heating power grows up. Thus, resonator with aberrational element is stable in wider range of pump pulse repetition rates than it would be predicted by the assumption of a proportional dependence of the optical power on the heating power.

4.6 Energy Gain and Saturation Condition

Realization of high peak-power and energy effective single picosecond pulse amplification using diode pumping schemes is very important for practical laser system designing. Regenerative amplification approach allows increasing picosecond pulse energy from pico- to millijoule level that is limited basically by optical damage of resonator and electro-optical elements at mode diameter of the order of 1 mm. In turn, this mode size roughly corresponds to the resonator length of 1–2 m which seems to be a good compromise in view of limited operation speed of the electro-optical switching, from one side, and of complexity and poor stability of long and wide-aperture resonators, from the other side. Laser pulse energy is growing up during 10–100 roundtrips while population inversion depletion is not still achieved. Gain saturation brings to higher pump efficiency and energy stability of the output pulses. Regenerative amplifier setup complexity is quite compensated by obtaining overall gain of 10^4–10^9 that is hard to obtain in another ways. Energy fluence of millijoule level pulse in about millimeter width beam is consistent with saturation fluence:

$$F_{\text{sat}} = \frac{h\nu_{\text{las}}}{\gamma\sigma_{\text{las}}} \tag{6.1}$$

which is approximately equal to some fractions of 1 J/cm^2 for commonly used Nd-doped laser crystals. Here ν_{las} and σ_{las} are laser frequency and stimulated emission cross-section respectively, γ is a factor of population inversion change 1 or 2 after one photon emission. It should be noted that parameter γ is equal to 2 due to finite lifetime of laser transition terminal level in Nd ion therefore its relaxation can be neglected during picosecond pulse amplification process.

Powerful end-amplifying cascades of over- and multi-millijoule levels may use multipass schemes providing effective depletion of population inversion.

Then let us consider single picosecond pulse double-pass amplification in a diode-end-pumped active crystal (see Fig. 4.12) and determine acceptable focusing the pump beam into the laser crystal so that the gain is at maximum, but the output energy density does not exceed the damage threshold of the laser crystal. Diode pump pulse with duration approximately equal to spontaneous luminescence time τ_{lum} and peak power P_{pump} is focused into active laser crystal in spot of radius r_{pump} and is absorbed on the length l_{abs}. For simplicity, we neglect losses and assume homogeneous laser intensity over beam section and density distribution of population inversion inside cylindrical volume $\pi r_{\text{pump}}^2 \times l_{\text{abs}}$:

$$n = \frac{\eta_{\text{lum}} P_{\text{pump}} \tau_{\text{lum}}}{h\nu_{\text{pump}} \pi r_{\text{pump}}^2 l_{\text{abs}}} \tag{6.2}$$

Factor η_{lum} takes into account luminescence losses and is equal to $1 - e^{-1}$ for amplification process at the end of pump pulse with duration τ_{lum} [45]. Beam radius is assumed to be equal to the pumped region radius. Laser beam goes through the pumped region, is reflected by the mirror and goes back. One may write the first pass gain using expression [82]:

$$F_{1\ \text{pass}} = F_{\text{sat}} \ln\left\{1 + \left[\exp\left(\frac{F_{\text{in}}}{F_{\text{sat}}}\right) - 1\right] \exp(\sigma_{\text{las}} n l_{\text{abs}})\right\} \tag{6.3}$$

where F_{in} and $F_{1\ \text{pass}}$ are the laser pulse fluencies of input and passed one-way trough the laser crystal radiations correspondingly. Parameters σ_{las} and τ_{lum} of mainly used Nd-doped materials are shown in Table 4.1.

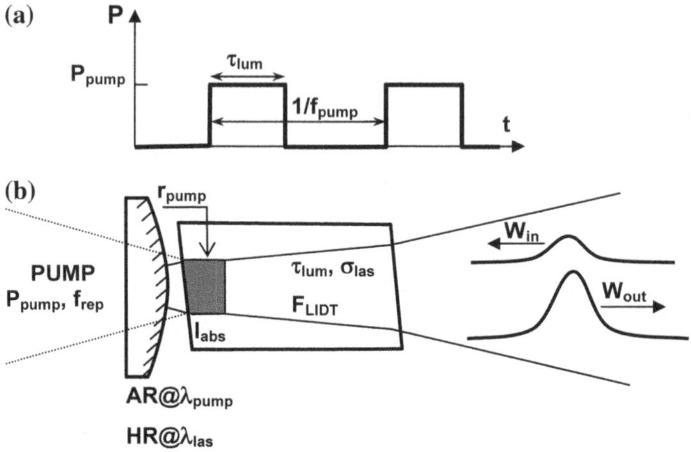

Fig. 4.12 Qcw (pulsed) pump pulses diagram (**a**) and double-pass diode-end-pumped scheme of picosecond pulse amplification (**b**)

Table 4.1 Active crystals characteristics [81]

	Fluorescence lifetime τ_{lum}, μs	Emission cross section σ, ×10^{19} cm^2	Thermal lens host factor, × 10^6	Thermal shock R, W/cm
Nd:YAG	230	2.8	0.9	7.9
Nd:YLF	485	π: 1.8 σ: 1.2	1.6–2.1	1.3
$Nd{:}YVO_4$	90	15	0.2–0.4	3.2

Then population inversion depletion may be taken into account as follows:

$$n_1 = n - \gamma \frac{F_{1\,\text{pass}} - F_{\text{in}}}{l_{\text{abs}} h \nu_{\text{las}}} \tag{6.4}$$

To describe the gain on the return pass, one may substitute n_1, $F_{1\,\text{pass}}$ and F_{LIDT} instead of n, F_{in} and $F_{1\,\text{pass}}$ correspondingly into (6.3), where maximum possible output energy fluence F_{LIDT} due to laser induced optical damage (LIDT) limitation is taken into account. Input and output pulse energies are expressed as $W_{\text{in}} = \pi r_{\text{pump}}^2 F_{\text{in}}$ and $W_{\text{out}} = \pi r_{\text{pump}}^2 F_{\text{LIDT}}$. Finally, implicit equation for finding r_{pump} may be written as:

$$\ln\left\{\exp\left(\frac{F_{\text{LIDT}}}{F_{\text{sat}}}\right) - 1\right\} - \ln\left\{\exp\left(\frac{F_{1\,\text{pass}}}{F_{\text{sat}}}\right) - 1\right\} = \sigma_{\text{las}} n_1 l_{\text{abs}} \tag{6.5}$$

On the basis of W_{in} and r_{pump} one can find overall gain $G = \frac{F_{\text{out}}}{F_{\text{in}}}$ and pump efficiency $\eta_{\text{eff}} = \frac{W_{\text{out}} - W_{\text{in}}}{P_{\text{pump}} \tau_{\text{lum}}}$. Considering commonly used qcw (pulsed) pump with

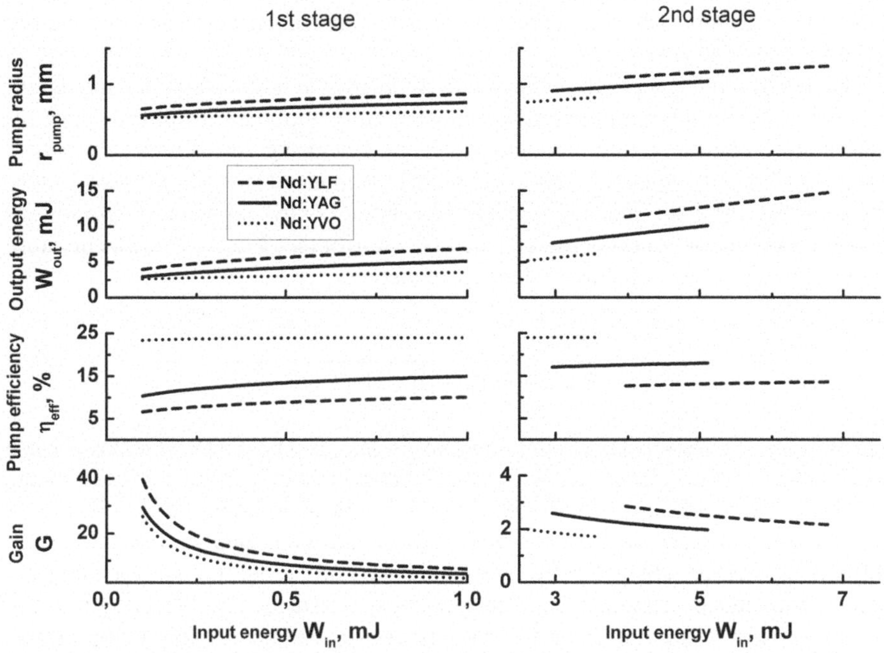

Fig. 4.13 Simulation of picosecond pulse amplification in two consequent double-pass diode-end-pumped active crystals. Input pulse with energy in the range of 0.1–1 mJ is amplified to LIDT energy fluence in first stage by appropriate choosing of pump beam radius at fixed peak power. Output of first stage is the input for second stage with the same limiting condition

diode fiber coupled laser modules with peak power e.g. of 120 W [e.g. 74] and LIDT value 3 J/cm^2 for picosecond pulses [83, 84], which we assume to be approximately the same for the crystals under consideration, one can simulate optimized pulse diode-end-pumped double-pass amplifier for the range within 0.1–1 mJ of input pulse energies.

Simulation results for optimized in the same way Nd:YLF, Nd:YAG and Nd:YVO_4 two-pass amplifiers are shown in the left column of Fig. 4.13 and indicated as the "1st stage". For Nd:YAG, they are in a good correspondence with the experimental results presented in the Sect. 4.2. In the right column of Fig. 4.13 we present calculation results for the optimized scheme of similar "2nd stage" amplifier operating near saturation and below damage threshold conditions with the same pump peak power and respectively enhanced pump beam diameter.

Maximum energy output at fixed peak pump power can be obtained with Nd:YLF active crystal due to longest lifetime of upper laser transition level. At the same time Nd:YVO_4 has highest pump efficiency due to largest stimulated emission cross-section. Nevertheless, maximal pump efficiency does not exceed 25% because of three factors: 37% spontaneous luminescence losses, 24% quantum defect losses and only 50% of stored energy available due to quasi-three-level scheme at picosecond pulse amplification. Nd:YAG active crystal is a compromise between pulse energy and pump efficiency.

When pump pulse repetition rates increase, heat generation starts playing essential role in achievable amplifier performance. Most important limiting factors are heat induced aberrative lens, depolarization and fracturing due to internal stresses. Amplified beam divergence is affected by the heat induced lens that brings to intensity increase and laser induced damage. Parabolic contribution into heat induced phase incursion can be compensated by convex mirror (see Fig. 4.12) with curvature radius approximately equal to the lens focal distance:

$$f_T = \frac{2\pi r_{pump}^2}{\eta_{heat} P_{abs}} \left[\frac{K_T}{\frac{dn}{dT} + (n-1)(1+\nu)\alpha_T} \right] \quad (6.6)$$

where P_{abs} is an absorbed average pump power, η_{heat} is a heating efficiency, K_T is a heat conductivity, dn/dT and α_T is a thermo-optic and thermal expansion coefficients correspondingly, n is a refraction index, and ν is a Poisson ratio.

Values for material factor in square brackets are presented in Table 4.1 [81]. Nd:YLF crystals are characterized by weak lensing properties, but its worst thermal shock parameter is accompanied with low stress resistance and high fragility that often results in destroying Nd:YLF crystals even at reasonably low pump power. Nd:YAG crystal has optimal combination of thermo-lensing and stress resistance at moderate (below 1 kHz) repetition rates. Nd:YVO_4 amplifier crystals are characterized by highest efficiency but at the same peak power provide lower single pulse energy due to lower fluorescence lifetime. At the same time, they may maintain effective pulsed operation regime at a higher repetition rates up to several kHz with the use of appropriate diode laser pump systems. The limiting factors associated with the thermal lens manifestation can be analyzed on the basis of the approaches presented in the paper.

4.7 Conclusions

Picosecond laser with single pulse energy of millijoule and multi-millijoule level and repetition rates from sub-kilohertz to multi-kilohertz are required in a numerous scientific and technological applications. We reviewed basic approaches to development of such high peak-power pulsed diode-pumped picosecond laser systems based on Nd-doped bulk active crystals and advanced mode-locking schemes. We developed approach utilizing active-passive mode-locked and negative feedback control oscillator scheme operating with Nd:YLF and Nd:YAG, that can provide generation of stable, closed to transform limited pulses with pulse duration of 25 ps (with Nd:YAG) and 16 ps (with Nd:YLF). Oscillator—regenerative amplifier scheme based on the common end-diode-pumped laser crystal generates pulses up to 1.2 mJ with Nd:YAG and up to 2 mJ with Nd:YLF crystals. Two-pass Nd:YAG diode end-pumped amplifier provides output radiation of 4 mJ single pulse energy at 300 Hz repetition rate. Output radiation is converted into the second harmonic with more than 60% efficiency. Numerical modeling which describes main factors determining picosecond pulse formation allows better understanding of the process. Installing Fabry-Perot etalons of 300 μm thickness and of different reflection coating inside oscillator provided generation of pulses with enhanced up to 120, 280 and 400 ps pulse width. Aberrative character of thermal lens and mode structure at end-pump geometry were analyzed using non-Gaussian mode decomposition into Laguerre-Gaussian beams with the width equal to the embedded beam value. We supposed that adaptive action of the aberration lens might enhance the resonator stability range. Comparison of optimized pulse diode-end-pumped double-pass amplifier schemes utilizing Nd:YLF, Nd:YAG and Nd:YVO_4 crystals was developed and discussed.

Funding The work was partly granted by M.V. Lomonosov Moscow State University Program of Development.

References

1. Y. Wang, K.B. Eisenthal, Picosecond laser studies of ultrafast processes in chemistry. J. Chem. Ed. **59**(6), 482–489 (1982)
2. S.-B. Zhu, J. Lee, G.W. Robinson, Effects of an intense picosecond laser on liquid carbon disulfide: a molecular dynamics study. J. Opt. Soc. Am. B **6**(2), 250–256 (1989)
3. V.G. Arakcheev, V.V. Kireev, V.B. Morozov, A.N. Olenin, V.G. Tunkin, A.A. Valeev, D.V. Yakovlev, Collisionally induced dephasing and rotational energy transfer in CO_2 Fermi dyad "blue" Q-branch 1388 cm^{-1}. J. Raman Spectr. **38**(8), 1046–1051 (2007)
4. A. Montello, M. Nishihara, J.W. Rich, I.V. Adamovich, W.R. Lempert, Picosecond CARS measurements of nitrogen rotational/translational and vibrational temperature in a nonequilibrium Mach 5 flow. Exp. Fluids **54**, 1422 (2013)
5. R. Knappe, Applications of picoseconds lasers and pulse-bursts in precision manufacturing. Proc. SPIE **8243**, 82430I (2012)
6. S. Brüning, G. Hennig, S. Eifel, A. Gillner, Ultrafast scan techniques for 3D micrometer structuring of metal surfaces with high repetitive ps-laser pulses, in *Lasers in Manufacturing* (Munich, 2011)
7. P. Likschat, A. Demba, S. Weissmantel, Ablation of steel using picoseconds laser pulses in burst mode. Appl. Phys. A **123**, 137 (2017)
8. G. Scotti, D. Trusheim, P. Kanninen, D. Naumenko, M. Shulz-Ruhtenberg, V. Snitka, T. Kallio, S. Franssila, Picosecond laser ablation for silicon micro fuel cell fabrication. J. Micromech. Microeng. **23**, 055021 (2013)
9. J. Albello, P. Piogovsky, J. O'Brien, B. Baird, Picosecond laser micromachining of advanced semiconductor logic devices. Proc. SPIE **6871**, 687122 (2008)
10. R. Moser, M. Kunzer, C. Gossler, K. Köhler, W. Pletschen, U.T. Schwarz, J. Wagner, Laser processing of gallium nitride-based light-emitting diodes with ultraviolet picoseconds laser pulses. Opt. Eng. **51**(11), 114301 (2012)
11. E. Markauskas, P. Gečys, I. Repins, C. Beall, G. Račiukaitis, Laser lift-off scribing of the CZTSe thin-film solar cells at different pulse durations. Sol. Energy **150**, 246–254 (2017)
12. M. Domke, G. Heise, I. Richter, S. Sarrach, H.P. Huber, Pump-probe investigations on the laser ablation of CIS thin film solar cells. Phys. Procedia **12**, 396–403 (2011)

13. G.J. Spuhler, R. Paschotta, U. Keller, M. Moser, M.J.P. Dymott, D. Kopf, J. Meyer, K.J. Weingarten, J.D. Kmetec, J. Alexander, G. Truong, Diode-pumped passively mode-locked Nd:YAG laser with 10-W average power in a diffraction-limited beam. Opt. Lett. **24**(8), 528–530 (1999)
14. J. Kleinbauer, R. Knappe, R. Wallenstein, A powerful diode-pumped laser source for micromachining with ps pulses in the infrared, the visible and the ultraviolet. Appl. Phys. B **80**, 315–320 (2005)
15. X. Wushouer, P. Yan, H. Yu, Q. Liu, X. Fu, X. Yan, M. Gong, High peak power picosecond hybrid fiber and solid-state amplifier system. Laser Phys. Lett. **7**(9), 644–649 (2010)
16. Z.G. Peng, M. Chen, C. Yang, L. Chang, G. Li, A cavity-dumped and regenerative amplifier system for generating high-energy, high-repetition-rate picosecond pulses. Jpn. J. Appl. Phys. **54**, 028001 (2015)
17. Z. Ma, D.-J. Li, P. Shi, P.-X. Hu, N.-L. Wu, K.-M. Du, Compact multipass Nd:YVO_4 slab laser amplifier. J. Opt. Soc. B. **24**(5), 1061–1065 (2007)
18. C.K. Nielsen, B. Ortac,, T. Schreiber, J. Limpert, R. Hohmuth,W. Richter, A. Tünnermann, Self-starting self-similar all-polarization maintaining Yb-doped fiber laser. Opt. Expr. **13**(23), 9346–9351 (2005)
19. L.A. Gomes, L. Orsila, T. Jouhti, O.G. Okhotnikov, Picosecond SESAM-based ytterbium mode-locked fiber lasers. IEEE J. Sel. Top. Quant. Electr. **10**(1), 129–136 (2004)
20. M.E. Fermann, I. Hartl, Ultrafast fiber laser technology. IEEE J. Sel. Top. Quant. Electr. **15**(1), 191–206 (2009)
21. P. Dupriez, A. Piper, A. Malinowski, J.K. Sahu, M. Ibsen, B.C. Thomsen, Y. Jeong, L.M.B. Hickey, M.N. Zervas, J. Nilsson, D.J. Richardson, High average power, high repetition rate, picosecond pulsed fiber master oscillator power amplifier source seeded by a gain-switched laser diode at 1060 nm. IEEE Photonics Technol. Lett. **18**(9), 1013–1015 (2006)
22. H.-Y. Chan, S. Alam, L. Xu, J. Bateman, D.J. Richardson, D.P. Shepherd, Compact, high-pulse-energy, high-power, picoseconds master oscillator power amplifier. Opt. Expr. **22**(18), 21938–21942 (2014)
23. S. Matsubara, M. Tanaka, M. Takama, H. Hitotsuya, T. Kobayashi, S. Kawato, A picosecond thin-rod Yb:YAG regenerative laser amplifier with the high average power of 20W. Laser Phys. Lett. **10**, 055810–055814 (2013)
24. K.-H. Hong, A. Siddiqui, J. Moses, J. Gopinath, J. Hybl, F.Ö. Ilday, T.Y. Fan, F.X. Kärtner, Generation of 287 W, 5.5 ps pulses at 78 MHz repetition rate from a cryogenically cooled Yb:YAG amplifier seeded by a fiber chirped-pulse amplification system. Opt. Lett. **33**(21), 2473–2475 (2008)
25. Z. Ma, D. Li, P. Shi, P. Hu, N. Wu, K. Du, Compact multipass Nd:YVO_4 slab laser amplifier based on a hybrid resonator. J. Opt. Soc. Am. B. **24**(5), 1061–1064 (2007)
26. K.K. Chen, J.H.V. Price, S. Alam, J.R. Hayes, D. Lin, A. Malinowski, D.J. Richardson, Polarisation maintaining 100 W Yb-fiber MOPA producing μJ pulses tunable in duration from 1 to 21 ps. Opt. Expr. **18**(14), 14385–1394 (2010)
27. T. Eidam, S. Hanf, E. Seise, T.V. Andersen, T. Gabler, C. Wirth, T. Schreiber, J. Limpert, A. Tünnermann, Femtosecond fiber CPA system emitting 830 W average output power. Opt. Lett. **35**(2), 94–96 (2010)
28. D.J. Richardson, J. Nilsson, W.A. Clarkson, High power fiber lasers: current status and future perspectives. J. Opt. Soc. Am. B **27**(11), B63–B92 (2010)
29. H. Lin, J. Li, X. Liang, 105 W, <10 ps, TEM 00 laser output based on an in-band pumped Nd: YVO_4 Innoslab amplifier. Opt. Lett. **37**(13), 2634–2636 (2012)
30. P. Russbueldt, D. Hoffmann, M. Höfer, J. Löhring, J. Luttmann, A. Meissner, J. Weitenberg, M. Traub, T. Sartorius, D. Esser, R. Wester, P. Loosen, R. Poprawe, Innoslab amplifiers. IEEE J. Sel. Top. Quant. Electr. **21**(1), 3100117 (2015)
31. D. Li, K. Du, Picosecond laser with 400 W average power and 1 mJ pulse energy. Proc. SPIE **7912**, 79120N (2011)
32. A. Giesen, J. Speiser, Fifteen years of work on thin-disk lasers: results and scaling laws. IEEE J. Sel. Top. Quant. Electron **13**(3), 598–609 (2007)

33. C.J. Saraceno, F. Emaury, C. Schriber, A. Diebold, M. Hoffmann, M. Golling, T. Südmeyer, U. Keller, Toward millijoule-level high-power ultrafast thin-disk oscillators. IEEE J. Sel. Top. Quant. Electron **13**(3), 598–609 (2007)
34. J.-P. Negel, A. Loescher, A. Voss, D. Bauer, D. Sutter, A. Killi, M.A. Ahmed, T. Graf, Ultrafast thin-disk multipass laser amplifier delivering 1.4 kW (4.7 mJ, 1030 nm) average power converted to 820 W at 515 nm and 234 W at 343 nm. Opt. Expr. **23**(16), 21064–21077 (2015)
35. J. Fischer, A.-C. Heinrich, S. Maier, J. Jungwirth, D. Brida, A. Leitenstorfer, 615 fs pulses with 17 mJ energy generated by an Yb:thin-disk amplifier at 3 kHz repetition rate. Opt. Lett. **41**, 246–249 (2016)
36. T. Seeger, J. Kiefer, A. Leipertz, B.D. Patterson, C.J. Kliewer, T.B. Settersten, Picosecond time-resolved pure-rotational coherent anti-Stokes Raman spectroscopy for N_2 thermometry. Opt. Lett. **34**(23), 3755–3757 (2009)
37. G. Seeber, *Satellite Geodesy* (Walter de Gruyter, Berlin, New York, 2003) (Chapter 8)
38. B. Gourine, French transportable laser ranging station: positioning campaigns for satellite altimeter calibration missions in occidental Mediterranean Sea. Larhyss J. **12**, 57–69 (2013)
39. J.O. Dickey, P.L. Bender, J.E. Faller, X.X. Newhall, R.L. Ricklefs, J.G. Ries, P.J. Shelus, C. Veillet, A.L. Whipple, J.R. Wiant, J.G. Williams, C.F. Yoder, Lunar laser ranging: a continuing legacy of the Apollo Program. Science **265**(5171), 482–490 (1994)
40. J.G. Williams, S.G. Turyshev, D.H. Boggs, Progress in Lunar laser ranging tests of relativistic gravity. Phys. Rev. Lett. **93**, 261101 (2004)
41. R. Intartaglia, K. Bagga, F. Brandi, Study on the productivity of silicon nanoparticles by picosecond laser ablation in water: towards gram per hour yield. Opt. Expr. **22**(3), 3117–3127 (2014)
42. E.I. Gacheva, A.K. Poteomkin, SYu. Mironov, V.V. Zelenogorskii, E.A. Khazanov, K.B. Yushkov, A.I. Chizhikov, V.Ya. Molchanov, Fiber laser with random-access pulse train profiling for a photoinjector driver. Photonics Res. **5**(4), 293–298 (2017)
43. M. Petrarca, M. Martyanov, M.S. Divall, G. Luchinin, Study of the powerful Nd:YLF laser amplifiers for the CTF3 photoinjectors. IEEE J. Quant. Electr. **47**, 306–313 (2011)
44. G. Mourou, C.V. Stancampiano, A. Antonetti, A. Orszag, Picosecond microwave pulses generated with a subpicosecond laser-driven semiconductor switch. Appl. Phys. Lett. **39**(4), 295–296 (1981)
45. W. Koechner, *Solid-State Laser Engineering* (Springer, 2014)
46. R.-Q. Xu, Y.-R. Song, Z.-K. Dong, K.-X. Li, J.-R. Tian, Compact Yb-doped mode-locked fiber laser with only one polarized beam splitter. Appl. Opt. **56**(6), 1674–1681 (2017)
47. D. Nodop, J. Limpert, R. Hohmuth, W. Richter, M. Guina, A. Tünnermann, High-pulse-energy passively Q-switched quasi-monolithic microchip lasers operating in the sub-100-ps pulse regime. Opt. Lett. **32**(15), 2115 (2015)
48. D. Derickson, R. Helkey, A. Mar, J. Karin, J. Wasserbauer, J. Bowers, Short pulse generation using multisegment mode-locked semiconductor lasers. IEEE J. Quant. Electr. **28**(10), 2186–2202 (1992)
49. J.C. Balzer, T. Schlauch, T. Hoffmann, A. Klehr, G. Erbert, M.R. Hofmann, Modelocked semiconductor laser system with pulse picking for variable repetition rate. Electron. Lett. **47**(25), 1387–1388 (2011)
50. P. Heinz, A. Laubereau, Feedback-controlled mode-locking operation of Nd-doped crystal lasers. J. Opt. Soc. Am. B **7**(2), 182–186 (1990)
51. A. Agnesi, C. Pennacchio, G.C. Reali, V. Kubecek, High-power diode-pumped picosecond Nd^{3+}:YVO_4 laser. Opt. Lett. **22**(21), 1645–1647 (1997)
52. A.V. Ramamurthi, K.P.J. Reddy, Theory of combined AM and high-harmonic FM mode-locked laser. Pramana-J. Phys. **52**(1), 19–24 (1999)
53. U. Keller, K.J. Weingarten, F.X. Kärtner, D. Kopf, B. Braun, I.D. Jung, R. Fluck, C. Honninger, N. Matuschek, J.A. der Au, Semiconductor saturable absorber mirrors (SESAM's) for femtosecond to nanosecond pulse generation in solid-state lasers. IEEE. Sel. Top. Quant. Electr. **2**(3), 1077–1079 (1996)

54. U. Keller, Recent development in compact ultrafast lasers. Nature **424**, 831–838 (2003); U. Keller, Ultrafast solid-state laser oscillators: a success story for the last 20 years with no end in sight. Appl. Phys. B. **100**(1), 15–28 (2010)
55. M.V. Gorbunkov, A.V. Konyashkin, P.V. Kostryukov, V.B. Morozov, A.N. Olenin, V.A. Rusov, L.S. Telegin, V.G. Tunkin, Y. Shabalin, D.V. Yakovlev, Pulsed-diode-pumped, all-solid-state, electro-optically controlled picosecond Nd:YAG lasers. Quant. Electron. **35**(1), 2–6 (2005)
56. A. Del Corno, G. Gabetta, G.C. Reali, V. Kubecek, J. Marek, Active-passive mode-locked Nd:YAG laser with passive negative feedback. Opt. Lett. **15**(13), 734–736 (1990)
57. A.A. Karnaukhov, V.B. Morozov, A.N. Olenin, D.V. Yakovlev, Precise synchronization of qcw pumped active-passive mode locked picosecond lasers. J. Phys. Conf. Ser. **414**, 012–027 (2013)
58. A. Sennaroglu, *Solid-State Lasers and Applications* (2007), pp. 1–76
59. P. Wang, S.-H. Zhou, K.K. Lee, Y.C. Chen, Picosecond laser pulse generation in a monolithic self-Q-switched solid-state laser. Opt. Comm. **114**, 439–441 (1995)
60. J.J. Zayhowski, Passively Q-switched Nd:YAG microchip lasers and applications. J. All. Comp. **303–304**, 393–400 (2000)
61. D. Nodop, J. Limpert, R. Hohmuth, W. Richter, M. Guina, A. Tünnermann, High-pulse-energy passively Q-switched quasi-monolithic microchip lasers operating in the sub-100-ps pulse regime. Opt. Lett. **32**(15), 2115–2117 (2007)
62. G.J. Spühler, R. Paschotta, R. Fluck, B. Braun, M. Moser, G. Zhang, E. Gini, U. Keller, Experimentally confirmed design guidelines for passively Q-switched microchip lasers using semiconductor saturable absorbers. J. Opt. Soc. Am. B. **16**(3), 376–388 (1999)
63. B. Ryvkin, E. Avrutin, J. Kostamovaara, Asymmetric-waveguide laser diode for high-power optical pulse generation by gain switching. IEEE J. Lightwave Technol. **27**(12), 2125–2131 (2009)
64. J.E. Murray, W.H. Lowdermilk, Nd:YAG regenerative amplifier. J. Appl. Phys. B. **51**(7), 3548–3555 (1980)
65. P. Bado, M. Bouvier, J.S. Coe, Nd:YLF mode-locked oscillator and regenerative amplifier. Opt. Lett. **12**(5), 319–321 (1987)
66. J.C. Postlewaite, J.B. Miers, C.C. Reiner, D.D. Dlott, Picosecond Nd:YAG regenerative amplifier with acoustooptic injection and electrooptic VFET pulse switchout. IEEE J. Quant. Electr. **24**(2), 411–417 (1988)
67. M.D. Dawson, W.A. Schroeder, D.P. Norwood, A.L. Smirl, J. Weston, R.N. Ettelbrick, R. Aubert, Characterization of a high-gain picosecond flash-lamp-pumped Nd:YAG regenerative amplifier. Opt. Lett. **13**(11), 990–992 (1988)
68. D.R. Walker, C.J. Flood, H.M. van Driel, U.J. Greiner, H.H. Klingenberg, High power diode-pumped Nd:YAG regenerative amplifier for picoseconds pulses. Appl. Phys. Lett. **65**(16), 1992–1994 (1994)
69. M. Siebold, M. Hornung, J. Hein, G. Paunescu, R. Sauerbrey, T. Bergmann, G. Hollemann, A high-average-power diode-pumped Nd:YVO_4 regenerative laser amplifier for picoseconds pulses. Appl. Phys. B **78**, 287–290 (2004)
70. J. Kleinbauer, R. Knappe, R. Wallenstein, Ultrashort pulse lasers and amplifiers based on Nd:YVO_4 and Yb:YAG bulk crystals, in: *Femtosecond Technology for Technical and Medical Applications,* ed. by F. Dausinger, F. Lichtner, H. Lubatschowski; Topics Appl. Phys. **96**, 17–34 (2004)
71. M. Lührmann, C. Theobald, R. Wallenstein, J.A. L'huillier, High energy cw-diode pumped Nd:YVO_4 regenerative amplifier with efficient second harmonic generation. Opt. Expr. **17**(25), 22761–22766 (2009)
72. M.V. Gorbunkov, P.V. Kostryukov, V.B. Morozov, A.N. Olenin, L.S. Telegin, V.G. Tunkin, D.V. Yakovlev, Spatial radiation intensity distribution of linear diode arrays and calculation of inversion in fiber-coupled end-pumped solid-state lasers. Quant. Electr. **35**(12), 1121–1125 (2005)
73. V.B. Morozov, A.N. Olenin, V.G. Tunkin, D.V. Yakovlev, Operation conditions for a picosecond laser with an aberration thermal lens under longitudinal pulsed diode pumping. Quant. Electr. **41**(6), 508–514 (2011)

74. https://www.jenoptik.us/products/lasers/high-power-diode-lasers/diode-laser-modules
75. H.A. Haus, Mode-locking of lasers. IEEE J. Sel. Top. Quant. Electr. **6**(6), 1173–1185 (2000)
76. N.G. Mikheev, V.B. Morozov, A.N. Olenin, D.V. Yakovlev, Picosecond lasers with the dynamical operation control. Proc. SPIE **9917**, 99170A (2016)
77. D.J. Kuizenga, A.E. Siegman, FM and AM mode locking of the homogeneous laser-part I: theory. IEEE J. Quant. Electr. **QE-6**(11), 694–708 (1970)
78. H. Roskos, T. Robl, A. Seilmeier, Pulse shortening to 25 ps in a cw mode-locked Nd:YAG laser by introducing an intracavity etalon. Appl. Phys. B **40**, 59–65 (1986)
79. Q.S. Panga, Y. Liub, L. Changa, L.Z. Xua, C. Yanga, M. Chena, G. Lia, Adjustable picosecond pulse duration in a LD end pumped SESAM passively mode-locked Nd:YVO_4 laser. Laser Phys. **21**(6), 1009–1012 (2011)
80. A.E. Siegman, How to (maybe) measure laser beam quality. OSA TOPS **17**, 184–199 (1998)
81. A.E. Siegman, Defining the effective radius of curvature for a nonideal optical beam. IEEE J. Quant. Electr. **27**(5), 1146–1148 (1991)
82. L.M. Frantz, J.S. Nodvik, Theory of pulse propagation in a laser amplifier. J. Appl. Phys. **34**, 2346 (1963)
83. C. Dolda, G. Eberleb, K. Jefimovsc, M. Axtnerd, F. Pudea, K. Wegenera, Analysis of damage thresholds of laser scanning mirrors using ultrashort laser pulses. Phys. Procedia **12**, 445–451 (2011)
84. D.E. Zelmon, K.L. Schepler, S. Guha, D.J. Rush, S.M. Hegde, L.P. Gonzalez, J. Lee, Optical properties of Nd-doped ceramic yttrium aluminum garnet. Proc. SPIE. **5647**; *Laser-Induced Damage in Optical Materials*, vol. 2004 (2005), pp. 255–264

Chapter 5
Gravitational-Wave Astronomy by Precision Laser Interferometry

Norikatsu Mio

Abstract Gravitational waves are the waves of gravitational interaction, which were predicted by Einstein's theory of general relativity. One hundred years after the theoretical prediction, in February 2016, the US LIGO (laser Interferometer Gravitational Wave Observatory) project reported that LIGO had detected a gravitational wave occurring at the coalescence of a black hole binary. Furthermore, the Nobel Prize in Physics in 2017 was awarded to three American physicists who made an outstanding contribution to the LIGO project. Since then, six gravitational wave events have been observed during two observation periods. Moreover, astronomy combined with the observation of electromagnetic waves in a wide wavelength range from radio waves to gamma rays is beginning. In this chapter we will outline the fundamental nature of gravitational waves, their detection method, and this new astronomy.

5.1 Introduction

Gravitational waves (GWs) are the waves of gravitational interaction, which were predicted by Einstein's theory of general relativity (GR)[1] more than 100 years ago. According to GR, a GW appears as strain propagating as a transverse wave with the speed of light. When the theory was proposed, the effect of GWs was considered to be too small to be detected. However, pioneers had started to experimentally investigate the possibility of GW detection in the 1960s. The most famous attempt to detect GWs was by Weber using a cylindrical aluminum body; this detector is known as a Weber bar. He observed the vibration of an elastic mode of the detector. When a GW is incident to the detector, this mode is excited. If the frequency of the GW is close to its resonance, the excitation will be enhanced. In 1969, Weber reported that he could detect GWs; he found coinciding events that were detected by two independent detectors operated at distant locations, one in Maryland and the other in Argonne

[1] The basics on GR are described in the well-known textbook "Gravitation" [1].

N. Mio (✉)
Institute for Photon Science and Technology, University of Tokyo, Bunkyo-ku, Tokyo, Japan
e-mail: mio@ipst.s.u-tokyo.ac.jp

K. Yamanouchi et al. (eds.), *Progress in Photon Science*, Springer Series in Chemical Physics 119, https://doi.org/10.1007/978-3-030-05974-3_5

[2]. After his report, several experiments to confirm his result were performed but no plausible events were detected. In addition, theoretically unreasonable points were found. Thus, Weber's events are not considered to have been GWs. We had to wait a further 50 years before their actual detection. In 2016, the LIGO project announced that they had detected a GW emitted upon the coalescence of a black hole (BH) binary [3].

Since the first detection, six events have been observed: five from BH binaries and one from a neutron star (NS) binary [3–8]. The information obtained from GWs is different from that obtained from observations using an electromagnetic wave (EMW). We are now strongly convinced that GW astronomy is a powerful tool for understanding our Universe.

5.2 Sources of GWs

What can we discover by observing GWs? To discuss this issue, we need to know the mechanism of GW generation. Although a strict discussion requires GR, here we will proceed with a discussion by analogy with EMWs.

EMWs are generated when an electric charge distribution varies with time. For example, charged particles emit an EMW when accelerated, which is derived from the radiation process called electric dipole radiation. An electric dipole moment is defined by

$$\mathbf{p}(t) = \int \mathbf{r}\rho_{\mathrm{e}}(\mathbf{r}, t)\mathrm{d}^3\mathbf{r}, \tag{5.1}$$

where $\rho_{\mathrm{e}}(\mathbf{r}, t)$ is the electric charge distribution. The electric power radiated from a dipole moment is calculated as

$$W = \frac{\mu_0}{6\pi c}\overline{\left|\frac{\mathrm{d}^2\mathbf{p}}{\mathrm{d}t^2}\right|^2}, \tag{5.2}$$

where μ_0 is the vacuum permeability, c is the light velocity [9], and the overline represents the time average.

According to GR, a GW is due to the time variation of the mass distribution $\rho_{\mathrm{m}}(\mathbf{r}, t)$. Since the mass distribution is always positive, the dipole moment always vanishes when we take a centroid coordinate system. Therefore, a GW is generated by the time-dependent mass-quadrupole moment defined by

$$Q_{ij}(t) = \int (x_i x_j - \frac{1}{3}\delta_{ij} r^2)\rho_{\mathrm{m}}(\mathbf{r}, t)\mathrm{d}^3\mathbf{r}, \tag{5.3}$$

where $i, j = 1, 2, 3$ and $x_1 = x, x_2 = y, x_3 = z, r = |\mathbf{r}| = \sqrt{x^2 + y^2 + z^2}$.

The radiation power of a GW can be calculated using the following quadrupole formula:

$$W_{\mathrm{GW}} = \frac{G}{5c^5}\sum_{i,j=1}^{3}\overline{\frac{\mathrm{d}^3 Q_{ij}}{\mathrm{d}t^3}\frac{\mathrm{d}^3 Q_{ij}}{\mathrm{d}t^3}}. \tag{5.4}$$

For example, for the system where two point masses of mass m are performing circular motion with radius a and angular velocity ω around their center of mass, the energy released per unit time is given by

$$W_{\mathrm{GW}} = \frac{128G}{5c^5}m^2a^4\omega^6. \tag{5.5}$$

Since gravitational interaction is weak compared with electromagnetic interaction, the energy emitted from artificial objects on the Earth never reaches the detectable level. The energy released from typical objects on the Earth ($a = 1\,\mathrm{m}$, $m = 100\,\mathrm{kg}$, $\omega/2\pi = 100\,\mathrm{Hz}$) is

$$W_{\mathrm{GW}} = 4.3 \times 10^{-31}\,\mathrm{W}. \tag{5.6}$$

The energy emitted during one rotation is calculated as

$$\Delta E = 4.3 \times 10^{-33}\,\mathrm{J} = 0.03\hbar(2\omega). \tag{5.7}$$

Here, $\hbar$ is the Planck constant divided by 2π, and if the gravitational field is quantized, $\hbar(2\omega)$ represents the energy of one quantum.[2] This result shows that it is impossible to effectively generate GWs on the Earth.

Thus, we expect detectable GWs from astronomical objects; the larger the scale concerned, the more dominant the gravitational interaction. As a model calculation of the GW energy from an astronomical phenomenon, we consider a binary system.

Two stars are bound by gravity and move around their center of mass. For simplicity, we consider the case where stars of equal mass are moving circularly. Since the rotation speed is determined by the balance between the centrifugal force and gravitational force, the relation

$$ma\omega^2 = G\frac{m^2}{(2a)^2} \tag{5.8}$$

must be satisfied. Using this formula, we obtain the radiated GW energy as

$$W_{\mathrm{binary}} = \frac{2G^4m^5}{5c^5}\frac{1}{a^5}. \tag{5.9}$$

This equation shows that the smaller the radius of the orbit the greater the energy released.

[2] The angular frequency of the generated GW is 2ω.

Since the radius is limited by the size of the star, systems that emit a lot of GWs are binaries composed of compact stars such as an NS and a BH.

NSs are thought to be formed by a supernova explosion that produces a core that is made of highly compressed materials mainly composed of neutrons. A typical mass of an NS is considered to be $1.4M_\odot$ and a typical radius is 10 km ($M_\odot$ is the solar mass, 2.0×10^{30} kg). Then, assuming $m = 1.4M_\odot$ and $a = 1000$ km, we obtain $W_{\text{binary}} = 5.6 \times 10^{38}$ W. Since the radiant energy of the Sun is 3.9×10^{26} W, it can be understood that an extremely large amount of energy is released. If this source is at the center of our galaxy (the distance from the earth is about 10 kpc[3]), the energy density on the earth is $4.7 \times 10^{-4}\,\mathrm{W/m^2}$. Sirius, known as a bright star, emits 20 times more energy than the Sun. Since its distance is 2.6 pc, only a photon energy of $9.8 \times 10^{-8}\,\mathrm{W/m^2}$ reaches the Earth.

Despite the very weak gravitational interactions, such a large energy is released because a GW is generated by the movement of the mass of a whole star. Indeed, in the above example, objects of the mass with the Sun move at a speed of 2 % of the light speed.

A GW carries information reflecting the state of motion of such a system. Since the waveform of a GW from a binary system can be predicted, the detection efficiency can be increased by data processing. By analysis of the waveform, it is possible to determine the mass of the original stars, the angular momentum of motion, the state of the star, and so forth, as astronomical information that cannot be obtained from EMWs.

In this system, energy is lost by the emission of GWs, and the state of the orbit changes. Since the total energy of the system is given by

$$E = 2 \times \frac{1}{2} m a^2 \omega^2 - G\frac{m^2}{2a} = -G\frac{m^2}{4a}, \tag{5.10}$$

and the rate of the change in energy is given by (5.9), we obtain

$$\frac{\mathrm{d}E}{\mathrm{d}t} = G\frac{m^2}{4a^2}\frac{\mathrm{d}a}{\mathrm{d}t} = -\frac{2G^4 m^5}{5c^5}\frac{1}{a^5}. \tag{5.11}$$

Solving the above equation, we obtain the equation for the change in a as

$$\frac{\mathrm{d}a}{\mathrm{d}t} = -\frac{\Lambda}{4a^3} \tag{5.12}$$

$$\Lambda = \frac{32G^3 m^3}{5c^5}. \tag{5.13}$$

Assuming that when $t = t_0$, $a = 0$, we obtain

$$a(t) = [\Lambda(t_0 - t)]^{1/4} \tag{5.14}$$

---

[3] 1 pc is a distance commonly used in astronomy. 1 pc $= 3.08 \times 10^{16}$ m $= 3.26$ light years.

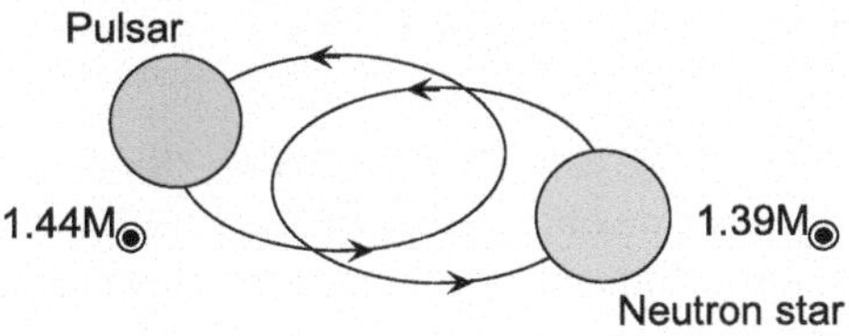

Fig. 5.1 Schematic view of the binary pulsar PSR1913+16

and

$$\omega(t) = \frac{\sqrt{Gm}}{2}[\Lambda(t_0 - t)]^{-3/8}. \tag{5.15}$$

These equations show that a is decreasing as a function of t, while the angular velocity of the rotation is increasing.

The existence of GWs was confirmed by observing such a celestial body. Figure 5.1 is a schematic representation of the NS binary system containing a pulsar called PSR1913+16.

A pulsar is a high-speed rotating NS that generates periodic pulses. However, a large fluctuation was observed in the signal cycle of this pulsar, and it turned out that it is a binary system. Then, by observing the state of the pulse signal, the state of the orbit of the binary system was determined very precisely. The orbital period is about 8 h and the radius of the orbit is approximately the diameter of the Sun (about 10^9 m). Furthermore, it was found that the orbital period of this NS binary is gradually decreasing at a rate of

$$\frac{dP_{obs}}{dt} = (-2.425 \pm 0.002) \times 10^{-12}\,\mathrm{s/s}. \tag{5.16}$$

When calculating the emission energy of the GW using the parameters obtained from the observation, the observed reduction in the orbital period is in good agreement with the theoretical prediction by GR of

$$\frac{dP_{GR}}{dt} = (-2.41958 \pm 0.005) \times 10^{-12}\,\mathrm{s/s}. \tag{5.17}$$

These values coincide with an accuracy of about 0.2% [10]. The Nobel Prize in Physics in 1993 was given to the discoverers of this NS binary.

A variety of relativistic theories of gravity have been proposed since GR was formulated. The observation of GWs is considered to be important for verifying the dynamical properties of theories of gravity [11].

The binary system is continuing to emit GWs. As a result, the two stars will finally merge and will coalesce in about 3×10^8 years.

Just before the coalescence, the orbital cycle will reach millisecond order and a large GW will be generated. All the events that have already been observed are due to this coalescence of the binary system.

5.3 Detection of GWs

How can we actually detect GWs? Since the equivalence principle holds, the effects of gravity on objects that are freely falling vanishes. For a person riding in an elevator freely falling, gravity disappears. However, an elevator has a finite size. If the gravity field is not uniform, the magnitude of the gravity force will be different at the ceiling and the floor. Therefore, we will feel as if we are stretched when we ride on a freely falling elevator. This effect is the same as the force that causes tides and is called the tidal force.

In Newtonian mechanics, two objects at positions $\mathbf{r}$ and $\mathbf{r} + \Delta\mathbf{r}$, they satisfy the following equations of motion:

$$\frac{\mathrm{d}^2\mathbf{r}}{\mathrm{d}t^2} = -\nabla\phi_\mathrm{g}(\mathbf{r}), \qquad \frac{\mathrm{d}^2(\mathbf{r} + \Delta\mathbf{r})}{\mathrm{d}t^2} = -\nabla\phi_\mathrm{g}(\mathbf{r} + \Delta\mathbf{r}), \tag{5.18}$$

where ϕ_g is the Newtonian gravitational potential. Thus, the relative acceleration is given by

$$\Delta\mathbf{a} = \frac{\mathrm{d}^2\Delta\mathbf{r}}{\mathrm{d}t^2} = -[\nabla\phi_\mathrm{g}(\mathbf{r} + \Delta\mathbf{r}) - \nabla\phi_\mathrm{g}(\mathbf{r})]. \tag{5.19}$$

The right term expresses the tidal force and can be expressed as

$$\Delta a_i = -\sum_{j=1}^{3} \frac{\partial^2\phi_\mathrm{g}}{\partial x_i \partial x_j}\Delta x_j \tag{5.20}$$

with the first-order approximation of $\Delta\mathbf{r}$.

If this tidal force is detected, the effect of gravity can be reliably detected. Since the influence of GWs is also known to be tidal, to detect GWs, it is necessary to measure the relative change between the two points.

Figure 5.2 shows the effect of a GW as field lines of the tidal force; these lines of force represent the relative force acting between two points. As mentioned above, a GW is a transverse wave and has two polarization components similarly to EMWs. The polarizations in the two figures have independent components, called the $+$ mode and $\times$ mode, which correspond to linear polarizations of EMWs. The polarization of GWs can be expressed as the superposition of $+$ and $\times$ modes, similarly to the way that any polarization state of EMWs can be expressed as the superposition of x and y polarizations.

Now considering a GW in the $+$ mode traveling along the z axis with amplitude h, the relative acceleration of two points separated by $(\Delta x, \Delta y)$ in the x–y plane is given by

$$\Delta a_x = \frac{1}{2}\frac{\partial^2 h}{\partial t^2}\Delta x, \qquad \Delta a_y = -\frac{1}{2}\frac{\partial^2 h}{\partial t^2}\Delta y. \tag{5.21}$$

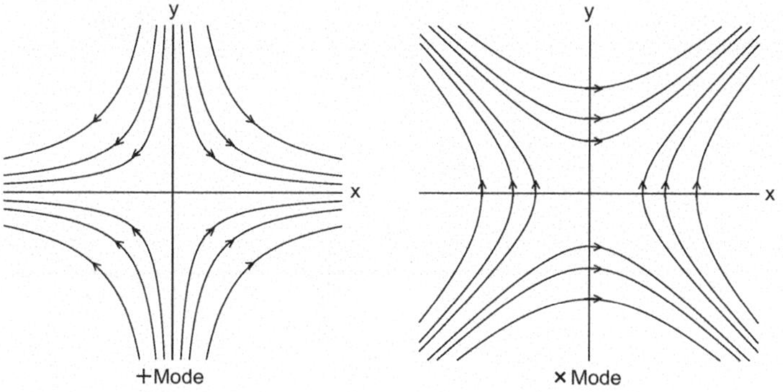

Fig. 5.2 Force lines of GW

For example, if two points separated by L in the x direction on the plane $z = 0$ change their distance by ΔL, $\Delta x = L + \Delta L(t)$ and

$$\Delta a_x = \frac{\mathrm{d}^2 \Delta L}{\mathrm{d}t^2} = \frac{1}{2}\frac{\mathrm{d}^2 h}{\mathrm{d}t^2}(L + \Delta L) \approx \frac{1}{2}\frac{\mathrm{d}^2 h}{\mathrm{d}t^2} L. \tag{5.22}$$

Integrating the above equation with respect to t, we obtain

$$\Delta L = \frac{1}{2} h L. \tag{5.23}$$

Also, in the y direction, the phases of expansion and contraction are inverted. By utilizing this property, we construct a GW detector.

According to a theoretical prediction, the expected value of h is on the order of 10^{-21} or less [12]. If we take L as the distance between the sun and the earth (1.5×10^{11} m), $\Delta L = 7.5 \times 10^{-11}$ m; this is almost the same as the Bohr radius. In other words, it is necessary to detect a relative change that is equivalent to the ratio of the size of a hydrogen atom to the distance between the sun and the earth.

To detect such a weak effect, various detection methods have been developed. Ultimately, laser interferometric detectors successfully detected GWs.

5.4 Laser Interferometric Detectors

As discussed in the previous section, a GW slightly changes the properties of space-time. To detect this effect, we accurately measure the distance between two distant points. For this purpose, a laser interferometer is a powerful tool.

A schematic view of an interferometer is shown in Fig. 5.3. We set two orthogonal optical paths as the x and y axes, When a GW is incident from the direction

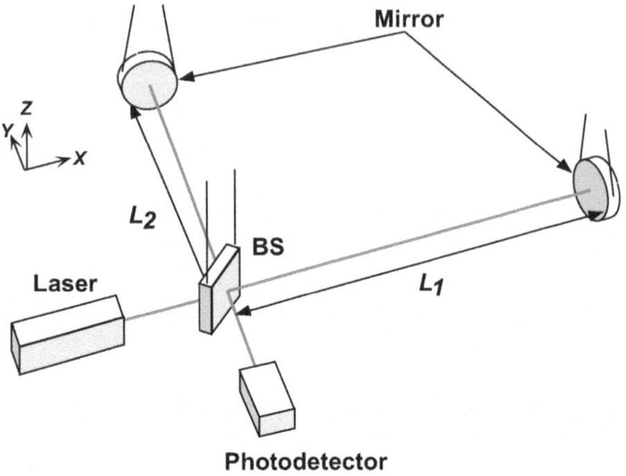

Fig. 5.3 Interferometric GW detector based on Michelson interferometer

perpendicular to the x–y lane (z axis), the x and y directions expand and contract in the x–y plane in opposite directions. A GW oscillates at a certain frequency, which can be observed as a change in the interference fringe of a Michelson interferometer.[4] Mirrors and a beam splitter, which define the optical path length, are suspended similarly to a pendulum. In such a setup, they behave as free masses. By doing so, the interferometer correctly responds to the GW. In addition, it is possible to reduce the influence of external disturbances.

The phase change of the interferometer $\Delta\phi$ is given in terms of the optical path difference $\Delta L_x - \Delta L_y$ as

$$\Delta\phi = \frac{4\pi}{\lambda}(\Delta L_x - \Delta L_y) = \frac{4\pi}{\lambda}hL \tag{5.24}$$

where (5.23) is used to calculate the optical path change as

$$\Delta L_x = -\Delta L_y = hL/2. \tag{5.25}$$

These equations show that the sensitivity increases as the value of L increases. However, if the time for light to travel back and forth through the optical path ($\tau = 2L/c$) becomes longer than a half period of the GW, the phase change due to its effect is integrated and reduced. When τ is on the same order as the period of the GW, the phase change in the Michelson interferometer should be calculated more rigorously using

$$\Delta\phi = \frac{4\pi c}{\lambda\omega}\sin(\omega\tau/2)\exp(-i\omega\tau/2)h_0\exp(i\omega t). \tag{5.26}$$

[4]Here, we assumed that the GW is polarized in the + mode. In this configuration, the detector has no sensitivity to GWs in the × mode.

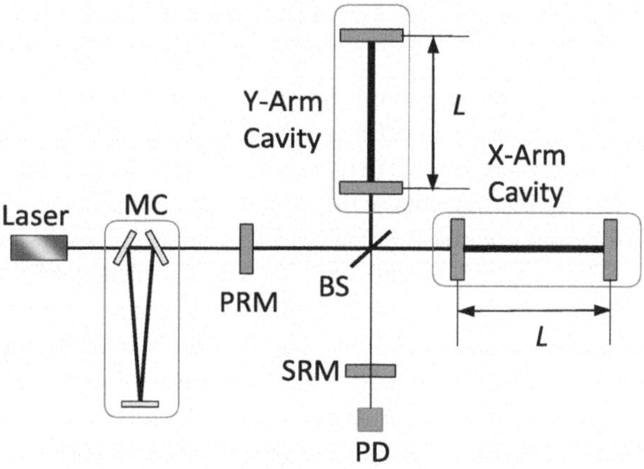

Fig. 5.4 Interferometric GW detector based on Michelson interferometer

Here, we assume that the GW is monochromatic and can be expressed by

$$h(t) = h_0 \exp(i\omega t), \tag{5.27}$$

where ω is the angular frequency of the GW. If $\omega\tau \ll 1$, (5.26) becomes (5.24). Also, from (5.26), $|\Delta\phi|$ becomes maximum if $\omega\tau = \pi$; this gives the optimal condition for the arm length of the interferometer. When we assume that the frequency of the GW is 1 kHz, the optimum value for L is 75 km. In an actual detector, as shown in Fig. 5.4, by incorporating Fabry-Perot optical cavities in both arms of a Michelson interferometer, the sensitivity can be enhanced even for short L of km order. In addition, it is necessary for the laser used as the light source to realize high power and high stability. There is an optical resonator called a mode cleaner (MC) between the interferometer and the laser that shapes the spatial mode of the emitted light and is also used as a reference for frequency stabilization [13]. A power-recycling mirror (PRM) between the MC and the interferometer returns the light to the interferometer again, increasing the effective optical power. It is possible to suppress the influence of shot noise by this technique (called power recycling). There is another recycling mirror at the output port of the beam splitter (BS) called the signal-recycling mirror (SRM), which controls the frequency dependence of the detector sensitivity to GWs. Since very small signals must be measured, to avoid disturbance, the interferometer is housed in a vacuum chamber. The reference mirrors are strictly vibration-isolated. Furthermore, the interferometer requires various control systems so that it can operate stably.

5.5 Noise of the Interferometric Detector

In this section, the noise sources that limit the sensitivity of the interferometer are discussed. In the interferometer, the displacement of the mirror is measured by the interference of light. There are various problems in reading small displacements; any noise source may mask the signal. The main noise sources are shot noise, thermal noise, and seismic noise (Fig. 5.5).

The shot noise determines the smallest detectable change in interference fringes. The shot noise originates from the quantum nature of light, which is a collection of photons; thus, this noise is one of the most fundamental noises. The intensity of light, which is calculated from the number of incident photons N during a unit time, does not become constant because N fluctuates in accordance with quantum effects. The fluctuation of N is given by the square root of N (Poisson process), namely $\delta N = \sqrt{N}$. Since the magnitude of a signal is proportional to N, the signal-to-noise ratio is determined by $N/\sqrt{N} = \sqrt{N}$. Since N is proportional to the power of the light source, a laser with a large power is necessary for the light source to read small changes.

According to detailed calculations, the shot-noise-limited sensitivity of the interferometer to a GW of frequency f is

$$h_{\rm shot} = 10^{-21}\left(\frac{f}{1\,{\rm kHz}}\right)\sqrt{\left(\frac{\lambda}{1\,\mu{\rm m}}\right)\left(\frac{1\,{\rm kW}}{P}\right)\left(\frac{\Delta f}{1\,{\rm kHz}}\right)}, \tag{5.28}$$

where Δf is the bandwidth of the measurement. Here, we assume that a laser with power P at wavelength λ is used with the optimized optical path length.

Ground vibration has a large influence at low frequencies. The ground is continuously shaking even when there is no earthquake. A device (referred to as a vibration

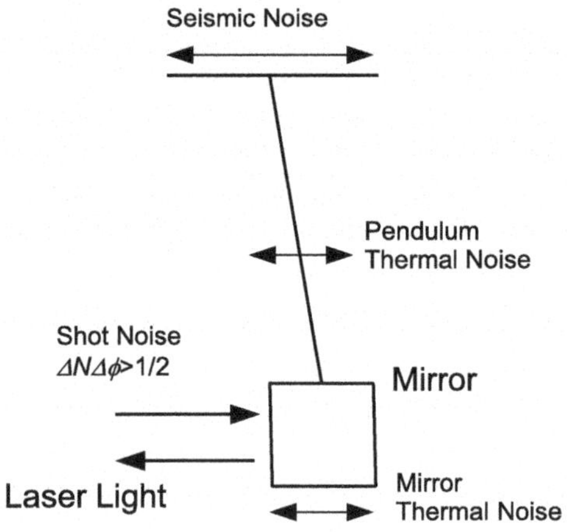

Fig. 5.5 Noise sources of the detector

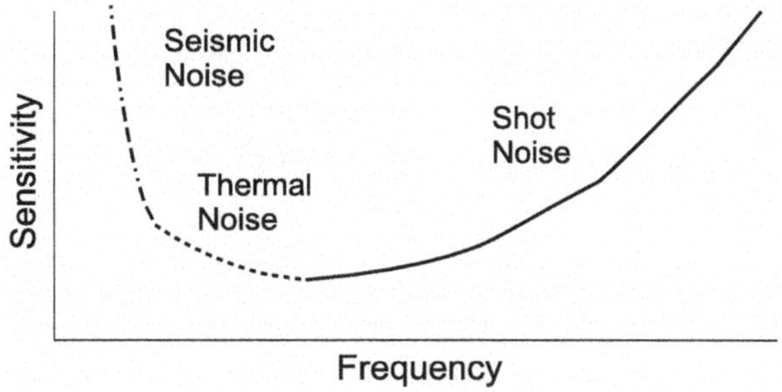

Fig. 5.6 Sensitivity of the detector shown as a function of frequency

isolation system) for blocking this noise can attenuate the magnitude of vibration in the measurement frequency band by a factor of about 10^8.

If the external disturbance is sufficiently removed, the thermal vibrations of the reflection surface of the mirror and the hanging pendulum become the dominant noise sources. These are due to thermal statistical mechanical fluctuation and are also considered also as a principal source of noise.

To reduce this noise, the most direct method is to lower the temperature. In addition, it is necessary to select materials with low mechanical loss. For this reason, high-purity synthetic silica is currently used. For the low-temperature environment, sapphire or other monocrystalline substrates must be used. The properties of single-crystal sapphire have been studied for use in the KAGRA interferometer being constructed.

The dominant noise changes with the frequency. At low frequencies (10 Hz or less), ground vibration is the largest noise. At intermediate frequencies (10–100 Hz order), thermal noise is high. At higher frequencies, shot noise becomes the main noise source. Figure 5.6 represents a schematic sensitivity curve limited by the noise source of the interferometer as a function of frequency.

5.6 Major Projects

An interferometer that can actually detect GWs is extremely large; such an interferometer is qualitatively different from ordinary optical devices used in laboratories. The construction of such an interferometer is a major project requiring a large amount of human resources, a huge budget and a very long time.

The LIGO project has two 4-km interferometers that have been constructed on the east coast (Livingstone, Louisiana) and west coast (Hanford, Washington) of the US. The budget was approved in 1990 and the construction began in 1992. In the autumn of 2000, the first interference fringe was observed. After that, the system, called initial LIGO, achieved its designed sensitivity in 2005. Advanced LIGO started in 2008, in which new vibration isolation systems, new mirrors, new lasers, and so forth

Table 5.1 Web sites of GW projects

LIGO	http://www.ligo.caltech.edu/
LSC	https://www.ligo.org/
VIRGO	http://www.virgo-gw.eu/
KAGRA	http://gwcenter.icrr.u-tokyo.ac.jp/en/

were developed and installed to the detectors. These efforts resulted in the successful detection of GWs [14].

In Europe, France and Italy are collaborating to build a detector named VIRGO, located near Pisa in Italy, with an interferometer of length 3 km. To detect low-frequency GWs, a special vibration isolation device was being introduced.

LIGO and VIRGO are now working together as LIGO-VIRGO collaboration. The first detection of a GW was simultaneously announced in the US and Europe.

In Japan, a 3-km interferometer named KAGRA is being constructed inside Kamioka mine. Many research institutes are participating.

The web URLs of these projects are shown in Table 5.1. For details, see these web pages.

5.7 KAGRA

Here, the Japanese project KAGRA is discussed [15]. This project was started in 2010 with the aim of constructing a 3-km-long interferometer at the underground site of Kamioka mine in Gifu prefecture, shown in Fig. 5.7.

To suppress thermal noise, a cryogenic technology to cool the arm mirrors has been introduced. Owing to their suitability for use in a cryogenic environment, large sapphire crystals are used as substrates for the arm mirrors. The target sensitivity is sufficient to detect GWs from the coalescence of an NS binary. When the KAGRA detector is operated, it will provide valuable information to the worldwide network of GW detectors because the distant location of KAGRA from the other detectors is advantageous in determining direction of the GW sources.

The construction of the tunnel of KAGRA was finished in 2014 and the huge vacuum system was installed. The first trial to obtain an interference signal was carried out in 2016, which was successful [16]. After that, the cryogenic system was installed. In 2018, an end test mass made of a single-crystal sapphire was installed in the KAGRA system. The first trial operation of the cryogenic interferometer without arm cavities has started in May 2018.

KAGRA requires many advanced technologies concerning optics, mechanics, cryogenics, control systems, and so forth. Our group is working on the development of the light source for KAGRA. In the following section, we describe the light source.

Fig. 5.7 Overview of KAGRA (offered by Shinji Miyoki at ICRR, the University of Tokyo)

5.8 Light Source for KAGRA

To realize the target sensitivity of KAGRA, the noise level must be controlled to the target value. For shot noise, the laser power limits the noise level and must be larger than 180 W with single-frequency and single-mode CW oscillation assuming a laser wave length of 1064 nm. There are two ways of realizing such laser oscillation: one is to use a master oscillator and power amplifier (MOPA) and the other is by injection locking. We have developed a laser system based on injection locking and achieved a 100 W single-frequency output [17, 18]. However, the injection-locking system requires a servo control system to maintain the injection-locked state and needs fine tuning in alignment. Since the complexity of the allover detector system should be minimized, we adopted a MOPA system for KAGRA. A schematic view of the KAGRA laser system is shown in Fig. 5.8; the system consists of a seed laser (NPRO), two fiber-laser amplifiers, and a three-stage solid-state laser amplifier.

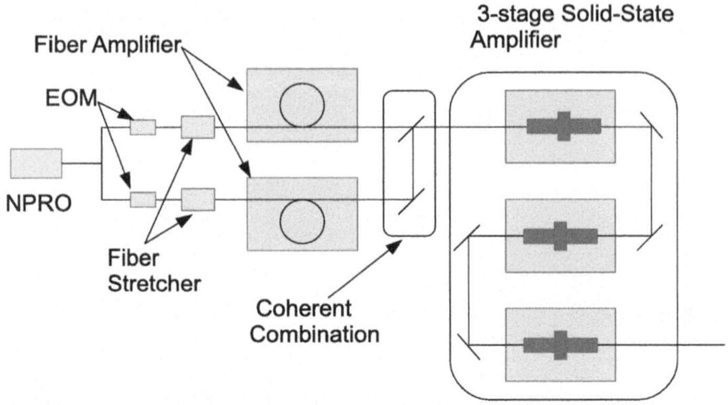

Fig. 5.8 Laser system for KAGRA

The seed laser emits a small power (about 400 mW) with very high frequency stability at 1064 nm. The output of the seed laser is divided into two optical paths using an optical-fiber-based system. An EOM and a fiber stretcher are placed on each optical path to control the phase of the laser light. The two outputs of the divided optical paths are introduced into two fiber-laser amplifiers with 40 W output power. These two laser beams are coherently combined to obtain a higher power than that necessary for the solid-state laser to work in the saturated amplification regime. After the coherent combination, the laser light is incident to the three-stage solid-state amplifier, which is composed of LD-pumped Nd:YAG laser modules. As mentioned above, by adopting the MOPA scheme, the operation of the laser becomes easier than that based on injection locking. Also, we can use fiber laser technology, which enables high-quality amplification. However, narrow-band amplification with a fiber laser sometimes becomes unstable because of nonlinear effects, mostly stimulated Brillouin scattering. Thus, single-frequency amplification to the level required for KAGRA is very difficult and we adopted a simple solid-state laser for the last stage of amplification. We have succeeded in high-efficiency coherent combination. As shown in Fig. 5.9, we obtained 78 W output from two 41 W outputs with a good optical mode. The efficiency of the coherent combination was 95 %.

Although we have also performed amplification using the solid-state laser, the output power with the fundamental optical mode was limited to about 150 W, which is not sufficient for use in KAGRA. We are currently considering how to improve the system to obtain the target laser power.

Regarding the frequency noise, we have confirmed that the noise level of the combined and amplified output was almost the same as that of the NPRO as shown in Fig. 5.10.

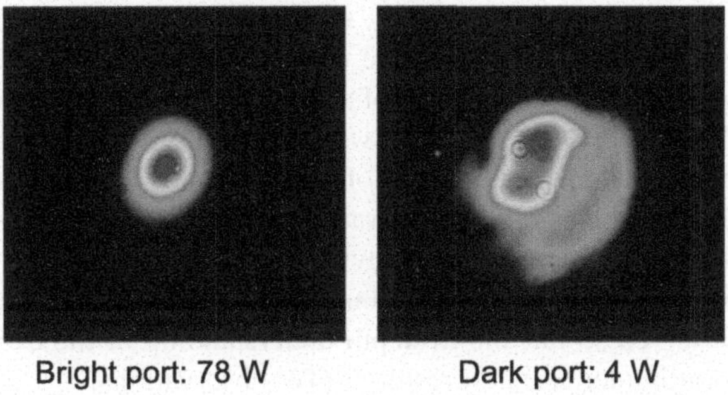

Fig. 5.9 Results of the coherent combination

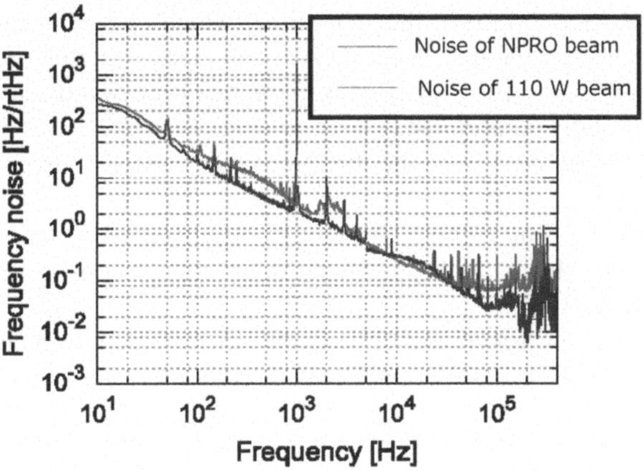

Fig. 5.10 Frequency noise of the laser system

5.9 GW Astronomy

Observation using GWs can clarify the state of motion of a global mass that cannot be determined by other means; this is the most advantageous feature of GW astronomy.

However, the detection of GWs is extremely rare. To obtain a sufficient observation rate (several times a year), we have to observe a wide area of the Universe, and sufficient sensitivity to observe very distant objects is necessary.

Actually, before the first detection of GWs, their most promising source was considered to be the coalescence of NS binaries. Furthermore, from the event rate estimated on basis of the observations of the NS binaries, GWs can be detected several times a year using detectors with a detection distance of about 200 Mpc.

However, the first event for which GWs were observed was BH binary coalescence. BHs of 36 $M_{\odot}$ and 29 $M_{\odot}$ located at a distance of 410 Mpc from the Earth merged into a single BH of 62 $M_{\odot}$. The energy of the emitted GW was about 3 $M_{\odot}c^2$; a very large amount of energy was carried by the GW. The detected amplitude of the GW at LIGO was about 10^{-21}. This was the first proof that BH binaries really exist. The masses of the BHs observed here were heavy; the masses of stellar-mass BHs observed before the first GW detection were smaller than 20 $M_{\odot}$.

Since the first detection, six events have been observed in two observation runs of LIGO and VIRGO. Five of the events were the coalescence of BH binaries. In addition, the coalescence of an NS binary has also been observed; this occurred at a distance of 44 Mpc, which was much closer than expected. From these findings, we consider that the prediction of the abundance of GW sources before the actual detector operation was very conservative; the population of possible sources of GWs seems to be much larger than previously estimated. Regarding the mass of the observed BHs, we found that the BH masses are extremely large among the stellar-mass BHs observed so far; the in-depth understanding of the evolution of BHs is expected to rapidly increase as a result of GW observations.

With regard to the observed coalescence of the NS binary, gamma rays were detected at almost the same time, which arrived at the Earth with 1.7 s delay; this indicates that the speed of GWs coincides with that of EMWs within an uncertainty of $(^{-3}_{+0.7}) \times 10^{-15}$ [19]. This is the first time that the speed of GWs has been experimentally determined.

Moreover, since the direction of a celestial source was first determined from the signal of a GW, observations of the source have been carried out with a wide range of EMWs. The Hubble constant H_0 has been estimated as $H_0 = 70.0^{+12.0}_{-7.0}$ km/s/Mpc on the basis of these observations; this is consistent with previously determined values [20].

A new astronomy, called multi-messenger astronomy involving the simultaneous observation of GWs and EMWs has been developed and is providing valuable results.

5.10 Summary

GW astronomy is in its infancy. In the period of over two years since the first observation of GWs, there have been six further observations, and the information obtained could not have been obtained by other means. Their simultaneous observation with EMWs has already been achieved, providing new knowledge on the Universe. Advanced photon science and technology have played an important role in the detection and will continue to contribute to improving detection sensitivity.

References

1. C.W. Misner, K.S. Thorne, J.A. Wheeler, Gravitation (W. H. Freeman, San Francisco, 1973), ISBN: 978-0-7167-0344-0
2. J. Weber, Evidence for discovery of gravitational radiation. Phys. Rev. Lett. **22**, 1320 (1969)
3. B.P. Abbott et al., (LIGO scientific collaboration and virgo collaboration) (11 Feb 2016), Observation of gravitational waves from a binary black hole merger. Phys. Rev. Lett. **116**, 061102 (2016)

4. B.P. Abbott et al., (LIGO scientific collaboration and virgo collaboration) (15 June 2016), GW151226: observation of gravitational waves from a 22-solar-mass binary black hole coalescence. Phys. Rev. Lett. **116**, 241103 (2016)
5. B.P. Abbott et al., (LIGO scientific collaboration and virgo collaboration) (1 June 2017). GW170104: observation of a 50-solar-mass binary black hole coalescence at redshift 0.2. Phys. Rev. Lett. **118**, 221101 (2017)
6. B.P. Abbott et al., (LIGO scientific collaboration and virgo collaboration) (18 Dec 2017). GW170608: observation of a 19-solar-mass binary black hole coalescence. Astrophys. J. Lett. **851**, L35 (2017)
7. B.P. Abbott et al., (LIGO scientific collaboration and virgo collaboration) (6 Oct 2017) GW170814: a three-detector observation of gravitational waves from a binary black hole coalescence. Phys. Rev. Lett. **119**, 141101 (2017)
8. B.P. Abbott et al., (LIGO scientific collaboration and virgo collaboration) (16 Oct 2017). GW170817: observation of gravitational waves from a binary neutron star inspiral. Phys. Rev. Lett. **119**, 161101 (2016)
9. J.D. Jackson, *Classical Electrodynamics*, Third Edition (Wiley, New York, 1999), ISBN: 0-471-30932-X, OCLC 925677836
10. J.M. Weisberg, J.H. Taylor, in *The Relativistic Binary Pulsar B1913+16: Thirty Years of Observations and Analysis*, ed. by F. A. Rasio, I.H. Stairs. Binary radio pulsars, ASP conference series, vol. 328, proceedings of conference held 11–17 Jan 2004, Aspen, Colorado, USA (Astronomical Society of the Pacific, San Francisco, 2005), p. 25
11. C.M. Will, The confrontation between general relativity and experiment. Living Rev. Relativ. (https://link.springer.com/article/10.12942/lrr-2014-4)
12. K.S. Thone, Gravitational radiation, in *Three Hundred Years of Gravitation*, ed. by S. Hawking, W. Israel (Cambridge University Press, Cambridge, New York, 1987), pp. 300–458
13. S. Reid, S. Rowan, J. Hough, Gravitational wave detection by interferometry (ground and space). Living Rev. Relativ. (https://link.springer.com/article/10.12942/lrr-2011-5)
14. J. Aasi et al., Advanced LIGO. Class. Quantum Grav. **32**, 074001 (2015)
15. T. Akutsu et al., Construction of KAGRA: an underground gravitational-wave observatory. Prog. Theor. Exp. Phys. **2018**, 013F01 (2018)
16. T. Akutsu, KAGRA Collaboration, Large-scale cryogenic gravitational-wave telescope in Japan: KAGRA., J. Phys. Conf. Ser. **610**, 012016 (2015)
17. K. Takeno, T. Ozeki, S. Moriwaki, N. Mio, 100 W, single-frequency operation of an injection-locked Nd:YAG laser. Opt. Lett. **30**, 2110 (2005)
18. N. Ohmae, S. Moriwaki, N. Mio, Wideband and high-gain frequency stabilization of a 100-W injection-locked Nd:YAG laser for second-generation gravitational wave detectors. Rev. Sci. Instrum. **81**, 073105 (2010)
19. B.P. Abbott et al., (LIGO scientific collaboration and virgo collaboration), Gravitational waves and gamma-rays from a binary neutron star merger: GW170817 and GRB 170817A. Astrophys. J. Lett. **848**, L13 (2017)
20. The LIGO Scientific Collaboration and The Virgo Collaboration, The 1M2H Collaboration, The Dark Energy Camera GW-EM Collaboration and the DES Collaboration, The DLT40 Collaboration, The Las Cumbres Observatory Collaboration, The VINROUGE Collaboration & The MASTER Collaboration, A gravitational-wave standard siren measurement of the Hubble constant. Nature **551**, 85 (2017). https://doi.org/10.1038/nature24471

Part II
Biological Applications

Chapter 6
Phosphorescence Lifetime Imaging (PLIM): State of the Art and Perspectives

Pavel S. Chelushkin and Sergey P. Tunik

Abstract This chapter reviews the status and perspectives of phosphorescence lifetime imaging (PLIM), an advanced imaging strategy that relies on phosphorescence lifetime measuring as a function of some particular biological microenvironment parameters. PLIM should be regarded as a functional imaging technique, as opposite to various forms of "localization" techniques, because it provides not only information on distribution pattern of the probe but also determines its "status" (via lifetime reporting).

6.1 Introduction

Luminescence imaging methods are among the most indispensable tools in the biosciences nowadays since they provide a unique combination of sensitivity and resolution in live experiments. The most common probes for luminescence imaging are fluorophores, singlet emitters featuring short lifetimes (0.1–10 ns) and small Stokes shifts (<100 nm) that results in strong interference of their emission with autofluorescence of the endogenous emitters. On the contrary, the triplet (phosphorescent) emitters based on luminescent transition metal complexes display both longer lifetimes (up to milliseconds) and larger Stokes shifts (>100 nm) that pave the way to efficient separation of their signals from background fluorescence. This improvement in signal-to-noise ratio is of particular importance for intensity measurement mode widely used in luminescent microscopy. However, application of this mode in quantitative measurements is considerably restricted because of uneven spatial distribution of the probe across the sample under study and uncertainty in emitting light intensity due to variations in media absorption characteristics.

P. S. Chelushkin · S. P. Tunik (✉)
Institute of Chemistry, St. Petersburg State University, Universitetskii pr., 26,
Saint Petersburg 198504, Russia
e-mail: sergey.tunik@spbu.ru

P. S. Chelushkin
e-mail: p.chelushkin@spbu.ru

K. Yamanouchi et al. (eds.), *Progress in Photon Science*, Springer Series
in Chemical Physics 119, https://doi.org/10.1007/978-3-030-05974-3_6

Application of lifetime-based techniques, referred to as fluorescence (FLIM) and phosphorescence (PLIM) lifetime imaging, makes possible to use the lifetime as a function of some particular biologically relevant microenvironment parameters. Thus, FLIM and PLIM should be regarded as "functional imaging" techniques, as opposite to various forms of "localization" imaging, because they provide not only information on distribution pattern of the probe in the sample but also determine its "status" (via reporting of lifetime).

While FLIM is a well-established part of modern biomedical studies [1, 2], its cognate, PLIM, is just an emerging area [3]. In this Perspective review, we will highlight the cutting edge research which use this method, without intention to cover exhaustively all the existing literature on PLIM, and present our vision of the prospectives of further development in this area with special focus on novel generations of PLIM probes and sensors.

6.2 State of the Art in PLIM Research

As for any other advanced bioimaging techniques, the overall progress in PLIM is a result of mutual synergistic stimulation of at least three areas of knowledge (Fig. 6.1), namely, physics (development of instrumentation and data treatment), chemistry (synthesis of advanced probes), and biomedicine (formulation of acute research directions and development of relevant models). Progress in a particular field opens up novel horizons and poses challenging requirements for the other fields thereby stimulating the overall progress. We will briefly describe in this section state of the art for each of the above fields; for historical survey of PLIM, the reader is referred to [4].

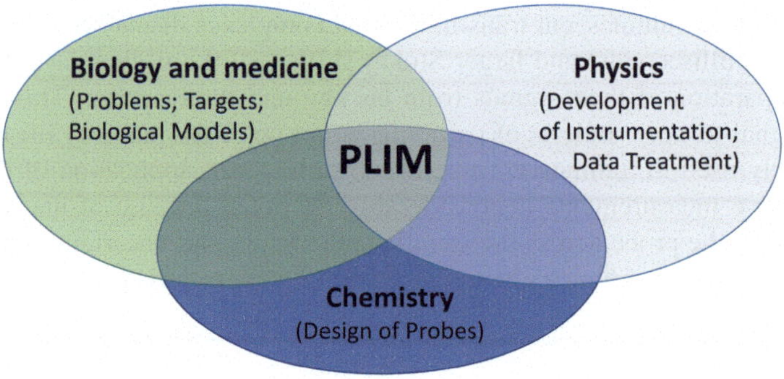

Fig. 6.1 Representation of PLIM as a multidisciplinary research area

6.2.1 PLIM Instrumentation: Confocal PLIM with TCSPC

To date, the most established approach for PLIM implementation is the use of raster scanning (confocal) technique accompanied with time-correlated single photon counting (TCSPC). Despite several promising alternatives (such as wide-field TCSPC PLIM [5] or frequency-domain PLIM [6] that can be performed on commercially available "LIFA-X" instrument from Lambert Instruments BV, Netherlands) this technique suggests at least two key advantages: (i) confocal scanning provides improved axial resolution inaccessible by wide-field techniques, and (ii) well established TCSPC data acquisition mode [7] makes possible robust and precise lifetime measurements compared to frequency-domain approaches although comes short in acquisition rate.

The up-to-date TCSPC FLIM/PLIM microscopy is described in detail elsewhere [8]; herein we briefly outline the essence of the routine. To collect the phosphorescence photons and emission decay of the probe molecules, the sample is scanned pixel-by-pixel, and during the pixel dwell time, the phosphors are first pumped by a high-frequency pulsed laser (the "laser ON" period); the laser is then switched off (the "laser OFF" period), phosphorescence photons are collected, and lifetime distribution is built up (Fig. 6.2a). Notably, the process described above allows simultaneous implementation of FLIM experiment by analyzing the distribution of the fluorescent photons during the "laser ON" period i.e. FLIM and PLIM data can be obtained simultaneously within the same acquisition time (Fig. 6.2b). Using of pulse train instead of single pulse for pumping the phosphorescence is highly advantageous since it maintains high sensitivity while decreasing photo damage and some undesirable processes (such as pile-up effect and detector overload) [8].

Thus, the combined TCSPC FLIM/PLIM experiment simultaneously provides information on the characteristics of fluorophores and phosphors presented in the object studied, it enables correlative mapping of metabolic information (so-called "optical redox ratio" [9], which measures ratio between flavin adenine dinucleotide, FAD, and nicotinamide adenine dinucleotide, NADH), and oxygen distribution derived from luminescence quenching of exogenous phosphorescent probe such as ruthenium tris-(2,2′-bipyridyl) dichloride [10], or covalent conjugates of human serum albumin with Pt(2-phenylpyridine)(triphenylphosphine)Cl complex that exerts "luminescence switch-on" effect via selective binding to histidine residues [11].

The raster scanning TCSPC approach can be extended to the macroscopic TCSPC PLIM. Moreover, confocal PLIM macro scanners (for example, "DCS-120 MACRO", Becker and Hickl, GmbH, Germany) have recently become commercially available. Nevertheless, there are only a few reports on the confocal macroscopic TCSPC PLIM to date [12, 13].

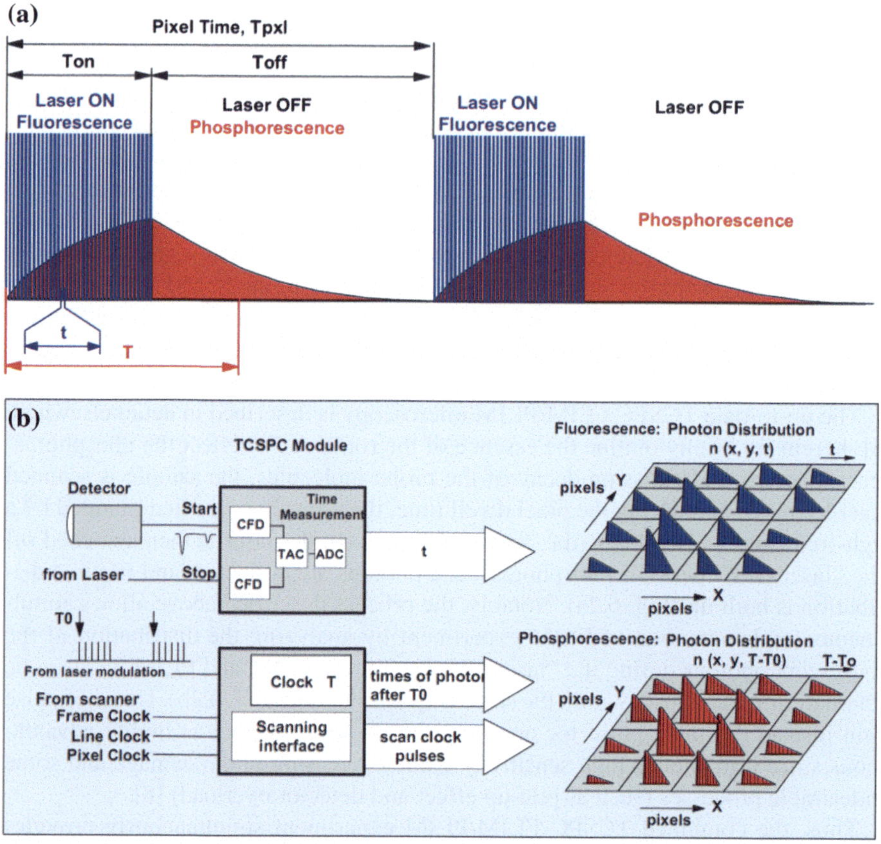

Fig. 6.2 Principle of simultaneous FLIM/PLIM with TCSPC. **a** Modulation scheme: a high-frequency pulsed laser is on-off modulated synchronously with the pixels. FLIM is recorded in the "Laser ON" phases, PLIM in the "Laser OFF" phases. **b** Scheme of simultaneous determination of fluorescence and phosphorescence lifetime distributions. Reprinted with permission from [8]. Copyright 2017 Springer

6.2.2 Biomedical Models

To date, PLIM was used for the variety of biomedical models, including monolayer cell cultures [10, 11], 3D tissue models (cell spheroids and organoids [14]), and living organisms (small animals [15]). In the case of 2D and 3D cell cultures (as well as in the case of in vivo microscopy) confocal TCSPC PLIM (or its FLIM/PLIM version) is the method of choice. In the case of macroscopic PLIM, to date the majority of studies were carried out using the wide-field macroscopic instruments that lack spatial resolution [15]. We believe that dissemination of confocal PLIM macro scanners mentioned above will change this area and eventually will lead to high-resolution in vivo PLIM.

6.2.3 Probes for Practical PLIM

Numerous phosphorescent probes designed for various PLIM applications have been described to date [3], but only one class of these compounds, namely, O_2 PLIM sensors, have been developed into commercial products applicable for oxygen mapping in biological samples. There are two major properties of the phosphorescent emitters, which made possible to use them as O_2 sensors in PLIM mode. The first one is that the excited state of phosphorescent compound and ground state of molecular oxygen are triplets that allows for energy transfer from excited state of the phosphor to O_2 molecule thus quenching the emission of triplet probes. Because the quenching is a dynamic process, it changes phosphorescence lifetime, making these probes excellent candidates for O_2 sensing by PLIM. The second reason consists in a relatively easy design of the probes with the combination of strong lifetime response onto oxygen concentration and minor dependence on other environment variables, including pH, temperature, interactions with proteins, etc.

To date, two approaches for the design of practically applicable PLIM O_2 sensors have been elaborated, which combine high lifetime sensitivity and selectivity with respect to oxygen. The first one was developed by Prof. D. Papkovsky's group and consists in solubilization of hydrophobic organometallic PLIM sensors (various platinum porphyrins, e.g. Pt(II)-*meso*-tetrakis(pentafluorophenyl)porphyrin, PtPFPP, Fig. 6.3) through their incorporation into nanoparticle-forming polymers (for example, cationic polymer Eudragit RL-100, Fig. 6.3) [16]. The resulting nanosensors (e.g. $NanO_2$, Fig. 6.3) easily penetrate into cells in 2D and 3D cell cultures, but are hardly applicable for whole-body in vivo imaging since (i) systemic administration of $NanO_2$ is doubtful because opsonization and phagocytosis caused by positive net charge of sensor nanoparticles [17–19] lead to accumulation cationic nanosensors in liver [20] and (ii) excitation wavelengths of $NanO_2$ (excitation maximum at 395 nm) are far from the first window of transparency of biological tissues (650–900 nm [21]).

The second approach was developed by Prof. Vinogradov's group and includes dendronization of Pd- (referred to as "Oxyphors" [15, 22], Fig. 6.4) or Pt-porphyrins [23, 24] with subsequent modification by polyethylene glycol (PEG) arms. PEGylation is prone to endow macromolecules with prolonged circulation in blood [25]. Though being cell-impermeable and thus inapplicable for intracellular oxygen mea-

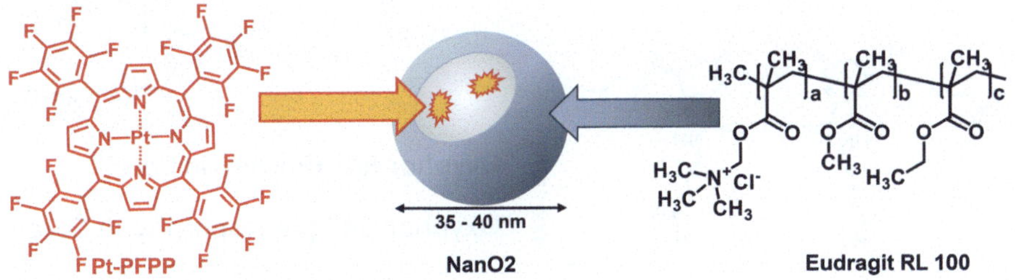

Fig. 6.3 Schematic structure of $NanO_2$ oxygen nanosensor described in [16]

surements, "Oxyphors" and related sensors are suitable for systemic administration and consequently promising for in vivo oxygenation measurements [15, 22].

The both approaches are based on essentially similar Pt/Pd-porphyrin emitters, which demonstrate phosphorescence in NIR area of the spectrum, display strong response to oxygen concentration and are rather insensitive to variations in temperature and pH. Additionally, the outer polymer "shell" (either polymer nanoparticle or branched dendrons) provides steric protection of the luminescent center from interaction with biomolecules, which may considerably and uncontrollably change the probe sensitivity to oxygen. However, this type of protection does not eliminate phosphor's sensitivity to oxygen since O_2 molecule is small enough to easily diffuse through the protective polymeric "shell" and get in contact with the triplet chromophore.

Nevertheless, in the context of increasing PLIM potential for in vivo imaging (*vide supra*) further tuning of photophysical and physico-chemical properties of prospective probes is highly desirable. Several key ideas have already been realized. The first one consists in variations in the structure and composition of metal porphyrins, e.g. in development of the porphyrine aromatic system, which may shift excitation and emission wavelengths into the first window of transparency of biological tissues (650–900 nm). As an example, the structure of Pd-tetrabenzoporphyrins [15] is presented in Fig. 6.4, which differs from standard porphyrin core (drawn by red color) by its conjugation with four benzo-groups (drawn by green color). This idea has led to several near-infrared sensors, including "NanO$_2$-IR" (Excitation/Emission wavelengths are 615/760 nm [26]) and "Oxyphor G4" (Excitation/Emission wavelengths are 637/813 nm [15]), and enabled whole-body tumor hypoxia imaging [15]. Unfortunately, excitation bands of the above probes still do not exactly fall into the window of transparency of biological tissues, and, to the best of our knowledge, there are no phosphors with one-photon excitation wavelength over 650 nm.

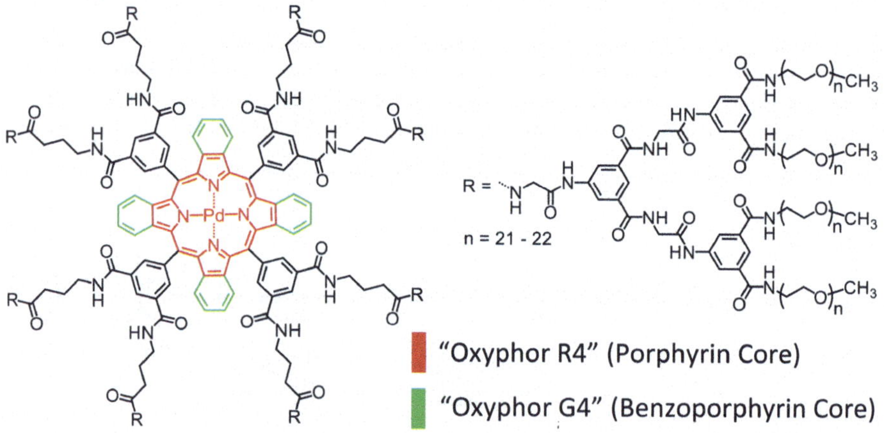

Fig. 6.4 Chemical structures of Oxyphors R4 and G4 described in [15]

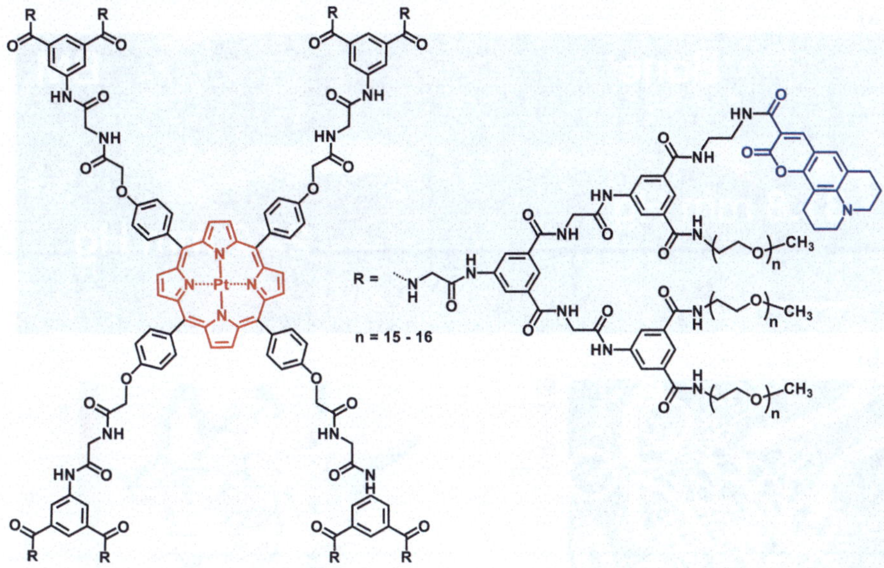

Fig. 6.5 Chemical structure of two-photon PLIM oxygen sensor—platinum porphyrin-coumarin-343 described in [23] and used in [28, 29] (see Fig. 6.6). Two-photon absorbing antenna (Coumarin 343) is depicted by blue, phosphorescent acceptor (Pt porphyrin core) is depicted by red color

The second improvement consists in the synthesis of molecular ensembles containing triplet chromophore with the excitation spectrum fit in the 350–500 nm range and two-photon absorbing antenna excited by femtosecond lasers operating at the wavelengths from 700 to 1000 nm, see Fig. 6.5. Two-photon antenna excitation followed by the resonance energy transfer to the triplet emitter gives resulting emission above 650 nm. In this case both excitation and emission fall into the first window of transparency of biological tissues. This approach requires extremely high density of photon flux, which is a major drawback of the probes of this type, because high light scattering in biological tissues makes it highly diffuse at the depths higher than 1 mm and inapplicable for multiphoton confocal experiments [21]. Consequently, in vivo use of this type of oxygen sensors is restricted to the measurements at depth not exceeding 1 mm [27].

Nevertheless, several two-photon O_2 sensors were developed either via covalent bonding of two-photon absorbing antenna (Coumarin 343, [23]; Coumarin 307, [24]) to PEGylated shell of Pt porphyrins (platinum analogs of "Oxyphors") or via simultaneous incorporation of two-photon antennas (polyfluorene) and Pt-porphyrins into either cationic nanoparticles made of Eudragit RL-100 (MM2 [30]) or anionic poly(methyl methacrylate-*co*-methacrylic acid) nanoparticles, referred to as PA2 [31]. Application of these sensors enabled direct in vivo measurements of local O_2 concentration in the bone marrow through the intact animals skull [28] (Fig. 6.6a), as well as high-resolution measurements of O_2 partial pressure in cerebral vasculature of animals (Fig. 6.6b) [29].

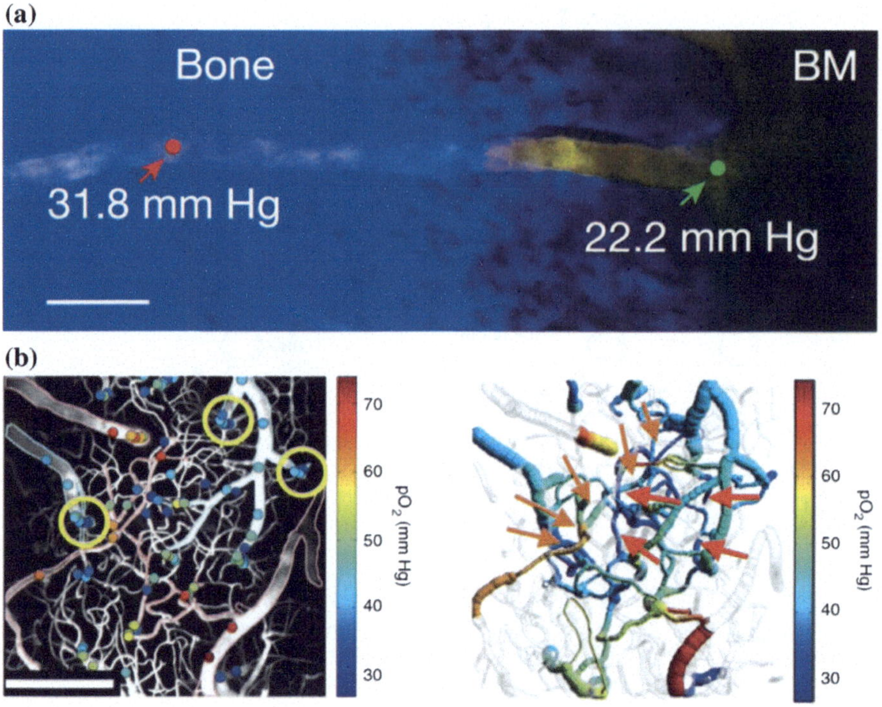

Fig. 6.6 Examples of in vivo two-photon PLIM oxygen sensing of platinum porphyrin-coumarin-343 (see Fig. 6.5 for structure). **a** Maximum intensity projection image montage of a blood vessel entering the bone marrow (BM) from the bone. Bone (blue) and blood vessels (yellow) are delineated with collagen second harmonic generation signal and Rhodamine B—dextran fluorescence, respectively. The two arrows point to locations of pO_2 measurements just before and after the vessel enters the BM. Scale bar: 100 μm. Reprinted with permission from [28]. Copyright 2014 Nature Publishing Group. **b** Measurement of pO_2 in cortical microvasculature. Left: measured pO_2 values in microvasculature at various depths (colored dots), overlaid on the maximum intensity projection image of vasculature structure (grayscale). Digital processing was performed to remove images of the dura vessels. Edges of the major pial arterioles and venules are outlined in red and blue, respectively. Right: composite image showing a projection of the imaged vasculature stack. Red arrows mark pO_2 measurement locations in the capillary vessels at 240 μm depth. Orange arrows point to the consecutive branches of the vascular tree, from pial arteriole (bottom left arrow) to the capillary and then to the connection with ascending venule (top right arrow). Scale bars: 200 μm. Reprinted with permission from [29]. Copyright 2010 Nature Publishing Group

Another prospective way to obtain the NIR excitation/emission probes consists in the use of up-converting antenna instead of two-photon one. This approach also allows shifting of both excitation and emission maxima into the first window of transparency, but, contrary to the previous case, excitation of up-converting antenna requires much less intense laser power and can be used for deep in vivo imaging. Despite the high promising potential of this approach, only a few publications have described this method [32, 33], with only one example of PLIM oxygen imaging by up-converting sensors [32].

In conclusion, the most advanced to date application of PLIM is sensing of oxygen and imaging of its distribution in biological samples. A limited amount of probes based on the Pt/Pd porphyrins have found practical application in biomedicine; their derivatives containing tetra-benzoporphyrins as emitters and either two-photon or up-converting antennae display excitation and emission wavelengths in the windows of transparency of biological tissues that makes them applicable for in vivo microscopic experiments.

6.3 Perspectives of PLIM

As mentioned above, the general progress of PLIM is a result of mutual synergistic influence of at least three major research areas, including development of instrumentation, biological models, and phosphorescent probes.

In the field of instrumentation, we believe that further progress will focus on implementation of more affordable instruments, which will provide shorter acquisition times (now typical PLIM experiment takes from minutes to tens of minutes) accompanied by enhanced spatial resolution. This could be achieved via optical schemes allowing more efficient photon harvesting, using multiple detectors, etc.

In the field of biological models, on the one hand, new animal models are being intensely developed, especially in the areas of neuro- and cancer imaging; so PLIM techniques must progress concurrently to match requirements of new objects to be studied. The ultimate goal in this area is performing of single cell visualization within living and intact body [34]. On the other hand, more sophisticated cell constructs are evolved, and increasing tendency of switching from 2D and “quasi-uniform” (i.e. consisting of one cell type) 3D models, to highly differentiated 3D tissue models including “organoids” and artificially engineered tissues. Reference [14] presents excellent recent overview of implementation and perspectives of PLIM and related imaging techniques in the field of 3D tissue modelling.

The progress in the above-mentioned fields is definitely posing novel challenges for the development of phosphorescent probes compatible with requirements of biomedical experiments (solubility and stability in physiological media, biocompatibility, retaining of sensitivity to analytes in biological samples) and best suitable for effective acquisition of the data in lifetime domain. Below we will highlight the main challenges and emerging areas of design and development of phosphorescent probes.

6.3.1 Non-oxygen Sensing

The most obvious idea in this field is application of PLIM for sensing of various microenvironment parameters (such as pH, temperature, ionic strength) and biologically relevant analytes other than oxygen (cations/anions, biothiols etc.). The latter area includes but not limited to development of sensors to low molecular ions (Zn^{2+}, Cl^-, F^-, ClO^-), biologically important molecules (including H_2S, cysteine as well as other biothiols, DNA, miRNA, etc.) and selective stains for different organelles (nucleus, nucleoli, mitochondria, lysosomes, and cell membrane). The key achievements in bioanalytics of this sort are exhaustively reviewed in the recent publication [3], however neither of these probes have become a commercial product yet. One of possible explanations (in addition to exceptional youth of this area) is that, contrary to oxygen sensor, other sensors and trackers do not possess such a robust and unambiguously interpretable lifetime correlations with appropriate analytes (as it was achieved in the case of oxygen sensing) and often display concomitant lifetime dependence on other microenvironment characteristics.

Two possible strategies may be used to decouple/avoid the cross-talk between target and obstructive parameters and to design effective non-oxygen PLIM sensors. The first strategy consists in rational design of the probe architecture (choice of appropriate luminescent center and ligand environment selectively responsive to the target analytes) while the other one should be based on embedding of phosphors into responsive matrix (various kinds of nanoparticles, conjugation to polymers, etc.). Below (in the rest of this Section) we describe both approaches in detail.

6.3.1.1 Rational Design of Phosphor Architecture

Molecular thermometers. One of the challenging tasks in functional bioimaging is preparation of PLIM molecular sensor with lifetime response onto temperature variations in physiological interval. The design of this type of probes relies on the recent findings of strong luminescence intensity [35] and lifetime [36] dependence of europium complexes on temperature within the 30–45 °C range. Moreover, the europium compounds luminescence was found to be nearly independent of oxygen concentration that allows avoiding cross-talking of these physiological parameters. Consequently, europium complexes are very attractive candidates for PLIM temperature sensing. For example, Eu-tris(dinaphthoylmethane)-bis-(trioctylphosphine oxide) complex (Fig. 6.7a) incorporated into poly(methyl methacrylate)-based nanoparticles demonstrated variations of luminescence lifetime from 230 to 170 μs in response to the temperature increase from 30 to 40 °C [36]. Though these PLIM experiments were restricted to investigation of sensor nanoparticles embedded into poly(vinyl alcohol) films [36], this example demonstrates high potential of Eu complexes in temperature tracking in biological samples, e.g. in living cells, in response to stimuli of various nature, using PLIM technique. Aside from Eu complexes, various nanomaterials, such as silicon nanoparticles [37] or gold nan-

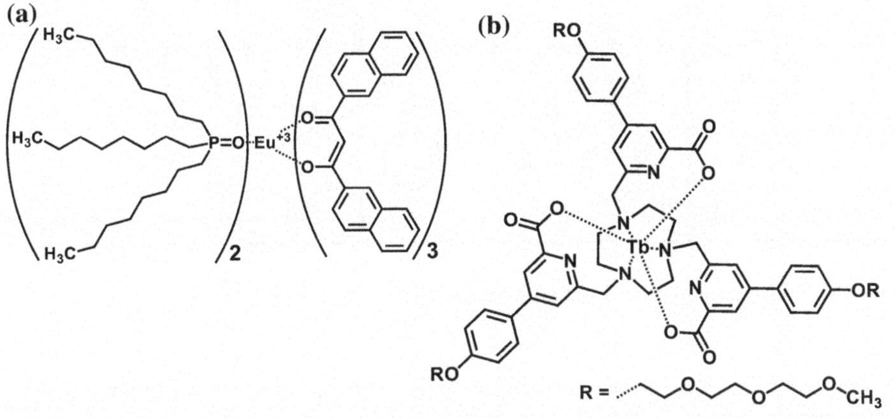

Fig. 6.7 Structures of **a** Eu(III) thermometer described in [36] and **b** Tb(III) viscometer described in [39]

oclusters [38], also demonstrate substantial lifetime temperature dependences and are promising for application as PLIM molecular thermometers.

Molecular viscometers. In the case of microviscosity measurements, an effective strategy consists in the development of probes bearing ligands with auxiliary rotors. In the case of correct design, the rotor's internal rotation can contribute into one of nonradiative relaxation channels. As a result, increase in microviscosity slows down internal rotation to give the decrease in nonradiative relaxation rate, and, ultimately, elongation of luminescence lifetimes and increase in luminescence intensity. This concept was exemplified by Tb(III) complex bearing biaryl rotors (Fig. 6.7b) that exhibited statistically significant increase in lifetimes within the nuclei compared to the cytoplasm [39].

Molecular pH meters. Incorporation of protonatable groups into ligands of phosphorescent complexes renders them pH sensitive. For example, two Ir(III) complexes bearing N,N ligands featuring β-carboline motif (Fig. 6.8) displayed both pH dependence and accumulation in lysosomes as a result of secondary amino group protonation [40]. The complexes are thus regarded as promising pH sensors though their sensitivity to oxygen quenching and interaction with lysosomal constituents should be investigated to avoid potential misinterpretations of lifetime changes.

6.3.1.2 Polymer-Based Environmentally Responsive Sensors

An alternative idea of sensors development consists in the use of phosphor conjugates with environmentally responsive polymers. This approach was successfully exemplified by molecular thermometer based on the conjugate of luminescent iridium complexes with thermally responsive polymer [41]. The polymer undergoes conformational coil-to-globule transition in physiological temperature range that results

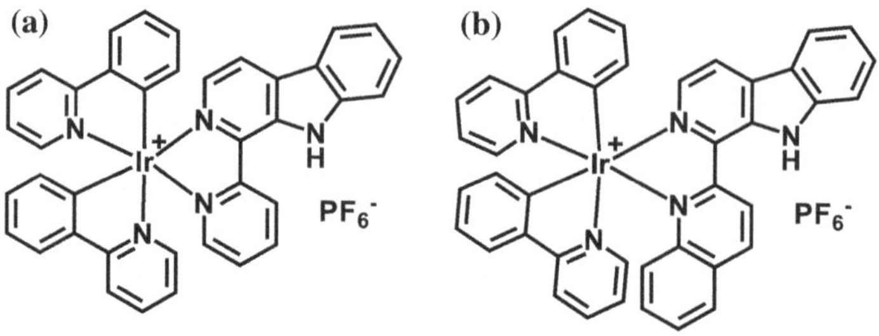

Fig. 6.8 Structures of pH-sensitive Ir(III) complexes described in [40]

in dramatic change in luminescence intensity. This approach was implemented in both ratiometric and PLIM imaging modalities for temperature measurements, but generally this idea can be broadened to other microenvironment parameters such as pH, ionic strength, and viscosity provided that at least two requirements were met. First, luminescent complex should be as little as possible environmentally sensitive by itself. Second, the core-shell architecture with environmentally responsive core conjugated with the phosphor and hydrophilic outer shell that sterically protects labels from unwanted interactions seems to be more preferable compared to linear polymers.

6.3.2 Towards Live Imaging of Cell/Tissue Metabolism

Another growth point in the PLIM research will be the switching from "concentration measurement" to "metabolic" imaging, i.e. obtaining of information on the status of organelle, cell or tissue rather than just quantitative imaging of the distribution of oxygen or other analytes. In fact, such metabolic imaging has already been implemented in FLIM, where the "optical redox ratio", the term encompassing ratios between free and protein-bound NADH (electron donor) and FAD (electron acceptor), directly calculated from FLIM, is intensely investigated and was proved to differentiate between normal and precancerous cells [9]. In analogy to the above term, we believe that it is possible to design PLIM probes, which would respond to some metabolic shifts or processes by corresponding changes in lifetimes. To date, there is no clear understanding of what kind of metabolic response could be used for such analysis but we would highlight several emerging applications that have the potential of transformation into metabolic imaging.

The first of these examples is the real-time sensing of newly synthesized proteins [42]. To realize this analytical instrument, the authors first forced cells to produce proteins with methionine substituted by L-azidohomoalanine (AHA), the methionine structural analog, by incubating them in methionine-free media supplemented with AHA (Fig. 6.9a) [43]. The newly synthesized proteins contained azido groups and thus were able to conjugate with iridium complex containing alkynyl groups via

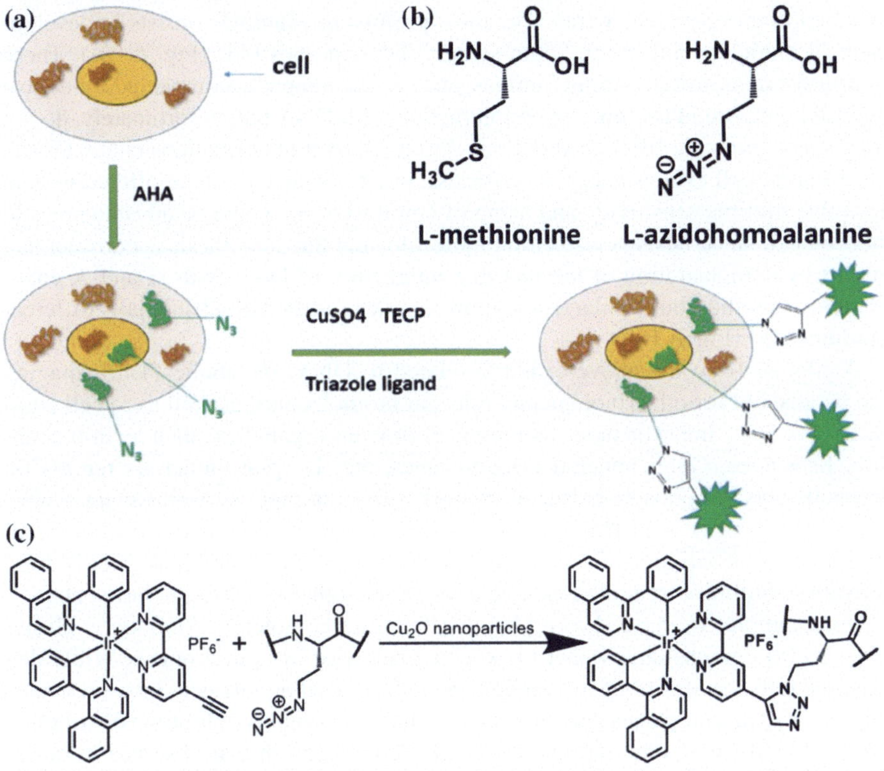

Fig. 6.9 Experimental scheme of bioorthogonal labelling of newly synthesized proteins for fluorescence visualization in cells. **a** Schematic diagram of metabolic labelling of newly synthesized proteins in mammalian cells using L-azidohomoalanine (AHA) incorporation and then over the Cu (I)-catalyzed [3+2] azide–alkyne cycloaddition to label fluorescence. Reprinted with permission from [43]. Copyright 2017 Elsevier. **b** Chemical structures of L-methionine and AHA. C. Scheme of click reaction between Ir-alkyne complex and AHA residue described in [42]

click reaction directly inside cells (Fig. 6.9b). Upon this conjugation the lifetime of iridium complex dramatically decreased from 600–800 (free complex) to 400–600 ns (conjugated complex) without variations in the energy of emission. As a result, monitoring of protein synthesis dynamics became possible. This example shows not only principal ability to perform selective bioconjugation inside living cells but also paves the way to the design of sensors with lifetime response onto changes in essential characteristics of biomolecules and organelles including complex formation or disintegration, dimerization, conformational changes, etc.

In this context, metabolic DNA imaging looks particularly attractive. DNA is the molecule that undergoes substantial conformational transitions during cell cycle, and elaboration of PLIM sensors changing their lifetimes in response to DNA compaction/unfolding would eventually lead to creation of "PLIM cell cycle clocks". To date, such complexes are not found yet, though there are some significant results obtained in closely related area. The most promising example consists in development of dinuclear ruthenium "light-switch" DNA probes [44] (Fig. 6.10a). These complexes demonstrate strong luminescence enhancement accompanied by almost two-fold increase in lifetimes upon interaction with DNA but, unfortunately, do not show any significant lifetime variations during DNA transformations characteristic for different cell cycle stages. Nevertheless, other chemistry can be offered to gain desirable lifetime sensitivity, and using of chiral isomers would be advantageous in this context, since DNA is also chiral molecule, and this hypothesis is corroborated in part by demonstration of the fact that interaction of DNA with Δ and Λ enantiomers of another (but similar) ruthenium complex resulted in substantially different lifetimes [45] (Fig. 6.10b).

Analogously to the above example related to DNA, metabolic PLIM imaging can be adapted for other biologically relevant biomolecules, as well as to cell compartments (e.g., inner or outer membranes) or even organelles, such as mitochondria, lisosomes, Golgi apparatus, adiposomes, etc. Despite numerous reports on organelle-specific phosphors, there are only a few attempts to evaluate applicability of these probes in PLIM. Potential metabolic sensors may respond to different stimuli such as variations on redox potential or degree of membrane polarization, activation/inhibition of proteases, esterases, or any other ferments, polarity, etc.

For example, we have shown [46] that lipohilic homoleptic Au-alkynyl cluster (Fig. 6.11a) demonstrates rather unexpected behavior in lipid droplets. While distributing rather uniformly between lipid droplets (based on intensity measurements; Fig. 6.11b), the complexes demonstrate two distinct domains in lifetime distribution (Fig. 6.11c–d), with one lifetime drastically different from that observed in model lipid media (vegetable oil) [46]. Though there is no experimentally supported explanation of the appearance of the additional lifetime mode, this observation clearly shows advantageous nature of PLIM functional imaging over usual "localization"

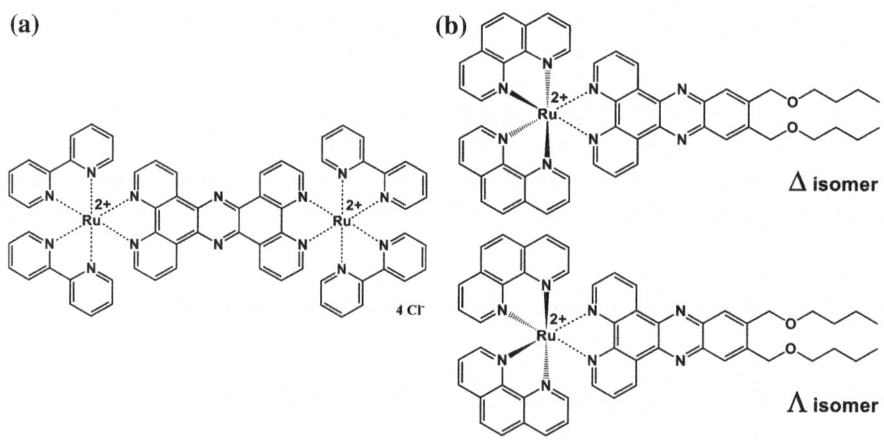

Fig. 6.10 Chemical structures of Ru complexes described in [44] (**a**) and [45] (**b**)

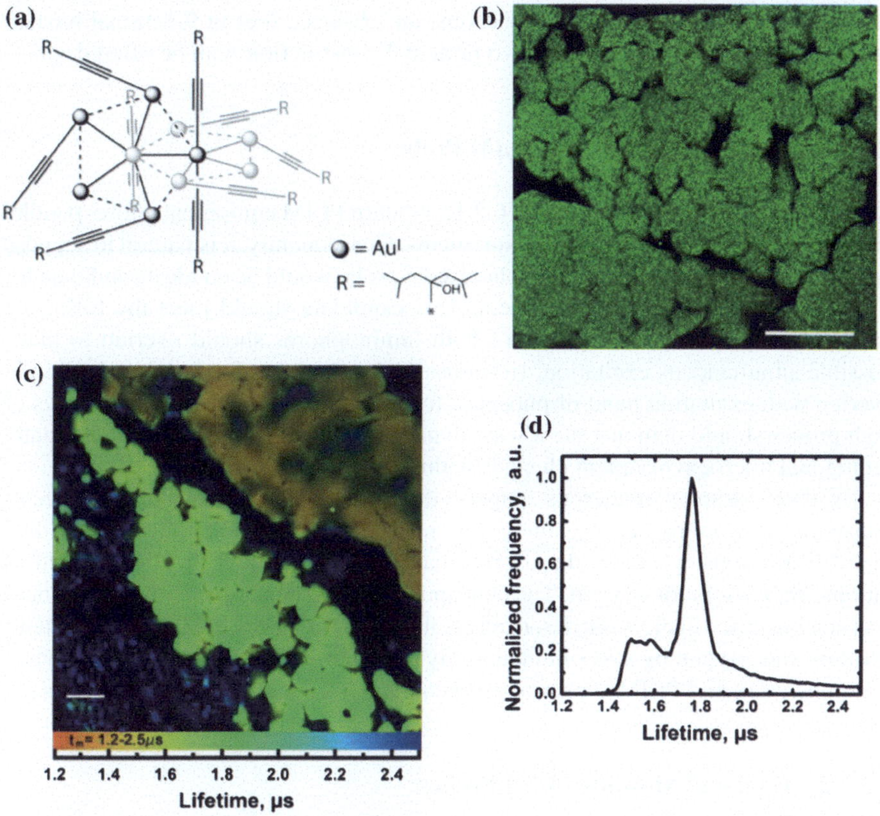

Fig. 6.11 **a** Schematic structure of lipophilic Au cluster. Visualization of adipocytes in the chicken subcutaneous adipose tissue (**b**, **c**): **b** Signal detection with spectral imaging. **c** Phosphorescence lifetime image. **d** Distribution of the phosphorescence lifetimes across the image (**c**). Two-photon excitation at 710 nm. Scale bar: 100 μm. Reproduced with permission from [46]. Copyright 2017 Elsevier

mode: in addition to visualization and "localization" of cell compartments provided by both modalities, functional imaging also provides information on the "status" of the region of interest, provided that there is an unambiguous correlation between the specific environmental property and lifetime.

6.3.3 Dual (Ultimately—Multiple) Sensors

Another approach that seems to be developed rapidly in the nearest future is elaboration of dual emission probes, which could perform simultaneous sensing of two independent parameters for correlative mapping. In a remote perspective multiple sensors have a clear prospective to become an advanced tool in functional bioanalytics. In the case of dual sensors, two principal constructions can be offered.

6.3.3.1 Dual-Modality FLIM/PLIM Probes

As it was demonstrated above (Sect. 6.2.1), modern PLIM equipment allows parallel performing of PLIM and FLIM measurements. Consequently, it is natural to suppose that dual emission fluorescent-phosphorescent probe would be an ideal candidate for simultaneous FLIM/PLIM experiment. This candidate should meet the following requirements: (i) excitation bands of both luminophores should overlap to make possible simultaneous excitation; (ii) emission band of the fluorophore should not overlap with excitation band of phosphor to avoid self-quenching; (iii) lifetimes of both probes should demonstrate a wide dynamic range in response to two mutually orthogonal microenvironment changes within physiological interval.

To date, despite numerous papers describing dual emission fluorescent-phosphorescent constructions, there was no reports on their application in dual FLIM/PLIM sensing mode: the overall majority of such dual probes are used as ratiometric sensors for oxygen. The conceptually closest example to the above ideas was demonstrated in the work describing simultaneous FLIM/PLIM mapping of temperature and oxygen by two simultaneously administered sensors: the FLIM sensor on temperature and PLIM sensor on oxygen [47].

6.3.3.2 Dual and Multiple PLIM Probes

Since lifetime range characteristic of phosphorescent emitters and potentially applicable in PLIM is at least of 5 orders of magnitude (from 10^{-8} to 10^{-3} s; Fig. 6.12), simultaneous acquisition of lifetime information from several phosphors is possible. For example, Lakowicz and co-workers clearly demonstrated PLIM differentiation of three Ru complexes displaying rather close lifetimes (7, 17, and 42 ns for tris(5-amino-1,10-phenanthroline)ruthenium(II), tris(2,2′-bipyridine)ruthenium(II), and tris(2,3-bis(2-pyridyl)pyrazine))ruthenium(II), correspondingly) individually incorporated into silver nanoshells (SiO_2 nanoparticles covered by Ag nanolayer) [48]. Hence, upon appropriate intracellular targeting, simultaneous multi-tracker imaging is possible, with the number of trackers sufficiently exceeding currently available range of fluorescent ones (up to 4 species in FLIM mode; up to 6 species in spectral imaging mode). In the case of phosphorescent sensors working in much broader lifetime interval (up to 10-fold change of lifetime upon variation of analyte concentration) differentiation of at least four phosphors with lifetimes differing by at least one order of magnitude (e.g., 10–100 ns, 0.1–1, 1–10 and 10–100 μs) is possible.

Regarding the combination of several phosphors into one construction, several different Eu–Ir dual phosphors were involved in assessment of their applicability in PLIM to date (Fig. 6.13) [49–51]. Unfortunately, PLIM experiments were described

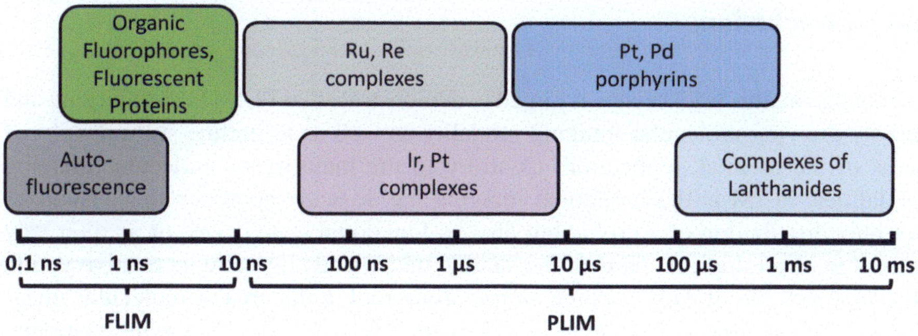

Fig. 6.12 Typical time intervals characteristic for FLIM and PLIM, and typical lifetimes of different classes of phosphors

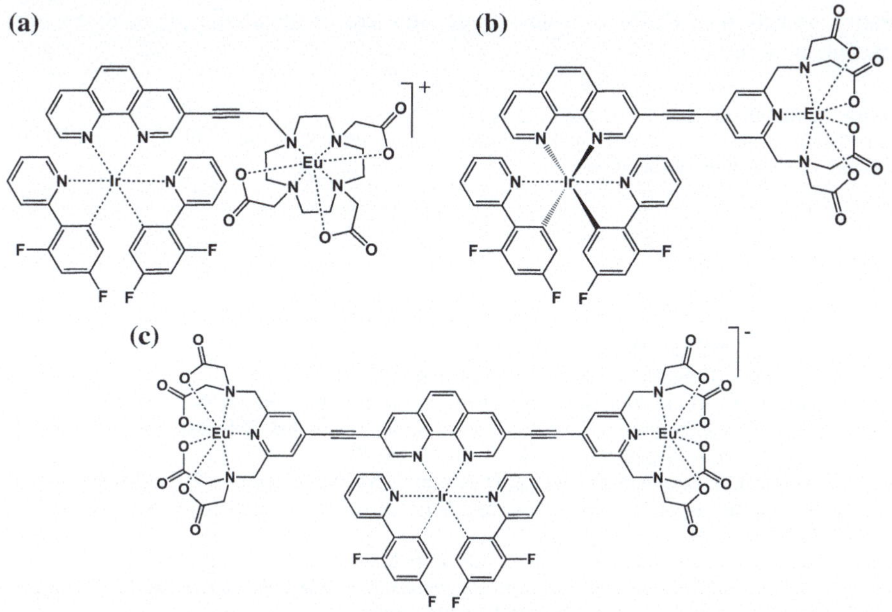

Fig. 6.13 Structures of dual Eu–Ir complexes described in [49] (**a**), [50] (**b**) and [51] (**c**), respectively

for Ir-based luminescence solely while Eu-based luminescence was used in time gating mode [49]. Most probably, this was because of a very long emission lifetimes of Eu chromophore that turned image acquisition time unacceptably long. Nevertheless, this approach seems to be very promising, and other luminophore combinations are anticipated.

6.4 Conclusions

In conclusion, this brief review is aimed to demonstrate that PLIM is an emerging and rapidly growing molecular imaging modality as well as to outline still unexplored areas of the method application. Contrary to the majority of molecular imaging modalities, PLIM is the functional imaging, i.e. it is the approach that visualizes not only distribution of a probe, but also its functional state in one or another way related to physiological status of the sample under study. Despite its relative youth, this approach has already become an important tool in the area of molecular imaging with the alternative and quite robust methodology of oxygen sensing/mapping. We envision that nearest future will see not only step-by-step progress in this area (i.e. improvement of equipment, sophistication of biomedical models, appearance of novel commercial probes and selective trackers suitable for PLIM) but also qualitative transition of PLIM to a new stage—the era of metabolic phosphorescence lifetime imaging.

Acknowledgements The authors greatly appreciate financial support from the Russian Science Foundation (grant 16-43-03003) and the organizing committee of the STEPS program for the support in publication of this review.

References

1. W. Becker, J. Microsc. **247**, 119 (2012)
2. M.Y. Berezin, S. Achilefu, Chem. Rev. **110**, 2641 (2010)
3. K.Y. Zhang, Q. Yu, H. Wei, S. Liu, Q. Zhao, W. Huang, Chem. Rev. **118**, 1770 (2018)
4. E. Baggaley, J.A. Weinstein, J.A.G. Williams, Struct. Bond. **165**, 205 (2015)
5. L.M. Hirvonen, K. Suhling, Meas. Sci. Technol. **28** (2017)
6. S.S. Howard, A. Straub, N.G. Horton, D. Kobat, C. Xu, Nat. Photonics **7**, 33 (2013)
7. W. Becker, *Advanced Time-Correlated Single Photon Counting Techniques* (Springer, Berlin, Heidelberg, 2005)
8. W. Becker, V. Shcheslavskiy, A. Rück, Adv. Exp. Med. Biol. 19–30, (2017)
9. M.C. Skala, K.M. Riching, A. Gendron-Fitzpatrick, J. Eickhoff, K.W. Eliceiri, J.G. White, N. Ramanujam, Proc. Natl. Acad. Sci. **104**, 19494 (2007)
10. S. Kalinina, J. Breymayer, P. Schäfer, E. Calzia, V. Shcheslavskiy, W. Becker, A. Rück, J. Biophotonics **9**, 800 (2016)
11. A.I. Solomatina, P.S. Chelushkin, D.V. Krupenya, I.S. Podkorytov, T.O. Artamonova, V.V. Sizov, A.S. Melnikov, V.V. Gurzhiy, E.I. Koshel, V.I. Shcheslavskiy, S.P. Tunik, Bioconjug. Chem. **28**, 426 (2017)
12. V. Shcheslavskiy, M. Shirmanova, V. Dudenkova, K. Lukyanov, A. Gavrina, A. Shumilova, E. Zagaynova, W. Becker, Opt. Lett. **43** (2018)
13. A.I. Solomatina, S.-H. Su, M.M. Lukina, V.V. Dudenkova, V.I. Shcheslavskiy, C.-H. Wu, P.S. Chelushkin, P.-T. Chou, I.O. Koshevoy, S.P. Tunik, RSC Adv. **8**, 17224 (2018)
14. R.I. Dmitriev (ed.), *Multi-Parametric Live Cell Microscopy of 3D Tissue Models* (Springer International Publishing, Cham, 2017)
15. T.V. Esipova, A. Karagodov, J. Miller, D.F. Wilson, T.M. Busch, S.A. Vinogradov, Anal. Chem. **83**, 8756 (2011)

16. A. Fercher, S.M. Borisov, A.V. Zhdanov, I. Klimant, D.B. Papkovsky, ACS Nano 5499–5508 (2011)
17. D.E. Owens, N.A. Peppas, Int. J. Pharm. **307**, 93 (2006)
18. S. Dufort, L. Sancey, J.L. Coll, Adv. Drug Deliv. Rev. **64**, 179 (2012)
19. F. Alexis, E. Pridgen, L.K. Molnar, O.C. Farokhzad, Mol. Pharm. 505–515 (2008)
20. R.I. Dmitriev, S.M. Borisov, H. Düssmann, S. Sun, B.J. Müller, J. Prehn, V.P. Baklaushev, I. Klimant, D.B. Papkovsky, ACS Nano **9**, 5275 (2015)
21. V. Ntziachristos, Nat. Methods **7**, 603 (2010)
22. I. Dunphy, S.A. Vinogradov, D.F. Wilson, Anal. Biochem. **310**, 191 (2002)
23. O.S. Finikova, A.Y. Lebedev, A. Aprelev, T. Troxler, F. Gao, C. Garnacho, S. Muro, R.M. Hochstrasser, S.A. Vinogradov, Chem. Phys. Chem. **9**, 1673 (2008)
24. E. Roussakis, J.A. Spencer, C.P. Lin, S.A. Vinogradov, Anal. Chem. **86**, 5937 (2014)
25. J. Milton Harris, R.B. Chess, Nat. Rev. Drug Discov. **2**, 214 (2003)
26. V. Tsytsarev, H. Arakawa, S. Borisov, E. Pumbo, R.S. Erzurumlu, D.B. Papkovsky, J. Neurosci. Methods **216**, 146 (2013)
27. F. Helmchen, W. Denk, Nat. Methods **2**, 932 (2005)
28. J.A. Spencer, F. Ferraro, E. Roussakis, A. Klein, J. Wu, J.M. Runnels, W. Zaher, L.J. Mortensen, C. Alt, R. Turcotte, R. Yusuf, D. Côté, S.A. Vinogradov, D.T. Scadden, C.P. Lin, Nature **508**, 269 (2014)
29. S. Sakadžić, E. Roussakis, M.A. Yaseen, E.T. Mandeville, V.J. Srinivasan, K. Arai, S. Ruvinskaya, A. Devor, E.H. Lo, S.A. Vinogradov, D.A. Boas, Nat. Methods **7**, 755 (2010)
30. A.V. Kondrashina, R.I. Dmitriev, S.M. Borisov, I. Klimant, I. O'Brien, Y.M. Nolan, A.V. Zhdanov, D.B. Papkovsky, Adv. Funct. Mater. **22**, 4931 (2012)
31. R.I. Dmitriev, S.M. Borisov, A.V. Kondrashina, J.M.P. Pakan, U. Anilkumar, J.H.M. Prehn, A.V. Zhdanov, K.W. McDermott, I. Klimant, D.B. Papkovsky, Cell. Mol. Life Sci. **72**, 367 (2015)
32. W. Lv, T. Yang, Q. Yu, Q. Zhao, K.Y. Zhang, H. Liang, S. Liu, F. Li, W. Huang, Adv. Sci. **2** (2015)
33. J. Liu, Y. Liu, W. Bu, J. Bu, Y. Sun, J. Du, J. Shi, J. Am. Chem. Soc. **136**, 9701 (2014)
34. T. Misgeld, M. Kerschensteiner, Nat. Rev. Neurosci. **7**, 449 (2006)
35. Y. Takei, S. Arai, A. Murata, M. Takabayashi, K. Oyama, S. Ishiwata, S. Takeoka, M. Suzuki, ACS Nano **8**, 198 (2014)
36. H. Peng, M.I.J. Stich, J. Yu, L.N. Sun, L.H. Fischer, O.S. Wolfbeis, Adv. Mater. **22**, 716 (2010)
37. Q. Li, Y. He, J. Chang, L. Wang, H. Chen, Y.W. Tan, H. Wang, Z. Shao, J. Am. Chem. Soc. **135**, 14924 (2013)
38. L. Shang, F. Stockmar, N. Azadfar, G.U. Nienhaus, Angew. Chemie Int. Ed. **52**, 11154 (2013)
39. A.T. Bui, A. Grichine, A. Duperray, P. Lidon, F. Riobé, C. Andraud, O. Maury, J. Am. Chem. Soc. **139**, 7693 (2017)
40. L. He, C.P. Tan, R.R. Ye, Y.Z. Zhao, Y.H. Liu, Q. Zhao, L.N. Ji, Z.W. Mao, Angew. Chemie Int. Ed. **53**, 12137 (2014)
41. Z. Chen, K.Y. Zhang, X. Tong, Y. Liu, C. Hu, S. Liu, Q. Yu, Q. Zhao, W. Huang, Adv. Funct. Mater. **26**, 4386 (2016)
42. J. Wang, J. Xue, Z. Yan, S. Zhang, J. Qiao, X. Zhang, Angew. Chemie Int. Ed. **56**, 14928 (2017)
43. J. Wang, L. Sheng, H. Zhao, X. Zhang, S. Zhang, Talanta **162**, 641 (2017)
44. E. Baggaley, M.R. Gill, N.H. Green, D. Turton, I.V. Sazanovich, S.W. Botchway, C. Smythe, J.W. Haycock, J.A. Weinstein, J.A. Thomas, Angew. Chemie Int. Ed. **53**, 3367 (2014)
45. F.R. Svensson, M. Abrahamsson, N. Strömberg, A.G. Ewing, P. Lincoln, J. Phys. Chem. Lett. **2**, 397 (2011)
46. E.I. Koshel, P.S. Chelushkin, A.S. Melnikov, P.Y. Serdobintsev, A.Y. Stolbovaia, A.F. Saifitdinova, V.I. Shcheslavskiy, O. Chernyavskiy, E.R. Gaginskaya, I.O. Koshevoy, S.P. Tunik, J. Photochem. Photobiol. A Chem. **332**, 122 (2017)
47. J. Jenkins, S.M. Borisov, D.B. Papkovsky, R.I. Dmitriev, Anal. Chem. **88**, 10566 (2016)

48. J. Zhang, Y. Fu, J.R. Lakowicz, J. Phys. Chem. C **115**, 7255 (2011)
49. E. Baggaley, D.K. Cao, D. Sykes, S.W. Botchway, J.A. Weinstein, M.D. Ward, Chem. A Eur. J. **20**, 8898 (2014)
50. A. Jana, E. Baggaley, A. Amoroso, M.D. Ward, Chem. Commun. **51**, 8833 (2015)
51. A. Jana, B.J. Crowston, J.R. Shewring, L.K. McKenzie, H.E. Bryant, S.W. Botchway, A.D. Ward, A.J. Amoroso, E. Baggaley, M.D. Ward, Inorg. Chem. **55**, 5623 (2016)

Chapter 7
Optical Control of G Protein-Coupled Receptor Activities in Living Cells

Hideaki Yoshimura and Takeaki Ozawa

Abstract Membrane receptors transmit external signals into cells in response to extracellular stimuli and their activities are controlled spatiotemporally. In recent years, it has become possible to control activity of a target membrane receptor by external light using a photoreceptor protein. It forms oligomers or interacts specifically with its binding partners after light absorption. Controlling receptor activities by external light is now a powerful approach to elucidating the role of receptor activities and its dynamics in various life phenomena. In this review, we describe a new technology of optically controllable receptor using a photoreceptor protein, CRY2, and its application to the interaction of GPCR with β-arrestin in living cells.

7.1 Introduction

Basic knowledge of biochemistry of bioluminescence becomes driving forces behind development of basic tools used for modern life science. A typical example is the discovery of the light source of firefly luciferase; the light source is composed of an enzymatic reaction of luciferase with its substrate, D-luciferin [1]. Since ATP and oxygen molecules are necessary for this reaction, luciferase has been used for highly sensitive luminescence analysis of ATP. Also, the luciferase enzyme is still one of the most versatile proteins as a reporter of gene expression, which is used for measuring transcriptional activities in living cells [2]. Another example is the green fluorescent protein (GFP) discovered in the 1962 [3], which is now indispensable as a tool to visualize target molecules and proteins in living cells and animals. Furthermore, various amino acid mutants of GFP and identification of analogues thereof have advanced the development of fluorescent proteins showing new optical properties [4], which support the current fluorescence imaging technologies. Thus,

H. Yoshimura · T. Ozawa (✉)
Department of Chemistry, School of Science, The University of Tokyo, 7-3-1 Hongo, Bunkyo-Ku, Tokyo 113-0033, Japan
e-mail: ozawa@chem.s.u-tokyo.ac.jp

H. Yoshimura
e-mail: hideaki@chem.s.u-tokyo.ac.jp

K. Yamanouchi et al. (eds.), *Progress in Photon Science*, Springer Series in Chemical Physics 119, https://doi.org/10.1007/978-3-030-05974-3_7

the technical basis of modern biological science is strongly dependent on the results of photobiological research.

It has been well known that plant cells and algae include many light-absorbing proteins. The proteins contain organic molecules as their chromophore originated from metabolites in the cells. When the chromophore in the proteins absorbs light in a visible region, the proteins undergo dynamical changes in their structure, and translate the light signal into activation of intracellular signaling. Retinal in an ion channel (Fig. 7.1a), for example, absorbs light with a specific wavelength region, and then the channel protein transduces the light signal into the activation of visual cells. Based on the knowledge of light-absorbing ion channel, new technologies called optogenetics have been emerged by artificially expressing such ion channels in neurons [5]; a specific nerve activity in living animals can be controlled by turning on and off of external light. Since the appearance of optogenetics technology, it has spread into various ion channel proteins and their applications [6]. In parallel with the ion channel technologies, various new photoreceptor proteins have also been utilized for directly controlling intracellular enzyme activity and gene expression in living cells and animals [7].

Receptor proteins embedded in the plasma membrane in living cells recognize extracellular ligand molecules and transmit signals into cells, thereby inducing appropriate cellular responses according to environmental differences. One of the important membrane receptor families is G protein-coupled receptors (GPCRs), which are widely distributed in different organs and mediate numerous hormones, neurotransmitters and odor molecules, etc., into intracellular signaling [8]. Although the elucidation of the structure of GPCR is progressing [9], most of the details of the mechanism such as binding with G protein or β-arrestin regulating its function,

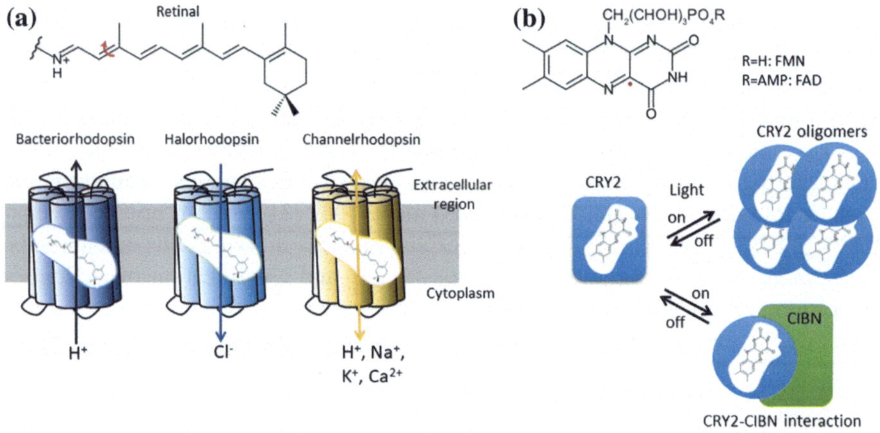

Fig. 7.1 Different types of optogenetics tools. **a** Structure of retinal and different types of ion channels. Arrows in the bottom figure indicate direction of each ion flow. **b** Structure of FMN and FAD and character of CRY2. Upon blue light absorption, CRY2 forms oligomers or interacts with its specific binding protein, CIBN

its accompanying intracellular signals, and dynamic process of the translocation to endosomes. Many studies using specific ligands and inhibitors for GPCRs have been carried out so far, it is difficult to manipulate these compounds in the spatiotemporal way. Therefore, it is desirable to develop new technologies for simultaneous analysis of GPCR and its related proteins in living cells. We herein focus on a new methodology of the analysis of GPCR and β-arrestin interactions using a photoreceptor proteins, CRY2 and new insight into the mechanism of intracellular trafficking of GPCRs.

7.2 Optogenetics Tool of CRY2

Many photoreceptor proteins undergo their structural changes upon light absorption by their chromophores. In optogenetics, the property of photoreceptor proteins is used for controlling activity of specific proteins of interest in living systems. In particular, photoreceptor proteins that form dimers and oligomers by light absorption are widely used as a module that induces protein-protein interactions and localization changes of a target protein in living cells.

Photoreceptor proteins responsive to blue light are often used as optogenetics tools. Of various photoreceptor proteins absorbing blue light, the light-oxygen-voltage sensing (LOV) domain is a representative: Aureochrome 1a derived from *Vaucheria frigida* which is a kind of yellow-green alga has a LOV domain (AU1) enveloping flavin mononucleotide (FMN) as a chromophore (Fig. 7.1b). Blue light irradiation induces the protein dimer formation reversibly, which was applied for manipulating gene expression by external blue light [10].

On the other hand, Cryptochrome is a family of photoreceptor proteins having flavin adenine dinucleotide (FAD) as a chromophore. Among them, Cryptochrome 2 (CRY2) derived from *Arabidopsis thaliana* is widely used as a light control module in response to blue light. CRY2 exhibits two important characteristics with blue light absorption; one is a reaction in which CRY2 forms an oligomer (Fig. 7.1b). Particularly in plant cells, it forms agglomerates in a microscopic observable sizes called Photobody in the nucleus. Until now, the CRY2 protein has been used as an optogenetics module to control activities of tyrosine kinase and membrane receptors such as Trk and DCC receptors [11, 12]. The other is a character of interaction between CRY2 and CIB1 protein, which is a target protein of CRY2. Because CRY2 interacts with the N-terminal domain of CIB1, the truncated domain (called CIBN) is used as a tool for light manipulation. The interaction of CRY2 with CIBN is reversible and their complex dissociates promptly upon turning off the light (Fig. 7.1b). Taking these advantages of CRY2-CIB1 interactions, some researchers have demonstrated their application to controlling kinase activities such as intracellular kinases Akt and Ras [13, 14].

7.3 Signal Transduction of GPCR and Downstream Molecules

GPCRs represent the largest protein family and mediate large number of external stimuli such as hormones, and neurotransmitters, which are therefore implicated in myriad of physiological functions and diseases in our body [15, 16]. Even though the GPCR family consists of nearly 1000 members, their structural features and the signal transduction pathways through them are highly conserved throughout the members. GPCRs consist of single polypeptide chain that spans 7 times across the cell membrane. The GPCRs activated with a specific agonist lead to interaction with heterotrimeric G proteins inside of the cells and stimulate signal transduction through G proteins. The activated GPCRs are then C-terminally phosphorylated by GPCR kinases (GRK), and then β-arrestin is recruited through interaction to the phosphorylated site on the GPCRs. The β-arrestin recruitment induces stimulation of the MAP/ERK signaling pathway, receptor desensitization, and clathrin-dependent endocytosis.

Hetero trimeric G proteins are typical downstream proteins of activated GPCRs, consisting of three subunits, Gα, Gβ, and Gγ. Upon an agonist binds to GPCR, the Gα subunit in heterotrimeric G protein, which are tethered to the inner leaflet of the plasma membrane are associated with the active GPCRs. Then, the signal is transduced from the GPCR to the heterotrimeric G protein, leading activation and dissociation of the heterotrimeric G protein into a Gα and a Gβγ dimer.

GPCRs in an active form undergo the function of G protein-coupled receptor kinase (GRK) and are phosphorylated at the C-terminal region [17]. Then, β-arrestin, another downstream molecule of GPCRs, interacts with the phosphorylation site and forms a complex with the GPCRs. β-arrestin attaching to a GPCR has different functions. First, β-arrestin inhibits the activity of GPCRs to transduce signal to G proteins by preventing GPCR-G protein interaction (desensitization) [18]. Second, upon attaching to GPCRs, β-arrestin transduce signal to its downstream molecules such as ERK1/2. In addition, β-arrestin provides a scaffold to concentrate clathrin and its related proteins, and induces endocytosis of the GPCRs [19]. Beside the importance of the G-protein signaling pathway, the roles of β-arrestin are thus also critical to regulate GPCR functions.

7.4 Light Regulation of GPCR-β-Arrestin Interactions

The interaction between GPCRs and β-arrestin is crucial for GPCR signaling, regulation of GPCR activity, and GPCR trafficking in the cells. Artificial control of GPCR-β-arrestin interaction is a promising approach to control the function of GPCRs comprehensively. We have established a method to manipulate GPCR-β-arrestin interaction through photo irradiation in living cells. We introduced briefly the work of photo regulation of GPCR-β-arrestin interaction in living cells [20].

β2-adrenergic receptor (ADRB2), which is one of the most investigated GPCRs, was selected as a target GPCR. ADRB2 was fused with CIBN whereas β-arrestin was connected to CRY2. Upon blue light irradiation, photo-induced interaction between the CRY and CIBN leaded association between ADRB2 and β-arrestin (Fig. 7.2) [21, 22]. In addition, a SNAP tag, which is a tag protein to conjugate an organic fluorescent dye, and a red fluorescent protein mCherry were attached to ADRB2-CIB and β-arrestin-CRY, respectively, to monitor their localization in cells. The SNAP tag was conjugated with SNAP-Cell 647-SiR, which is a far red fluorescent dye, in microscope observation, allowing simultaneous localization analysis of ADRB2-CIB and β-arrestin-CRY in the same living cells.

Before blue light irradiation, ADRB2-CIB and β-arrestin-CRY were localized on the plasma membrane and in the cytosol, respectively. This localization property was consistent with those of ADRB2 and β-arrestin in the resting state without stimulation. Upon blue light irradiation, interaction of CRY2 and CIBN led to β-arrestin recruitment to the plasma membrane within 2 min, then particle-like structures that include ADRB2 and β-arrestin appeared in the cytoplasm 15 min after the light irradiation (Fig. 7.3). Treatment of the sample cells with a dynamin inhibitor, Dingo4a, before the blue light irradiation hampered the generation of the particle-like structures in the cytoplasm, indicating that the particle-like structures were endosomes that form through clathrin-dependent endocytosis of ADRB2-β-arrestin complexes.

In endocytosis on various receptor molecules including ADRB2, some cytosolic proteins were recruited to endosomes to initiate, compose, and/or maintain the endosomes. To assess the recruitment of the endocytosis-related proteins to the photo-induced endosomes of ADRB2-β-arrestin complexes, localization of one of such proteins, Mdm2 [23], was monitored during the photo-induced ADRB2 endocytosis under a fluorescence microscope. Upon blue light irradiation of the sample cells, Mdm2 was recruited to the photo-induced endosomes. After the blue light irradiation stopped, Mdm2 relocated to the cytosol homogenously as the endosomes were decomposed. This result showed that the photo-induced interaction between ADRB2 and β-arrestin induced formation of endosomes of the same properties as those generated upon physiological stimulations.

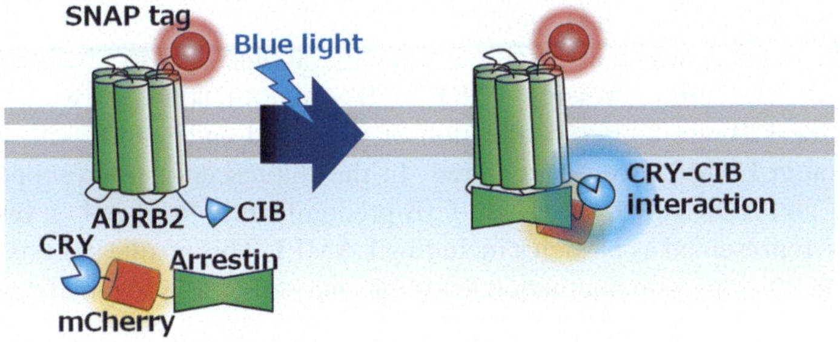

Fig. 7.2 Schematic of the optogenetic tool to control ADRB2-β-arrestin interaction

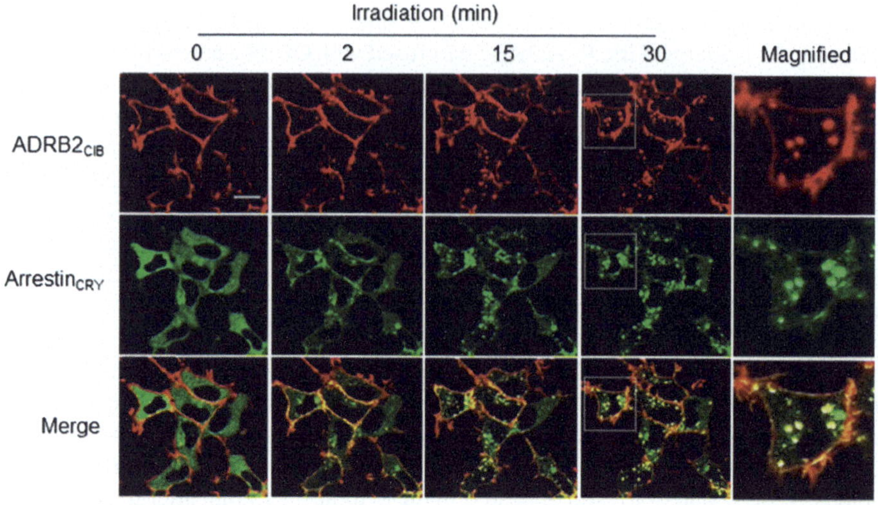

Fig. 7.3 Light-induced endocytosis of ADRB2-CIB. Scale bar, 20 μm. Blue light was irradiated for 30 min to the cells expressing ADRB2-CB and Arrestin-CRY under a confocal fluorescence microscope

Considering the reversibility of CRY-CIB interaction, dissociation of photo-induced ADRB2-β-arrestin complex after stopping light irradiation was tested. The cells expressing ADRB2-CIB and β-arrestin-CRY were irradiated with blue light for 30 min, which was sufficient to form endosomes in the cytosol. Then the light irradiation was quitted and the cells were kept in dark. In the dark condition, the endosomes disappeared in 20 min, and ADRB2-CIB and β-arrestin-CRY relocated to the plasma membrane and the cytosol, respectively (Fig. 7.4). The same endosome disappearing and relocation of ADRB2 and β-arrestin occurred in case of stimulation with an agonist, isoproterenol (ISO). ISO treatment for 60 min led to reduction of ADRB2 on the plasma membrane, whereas ADRB2 recovered on the plasma membrane over time after washing ISO out. Thus the photo-regulation system on ADRB2-β-arrestin interaction reproduces the trafficking dynamics of ADRB2 and β-arrestin upon agonist stimulation.

Next, the dependency of endosome trafficking on blue light irradiation time was assessed [20]. The cells expressing ADRB2-CIB and β-arrestin-CRY were irradiated blue light for 120 min and colocalization of generated endosomes with lysosome marker protein LAMP1 were monitored. In the fluorescence microscope images of LAMP1 and ADRB2-CIB, ADRB2-CIB predominantly localized on lysosomes, which were represented as particles including LAMP1. This result suggests that prolonged ADRB2-β-arrestin interaction leads the endosomes to lysosomal degradation pathway.

Finally, biochemical activity of ADRB2-CIB as a GPCR upon blue light irradiation was estimated. In the assay of ADRB2 activity to transduce the signal to G protein pathways, nanobody 80 (Nb80), which selectively bind to the active state

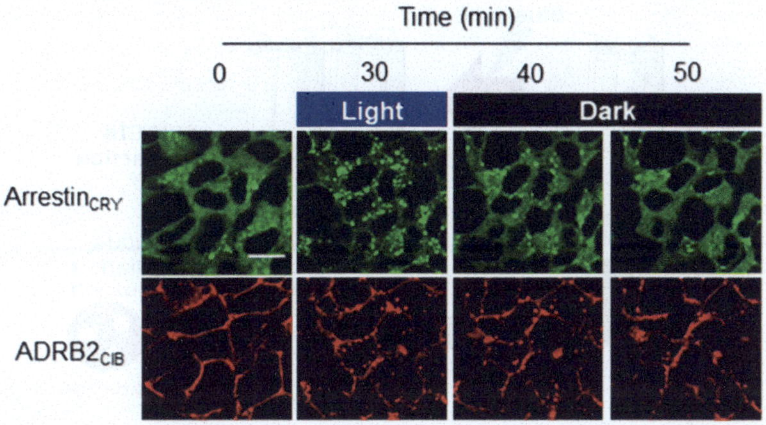

Fig. 7.4 Recycling of ADRB2-CIB after quitting light irradiation. Scale bar, 20 μm. The cells expressing ADRB2-CIB and β-arrestin-CRY was illuminated with blue light for 30 min and kept in dark for further 20 min under a confocal fluorescence microscope

ADRB2 interacting to G proteins [24], was used. The gene of GFP-fused Nb80 (GFP-NB80) was introduced and expressed in the cells stably expressing ADRB2-CIB and β-arrestin-CRY, and the localization of GFP-Nb80 was monitored upon blue light or ISO stimulation under a fluorescence microscope. Whereas 30 min ISO stimulation led to formation of endosomes that contained GFP-Nb80, the endosomes formed after blue light irradiation did not include GFP-Nb80. In addition, the signal transduction activity through the β-arrestin pathway was assessed by Western blotting analysis to detect ERK1/2 phosphorylation, which is a downstream of the β-arrestin signal pathway. Different from ISO stimulation, blue light irradiation of ADRB2-CIB and β-arrestin-CRY-expressing cells failed to induce ERK1/2 phosphorylation. Phosphorylation on the C-terminal region of ADRB2, which occurred upon ISO stimulation, was also assessed by Western Blotting analysis. Even upon blue light irradiation, the C-terminal region did not phosphorylated. These results indicates that the photo activation system of ADRB2-CIB and β-arrestin-CRY induces endocytosis of ADRB2-β-arrestin complex, even though such biochemical activities of ADRB2 as signal transduction through G-protein pathway, ERK1/2 pathway stimulation, and phosphorylation of the ADRB2 C-terminal region were not introduced.

The study described above shows a potential of the optogenetics tool based on CRY and CIB to induce a particular physiological event. In the case of ADRB2 and β-arrestin, CRY-CIB interaction upon blue light irradiation induces ADRB2-β-arrestin complexes formation, resulting in endocytosis of the complexes (Fig. 7.5). It is noteworthy that the light irradiation triggers complex formation of ADRB2 and β-arrestin and following endocytosis, a downstream event of ADRB2—β-arrestin complex formation. On the other hand, upstream phenomena of the endocytosis, such as G-protein pathway signaling, the C-terminal phosphorylation, or ERK1/2 pathway signaling, were not induced. These results imply two aspects of GPCR endocytosis.

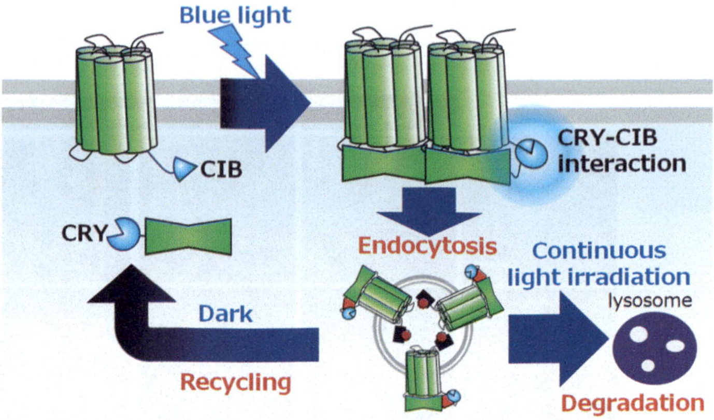

Fig. 7.5 Schematic on trafficking of ADRB2-CIB and β-arrestin-CRY upon photo irradiation

First, formation of endosomes is induced upon GPCR-β-arrestin complex formation, which would provide a scaffold to recruit various endocytosis related proteins to the endosomes. Second, endocytosis of GPCR does not require other upstream biological activation, such as signaling activity through G-proteins and ERK1/2, and phosphorylation of the GPCR. These finding are newly provided as a GPCR functioning mechanism, and would not be revealed without a photo-activation system based on CRY and CIB.

7.5 Perspective

Upon an extracellular signal is induced into a cell, the signal is transmitted through many molecules and also branched multiple pathways. In addition, temporal pattern of signal input causes a variety of signal outputs. Analyses of the function and mechanism of each signaling molecule are still difficult in conventional methods. Recent development of optogenetic tools allows for spatiotemporally-specific signal input, as well as stimulation of a particular target molecule in the middle of a signaling pathway. As shown in the case of optogenetic control of ADRB2-β-arrestin interaction, different duration time of molecular-molecular interaction results in various outputs; ADRB2 was recycled after β-arrestin dissociation whereas continuous ADRB2-β-arrestin interaction for 60 min or longer led them to a degradation pathway in lysosomes. Activation of ADRB2-β-arrestin interaction, which is a mid-point in whole signaling cascades starting from ADRB2, revealed that upstream signaling events such as activation of G-protein signaling and ERK1/2 signaling are not required to initiate and control ADRB2 trafficking. Such analyses were realized by artificial manipulation of ADRB2-β-arrestin interaction by the CRY-CIB-based optogenetic tool.

Further development of optogenetic tools and their practical application will provide much suggestive information in biological studies. Variation of spatial and temporal patterns in signal input generated by an optogenetic technique and quantitative analysis of signal output will provide clues to establish a mathematical model on a target signaling cascade, which will highly contribute to works on systems biology. Optogenetic signal input to a particular cell in living organ will also be useful to investigate cell-cell communication and intercellular signal transduction. Thus, optogenetic techniques will provide new aspects in investigation on signal transduction and mechanisms in biological phenomena of living animals.

Acknowledgements This work was supported by CREST (JPMJCR1752 to T.O.) from Japan Science and Technology (JST) and the Japan Society for the Promotion of Science (JSPS) KAKENHI (Grants-in-Aid for Scientific Research (S) 26220805 to T.O.).

References

1. E.N. Harvey (1916) The Mechanism of light production in animals. Science (New York, NY) **44** (1128), 208–209. https://doi.org/10.1126/science.44.1128.208
2. T. Ozawa, H. Yoshimura, S.B. Kim, Advances in fluorescence and bioluminescence imaging. Anal. Chem. **85**(2), 590–609 (2013). https://doi.org/10.1021/ac3031724
3. O. Shimomura, F.H. Johnson, Y. Saiga, Extraction, purification and properties of aequorin, a bioluminescent protein from the luminous hydromedusan, Aequorea. J. Cellular Comparat. Physiol. **59**, 223–239 (1962)
4. J.R. Enterina, L. Wu, R.E. Campbell, Emerging fluorescent protein technologies. Curr. Opin. Chem. Biol. **27**, 10–17 (2015). https://doi.org/10.1016/j.cbpa.2015.05.001
5. K. Deisseroth, G. Feng, A.K. Majewska, G. Miesenbock, A. Ting, M.J. Schnitzer, Next-generation optical technologies for illuminating genetically targeted brain circuits. J Neurosci. Off. J. Soc. Neurosci. **26**(41), 10380–10386 (2006). https://doi.org/10.1523/jneurosci.3863-06.2006
6. G. Aston-Jones, K. Deisseroth, Recent advances in optogenetics and pharmacogenetics. Brain Res. **1511**, 1–5 (2013). https://doi.org/10.1016/j.brainres.2013.01.026
7. M. Endo, T. Ozawa, Strategies for development of optogenetic systems and their applications. J. Photochem. Photobiol. C **30**, 10–23 (2017). https://doi.org/10.1016/j.jphotochemrev.2016.10.003
8. L.M. Luttrell, Transmembrane signaling by G protein-coupled receptors. Methods Mol. Biol. (Clifton, NJ) **332**, 3–49 (2006). https://doi.org/10.1385/1-59745-048-0:1
9. A. Manglik, A.C. Kruse, Structural basis for G protein-coupled receptor activation. Biochemistry **56**(42), 5628–5634 (2017). https://doi.org/10.1021/acs.biochem.7b00747
10. L.B. Motta-Mena, A. Reade, M.J. Mallory, S. Glantz, O.D. Weiner, K.W. Lynch, K.H. Gardner, An optogenetic gene expression system with rapid activation and deactivation kinetics. Nat. Chem. Biol. **10**(3), 196–202 (2014). https://doi.org/10.1038/nchembio.1430
11. K.Y. Chang, D. Woo, H. Jung, S. Lee, S. Kim, J. Won, T. Kyung, H. Park, N. Kim, H.W. Yang, J.W. Park, E.M. Hwang, D. Kim, W.D. Heo, Light-inducible receptor tyrosine kinases that regulate neurotrophin signalling. Nat. Commun. 5, 4057. https://doi.org/10.1038/ncomms5057
12. Endo M, Hattori M, Toriyabe H, Ohno H, Kamiguchi H, Iino Y, Ozawa T (2016) Optogenetic activation of axon guidance receptors controls direction of neurite outgrowth. Scientific reports 6:23976. https://doi.org/10.1038/srep23976
13. K. Aoki, Y. Kumagai, A. Sakurai, N. Komatsu, Y. Fujita, C. Shionyu, M. Matsuda, Stochastic ERK activation induced by noise and cell-to-cell propagation regulates cell density-dependent proliferation. Mol. Cell **52**(4), 529–540. https://doi.org/10.1016/j.molcel.2013.09.015

14. Y. Katsura, H. Kubota, K. Kunida, A. Kanno, S. Kuroda, T. Ozawa, An optogenetic system for interrogating the temporal dynamics of Akt. Scientif. Rep. **5**, 14589 (2015). https://doi.org/10.1038/srep14589
15. A. Musnier, B. Blanchot, E. Reiter, P. Crepieux, GPCR signalling to the translation machinery. Cell. Signal. **22**(5), 707–716 (2010). https://doi.org/10.1016/j.cellsig.2009.10.012
16. J. Bockaert, J.P. Pin, Molecular tinkering of G protein-coupled receptors: an evolutionary success. EMBO J. **18**(7), 1723–1729 (1999). https://doi.org/10.1093/emboj/18.7.1723
17. Zhou XE, He Y, de Waal PW, Gao X, Kang Y, Van Eps N, Yin Y, Pal K, Goswami D, White TA, Barty A, Latorraca NR, Chapman HN, Hubbell WL, Dror RO, Stevens RC, Cherezov V, Gurevich VV, Griffin PR, Ernst OP, Melcher K, Xu HE (2017) Identification of Phosphorylation Codes for Arrestin Recruitment by G Protein-Coupled Receptors. Cell 170 (3):457–469.e413. https://doi.org/10.1016/j.cell.2017.07.002
18. T.J. Cahill, 3rd, A.R. Thomsen, JT Tarrasch, B. Plouffe, A.H. Nguyen, F. Yang, L.Y. Huang, A.W. Kahsai, D.L. Bassoni, B.J. Gavino, J.E. Lamerdin, S. Triest, A.K. Shukla, B. Berger, J.T. Little, A. Antar, A. Blanc, C.X. Qu, X. Chen, K. Kawakami, A. Inoue, J. Aoki, J. Steyaert, J.P. Sun, M. Bouvier, G. Skiniotis, R.J. Lefkowitz, Distinct conformations of GPCR-beta-arrestin complexes mediate desensitization, signaling, and endocytosis. Proc. Natl. Acad. Sci. USA **114**(10), 2562–2567 (2017). https://doi.org/10.1073/pnas.1701529114
19. J. Zhang, L.S. Barak, K.E. Winkler, M.G. Caron, S.S. Ferguson, A central role for beta-arrestins and clathrin-coated vesicle-mediated endocytosis in beta2-adrenergic receptor resensitization. Differential regulation of receptor resensitization in two distinct cell types. J. Biol. Chem. **272**(43), 27005–27014 (1997)
20. O. Takenouchi, H. Yoshimura, T. Ozawa, Unique roles of beta-arrestin in GPCR trafficking revealed by photoinducible dimerizers. Scientif. Rep. **8**(1), 677 (2018). https://doi.org/10.1038/s41598-017-19130-y
21. M. Weitzman, K.M. Hahn, Optogenetic approaches to cell migration and beyond. Curr. Opin. Cell Biol. **30**, 112–120 (2014). https://doi.org/10.1016/j.ceb.2014.08.004
22. M.J. Kennedy, R.M. Hughes, L.A. Peteya, J.W. Schwartz, M.D. Ehlers, C.L. Tucker, Rapid blue-light-mediated induction of protein interactions in living cells. Nat. Methods **7**(12), 973–975 (2010). https://doi.org/10.1038/nmeth.1524
23. S.K. Shenoy, A.S. Modi, A.K. Shukla, K. Xiao, M. Berthouze, S. Ahn, K.D. Wilkinson, W.E. Miller, R.J. Lefkowitz, Beta-arrestin-dependent signaling and trafficking of 7-transmembrane receptors is reciprocally regulated by the deubiquitinase USP33 and the E3 ligase Mdm2. Proc. Natl. Acad. Sci. USA **106**(16), 6650–6655 (2009). https://doi.org/10.1073/pnas.0901083106
24. R. Nygaard, Y. Zou, R.O. Dror, T.J. Mildorf, D.H. Arlow, A. Manglik, A.C. Pan, C.W. Liu, J.J. Fung, M.P. Bokoch, F.S. Thian, T.S. Kobilka, D.E. Shaw, L. Mueller, R.S. Prosser, B.K. Kobilka, The dynamic process of beta(2)-adrenergic receptor activation. Cell **152**(3), 532–542 (2013). https://doi.org/10.1016/j.cell.2013.01.008

Chapter 8
Perspective Tools for Optogenetics and Photopharmacology: From Design to Implementation

Dmitrii M. Nikolaev, Maxim S. Panov, Andrey A. Shtyrov, Vitaly M. Boitsov, Sergey Yu. Vyazmin, Oleg B. Chakchir, Igor P. Yakovlev and Mikhail N. Ryazantsev

Abstract Optogenetics and photopharmacology are two perspective modern methodologies for control and monitoring of biological processes from an isolated cell to complex cell assemblies and organisms. Both methodologies use optically active components that being introduced into the cells of interest allow for optical control or monitoring of different cellular processes. In optogenetics, genetic materials are introduced into the cells to express light-sensitive proteins or protein constructs. In photopharmacology, photochromic compounds are delivered into a cell directly but not produced inside the cell from a genetic material. The development of both optogenetics and photopharmacology is inseparable from the design of improved tools (protein constructs or organic molecules) optimized for specific applications. Herein, we review the main tools that are used in modern optogenetics and photopharmaclogy and describe the types of cellular processes that can be controlled by these tools. Although a large number of different kinds of optogenetic tools exist, their performance can be evaluated with a limited number of metrics that have to be optimized for specific applications. We classify these metrics and describe the ways of their improvement.

D. M. Nikolaev · A. A. Shtyrov · V. M. Boitsov · S. Yu. Vyazmin · O. B. Chakchir
Saint-Petersburg Academic University – Nanotechnology Research and Education Centre RAS, 8/3 Khlopina street, Saint Petersburg 194021, Russia

M. S. Panov · M. N. Ryazantsev (✉)
Institute of Chemistry, Saint Petersburg State University, 26 Universitetskii pr., Saint Petersburg 198504, Russia
e-mail: mikhail.n.ryazantsev@gmail.com

I. P. Yakovlev
St. Petersburg State Chemical Pharmaceutical Academy, 14A Professora Popova street, Saint Petersburg 197376, Russia

M. N. Ryazantsev
Saint-Petersburg Scientific Center of the Russian Academy of Sciences, Universitetskaya nab. 5, Saint Petersburg 199034, Russia

K. Yamanouchi et al. (eds.), *Progress in Photon Science*, Springer Series in Chemical Physics 119, https://doi.org/10.1007/978-3-030-05974-3_8

8.1 Introduction

Modern photonics provides versatile tools for control and monitoring of biological processes from an isolated cell to complex cell assemblies and organisms. The most straightforward approach is to use intrinsic optical properties of molecules and molecular aggregates that are produced naturally in the cell. Several implementations of this approach have been developed and used for biological and medical applications [1–3]. However, the real progress was made by using specially designed tools with desired optical, chemical, and biological properties that, being incorporated into the cellular processes, allow changing them after absorption of a photon at the specific wavelength.

Two main methodologies have been proposed for such optical monitoring and control of cellular processes: optogenetics and photopharmacology. Their key difference is in the tools employed. In optogenetics, light-sensitive proteins or protein constructs are genetically introduced into the cells of interest becoming an integral part of the cellular machinery [4]. Illumination of cells modified in such way allows for the change of various cellular properties and processes, for example, the membrane potential or gene transcription. In photopharmacology, photochromic compounds are delivered into the cell or to its surface [5, 6]. Such compound has to satisfy two requirements: the compound can be transferred from one form to another form by a light stimulus, and physiological activities of these two forms must differ. One of the interesting examples is photochromic ion channel blockers—compounds that block a channel in one form and open in another one, allowing for the change in ion transport from extracellular to intracellular regions.

The development of both optogenetics and photopharmacology is inseparable from the design of improved tools (protein constructs or organic molecules) optimized for specific applications. These improvements regard either to the biological features, such as expression levels, toxicity, an efficiency of localization at the target sites, either to the properties of the light-induced response. In this review, we will cover the main factors determining the functionality of both optogenetic and photopharmacological tools and consider the commonly applied methods for their improvement.

This review is organized as follows. First, we will introduce the types of cellular processes that can be controlled with modern optogenetic and photopharmacological tools. For each target process, the applicable tools will be described. Although a large number of different kinds of optogenetic tools exist, their performance can be evaluated with a limited number of metrics. In the following subsections, we will classify these metrics in the Tables 8.1 and 8.2. For each metric, we will consider challenges and proposed solutions.

Table 8.1 The performance metrics of optogenetic tools and possible approaches for development of optogenetic tools with improved performance

Performance metrics	Optogenetic tools	Possible solution
Signal intensity (increasing)	Microbial rhodopsins as neural actuators or inhibitors	• Conversion of proton pumps into anion channels [21, 22] • Finding natural anion-conducting microbial rhodopsins [23, 24] • Shifting ion selectivity to divalent ions [86] • Rational mutagenesis near the active site [20, 87, 88]
	Microbial rhodopsins as fluorescent reporters	• Subject protein to the directed evolution [89] • High-throughput screening of mutants [90] • Combining microbial rhodopsins with bright fluorescent proteins in FRET system [75]
	LOV domains	• Rational mutagenesis near the active site [91, 92]
	Cryptochromes	• Rational mutagenesis near the active site [93, 94]
	UVR8 receptors	• Coupling two UVR8 receptors [95]
	Sensors based on fluorescent proteins	• Finding brighter FPs [96] • Increasing the expression levels and plasma membrane localization [97] • Mutation of existing FP [98–100]
Kinetics (acceleration or deceleration)	Microbial rhodopsins (ms timescale)	Acceleration of kinetics: • Rational mutagenesis of the residues involved into the photocycle [90, 101, 102] • Using faster rhodopsins from other organisms [103] • High-throughput screening of mutants [90] Deceleration of kinetics: • Site-directed mutagenesis of residues involved in the photocycle [104, 105]

(continued)

Table 8.1 (continued)

Performance metrics	Optogenetic tools	Possible solution
	GPCRs (second-minutes timescale)	Acceleration of kinetics: • Using GPCRs from vertebrate cone opsins [28, 106] Deceleration of kinetics: • Using bi-stable GPCRs with long-living open states [30, 107] • Expression of high levels of GPCRs leading to the saturation [108]
	LOV domains (minutes–hours timescale)	Acceleration of kinetics: • Rational mutagenesis altering hydrogen-bonding network near the active site [49, 109–112]
	Sensors based on fluorescent proteins	Acceleration of kinetics: • Finding faster functional domains [113–115] • Change the position of fluorescent protein [97] • Optimize linker between fluorescent protein and functional domain [116] • Rational mutagenesis [98] • Random mutagenesis [100]
Activation wavelength (red-shifting or obtaining tools with substantially different on/off activation wavelengths)	Microbial rhodopsins	• Natural variation of wavelengths [103, 117, 118] • Rational mutagenesis near the active site [89, 119, 120] • Creating protein chimeras [121, 122] • Subject protein to the directed evolution [68]
	GPCRs	• Applying OptoXR approach, choosing visual rhodopsin with required absorption maximum [31, 34, 123]
	LOV, BLUF, Cryptochromes	• Activation wavelength is not altered • Switching to phytochromes that perform same functions [60]

(continued)

Table 8.1 (continued)

Performance metrics	Optogenetic tools	Possible solution
	Phytochromes	• Using a different bilin chromophore [124, 125]
	Tools based on fluorescent proteins	• Mutations of GFP [126] • Choosing FP with different absorption wavelength [127]
Stability of the signal (increasing)	Microbial rhodopsins as neural actuators or inhibitors	• Rational mutagenesis [123] • Shifting ion selectivity to divalent ions [86]
	Microbial rhodopsins as fluorescent reporters	• Rational mutagenesis [90] • Decreasing illumination intensity [128]
	GPCRs	• Switching to opsins from invertebrates [106, 123]
	LOV domains	• Very stable optogenetic sensors [85, 129, 130]
	Tools based on fluorescent proteins	• Decreasing illumination brightness [131]
Sensitivity of the signal (increasing)	Microbial rhodopsins as neural actuators or inhibitors	• Using rhodopins with slower kinetics [87, 105] • Shifting ion selectivity to divalent ions [86] • Using more sensitive GPCRs [106, 132]
	Microbial rhodopsins as fluorescent reporters	• Mutating residues involved in the photocycle [90, 102]
	Tools based on fluorescent proteins	• Mutating fluorescent protein near the active site [98, 133, 134] • Mutating functional domain [72, 100] • Optimizing linker between fluorescent protein and the functional domain [100] • Replacing fluorescent protein or functional domain with homologs [97, 113, 127]
Measurement of absolute values	Tools based on fluorescent proteins	• Ratiometric measurements [135, 136] • Preliminary calibration [137]

(continued)

Table 8.1 (continued)

Performance metrics	Optogenetic tools	Possible solution
	Microbial rhodopsins	• Monitoring the change of signal upon alteration of illumination wavelengths [138]
Compatibility of optogenetic tools	Two microbial rhodopsins	• Using two independent actuators [122, 139] • Using actuator/inhibitor system in a single neuron [140] • "Closed-loop optogenetics": combination of the actuator and the reporter of neural electrical activity [90, 141] • Using outward proton pump and inward chloride pump for pumping chloride ions from neurons [142]
	Two fluorescent proteins	• Using several fluorescent sensors for simultaneous monitoring of different processes [143]
Expression and plasma membrane localization levels	Microbial rhodopsins	• Finding natural analogs with higher expression and localization levels [66, 140, 144] • Creating chimeras with highly expressing protein [122] • Using additional sequences in expression system [145, 146]
	Tools based on fluorescent proteins	• Using smaller functional domains [72, 73] • Optimization of coupling between a fluorescent protein and a functional domain [97] • Replacing a fluorescent protein with analogs [147]
Availability of the chromophore	Microbial rhodopsins and GPCRs	• Retinal chromophore is available in animals or can be delivered into their organisms with food [10]
	LOV, BLUF, cryptochromes	• Flavin chromophore is available in animals or can be delivered into their organisms with food [130]

(continued)

Table 8.1 (continued)

Performance metrics	Optogenetic tools	Possible solution
	Phytochromes	• Bilin chromophores need extra synthesis in animals [148, 149]
	UVR8 receptors	• Do not use chromophores for light absorption
Toxicity	Microbial rhodopsins	For ion pumps: • Shifting to potassium or sodium ions [16–18] • Using anion channels [21, 24] For sensors: • Rational mutagenesis to eliminate residual photocurrents [66]
	LOV, BLUF, cryptochromes	• Using natural analogs with reduced dark activity [46, 47] • Expressing sensors with additional constructs which will suppress dark activity [150]
	Tools based on fluorescent proteins	• Reducing expression levels [151]

Table 8.2 The performance metrics of photopharmacological tools and possible approaches for development of photopharmacological tools with improved performance

Performance metrics	Possible solution
Efficiency of the pharmaceutical (increasing)	• Increasing the isomerization efficiency of a molecular photoswitch [181] • Altering the conformational change upon isomerization [182]
Activation wavelength (red-shifting)	• Extending π-conjugated system [183] • Introducing electron-donating or electron-withdrawing groups [184, 185] • Adding bridge between the ring moieties of the molecular photoswitch [186, 187]
Kinetics of thermal relaxation (acceleration or deceleration)	Acceleration: • Introducing electron-donating or electron-withdrawing groups, creating "push-pull" molecular systems [184, 185] Deceleration: • Adding bridge between the ring moieties of the molecular photoswitch [186] • Functionalizing molecule with ortho-fluorine [190]
Stability (increasing)	• Inserting electron-donating moieties [193–195]

8.2 Optogenetics. Properties to Control or Monitor and Applied Tools

In this section we will consider cellular properties and processes that can be controlled or monitored with optogenetic tools. For each application, we will describe the classes of tools and briefly describe the molecular mechanisms of their functioning.

Control of electrical activity of excitable cells (neurons and cardiomyocytes). In order to achieve this goal, one has to activate ion currents through the cell membrane [7, 8]. Two main protein families are used for this purpose.

Microbial rhodopsins. Microbial rhodopsins are heptahelical transmembrane proteins naturally found in archaea, algae, fungi, and bacteria (Fig. 8.1a). Working as ion pumps or ion channels, microbial rhodopsins allow effective activation and inhibition of neural electrical activity by changing ion currents through the membrane.

Activation of neural excitability is achieved by intracellular transport of cations, leading to the neural depolarization. Proton-pumping channelrhodopsins are the most commonly used actuators of neural activity [9–12].

The detailed molecular mechanism of proton pumping in channelrhodopsin has not been determined yet. However, it is known that at some stages of the photocycle the retinal chromophore becomes deprotonated and the opsin converts into an open-gate state, allowing for proton transfer from the chromophore into the cell (neuron). Afterward, the retinal chromophore can be reprotonated with a proton from the intracellular medium and the opsin converts back into a closed-gate state [13]. Several studies demonstrated that channelrhodopsins can function not only as proton pumps but also as cation channels, allowing for cation transfer inside a cell [12]. The pumped protons (or other cations) change the membrane potential of the neuron, leading to its depolarization.

Inhibition of neural activity with other microbial rhodopsins can be achieved via three mechanisms.

(1) Neural hyperpolarization achieved by outward cation transport or inward anion transport. Tools for outward cation transport: proton-pumping archaerhodopsins [14, 15], natural and engineered sodium pumps [16, 17], and potassium pumps [18]. Tools for inward anion transport: chloride-pumping halorhodopsins and cruxhalorhodopsins [19, 20]. Here, the conformational changes in proteins triggered by light absorption lead to the formation of an ion pore, which is highly selective to a certain ion type [18–20]. This selectivity is determined by the amino acids forming the pore, however, exact mechanisms are not always clear.

(2) Blocking neural depolarization by triggering "shunting" currents of chloride ions through light-gated chloride channels. Initially, light-activated chloride channels were engineered on the basis of proton-pumping channelrhodopsins [21, 22]. A single replacement of a negative residue on a positive one led to the change of protein ion specificity. Later natural chloride channels with increased conductance were found [23, 24]. While the exact working principle of these anion-conducting channelrhodopsins is not clear yet, recent site-directed mutagenesis studies support the idea that two mechanisms are involved.

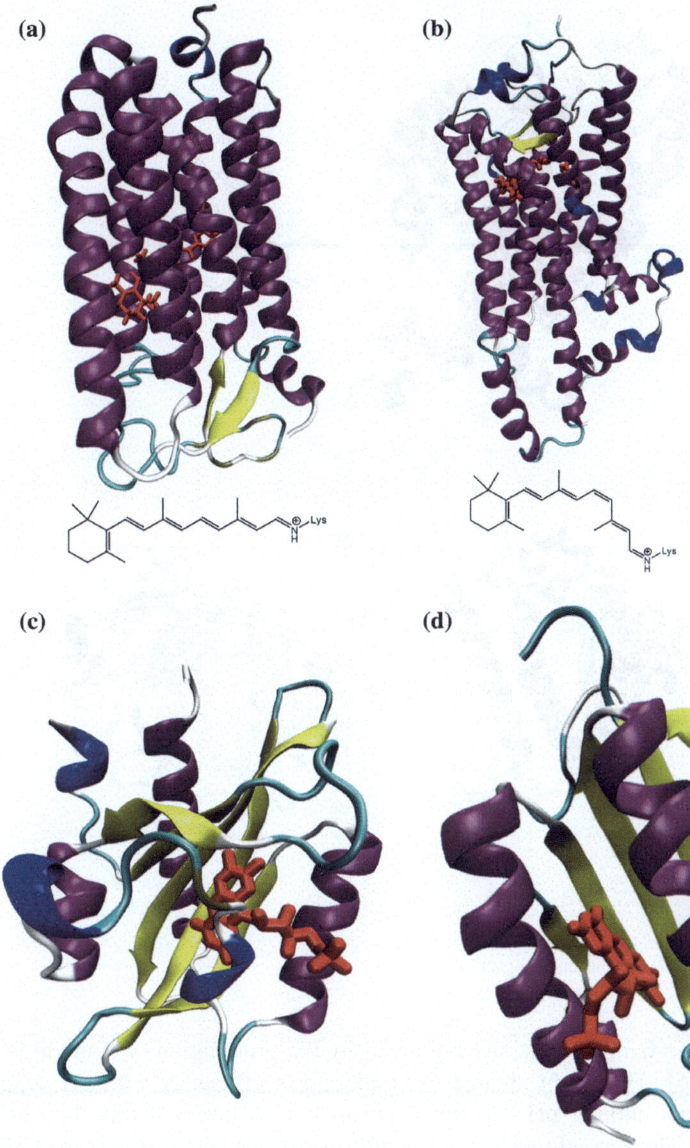

Fig. 8.1 Structure of main classes of optogenetic tools. **a** Microbial rhodopsin N. pharaonis halorhodopsin (PDB entry 3A7K) with an all-trans retinal chromophore, **b** Visual rhodopsin from T. Pacificus (PDB entry 2Z73) with an 11-cis retinal chromophore, **c–e** LOV domain (PDB entry 5EFW) (**c**), BLUF domain (PDB entry 2IYG) (**d**) and cryptochrome (PDB entry 2J4D) (**g**) are optogenetic tools with a flavin-based chromophore, flavin mononucleotide is presented in the figure, **f** Phytochrome (PDB entry 4OUR) with bilin chromophore, **g** UVR8 receptor (PDB entry 4DNW) utilizing tryptophan clusters for light absorption

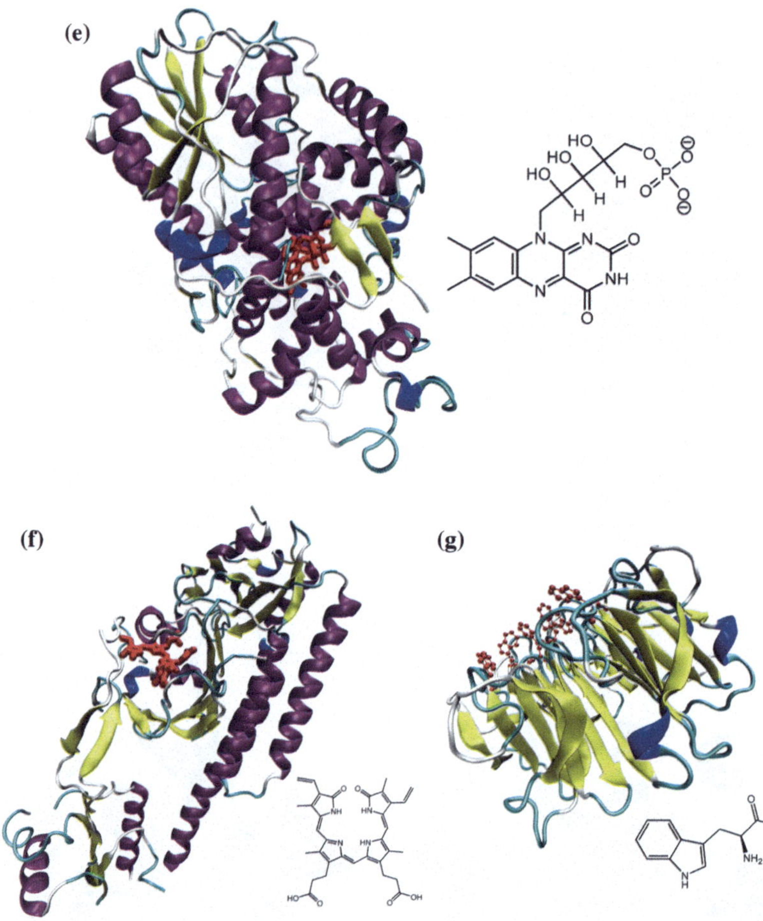

Fig. 8.1 (continued)

The first mechanism is mostly regulated by the protonation of a single Glu68 residue [25]. The slight change of conductance efficiency with pH supports the idea that another pH-dependent group is also involved into the channel dynamics. The second mechanism is fully regulated with a conserved Cys102 residue. Probably, Cys102 residue forms a hydrogen bond with a conserved Asp156 residue in a closed-gate state. Upon light absorption, the hydrogen bond dissociates, and the subsequent conformational changes lead to the formation of an open-gate state [25]. The channel opening leads to the occurrence of "extra" flow of cations, which decreases the local electrical resistance of the cellular system and, subsequently, decreases the excitatory postsynaptic potential.

(3) Excitation of inhibitory interneurons with proton-pumping channelrhodopsins [26]. The activation of interneurons leads to the silencing of local areas of the brain [27].

GPCRs. G protein-coupled receptors constitute a large class of transmembrane proteins found in eukaryotes (Fig. 8.1b). Upon activation with photon absorption or binding to a specific signaling molecule, GPCRs activate signaling transduction by interacting with a specific G protein. GPCRs are most commonly used for the inhibition of neural activity through intrinsic $G_{i/o}$ and G_s pathways. Usually, visual rhodopsins from vertebrates and invertebrates are used [28, 29].

Specifically, light activation triggers the interaction of visual rhodopsin with G proteins. G proteins are involved in the regulation of G protein inward rectifying potassium (GIRK) channels and presynaptic calcium channels. Thus, the light-activated interaction between rhodopsins and G proteins leads to the light-activated ion flow through GIRK and presynaptic calcium channels that induces neural inhibition [28].

In other studies bi-stable melanopsins, neuropsins, and parapinopsins that can be switched on and off with brief pulses of light with different wavelengths are used [30].

In order to change the G protein-coupling specificity of GPCR, an OptoXR approach is used. In the OptoXR method, rhodopsin intracellular or extracellular loops are replaced with corresponding parts of the ligand-activated GPCR with required specificity [31–33]. Alternatively, the ligand-activated GPCR can be attached to the C- or N-terminus of the light-activated rhodopsin [34].

Control of synaptic transmission between neurons. In order to achieve this goal, modern optogenetic tools that can change pH levels in the presynaptic region of neurons are used. Up to date, only proteins from the family of microbial rhodopsins were utilized for this purpose.

Microbial rhodopsins. Microbial rhodopsin archaerhodopsin-3 expressed in presynaptic terminals demonstrated light-activated inhibition of neurotransmitter release via increasing intracellular pH levels [35, 36]. The substantial increase of pH levels at the presynaptic terminals is induced by the light-activated proton pumping activity of archaerhodopsin-3. However, the exact mechanism of neurotransmitter release inhibition via pH increase is not determined yet, even though several speculations have been made [35].

Control of enzyme activity. In order to achieve this goal, modern optogenetic tools mediate a light-induced structural reorganization of a target enzyme. This allows controlling intracellular concentrations of second messengers, such as cyclic adenosine monophosphate (cAMP). Proteins from five protein families are used for this purpose.

LOV proteins. LOV proteins, naturally found in plants, bacteria, algae and fungi, absorb blue light with a flavin mononucleotide (FMN) chromophore (Fig. 8.1c). For optogenetic applications they are attached to specific functional domains. Light activation triggers structural reorganization in the LOV protein.

Specifically, this reorganization starts from the formation of a covalent bond between the FMN chromophore and the conserved cysteine residue [37]. The reorganization is then transmitted via hydrogen bonds and salt bridges to the alpha-helical linker, leading to its rotation. The rotation of the linker in its turn causes the reorientation of the functional domain, triggering its activation or inactivation. The induced effect depends on the number of coils in the alpha-helical linker, but does not critically depend on the linker length. Thus, addition of seven coils conserves the rotational angle of the linker and has a little effect on the signal transmission from the LOV domain to the functional domain.

In one of the examples, LOV domain was attached to histidine kinase (HisK) instead of its natural chemosensory domain, which allowed light activation of the enzyme [38]. Here, LOV-HisK constructs were paired in order to achieve the effective light modulation of HisK activity. It was demonstrated, that alteration of alpha-helical linker length led to three different cases: activation, deactivation or independence of HisK activity upon blue light illumination, supporting the assumption of crucial dependence of LOV-triggered effects on the linker rotational angle.

In other applications to modulate calcium concentrations, LOV proteins were attached to calcium-binding proteins, altering their calcium-binding capability [39]. The same strategy was used to activate Rac1, regulating cell protrusions [40].

Blue Light Utilizing Flavine adenine dinucleotide (FAD) (BLUF) domains. The action mechanism of BLUF domains is similar to that of LOV proteins (Fig. 8.1d). Photon absorption by the flavin chromophore also triggers structural reorganization of these proteins, which is transferred to distinct functional domains. It is proposed that the first steps of this photoactivated reorganization involve the electron transfer between FAD and the conserved tyrosine amino acid, leading to the formation of two radicals.

Subsequently, the formation of the radicals leads to the alteration of a hydrogen-bonding network of the protein and induces significant conformational alterations. However, the mechanism describing how this structural reorganization is transmitted to the linker between BLUF and functional domains remains is completely unclear [41].

In one of the experiments considering blue light triggered enzyme activation, BLUF domain was coupled to the catalytic domain that produces cAMP second messenger, which allowed blue-light mediated cAMP production [42, 43].

Cryptochromes. Cryptochromes are the third class of optogenetic tools that use the flavine chromophore (FAD) for blue light absorption (Fig. 8.1e). Cryptochromes are naturally found in plants and animals. Their functioning is based on the same principles as the functioning of LOV proteins and BLUF domains. Thus, attachment of cryptochromes to specific enzymes allows for light-induced activation of these proteins. For example, the light-inducible region of *Arabidopsis thaliana* cryptochrome attached to its binding partner allowed for the activation of the tropomyosin-related kinase, which regulates neurotrophin signaling [44].

Phytochromes. Phytochromes are photoreceptors naturally found in plants. They use bilin chromophore for light absorption and are attached to distinct functional domains in optogenetic experiments (Fig. 8.1f). In a recent study, a red light-activated bacterial phytochrome was linked via an alpha-helical linker to an effector module of *Homo sapiens* phosphodiesterase 2A. Resulting light-activated phosphodiesterase (LAPD) modulated the hydrolysis of cAMP and cGMP [45]. The mechanism of signal transduction here is similar to that of LOV domains; however, it is not completely described yet.

Microbial rhodopsins. The microbial rhodopsin from fungus *Blastocladiella emersonii* fused to the guanylyl cyclase catalytic domain allowed green light-activated modulation of cGMP levels [46, 47]. These unique eight alpha-helical rhodopsins allowed for the light-modulated control of cyclic nucleotide gated channels opening, modulating the intracellular cGMP levels.

Recruitment of protein to plasma membrane or organelles. In order to achieve this goal, one has to alter protein binding to specific functional proteins that are attached to the membrane or an organelle in a light-dependent manner. This allows controlling protein availability to its binding partners, modulating its activity. Three main protein families are used for this purpose.

LOV proteins. Light-activated interaction of LOV domain from *Arabidopsis thaliana* with GIGANTEA protein mediated the recruitment of yellow fluorescent protein to the plasma membrane, which led to the dimerization of the target protein [48]. In other works light-induced protein binding was used for activating scaffold proteins, kinases, and nucleotide-exchange factors [49].

Phytochromes. Plant photoreceptor phytochrome B from *Arabidopsis thaliana* coupled with phytochrome interaction factor 3 (PIF3) was attached to the cellular membrane. Red-light illumination regulated the binding of PIF3 to target proteins, in this way regulating attachment of target proteins to the membrane. This approach was used to modulate intracellular signaling by changing protein availability to its binding partners and to control nuclear localization of proteins [50–52].

Cryptochromes. The activation of Raf kinase occurs when the protein is bound to the cellular membrane. In a recent study, Raf proteins were labeled with cryptochromes. Upon blue-light absorption, cryptochromes attached to their interaction partners, which were recruited to the plasma membrane. Thus, Raf kinase recruitment and activation occurred [53].

Light modulated organelle relocalization. In order to achieve this goal, one has to induce organelle binding to specific functional proteins that are attached to mobile intracellular structures. Control of organelle relocalization allows the modulation of cellular signaling, cell growth, and other vital cellular activities. Up to date, only LOV proteins and cryptochromes were used for this purpose.

Light-oxygen-voltage (LOV) proteins. In order to achieve blue-light induced relocalization of peroxisomes, they were fused to the LOV domain from *Avena sativa* phototropin 1. Upon blue-light absorption, LOV domain attached to the cytoskeletal motor protein kinesin, which moved the peroxisome [54].

Cryptochromes. In one of the experiments, organelles were labeled with cryptochrome 2. Upon blue-light activation, cryptochrome bound to its interaction partner CIB1. CIB1 in its turn was fused to intracellular molecular motor kinesin. Thus, blue-light activation bound organelles to kinesin, which resulted in their translocation [55].

Control of protein-protein interactions. In order to alter protein-protein interactions, one has to change the position of the target protein relative to its interaction partner. This allows for modulation of protein activation and deactivation, and can be used for light-modulated opening of ion channels. Two main protein classes are used for this purpose.

LOV proteins. In one of the experiments, the photoswitch consisted of a LOV domain bound to a peptide toxin that blocks ion channels. Upon light activation, the photoswitch unfolded, lowering toxin concentration near the cell membrane. The lack of toxin near the cell membrane led to unblocking of ion channels [56]. In another study, target protein was bound to the localized LOV domain. Upon blue-light absorption, target protein detached from LOV domain and moved to the site of action [49]. LOV domains bound to peptides allowed for blue-light mediated control of peptide binding affinity [57].

Phytochromes. Phytochrome B from *Arabidopsis thaliana* with its interaction partner PIF3 was combined into a conditional protein splicing system. Red light absorption led to the attachment of phytochrome to the specific intein, which triggered protein splicing processes [58].

Control of gene transcription and expression. In order to induce gene transcription or expression in a light-dependent manner, modern optogenetic tools activate specific DNA binding proteins. Four main protein classes are used for this purpose.

LOV domains. Fusion of a LOV domain with DNA-binding protein EL222 prevented the dimerization of the latter protein. Upon blue-light absorption EL222 detached from the LOV domain, underwent dimerization, which led to DNA-binding and triggering of the gene transcription [59]. Light-induced interaction of LOV domain from *Arabidopsis thaliana* with GIGANTEA protein bound to the DNA-binding domain also allowed to regulate gene transcription [48].

Phytochromes. Red light absorption triggers interaction of phytochrome B with its interaction partner PIF6, reconstituting a factor mediating gene transcription. Upon far-red light absorption, the interaction breaks, silencing gene transcription [60].

Cryptochromes. Plant cryptochrome 2 binds *Arabidopsis* CIB1 protein upon blue-light absorption. This system allows for light-activated DNA binding and activation of gene expression and DNA transcription [61].

UV Resistance locus 8 (UVR8) receptors. UVR8 receptors are naturally found in plants. These proteins utilize tryptophan clusters for UV-light absorption (Fig. 8.1f). UVR8 receptors from *A. thaliana* in the dark-adapted state are organized in homodimers. Light activation leads to UVR8 monomerization.

As it was demostrated in recent mutational studies supported by dynamic crystallography and quantum chemical calculations, in the dark adapted state the two UVR8 monomers are coupled together with a network of hydrogen bonds and salt bridges [62, 63]. Upon light absorption, the charge separation in the conserved Trp233/Trp285 residues occurs, which leads to the disruption of salt bridges between arginine and aspartic acid residues that connected the two monomers. The molecular dynamics simulations also support the idea that upon light absorption the water molecule located at the interface of two UVR8 monomers is released, which also results in the dissociation of hydrogen bonds between the monomers. However, these assumptions have to be proved experimentally in further investigations [62].

In a construct for UV-activated gene expression, the monomeric form of UVR8 receptor interacts with the E3-ubiquitin ligase COP1 factor activating gene expression [64]. Specifically, upon light absorption the monomeric UVR8 receptor binds to the WD40 domain of the COP1 factor, which activates the target promoter and starts the gene expression [64].

Monitoring the electrical activity of excitable cells (neurons). In order to detect the changes of cellular membrane potential, modern optogenetics uses tools that alter their fluorescent signal upon the change of external electric potential.

Microbial rhodopsins. Several microbial rhodopsins—archaerhodopsin-3, green light-absorbing proteorhodopsin, and *Gloebacter violaceus* rhodopsin demonstrate the voltage-dependent intensity of fluorescence [65–68]. They absorb light with the retinal chromophore that can be either in all-trans or 13-cis conformations in the dark-adapted state (corresponding to two coexisting conformations of proteins).

The fluorescence spectroscopy and near-IR resonance Raman confocal microscopy studies of archaerhodopsin-3 suggest that the fluorescent Q-form of the protein is an intermediate that is generated after the photoexcitation of the 13-cis archaerhodopsin-3 conformation [69, 70]. Specifically, during the photocycle of archaerhodopsin-3 the deprotonation of the retinal chromophore occurs (M-state). Subsequently, the chromophore is reprotonated (N-state), and the photoexcitation of the N-state generates the Q-state. In its turn the photoexcitation of the Q-state results in the fluorescent signal.

It is assumed that membrane voltage regulates the equilibrium between the deprotonated M and the protonated N-states, thus regulating the concentration of the Q-form. However, the exact mechanism of potential-dependent fluorescence of archaerhodopsin-3 and other similar proteins is not clear yet [69].

Sensors based on fluorescent proteins. A fluorescent protein or two fluorescent proteins (FPs) are attached to a functional voltage-sensing domain. Membrane voltage triggers structural reorganization in the functional domain, which is transferred to FPs via a connecting linker, altering the fluorescent signal [71–73]. For example, in VSFP1 voltage sensor, two fluorescent proteins (CFP and YFP) connected with a flexible linker were fused to the fourth alpha-helix of the *Ciona intestinalis* voltage-sensing domain [74]. The alteration of membrane voltage induced the rotation of the fourth alpha-helix of the domain and consequently changed the relative orientation of the two fluorescent proteins. In its turn the change of orientation altered the Fluorescence Resonance Energy Transfer (FRET) efficiency between the proteins, and caused the alteration of the output fluorescent signal [74].

In recent studies, GFP was introduced into the neuron in combination with the fluorescent voltage-dependent microbial rhodopsin archaerhodopsin-3. Membrane voltage altered the intensity of the fluorescent signal of archaerhodopsin-3 that was absorbed with the GFP via FRET, which caused a more substantial alteration in the GFP fluorescence [75].

Monitoring the intracellular pH levels. In order to detect the changes of pH levels, modern optogenetics uses fluorescent proteins that alter their signal properties upon the change of intracellular or extracellular pH level.

Sensors based on fluorescent proteins. Specific mutations in the vicinity of fluorescent protein active site shift the pKa values of the protein chromophore and endow the GFP and its analogs (such as YFP, RFP) with the pH sensitivity [76].

Specifically, the intensities of absorption peaks at two wavelengths, corresponding to the protonated and deprotonated states of the protein chromophore, change with the alteration of the intracellular pH value [77]. Most probably, intracellular pH value regulates the relative concentration of the protonated and deprotonated states of the fluorescent protein.

Optical detection of redox reactions. In order to detect redox reaction, modern optogenetics uses tools that change the properties of their fluorescent signal upon oxidation or reduction.

Sensors based on fluorescent proteins. Redox reactions are indicated by H_2O_2 concentrations. Fluorescent reporters either consist of a fluorescent protein and a distinct H_2O_2 sensor, which undergoes structural reorganization upon oxidation, either is represented by mutated fluorescent proteins [78, 79]. In the latter case, additional peptides for the structural stabilization of the fluorescent protein after oxidation are required [79].

Specifically, the first H_2O_2 sensor consisted of a circularly permutated yellow fluorescent protein (cpYFP) that was inserted into the regulatory domain (RD) of the H_2O_2 sensitive protein (OxyR). Upon oxydation, the Cys199 converted into a sulfenic acid derivative and formed a disulfide bond with the Cys208 of the RD. The formation of the disulfide bridge dramatically changed the conformation of the RD, leading to the increase of the cpYFP fluorescence [78].

In further investigations, the cpRFP was modified in such a way that the oxydation led to the disulfide bridge formation between cysteines located at the N- and C-termini of the protein, and the induced conformational change led to the alteration of the protein fluorescent signal [79].

Monitoring the intracellular ion concentrations. In order to achieve this goal, modern optogenetics uses fluorescent proteins that change their signal properties upon binding to specific ions.

Sensors based on fluorescent proteins. Specific mutations created fluorescent proteins with conformational changes of their chromophore upon ion binding. This conformational change leads to the change of the fluorescent signal. For example, the chromophore of the GFP S65T/T203Y mutant in chloride-unbound form forms a hydrogen bond with the Tyr203 residue of the apoprotein [80]. The bond dissociation upon chloride binding to Tyr203 leads to the increased distance between the GFP chromophore and the Tyr203 residue and causes the increase of absorbance at 400 nm.

In other constructs, more complex systems involving several fluorescent proteins were developed. For example, in one of the experiments, increase of chloride concentration reduced the excitation efficiency of YFP which was connected with CFP in a FRET system. Thus, the increase of chloride concentration led to the increase of CFP fluorescence relative to the YFP fluorescence [81].

Sensors for a number of ions were developed, including calcium, cadmium, zinc ions [80–82]. Here, two general types of sensors exist. In the sensors of first type, ions interact not with fluorescent proteins, but rather with distinct modules. For example, in a calcium sensor the CFP-YFP FRET system was linked to the calmodulin-M13 (CaM-M13) system [81, 83]. In the absence of calcium ions CaM was not bound to M13 and the two fluorescent proteins were located far away from each other, not allowing FRET to occur. Interaction of CaM with calcium ions led to the CaM-M13 binding, decrease of the distance between CFP and YFP and the occurrence of FRET between the two proteins [81].

The second type of sensors does not involve distinct domains for ion binding. For example, in one of zinc sensors, specific mutations were introduced into CFP and YFP, creating zinc-binding sites at their surface. These two fluorescent proteins were linked in a single FRET system. Upon zinc binding, the two proteins stuck together via zinc bridges, which led to the sufficient change of FRET efficiency between the two proteins [84].

Optical monitoring protein expression and protein-protein interactions. In order to achieve this goal, the tool must to change the properties of its fluorescent signal upon binding or interacting with specific proteins.

LOV proteins. LOV proteins have intrinsic dim fluorescence. Rational mutagenesis of the protein active site increased this fluorescence and allowed using LOV proteins as fluorescent reporters of protein-protein interactions [85].

8.3 Optogenetics. Main Properties to Optimize

In this section, we will consider the main properties that define the performance of optogenetic tools. For each property, we will briefly describe the main directions of improvement and the approaches that are usually used for the optimization of this property to specific experimental conditions. This information is summarized in Table 8.1.

Signal intensity. A lot of studies were aimed at increasing the signal intensity of optogenetic tools. For tools that are aimed at the control of cellular processes, the increase of signal intensity is required to achieve the desired effect upon moderate illumination. For fluorescent reporters of cellular processes or properties, the increase of fluorescent signal intensity allows for more reliable measurements.

Microbial rhodopsins for control of neural activity. Ion pumping microbial rhodopsins usually demonstrate suboptimal currents, because a single ion is conducted upon absorption of a single photon. Weak ion currents cannot trigger strong action potentials or completely inhibit neural electrical activity.

One of the solutions is increasing the conductivity of microbial rhodopsins. For example, proton-pumping channelrhodopsins were converted into light-gated anion channels. In an open-gate state, these constructs allowed chloride ions to move freely through the pore, shunting excitatory ion currents.

Substitution of a single acidic residue with a basic residue in the ion-pumping region led to the halide selectivity of channelrhodopsin-2 [21, 22]. While the first variants demonstrated residual proton currents, additional mutagenesis solved this problem, leading to variants with even more intense photocurrents [119]. Natural anion selective channelrhodopsins were even more efficient [23, 24]. Another approach to obtain large ion currents is shifting the ion selectivity of ion pumps from monovalent to divalent ions [86].

Site-directed mutagenesis combined with screening assays increased the channelrhodopsin-2 photocurrents by introducing T159C point mutation [87]. In other studies, rational mutagenesis in the vicinity of protein active site also led to the increased signal [20, 88].

Microbial rhodopsins as sensors of membrane potential. Wild-type microbial rhodopsins used as fluorescent reporters of membrane potential are very dim and unsuitable for in vivo experiments [66]. Application of directed evolution approach [89], high-throughput screening of mutants in combination with site-directed mutagenesis [90] allowed to increase the fluorescence of archaerhodopsin-3. Red-shifting the absorption spectrum also led to the increase of fluorescence intensity; however, the mechanism of such improvement is unclear [68].

Another approach is combining microbial rhodopsin with bright fluorescent protein in a FRET system, monitoring only the change of FP fluorescence intensity [75]. In future studies, computational methodologies can help to develop a protein with increased signal intensity. Such development can be based on the rational mutagenesis or on application of more efficient artificial chromophores with higher quantum yields [152–163].

LOV proteins and cryptochromes. The intensity of LOV signal is defined by the change of the regulatory functions, e.g. of target protein binding affinity, upon light illumination. Rational structure-based mutagenesis of the LOV2 domain from *Avena sativa* increased the change of binding affinity of the corresponding functional domain to DNA upon light illumination in 14 times. Increased change of binding affinity allowed better control of gene transcription and other DNA-related processes [91].

Structure-guided mutagenesis in combination with screening assays led to LOV domain with increased control of interaction between to peptides [92]. The rational design also increased the light-induced oligomerization of proteins bound to *Arabidopsis* photoreceptor cryptochrome 2 [93, 94].

UVR8 receptors. The enhancement of UVR8 receptor activity was achieved by coupling two UVR8 receptors via a flexible peptide linker [95].

Sensors based on fluorescent proteins. The signal intensity of fluorescent sensors must also be high enough for reliable monitoring of cellular processes. Finding brighter fluorescent proteins, inserting mutations or circular permutating GFP are possible solutions [96, 98–100], along with finding sensors with higher levels of expression and plasma membrane localization [51, 97].

Photoreceptor kinetics. The desired on/off kinetics of an optogenetic tool greatly depends on the application of interest. Generation of frequent neural spikes or monitoring action potentials requires tools with on/off kinetics faster than 1 ms. When the inhibition of neural activity is concerned, long off-kinetics is preferable. Long-living open states of neural silencers allow maintaining neural inhibition with brief pulses of light, reducing the photodamage and increasing the stability of the tool.

Microbial rhodopsins. Natural microbial rhodopsins have millisecond-timescale photocycle kinetics. However, generation of frequent spikes can require additional protein engineering. For example, channelrhodopsin-2 could not stably generate action potentials with frequencies higher than 40 Hz. Rational mutagenesis of the channelrhodopsin-2 active site allowed stable generation of action potentials with frequencies up to 200 Hz [87, 101]. Finding faster natural analogs in another possible approach [103].

Considering microbial rhodopsins used for sensing membrane potential, site-directed mutagenesis of residues involved in the photocycle along with screening assays generated variants with submillisecond kinetics [90, 102]. The mutant of archaerhodopsin-3, QuasAr1, demonstrated response time constants of only 0.05 ms, which is close to the limitations imposed by electronics [90].

Site-directed mutagenesis of C128 and D156 positions of channelrhodopsin-2, which are involved in the protein photocycle, led to the variants with substantially decreased off-kinetics [104, 105, 164]. Thus, D156A mutant had open-gate state lasting for several minutes, compared to about 19 ms of the wild-type protein. Such great elongation of the protein photocycle is supposed to be related to the alteration of hydrogen-bonding network between protein helices [104]. Rhodopsins with long-living open states allowed the engineering of step-function rhodopsins, which are rapidly turned on and off with brief pulses of light [104, 105].

GPCRs. GPCRs have slower kinetics compared to microbial rhodopsins, with photocycle lasting for seconds or minutes. Among them, vertebrate cone opsins have faster responses compared to other types of GPCRs [28, 106]. On the other hand, melanopsins, neuropsins and parapinopsins have very long open states.

Moreover, these three classes of GPCRs are bistable, i.e. can be turned on and off with brief pulses of light [30, 107]. In order to obtain even more stable inhibition of neural activity, sufficient levels of GPCRs were expressed in the target tissue. High expression levels led to the saturation—there always were proteins in active state, which led to constant neural inhibition [108].

LOV and BLUF sensors. Activation kinetics of LOV and BLUF sensors is relatively slow and can last from minutes to hours [48, 165]. The slow off-kinetics of these sensors is related to the long thermal decay of their flavin chromophore to the dark-adapted state [166]. Rational mutagenesis, e.g. altering the protein hydrogen-bonding network in the vicinity of the flavin chromophore, allowed altering the signaling lifetimes of LOV and BLUF sensors by orders of magnitude [109, 110]. It allowed increasing the activation kinetics of protein recruitment to cellular compartments [49, 111], gene expression [112].

Sensors based on fluorescent proteins. If a fluorescent protein is attached to a distinct functional, e.g. voltage-sensing, domain, finding faster domains [72, 113–115], changing the position of fluorescent protein relative to the functional domain and the cellular membrane [97], and optimizing the linker between fluorescent protein and the functional domain [116] are possible solutions. Different mutation techniques, such as random [100] and rational [98] mutagenesis are also applicable here.

The activation wavelength. A large number of works were aimed at shifting the activation wavelength of optogenetic tools towards the IR region. This direction is related to two issues. First of all, the ability of deep tissue imaging requires the red-shifted activation wavelength, ideally in the IR spectral range, because biological tissues are almost transparent for radiation in the 700–900 nm spectral range [167]. While the modern solutions to this problem include using highly invasive optical fibers or application of a two-photon microscopy technique [168, 169] the availability of deep tissue single-photon activation could be much more effective.

Another reason for obtaining red-shifted optogenetic tools is related to the reduction of phototoxicity—photons with longer wavelength have lower energy and less affect the biological tissues.

Simultaneous applications of two different optogenetic tools also can require tuning of their absorption properties. In this case, it is very important to use variants with substantially different activation wavelengths in order to prevent the spectral cross-talk (see *Compatibility* of *optogenetic tools*).

Microbial rhodopsins. Activation wavelengths of microbial rhodopsins alter in a wide range even without additional protein engineering. For example, choosing channelrhodopsin from different bacteria allows varying the activation wavelength from 436 to 590 nm [103, 117, 118]. Rational mutagenesis of the active site and engineering protein chimeras allowed obtaining red-shifted variants of different microbial rhodopsins [89, 120–122]. Rational mutagenesis was also used to obtain bi-stable channelrhodopsin with sufficiently different on/off wavelengths (488/600 nm) [119]. Computational methodologies can be a prospective approach for the rational spectral tuning of rhodopsins [153, 160, 170–174].

Directed evolution demonstrated itself as a powerful method for shifting protein absorption maximum. Thus, directed evolution of *Gloebacter violaceus* rhodopsin generated mutants with absorption maximum red-shifted at 80 nm relative to the wild-type protein [68].

GPCRs. Engineering of GPCRs with required G protein specificity and activation wavelength exploits OptoXR approach. As described above, in the OptoXR approach visual opsin with required absorption maximum wavelength is combined with parts of ligand-activated GPCR of interest. Thus, an independent variation of neural inhibition pathway and activation wavelength is possible [31, 34, 123].

LOV proteins, BLUF, cryptochromes. Color tuning of optogenetic tools with flavin chromophore is impossible. An alternative is using red light absorbing phytochromes that perform the same functions [60].

Phytochromes. Modulation of activation wavelength of phytochromes can be achieved by changing the protein chromophore or finding a natural phytochrome with different absorption [124, 125].

Sensors based on fluorescent proteins. For tools based on fluorescent proteins, choosing FPs with naturally different absorption wavelength or insertion mutations in the current FPs are the two possible options [126, 127].

Stability. There are two important parameters of the stability of optogenetic tools that need to be considered: the stability of the signal during long-lasting experiments and the independence from the cellular environment.

Microbial rhodopsins. Signal intensity of all microbial rhodopsins used for the control of the neural electrical activity, decrease upon prolonged illumination. This instability has two reasons. First, the photocycle of microbial rhodopsins has inactive intermediates with slow recovery kinetics, and these intermediates accumulate with time [175]. Second, because microbial rhodopsins work as intracellular or extracellular ion pumps or ion channels, the change of corresponding ion concentrations leads to the decrease of the current [142].

Finding more stable natural variants or introducing mutations in the existing ones are possible solutions. For example, a channelrhodopsin-1/channelrhodopsin-2 chimera with crossover site at E-F loop exhibited significantly less inactivation upon prolonged light stimulation [176]. Switching to divalent ions also led to the increase of signal stability [86].

Fluorescence of microbial rhodopsins used as sensors of membrane potential also decreases very fast [66, 90]. The speed of photobleaching depends on the illumination intensity and wavelength, and time-varying modulation of the illumination [128]. Less photobleaching variants can be obtained by insertion of specific mutations [90], however, even the most stable mutants available up to date are unsuitable for long-term experiments.

GPCRs. GPCRs demonstrate unstable signal, which adapts under repetitive stimulation [28]. Switching to opsins from invertebrates, e.g. box jellyfish opsin or using specific cone opsins partially solved this problem [106, 123].

LOV proteins. LOV protein can be used as stable fluorescent reporters of protein-protein interactions. LOV proteins have very low photobleaching levels, and their signal is stable over a wide range of pH, temperatures and does not depend on the availability of molecular oxygen, i.e. they can work in the hypoxia conditions [85, 129, 130].

Sensors based on fluorescent proteins. In case of tools based on fluorescent proteins, using brighter FPs and thus decreasing illumination brightness is a possible solution [131].

Sensitivity. The sensitivity of an optogenetic control tool is defined by the change of the signal upon the change of illumination intensity. On the other hand, the sensitivity of fluorescent reporters is defined by the change of fluorescent intensity upon the change of the property of interest. The sensitivity should be maximal in the desired physiological range of the target property [143].

Microbial rhodopsins. The sensitivity of microbial rhodopsins used as actuators/inhibitors of neural activity was achieved by the accumulation of rhodopsins with slow kinetics at the site of interest [87, 105]. Shifting to divalent ions also increased the sensitivity in 70 times compared to the corresponding monovalent ion-pumping rhodopsins [86].

Mutating the residues of fluorescent voltage-dependent microbial rhodopsins, which are involved in the photocycle, increased their sensitivity [90, 102].

GPCRs. GPCRs as a tool for the control of neural activity are on orders of magnitude more sensitive than channelrhodopsins [106, 132].

Sensors based on fluorescent proteins. Considering sensors based on fluorescent proteins the following approaches are applied to increase the sensitivity of the tools.

1. Mutation of fluorescent protein at the periphery of the chromophore or circular permutation of the protein [98, 133, 134].
2. Mutation of the distinct functional domain, e.g. voltage-sensing domain [72, 100].
3. Changing the linker between fluorescent protein and the functional domain [100, 151].
4. Replacing fluorescent protein or functional domain with more sensitive homologs [97, 113, 127].

For example, mutations that alter the chromophore pKa values change the pH sensitivity of the protein [177, 178]. Mutations that alter the affinity of chloride ions to the chromophore change the values of halide sensitivity [80, 179].

Measurement of absolute values. Most of the modern fluorescent sensors are capable of monitoring the changes of the target property, being incapable of measuring the absolute values. A commonly used technique, applied both for measuring absolute pH and membrane potential values, is ratiometric measurements. Here, the relative response of the sensor to illumination with two different wavelengths is measured [135, 136]. However, this approach is quite unstable and has slow signal detection fidelity.

Attempts to measure absolute values of membrane potential using preliminary calibration of sensors were performed, however, such calibration is very unstable and is extremely challenging for in vivo applications [137].

Finally, a complicated reliable approach for measuring absolute values of membrane potential was presented for microbial rhodopsin sensors [138]. The response of archaerhodopsin-3 mutant upon alteration of two wavelengths was measured. Such measurements were possible because of a complex nature of archaerhodopsin-3, which has two dark-adapted conformations.

Compatibility of optogenetic tools. Several studies were aimed at combining optogenetic tools for a more complex interaction with the target cells. Five different combinations are used in the experiments.

Two actuators of neural activity. First of all, two kinds of actuators of neural electrical activity can be simultaneously introduced into different cell types. This allows independent actuation of distinct neural subpopulations and can be applied for studying interactions between these subpopulations [122, 139].

Actuator and inhibitor of neural activity. Second, actuator and inhibitor can be simultaneously introduced into a single neuron. This allows the researchers to easily turn on and off neural activity with the light of different wavelengths as it was demonstrated for the channelrhodopsin/halorhodopsin system [140].

Actuator and sensor of neural activity. The third approach, so-called "closed-loop optogenetics", combines an optogenetic actuator and a fluorescent voltage sensor in a single system [90, 141]. Ideally, the activity of the actuator must depend on the output signal from the fluorescent sensor.

Several fluorescent sensors. Fourth, several fluorescent sensors can be combined for simultaneous monitoring of different cellular processes [143].

Combination of ion pumps for a complex effect. Fifth, two tools can be combined in order to achieve some complex effect. For example, outward proton-pumping archaerhodopsin was combined with inward chloride pumping halorhodopsin in order to pump chloride ions out from the cell [142].

Requirements for the combined optogenetic tools. The combined optogenetic tools must satisfy certain requirements. First of all, the activation spectrum of the combined tools must be sufficiently different in order to prevent their spectral cross-talk [90, 121]. The only exclusion here is when two tools must be activated simultaneously in order to achieve the desired effect, as in the case of archaerhodopsin/halorhodopsin outward chloride pump [142].

Second, while illumination of a tool with red-shifted activation wavelength cannot trigger the blue-shifted tool, the reverse effect is often observed. Such residual activation can be suppressed by mutations [141].

Finally, when two different tools are expressed into a single cell, the question of using single or separate expression systems occurs. In a recent study, it was demonstrated that a single expression system leads to more stable expression and plasma membrane localization [180].

Biological problems. In this section we will consider four main classes of problems that are related with the functioning of optogenetic tools inside the cell.

(a) ***Expression levels and plasma membrane localization***. High expression levels and excellent plasma localization are of key importance for the efficient and intensive work of the optogenetic tools. Improving these characteristics is an especially acute problem for in vivo applications [114].
    Microbial rhodopsins. The first solution is finding natural analogs with higher expression and membrane localization levels. For example, *N. pharaonis* halorhodopsin has superior expression levels compared to archaerhodopsins from different organisms. Both protein types are used for the inhibition of membrane potential [140, 144]. Between fluorescent sensors of membrane potential, archaerhodopsin-3 demonstrated better membrane localization than green-absorbing proteorhodopsin in eukaryotic cells [66].
    The second approach is creating chimeras with highly-expressing proteins. For example, replacing the first two helices of green light-absorbing *Volvox carteri* rhodopsin with the corresponding part of highly expressing channelrhodopsin-1 resulted in chimeric rhodopsin with superior expression levels compared to wild-type *Volvox carteri* rhodopsin [122].
    Using additional sequences in an expression system, e.g. Golgi export sequences or plasma membrane targeting motifs, can also increase the membrane localization levels [145, 146].

Sensors based on fluorescent proteins. In case of optogenetic sensors based on fluorescent proteins, which utilize distinct functional domains, using smaller domains [72, 73], optimization of coupling between the functional domain and the fluorescent protein [97], or replacing the fluorescent protein with analog that do not lead to accumulation of intracellular aggregates [147] are possible solutions.

(b) ***Availability of the chromophore***. Except for UVR8 receptors, which absorb light via intrinsic tryptophan clusters, optogenetic tools require a chromophore cofactor for functioning. This cofactor cannot be encoded in the expression system of the protein-based tool and binds to the protein inside the cells. For this reason, the availability of chromophore molecules in the cells of interest is of critical importance for the tool functioning.

The retinal chromophore, which is required for the functioning of microbial rhodopsins and GPCRs, and the flavin chromophore required for LOV-based tools and cryptochromes, are available in animal tissues or can be easily delivered into their organisms with food [10, 130].

On the other hand, phytochromes, one of the common tools for altering protein-protein interactions, exploit bilin chromophores, which are absent in animals and require additional two-stage synthesis [148, 149]. Because of this obstacle, phytochromes were rarely used for the modulation of protein-protein interactions in animals, giving the preference to LOV-based protein constructs and cryptochromes.

(c) ***Undesirable altering the cellular physiology***. Another challenge is related to minimizing the side effects imposed by optogenetic tool functioning on the cell.

Microbial rhodopsins. Proton and chloride pumping microbial rhodopsins, such as archaerhodopsins and halorhodopsins, alter intracellular pH level, proton, and chloride concentrations during functioning [14, 15]. Changing of ion specificity to sodium or potassium is a possible solution [16].

Anion conducting channelrhodopsins, which inhibit neural excitations via shunting (decreasing ionic flux) rather than hyperpolarization of the cell membrane, also do not cause chloride concentration changes [119, 181].

Light illumination of microbial rhodopsins which are used as fluorescent reporters of membrane potential can cause the emergence of residual photocurrents. These photocurrents can be blocked by mutating a single residue involved in the photocycle of the rhodopsin [66].

LOV, BLUF, cryptochromes. Optogenetic tools with a flavin chromophore, i.e. LOV, BLUF domains, and cryptochromes, demonstrate significant dark activity [43]. Using alternative BLUF domains with reduced dark activity [46, 47] and truncated versions of cryptochromes [94] can help to solve the problem. Another approach is expressing these sensors with additional constructs which suppress dark activity [150].

Sensors based on fluorescent proteins. Fluorescent sensors of membrane voltage increase the capacitance of cellular membrane and alter the properties of the cellular electrical activity [182]. Solving this problem can be achieved by reducing the expression levels of the tool, which should be compensated by increased fluorescence intensity.

(d) ***Phototoxicity***. Another source of cellular damage during optogenetic experiments is light. Phototoxicity involves altering cell morphology and physiology, DNA damage [183, 184]. Light photons of longer wavelength contain less energy, and for this reason, cause less damage [185]. Unfortunately, red-shifting the activation wavelength is not applicable to all kinds of tools, but only to microbial rhodopsins, GPCRs, phytochromes, and tools based on fluorescent proteins.
Other approaches for decreasing phototoxicity include the decrease of illumination time and the decrease of illumination intensity [87, 105].

8.4 Photopharmacology. Some Tools and Ways for Their Improvement

Photopharmacology is a new and fast developing area of medicine [6]. The workhorses of photopharmacology are different types of organic molecules, which change their conformation upon light illumination, such as azobenzenes, diarylethenes, azonaphthalenes [186–188]. In order to control specific functions in a light-dependent manner, these compounds are attached to different pharmaceuticals. Two types of constructs are most commonly used. In the first variant, the photoswitch is used as a linker between two distinct parts of the drug. Upon light absorption, the photoswitch isomerizes, and the distance between the pharmacores change leading to drug activation [187]. For example, in several experiments, azobenzene-based switches were inserted between acrylamide moiety and quaternary ammonium ions (AAQ) (Fig. 8.2). In the trans-conformation of the photoswitch, the AAQ conformation allows the quaternary ammonium ions to effectively block neural potassium channels. Upon light illumination, trans-to-cis conversion of azobenzene takes place, leading to the unblocking of potassium channels [189, 190]. Such optical control of neurons can be used for the treatment of vision loss caused by neural degradation [190–193]. In the second type of constructs, molecular photoswitch is bound to the one-piece drug, altering its affinity to the receptor upon light illumination. For example, this strategy was used for the control of drugs acting on glutamate and GABAA receptors [194, 195]. The variation of molecular photoswitches is obtained by chemical modification of various molecules, which have already proved their effectiveness. These modifications are aimed at optimization of key characteristics of the photoswitches for different experimental requirements. Below we will consider these characteristics and run over the standard methods used for their optimization (Table 8.2).

The efficiency of the photopharmaceutical. The efficiency of the photopharmaceutical depends on two key parameters: the efficiency of the photoswitch isomerization and the change in the drug activity upon light activation. The efficiency of isomer-

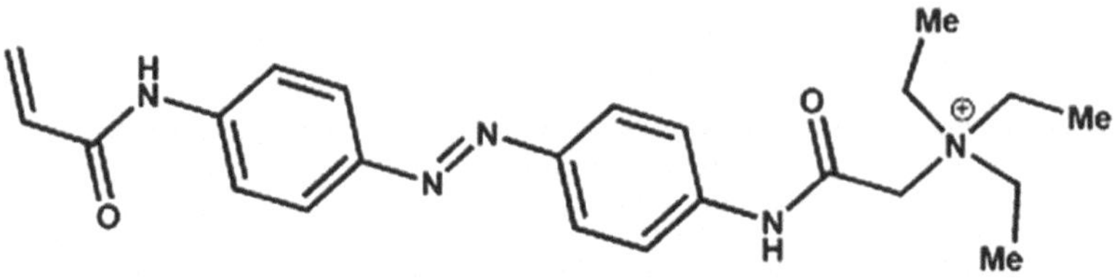

Fig. 8.2 Structural formula of a commonly used in photopharmacology azobenzene derivative, AAQ

ization is defined by the difference between relative concentrations of active/inactive states of the molecular photoswitch in the dark and illuminated conditions. Ideally, only non-active form must be present in the dark-adapted state, and only active form in the light-adapted state. However, such situation is prevented by the laws of thermodynamics. The presence of a small fraction of the active state in the darkness is not critical if this fraction is below a certain biological threshold [5].

Studies are aimed at increasing the isomerization efficiency of molecular photoswitches because this will allow increasing the activity of the drug along with decreasing the required illumination intensity. Up to date, chemical modification of azobenzene photoswitches allowed obtaining more than 95% cis-isomer upon light activation [196]. The magnitude of activity change of the drug can be varied by modification of the photochromic compound. For example, in case of AAQ and the similar switches used for blocking of potassium channels, their efficiency can be altered by increasing the geometry change upon trans-to-cis isomerization. This alteration can be obtained by adding proper moieties into azobenzene linkers [197].

Activation wavelength. As in the case of optogenetic tools, numerous works were aimed at red-shifting the absorption maximum of photochromic switches, which is related to less phototoxicity and the possibility of illuminating deeper tissues. Unfortunately, most of the switches used nowadays, such as azobenzenes, are activated by UV light that is extremely harmful to cells [167].

Activation of the drug prior injecting it into the organism is one of the possible solutions. However, this methodology is not always applicable. Extending π-conjugated system of the molecule is one of the mechanisms that leads to the red-shift of the absorption [198]. Thus, azobenzenes para-substituted with phenylacetylene absorb in the visible spectrum range. Another mechanism, which is commonly applied to azobenzenes, is introducing electron-donating (ED) or electron-withdrawing (EW) groups at ortho or para positions relative to the double N-N bond [199]. Two variants of azobenzene derivatives are generated with these methods—"amino" azobenzenes with one or both rings functionalized with ED groups or "push-pull" azobenzenes with AD group at one ring and AW group at another one. Both modifications lead to substantial red-shift of absorption spectrum to the visible region [200]. Another approach is providing an additional bridge between the ring moieties of the molecule [201, 202]. Finally, ortho-methoxy and ortho-fluoro azobenzenes also demonstrated absorption in the visible spectrum range [203].

Red-shifting the absorption spectrum is usually related to alteration of other characteristics. Thus, an extension of a π-conjugated system or adding ED and EW groups leads to faster thermal relaxation times [200]. On the other hand, adding a ring strain into the molecule led to the inverse thermal stability of the isomers [201]. Finally, ortho-fluorinating azobenzenes led to both isomers with approximately equal stability [203].

Kinetics of thermal relaxation. The optimal kinetics of the thermal relaxation of molecular photoswitches depends on the application of interest. For example, control of neural excitability, e.g. applied for the vision restoration, requires very fast (of millisecond timescale) thermal relaxation time constants [204]. On the other hand, obtaining active conformations with very long relaxation kinetics allows creating bi-stable molecular photoswitches, which can be turned on and off with the light of different wavelengths [161, 201, 205].

The kinetics of thermal relaxation is directly related to the energy difference between the two conformations of the molecular photoswitch. Decreasing the energy gap leads to slower relaxation kinetics. If the energy of two conformations is equal, both forms remain stable. The variation of thermal isomerization time constants can be controlled by adding different functional moieties, which alter electronic and steric nature of the molecules. For example, adding electron donating and electron withdrawing groups, creating "push-pull" azobenzenes, led to a much faster kinetics of thermal relaxation [199, 200]. Modern quantum chemical methodologies allow for high quality investigation of different chemical reactions [206–208] and can greatly facilitate the development of molecular photoswitches with altered kinetics.

Other issues. Other issues related to the use of synthetic photochromic compounds as light-activated drugs are the stability of their work in living organisms and their toxicity. Studying molecular mechanisms of the photoswitch degradation allows finding modifications that prevent the very first steps in the degradation process. For example, the mechanism of azobenzene degradation under the influence of different enzymes and glutathione were thoroughly investigated. Subsequently, insertion of specific electron-donating moieties prevented the degradation initiation reactions [209–211].

If the application of interest requires long-term work of the photopharmaceutical, the intrinsic stability of the photoswitch must be considered. In this case, the molecule should not lose the efficiency of conversion even after billions of transitions. For example, in visual restoration single photocycle of a photoswitch corresponds to the single action of photon acquisition by the eye retina [69]. In this case, millisecond timescale of the photocycle is required and efficient work of the drug for at least several days is desirable.

The toxicity of various azobenzene photoswitches was tested, and the main sources of their toxicity were defined [212]. In order to prevent the toxicity of azobenzenes, different chemical modifications are performed.

8.5 Conclusions

Large progress of optogenetics and photopharmacology achieved since the beginning of the XXI century is inseparably related with the development of improved tools. This improvement, guided by the experiment requirements, applied numerous approaches for achieving the goal. Unfortunately, in most cases these were solely experimental techniques, usually not based on solid surface of understanding the molecular mechanisms, which define the certain property of the tool. For example, in a number of experiments structure-guided rational design of channelrhodopsin-2, which applied as an actuator of membrane potential, was based on the crystallographic structure of channelrhodopsin-1/channelrhodopsin-2 chimeric protein [160]. However, FTIR and electrophysiological experiments demonstrated that this chimeric protein has different light-induced responses and biophysical properties [12]. Luckily, in November 2017 the high-quality crystallographic structure of channelrhodopsin-2 was published, and its analysis has already shed light on the molecular mechanisms of channelhrodopsin-2 functioning [213].

Extensive computational techniques can also facilitate the progress in this area. Thus, modern homology modeling techniques can be applied for the prediction of the proteins of interest if their crystallographic structure is not available, and for prediction of structural changes caused by mutations. On the other hand, techniques for the accurate calculation of protein absorption properties also exist. These two methodologies can be combined for in silico search for new protein-based tools with altered activation wavelength. Moreover, complex computational simulations, including molecular dynamics and hybrid quantum-mechanics/molecular-mechanics calculations can help to understand the molecular mechanisms of protein functioning and greatly facilitate the rational design approach.

Acknowledgements This work was supported by the Russian Foundation for Basic Research (grant numbers 15-29-03872 ofi_m and 16-04-00494 A) and by the program No. 32 of the Basic Research of Presidium RAS "Nanostructures: physics, chemistry, biology, basics of technology". The work was supported by the Skolkovo Foundation (grant agreement for Russian educational and scientific organization No. 7 dd 19.12.2017) and the Skolkovo Institute of Science and Technology (General agreement No. 3663-MRA dd. 25.12.2017). The work was also supported by Ministry of Education and Science of Russian Federation.

References

1. A.B. Ghisaidoobe, S.J. Chung, Int. J. Mol. Sci. **15**, 22518 (2014)
2. M. Monici, Biotechnol. Ann. Rev. **11**, 227 (2005)
3. S. Schmitz-Valckenberg, F.G. Holz, A.C. Bird, R.F. Spaide, Retina **28**, 385 (2008)
4. B.R. Rost, F. Schneider-Warme, D. Schmitz, P. Hegemann, Neuron **96**, 572 (2017)
5. W.A. Velema, W. Szymanski, B.L. Feringa, J. Am. Chem. Soc. **136**, 2178 (2014)
6. M.M. Lerch, M.J. Hansen, G.M. van Dam, W. Szymanski, B.L. Feringa, Angew. Chem. Int. Ed. **55**, 10978 (2016)
7. J.S. Wiegert, M. Mahn, M. Prigge, Y. Printz, O. Yizhar, Neuron **95**, 504 (2017)

8. A.B. Arrenberg, D.Y. Stainier, H. Baier, J. Huisken, Science **330**, 971 (2010)
9. F. Zhang, L.-P. Wang, E.S. Boyden, K. Deisseroth, Nat. Methods **3**, 785 (2006)
10. G. Nagel, M. Brauner, J.F. Liewald, N. Adeishvili, E. Bamberg, A. Gottschalk, Curr. Biol. **15**, 2279 (2005)
11. E.S. Boyden, F. Zhang, E. Bamberg, G. Nagel, K. Deisseroth, Nat. Neurosci. **8**, 1263 (2005)
12. F. Schneider, C. Grimm, P. Hegemann, Annu. Rev. Biophys. **44**, 167 (2015)
13. V.A. Lórenz-Fonfría, J. Heberle, Biochim. Biophys. Acta (BBA)-Bioenerg. **1837**, 626 (2014)
14. B.Y. Chow et al., Nature **463**, 98 (2010)
15. X. Han et al., Front Syst. Neurosci. **5**, 18 (2011)
16. H.E. Kato et al., Nature **521**, 48 (2015)
17. M.R. Hoque, T. Ishizuka, K. Inoue, R. Abe-Yoshizumi, H. Igarashi, T. Mishima, H. Kandori, H. Yawo, PLoS ONE **11**, e0166820 (2016)
18. I. Gushchin et al., Nat. Struct. Mol. Biol. **22**, 390 (2015)
19. V. Gradinaru, K.R. Thompson, K. Deisseroth, Brain Cell Biol. **36**, 129 (2008)
20. A.S. Chuong et al., Nat. Neurosci. **17**, 1123 (2014)
21. J. Wietek et al., Science **344**, 409 (2014)
22. A. Berndt, S.Y. Lee, C. Ramakrishnan, K. Deisseroth, Science **344**, 420 (2014)
23. E.G. Govorunova, O.A. Sineshchekov, R. Janz, X. Liu, J.L. Spudich, Science, aaa7484 (2015)
24. E.G. Govorunova, O.A. Sineshchekov, J.L. Spudich, Photochem. Photobiol. **92**, 257 (2016)
25. O.A. Sineshchekov, E.G. Govorunova, H. Li, J.L. Spudich, Proc. Natl. Acad. Sci. **112**, 14236 (2015)
26. L. Madisen et al., Nat. Neurosci. **15**, 793 (2012)
27. J.-Z. Guo et al., Elife **4** (2015)
28. X. Li, D.V. Gutierrez, M.G. Hanson, J. Han, M.D. Mark, H. Chiel, P. Hegemann, L.T. Landmesser, S. Herlitze, Proc. Natl. Acad. Sci. U.S.A. **102**, 17816 (2005)
29. H.J. Bailes, L.-Y. Zhuang, R.J. Lucas, PLoS ONE **7**, e30774 (2012)
30. K. Spoida et al., Curr. Biol. **26**, 1206 (2016)
31. R.D. Airan, K.R. Thompson, L.E. Fenno, H. Bernstein, K. Deisseroth, Nature **458**, 1025 (2009)
32. P.A. Barish, Y. Xu, J. Li, J. Sun, Y.P. Jarajapu, W.O. Ogle, Eur. J. Pharmacol. **705**, 42 (2013)
33. E.R. Siuda et al., Nat. Commun. **6**, 8480 (2015)
34. E. Oh, T. Maejima, C. Liu, E. Deneris, S. Herlitze, J. Biol. Chem. **285**, 30825 (2010)
35. M. El-Gaby, Y. Zhang, K. Wolf, C.J. Schwiening, O. Paulsen, O.A. Shipton, Cell Rep. **16**, 2259 (2016)
36. M. Mahn, M. Prigge, S. Ron, R. Levy, O. Yizhar, Nat. Neurosci. **19**, 554 (2016)
37. E. Peter, B. Dick, S.A. Baeurle, Nat. Commun. **1**, 122 (2010)
38. A. Möglich, R.A. Ayers, K. Moffat, J. Mol. Biol. **385**, 1433 (2009)
39. N. Fukuda, T. Matsuda, T. Nagai, ACS Chem. Biol. **9**, 1197 (2014)
40. Y.I. Wu, D. Frey, O.I. Lungu, A. Jaehrig, I. Schlichting, B. Kuhlman, K.M. Hahn, Nature **461**, 104 (2009)
41. S. Masuda, Plant Cell Physiol. **54**, 171 (2012)
42. T. Nagahama, T. Suzuki, S. Yoshikawa, M. Iseki, Neurosci. Res. **59**, 81 (2007)
43. S. Schröder-Lang et al., Nat. Methods **4**, 39 (2007)
44. K.-Y. Chang et al., Nat. Commun. **5**, 4057 (2014)
45. C. Gasser, S. Taiber, C.-M. Yeh, C.H. Wittig, P. Hegemann, S. Ryu, F. Wunder, A. Möglich, Proc. Natl. Acad. Sci. U.S.A. **111**, 8803 (2014)
46. G.M. Avelar, R.I. Schumacher, P.A. Zaini, G. Leonard, T.A. Richards, S.L. Gomes, Curr. Biol. **24**, 1234 (2014)
47. S. Gao, J. Nagpal, M.W. Schneider, V. Kozjak-Pavlovic, G. Nagel, A. Gottschalk, Nat. Commun. **6**, 8046 (2015)
48. M. Yazawa, A.M. Sadaghiani, B. Hsueh, R.E. Dolmetsch, Nat. Biotechnol. **27**, 941 (2009)
49. H. Wang et al., Nat. Methods **13**, 755 (2016)
50. A. Levskaya, O.D. Weiner, W.A. Lim, C.A. Voigt, Nature **461**, 997 (2009)
51. X. Yang, A.P.-T. Jost, O.D. Weiner, C. Tang, Mol. Biol. Cell **24**, 2419 (2013)

52. H.M. Beyer et al., ACS Synth. Biol. **4**, 951 (2015)
53. K. Zhang, L. Duan, Q. Ong, Z. Lin, P.M. Varman, K. Sung, B. Cui, PLoS ONE **9**, e92917 (2014)
54. P. Van Bergeijk, M. Adrian, C.C. Hoogenraad, L.C. Kapitein, Nature **518**, 111 (2015)
55. L. Duan, D. Che, K. Zhang, Q. Ong, S. Guo, B. Cui, Chem. Biol. **22**, 671 (2015)
56. D. Schmidt, P.W. Tillberg, F. Chen, E.S. Boyden, Nat. Commun. **5**, 3019 (2014)
57. O.I. Lungu, R.A. Hallett, E.J. Choi, M.J. Aiken, K.M. Hahn, B. Kuhlman, Chem. Biol. **19**, 507 (2012)
58. A.B. Tyszkiewicz, T.W. Muir, Nat. Methods **5**, 303 (2008)
59. B.D. Zoltowski, L.B. Motta-Mena, K.H. Gardner, Biochemistry **52**, 6653 (2013)
60. K. Müller et al., Nucleic Acids Res. **41**, e77 (2013)
61. M.J. Kennedy, R.M. Hughes, L.A. Peteya, J.W. Schwartz, M.D. Ehlers, C.L. Tucker, Nat. Methods **7**, 973 (2010)
62. X. Yang, S. Montano, Z. Ren, Photochem. Photobiol. **91**, 993 (2015)
63. K. Tilbrook, A.B. Arongaus, M. Binkert, M. Heijde, R. Yin, R. Ulm, The Arabidopsis book. Am. Soc. Plant Biol. **11** (2013)
64. K. Müller et al., Nucleic Acids Res. **41**, e124 (2013)
65. J.M. Kralj, D.R. Hochbaum, A.D. Douglass, A.E. Cohen, Science **333**, 345 (2011)
66. J.M. Kralj, A.D. Douglass, D.R. Hochbaum, D. Maclaurin, A.E. Cohen, Nat. Methods **9**, 90 (2012)
67. N.C. Flytzanis, C.N. Bedbrook, H. Chiu, M.K. Engqvist, C. Xiao, K.Y. Chan, P.W. Sternberg, F.H. Arnold, V. Gradinaru, Nat. Commun. **5**, 4894 (2014)
68. M.K. Engqvist, R.S. McIsaac, P. Dollinger, N.C. Flytzanis, M. Abrams, S. Schor, F.H. Arnold, J. Mol. Biol. **427**, 205 (2015)
69. D. Maclaurin, V. Venkatachalam, H. Lee, A.E. Cohen, Proc. Natl. Acad. Sci. U. S. A. **110**, 5939 (2013)
70. E.C. Saint Clair, J.I. Ogren, S. Mamaev, D. Russano, J.M. Kralj, K.J. Rothschild, J. Phys. Chem. B **116**, 14592 (2012)
71. M.S. Siegel, E.Y. Isacoff, Neuron **19**, 735 (1997)
72. D. Dimitrov, Y. He, H. Mutoh, B.J. Baker, L. Cohen, W. Akemann, T. Knöpfel, PLoS ONE **2**, e440 (2007)
73. B.E. Kang, B.J. Baker, Sci. Rep. **6**, 23865 (2016)
74. R. Sakai, V. Repunte-Canonigo, C.D. Raj, T. Knöpfel, Eur. J. Neurosci. **13**, 2314 (2001)
75. Y. Gong, M.J. Wagner, J.Z. Li, M.J. Schnitzer, Nat. Commun. **5**, 3674 (2014)
76. G.H. Patterson, S.M. Knobel, W.D. Sharif, S.R. Kain, D.W. Piston, Biophys. J. **73**, 2782 (1997)
77. M. Kneen, J. Farinas, Y. Li, A. Verkman, Biophys. J. **74**, 1591 (1998)
78. V.V. Belousov, A.F. Fradkov, K.A. Lukyanov, D.B. Staroverov, K.S. Shakhbazov, A.V. Terskikh, S. Lukyanov, Nat. Methods **3**, 281 (2006)
79. Y. Fan, Z. Chen, H.-W. Ai, Anal. Chem. **87**, 2802 (2015)
80. R.M. Wachter, S.J. Remington, Curr. Biol. **9**, R628 (1999)
81. A. Germond, H. Fujita, T. Ichimura, T.M. Watanabe, Biophys. Rev. **8**, 121 (2016)
82. A. Miyawaki, J. Llopis, R. Heim, J.M. McCaffery, J.A. Adams, M. Ikura, R.Y. Tsien, Nature **388**, 882 (1997)
83. M. Ikura, G.M. Clore, A.M. Gronenborn, G. Zhu, C.B. Klee, A. Bax, Science **256**, 632 (1992)
84. K.K. Jensen, L. Martini, T.W. Schwartz, Biochemistry (Moscow) **40**, 938 (2001)
85. S. Chapman, C. Faulkner, E. Kaiserli, C. Garcia-Mata, E.I. Savenkov, A.G. Roberts, K.J. Oparka, J.M. Christie, Proc. Natl. Acad. Sci. U.S.A. **105**, 20038 (2008)
86. F. Schneider, D. Gradmann, P. Hegemann, Biophys. J. **105**, 91 (2013)
87. A. Berndt, P. Schoenenberger, J. Mattis, K.M. Tye, K. Deisseroth, P. Hegemann, T.G. Oertner, Proc. Natl. Acad. Sci. U.S.A. **108**, 7595 (2011)
88. Y. Sudo et al., J. Biol. Chem. **288**, 20624 (2013)
89. R.S. McIsaac et al., Proc. Natl. Acad. Sci. U.S.A. **111**, 13034 (2014)
90. D.R. Hochbaum et al., Nat. Methods **11**, 825 (2014)

91. D. Strickland, X. Yao, G. Gawlak, M.K. Rosen, K.H. Gardner, T.R. Sosnick, Nat. Methods **7**, 623 (2010)
92. G. Guntas, R.A. Hallett, S.P. Zimmerman, T. Williams, H. Yumerefendi, J.E. Bear, B. Kuhlman, Proc. Natl. Acad. Sci. U.S.A. **112**, 112 (2015)
93. A. Taslimi, J.D. Vrana, D. Chen, S. Borinskaya, B.J. Mayer, M.J. Kennedy, C.L. Tucker, Nat. Commun. **5**, 4925 (2014)
94. A. Taslimi, B. Zoltowski, J.G. Miranda, G.P. Pathak, R.M. Hughes, C.L. Tucker, Nat. Chem. Biol. **12**, 425 (2016)
95. R.P. Crefcoeur, R. Yin, R. Ulm, T.D. Halazonetis, Nat. Commun. **4**, 1779 (2013)
96. H. Tsutsui, S. Karasawa, Y. Okamura, A. Miyawaki, Nat. Methods **5**, 683 (2008)
97. F. St-Pierre, J.D. Marshall, Y. Yang, Y. Gong, M.J. Schnitzer, M.Z. Lin, Nat. Neurosci. **17**, 884 (2014)
98. L. Jin, Z. Han, J. Platisa, J.R. Wooltorton, L.B. Cohen, V.A. Pieribone, Neuron **75**, 779 (2012)
99. G.S. Baird, D.A. Zacharias, R.Y. Tsien, Proc. Natl. Acad. Sci. U.S.A. **96**, 11241 (1999)
100. A.S. Abdelfattah et al., J. Neurosci. **36**, 2458 (2016)
101. L.A. Gunaydin, O. Yizhar, A. Berndt, V.S. Sohal, K. Deisseroth, P. Hegemann, Nat. Neurosci. **13**, 387 (2010)
102. Y. Gong, J.Z. Li, M.J. Schnitzer, PLoS ONE **8**, e66959 (2013)
103. N.C. Klapoetke et al., Nat. Methods **11**, 338 (2014)
104. C. Bamann, R. Gueta, S. Kleinlogel, G. Nagel, E. Bamberg, Biochemistry **49**, 267 (2009)
105. A. Berndt, O. Yizhar, L.A. Gunaydin, P. Hegemann, K. Deisseroth, Nat. Neurosci. **12**, 229 (2009)
106. O.A. Masseck, K. Spoida, D. Dalkara, T. Maejima, J.M. Rubelowski, L. Wallhorn, E.S. Deneris, S. Herlitze, Neuron **81**, 1263 (2014)
107. T. Yamashita et al., J. Biol. Chem. **289**, 3991 (2014)
108. B.L. Roth, Neuron **89**, 683 (2016)
109. J.M. Christie, Annu. Rev. Plant Biol. **58**, 21 (2007)
110. S. Raffelberg, M. Mansurova, W. Gärtner, A. Losi, J. Am. Chem. Soc. **133**, 5346 (2011)
111. D. Strickland, Y. Lin, E. Wagner, C.M. Hope, J. Zayner, C. Antoniou, T.R. Sosnick, E.L. Weiss, M. Glotzer, Nat. Methods **9**, 379 (2012)
112. L.B. Motta-Mena, A. Reade, M.J. Mallory, S. Glantz, O.D. Weiner, K.W. Lynch, K.H. Gardner, Nat. Chem. Biol. **10**, 196 (2014)
113. B.J. Baker, L. Jin, Z. Han, L.B. Cohen, M. Popovic, J. Platisa, V. Pieribone, J. Neurosci. Methods **208**, 190 (2012)
114. Y. Gong, C. Huang, J. Z. Li, B.F. Grewe, Y. Zhang, S. Eismann, M.J. Schnitzer, Science, aab0810 (2015)
115. Y. Mishina, H. Mutoh, C. Song, T. Knöpfel, Front. Mol. Neurosci. **7**, 78 (2014)
116. U. Sung, M. Sepehri-Rad, H.H. Piao, L. Jin, T. Hughes, L.B. Cohen, B.J. Baker, PLoS ONE **10**, e0141585 (2015)
117. E.G. Govorunova, O.A. Sineshchekov, H. Li, R. Janz, J.L. Spudich, J. Biol. Chem. **288**, 29911 (2013)
118. O. Yizhar et al., Nature **477**, 171 (2011)
119. A. Berndt et al., Proc. Natl. Acad. Sci. U.S.A. **113**, 822 (2016)
120. J.Y. Lin, P.M. Knutsen, A. Muller, D. Kleinfeld, R.Y. Tsien, Nat. Neurosci. **16**, 1499 (2013)
121. K. Erbguth, M. Prigge, F. Schneider, P. Hegemann, A. Gottschalk, PLoS ONE **7**, e46827 (2012)
122. M. Prigge, F. Schneider, S.P. Tsunoda, C. Shilyansky, J. Wietek, K. Deisseroth, P. Hegemann, J. Biol. Chem. **287**, 31804 (2012)
123. W.A. Karunarathne, L. Giri, V. Kalyanaraman, N. Gautam, Proc. Natl. Acad. Sci. U.S.A. **110**, E1565 (2013)
124. N.C. Rockwell, D. Duanmu, S.S. Martin, C. Bachy, D.C. Price, D. Bhattacharya, A.Z. Worden, J.C. Lagarias, Proc. Natl. Acad. Sci. U.S.A. **111**, 3871 (2014)
125. K.D. Piatkevich, F.V. Subach, V.V. Verkhusha, Chem. Soc. Rev. **42**, 3441 (2013)

126. M. Ormö, A.B. Cubitt, K. Kallio, L.A. Gross, R.Y. Tsien, S.J. Remington, Science **273**, 1392 (1996)
127. N.C. Shaner, R.E. Campbell, P.A. Steinbach, B.N. Giepmans, A.E. Palmer, R.Y. Tsien, Nat. Biotechnol. **22**, 1567 (2004)
128. K.M. Dean, J.L. Lubbeck, J.K. Binder, L.R. Schwall, R. Jimenez, A.E. Palmer, Biophys. J. **101**, 961 (2011)
129. T.E. Swartz, S.B. Corchnoy, J.M. Christie, J.W. Lewis, I. Szundi, W.R. Briggs, R.A. Bogomolni, J. Biol. Chem. **276**, 36493 (2001)
130. J.M. Christie, J. Gawthorne, G. Young, N.J. Fraser, A.J. Roe, Mol. Plant **5**, 533 (2012)
131. N.C. Shaner et al., Nat. Methods **10**, 407 (2013)
132. S. Kleinlogel, K. Feldbauer, R.E. Dempski, H. Fotis, P.G. Wood, C. Bamann, E. Bamberg, Nat. Neurosci. **14**, 513 (2011)
133. S. Jayaraman, P. Haggie, R.M. Wachter, S.J. Remington, A. Verkman, J. Biol. Chem. **275**, 6047 (2000)
134. M.J. Mahon, Adv. Biosci. Biotechnol. (Print) **2**, 132 (2011)
135. T. Kuner, G.J. Augustine, Neuron **27**, 447 (2000)
136. B.A. Wilt, J.E. Fitzgerald, M.J. Schnitzer, Biophys. J. **104**, 51 (2013)
137. M.B. Hoppa, G. Gouzer, M. Armbruster, T.A. Ryan, Neuron **84**, 778 (2014)
138. J.H. Hou, V. Venkatachalam, A.E. Cohen, Biophys. J. **106**, 639 (2014)
139. F. Zhang, M. Prigge, F. Beyrière, S.P. Tsunoda, J. Mattis, O. Yizhar, P. Hegemann, K. Deisseroth, Nat. Neurosci. **11**, 631 (2008)
140. X. Han, E.S. Boyden, PLoS ONE **2**, e299 (2007)
141. S. Lou et al., J. Neurosci. **36**, 11059 (2016)
142. H. Alfonsa, E.M. Merricks, N.K. Codadu, M.O. Cunningham, K. Deisseroth, C. Racca, A.J. Trevelyan, J. Neurosci. **35**, 7715 (2015)
143. M. Tantama, Y.P. Hung, G. Yellen, *Progress in Brain Research* (Elsevier, 2012), p. 235
144. F. Zhang et al., Nature **446**, 633 (2007)
145. V. Gradinaru, F. Zhang, C. Ramakrishnan, J. Mattis, R. Prakash, I. Diester, I. Goshen, K.R. Thompson, K. Deisseroth, Cell **141**, 154 (2010)
146. W. Akemann, H. Mutoh, A. Perron, Y.K. Park, Y. Iwamoto, T. Knöpfel, J. Neurophysiol. **108**, 2323 (2012)
147. K. Tsutsui, T. Ubuka, G.E. Bentley, L. Kriegsfeld, Front. Neurosci. **7**, 60 (2013)
148. K. Müller, R. Engesser, J. Timmer, F. Nagy, M.D. Zurbriggen, W. Weber, Chem. Commun. **49**, 8970 (2013)
149. C.E. Buckley, R.E. Moore, A. Reade, A.R. Goldberg, O.D. Weiner, J.D. Clarke, Dev. Cell **36**, 117 (2016)
150. Y.I. Wu, X. Wang, L. He, D. Montell, K.M. Hahn, *Methods in Enzymology* (Elsevier, 2011), p. 393
151. H.H. Yang, F. St-Pierre, J. Neurosci. **36**, 9977 (2016)
152. L. Herwig et al., Cell Chem. Biol. **24**, 415 (2017)
153. I. Schapiro, M.N. Ryazantsev, L.M. Frutos, N. Ferré, R. Lindh, M. Olivucci, J. Am. Chem. Soc. **133**, 3354 (2011)
154. A. Melloni et al., J. Am. Chem. Soc. **132**, 9310 (2010)
155. A. Sinicropi et al., Proc. Natl. Acad. Sci. U.S.A. **105**, 17642 (2008)
156. I.L. Zheldakov, M.N. Ryazantsev, A.N. Tarnovsky, J. Phys. Chem. Lett. **2**, 1540 (2011)
157. P.Z. El-Khoury, A.N. Tarnovsky, I. Schapiro, M.N. Ryazantsev, M. Olivucci, J. Phys. Chem. A **113**, 10767 (2009)
158. J.C. Williams, J. Xu, Z. Lu, A. Klimas, X. Chen, C.M. Ambrosi, I.S. Cohen, E. Entcheva, PLoS Comp. Biol. **9**, e1003220 (2013)
159. J. Kuhne, K. Eisenhauer, E. Ritter, P. Hegemann, K. Gerwert, F. Bartl, Angew. Chem. Int. Ed. **54**, 4953 (2015)
160. Y. Hontani, M. Marazzi, K. Stehfest, T. Mathes, I.H. Stokkum, M. Elstner, P. Hegemann, J.T. Kennis, Sci. Rep. **7**, 7217 (2017)
161. M. Sumita, M.N. Ryazantsev, K. Saito, PCCP **11**, 6406 (2009)

162. M.S. Panov, V.D. Voskresenska, M.N. Ryazantsev, A.N. Tarnovsky, R.M. Wilson, J. Am. Chem. Soc. **135**, 19167 (2013)
163. A. Filatov, N. Knyazev, M. Ryazantsev, V. Suslonov, A. Larina, A. Molchanov, R. Kostikov, V. Boitsov, A. Stepakov, Org. Chem. Front. **5**, 595 (2018)
164. A. Dawydow et al., Proc. Natl. Acad. Sci. U.S.A. **111**, 13972 (2014)
165. R.E. Dixon, C. Yuan, E.P. Cheng, M.F. Navedo, L.F. Santana, Proc. Natl. Acad. Sci. U.S.A. **109**, 1749 (2012)
166. A. Losi, W. Gärtner, Photochem. Photobiol. **87**, 491 (2011)
167. I. Yoon, J.Z. Li, Y.K. Shim, Clin. Endosc. **46**, 7 (2013)
168. B.A. Wilt, L.D. Burns, E.T. Wei Ho, K.K. Ghosh, E.A. Mukamel, M.J. Schnitzer, Annu. Rev. Neurosci. **32**, 435 (2009)
169. K. Svoboda, R. Yasuda, Neuron **50**, 823 (2006)
170. K. Welke, J.S. Frähmcke, H.C. Watanabe, P. Hegemann, M. Elstner, J. Phys. Chem. B **115**, 15119 (2011)
171. M.N. Ryazantsev, A. Altun, K. Morokuma, J. Am. Chem. Soc. **134**, 5520 (2012)
172. D.M. Nikolaev, A. Emelyanov, V.M. Boitsov, M.S. Panov, M.N. Ryazantsev, F1000Research **6** (2017)
173. I. Schapiro, M.N. Ryazantsev, W.J. Ding, M.M. Huntress, F. Melaccio, T. Andruniow, M. Olivucci, Aust. J. Chem. **63**, 413 (2010)
174. D.M. Nikolaev, A.A. Shtyrov, M.S. Panov, A. Jamal, O.B. Chakchir, V.A. Kochemirovsky, M. Olivucci, M.N. Ryazantsev, ACS Omega **3**, 7555 (2018)
175. J. Mattis et al., Nat. Methods **9**, 159 (2012)
176. J.Y. Lin, M.Z. Lin, P. Steinbach, R.Y. Tsien, Biophys. J. **96**, 1803 (2009)
177. N.C. Shaner, M.Z. Lin, M.R. McKeown, P.A. Steinbach, K.L. Hazelwood, M.W. Davidson, R.Y. Tsien, Nat. Methods **5**, 545 (2008)
178. D.E. Johnson, H.-W. Ai, P. Wong, J.D. Young, R.E. Campbell, J.R. Casey, J. Biol. Chem. **284**, 20499 (2009)
179. L.J. Galietta, P.M. Haggie, A. Verkman, FEBS Lett. **499**, 220 (2001)
180. S. Kleinlogel, U. Terpitz, B. Legrum, D. Gökbuget, E.S. Boyden, C. Bamann, P.G. Wood, E. Bamberg, Nat. Methods **8**, 1083 (2011)
181. S.M. Iyer et al., Sci. Rep. **6**, 30570 (2016)
182. W. Akemann, A. Lundby, H. Mutoh, T. Knöpfel, Biophys. J. **96**, 3959 (2009)
183. M. Pflaum, C. Kielbassa, M. Garmyn, B. Epe, Mutat. Res. DNA Repair **408**, 137 (1998)
184. R. Dixit, R. Cyr, PlJ **36**, 280 (2003)
185. V. Magidson, A. Khodjakov, *Methods in Cell Biology* (Elsevier, 2013), p. 545
186. M. Irie, T. Fukaminato, K. Matsuda, S. Kobatake, Chem. Rev. **114**, 12174 (2014)
187. C. Chittasupho, Ther. Deliv. **3**, 1171 (2012)
188. M. Irie, Chem. Rev. **100**, 1685 (2000)
189. R.O. Blaustein, P.A. Cole, C. Williams, C. Miller, Nat. Struct. Mol. Biol. **7**, 309 (2000)
190. L. Laprell et al., J. Clin. Invest. **127**, 2598 (2017)
191. I. Tochitsky, J. Trautman, N. Gallerani, J.G. Malis, R.H. Kramer, Sci. Rep. **7**, 45487 (2017)
192. A. Polosukhina et al., Neuron **75**, 271 (2012)
193. A.Y. Rotov, L.A. Astakhova, V.S. Sitnikova, A.A. Evdokimov, V.M. Boitsov, M.V. Dubina, M.N. Ryazantsev, M.L. Firsov, Acta Naturae **10**, 75 (2018)
194. M. Volgraf, P. Gorostiza, R. Numano, R.H. Kramer, E.Y. Isacoff, D. Trauner, Nat. Chem. Biol. **2**, 47 (2006)
195. M. Stein, S.J. Middendorp, V. Carta, E. Pejo, D.E. Raines, S.A. Forman, E. Sigel, D. Trauner, Angew. Chem. Int. Ed. **51**, 10500 (2012)
196. W. Szymański, B. Wu, C. Poloni, D.B. Janssen, B.L. Feringa, Angew. Chem. Int. Ed. **52**, 2068 (2013)
197. J. Broichhagen, J.A. Frank, D. Trauner, Acc. Chem. Res. **48**, 1947 (2015)
198. G.M. Tsivgoulis, J.M. Lehn, Adv. Mater. **9**, 627 (1997)
199. S. Samanta, A.A. Beharry, O. Sadovski, T.M. McCormick, A. Babalhavaeji, V. Tropepe, G.A. Woolley, J. Am. Chem. Soc. **135**, 9777 (2013)

200. J. Garcia-Amorós, A. Bučinskas, M. Reig, S. Nonell, D. Velasco, J. Mater. Chem. C **2**, 474 (2014)
201. S. Samanta, C. Qin, A.J. Lough, G.A. Woolley, Angew. Chem. Int. Ed. **51**, 6452 (2012)
202. A.A. Beharry, O. Sadovski, G.A. Woolley, J. Am. Chem. Soc. **133**, 19684 (2011)
203. C. Knie, M. Utecht, F. Zhao, H. Kulla, S. Kovalenko, A.M. Brouwer, P. Saalfrank, S. Hecht, D. Bléger, Chem. Eur. J. **20**, 16492 (2014)
204. R.H. Kramer, A. Mourot, H. Adesnik, Nat. Neurosci. **16**, 816 (2013)
205. D. Bléger, J. Schwarz, A.M. Brouwer, S. Hecht, J. Am. Chem. Soc. **134**, 20597 (2012)
206. M.N. Ryazantsev, A. Jamal, S. Maeda, K. Morokuma, PCCP **17**, 27789 (2015)
207. D.S. Parker, B.B. Dangi, R.I. Kaiser, A. Jamal, M. Ryazantsev, K. Morokuma, J. Phys. Chem. A **118**, 12111 (2014)
208. L.G. Muzangwa, T. Yang, D.S. Parker, R.I. Kaiser, A.M. Mebel, A. Jamal, M. Ryazantsev, K. Morokuma, PCCP **17**, 7699 (2015)
209. C. Boulegue, M. Löweneck, C. Renner, L. Moroder, ChemBioChem **8**, 591 (2007)
210. E.M. Kosower, H. Kanety-Londner, J. Am. Chem. Soc. **98**, 3001 (1976)
211. S. Zbaida, Drug Metab. Rev. **27**, 497 (1995)
212. M.A. Brown, S.C. De Vito, Crit. Rev. Environ. Sci. Technol. **23**, 249 (1993)
213. O. Volkov et al., Science **358**, eaan8862 (2017)

Chapter 9
Molecular-Plasmon Nanostructures for Biomedical Application

Alexey Povolotskiy, Marina Evdokimova, Alexander Konev, Ilya Kolesnikov, Anastasia Povolotckaia and Alexey Kalinichev

Abstract The development of modern nanotechnology opens new opportunities in the design of hybrid structures. This review is devoted to the general view on hybrids, which are based on metal nanoparticles and molecules. Functional properties of nanostructures that have a biomedical application are presented, including singlet oxygen generation for photodynamic therapy, photo-induced heating for photothermal therapy, photo-induced reactions for chemotherapy, luminescent thermometry and surface enhanced Raman scattering for drug delivery control etc. The association of nanostructures into hybrids allows to combine their functional properties and create universal preparations for controlled complex therapy.

9.1 Introduction

Over the years, the development of new cancer treatment strategies has been a priority for health care system. There are a number of issues and problems of oncology, the solution of which will significantly improve therapy methods. The central concern of scientists is to overcome multiple drug resistance, reduce anti-cancer drugs toxicity, and develop non-invasive methods. In order to demonstrate the effectiveness of the abovementioned strategy, materials should have specific physicochemical properties and ability to integrate different anti-cancer modalities, such as chemotherapy, photothermal therapy (PTT) and photodynamic therapy (PDT) into a single nanoplatform [1, 2]. Currently, latest publications and studies have demonstrated successful synthesis and in vivo experiments of these multifunctional systems based on hybrid metallic nanostructures in biomedical applications [3–7]. To create hybrid nanostructures, noble nanoparticles are usually used as nanocarrier forming the basis of surface coatings for various functionalities [8–10]. Different components have been used to modify nanomaterial surfaces for reaching desired properties and applications: small molecule drugs, biologic medical products, luminescent and photosensitive components. According to this approach, such class of nanohybrids is attracting

A. Povolotskiy (✉) · M. Evdokimova · A. Konev · I. Kolesnikov · A. Povolotckaia · A. Kalinichev
Saint-Petersburg State University, Saint-Petersburg, Russia
e-mail: alexey.povolotskiy@spbu.ru

K. Yamanouchi et al. (eds.), *Progress in Photon Science*, Springer Series in Chemical Physics 119, https://doi.org/10.1007/978-3-030-05974-3_9

considerable attention because of the possibility for controlling the incorporation of light-activated nanoparticles into tumors, minimizing the damage in the surrounding healthy tissue, combining both diagnostic and therapeutic capabilities in one structure, known as theranostics approach [11].

9.2 Photothermal Therapy

Photothermal therapy (PTT) is a growing therapeutic method which leads to cell death in malignant tumor by an increase in temperature caused by hybrid metallic nanostructures upon NIR laser irradiation. The fast thermalization of gold nanoparticles (AuNPs) when selective absorption of light occurs, combined with the NIR plasmon resonance makes them ideal candidate as contrast agents for photothermal therapy (Fig. 9.1) [12, 13].

At present, four major types of AuNPs demonstrate efficient NIR photothermal heat conversion, including gold nanospheres [14, 15], nanorods [16, 17], nanoshells [18, 19], and nanocages [20, 21]. When the works deal with in vitro photothermal experiments, the effective destruction of the diseased tissue can be achieved by changing treatment conditions such as the power density, laser exposure duration, and the response time after irradiation. For photothermal treatment, typical intensities are considered to range from 1 to 100 W/cm^2 using continuous wave laser [22]. Kang's group synthesized novel GNRs-porphyrin-trastuzumab complexes (TGNs) using gold nanorods (GNRs) conjugated with porphyrin and trastuzumab (anti-HER2 antibody) to target to HER2-positive breast cancer for NIR light-activatable photothermal therapy [16]. The nanohybrids had an excellent photothermal effect in vitro under 808 nm (with an intensity of 6.07 W/cm^2) laser exposure for 12 min. A significant photothermal effect had induced tumor reduction, which was reached after TGNs-injection into tumor-bearing mice models. Compared with alone GNRs or porphyrin, the hybrid TGNs with lower cytotoxicity showed durable elevated temperature to around 56 °C, high enough to cause tumor ablation. As expected, the

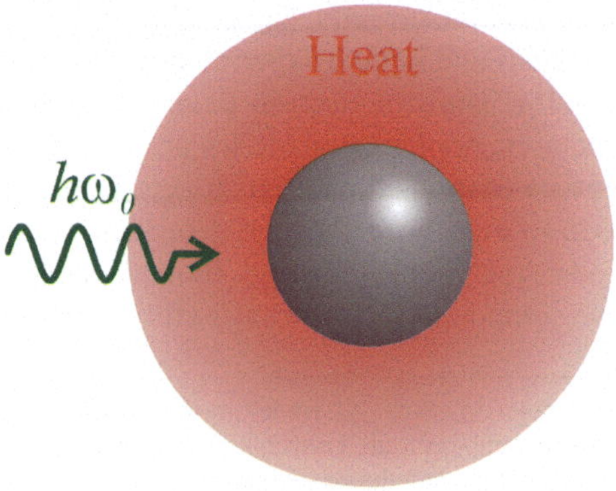

Fig. 9.1 Photo-induced thermalization of gold nanoparticles

in vivo experiments showed different effects for the following control groups such as TGNs nanohybrids only, TGNs + laser, and GNR + laser. The breast cancer-bearing nude mice in TGNs + laser group were nearly healed at 12 days. Moreover, the pathological changes in tumors indicated the significantly higher apoptosis index of cancer cells when compared with the control group. Therefore, the TGNs nanohybrids represented a great agent for photothermal treatment of tumor in vivo.

In addition, gold nanoparticles coated over a silver core exhibited even better photothermal conversion properties than the free gold nanoparticles. Shi et al. [23] synthesized Au@Ag/Au nanoplatform assembled with activatable probes containing thiolated aptamer and fluorophore-labeled complementary DNA for image-guided cancer therapy in vitro and in vivo. In these nanohybrids, Au@Ag/Au nanoparticles manifested dual functionality as both fluorescence quenchers and optical heaters. They obtained excellent results in selective activation of fluorescence signal during conformational reorganization of aptamer and targeting activatable theranostic nanoprobe to A549 lung cancer cells. Recently, in addition to selective destroy of targeted cancer cells through photothermal therapy, other mechanisms of AgNP-induced cancer cell death have extensively been proposed, such as endoplasmic reticulum stress (ER), reactive oxygen species (ROS), upregulation of autophagy genes [24–26]. Platinum-based nanomaterials could also be applied for PTT. For example, Chen's group realized green one-step synthesis of biocompatible fluorescent platinum nanoclusters (Pt NCs) by reducing chloroplatinic acid with ascorbic and glutathione acid for the bio-imaging and photothermal treatment of target cancers [27]. In other work, Chen et al. established a new effect of spontaneous synthesis of platinum nanoclusters by cancer cells and tissues via their propensity to generate both H_2O_2 and ROS through dioxygen reductions. Also, they improved image-guided PTT by combining porphyrin derivative 5,10,15,20-tetrakis(4-sulfonatophenyl)porphyrin (TSPP) with Pt NCs [28]. However, in most cases, systemic toxicity and large size of platinum nanoparticles doesn't allow them to be used for anti-cancer therapy [29].

9.3 Photodynamic Therapy

The integration of photoactive molecules (photosensitizers) in hybrid metallic nanostructures allow to significantly improve therapeutic efficacy of nanoparticles and apply them not only in photothermal (PTT), but also in photodynamic therapy (PDT). Therefore, photosensitizer-coated nanoparticles (NPs) have been reported to achieve highly promising results for PDT in several animal models [30]. This approach requires selective absorption of the photosensitizers (PSs) by the cancer cell and its further laser irradiation. As a result, photoactive molecule goes into excited triplet state, transfers accumulated energy to the surrounding oxygen molecules for producing reactive oxygen species (ROS) such as singlet oxygen ($^1O_2$) (Fig. 9.2) or free radicals, thus killing cancer cells [31].

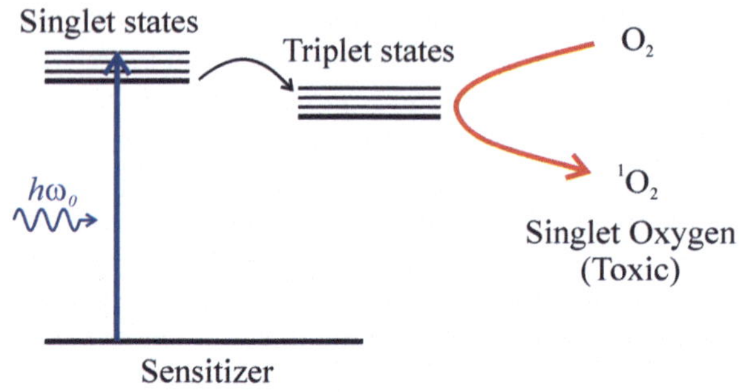

Fig. 9.2 Photo-induced generation of singlet oxygen

Moreover, introduction of some heavy atoms (e.g. Br and I) on a PS has been reported to increase intersystem crossing rates, leading to higher ROS generation [32, 33]. It should be noted that many examples combine NPs and photosensitizers, especially porphyrins and phtalocyanines. Several groups [34–36] demonstrated both PDT efficacy of porphyrin-functionalized GNPs and significant influence on the viability of different class of cancer cell line in vitro and in vivo in comparison with the free ligands. In these published works, protoporphyrin, hematoporphyrin, and brucine–porphyrin derivatives were used to functionalize GNPs. Dissymmetrical porphyrin derivatives having one alkyl chain with a thiol end group incorporating to phenyl group is proposed in the search for new photosensitizers. Penon's group synthesized novel water-soluble multifunctional nanosystem with low intrinsic toxicity comprising of gold nanoparticles (GNP-PR/PEG) immobilized 5-[4(11-mercaptoundecyloxy)-phenyl-10,15,20-triphenylporphyrin (PR-SH) and thiolated polyethylene glycol, which were used as advanced nanotheranostic agents for PDT [37]. Thus, they proved the capability to incorporate a lipophilic PS onto the GNP surface that increases the water solubility of the nanosystem. To evaluate the photodynamic activity nanohybrids, various chemical quenchers are used to characterize the amount of ROS after laser irradiation by monitoring fluorescence intensity of dyes [38, 39].

For enhancing the photodynamic therapy (PDT) efficacy in the near infrared (NIR) range against human breast cancer cells (MDA-MB-231), Hua and co-workers reported a layer-by-layer (LbL) assembly method to realize precise microfabrication of nanostructures consisted of gold nanorings (AuNRs) coated with two layers of Al(III) phthalocyanine chloride tetrasulfonic acid ($AlPcS_4$) [40]. It is revealed that deposition of two $AlPcS_4$ layers had increased the PDT of cancer cells in the NIR by a factor of 8 (is about ~85% cancer cells) compared with $AlPcS_4$ only or the Au NR-$AlPcS_4$ mixture. In absence of NIR radiation, the porphyrin photosensitivity adsorbed on the Au NR surface is inhibited due to charge-transfer induced quenching. During NIR irradiation, the dye is highly activated upon release of the PS from and in the immediate vicinity of the Au NRs resulting in field-enhanced ROS generation. Thus, this strategy, recommended by other researchers, has been successfully employed for constructing interesting nanostructured assemblies that display a high ROS yield [41, 42].

9.4 Combined Photothermal and Photodynamic Therapies

The integrated PTT and PDT nanoplatform is designed to achieve moderate hyperthermia (<45 °C) for less invasive cancer cell death and reactive oxygen species (ROS)-mediated intracellular damage, thus obtaining improved anti-cancer efficacy [43]. The treatment efficiency and the synergistic effect of PDT and PTT can be reached, if the photosenstisizer and photothermal agent are simultaneously delivered to cancer cells in a specific location. Thus, recent investigations are primarily focused on the PS-photothermal agent hybrid nanostructures for collaborative PDT/PTT therapy triggered by a single NIR laser [43, 44]. Using abovementioned strategy, Kumar et al. developed hybrid core-petal nanostructures (CPNs) for PDT–PTT-based apoptotic therapeutics realized through the gold chloride-induced oxidative disassembly and rupture of the polydopamine corona around Au nanoparticles and subsequent anisotropic growth of Au nanopetals with various protrusion lengths and densities. It was reached the efficient killing of cancer cell with CPN-4 (with maximum density of nanopetals) at mild increase in temperature (~42 °C) under 785 nm laser power density of 2 W/cm^2 for 6 min [45]. Photosensitizer-conjugated gold nanostars have been prepared by Wang's group, showing combined PDT and PTT effects using single wavelength NIR laser and improved cancer therapy efficacy in MDA-MB-435 tumour-bearing mice [44]. Nevertheless, the simultaneous synergistic therapy directly depends on the overlap both optical absorption of photosenstisizer and photothermal agent. Unfortunately, there are almost no articles conceptualizing the PDT/PTT combination treatment because of the particular requirement and complicated synthesis [46, 47].

Many NIR dyes from porphyrin family are clinically approved by the U.S. Food and Drug Administration (FDA). Their molecules can convert the absorbed light energy to local hyperthermia for PDT and PTT and ROS generation. Therefore, porphyrin derivatives could be considered as a kind of ideal theranostic platform for biomedical applications. Recently, several research groups have applied gold nanoparticles as nanocarriers to prepare porphyrin-coated NPs for improving stability and tumour-specificity.

For example, gold-nanoclustered hyaluronan nanoassembly (GNc-HyNA) has been developed as a nanomedicine platform. For the synthesis of the hybrid nanomaterials were employed amphiphilic hyaluronan-polycaprolactone (Hy-PCL) conjugates as a drug carrier for a hydrophobic photodynamic therapy agent verteporfin (V_p), a polymeric reducing agent and an organic nanoscaffold, into a single nanoplatform [48]. Both in vitro and in vivo experiments demonstrated that GNc-HyNA loaded verteporfin exhibited excellent stability in the bloodstream and exerted a great potential to treat tumors with a 100% survival rate. Thus, it is proved that V_p-GNc-HyNA can potentially be applied for photothermally boosted photodynamic tumor ablation.

In addition, Zeng's and Lokesh's group reported that Au NPs coated with cobalt and manganese porphyrin derivatives can be used for the development of the artificial photosynthetic and catalytic materials [49, 50]. Recently, it has been reported that combination of meso-tetrakis(4-sulphonatophenyl)porphyrin (TPPS) and chitosan-coated Au NPs (TPPS/QCS-SH/Au NPs) contributed to development novel nanoplatform system with improved photoproperties, including excellent biocompatibility, stability, high $^{1}O_2$ generation and photothermal conversion efficiency, due to synergistic effect of PDT and PTT [51]. Lin's group synthesized Chlorin e6 (C6)-encapsulatedplasmonic gold vesicles (GV-Ce6) for dual-modality PTT/PDT cancer treatment. The tumor treatment efficiency of the GV-Ce6 nanohybrids were demonstrated be higher than the sum effect of individual GVs and Ce6 components, representing a more effective tumor localized therapy [52].

9.5 Combined Chemo-photothermal Therapy

In contrast to traditional chemotherapy, in which drugs spread freely in the bloodstream resulting in the cytotoxicity both cancer and healthy cells [53, 54], a new approach enable the optically controlled delivery of drugs based on photothermal reaction mechanism. It means that drug delivery could be achieved through the light-heat activation of nanoparticles after they absorb light. The activation causes the local temperature increase and the release of drugs from the drug carrier along with photothermal ablation treatment due to thermal energy production. This approach can not only potentially increase drug bioavailability and capacity to overcome physical barriers, but also reduce the dosage to be used for treatment, thus reducing the toxic side effects [55].

Several nanocarrier options have been pursued, depending on the type of tumor cells and desired functions [10, 56, 57]. Khandelia's group studied the possibility to use AuNS–protein agglomerate-based nanohybrids as multimodal drug-delivery vehicles against cancer cells. Albumin stabilized agglomerated structures of gold nanoparticle (Au NP)–lysozyme (Lyz) have been fabricated, showing high drug-loading and-releasing capabilities for both hydrophilic doxorubicin (DOX) and hydrophobic pyrene (PYR) molecules [10]. DOX release was done via changes in pH value of the media inside the cell and was 35% at pH 4.0 and 27% at pH 7.4 up to 24 h, while any release of PYR wasn't observed in both the buffers due to strong binding of PYR in the hydrophobic regions of agglomerates. In another study, Lee and co-workers proposed AuNPs as carriers in target-specific systemic treatment of hepatitis C virus (HCV) infection [58]. New developed HA-AuNPs/IFNα complex for targeted systemic treatment based on interferon α (IFNα) loaded on thiolated HA-modified AuNPs (HA-AuNPs) have demonstrated an enhanced serum stability in human serum and prolonged efficacy in liver tissue for the treatment of chronic HCV infection.

DOX is one of the most widely used anti-cancer agents, causing inhibition of the progression of the enzyme topoisomerase II which leads to inhibition of cell growth and reproduction by interferencing DNA replication. The mechanism of drug release to cause DNA damage is the key of combined chemo-photothermal therapy. Chen et al. developed the tumor-targeting nanoplatform AuNS-pep/DOX@HA by using gold nanostar (AuNS) bearing peptide TPP-KLA and DOX encapsulated in hyaluronic acid (HA) protective shell [56]. It was proved that nanoplatform selectively accumulates at tumor site and the internalization mechanism into tumor cells occurs via CD44 receptor-mediated recognition. Followed degradation of HA by hyaluronidase (HAase), the therapeutic AuNS-pep/DOX@HA system demonstrated the ability to release DOX for chemotherapy and mitochondria-targeting/anchoring heat generator AuNS-pep for NIR light triggered subcellular photothermal therapy (PTT). Both in vitro and in vivo experiments showed that the AuNS-pep/DOX@HA nanoplatform has an ability to be a prominent non-resistant or resistant tumor inhibition. New nanotheranostic agent of DOX@PLA@Au-PEG-MnP has been developed by Jing et al. [7] integrating different functionalities in single nanoplatform. Hybrid drug delivery system made of various components, including: (1) poly(lactic acid) (PLA) as biodegradable drug carrier; (2) gold nanoshell as NIR photo-absorber to perform photothermal therapy and trigger an instant drug release; (3) Mn-porphyrin (MnP) as T_1 contrast agent to enhance MR imaging, and PEG to extend the circulation time in vivo experiments. This multifunctional nanoparticle manifested its photothermal therapeutic potential and triggered DOX release activated during NIR irradiation of Au nanoshells on both cellular experiments and HT-29 tumor-bearing nude mice. Systemic toxicity studies conducted in mice revealed that hybrid DOX@PLA@Au-PEG-MnP nanoparticle combined DOX and photothermal treatment were more cytotoxic than either agent alone in the treatment. Wang's group used the same chemotherapeutic agent in multi-stimuli responsive nanoplatform for targeted, non-invasive and pinpointed intracellular DOX release, which consisted of Au nanocages, hyaluronic acid (HA) and DOX encapsulated in gold nanocages@hyaluronic acid (AuNCs-HA) [59]. In vitro results confirmed efficiently drug-loaded nanohybrids uptake by cancer cells via interactions of hyaluronic acid with CD44 active sites holding significant importance in modulating targeting ability of hyaluronic acid to different cancer types and subsequently intracellular degradation into small fragments, releasing DOX. On the other hand, in vivo experiments have further revealed that AuNCs-HA nanohybrids demonstrate biocompatibility, CD44-targetability, multi-stimuli responsiveness, pinpointed drug release and chemo-photothermal synergistic effects and thus, have a great potential application in cancer therapy.

9.6 Triple Combination Nanotherapeutics

Although combined therapy of PTT/PDT, PDT/chemotherapy, and PTT/chemotherapy has been reported to enhance therapeutic efficiencies, it remains in the early stages and certain limitations still exist, including of the residual cancer cells survival after photothermal damage, the PDT-induced hypoxia microenvironment and hypermethylated cancer cell resistance to chemotherapeutic drugs [60, 61]. Therefore, an alternative approach was created for further improving therapeutic efficacy. Zeng's group used core–shell gold nanorod@metal–organic frameworks (AuNR@MOFs) made up of gold nanorods functionalized with lipoic acid and poly(ethylene glycol) (PEG)-SH polymer to develop organic nanotheranostic agents with triple-modal therapy. As a photodynamic agent was used tetrakis (4-carboxyphenyl)porphyrin) consisting of 6-connected Zr_6 cluster ($Zr_6O_4(OH)_4(H_2O)_6(OH)_6(COO)_6$) and tetratopic linker (TCPP) and drug (CPT: camptothecin) through π–π stacking and electrostatic interaction were loaded [6]. In this as-prepared nanoparticle, not only the loading of porphyrinic MOF on AuNR@(PEG)-SH could accelerate cellular uptake, but also the photothermal effect of porphyrinic MOF could promote drug delivery, enabling combined photodynamic, photothermal and chemotherapy for synergistic damaging the tumor cells in vitro and in vivo.

Park and co-workers prepared doxorubicin-loaded hollow gold nanoparticles (DOX-HGNPs) that demonstrated the effects of combined chemotherapy, thermal and radiotherapy [62]. The DOX release mechanism was triggered by an NIR laser and increased with irradiation. The triple-combination therapy strategy reduced tumor's weight by 6.8-fold and showed delayed tumor growth by 4.3-fold for A549 bearing mice compared to control groups. The radio-enhancing effect of DOX-HGNPs was confirmed by the high level of γ-H2AX (phosphorylated histone) foci than before the irradiation. CT imaging studies obtained were compared to the clinically available Ultravist 300 and HGNPs, showing a linear dependence of the absorption on the concentration and an attenuation coefficient higher than that of Ultravist 300.

9.7 Luminescent Probes and Sensors for Temperature

Temperature is one of the most fundamental parameter in all kinds of science. Respective sensors [63, 64] are widely used in daily life, in metrology, aerodynamics, climate and marine research, in chemistry, medicine, biology, military technology, air conditioning, in practically all devices for heating and cooling. The share of thermal sensors can be estimated as 75–80% of the world's sensor market [65]. Types of thermometers include liquid-filled glass thermometers based on the thermal expansion of materials [66]; thermocouples based on the Seebeck effect [67, 68]; and optical sensors [69]. All contact measurements require convective heat transfer and thus need to reach equilibrium between the sensor and the object. This procedure can alter the actual temperature of the object during the measurement, especially if the size of the sample is small. Moreover, traditional liquid-filled and bimetallic thermometers, thermocouples, pyrometers, and thermistors cannot be easily miniaturized and therefore are not suitable for temperature measurement with a spatial resolution of <10 μm which is desirable for micro/nanoelectronics, photonic devices and especially for various biological application.

Thermometry at the nanoscale requires, therefore, a new paradigm in the use of both materials and thermometric properties. Moreover, new synthetic techniques are helping to reduce materials limitation for sensing temperature at the nanoscale by either improving qualitatively inherent materials properties, e.g., size dispersion,

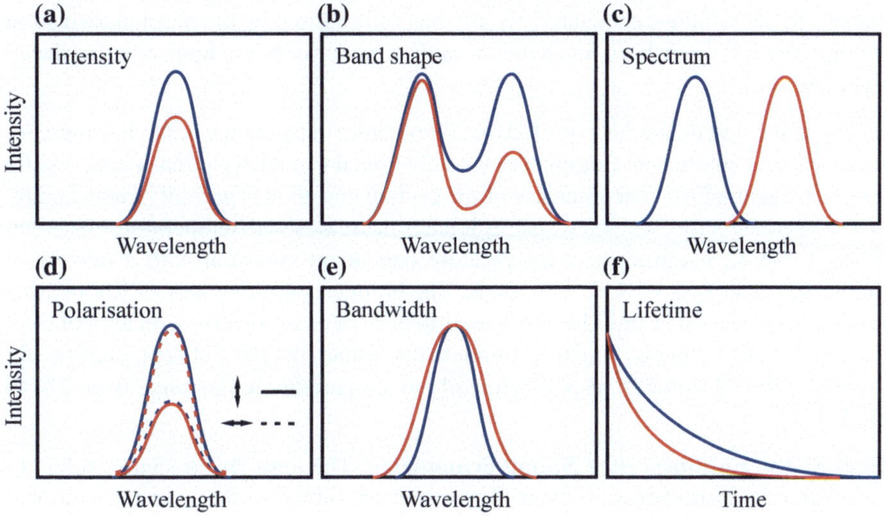

Fig. 9.3 Schematic representation of the possible effects caused by a temperature increment on the luminescence. Red lines correspond to higher temperatures

surface roughness, or by opening entirely new possibilities based on new materials with new properties. Recently luminescence thermometry becomes one of the most popular approaches to thermal sensing at nanoscale. This technique possesses unique properties including noninvasive, contactless, and simplicity of read-out [70]. Luminescence is the emission of light from a given substance, occurring from electronically excited states that have been populated by an external excitation source. The properties of the emitted photons depend on the characteristics of the electronic states involved in photon emission [71, 72]. These, in turn, depend on the local temperature and thus luminescence nanothermometry exploits the relationship between temperature and luminescence properties to achieve thermal sensing from the spatial and spectral analysis of the light generated from the object to be thermally imaged. Luminescence thermal sensors can be grouped into different classes based on the particular luminescent parameter which is analyzed and from which the thermal reading is ultimately extracted. Figure 9.3 schematically depicts the six parameters that characterize the emission of a given material: intensity, band-shape, spectral position, polarization, lifetime and bandwidth.

Figure 9.3 also qualitatively demonstrates how the luminescence emission spectrum is modified when each of these parameters is varied. Thus, based on these variations it is possible to define the following luminescence nanothermometry subclasses.

Intensity Luminescence Nanothermometry. In this case, thermal sensing is achieved through the analysis of the luminescence intensity. When temperature changes, there is an overall change in the number of emitted photons per second such that the emission spectrum becomes less (or more) intense. Temperature induced changes in the luminescence intensity are generally caused by the thermal activation of luminescence quenching mechanisms and/or increases in the non-radiative decay probabilities.

One of the first examples reported the temperature dependence of the luminescent intensity of a naphthalene fragment covalently linked to a Ni(II) tetraazamacrocyclic complex, cyclam [73]. The intensity of the cyclam emission is partially quenched by an energy transfer mechanism whose efficiency decreases with temperatures between 300 and 338 K, resulting in a temperature dependent emission with a maximum relative sensitivity of 3.6% K^{-1} at 300 K. The luminescence intensity of Rhodamine B, which can be used as liquid solution or thin film, reduces linearly with temperature at a rate close to 2% per C, which is similar to that found for QDs [74–76]. Fluorescein covalently linked to starch responds linearly to temperature in the range from 273 to 333 K [77].

Band-Shape Luminescence Nanothermometry. The term "band-shape" refers to the relative intensity between the different spectral lines that make up the luminescence spectrum. Thermally induced variations in the band-shape usually take place when the electronic states from which emission is generated are very close in energy such that they are thermally coupled. It can be also present in mixed systems, i.e. systems containing more than one class of emitting centers. There are a great variety of examples in the literature that report on ratiometric luminescent systems including lanthanide complexes, quantum dots and organic dyes.

$[Tb_{0.99}Eu_{0.01}(hfa)_3(dpbp)]_n$ (hfa: hexafluoroacetylacetonato, dpbp: 4,4′-bis(diphenylphosphoryl) coordination polymer is a ratiometric probe (emission at 543 nm is sensitive to temperature, but the emission at 613 is not) with relative sensitivity of 0.83% K^{-1}, which can be utilized in temperature range of 200–500 K [78]. An example of a ratiometric highly sensitive molecular thermometer based on a platinum octaethyl porphyrin was described by Lupton [79]. The intensity ratio uses two transitions of the PtOEP (2,3,7,8,12,13,17,18-octaethyl-21H,23H-porphyrin platinum(II)): the first excited triplet level, at 650 nm, and one band, at 540 nm, which origin is not entirely clear. PtOEP is able to monitor temperature changes in the range 290–320 K with a maximum relative sensitivity of 4.6% K^{-1} at 305 K. A trizwitterionic dicationic Cu5 cluster exhibits an excellent thermochromic temperature-dependent luminescence in the range between 228 and 353 K with a high sensitivity and temporal (sub-millisecond) as well as spatial (sub-micrometer) resolution [80]. Chromium(III)-based dye was used an unprecedented molecular ratiometric thermometer in the 210–373 K temperature range in organic and in aqueous media [81].

Spectral Luminescence Nanothermometry. It is based on the analysis of the spectral positions of the emission lines, which are unequivocally determined by the energy separation between the two electronic levels involved in the emission. In turn, this depends on a large variety of temperature dependent parameters of the emitting material including refractive index and inter-atomic distances (density). Thus, in any emitting material the spectral positions of the luminescence lines are expected to be temperature dependent, and this is exploited by spectral luminescence nanothermometry to translate spectral shifts into temperature. It should be noted that Spectral Luminescence Nanothermometry is not based on the analysis of absolute or relative intensities but on the determination of the spectral position of the luminescence lines. Consequently, temperature reading is not affected by luminescence intensity fluctuations caused by variations in the local concentration of emitting centers.

A pyrene-containing triarylboron molecule, DPTB (dipyren-1-yl(2,4,6-triisopropylphenyl)borane), dissolved in 2-methoxyethyl ether, shows temperature-dependent green to blue luminescence over the temperature range 223–373 K. The temperature can be determined by the blue-shift of the broad emission spectra, ascribed to the thermal equilibrium between twisted intramolecular charge transfer and local excited state of the DPTB molecule with 40 μm resolution [82]. Copper(I) cluster complexes give visually observable thermochromic shifts in luminescence [83–85]. Such probes cover a whole range from 8 to 290 K.

Polarization Luminescence Nanothermometry. In anisotropic media, the emitted radiation is generally non-isotropically polarized and consequently, the shape and intensity of emitted radiation are strongly dependent on its polarization. This allows the definition of the "polarization anisotropy" parameter, which is the ratio between the luminescence intensities emitted at two orthogonal polarization states. As a result, polarization luminescence nanothermometry is based on the influence of temperature on this polarization anisotropy.

One of the molecules used in the past for Polarization Luminescence Nanothermometry is fluorescein [86]. Its polarization anisotropy is modified by more than a 100% making it a highly sensitive temperature sensor. The green fluorescent protein (GFP) can also act as a thermal sensitive intracellular nanoprobe [87], because its fluorescence polarization anisotropy (FPA) depends on temperature. The method was applied to GFP-transfected HeLa and other cancer cell lines to monitor the heat generated after photothermal heating using gold nanorods surrounding the cells. A spatial resolution of 300 nm and a thermal resolution of about 0.4 K were achieved.

Bandwidth Luminescence Nanothermometry. The width of the various emission lines that make up any luminescence spectrum is determined by the properties of the material (such as the degree of disorder) and temperature. It is well-known that as the temperature of a luminescent material is increased, a corresponding increase in the density of phonons occurs resulting from the spectral contribution of homogeneous line broadening. Generally, in the vicinity of room temperature, homogeneous line broadening leads to a linear relationship between bandwidth and temperature. The change in the bandwidth of the luminescence spectra is exploited in bandwidth luminescence nanothermometry to achieve a thermal reading.

The magnitude of the temperature induced luminescence line broadening is in general small and thus, can be only observed in systems showing inherent narrow emission lines and in which homogeneous line broadening dominates over the inhomogeneous one. This is the case of rare earth ions incorporated in some crystalline hosts [88–91].

Lifetime Luminescence Nanothermometry. Luminescence lifetime is defined as the time that the emitted luminescence intensity decays down to 1/e of its initial value after a pulsed excitation. This is an indication of the total decay probability of the emitted intensity (indeed this probability is defined as the inverse of the luminescence lifetime). Decay probabilities from electronic levels depend on a great variety of factors and many of them are related to temperature (such as phonon assisted energy transfer processes and multiphonon decays). This temperature dependence makes it possible to extract temperature readings from the determination of the luminescence lifetime.

The temperature-induced reduction of the observed lifetime of luminescent organic dyes is a well-known phenomenon and is especially noticeable in the case of the Rhodamine B dye [92–94]. A luminescence lifetime reduction close to 75% is observed in the 10–70 °C range. The terbium(III) tris[(2-hydroxybenzoyl)-2-aminoethyl]amine chelate was found to be a viable probe for lifetime temperature sensing [95]. Molecular thermometer consists of [Eu(btfa)$_3$ (MeOH)(bpeta)] and [Tb(btfa)$_3$(MeOH)(bpeta)] β-diketonate chelates (btfa: 4,4,4-trifluoro-l-phenyl-1,3-butanedione, bpeta: bis(4–pyridyl)ethane) embedded into organic-inorganic hybrid nanoclusters formed by a maghemite (γ-Fe_2O_3) magnetic core coated with a tetraethyl orthosilicate/aminopropyltriethoxysilane organosilica shell demonstrated temperature dependence of luminescence lifetime in 14–300 K range [96]. A tetradecanuclear copper(I) cluster compound demonstrated temperature dependence of the emission decay lifetimes with Mott–Seitz model fitting [97].

Thus, we can conclude that luminescence nanothermometry provides several options to achieve thermal sensing from the analysis of the emission spectrum generated by the system under study.

One of the most promising thermal sensing techniques is based on Band-Shape Luminescence. Ratiometric temperature sensing can be achieved with three design concepts: (i) a combination of two emissive dyes (temperature-responsive probe and temperature-inert reference); (ii) a combination of two dyes which interact by a distance-dependent process, such as fluorescence resonance energy transfer (EnT) in systems, in which distance between dyes is modified by temperature; and (iii) a single dye displaying dual emission (Fig. 9.4) [72, 98–100].

First concept is commonly achieved by combining two phosphors in a fixed ratio (Fig. 9.4a). For example, both dyes must be excitable at the same wavelength and show spectrally distinguishable emission bands to allow separating and integrating the luminescence signals for the calculation of the temperature sensitive ratio I_{probe}/I_{ref}.

Fluorescent temperature sensors, which exploit temperature-dependent structural features connected with changes in fluorescence intensity of the reporter dyes are fluorophore-labeled molecular beacons (MBs) (Fig. 9.4b) [101–104]. These flexible single-stranded oligonucleotides are either dually labeled at their 5′- and 3′-ends with a fluorophore and a nonemissive quencher (nonratiometric MBs) or with two spectrally distinguishable emitters (ratiometric MBs). The fluorescence properties of these MBs are determined by the temperature-dependent conformation of the

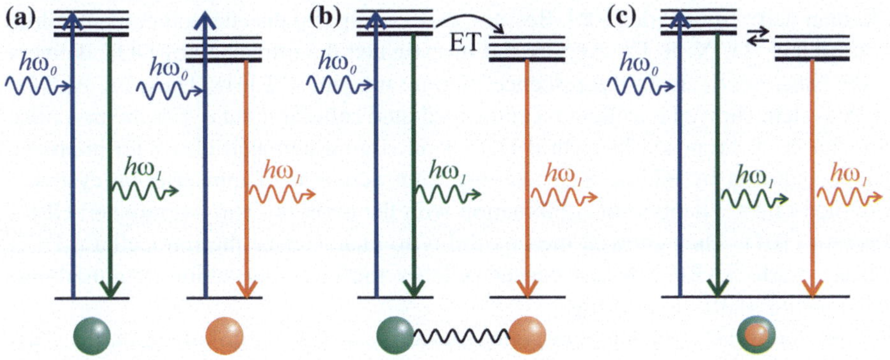

Fig. 9.4 Principles of dual-emission optical thermometers: **a** two distinct dyes; **b** two fluorophores interacting through energy transfer EnT; and **c** single dye with two equilibrating excited states

stem region, which defines the label distance. Because large molecular motions are involved, thermometers based on MBs are only operative in solution. The third scenario, internal referencing of the temperature-sensitive optical signal of a single dual emissive fluorophore, is principally more straightforward approach than the use of two separate emitters (Fig. 9.4c). Herein, nonspecific environmental factors affecting the dye luminescence differently, concentration variations, and a different photostability are eliminated [98]. This design concept is seldom realized, particularly for small molecular sensors, because most emitters display only a single-luminescence band.

9.8 Surface-Enhanced Raman Scattering Diagnostics of Cancer

Currently, surface-enhanced Raman scattering (SERS) has been proved to be an ultrasensitive tool for noninvasively cancer cell detection and imaging in vitro and in vivo [105]. Successful realization of SERS experiments depends on interaction between adsorbed molecules and nanoparticles surface. Development of new synthesis methods of hybrid SERS-active nanostructures have the potential to improve the visualization, diagnostics of various diseases, therapy and drug delivery. SERS labels consisted of a hybrid material with Raman-active probe and a molecule capable of specific binding to the studied biological objects (antigen-antibody, ligand-receptor, enzyme-inhibitor), have a number of advantages over fluorescent labels, as they have greater sensitivity when detecting biological molecules.

For Raman tracking and imaging of drug release in tumors after nanodelivery, some of the most significant investigations have employed doxorubicin. For instance, Wang and co-workers synthesized folic acid (FA)-coated AgNPs in which FA acted as a specific targeting agent of the folate receptor expressed in the KB cancer cell and a Raman-active molecule [106]. Besides, they conjugated the chemotherapeutic drug doxorubicin (DOX) to FA-AgNPs and investigated the efficiency of DOX delivery to the cancer cells using fluorescence lifetime imaging (FLIM). Thus, FA-AgNPs-DOX system showed excellent receptor-mediated cellular uptake. For the first time, Hossain et al. proposed biohybrid DOX-loaded gold nanoparticles with improved cell recognition by adding a penetrating trans-activator of transcription cysteine-modified (Tat-C) peptide in combination with the antibody-antigen targeted HER2 receptor [107]. They showed the possibility to characterize the intracellural DOX release inside SKBR-3 breast cancer cells by the releasing action of glutathione using in situ SERS monitoring.

Taking into account active targeting concerns, instead of doxorubicin, the other drugs have been suggested by other researches based on the principle of antibody-antigen recognition [108–110]. For Raman spectroscopy, one such drug is cetuximab, an epidermal growth factor receptor (EGFR) inhibitor used as anti-cancer agents. Molecular-plasmon hybrids using for SERS diagnostics of cancer have similar structures and integrate into single nanoplatform components having defined functional properties: (1) metallic nanoparticles as a core to perform photothermal therapy (PTT) and control a drug release; (2) PEG as biologically inert shell, protecting nanoparticles from immune system; (3) Raman-active molecule to enhance the intensity of SERS signal; and (4) anti-cancer agents for chemotherapy. For example, Conde et al. prepared gold nanoparticles coated by Raman-active molecule DTTC (3,3′-Diethylthiatricarbocyanine iodide) inside a polymer shell (HS-PEG) and after that covered with FDA approved cetuximab bonded via EDC (1-Ethyl-3-(3-dimethylaminopropyl) carbodiimide)/NHS (*N*-hydroxysuccinimide) coupling reaction, occurring between the carboxylated PEG and the terminal NH_2 group of the antibody. In such system they investigated in vivo ability of cetuximab attached at the surface of gold nanoprobes to target specific markers at the tumor surface [111]. Characteristic Raman spectra of drug–Raman NPs in HT-adenocarcinoma cells from colorectal cancer revealed a 4.5-fold higher signal of the characteristic Raman "fingerprint" of DTTC from the nanoparticles on the cell membrane than Raman-NPs only, showing a characteristic SERS peak at 518 cm^{-1}.

Also, hybrid nanostructures can be used in the construction of Raman scattering (SERS) nanotags for the photosensitizer's distribution in biopsied human cancerous and non-cancerous tissues. Using SERS technique, it may be possible in photosensitizer-functionalized GNPs to realize in real-time the following options, such as: (1) detection of the disease; (2) delivery the ROS agent directly to the tumor site; and (3) the quantitative dosage of photosensitizer required for efficient ROS conversion and damage [112, 113]. It has been reported that porphyrin derivatives are widely used as photosensitizers [114]. For example, Farhadi's group studied the palladium-doped photosensitizer pheophorbide A (Pd-pyrolipid) encapsulated at the surface of spherical AuNPs for monitoring the PDT by SERS [115]. Without using a Raman reporter, the researchers achieved the enhanced signal of the Pd-pyrolipid spectrum upon 638 nm wavelength laser excitation and generation of ROS species for PDT as well. SERS-active gold nanochains (AuNCs) consisted of AuNPs functionalized with a Raman reporter (2-naphthalenethiol) and a photosensitizer (Pheophorbide

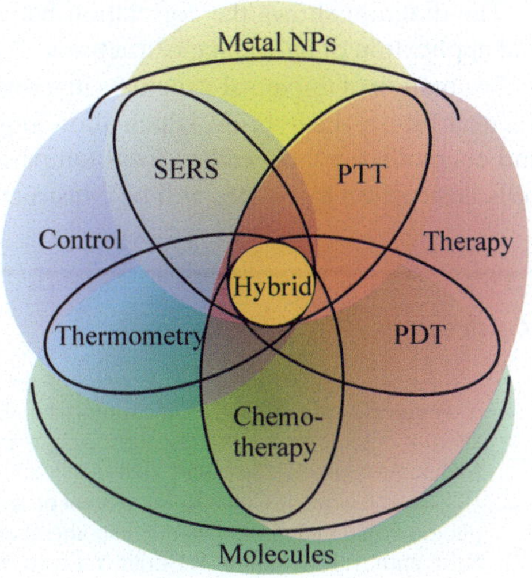

Fig. 9.5 The diagram of bio- and medicine hybrid nanostructures application

A) covalently attached to a polymeric chain of hyaluronic acid and hydrocaffeic acid (HA–HCA) has been synthesized [116]. Compared with the laser irradiated free photosensitizer, conjugated AuNCs exhibited the best absorption and SERS properties in the NIR spectral region as well as 99% cellular uptake and excellent phototoxicity, even at low photosensitizer concentrations.

Nowadays, newly synthesized analogues of heptamethine dyes such as 2-[2-[2-Chloro-3-[(1,3-dihydro-3,3-dimethyl-1-propyl-2*H*-indol-2-ylidene) ethylidene]-1-cyclohexen-1-yl]ethenyl]-3,3-dimethyl-1-propylindolium (IR780), 2-[2-[2-Chloro-3-[[1,3-dihydro-1,1-dimethyl-3-(4-sulfobutyl)-2*H*-benzo[e]indol-2-ylidene]-ethylidene]-1-cyclohexen-1-yl]-ethenyl]-1,1-dimethyl-3-(4-sulfobutyl)-1*H*-benzo[e]indolium (IR820), and 2-[2-[2-Chloro-3-[2-[1,3-dihydro-3,3-dimethyl-1-(5-carboxypentyl)-2*H*-indol-2-ylidene]-ethylidene]-1-cyclohexen-1-yl]-ethenyl]-3,3-dimethyl-1-(5-carboxypentyl)-3*H*-indolium bromide (MHI-148), which presents an opportunity in photothermal therapy (PTT), photodynamic therapy (PDT), and other combinatorial therapeutic methods, are promising materials for fabrication of hybrid nanoplatform due to its ability to generate heat upon laser irradiation, and generation of reactive oxygen species (ROS), such as singlet oxygen, acting as excellent photothermal and photodynamic agent simultaneously [117–119]. New theranostic nanoplatforms based on gold nanoparticles covered by different stabilizing agents and hydrophobic near infrared (NIR) dye, IR780 iodide, able to perform live cell imaging through surface-enhanced resonance Raman scattering (SERRS) microscopy [120, 121]. Subsequently, many researchers demonstrated the possibility to provide an accurate mapping of the cell using hybrid nanostructures due to strong, distinct intensity of SERS signal inside cells upon laser (785–808 nm) excitation.

All the examples considered can be summarized in the form of the diagram shown in the Fig. 9.5.

The diagram shows the correlation between composition, functional properties and application of hybrid nanostructures. The modern trends in hybrids design leads to the creation of universal structures in terms of controlled complex therapy. Thus, in the near future, the hybrids, which allow simultaneous implementation of PDT, PTT and chemotherapy, as well as local temperature control and tracking of the cancer cells destruction by SERS, will be constructed.

References

1. Y. Wang, F. Wang, Y. Shen et al., Tumor-specific disintegratable nanohybrids containing ultrasmall inorganic nanoparticles: from design and improved properties to cancer applications. Mater. Horizons **5**, 184–205 (2018)
2. H. Zhu, P. Cheng, P. Chen, K. Pu, Recent progress in the development of near-infrared organic photothermal and photodynamic nanotherapeutics. Biomater. Sci. **6**, 746–765 (2018)
3. S. Sreejith, T.T.M. Huong, P. Borah, Y. Zhao, Organic–inorganic nanohybrids for fluorescence, photoacoustic and Raman bioimaging. Sci. Bull. **60**, 665–678 (2015)
4. J. Zhou, P. Wang, C. Wang et al., Versatile core-shell nanoparticle@metal–organic framework nanohybrids: exploiting mussel-inspired polydopamine for tailored structural integration. ACS Nano **9**, 6951–6960 (2015). https://doi.org/10.1021/acsnano.5b01138
5. J. Song, X. Yang, O. Jacobson et al., Ultrasmall gold nanorod vesicles with enhanced tumor accumulation and fast excretion from the body for cancer therapy. Adv. Mater. **27**, 4910–4917 (2015)
6. J.-Y. Zeng, M.-K. Zhang, M.-Y. Peng et al., Porphyrinic metal-organic frameworks coated gold nanorods as a versatile nanoplatform for combined photodynamic/photothermal/chemotherapy of tumor. Adv. Funct. Mater. **28**, 1705451 (2018)
7. L. Jing, X. Liang, X. Li et al., Mn-porphyrin conjugated Au nanoshells encapsulating doxorubicin for potential magnetic resonance imaging and light triggered synergistic therapy of cancer. Theranostics **4**, 858–871 (2014)
8. D. Yang, J. Xu, G. Yang et al., Metal-organic frameworks join hands to create an anti-cancer nanoplatform based on 808 nm light driving up-conversion nanoparticles. Chem. Eng. J. **344**, 363–374 (2018)
9. Q. Mu, G. Jiang, L. Chen et al., Chemical basis of interactions between engineered nanoparticles and biological systems. Chem. Rev. **114**, 7740–7781 (2014)
10. R. Khandelia, A. Jaiswal, S.S. Ghosh, A. Chattopadhyay, Gold nanoparticle-protein agglomerates as versatile nanocarriers for drug delivery. Small **9**, 3494–3505 (2013)
11. Y. Ma, J. Huang, S. Song et al., Cancer-targeted nanotheranostics: recent advances and perspectives. Small **12**, 4936–4954 (2016)
12. C. Yi, S. Zhang, K.T. Webb, Z. Nie, Anisotropic self-assembly of hairy inorganic nanoparticles. Acc. Chem. Res. **50**, 12–21 (2017)
13. P.K. Jain, X. Huang, I.H. El-Sayed, M.A. El-Sayed, Noble metals on the nanoscale: optical and photothermal properties and some applications in imaging, sensing, biology, and medicine. Acc. Chem. Res. **41**, 1578–1586 (2008)
14. M. Kodiha, E. Hutter, S. Boridy et al., Gold nanoparticles induce nuclear damage in breast cancer cells, which is further amplified by hyperthermia. Cell. Mol. Life Sci. **71**, 4259–4273 (2014)
15. P. Mishra, S. Ray, S. Sinha et al., Facile bio-synthesis of gold nanoparticles by using extract of *Hibiscus sabdariffa* and evaluation of its cytotoxicity against U87 glioblastoma cells under hyperglycemic condition. Biochem. Eng. J. **105**, 264–272 (2016)
16. X. Kang, X. Guo, W. An et al., Photothermal therapeutic application of gold nanorods-porphyrin-trastuzumab complexes in HER2-positive breast cancer. Sci. Rep. **7**, 42069 (2017)

17. M.A. Mackey, M.R.K. Ali, L.A. Austin et al., The most effective gold nanorod size for plasmonic photothermal therapy: theory and in vitro experiments. J. Phys. Chem. B **118**, 1319–1326 (2014)
18. A. Ahmadi, S. Arami, Potential applications of nanoshells in biomedical sciences. J. Drug Target. **22**, 175–190 (2014)
19. M.E. Khosroshahi, Z. Hassannejad, M. Firouzi, A.R. Arshi, Nanoshell-mediated targeted photothermal therapy of HER2 human breast cancer cells using pulsed and continuous wave lasers: an in vitro study. Lasers Med. Sci. **30**, 1913–1922 (2015)
20. S.E. Skrabalak, J. Chen, L. Au et al., Gold nanocages for biomedical applications. Adv. Mater. **19**, 3177–3184 (2007)
21. C.M. Cobley, L. Au, J. Chen, Y. Xia, Targeting gold nanocages to cancer cells for photothermal destruction and drug delivery. Expert Opin. Drug Deliv. **7**, 577–587 (2010)
22. D. Jaque, L. Martínez Maestro, B. del Rosal et al., Nanoparticles for photothermal therapies. Nanoscale **6**, 9494–9530 (2014)
23. H. Shi, X. Ye, X. He et al., Au@Ag/Au nanoparticles assembled with activatable aptamer probes as smart "nano-doctors" for image-guided cancer thermotherapy. Nanoscale **6**, 8754 (2014)
24. J. Lin, Z. Huang, H. Wu et al., Inhibition of autophagy enhances the anticancer activity of silver nanoparticles. Autophagy **10**, 2006–2020 (2014)
25. R. Foldbjerg, E.S. Irving, Y. Hayashi et al., Global gene expression profiling of human lung epithelial cells after exposure to nanosilver. Toxicol. Sci. **130**, 145–157 (2012)
26. P. AshaRani, S. Sethu, H. Lim et al., Differential regulation of intracellular factors mediating cell cycle, DNA repair and inflammation following exposure to silver nanoparticles in human cells. Genome Integr **3**, 2 (2012)
27. D. Chen, S. Gao, W. Ge et al., One-step rapid synthesis of fluorescent platinum nanoclusters for cellular imaging and photothermal treatment. RSC Adv. **4**, 40141 (2014)
28. D. Chen, C. Zhao, J. Ye et al., In situ biosynthesis of fluorescent platinum nanoclusters: toward self-bioimaging-guided cancer theranostics. ACS Appl. Mater. Interfaces **7**, 18163–18169 (2015)
29. T.C. Johnstone, G.Y. Park, S.J. Lippard, Understanding and improving platinum anti-cancer drugs—phenanthriplatin. Anticancer Res. **34**, 471–476 (2014)
30. Z.S. Silva, S.K. Bussadori, K.P.S. Fernandes et al., Animal models for photodynamic therapy (PDT). Biosci. Rep. **35**, e00265–e00265 (2015)
31. G. Calixto, J. Bernegossi, L. de Freitas et al., Nanotechnology-based drug delivery systems for photodynamic therapy of cancer: a review. Molecules **21**, 342 (2016). https://doi.org/10.3390/molecules21030342
32. X. Li, S. Kolemen, J. Yoon, E.U. Akkaya, Activatable photosensitizers: agents for selective photodynamic therapy. Adv. Funct. Mater. **27**, 1604053 (2017)
33. X. Li, B.-D. Zheng, X.-H. Peng et al., Phthalocyanines as medicinal photosensitizers: developments in the last five years. Coord. Chem. Rev. (2017)
34. H. Eshghi, A. Sazgarnia, M. Rahimizadeh et al., Protoporphyrin IX–gold nanoparticle conjugates as an efficient photosensitizer in cervical cancer therapy. Photodiagnosis Photodyn. Ther. **10**, 304–312 (2013)
35. S.C. Hayden, L.A. Austin, R.D. Near et al., Plasmonic enhancement of photodynamic cancer therapy. J. Photochem. Photobiol. A Chem. **269**, 34–41 (2013)
36. K. Záruba, J. Králová, P. Řezanka et al., Modified porphyrin–brucine conjugated to gold nanoparticles and their application in photodynamic therapy. Org. Biomol. Chem. **8**, 3202 (2010)
37. O. Penon, T. Patiño, L. Barrios et al., A new porphyrin for the preparation of functionalized water-soluble gold nanoparticles with low intrinsic toxicity. ChemistryOpen **4**, 127–136 (2015)
38. Y. Guo, M. Kumar, P. Zhang, Nanoparticle-based photosensitizers under CW infrared excitation. Chem. Mater. **19**, 6071–6072 (2007)

39. K. Knop, A.-F. Mingotaud, N. El-Akra et al., Monomeric pheophorbide(a)-containing poly(ethyleneglycol-b-ε-caprolactone) micelles for photodynamic therapy. Photochem. Photobiol. Sci. **8**, 396 (2009)
40. Y. Hu, Y. Yang, H. Wang, H. Du, Synergistic integration of layer-by-layer assembly of photosensitizer and gold nanorings for enhanced photodynamic therapy in the near infrared. ACS Nano **9**, 8744–8754 (2015)
41. T.G. Shutava, S.S. Balkundi, P. Vangala et al., Layer-by-layer-coated gelatin nanoparticles as a vehicle for delivery of natural polyphenols. ACS Nano **3**, 1877–1885 (2009)
42. Z. Poon, D. Chang, X. Zhao, P.T. Hammond, Layer-by-layer nanoparticles with a pH-sheddable layer for in vivo targeting of tumor hypoxia. ACS Nano **5**, 4284–4292 (2011)
43. Z. Sheng, D. Hu, M. Zheng et al., Smart human serum albumin-indocyanine green nanoparticles generated by programmed assembly for dual-modal imaging-guided cancer synergistic phototherapy. ACS Nano **8**, 12310–12322 (2014)
44. S. Wang, P. Huang, L. Nie et al., Single continuous wave laser induced photodynamic/plasmonic photothermal therapy using photosensitizer-functionalized gold nanostars. Adv. Mater. **25**, 3055–3061 (2013)
45. A. Kumar, S. Kumar, W.-K. Rhim et al., Oxidative nanopeeling chemistry-based synthesis and photodynamic and photothermal therapeutic applications of plasmonic core-petal nanostructures. J. Am. Chem. Soc. **136**, 16317–16325 (2014)
46. M.E. Alea-Reyes, J. Soriano, I. Mora-Espí et al., Amphiphilic gemini pyridinium-mediated incorporation of Zn(II)meso-tetrakis(4-carboxyphenyl)porphyrin into water-soluble gold nanoparticles for photodynamic therapy. Colloids Surf. B Biointerfaces **158**, 602–609 (2017)
47. O. Penon, M.J. Marín, D.A. Russell, L. Pérez-García, Water soluble, multifunctional antibody-porphyrin gold nanoparticles for targeted photodynamic therapy. J. Colloid Interface Sci. **496**, 100–110 (2017)
48. H.S. Han, K.Y. Choi, H. Lee et al., Gold-nanoclustered hyaluronan nano-assemblies for photothermally maneuvered photodynamic tumor ablation. ACS Nano **10**, 10858–10868 (2016)
49. Y. Zheng, Y. Yuan, Y. Chai, R. Yuan, l-cysteine induced manganese porphyrin electrocatalytic amplification with 3D DNA-Au@Pt nanoparticles as nanocarriers for sensitive electrochemical aptasensor. Biosens. Bioelectron. **79**, 86–91 (2016)
50. K.S. Lokesh, A. Shambhulinga, N. Manjunatha et al., Porphyrin macrocycle-stabilized gold and silver nanoparticles and their application in catalysis of hydrogen peroxide. Dye Pigment **120**, 155–160 (2015)
51. J. Zeng, W. Yang, D. Shi et al., Porphyrin derivative conjugated with gold nanoparticles for dual-modality photodynamic and photothermal therapies in vitro. ACS Biomater. Sci. Eng. **4**, 963–972 (2018)
52. J. Lin, S. Wang, P. Huang et al., Photosensitizer-loaded gold vesicles with strong plasmonic coupling effect for imaging-guided photothermal/photodynamic therapy. ACS Nano **7**, 5320–5329 (2013)
53. Z. Wang, S. Sau, H.O. Alsaab, A,K, Iyer, CD44 directed nanomicellar payload delivery platform for selective anticancer effect and tumor specific imaging of triple negative breast cancer. Nanomed. Nanotechnol. Biol. Med. (2018)
54. X. Chen, W. Zhang, Diamond nanostructures for drug delivery, bioimaging, and biosensing. Chem. Soc. Rev. **46**, 734–760 (2017)
55. C. Ding, Z. Li, A review of drug release mechanisms from nanocarrier systems. Mater. Sci. Eng. C **76**, 1440–1453 (2017)
56. S. Chen, Q. Lei, W.-X. Qiu et al., Mitochondria-targeting "nanoheater" for enhanced photothermal/chemo-therapy. Biomaterials **117**, 92–104 (2017)
57. J. Nam, S. Son, L.J. Ochyl et al., Chemo-photothermal therapy combination elicits anti-tumor immunity against advanced metastatic cancer. Nat. Commun. **9**, 1074 (2018)
58. M.-Y. Lee, J.-A. Yang, H.S. Jung et al., Hyaluronic acid-gold nanoparticle/interferon α complex for targeted treatment of hepatitis C virus infection. ACS Nano **6**, 9522–9531 (2012)
59. Z. Wang, Z. Chen, Z. Liu et al., A multi-stimuli responsive gold nanocage–hyaluronic platform for targeted photothermal and chemotherapy. Biomaterials **35**, 9678–9688 (2014)

60. W. Chen, S. Zhang, Y. Yu et al., Structural-engineering rationales of gold nanoparticles for cancer theranostics. Adv. Mater. **28**, 8567–8585 (2016)
61. D. Luo, K.A. Carter, D. Miranda, J.F. Lovell, Chemophototherapy: an emerging treatment option for solid tumors. Adv. Sci. **4**, 1600106 (2017)
62. J. Park, J. Park, E.J. Ju et al., Multifunctional hollow gold nanoparticles designed for triple combination therapy and CT imaging. J. Control Release **207**, 77–85 (2015)
63. J.B. Weaver, Hot nanoparticles light up cancer. Nat. Nanotechnol. **5**, 630–631 (2010)
64. E.F.J. Ring, The historical development of temperature measurement in medicine. Infrared Phys. Technol. **49**, 297–301 (2007)
65. P.R.N. Childs, J.R. Greenwood, C.A. Long, Review of temperature measurement. Rev. Sci. Instrum. **71**, 2959–2978 (2000)
66. L. Michalski, K. Eckersdorf, J. Kucharski, J. McGhee, *Temperature Measurement*, 2nd ed. (Wiley, New York, 2001)
67. S. Maekawa, T. Tohyama, S.E. Barnes et al., *Physics of Transition Metal Oxides* (Springer, Berlin, 2004)
68. N.A.N. Mermin, *Solid State Physics* (Saunders College, Philadelphia, 1976)
69. M. Tabib-Azar, Optical temperature sensors, in *Integrated Optics, Microstructures, and Sensors* (Springer, Berlin, 1995), pp. 285–313
70. C.D.S. Brites, P.P. Lima, N.J.O. Silva et al., Thermometry at the nanoscale. Nanoscale **4**, 4799 (2012)
71. G.F. Imbusch, B. Henderson, *Optical Spectroscopy of Inorganic Solids* (Oxford Science Publications, London, 2006)
72. D. Jaque, F. Vetrone, Luminescence nanothermometry. Nanoscale **4**, 4301 (2012)
73. M. Engeser, L. Fabbrizzi, M. Licchelli, D. Sacchi, A fluorescent molecular thermometer based on the nickel(II) high-spin/low-spin interconversion. Chem. Commun. 1191–1192 (1999)
74. R. Samy, T. Glawdel, C.L. Ren, Method for microfluidic whole-chip temperature measurement using thin-film poly(dimethylsiloxane)/rhodamine B. Anal. Chem. **80**, 369–375 (2008)
75. V.M. Chauhan, R.H. Hopper, S.Z. Ali et al., Thermo-optical characterization of fluorescent rhodamine B based temperature-sensitive nanosensors using a CMOS MEMS micro-hotplate. Sens. Actuators B Chem. **192**, 126–133 (2014)
76. D. Ross, M. Gaitan, L.E. Locascio, Temperature measurement in microfluidic systems using a temperature-dependent fluorescent dye. Anal. Chem. **73**, 4117–4123 (2001)
77. F.-L. Mi, Synthesis and characterization of a novel chitosan–gelatin bioconjugate with fluorescence emission. Biomacromol **6**, 975–987 (2005)
78. K. Miyata, Y. Konno, T. Nakanishi et al., Chameleon luminophore for sensing temperatures: control of metal-to-metal and energy back transfer in lanthanide coordination polymers. Angew Chem. Int. Ed. **52**, 6413–6416 (2013)
79. J.M. Lupton, A molecular thermometer based on long-lived emission from platinum octaethyl porphyrin. Appl. Phys. Lett. **81**, 2478–2480 (2002)
80. D. Cauzzi, R. Pattacini, M. Delferro et al., Temperature-dependent fluorescence of Cu5 metal clusters: a molecular thermometer. Angew Chem. Int. Ed. **51**, 9662–9665 (2012)
81. S. Otto, N. Scholz, T. Behnke et al., Thermo-chromium: a contactless optical molecular thermometer. Chem. A Eur. J. **23**, 12131–12135 (2017)
82. J. Feng, K. Tian, D. Hu et al., A Triarylboron-based fluorescent thermometer: sensitive over a wide temperature range. Angew Chem. Int Ed. **50**, 8072–8076 (2011)
83. B. Huitorel, Q. Benito, A. Fargues et al., Mechanochromic luminescence and liquid crystallinity of molecular copper clusters. Chem. Mater. **28**, 8190–8200 (2016)
84. F. Parmeggiani, A. Sacchetti, Preparation and luminescence thermochromism of tetranuclear copper(I)–pyridine–iodide clusters. J. Chem. Educ. **89**, 946–949 (2012)
85. A. Yadav, A.K. Srivastava, A. Balamurugan, R. Boomishankar, A cationic copper(I) iodide cluster MOF exhibiting unusual ligand assisted thermochromism. Dalt. Trans. **43**, 8166–8169 (2014)
86. G. Baffou, M.P. Kreuzer, F. Kulzer, R. Quidant, Temperature mapping near plasmonic nanostructures using fluorescence polarization anisotropy. Opt. Express **17**, 3291 (2009)

87. J.S. Donner, S.A. Thompson, M.P. Kreuzer et al., Mapping intracellular temperature using green fluorescent protein. Nano Lett. **12**, 2107–2111 (2012)
88. I.E. Kolesnikov, A.A. Kalinichev, M.A. Kurochkin et al., YVO4:Nd^{3+} nanophosphors as NIR-to-NIR thermal sensors in wide temperature range. Sci. Rep. **7**, 18002 (2017)
89. I.E. Kolesnikov, E.V. Golyeva, A.A. Kalinichev et al., Nd^{3+} single doped YVO 4 nanoparticles for sub-tissue heating and thermal sensing in the second biological window. Sens. Actuators B Chem. **243**, 338–345 (2017)
90. A. Benayas, E. Escuder, D. Jaque, High-resolution confocal fluorescence thermal imaging of tightly pumped microchip Nd:YAG laser ceramics. Appl. Phys. B **107**, 697–701 (2012)
91. A.A. Kaminskii, Laser crystals and ceramics: recent advances. Laser Photon. Rev. **1**, 93–177 (2007)
92. D. Moreau, C. Lefort, R. Burke et al., Rhodamine B as an optical thermometer in cells focally exposed to infrared laser light or nanosecond pulsed electric fields. Biomed. Opt. Express **6**, 4105 (2015)
93. T. Karstens, K. Kobs, Rhodamine B and rhodamine 101 as reference substances for fluorescence quantum yield measurements. J. Phys. Chem. **84**, 1871–1872 (1980)
94. D.A. Mendels, E.M. Graham, S. Magennis et al., Quantitative comparison of thermal and solutal transport in a T-mixer by FLIM and CFD. Microfluid. Nanofluid. **5**, 603–617 (2008)
95. L.-N. Sun, J. Yu, H. Peng et al., Temperature-sensitive luminescent nanoparticles and films based on a terbium (III) complex probe. J. Phys. Chem. C **114**, 12642–12648 (2010)
96. C.D.S. Brites, P.P. Lima, N.J.O. Silva et al., A luminescent molecular thermometer for long-term absolute temperature measurements at the nanoscale. Adv. Mater. **22**, 4499–4504 (2010)
97. J.-H. Wang, M. Li, J. Zheng et al., A dual-emitting Cu 6-Cu 2-Cu 6 cluster as a self-calibrated, wide-range luminescent molecular thermometer. Chem. Commun. **50**, 9115–9118 (2014)
98. X. Wang, O.S. Wolfbeis, R.J. Meier, Luminescent probes and sensors for temperature. Chem. Soc. Rev. **42**, 7834 (2013)
99. H. Zhou, M. Sharma, O. Berezin et al., Nanothermometry: from microscopy to thermal treatments. ChemPhysChem **17**, 27–36 (2016)
100. M. Schäferling, The art of fluorescence imaging with chemical sensors. Angew Chem. Int. Ed. **51**, 3532–3554 (2012)
101. C. Nellaker, U. Wallgren, H. Karlsson, Molecular beacon-based temperature control and automated analyses for improved resolution of melting temperature analysis using SYBR I green chemistry. Clin. Chem. **53**, 98–103 (2006)
102. G. Ke, C. Wang, Y. Ge et al., L-DNA molecular beacon: a safe, stable, and accurate intracellular nano-thermometer for temperature sensing in living cells. J. Am. Chem. Soc. **134**, 18908–18911 (2012)
103. S. Ebrahimi, Y. Akhlaghi, M. Kompany-Zareh, Å. Rinnan, Nucleic acid based fluorescent nanothermometers. ACS Nano **8**, 10372–10382 (2014)
104. Z. Fidan, A. Wende, U. Resch-Genger, Visible and red emissive molecular beacons for optical temperature measurements and quality control in diagnostic assays utilizing temperature-dependent amplification reactions. Anal. Bioanal. Chem. **409**, 1519–1529 (2017)
105. E. Darrigues, V. Dantuluri, Z.A. Nima et al., Raman spectroscopy using plasmonic and carbon-based nanoparticles for cancer detection, diagnosis, and treatment guidance. Part 2: treatment. Drug Metab. Rev. **49**, 253–283 (2017)
106. Y. Wang, B.B. Newell, J. Irudayaraj, Folic acid protected silver nanocarriers for targeted drug delivery. J. Biomed. Nanotechnol. **8**, 751–759 (2012)
107. M.K. Hossain, H.-Y. Cho, K.-J. Kim, J.-W. Choi, In situ monitoring of doxorubicin release from biohybrid nanoparticles modified with antibody and cell-penetrating peptides in breast cancer cells using surface-enhanced Raman spectroscopy. Biosens. Bioelectron. **71**, 300–305 (2015)
108. C. Peters, S. Brown, Antibody-drug conjugates as novel anti-cancer chemotherapeutics. Biosci. Rep. **35**, e00225–e00225 (2015)
109. S. Nussbaumer, P. Bonnabry, J.-L. Veuthey, S. Fleury-Souverain, Analysis of anticancer drugs: a review. Talanta **85**, 2265–2289 (2011)

110. H.-W. Kao, Y.-Y. Lin, C.-C. Chen et al., Biological characterization of cetuximab-conjugated gold nanoparticles in a tumor animal model. Nanotechnology **25**, 295102 (2014)
111. J. Conde, C. Bao, D. Cui et al., Antibody–drug gold nanoantennas with Raman spectroscopic fingerprints for in vivo tumour theranostics. J. Control Release **183**, 87–93 (2014)
112. J. Feng, L. Chen, Y. Xia et al., Bioconjugation of gold nanobipyramids for SERS detection and targeted photothermal therapy in breast cancer. ACS Biomater. Sci. Eng. **3**, 608–618 (2017)
113. J. Song, L. Pu, J. Zhou et al., Biodegradable theranostic plasmonic vesicles of amphiphilic gold nanorods. ACS Nano **7**, 9947–9960 (2013)
114. A.M. Fales, H. Yuan, T. Vo-Dinh, Cell-penetrating peptide enhanced intracellular raman imaging and photodynamic therapy. Mol. Pharm. **10**, 2291–2298 (2013)
115. A. Farhadi, Á. Roxin, B.C. Wilson, G. Zheng, Nano-enabled SERS reporting photosensitizers. Theranostics **5**, 469–476 (2015)
116. L. Zhao, T.-H. Kim, H.-W. Kim et al., Surface-enhanced Raman scattering (SERS)-active gold nanochains for multiplex detection and photodynamic therapy of cancer. Acta Biomater. **20**, 155–164 (2015)
117. C. Pais-Silva, D. de Melo-Diogo, I.J. Correia, IR780-loaded TPGS-TOS micelles for breast cancer photodynamic therapy. Eur. J. Pharm. Biopharm. **113**, 108–117 (2017)
118. S. Harmsen, M.A. Wall, R. Huang, M.F. Kircher, Cancer imaging using surface-enhanced resonance Raman scattering nanoparticles. Nat. Protoc. **12**, 1400–1414 (2017)
119. A. Yuan, J. Wu, X. Tang et al., Application of near-infrared dyes for tumor imaging, photothermal, and photodynamic therapies. J. Pharm. Sci. **102**, 6–28 (2013)
120. C.G. Alves, R. Lima-Sousa, D. de Melo-Diogo et al., IR780 based nanomaterials for cancer imaging and photothermal, photodynamic and combinatorial therapies. Int. J. Pharm. **542**, 164–175 (2018)
121. T. Nagy-Simon, M. Potara, A.-M. Craciun et al., IR780-dye loaded gold nanoparticles as new near infrared activatable nanotheranostic agents for simultaneous photodynamic and photothermal therapy and intracellular tracking by surface enhanced resonant Raman scattering imaging. J. Colloid Interface Sci. **517**, 239–250 (2018)

Part III
Atoms and Molecules

Chapter 10
Methods for the Simulation of Coupled Electronic and Nuclear Motion in Molecules Beyond the Born-Oppenheimer Approximation

Erik Lötstedt, Tsuyoshi Kato and Kaoru Yamanouchi

Abstract We review theoretical methods which can be used for the simulation of time-dependent electronic and nuclear dynamics of gas-phase molecules beyond the Born-Oppenheimer approximation. We concentrate on methods which allow for a description of extensive electronic excitation and ionization. Particular emphasis is placed on the extended multiconfiguration time-dependent Hartree-Fock (Ex-MCTDHF) method. We provide a derivation of the equations of motion of the Ex-MCTDHF method, and discuss its advantages and disadvantages over the methods based on the Born-Huang expansion.

10.1 Introduction

When molecules are irradiated with intense laser light, one or more electrons in the molecule absorb energy from the laser pulse. The motion of electrons is excited and the electrons can be ejected from molecules, and the molecules as well as the resulting molecular ions can be electronically and/or vibrationally excited. The acquired energy is subsequently transferred to the nuclei, which triggers a variety of nuclear motion within the molecule: vibrational motion, structural deformation, and dissociation. An interesting example is a process called hydrogen migration, in which a hydrogen atom or a proton in an excited molecule moves in the wide spatial range within the molecule on a fast time scale. In [1, 2], the hydrogen migration process in methanol was studied. In [1], it was revealed that, after the ionization $CH_3OH \rightarrow CH_3OH^+$ of a methanol molecule by an 800 nm, 60 fs laser pulse, a hydrogen atom moves within a molecule, so that CH_2^+ and OH_2^+ fragments are formed af-

E. Lötstedt (✉) · T. Kato · K. Yamanouchi
Department of Chemistry, School of Science, The University of Tokyo, 7-3-1 Hongo, Bunkyo-ku, Tokyo 113-0033, Japan
e-mail: lotstedt@chem.s.u-tokyo.ac.jp

T. Kato
e-mail: tkato@chem.s.u-tokyo.ac.jp

K. Yamanouchi
e-mail: kaoru@chem.s.u-tokyo.ac.jp

K. Yamanouchi et al. (eds.), *Progress in Photon Science*, Springer Series in Chemical Physics 119, https://doi.org/10.1007/978-3-030-05974-3_10

ter a double ionization. The hydrogen migration was found to proceed within the laser pulse, that is, on a time scale shorter than 60 fs. By employing a pump-and-probe scheme, the time scale of the hydrogen migration was found to be shorter than 25 fs [2].

Another example is the experiments on the asymmetric dissociation of H_2 into H + H^+ or H^+ + H [3, 4], in which H_2 is first ionized to H_2^+ by a short laser pulse, and then, in the course of the dissociation of H_2^+, the remaining electron is driven by the later part of the laser field so that it ends up with being caught by either one of the two protons. It was shown that the final position of the electron (on the left proton or on the right proton) could be controlled by varying the carrier-envelope phase of the driving laser field. We stress that ionization, electronic excitation and dissociation in this case proceed within the same laser pulse.

The complex interplay between electronic excitation, ionization, and nuclear motion has also been shown in experiments on other molecular species, such as I_2 [5], N_2 [6], H_3^+ [7], and C_4H_6 [8, 9].

In order to interpret the experimental results showing that laser-driven molecules undergo complex dynamical processes such as electronic excitation, ionization, nuclear vibration and dissociation, we need to develop a theoretical framework in which both electronic and nuclear motion are included in a general way without imposing any constraints on the electronic motion and the nuclear motion, so that arbitrary electronic excitation (including ionization) and nuclear motion (including dissociation) can be simulated. Presently, an efficient quantum mechanical method fulfilling fully these requirements has not been known. However, a few attempts have been made along this direction in these years. In this article, we first review in Sect. 10.2 the standard theoretical method for dealing with laser-molecule interaction based on the Born-Oppenheimer (BO) approximation. After discussing the advantages and the disadvantages of the BO approximation, we will review several attempts at going beyond the BO approximation. In Sect. 10.3, we give a detailed account of the extended multiconfiguration time-dependent Hartree-Fock (Ex-MCTDHF) method, including a derivation of the equations of motion. In the remaining sections, we provide brief overviews of three related methods, i. e., the multiconfiguration time-dependent Hartree (MCTDH) method (Sect. 10.4.1), the multi-configuration electron-nuclear dynamics (MCEND) method (Sect. 10.4.2), and the MCTDHF method for diatomic molecules (Sect. 10.4.3).

10.2 Born-Oppenheimer Approximation

The BO approximation is the standard method for simulating laser-molecule interaction. In order to derive the working equations for the BO approximation [10], we start with the Hamiltonian of a general molecule consisting of N_e electrons with mass m_e and N_N nuclei with masses M_k and charge numbers Z_k ($k = 1, \ldots, N_\mathrm{N}$),

$$H = T_\mathrm{N} + V_\mathrm{NN} + U_\mathrm{N}(t) + H_\mathrm{e} + U_\mathrm{e}(t), \tag{10.1}$$

where

$$T_{\mathrm{N}} = \sum_{k=1}^{N_{\mathrm{N}}} \frac{-\hbar^2}{2M_k} \nabla^2_{\mathbf{R}_k} \tag{10.2}$$

is the nuclear kinetic energy operator with $\mathbf{R}_k$ denoting the spatial coordinate of nucleus k,

$$V_{\mathrm{NN}} = \sum_{k=1}^{N_{\mathrm{N}}} \sum_{l<k} \frac{Z_k Z_l e^2}{4\pi\varepsilon_0 |\mathbf{R}_k - \mathbf{R}_l|} \tag{10.3}$$

is the nuclear-nuclear repulsive Coulomb potential,

$$U_{\mathrm{N}}(t) = -\sum_{k=1}^{N_{\mathrm{N}}} Z_k e \mathbf{E}(t) \cdot \mathbf{R}_k \tag{10.4}$$

is the nuclear-laser interaction expressed in the dipole approximation with the laser field $\mathbf{E}(t)$,

$$H_{\mathrm{e}} = \sum_{k=1}^{N_{\mathrm{e}}} \left(\frac{-\hbar^2}{2m_{\mathrm{e}}} \nabla^2_{\mathbf{r}_k} - \sum_{l=1}^{N_{\mathrm{N}}} \frac{Z_l e^2}{4\pi\varepsilon_0 |\mathbf{r}_k - \mathbf{R}_l|} + \sum_{l<k} \frac{e^2}{4\pi\varepsilon_0 |\mathbf{r}_k - \mathbf{r}_l|} \right) \tag{10.5}$$

is the electronic Hamiltonian with $\mathbf{r}_k$ denoting the spatial coordinate of electron k, and

$$U_{\mathrm{e}}(t) = \sum_{k=1}^{N_{\mathrm{e}}} e\mathbf{E}(t) \cdot \mathbf{r}_k \tag{10.6}$$

is the electron-laser interaction. In the following discussion, we introduce for convenience a collective coordinate $\overline{\mathbf{R}} = (\mathbf{R}_1, \mathbf{R}_2, \ldots, \mathbf{R}_{N_{\mathrm{N}}})$ to denote the spatial coordinates of all the nuclei, and similarly $\overline{\mathbf{r}} = (\mathbf{r}_1, \mathbf{r}_2, \ldots, \mathbf{r}_{N_{\mathrm{e}}})$ for the electrons. We also define the spin coordinate of electron k as s_k, the combined spatial-spin coordinate as $x_k = (\mathbf{r}_k, s_k)$, and the collective spin-spatial coordinate as $\overline{x} = (x_1, x_2, \ldots, x_{N_{\mathrm{e}}})$

In the BO approximation, we first construct a set of L_{BO} electronic states $\Phi_j(\overline{x}; \overline{\mathbf{R}})$, $j = 1, 2, \ldots, L_{\mathrm{BO}}$, which parametrically depends on the nuclear coordinate $\overline{\mathbf{R}}$. The electronic states are calculated by solving the eigenvalue equation

$$\mathscr{E}_j(\overline{\mathbf{R}}) \Phi_j(\overline{x}; \overline{\mathbf{R}}) = H_{\mathrm{e}}(\overline{\mathbf{R}}) \Phi_j(\overline{x}; \overline{\mathbf{R}}), \tag{10.7}$$

where $\overline{\mathbf{R}}$ is treated as a set of parameters $(\mathbf{R}_1, \ldots, \mathbf{R}_{N_{\mathrm{N}}})$ taking fixed values. Because H_{e} is Hermitian, the electronic states are orthogonal. We have $\int \Phi_j^*(\overline{x}; \overline{\mathbf{R}}) \Phi_k(\overline{x}; \overline{\mathbf{R}}) \mathrm{d}\overline{x} = \delta_{jk}$ after appropriate normalization. Equation (10.7) has to be solved at each value of the nuclear coordinate $\overline{\mathbf{R}}$, which may become a difficult task for large molecules. This point is discussed more in detail later at the end of this section.

We make the ansatz here

$$\Psi(\overline{\mathbf{R}}, \overline{x}, t) = \sum_{j=1}^{L_{\mathrm{BO}}} \chi_j(\overline{\mathbf{R}}, t)\Phi_j(\overline{x}; \overline{\mathbf{R}}) \tag{10.8}$$

for the total wave function. The expansion (10.8) is commonly referred to as the Born-Huang (BH) expansion [11]. The time-dependence of the wave function is contained in the nuclear wave functions $\chi_j(\overline{\mathbf{R}}, t)$. Inserting the ansatz (10.8) into the time-dependent Schrödinger equation (TDSE)

$$\mathrm{i}\hbar\frac{\partial\Psi(\overline{\mathbf{R}}, \overline{x}, t)}{\partial t} = H\Psi(\overline{\mathbf{R}}, \overline{x}, t) \tag{10.9}$$

and integrating out the electronic degrees of freedom leads to a set of coupled equations for the nuclear wave functions,

$$\begin{aligned} \mathrm{i}\hbar\frac{\partial\chi_j(\overline{\mathbf{R}}, t)}{\partial t} &= \left(T_{\mathrm{N}} + U_{\mathrm{N}}(t) + V_j^{\mathrm{BO}}(\overline{\mathbf{R}})\right)\chi_j(\overline{\mathbf{R}}, t) \\ &+ \sum_{k=1}^{L_{\mathrm{BO}}}\left(-\mathbf{E}(t)\cdot\boldsymbol{\mu}_{jk}(\overline{\mathbf{R}}) + A_{jk} + B_{jk}\right)\chi_k(\overline{\mathbf{R}}, t), \end{aligned} \tag{10.10}$$

where

$$V_j^{\mathrm{BO}}(\overline{\mathbf{R}}) = \mathscr{E}_j(\overline{\mathbf{R}}) + V_{\mathrm{NN}}(\overline{\mathbf{R}}) \tag{10.11}$$

is a BO potential energy surface defined using the electronic eigenenergy in (10.7),

$$\boldsymbol{\mu}_{jk}(\overline{\mathbf{R}}) = -e\int \mathrm{d}\overline{x}\Phi_j^*(\overline{x}; \overline{\mathbf{R}})\left(\sum_{l=1}^{N_{\mathrm{e}}}\mathbf{r}_l\right)\Phi_k(\overline{x}; \overline{\mathbf{R}}) \tag{10.12}$$

is the transition dipole matrix element, and

$$A_{jk} = -\hbar^2\int \mathrm{d}\overline{x}\Phi_j^*(\overline{x}; \overline{\mathbf{R}})\left(\sum_{l=1}^{N_{\mathrm{N}}}\frac{\nabla_{\mathbf{R}_l}}{M_l}\right)\Phi_k(\overline{x}; \overline{\mathbf{R}})\cdot\nabla_{\mathbf{R}_l} \tag{10.13}$$

and

$$B_{jk} = -\frac{\hbar^2}{2}\int \mathrm{d}\overline{x}\Phi_j^*(\overline{x}; \overline{\mathbf{R}})\left(\sum_{l=1}^{N_{\mathrm{N}}}\frac{\nabla^2_{\mathbf{R}_l}}{M_l}\right)\Phi_k(\overline{x}; \overline{\mathbf{R}}) \tag{10.14}$$

are non-adiabatic coupling terms.

The BO approximation consists in neglecting the non-adiabatic coupling terms A_{jk} and B_{jk}. This neglection can be rationalized by (i) the small value of the $1/M_l$ nuclear-mass factors in the expressions (10.13) and (10.14), and (ii) the small value of $\nabla_{\mathbf{R}_l}\Phi_k(\overline{x}; \overline{\mathbf{R}})$, which is assured as long as the electronic states $\Phi_k(\overline{x}; \overline{\mathbf{R}})$ change

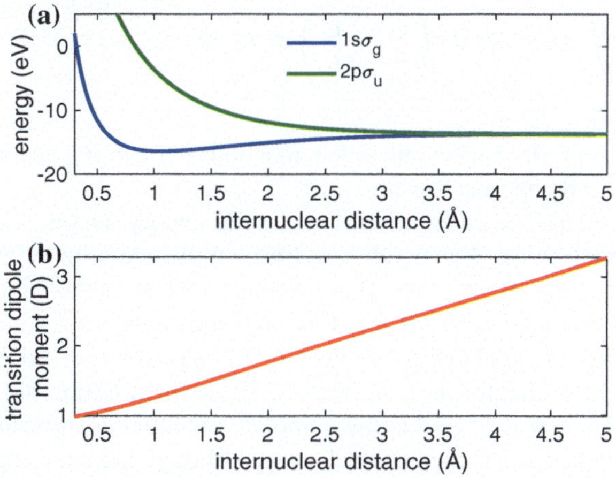

Fig. 10.1 **a** Potential energy curves for the ground and first excited state of H_2^+. **b** Transition dipole moment between the $1s\sigma_g$ state and the $2p\sigma_u$ state

slowly with $\overline{\mathbf{R}}$. If we also neglect the term $U_{\mathrm{N}}(t)$ in (10.10) by considering that it only affects the center-of-mass motion, we arrive at the BO TDSE,

$$\mathrm{i}\hbar\frac{\partial \chi_j(\overline{\mathbf{R}},t)}{\partial t} = \left(T_{\mathrm{N}} + V_j^{\mathrm{BO}}(\overline{\mathbf{R}})\right)\chi_j(\overline{\mathbf{R}},t) - \mathbf{E}(t)\cdot\sum_{k=1}^{L_{\mathrm{BO}}}\boldsymbol{\mu}_{jk}(\overline{\mathbf{R}})\chi_k(\overline{\mathbf{R}},t). \quad (10.15)$$

The physical picture of (10.15) is that each nuclear wave packet $\chi_j(\overline{\mathbf{R}}, t)$ moves on a potential energy surface $V_j^{\mathrm{BO}}(\overline{\mathbf{R}})$. Transitions between different electronic states are described by the $\overline{\mathbf{R}}$-dependent transition dipole matrix elements $\boldsymbol{\mu}_{jk}(\overline{\mathbf{R}})$. In Fig. 10.1, we show an example of the BO potential energy curves and transition dipole moment for a hydrogen molecular ion, H_2^+. In H_2^+, there is only one internal nuclear coordinate, the internuclear distance. The potential energy curves shown in Fig. 10.1 were calculated by solving (10.7) with the finite-difference method at each value of the internuclear distance.

It is true that the BH expansion together with the BO approximation works well for many cases of laser-molecule interaction, and is a standard method for the simulation and interpretation of experimental results, but there are of course some limits. We would like to point out a few of the disadvantageous points of the BO approximation, which serve as a motivation for developing methods which go beyond the BO approximation.

(i) The magnitude of the A_{jk} and B_{jk} operators may become unrealistically large at a nuclear geometry $\overline{\mathbf{R}}$ where j-th and k-th BO potential energy surfaces are close in energy to satisfy $\mathscr{E}_j(\overline{\mathbf{R}}) \approx \mathscr{E}_k(\overline{\mathbf{R}})$. By using (10.7), we can derive

$$\int d\overline{x}\Phi_j^*(\overline{x};\overline{\mathbf{R}})\left(\sum_{l=1}^{N_N}\frac{\nabla_{\overline{\mathbf{R}}_l}}{M_l}\right)\Phi_k(\overline{x};\overline{\mathbf{R}}) = \frac{\int d\overline{x}\Phi_j^*(\overline{x};\overline{\mathbf{R}})\left(\sum_{l=1}^{N_N}\frac{(\nabla_{\overline{\mathbf{R}}_l}H_e)}{M_l}\right)\Phi_k(\overline{x};\overline{\mathbf{R}})}{\mathscr{E}_k(\overline{\mathbf{R}}) - \mathscr{E}_j(\overline{\mathbf{R}})}, \tag{10.16}$$

from which it becomes clear that both A_{jk} and B_{jk} become extremely large if $\mathscr{E}_k(\overline{\mathbf{R}}) - \mathscr{E}_j(\overline{\mathbf{R}})$ approaches zero.

(ii) The calculation of potential energy curves is a very difficult task for large molecules. A general, non-linear molecule with N atoms has $3N - 6$ vibrational degrees of freedom. If we assume that K points along the coordinate of each vibrational degree of freedom are required for a sufficiently accurate description of the potential energy surface, (10.7) has to be solved K^{3N-6} times, which increases exponentially as N increases. Even for a tetratomic molecule ($N = 4$), we have $3N - 6 = 6$, so that the complete potential energy surface would have to be represented as a 6-dimensional array. Although there are sophisticated methods for fitting high-dimensional potential energy surfaces (see for example [12]), the accurate representation of high-dimensional potential energy surfaces is a difficult problem.

(iii) Omission of electron excitation and ionization. In the BH expansion, we always have to limit the number of electronic states L_{BO} included in the expansion of the wave function (see (10.8)). Typically, L_{BO} is set to be $L_{\mathrm{BO}} < 5$, meaning that only the electronic states with the lowest energy (ground state + a few excited states) are included. Extensive excitation to Rydberg states, doubly excites states, continuum electronic states, which would be required to describe ionization, are difficult to be treated. The omission of such highly excited states and continuum states becomes particularly problematic when we describe the interaction of molecules with ultrashort and intense laser pulses, because ionization and excitation occur with a high probability. In the simplest case of a hydrogen molecule H_2, there exist models by which potential energy curves are calculated for quasi-stable doubly excited electronic states as well as for electronic states corresponding to the ionization [13, 14], but for a general, polyatomic molecule, this approach is not practically applicable.

10.3 Extended Multiconfiguration Time-Dependent Hartree-Fock Method

10.3.1 Basic Concepts

In this section, we describe one promising attempt to go beyond the BO approximation, called the extended multiconfiguration time-dependent Hartree-Fock (Ex-MCTDHF) method [15]. The Ex-MCTDHF method is an extension of the multiconfiguration time-dependent Hartree-Fock (MCTDHF) method for electron dynamics [16, 17] (see also [18] for an overview of the method), which is a natural extension of the multiconfiguration time-dependent Hartree (MCTDH) method for nuclear dynamics [19, 20].

The basic idea of all the methods mentioned above is to introduce time-dependent single-particle (or single-mode) functions for the description of the time-dependent dynamics. This means that each degree of freedom is described by its own time-dependent basis set. In this way, any kind of excitation can be described in a flexible manner.

In the Ex-MCTDHF method, the total time-dependent wave function for both nuclei and electrons is written as

$$\Psi(\overline{\mathbf{R}}, \overline{x}, t) = \sum_{J=1}^{L_e} \chi_J(\overline{\mathbf{R}}, t)\Phi_J(\overline{x}, t), \tag{10.17}$$

using the same notation for the nuclear and electronic coordinates as adopted in Sect. 10.2. In (10.17), $\chi_J(\overline{\mathbf{R}}, t)$ is a time-dependent nuclear wave function, $\Phi_J(\overline{x}, t)$ is a time-dependent electronic wave function, and L_e is the number of terms included in the wave function expansion. In order to make the total wave function antisymmetrized with respect to an exchange of electrons, $\Phi_J(\overline{x}, t)$ is represented as a Slater determinant, constructed from a set of time-dependent electronic spin-orbitals $\phi_k(x, t)$ ($k = 1, \ldots, K_e$),

$$\Phi_J(\overline{x}, t) = |\phi_{J_1}(t) \ldots \phi_{J_{N_e}}(t)|, \tag{10.18}$$

where the label J is a composite index $J = (J_1, \ldots, J_{N_e})$. We should have at least as many spin-orbitals as the number of electrons in the molecule, that is, $K_e \geq N_e$. It is assumed that at all times, the spin-orbitals are orthonormal, $\langle \phi_j(t) | \phi_k(t) \rangle = 0$.

We point out here that the meanings of the electronic wave function $\Phi_J(\overline{x}, t)$ in (10.17) and the electronic wave function $\Phi_j(\overline{x}; \overline{\mathbf{R}})$ in the Born-Huang expansion (10.8) are different. In (10.8), $\Phi_j(\overline{x}; \overline{\mathbf{R}})$ represents an electronic state, an eigenfunction of the electronic Hamiltonian $H_e(\overline{\mathbf{R}})$ at fixed $\overline{\mathbf{R}}$, which is usually expanded into a linear combination of several Slater determinants, but in (10.17), $\Phi_J(\overline{x}, t)$ is a single, time-dependent Slater determinant.

It is frequently assumed that the same spatial orbitals are used to construct spin-orbitals of both spin types. That is, we assume a set of M_e spatial orbitals $\varphi_k(\mathbf{r}, t)$, $k = 1, \ldots, M_e$, and construct $K_e = 2M_e$ spin-orbitals according to

$$\phi_k(x, t) = \begin{cases} \varphi_k(\mathbf{r}, t)\alpha(s) & \text{if } k \leq M_e, \\ \varphi_{k-M_e}(\mathbf{r}, t)\beta(s) & \text{if } k > M_e. \end{cases} \tag{10.19}$$

In (10.19), $\alpha(s)$ and $\beta(s)$ denote spin-up and spin-down spin functions, respectively. In this case, each Slater determinant is labeled by a double index $J = (J^\alpha, J^\beta)$, with $J^\alpha = (J_1^\alpha, \ldots, J_{N_e^\alpha}^\alpha)$ and $J^\beta = (J_1^\beta, \ldots, J_{N_e^\beta}^\beta)$. The numbers N_e^α of α electrons and N_e^β of β electrons satisfy $N_e^\alpha + N_e^\beta = N_e$, and we assume that all Slater determinants have the same value of N_e^α and N_e^β. A determinant is written as

$$\Phi_J(\overline{x},t) = |\varphi_{J_1^\alpha}(t)\alpha \ldots \varphi_{J_{N_e^\alpha}^\alpha}(t)\alpha\, \varphi_{J_1^\beta}(t)\beta \ldots \varphi_{J_{N_e^\beta}^\beta}(t)\beta|. \tag{10.20}$$

The total number of electronic Slater determinants that can be constructed in this way is $L_e = \binom{M_e}{N_e^\alpha}\binom{M_e}{N_e^\beta} = M_e!^2/[N_e^\alpha!N_e^\beta!(M_e - N_e^\alpha)!(M_e - N_e^\beta)!]$.

The wave function defined by (10.17) looks very similar to the BO wave function in (10.8). The important difference is that the electronic wave function $\Phi_J(\overline{x},t)$ in (10.17) does not depend on the nuclear coordinates $\overline{\mathbf{R}}$, but instead depends on time t. This means that each electronic wave function $\Phi_J(\overline{x},t)$ is no longer associated with a certain electronic state with a certain energy. At some moment in time, $\Phi_J(\overline{x},t)$ may be a superposition of a bound part (consisting of both ground-state and excited-state components) and a continuum part. This feature makes it possible for a wave function of the type (10.17) to describe arbitrary electronic excitation, including ionization, with a limited number of terms L_e in the sum over J in (10.17). Moreover, because the electronic wave functions do not depend on $\overline{\mathbf{R}}$, there is no need to calculate potential energy surfaces in the Ex-MCTDHF method. The major drawback of an ansatz like the one in (10.17) is that the equations governing the time-evolution of the time-dependent orbitals become non-linear, as we will see below in (10.31), (10.46), and (10.54).

In [15], the ansatz (10.17) was further adapted to describe "diatomic-like" molecules, which refers to molecules consisting of two heavy atoms like O or C and N_p light hydrogen atoms. Explicit examples are C_2H_2 and CH_3OH. The idea proposed in [15] was that the protonic part of the wave function can be expanded in terms of Slater determinants, because protons are also fermions whose wave function needs to be properly antisymmetrized. The motivation of using Slater determinants also for the description of the protonic motion is to make the structure of the wave function as flexible as possible, which can allow us to simulate highly distorted molecular structures such as those appearing in the course of hydrogen migration.

In order to describe the protonic part of the wave function with Slater determinants, the wave function $\chi_J(\overline{\mathbf{R}},t)$ for the nuclear coordinate is expanded as

$$\chi_J(\overline{\mathbf{R}},t) = \sum_{I=1}^{L_p} C_{IJ}(\overline{\mathbf{R}}_h,t)\Lambda_I(\overline{X},t), \tag{10.21}$$

where $\overline{\mathbf{R}}_h$ is the collective coordinate for the two heavy atoms, $\overline{X}$ is the collective spatial-spin coordinate for the N_p protons, and L_p is the number of terms included in the expansion of the nuclear wave function. The proton coordinate $\overline{X}$ is defined as $\overline{X} = (X_1,\ldots,X_{N_p})$ with $X_k = (\mathbf{R}_{pk}, S_k)$ expressed in terms of the spatial coordinate $\mathbf{R}_{pk}$ and the spin coordinate S_k of the proton k. Similarly to the electronic part of the wave function, $\Lambda_I(\overline{X},t)$ is taken to be a Slater determinant constructed from the time-dependent protonic orbitals $\lambda_i(X,t)$,

$$\Lambda_I(\overline{X},t) = |\lambda_{I_1}(t)\ldots\lambda_{I_{N_p}}(t)|, \tag{10.22}$$

with $I = (I_1, \ldots, I_{N_\mathrm{p}})$. In the same way as for the electronic Slater determinants, it is convenient to assume that the same spatial orbitals are used both for α and β spin-orbitals, that is,

$$\lambda_j(X,t) = \begin{cases} \kappa_j(\mathbf{R},t)\alpha(S) & \text{if } j \leq M_\mathrm{p}, \\ \kappa_{j-M_\mathrm{p}}(\mathbf{R},t)\beta(S) & \text{if } j > M_\mathrm{p}, \end{cases} \tag{10.23}$$

where we have assumed that there are M_p spatial protonic orbitals. The protonic Slater determinants are in this case written as

$$\Lambda_I(\overline{X},t) = |\kappa_{I_1^\alpha}(t)\alpha \ldots \kappa_{I^\alpha_{N_\mathrm{p}^\alpha}}(t)\alpha\, \kappa_{I_1^\beta}(t)\beta \ldots \kappa_{I^\beta_{N_\mathrm{p}^\beta}}(t)\beta|, \tag{10.24}$$

where there are N_p^α α-spin protons and N_p^β β-spin protons in each determinant. The total number of protons is $N_\mathrm{p} = N_\mathrm{p}^\alpha + N_\mathrm{p}^\beta$. Similarly to the electronic determinants, we have a total number of $L_\mathrm{p} = \binom{M_\mathrm{p}}{N_\mathrm{p}^\alpha}\binom{M_\mathrm{p}}{N_\mathrm{p}^\beta} = M_\mathrm{p}!^2/[N_\mathrm{p}^\alpha! N_\mathrm{p}^\beta!(M_\mathrm{p} - N_\mathrm{p}^\alpha)!(M_\mathrm{p} - N_\mathrm{p}^\beta)!]$ protonic Slater determinants.

Substituting (10.21) into (10.17), we find for the total wave function,

$$\Psi(\overline{\mathbf{R}}_\mathrm{h}, \overline{X}, \overline{x}, t) = \sum_{J=1}^{L_\mathrm{e}} \sum_{I=1}^{L_\mathrm{p}} C_{IJ}(\overline{\mathbf{R}}_\mathrm{h}, t)\Lambda_I(\overline{X},t)\Phi_J(\overline{x},t). \tag{10.25}$$

Due to the similarity of (10.25) to the configuration-interaction expansion of a many-electron wave function [21], we refer to the $C_{IJ}(\overline{\mathbf{R}}_\mathrm{h}, t)$ as time-dependent configuration-interaction (CI) coefficients.

Because all factors $[C_{IJ}(\overline{\mathbf{R}}_\mathrm{h}, t)$, $\Lambda_I(\overline{X},t)$, and $\Phi_J(\overline{x},t)]$ depend on t, we have to derive appropriate evolution equations. In order to derive the evolution equations, also referred to as the equations of motion, we employ the time-dependent variational principle [22, 23],

$$\langle \delta\Psi(t)|H - i\hbar\frac{\partial}{\partial t}|\Psi(t)\rangle = 0, \tag{10.26}$$

where $\delta\Psi(t)$ is the variation of the total wave function with respect to parameters $C_{IJ}(\overline{\mathbf{R}}_\mathrm{h}, t)$, $\Lambda_I(\overline{X},t)$, and $\Phi_J(\overline{x},t)$, and the bra-ket in (10.26) implies the integration over all variables $\overline{\mathbf{R}}_\mathrm{h}$, $\overline{X}$, and $\overline{x}$. Orthogonality of the electronic and protonic orbitals, $\langle\varphi_j(t)|\varphi_k(t)\rangle = \langle\kappa_j(t)|\kappa_k(t)\rangle = \delta_{jk}$, is assumed by introducing suitable Lagrange multipliers in (10.26).

Before we derive the equations of motion for the Ex-MCTDHF method, we first define the single-proton Hamiltonian,

$$h_\mathrm{p}(\mathbf{R},t) = -\frac{\hbar^2}{2m_\mathrm{p}}\nabla^2_\mathbf{R} - e\mathbf{E}(t)\cdot\mathbf{R}, \tag{10.27}$$

where m_p is the mass of the proton, the single-electron Hamiltonian,

$$h_{\rm e}(\mathbf{r},t) = -\frac{\hbar^2}{2m_{\rm e}}\nabla_{\mathbf{r}}^2 + e\mathbf{E}(t)\cdot\mathbf{r}, \tag{10.28}$$

and the Hamiltonian for the $N_{\rm h} = N_{\rm N} - N_{\rm p}$ heavy particles,

$$H_{\rm h}(\overline{\mathbf{R}}_{\rm h},t) = \sum_{k=1}^{N_{\rm h}}\left(-\frac{\hbar^2}{2M_k}\nabla_{\mathbf{R}_{{\rm h}k}}^2 - e\mathbf{E}(t)\cdot\mathbf{R}_{{\rm h}k} + \sum_{l<k}\frac{Z_l Z_k e^2}{4\pi\varepsilon_0|\mathbf{R}_{{\rm h}k} - \mathbf{R}_{{\rm h}l}|}\right), \tag{10.29}$$

where Z_k here represents the charge number of the heavy nucleus k. In [15], it was assumed that $N_{\rm h} = 2$. Then, the total Hamiltonian can now be written as

$$\begin{aligned} H =& \sum_{k=1}^{N_{\rm p}} h_{\rm p}(\mathbf{R}_{{\rm p}k},t) + \sum_{k=1}^{N_{\rm p}}\sum_{l<k}\frac{e^2}{4\pi\varepsilon_0|\mathbf{R}_{{\rm p}k} - \mathbf{R}_{{\rm p}l}|} \\ &+ \sum_{k=1}^{N_{\rm e}} h_{\rm e}(\mathbf{r}_k,t) + \sum_{k=1}^{N_{\rm e}}\sum_{l<k}\frac{e^2}{4\pi\varepsilon_0|\mathbf{r}_k - \mathbf{r}_l|} \\ &- \sum_{k=1}^{N_{\rm p}}\sum_{l=1}^{N_{\rm e}}\frac{e^2}{4\pi\varepsilon_0|\mathbf{R}_{{\rm p}k} - \mathbf{r}_l|} - \sum_{k=1}^{N_{\rm h}}\sum_{l=1}^{N_{\rm e}}\frac{Z_k e^2}{4\pi\varepsilon_0|\mathbf{R}_{{\rm h}k} - \mathbf{r}_l|} + \sum_{k=1}^{N_{\rm h}}\sum_{l=1}^{N_{\rm p}}\frac{Z_k e^2}{4\pi\varepsilon_0|\mathbf{R}_{{\rm h}k} - \mathbf{R}_{{\rm p}l}|} \\ &+ H_{\rm h}(\overline{\mathbf{R}}_{\rm h},t). \end{aligned} \tag{10.30}$$

We note that the Hamiltonian (10.30) is the same as that in (10.1), but the form is rewritten so that the different interaction terms appear more clearly.

10.3.2 Equations of Motion

In order to obtain the equation of motion for the electronic orbitals, we take the variation $\delta\Psi$ with respect to one spatial orbital φ_k in (10.26). The result is

$$\begin{aligned} {\rm i}\hbar\frac{\partial\varphi_k(\mathbf{r},t)}{\partial t} =& Q_{\rm e}(t)\sum_{lm} D_{kl}^{{\rm e}-1}(t)\Big(D_{lm}^{\rm e}(t)h_{\rm e}(\mathbf{r},t) + W_{lm}^{\rm ee}(\mathbf{r},t) + W_{lm}^{\rm ep}(\mathbf{r},t) \\ &+ W_{lm}^{\rm eh}(\mathbf{r},t)\Big)\varphi_m(\mathbf{r},t), \end{aligned} \tag{10.31}$$

where

$$D_{lm}^{\rm e}(t) = \sum_{IPQ}\int {\rm d}\overline{\mathbf{R}}_{\rm h} C_{IP}^*(\overline{\mathbf{R}}_{\rm h},t)C_{IQ}(\overline{\mathbf{R}}_{\rm h},t)\mathscr{E}_{PQlm}^{\rm e} \tag{10.32}$$

is the spin-summed electronic first-order density matrix,

$$W^{\rm ee}_{lm}(\mathbf{r},t) = \sum_{pq} d^{\rm e}_{lmpq}(t) g^{\rm ee}_{pq}(\mathbf{r},t) \tag{10.33}$$

is the electron-electron interaction defined using the spin-summed electronic second-order density matrix,

$$d^{\rm e}_{klmn}(t) = \sum_{IPQ} \int {\rm d}\overline{\mathbf{R}}_{\rm h} C^*_{IP}(\overline{\mathbf{R}}_{\rm h},t) C_{IQ}(\overline{\mathbf{R}}_{\rm h},t) \mathscr{F}^{\rm e}_{PQklmn}, \tag{10.34}$$

and the repulsive electron-electron Coulomb potential

$$g^{\rm ee}_{pq}(\mathbf{r},t) = \frac{e^2}{4\pi\varepsilon_0} \int {\rm d}\mathbf{r}' \frac{\varphi^*_p(\mathbf{r}',t)\varphi_q(\mathbf{r}',t)}{|\mathbf{r}-\mathbf{r}'|}, \tag{10.35}$$

the electron-proton interaction is

$$W^{\rm ep}_{lm}(\mathbf{r},t) = \sum_{IJPQrs} \int {\rm d}\overline{\mathbf{R}}_{\rm h} C^*_{IP}(\overline{\mathbf{R}}_{\rm h},t) C_{JQ}(\overline{\mathbf{R}}_{\rm h},t) \mathscr{E}^{\rm e}_{PQlm} \mathscr{E}^{\rm p}_{IJrs} g^{\rm ep}_{rs}(\mathbf{r},t) \tag{10.36}$$

defined with the attractive electron-proton Coulomb potential

$$g^{\rm ep}_{pq}(\mathbf{r},t) = -\frac{e^2}{4\pi\varepsilon_0} \int {\rm d}\mathbf{R} \frac{\kappa^*_p(\mathbf{R},t)\kappa_q(\mathbf{R},t)}{|\mathbf{r}-\mathbf{R}|}, \tag{10.37}$$

and the attractive electron-heavy nuclei interaction is

$$W^{\rm eh}_{lm}(\mathbf{r},t) = -\frac{e^2}{4\pi\varepsilon_0} \sum_{IPQ} \mathscr{E}^{\rm e}_{PQlm} \int {\rm d}\overline{\mathbf{R}}_{\rm h} C^*_{IP}(\overline{\mathbf{R}}_{\rm h},t) C_{IQ}(\overline{\mathbf{R}}_{\rm h},t) \left(\sum_{k=1}^{N_{\rm h}} \frac{Z_k}{|\mathbf{R}_{{\rm h}k} - \mathbf{r}|} \right). \tag{10.38}$$

In (10.32), (10.34), (10.36) and (10.38), we have used the following matrix elements of the spin-summed excitation operators $\hat{\mathscr{E}}_{pq}$ and $\hat{\mathscr{F}}_{pqrs}$ [21],

$$\mathscr{E}^{\rm e}_{PQlm} = \langle \Phi_P(t) | \hat{\mathscr{E}}_{lm} | \Phi_Q(t) \rangle, \tag{10.39}$$

$$\mathscr{E}^{\rm p}_{IJrs} = \langle \Lambda_I(t) | \hat{\mathscr{E}}_{rs} | \Lambda_J(t) \rangle, \tag{10.40}$$

and

$$\mathscr{F}^{\rm e}_{PQklmn} = \langle \Phi_P(t) | \hat{\mathscr{F}}_{klmn} | \Phi_Q(t) \rangle. \tag{10.41}$$

For later use, we additionally define

$$\mathscr{F}^{\rm p}_{IJklmn} = \langle \Lambda_I(t) | \hat{\mathscr{F}}_{klmn} | \Lambda_J(t) \rangle. \tag{10.42}$$

The spin-summed excitation operators are defined using the creation and annihilation operators $\hat{a}^{\dagger}_{p\sigma}$ and $\hat{a}_{q\sigma}$ as [21]

$$\hat{\mathscr{E}}_{pq} = \sum_{\sigma=\alpha,\beta} \hat{a}^{\dagger}_{p\sigma}\hat{a}_{q\sigma} \tag{10.43}$$

and

$$\hat{\mathscr{F}}_{pqrs} = \hat{\mathscr{E}}_{pq}\hat{\mathscr{E}}_{rs} - \delta_{qr}\hat{\mathscr{E}}_{ps}. \tag{10.44}$$

Upon the operation of the creation operator $\hat{a}^{\dagger}_{p\sigma}$ on a Slater determinant (either electronic or protonic), the spatial orbital p with spin σ (=α or β) is created. On the other hand, upon the operation of the annihilation operator $\hat{a}_{q\sigma}$, the spatial orbital q with spin σ is annihilated if it exists in the determinant. The matrix elements defined in (10.39)–(10.42) are equal to either 0, -1, or 1. We can derive their explicit values for different combinations of indexes by using the orthonormality of the spatial orbitals and by taking into account the sign change of a determinant upon the permutation of the order of the orbitals. The matrix elements, $\mathscr{E}^{\mathrm{e}}_{PQlm}$ and $\mathscr{E}^{\mathrm{p}}_{PQlm}$, can take non-zero values only when the two determinants involved differ by at most one orbitals, and the matrix elements, $\mathscr{F}^{\mathrm{e}}_{PQklmn}$ and $\mathscr{F}^{\mathrm{p}}_{PQklmn}$, are non-zero only when the two determinants differ by at most two orbitals. We also note that $\mathscr{E}^{\mathrm{e}}_{PQlm}$, $\mathscr{E}^{\mathrm{p}}_{PQlm}$, $\mathscr{F}^{\mathrm{e}}_{PQklmn}$ and $\mathscr{F}^{\mathrm{p}}_{PQklmn}$ defined (in (10.39)–(10.42)) as matrix elements composed of time-dependent determinants, are independent of time.

The symbol Q_{e} in the equation of motion (10.31) is a projection operator whose action on an arbitrary function $f(\mathbf{r})$ is defined as

$$Q_{\mathrm{e}}(t)f(\mathbf{r}) = f(\mathbf{r}) - \sum_{k} f(\mathbf{r})\langle f|\varphi_k(t)\rangle. \tag{10.45}$$

The projection operator $Q_{\mathrm{e}}(t)$ appears in the equation of motion (10.31) because of the restriction of $\langle\varphi_j(t)|\varphi_k(t)\rangle = \delta_{jk}$ imposed by the Lagrange multipliers. By multiplying the equation of motion (10.31) by $\varphi^*_j(\mathbf{r},t)$ and integrating over $\mathbf{r}$, we may confirm that $\langle\varphi_j|(\partial/\partial t)|\varphi_k\rangle = 0$ holds for arbitrary j and k if $\langle\varphi_j(t)|\varphi_k(t)\rangle = \delta_{jk}$ because of the presence of $Q_{\mathrm{e}}(t)$, and therefore, $\langle\varphi_j(t)|\varphi_k(t)\rangle = \delta_{jk}$ is satisfied for all t provided that the orbital set $\{\varphi_j(t)\}$ is orthonormal at $t = 0$.

The equation of motion for the protonic orbitals $\kappa_j(\mathbf{R},t)$ is calculated by taking the variation $\delta\Psi$ in (10.26) with respect to κ_j on the condition that $\langle\kappa_j(t)|\kappa_k(t)\rangle = \delta_{jk}$. We obtain the following equation of motion, similar to that for the electronic orbitals, (10.31),

$$\begin{aligned} \mathrm{i}\hbar\frac{\partial\kappa_k(\mathbf{R},t)}{\partial t} = Q_{\mathrm{p}}(t)\sum_{lm} D^{\mathrm{p}-1}_{kl}(t)\Big(&D^{\mathrm{p}}_{lm}(t)h_{\mathrm{p}}(\mathbf{R},t) + W^{\mathrm{pp}}_{lm}(\mathbf{R},t) + W^{\mathrm{pe}}_{lm}(\mathbf{R},t) \\ &+ W^{\mathrm{ph}}_{lm}(\mathbf{R},t)\Big)\kappa_m(\mathbf{R},t), \end{aligned} \tag{10.46}$$

where

$$D^{\mathrm{p}}_{lm}(t) = \sum_{IJP} \int \mathrm{d}\overline{\mathbf{R}}_{\mathrm{h}} C^*_{IP}(\overline{\mathbf{R}}_{\mathrm{h}}, t) C_{JP}(\overline{\mathbf{R}}_{\mathrm{h}}, t) \mathscr{E}^{\mathrm{p}}_{IJlm} \tag{10.47}$$

is the spin-summed protonic first-order density matrix,

$$W^{\mathrm{pp}}_{lm}(\mathbf{R}, t) = \sum_{pq} d^{\mathrm{p}}_{lmpq}(t) g^{\mathrm{pp}}_{pq}(\mathbf{R}, t) \tag{10.48}$$

is the proton-proton interaction with the spin-summed protonic second-order density matrix,

$$d^{\mathrm{p}}_{klmn}(t) = \sum_{IJP} \int \mathrm{d}\overline{\mathbf{R}}_{\mathrm{h}} C^*_{IP}(\overline{\mathbf{R}}_{\mathrm{h}}, t) C_{JP}(\overline{\mathbf{R}}_{\mathrm{h}}, t) \mathscr{F}^{\mathrm{p}}_{IJklmn}, \tag{10.49}$$

and the proton-proton Coulomb potential,

$$g^{\mathrm{pp}}_{pq}(\mathbf{R}, t) = \frac{e^2}{4\pi\varepsilon_0} \int \mathrm{d}\mathbf{R}' \frac{\kappa^*_p(\mathbf{R}', t)\kappa_q(\mathbf{R}', t)}{|\mathbf{R} - \mathbf{R}'|}. \tag{10.50}$$

Furthermore,

$$W^{\mathrm{pe}}_{lm}(\mathbf{R}, t) = \sum_{IJPQrs} \int \mathrm{d}\overline{\mathbf{R}}_{\mathrm{h}} C^*_{IP}(\overline{\mathbf{R}}_{\mathrm{h}}, t) C_{JQ}(\overline{\mathbf{R}}_{\mathrm{h}}, t) \mathscr{E}^{\mathrm{e}}_{PQlm} \mathscr{E}^{\mathrm{p}}_{IJrs} g^{\mathrm{pe}}_{rs}(\mathbf{R}, t) \tag{10.51}$$

is the proton-electron interaction defined using the attractive proton-electron Coulomb potential,

$$g^{\mathrm{pe}}_{pq}(\mathbf{R}, t) = -\frac{e^2}{4\pi\varepsilon_0} \int \mathrm{d}\mathbf{r} \frac{\varphi^*_p(\mathbf{r}, t)\varphi_q(\mathbf{r}, t)}{|\mathbf{r} - \mathbf{R}|}, \tag{10.52}$$

and

$$W^{\mathrm{ph}}_{lm}(\mathbf{R}, t) = \frac{e^2}{4\pi\varepsilon_0} \sum_{IJP} \mathscr{E}^{\mathrm{p}}_{IJlm} \int \mathrm{d}\overline{\mathbf{R}}_{\mathrm{h}} C^*_{IP}(\overline{\mathbf{R}}_{\mathrm{h}}, t) C_{JP}(\overline{\mathbf{R}}_{\mathrm{h}}, t) \left(\sum_{k=1}^{N_{\mathrm{h}}} \frac{Z_k}{|\mathbf{R}_{\mathrm{h}k} - \mathbf{R}|} \right) \tag{10.53}$$

is the repulsive proton-heavy nuclei interaction.

Finally, we take the variation with respect to the coefficients $C_{IJ}(\overline{\mathbf{R}}_{\mathrm{h}}, t)$, which results in the equation of motion,

$$\begin{aligned} \mathrm{i}\hbar \frac{\partial C_{IJ}(\overline{\mathbf{R}}_{\mathrm{h}}, t)}{\partial t} &= H_{\mathrm{h}}(\overline{\mathbf{R}}_{\mathrm{h}}, t) C_{IJ}(\overline{\mathbf{R}}_{\mathrm{h}}, t) \\ &+ \sum_{KL} \left(W^{\mathrm{he}}_{IJKL}(\overline{\mathbf{R}}_{\mathrm{h}}, t) + W^{\mathrm{hp}}_{IJKL}(\overline{\mathbf{R}}_{\mathrm{h}}, t) + W^{0}_{IJKL}(t) \right) C_{KL}(\overline{\mathbf{R}}_{\mathrm{h}}, t). \end{aligned} \tag{10.54}$$

In (10.54), $H_\mathrm{h}(\overline{\mathbf{R}}_\mathrm{h}, t)$ is given by (10.29),

$$W^{\mathrm{he}}_{IJKL}(\overline{\mathbf{R}}_\mathrm{h}, t) = -\delta_{IK}\frac{e^2}{4\pi\varepsilon_0}\sum_{lm}\mathscr{E}^{\mathrm{e}}_{JLlm}\left(\sum_{k=1}^{N_\mathrm{h}} Z_k\int \mathrm{d}\mathbf{r}\frac{\varphi_l^*(\mathbf{r}, t)\varphi_m(\mathbf{r}, t)}{|\mathbf{R}_{\mathrm{h}k} - \mathbf{r}|}\right) \tag{10.55}$$

is the heavy nuclei-electron interaction,

$$W^{\mathrm{hp}}_{IJKL}(\overline{\mathbf{R}}_\mathrm{h}, t) = \delta_{JL}\frac{e^2}{4\pi\varepsilon_0}\sum_{lm}\mathscr{E}^{\mathrm{p}}_{IKlm}\left(\sum_{k=1}^{N_\mathrm{h}} Z_k\int \mathrm{d}\mathbf{R}\frac{\kappa_l^*(\mathbf{R}, t)\kappa_m(\mathbf{R}, t)}{|\mathbf{R}_{\mathrm{h}k} - \mathbf{R}|}\right) \tag{10.56}$$

is the heavy nuclei-proton interaction, and

$$\begin{aligned} W^0_{IJKL}(t) &= \delta_{IK}\left(\sum_{kl}\mathscr{E}^{\mathrm{e}}_{JLkl}\langle\varphi_k(t)|h_\mathrm{e}(t)|\varphi_l(t)\rangle + \frac{1}{2}\sum_{klmn}\mathscr{F}^{\mathrm{e}}_{JLklmn}\langle\varphi_m(t)|g^{\mathrm{ee}}_{kl}(t)|\varphi_n(t)\rangle\right) \\ &+ \delta_{JL}\left(\sum_{kl}\mathscr{E}^{\mathrm{p}}_{IKkl}\langle\kappa_k(t)|h_\mathrm{p}(t)|\kappa_l(t)\rangle + \frac{1}{2}\sum_{klmn}\mathscr{F}^{\mathrm{p}}_{IKklmn}\langle\kappa_m(t)|g^{\mathrm{pp}}_{kl}(t)|\kappa_n(t)\rangle\right) \\ &+ \sum_{lmrs}\mathscr{E}^{\mathrm{e}}_{JLlm}\mathscr{E}^{\mathrm{p}}_{IKrs}\langle\varphi_l(t)|g^{\mathrm{ep}}_{rs}(t)|\varphi_m(t)\rangle \end{aligned} \tag{10.57}$$

is an $\overline{\mathbf{R}}_\mathrm{h}$-independent matrix. The CI coefficients $C_{IJ}(\overline{\mathbf{R}}_\mathrm{h}, t)$ are not orthonormal in general, that is, $\int \mathrm{d}\overline{\mathbf{R}}_\mathrm{h}C^*_{IJ}(\overline{\mathbf{R}}_\mathrm{h}, t)C_{KL}(\overline{\mathbf{R}}_\mathrm{h}, t) \neq 0$ for $IJ \neq KL$. However, the total wave function is normalized as

$$\sum_{I=1}^{L_\mathrm{e}}\sum_{J=1}^{L_\mathrm{p}}\int \mathrm{d}\overline{\mathbf{R}}_\mathrm{h}C^*_{IJ}(\overline{\mathbf{R}}_\mathrm{h}, t)C_{IJ}(\overline{\mathbf{R}}_\mathrm{h}, t) = 1. \tag{10.58}$$

We remark here that the equations of motion (10.31), (10.46), and (10.54) presented above are the same as those given in the original publication [15] even though the notations are different.

The interaction potentials W^{xy}_{lm} (where $x, y = \mathrm{e}, \mathrm{p}, \mathrm{h}$) in general arise from the inter-particle Coulomb potentials (repulsive or attractive), but the Coulomb potentials always appear as those averaged over the particle distributions. For example, in the case of the electron-proton interaction term $W^{\mathrm{ep}}_{lm}(\mathbf{r}, t)$ defined in (10.36), the Coulomb potential $g^{\mathrm{ep}}_{pq}(\mathbf{r}, t)$ (defined in (10.37)) is not the bare Coulomb potential $-e^2/(4\pi\varepsilon_0|\mathbf{r} - \mathbf{R}|)$, but is that averaged over the protonic orbitals $\kappa_p(\mathbf{R}, t)$ and $\kappa_q(\mathbf{R}, t)$.

Even though we have presented the general equations of motion for $\overline{\mathbf{R}}_\mathrm{h}$-dependent CI coefficients as seen in (10.54), we can assume that the heavy nuclei, i.e., all nuclei except for the protons, in the molecule are immobile, which corresponds to the clamped-nuclei approximation. In this case, the coordinates of the heavy nuclei should be treated as a parameter, and the heavy-nuclei Hamiltonian $H_\mathrm{h}(\overline{\mathbf{R}}_\mathrm{h}, t)$

in (10.30) should be omitted. As a result, the first line of (10.54) involving the term $H_h(\overline{\mathbf{R}}_h, t)C_{IJ}(\overline{\mathbf{R}}_h, t)$ disappears, and the equation of motion for the CI coefficients becomes an ordinary differential equation (in t) instead of a partial differential equation. Moreover, all integrations over the heavy nuclei coordinates $\overline{\mathbf{R}}_h$ should be dropped, so that the electron-heavy nuclei potential (10.38) and the proton-heavy-nuclei potential (10.53) depend on a parameter $\overline{\mathbf{R}}_h$.

We can find the ground state wave function, that is, the Ex-MCTDHF wave function Ψ that minimizes the total energy $\mathscr{E} = \langle\Psi|H|\Psi\rangle$, by integrating the equations of motion (10.31), (10.46), and (10.54) in imaginary time [17]. This means that the time t is replaced by $-i\tau$, resulting in $i\partial/\partial t \rightarrow -\partial/\partial\tau$, so that the TDSE takes the form of a diffusion equation.

10.3.3 Applications of the Ex-MCTDHF Method

Thus far, only two reports [24, 25] have been published on the application of the Ex-MCTDHF method. Both of these applications deal with the stationary properties of ground-state wave functions. The equations of motion derived in the preceding Sect. 10.3.2 are used in this section to obtain the optimal ground state via the imaginary time propagation.

10.3.3.1 CH_3OH

In [24], the Ex-MCTDHF method was applied to the calculation of the electro-protonic ground state wave function of methanol, CH_3OH. The oxygen atom and the carbon atom were treated as heavy nuclei, and rotation of the molecule was neglected, which means that $\overline{\mathbf{R}}_h = R_{CO}$, the C–O internuclear distance. Furthermore, the clamped nuclei approximation was assumed for C and O, so that R_{CO} was treated as a parameter. Since the C atom and the O atom define a molecular axis, and the total Hamiltonian for the electrons and the protons is symmetric under rotations around this axis, cylindrical symmetry around the C–O axis can be assumed for the electronic and protonic orbitals. If we take the C–O axis to be the z-axis in the cylindrical coordinate system, we have $\mathbf{r}_k = (z_{ek}, \rho_{ek}, \phi_{ek})$ for the coordinate of the electron k, and $\mathbf{R}_{pl} = (z_{pl}, \rho_{pl}, \phi_{pl})$ for the coordinate of the proton l. We have

$$\varphi_k(\mathbf{r}_e, t) = f_k(z_e, \rho_e, t)e^{im_{ek}\phi_e} \tag{10.59}$$

for the electronic spatial orbitals, and

$$\kappa_k(\mathbf{R}_p, t) = g_k(z_p, \rho_p, t)e^{im_{pk}\phi_p} \tag{10.60}$$

for the protonic spatial orbitals. Both the electronic orbitals and the protonic orbitals were discretized using the grid method. In (10.59) and (10.60), the quantum numbers

m_{ek} and m_{pk} determine the angular momentum around the molecular axis of the respective orbital. In [24], the values $|m_{ek}| \leq 1$ and $|m_{pk}| \leq 3$ were adopted. For the electronic structure, a single closed-shell Slater determinant was adopted, meaning that $L_e = 1$ in (10.25). For the protonic wave function, a total of $M_p = 16$ spatial orbitals were employed, and determinants with the highest possible protonic spin, $S = 2$, were constructed. This means that $N_p^\alpha = 4$ and $N_p^\beta = 0$ in (10.24), and $L_p = \binom{16}{4} = 1820$ in (10.25).

The main result obtained in [24] is that the molecular structure of a polyatomic molecule such as CH_3OH in the absence of a laser field can indeed be described by the Ex-MCTDHF ansatz given in (10.25). This is a remarkable result, because the spatial distribution of the protons is not determined from the energy minimum of a potential energy surface, but is the result of the optimization of the Ex-MCTDHF ground state wave function.

The protonic structure of CH_3OH was elucidated first by calculating the 2-proton spatial distribution,

$$\Gamma_p(\mathbf{R}_{p1}, \mathbf{R}_{p2}) = \frac{1}{2} \sum_{klmn} d^p_{klmn} \kappa_k^*(\mathbf{R}_{p1}) \kappa_l(\mathbf{R}_{p1}) \kappa_m^*(\mathbf{R}_{p2}) \kappa_n(\mathbf{R}_{p2}), \tag{10.61}$$

expressed in terms of the spin-summed protonic second-order density matrix d^p_{klmn} defined in (10.49). $\Gamma_p(\mathbf{R}_{p1}, \mathbf{R}_{p2})$ represents the probability distribution of finding one proton at $\mathbf{R}_{p1}$ and another one at $\mathbf{R}_{p2}$. We can now define a conditional 1-proton distribution $D_p(\mathbf{R}_p|\mathbf{R}_0) = \Gamma_p(\mathbf{R}_p, \mathbf{R}_0)$ by fixing one of the proton positions in $\Gamma_p(\mathbf{R}_{p1}, \mathbf{R}_{p2})$ to $\mathbf{R}_0$. $D_p(\mathbf{R}_p|\mathbf{R}_0)$ represents the spatial proton distribution of the three remaining protons, given that the position of the fourth proton is fixed to $\mathbf{R}_0$.

In Fig. 10.2, we show the conditional 1-proton distribution $D_p(x_p, y_p|\mathbf{R}_0) = \int dz_p D_p(\mathbf{R}_p|\mathbf{R}_0)$ in the xy-plane, integrated along the z-axis (the C–O axis). The fixed position $\mathbf{R}_0$ of one of the protons is taken to be the most probable position of the proton in the hydroxyl group, $\mathbf{R}_0 = (z_0, \rho_0, \phi_0) = (-2.02\,a_0, 1.78\,a_0, 0)$, where $a_0 \approx 0.53$ Å denotes Bohr's radius. As expected, the proton distribution in the xy-plane exhibits three maxima, consistent with the protonic structure of the methyl group in CH_3OH. For comparison, we also show in Fig. 10.2 the conventional ball-and-stick representation of CH_3OH, corresponding to the positions of the nuclei yielding the lowest total energy on the BO potential energy surface.

The results presented in [24] show that the spatial correlation among protons in a molecule containing several protons can be correctly reproduced with an Ex-MCTDHF wave function, provided that the number of protonic orbitals included in the wave function expansion is sufficiently large. It was shown in [24] that the spatial correlations among the four protons in CH_3OH were properly described when $M_p = 16$ and $|m_{pk}| \leq 3$ as shown in Fig. 10.2a, and that the three distinct peaks seen in Fig. 10.2a, corresponding to the methyl-group protons, were not reproduced when a smaller set of $M_p = 12$ protonic orbitals including orbital angular momentum $|m_{pk}| \leq 2$ was adopted.

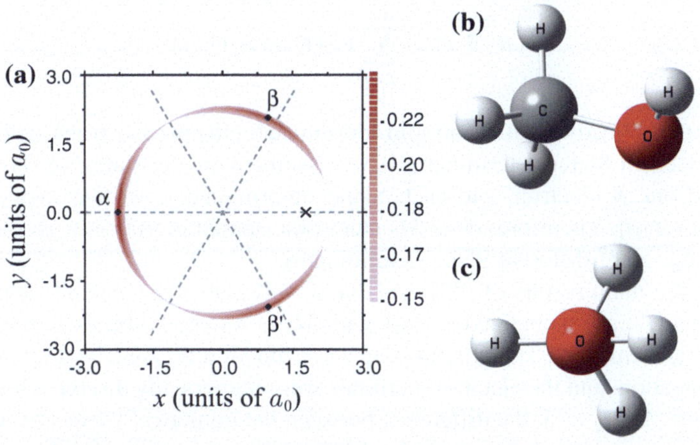

Fig. 10.2 **a** Conditional probability distribution $D_\mathrm{p}(x_\mathrm{p}, y_\mathrm{p}|\mathbf{R}_\mathrm{O})$, taken from [24]. One proton is fixed at the point marked with a cross ×, corresponding to the most probable position of the proton in the OH group. The peaks in the distribution labeled with α, β, and β′ correspond to the protons in the methyl group on the side of the C atom. **b** Ball-and-stick model of CH_3OH. The C–O axis defines the z-axis, while the xy-plane is perpendicular to the C–O axis. **c** Ball-and-stick model of CH_3OH oriented so that the C–O axis (z-axis) becomes perpendicular to the plane of the paper. The rightmost proton is the proton on the O atom side, corresponding to the position of the cross × in panel (**a**)

10.3.3.2 H_2

In [25], the Ex-MCTDHF method was applied to a one-dimensional model of an H_2 molecule. The one-dimensional model means that all particles, that is, two protons and two electrons, are restricted to move along one spatial dimension. A soft-core potential

$$V_\mathrm{SC}(r) = \pm\frac{e^2}{4\pi\varepsilon_0}\frac{1}{\sqrt{r^2+a^2}} \tag{10.62}$$

with soft-core parameter a is used for describing the attractive and repulsive potentials instead of the Coulomb potential. This model of H_2 contains only three degrees of freedom, that is, one for the vibrational motion, and two for the motion of the two electrons. Therefore, it is feasible to compute the total wave function of the system directly in a numerically exact way without making a product expansion as in the Ex-MCTDHF method. We may therefore compare the results of the numerically exact calculation and the results obtained by the Ex-MCTDHF method and examine the accuracy of the Ex-MCTDHF method in a rigorous way.

In the case of H_2, it is not necessary to use Slater determinants for the description of the protonic motion because there is only one vibrational degree of freedom represented by the internuclear distance R. Therefore, we can set $L_\mathrm{p} = 1$, $\Lambda_I = 1$, $\overline{\mathbf{R}}_\mathrm{h} = R$, and $C_{IJ}(\overline{\mathbf{R}}_\mathrm{h}, t) = C_I(R, t)$ in (10.25), so that the total wave function for H_2 is written as

$$\Psi(R,\overline{x},t) = \sum_{I=1}^{L_e} C_I(R,t)\Phi_I(\overline{x},t), \tag{10.63}$$

where $\overline{x}$ is the collective spatial and spin coordinate for the two electrons, and $\Phi_I(\overline{x},t)$ is a two-electron Slater determinant. If we assume a singlet state, we should have one α and one β electron, and each Slater determinant is written as $\Phi_I(\overline{x},t) = |\varphi_{I^\alpha}(t)\alpha\, \varphi_{I^\beta}(t)\beta|$. Therefore, once M_e electronic spatial orbitals are given, we can construct $L_e = M_e^2$ different Slater determinants.

One of the main results of [25] was that a well-converged ground state of one-dimensional H_2 can be obtained already with $M_e = 3$ spatial orbitals, corresponding to $L_e = 9$ determinants. The ground state wave function was obtained by imaginary time-propagation, and the electronic orbitals were numerically discretized using the grid method. At $M_e = 3$, the difference between the numerically exact ground state energy $\mathscr{E}_0^{(\text{exact})} = -39.34$ eV and the Ex-MCTDHF energy $\mathscr{E}_0^{(\text{Ex-MCTDHF})}$ was found to be $\mathscr{E}_0^{(\text{Ex-MCTDHF})} - \mathscr{E}_0^{(\text{exact})} \approx 10$ meV, and at $M_e = 5$, $\mathscr{E}_0^{(\text{Ex-MCTDHF})} - \mathscr{E}_0^{(\text{exact})} \approx 0.5$ meV.

Another feature of the Ex-MCTDHF wave function pointed out in [25] is that the expansion (10.63) allows us to represent the total wave function in a compact form, in the sense that the number of parameters needed to represent the wave function can be smaller than the number of parameters needed for describing a wave function using the Born-Huang expansion. If we assume for simplicity that both an electronic spatial orbital $\varphi_k(r,t)$ and a CI coefficient $C_I(R,t)$ are discretized using $\mathscr{N}$ grid points, the total number of parameters needed to specify the Ex-MCTDHF wave function becomes $\mathscr{N}_{\text{tot}} = L_e\mathscr{N} + M_e\mathscr{N} = M_e\mathscr{N}(M_e+1)$. On the other hand, in the case of one-dimensional H_2, when we employ a Born-Huang expansion to describe the total wave function,

$$\Psi_{\text{BH}}(R,\overline{x},t) = \sum_{I=1}^{L_e} C_I^{\text{BH}}(R,t)\Phi_I^{\text{BH}}(\overline{x};R), \tag{10.64}$$

we would need $\mathscr{N}_{\text{tot}}^{\text{BH}} = L_e\mathscr{N} + M_e\mathscr{N}^2 = M_e\mathscr{N}(M_e+\mathscr{N})$ parameters, because each spatial orbital $\varphi_k(r;R)$ is a function of both r and R, and therefore needs $\mathscr{N}^2$ grid points for the discretization. In practical calculations, the number of grid points is typically $\mathscr{N} > 10^2$, and the number of orbitals is typically $M_e \leq 10$. Therefore, the number of parameters $\mathscr{N}_{\text{tot}}^{\text{BH}}$ required in the Born-Huang expansion (10.64) is much larger than the number of parameters $\mathscr{N}_{\text{tot}}$ required in the Ex-MCTDHF expansion (10.63). If we assume that $\mathscr{N} \gg 1$ and $M_e \ll \mathscr{N}$, then $\mathscr{N}_{\text{tot}}^{\text{BH}}/\mathscr{N}_{\text{tot}} \approx \mathscr{N} \gg 1$.

10.4 Related Methods

In this section, we give a brief account on methods similar to the Ex-MCTDHF method, describing the coupled, time-dependent motion of both electrons and nuclei in a molecule in order that electronic and vibrational excitation are treated with a larger extent of flexibility than the BO approximation.

10.4.1 Multiconfiguration Time-Dependent Hartree Method

The multiconfiguration time-dependent Hartree (MCTDH) method [19, 20, 26] is a method originally invented for the simulation of vibrational motion of polyatomic molecules. The total vibrational wave function is written as

$$\Psi(q_1, \ldots, q_n, t) = \sum_{i_1,\ldots,i_n} C_{i_1\ldots i_n}(t)\eta_{i_1}^{(1)}(q_1, t)\ldots\eta_{i_n}^{(n)}(q_n, t), \tag{10.65}$$

where q_j represents the coordinate for the vibrational mode j, and there is a set of time-dependent basis functions $\{\eta_k^{(j)}(q_k, t)\}$ for each mode. We have assumed that there are a total of n modes. The equations of motion for the coefficients $C_{i_1\ldots i_n}(t)$ and the basis functions $\eta_k^{(j)}(q_k, t)$ can be derived using the time-dependent variational principle [20], similarly to the procedure described in Sect. 10.3.2. The MCTDH method can be used to simulate the vibrational motion of large, many-dimensional systems, as has been demonstrated in the simulation of the 15-dimensional vibrational motion of $H_5O_2^+$ [27] and the 21-dimensional vibrational motion of $C_3H_4O_2$ [28].

An extension of the MCTDH method is called the multi-layer MCTDH method [26, 29, 30]. In the multi-layer MCTDH method, the vibrational coordinates are combined into K groups of combined coordinates $\overline{\mathbf{Q}}_j$ as [26]

$$\overline{\mathbf{Q}}_1 = (q_1, \ldots, q_{k_1}),\ \overline{\mathbf{Q}}_2 = (q_{k_1+1}, \ \ldots \ , q_{k_1+k_2}), \ldots, \overline{\mathbf{Q}}_K = (q_{n-k_K+1}, \ldots, q_n), \tag{10.66}$$

where k_j is the number of coordinates in group j. The wave function is written in the same way as in the original MCTDH method, but in terms of the combined coordinates $\overline{\mathbf{Q}}_j$,

$$\Psi(q_1, \ldots, q_n, t) = \sum_{i_1,\ldots,i_K} C_{i_1\ldots i_K}(t)\zeta_{i_1}^{(1)}(\overline{\mathbf{Q}}_1, t)\ldots\zeta_{i_K}^{(K)}(\overline{\mathbf{Q}}_K, t). \tag{10.67}$$

The idea of the multi-layer MCTDH method is to express each time-dependent basis function $\zeta_m^{(j)}(\overline{\mathbf{Q}}_j, t)$ as a time-dependent multiconfiguration expansion,

$$\zeta_m^{(j)}(\overline{\mathbf{Q}}_j, t) = \sum_{l_1,\ldots,l_{k_j}} C_{m,l_1\ldots l_{k_j}}^{(j)}(t)\eta_{l_1}^{(1)}(q_{a_j+1}, t)\ldots\eta_{l_{k_j}}^{(k_j)}(q_{a_j+k_j}, t), \tag{10.68}$$

where $a_j = \sum_{l=1}^{j-1} k_l$. Using the multi-layer MCTDH method, we can simulate the vibrational motion of very large systems. For example, in [31], it was shown for CH_3I embedded in calix[4]resorcinarene ($C_{28}H_{24}O_8$) that the full 189-dimensional vibrational wave function as well as its electronic absorption spectra can be obtained by the multi-layer MCTDHF method.

In [32], it was shown that the expansion (10.65) can be applied to the coupled electro-nuclear motion of H_2^+, if q_1 represents the internuclear distance R and q_2 represents the coordinate r of the electron. If we assume a one-dimensional model like that described in Sect. 10.3.3.2, the total wave function becomes

$$\Psi(R, r, t) = \sum_{IJ} c_{IJ}(t)\chi_I(R, t)\varphi_J(r, t), \tag{10.69}$$

where $\chi_I(R, t)$ and $\varphi_J(r, t)$ are orbitals describing the nuclear and the electronic motion, respectively, and $c_{IJ}(t)$ is a time-dependent expansion coefficient. If we define an R-dependent CI coefficient by

$$C_J(R, t) = \sum_{I} c_{IJ}(t)\chi_I(R, t), \tag{10.70}$$

(10.69) can take the same form as (10.63),

$$\Psi(R, r, t) = \sum_{J} C_J(R, t)\varphi_J(r, t). \tag{10.71}$$

The difference of (10.71) and (10.63) is that the electronic wave function $\varphi_J(r, t)$ is a single-particle orbital in (10.71), while $\Phi_I(\overline{x}, t)$ is a two-electron Slater determinant in (10.63).

In [32], it was concluded that an accurate time-dependent wave function of H_2^+ could be obtained when $M_e \geq 8$ electronic spatial orbitals with the same number of protonic orbitals were included in the expansion (10.69). By comparing with numerically exact wave functions obtained by a direct solution of the TDSE, it was confirmed that both the time-dependent electronic and nuclear densities as well as the high-harmonic spectra were well reproduced by the MCTDH method. This conclusion was independently confirmed in [33, 34], in which methods based on the same type of multiconfiguration expansion of the wave function shown in (10.69) were used for investigating the strong-field induced dynamics in H_2^+.

10.4.2 Multi-Configuration Electron-Nuclear Dynamics Method

The multi-configuration electron-nuclear dynamics (MCEND) method was proposed by Nest in [35] for describing time-dependent coupled electron-nuclear motion. The ansatz for the total wave function is written as

$$\Psi(\overline{\mathbf{R}}, \overline{x}, t) = \sum_{IJ} C_{IJ}(t)\xi_I(\overline{\mathbf{R}}, t)\Phi_J(\overline{x}, t), \tag{10.72}$$

where $\overline{\mathbf{R}} = (q_1, \ldots, q_n)$ is a collective coordinate for the n vibrational modes of the molecule, where q_j represents the coordinate of one vibrational mode, and $\Phi_J(\overline{x}, t)$ is a Slater determinant described as a function of the collective electronic coordinate $\overline{x}$ and t. The Slater determinant is constructed from time-dependent electronic orbitals $\varphi_k(\mathbf{r}, t)$ as in (10.20). Similarly to the MCTDH method introduced in Sect. 10.4.1, the vibrational wave functions $\xi_I(\overline{\mathbf{R}}, t)$ are written as a product of time-dependent basis functions for representing the respective vibrational modes as

$$\xi_I(\overline{\mathbf{R}}, t) = \eta^{(1)}_{I_1}(q_1, t) \ldots \eta^{(n)}_{I_n}(q_n, t). \tag{10.73}$$

The MCEND wave function ansatz (10.72) is similar to the Ex-MCTDHF ansatz (10.25) if we regard the coordinate $\overline{\mathbf{R}}_{\text{h}}$ of the heavy nuclei as a constant. However, it should be noted that the nuclear motion is described in terms of vibrational modes in the MCEND method, whereas the motion of the protons is described by protonic orbitals in the Ex-MCTDHF method. For this reason, the Ex-MCTDHF method is considered to be suited for the simulation of molecules containing many (>3) protons as well as for the discussion of quantum effects arising from the fermionic nature of the protons, while the MCEND method could be suited for the simulation of small molecules having only a few vibrational modes. For diatomic molecules with only one vibrational degree of freedom, the nuclear motion is treated in exactly the same manner in the MCEND method and the Ex-MCTDHF method. In [36], the MCEND method was applied to investigate the time-dependent dynamics of LiH in an ultrashort laser pulse.

10.4.3 MCTDHF Method for Diatomic Molecules

Haxton et al. [37] presented a modified version of the MCTDHF method, in which the vibrational motion in a diatomic molecule is treated quantum mechanically in addition to the electronic degrees of freedom. The total wave function is written in a form similar to the Born-Huang expansion (see (10.8)) as

$$\Psi(R, \overline{x}, t) = \sum_{J} \chi_J(R, t)\Phi_J(\overline{x}, t; R), \tag{10.74}$$

where R is the internuclear distance, $\chi_J(R, t)$ is a nuclear wave function, and $\Phi_J(\overline{x}, t; R)$ is a time-dependent Slater determinant, which depends parametrically on the internuclear distance R. This parametric dependence on R makes this method different from the Ex-MCTDHF method, as can be seen from the comparison of (10.74) with (10.63). The Slater determinants are constructed from time-dependent electronic orbitals as in (10.20), but the spatial orbitals $\varphi_k(\mathbf{r}, t; R)$ depend parametrically on R in this case. However, differently from the Born-Huang expansion, where the electronic orbitals are used to construct electronic eigenfunctions of the electronic Hamiltonian at each R (see (10.7)), in the method presented in [37], the R-dependence of the electronic spatial orbitals arises from the R-dependent basis set adopted in the expansion of the orbitals. This means that each spatial electronic orbital can be written as

$$\varphi_k(\mathbf{r}, t; R) = \sum_i c_{ki}(t) F_i(\mathbf{r}; R), \tag{10.75}$$

where $c_{ki}(t)$ is a time-dependent coefficient which is independent of R, and $F_i(\mathbf{r}; R)$ is an R-dependent basis function. The finite element method and the discrete variable representation [38] in the prolate spheroidal coordinate system were used to construct the basis functions $F_i(\mathbf{r}; R)$ in [37].

The MCTDHF method for diatomic molecules was employed in [37] to calculate accurate vibronic eigenstates of HD^+, HD, H_2, and LiH. In [39], this method was applied to the calculation of the cross section of dissociative photoionization of H_2^+. It was found that the cross section at photon energies around 30 eV, corresponding to vertical ionization, could not be well reproduced, although the cross section for large photon energies was reproduced well. To the best of our knowledge, this method has not yet been applied to coupled electro-nuclear motion in molecules in strong laser fields.

10.5 Summary

We have presented several methods that have been developed for the simulation of the coupled time-dependent motion of electrons and nuclei in molecules. In the case of the Ex-MCTDHF method, the equations of motion were derived and presented in a compact form. The two examples to which the Ex-MCTDHF method was applied have been introduced, that is, the calculations of the ground state electro-protonic wave functions of CH_3OH and H_2. In the case of CH_3OH, it was shown that the spatial proton distribution corresponding to three protons around the C atom and one proton around the O atom was reproduced well by the Ex-MCTDHF method without using a potential energy surface. Brief overviews were also given on the three related methods, the MCTDH method, the MCEND method, and the MCTDHF method for diatomic molecules.

The real advantage of the Ex-MCTDHF method is expected to appear in the real-time propagation of molecular wave functions under the influence of short and strong laser pulses. Because the Ex-MCTDHF ansatz provides a very flexible form of the total wave function, this method is suited for the simulation of extensive structural change and dissociation of a molecule in the time domain. It is expected that the Ex-MCTDHF method will be a powerful and general method for simulating ultrafast dynamics of polyatomic molecules in intense laser fields.

Acknowledgements This work was supported by JSPS KAKENHI grants no. JP15K17805, no. JP18K05024, and no. JP15H05696.

References

1. T. Okino, Y. Furukawa, P. Liu, T. Ichikawa, R. Itakura, K. Hoshina, K. Yamanouchi, H. Nakano, Coincidence momentum imaging of ultrafast hydrogen migration in methanol and its isotopomers in intense laser fields. Chem. Phys. Lett. **423**, 220 (2006)
2. T. Ando, A. Shimamoto, S. Miura, K. Nakai, H. Xu, A. Iwasaki, K. Yamanouchi, Wave packet bifurcation in ultrafast hydrogen migration in CH_3OH^+ by pump-probe coincidence momentum imaging with few-cycle laser pulses. Chem. Phys. Lett. **624**, 78 (2015)
3. M.F. Kling, C. Siedschlag, A.J. Verhoef, J.I. Khan, M. Schultze, T. Uphues, Y. Ni, M. Uiberacker, M. Drescher, F. Krausz, M.J.J. Vrakking, Control of electron localization in molecular dissociation. Science **312**, 246 (2006)
4. M. Kremer, B. Fischer, B. Feuerstein, V.L.B. de Jesus, V. Sharma, C. Hofrichter, A. Rudenko, U. Thumm, C.D. Schröter, R. Moshammer, J. Ullrich, Electron localization in molecular fragmentation of H_2 by carrier-envelope phase stabilized laser pulses. Phys. Rev. Lett. **103**, 213003 (2009)
5. V. Tagliamonti, H. Chen, G.N. Gibson, Multielectron effects in charge asymmetric molecules induced by asymmetric laser fields. Phys. Rev. Lett. **110**, 073002 (2013)
6. X. Gong, M. Kunitski, K.J. Betsch, Q. Song, L.P.H. Schmidt, T. Jahnke, N.G. Kling, O. Herrwerth, B. Bergues, A. Senftleben, J. Ullrich, R. Moshammer, G.G. Paulus, I. Ben-Itzhak, M. Lezius, M.F. Kling, H. Zeng, R.R. Jones, J. Wu, Multielectron effects in strong-field dissociative ionization of molecules. Phys. Rev. A **89**, 043429 (2014)
7. J. McKenna, A.M. Sayler, B. Gaire, N.G. Johnson, K.D. Carnes, B.D. Esry, I. Ben-Itzhak, Benchmark measurements of H_3^+ nonlinear dynamics in intense ultrashort laser pulses. Phys. Rev. Lett. **103**, 103004 (2009)
8. H. Xu, T. Okino, K. Nakai, K. Yamanouchi, S. Roither, X. Xie, D. Kartashov, M. Schöffler, A. Baltuska, M. Kitzler, Hydrogen migration and C-C bond breaking in 1,3-butadiene in intense laser fields studied by coincidence momentum imaging. Chem. Phys. Lett. **484**, 119 (2010)
9. S. Roither, X. Xie, D. Kartashov, L. Zhang, M. Schöffler, H. Xu, A. Iwasaki, T. Okino, K. Yamanouchi, A. Baltuska, M. Kitzler, High energy proton ejection from hydrocarbon molecules driven by highly efficient field ionization. Phys. Rev. Lett. **106**, 163001 (2011)
10. M. Born, R. Oppenheimer, Zur Quantentheorie der Molekeln. Ann. Phys. **389**, 457 (1927)
11. M. Born, K. Huang, *Dynamical Theory of Crystal Lattices* (Oxford University Press, Oxford, 1954)
12. B. Jiang, J. Li, H. Guo, Potential energy surfaces from high fidelity fitting of ab initio points: the permutation invariant polynomial—neural network approach. Int. Rev. Phys. Chem. **35**, 479 (2016)
13. F. Martín, J. Fernández, T. Havermeier, L. Foucar, T. Weber, K. Kreidi, M. Schöffler, L. Schmidt, T. Jahnke, O. Jagutzki, A. Czasch, E.P. Benis, T. Osipov, A.L. Landers, A. Belkacem, M.H. Prior, H. Schmidt-Böcking, C.L. Cocke, R. Dörner, Single photon-induced symmetry breaking of H_2 dissociation. Science **315**, 629 (2007)
14. G. Sansone, F. Kelkensberg, J.F. Pérez-Torres, F. Morales, M.F. Kling, W. Siu, O. Ghafur, P. Johnsson, M. Swoboda, E. Benedetti, F. Ferrari, F. Lépine, J.L. Sanz-Vicario, S. Zherebtsov, I. Znakovskaya, A. L'Huillier, M.Y. Ivanov, M. Nisoli, F. Martín, M.J.J. Vrakking, Electron localization following attosecond molecular photoionization. Nature (London) **465**, 763 (2010)
15. T. Kato, K. Yamanouchi, Time-dependent multiconfiguration theory for describing molecular dynamics in diatomic-like molecules. J. Chem. Phys. **131**, 164118 (2009)
16. J. Zanghellini, M. Kitzler, C. Fabian, T. Brabec, A. Scrinzi, An MCTDHF Approach to multielectron dynamics in laser fields. Laser Phys. **13**, 1064 (2003)

17. T. Kato, H. Kono, Time-dependent multiconfiguration theory for electronic dynamics of molecules in an intense laser field. Chem. Phys. Lett. **392**, 533 (2004)
18. E. Lötstedt, T. Kato, K. Yamanouchi, in *Ultrafast Intense Laser Science XIII, Vol. 116 of Springer Series in Chemical Physics*, ed. by K. Yamanouchi, W.T. Hill III, G.G. Paulus (Springer International Publishing, Switzerland, 2017), pp. 15–40 (in progress)
19. H.-D. Meyer, U. Manthe, L. Cederbaum, The multi-configurational time-dependent Hartree approach. Chem. Phys. Lett. **165**, 73 (1990)
20. M. Beck, A. Jäckle, G. Worth, H.-D. Meyer, The multiconfiguration time-dependent Hartree (MCTDH) method: a highly efficient algorithm for propagating wavepackets. Phys. Rep. **324**, 1 (2000)
21. T. Helgaker, P. Jørgensen, J. Olsen, *Molecular Electronic-Structure Theory* (Wiley, Hoboken, NJ, 2000)
22. P.A.M. Dirac, Note on exchange phenomena in the Thomas atom. Math. Proc. Camb. Phil. Soc. **26**, 376 (1930)
23. P.-O. Löwdin, P.K. Mukherjee, Some comments on the time-dependent variation principle. Chem. Phys. Lett. **14**, 1 (1972)
24. T. Kato, K. Yamanouchi, Protonic structure of CH_3OH described by electroprotonic wave functions. Phys. Rev. A **85**, 034504 (2012)
25. Y. Ide, T. Kato, K. Yamanouchi, Non-Born-Oppenheimer molecular wave functions of H_2 by extended multi-configuration time-dependent Hartree-Fock method. Chem. Phys. Lett. **595–596**, 180 (2014)
26. U. Manthe, Wavepacket dynamics and the multi-configurational time-dependent Hartree approach. J. Phys. Cond. Matter **29**, 253001 (2017)
27. V. Oriol, G. Fabien, M. Hans-Dieter, Dynamics and infrared spectroscopy of the protonated water dimer. Angew. Chem. Int. Ed. **46**, 6918 (2007)
28. M.D. Coutinho-Neto, A. Viel, U. Manthe, The ground state tunneling splitting of malonaldehyde: accurate full dimensional quantum dynamics calculations. J. Chem. Phys. **121**, 9207 (2004)
29. H. Wang, M. Thoss, Multilayer formulation of the multiconfiguration time-dependent Hartree theory. J. Chem. Phys. **119**, 1289 (2003)
30. U. Manthe, A multilayer multiconfigurational time-dependent Hartree approach for quantum dynamics on general potential energy surfaces. J. Chem. Phys. **128**, 164116 (2008)
31. T. Westermann, R. Brodbeck, A.B. Rozhenko, W. Schoeller, U. Manthe, Photodissociation of methyl iodide embedded in a host-guest complex: a full dimensional (189D) quantum dynamics study of CH_3I@resorc[4]arene. J. Chem. Phys. **135**, 184102 (2011)
32. C. Jhala, M. Lein, Multiconfiguration time-dependent Hartree approach for electron-nuclear correlation in strong laser fields. Phys. Rev. A **81**, 063421 (2010)
33. A. Hanusch, J. Rapp, M. Brics, D. Bauer, Time-dependent renormalized-natural-orbital theory applied to laser-driven H_2^+. Phys. Rev. A **93**, 043414 (2016)
34. R. Anzaki, T. Sato, K.L. Ishikawa, A fully general time-dependent multiconfiguration self-consistent-field method for the electron-nuclear dynamics. Phys. Chem. Chem. Phys. **19**, 22008 (2017)
35. M. Nest, The multi-configuration electron-nuclear dynamics method. Chem. Phys. Lett. **472**, 171 (2009)
36. I.S. Ulusoy, M. Nest, The multi-configuration electron-nuclear dynamics method applied to LiH. J. Chem. Phys. **136**, (2012)
37. D.J. Haxton, K.V. Lawler, C.W. McCurdy, Multiconfiguration time-dependent Hartree-Fock treatment of electronic and nuclear dynamics in diatomic molecules. Phys. Rev. A **83**, 063416 (2011)
38. L. Tao, C.W. McCurdy, T.N. Rescigno, Grid-based methods for diatomic quantum scattering problems: a finite-element discrete-variable representation in prolate spheroidal coordinates. Phys. Rev. A **79**, 012719 (2009)
39. D.J. Haxton, K.V. Lawler, C.W. McCurdy, Qualitative failure of a multiconfiguration method in prolate spheroidal coordinates in calculating dissociative photoionization of H_2^+. Phys. Rev. A **91**, 062502 (2015)

Chapter 11
Separable Potentials Model for Atoms and Molecules in Strong Ultrashort Laser Pulses

Yu. V. Popov, A. Galstyan, B. Piraux, P. F. O'Mahony, F. Mota-Furtado, P. Decleva and O. Chuluunbaatar

Abstract In this contribution, we discuss a model based on the replacement of the potential describing the interaction of a single active electron with the nucleus or the nuclei of atoms or molecules, with a potential, separable in momentum space and consisting of several terms. Each term supports only one single electron bound state of the system. We apply this model to the description of the interaction of atomic and molecular hydrogen, hydrogen anion and water molecule with an external ultrashort laser pulse. As expected, this short range separable potential model works very well for the hydrogen negative ion due to the short range nature of it real potential. In the case of other systems, we show that, at high frequency, taking into account the

Yu. V. Popov
Skobeltsyn Institute of Nuclear Physics, Lomonosov Moscow State University, Moscow, Russia
e-mail: popov@srd.sinp.msu.ru

Yu. V. Popov · O. Chuluunbaatar
Joint Institute for Nuclear Research, Dubna, Moscow 141980, Russia
e-mail: chuka@jinr.ru

A. Galstyan (✉) · B. Piraux
Institute of Condensed Matter and Nanosciences, Université Catholique de Louvain, 2 Chemin du Cyclotron, L7.01.07, 1348 Louvain-la-Neuve, Belgium
e-mail: alexander.galstyan@uclouvain.be

B. Piraux
e-mail: bernard.piraux@uclouvain.be

P. F. O'Mahony · F. Mota-Furtado
Department of Mathematics, Royal Holloway, University of London, Egham, Surrey TW20 0EX, UK
e-mail: p.omahony@rhul.ac.uk

F. Mota-Furtado
e-mail: f.motafurtado@rhul.ac.uk

P. Decleva
Dipartimento di Scienze Chimiche e Farmaceutiche, Universita' di Trieste, Trieste, Italy
e-mail: decleva@units.it

O. Chuluunbaatar
Institute of Mathematics, National University of Mongolia, UlaanBaatar, Mongolia

K. Yamanouchi et al. (eds.), *Progress in Photon Science*, Springer Series in Chemical Physics 119, https://doi.org/10.1007/978-3-030-05974-3_11

long range interaction in the final state is equivalent to multiplying the ionisation probability by a constant factor independent of the laser parameters.

11.1 Introduction

Nonlocal two-particle separable potentials are widely used in nuclear physics. For more than two particle scattering processes, they might significantly simplify the calculation of matrix elements, in particular if it involves the evaluation of the two-body scattering amplitudes (see e.g. [1]). In some cases, one can even solve the problem analytically, e. g. when a light particle interacts with two heavy ones (see Takibayev [2]). This model was applied to study the structure of H_2^+ by the same author, but to our regret this paper exists only in Russian and does not have an online version [3]. It is important to note at this stage that in the configuration space the separable potential has a short range.

Given the simplicity of the calculations, the separable potentials can be applied to treat the laser-matter interactions. In this case, the main difficulty is the long range of the Coulomb interaction, which is usually dominating. Besides this problem, the use of separable potentials raises additional questions:

- Can a separable potential be uniquely defined?
- The observables in the presence of the electromagnetic field should be gauge invariant. In quantum physics the gauge invariance is assured by a property of locality of the potentials. What to do with separable non-local potentials?
- What is the predictive power of this model? Does it allow to find new mechanisms and processes in the domain of its validity?

These are the questions we are trying to answer in this chapter.

The model we consider here (we abbreviate it below by SPAM for Separable Potentials for Atoms and Molecules), was first introduced in [4] to describe the interaction of atomic hydrogen with a laser pulse. The separable potentials are defined in terms of the bound state wave functions, and the continuum functions are obtained by solving the stationary Schrödinger equation. Subsequently in [5–8], more complicated potentials have been considered and the calculation of different observables in the case of atomic hydrogen interacting with a laser pulse has been discussed in detail.

Electron energy and angular spectra in the case of atomic hydrogen, interacting with an external electromagnetic field of fixed frequency, have been measured [9–16]. On the other hand, a robust numerical code exists that is solving the time dependent Schrödinger equation (TDSE) by decomposing the initial wave function in a Sturmian basis and propagating it during its interaction with a laser pulse [17, 18]. Thus it is possible to compare our SPAM results to both experimental data and results of "exact" numerical calculations, such as TDSE based simulations, which may be considered as "numerical experiments". Finally, separable potentials have been also used in [19, 20], focusing, in particular, on the gauge invariance problem.

It is worth noting that the SPAM model is related to the strong field approximation (SFA) [21]. We remind here that SFA neglects the Coulomb potential after the emission of the electron when the electric field is sufficiently intense. In this case, the solution of the TDSE with a time dependent dipole interaction potential is a Volkov wave, so that the transition matrix elements in the length gauge and in the velocity gauge can be easily calculated. We can generalise it and build an iterative series in the Coulomb potential, which allows one to calculate the SFA wave packet at different orders in the Coulomb potential [22]. This approximation has, however, serious drawbacks:

- it is gauge dependent;
- the series has been shown to diverge in some situations [23].

Finally, within the single active electron (SAE) approximation, the SPAM can be applied to many-electron systems. The SAE approximation is used in the case of a single electron ionisation of a complex target by different projectiles, in particular by a laser pulse. The main idea behind the SAE approximation is that only one electron in the atomic (molecular) target is interacting with the projectile, and this is the electron which leaves the target. All other electrons are frozen. This model is working pretty well in the case of high energy projectiles, when the momentum of the ejected particle is much higher than a characteristic velocity of the target electrons, meaning that the transferred momentum has to be big. Conversely, if it is not the case, then the slow electron inelastically interacts with other electrons, and the SAE model is no longer applicable. For example, if the core electron is ejected from the atom, this vacancy will be filled with outer electrons, which could lead to cascade processes and, possibly, auto-ionisation. The SAE model would work well only if the ejection of the electron is much faster than these subsequent processes. We therefore expect that the SAE model can be successfully applied for the outer orbital ionisation.

All these features will be discussed in more detail below. Atomic units (a.u.), in which $\hbar = e = m_e = 1$, are used throughout unless otherwise specified.

11.2 Atomic Hydrogen

11.2.1 Preliminary Remarks on Gauge Invariance

The property of gauge invariance of theories which include the electromagnetic field is a central problem of such theories. This property allows one to obtain various equivalent forms of the Schrödinger equation related to each other through unitary transformations: this leads to the invariance of the physical observables. Let us recall that the Maxwell equations can be written in terms of the scalar and vector potentials, $U(\mathbf{r}, t)$ and $\mathbf{A}(\mathbf{r}, t)$. These potentials determine unambiguously the electric and magnetic fields. However, the potentials themselves are not uniquely defined. For example, two sets of potentials ($\mathbf{A}'$, U') and ($\mathbf{A}$, U), where

$$\mathbf{A}' = \mathbf{A} + \nabla f, \qquad U' = U - \frac{1}{c}\frac{\partial f}{\partial t},$$

give the same electric and magnetic fields for an arbitrary scalar function $f(\mathbf{r}, t)$.

The electric field $\mathbf{E}$ is related to the potentials as follows:

$$\mathbf{E}(\mathbf{r}, t) = -\frac{1}{c}\frac{\partial}{\partial t}\mathbf{A}(\mathbf{r}, t) - \nabla U(\mathbf{r}, t). \tag{11.1}$$

In the so-called Coulomb gauge it is assumed that

$$\mathrm{div}\mathbf{A} = 0.$$

Furthermore, one makes a physical assumption about weak dependence of the scalar potential on the coordinate within the atom, i.e.

$$U(\mathbf{r}, t) \simeq U(0, t),$$

which allows one to neglect the gradient of the scalar potential in (11.1). In addition, we write

$$\mathbf{A}(\mathbf{r}, t) \simeq \mathbf{A}(0, t) = \mathbf{A}(t).$$

which is the well known dipole approximation.

In the following, we consider a linearly polarised pulsed field which means that the vector potential can be written as $\mathbf{A}(t) = \mathbf{e}A(t)$, where $\mathbf{e}$ is the unit polarisation vector[1] and

$$A(t) = \sqrt{\frac{I}{I_0}}\sin^2\left(\pi\frac{t}{T}\right)\sin(\omega t), \quad 0 \le t \le T.$$

Here ω is field frequency, $T = \frac{2\pi N}{\omega}$ is the total pulse duration where N is the number of optical cycles and I is the peak intensity of the pulse with $I_0 = 3.5 \times 10^{16}$ W/cm^2. It is important to note that $A(t)$ must be zero before and after the laser pulse.

Consider the TDSE in velocity form (V-form) describing the interaction between an electric pulse and a hydrogen-like atom which is initially in its ground state,

$$\left[\mathrm{i}\frac{\partial}{\partial t} - \frac{1}{2}\left(-\mathrm{i}\nabla + \frac{1}{c}\,\mathbf{e}A(t)\right)^2 + \frac{Z}{r}\right]\Phi_V(\mathbf{r}, t) = 0, \quad \Phi_V(\mathbf{r}, 0) = \varphi_0(r) = \sqrt{\frac{Z^3}{\pi}}e^{-Zr},$$

$$\int \mathrm{d}^3r|\Phi_V(\mathbf{r}, t)|^2 = 1, \tag{11.2}$$

[1]The approach is not limited to linear polarisation of the external field or a particular form of the envelope.

where Z designates the nuclear charge. By using the well known unitary Göppert-Mayer transformation that relates the full wave packet in velocity and length gauges,

$$\Phi_V(\mathbf{r}, t) = \exp\left[-\mathrm{i}\frac{1}{c}A(t)(\mathbf{er})\right]\Phi_L(\mathbf{r}, t), \tag{11.3}$$

we obtain from (11.2) the length form (L-form) of the TDSE:

$$\left[\mathrm{i}\frac{\partial}{\partial t} + \frac{1}{2}\Delta - E(t)(\mathbf{er}) + \frac{Z}{r}\right]\Phi_L(\mathbf{r}, t) = 0, \quad \Phi_L(\mathbf{r}, 0) = \varphi_0(r).$$

Here $E(t) = -\frac{1}{c}\frac{\partial A(t)}{\partial t}$.

Usually one requires a good numerical algorithm to obtain a good agreement of the computed observables (level populations, electron angular and energy distributions etc.) in the L- and V- forms. In the exact theory they must be identical. In the momentum space the V-form TDSE takes the form

$$\left[\mathrm{i}\frac{\partial}{\partial t} - \frac{1}{2}(\mathbf{p} + \frac{1}{c}A(t)\mathbf{e})^2\right]\tilde{\Phi}_V(\mathbf{p}, t) + \int \frac{\mathrm{d}^3u}{(2\pi)^3}\frac{4\pi Z}{|\mathbf{p}-\mathbf{u}|^2}\tilde{\Phi}_V(\mathbf{u}, t) = 0, \tag{11.4}$$

$$\tilde{\Phi}_V(\mathbf{p}, 0) = \tilde{\varphi}_0(p) = \frac{8\sqrt{\pi Z^5}}{(p^2 + Z^2)^2}.$$

In (11.4), the function $\tilde{\Phi}_V(\mathbf{p}, t)$ designates the Fourier transform of the function $\Phi_V(\mathbf{r}, t)$. Similarly, the L-form TDSE becomes

$$\left[\mathrm{i}\frac{\partial}{\partial t} - \frac{p^2}{2} - \mathrm{i}E(t)(\mathbf{e}\cdot\nabla_p)\right]\tilde{\Phi}_L(\mathbf{p}, t) + \int \frac{\mathrm{d}^3u}{(2\pi)^3}\frac{4\pi Z}{|\mathbf{p}-\mathbf{u}|^2}\tilde{\Phi}_L(\mathbf{u}, t) = 0,$$

$$\tilde{\Phi}_L(\mathbf{p}, 0) = \tilde{\varphi}_0(p), \tag{11.5}$$

and the gauge transformation (11.3) becomes

$$\tilde{\Phi}_V(\mathbf{p}, t) = \tilde{\Phi}_L(\mathbf{p} + \frac{1}{c}A(t)\mathbf{e}, t).$$

Equations (11.4) and (11.5) are the basic ones that we use in our SPAM model.

11.2.2 Definition of Separable Potentials

First of all and based on [24, 25], it is possible to show that the kernel of the Coulomb potential may be expanded as follows

$$\frac{4\pi Z}{|\mathbf{p}-\mathbf{u}|^2} = \frac{(2\pi)^2 Z}{pu} \sum_{l=0}^{\infty} (l!)^2 \left(\frac{4qp}{q^2+p^2}\right)^{l+1} \left(\frac{4qu}{q^2+u^2}\right)^{l+1} \times$$

$$\sum_{m=-l}^{l} Y^*_{l,m}(\theta_p,\phi_p) Y_{l,m}(\theta_u,\phi_u) \sum_{n=0}^{\infty} \frac{n!}{(n+2l+1)!} C_n^{l+1}\left[\frac{q^2-p^2}{q^2+p^2}\right] C_n^{l+1}\left[\frac{q^2-u^2}{q^2+u^2}\right]. \tag{11.6}$$

In other words, this Coulomb kernel may be expanded as a sum of products of separable potentials, given by Gegenbauer polynomials $C_n^{l+1}(x)$. Note that this expansion is not unique since it depends on the arbitrary parameter q. In practice, we have to truncate the infinite expansion. In this case we deal again with a sum of the short range separable potentials, and q determines their range in the configuration space [8]. Equation (11.6) presents one way to define non-local separable potentials. Here, however, we focus on a different method.

Before describing this method in more detail, let us discuss a few important points. The approximation of a singular local potential by a non-singular non-local potential raises some questions about the validity of this approximation. Instead of the long range Coulomb potential that gives rise to an infinite number of bound states, we obtain a short range potential with a finite number of bound states. In fact, this problem exists only in theory, as in the reality the potential is always truncated. Moreover, the locality of the potential is an approximation as well. Finally, the gauge invariance is a feature of a theory with a local potential only. Separable potential theories cannot be gauge invariant. However, as we see below, separable potentials are very useful for calculations, and they allow one to investigate the dominating mechanisms. In the second method, we approximate the Coulomb potential in the following way

$$\frac{4\pi Z}{|\mathbf{p}-\mathbf{u}|^2} \approx \sum_{n=0}^{N}\sum_{l=0}^{L}\sum_{m=-l}^{l} v_{nl}(p) Y_{l,m}(\theta_p,\phi_p)\, v^*_{nl}(u) Y^*_{l,m}(\theta_u,\phi_u).$$

The components $v_{nl}(p)$ of the separable potential correspond to some eigenfunctions of the Hamiltonian with the Coulomb potential

$$\left(\varepsilon_j - \frac{p^2}{2}\right)\tilde{\varphi}_j(\mathbf{p}) + \sum_{q=1}^{N}\sum_{l=0}^{L}\sum_{m=-l}^{l} v_{ql}(p) Y_{l,m}(\theta_p,\phi_p) \times \left[\int \frac{d^3u}{(2\pi)^3} v^*_{ql}(u) Y^*_{l,m}(\theta_u,\phi_u)\tilde{\varphi}_j(\mathbf{u})\right] = 0, \tag{11.7}$$

where $\tilde{\varphi}_j(\mathbf{p}) = \tilde{\varphi}_{jl_j}(p) Y_{l_j,m_j}(\theta_p,\phi_p)$ are the Coulomb bound state eigenfunctions. By substituting in (11.7), we get

$$\left(\varepsilon_j - \frac{p^2}{2}\right)\tilde{\varphi}_{jl_j}(p) + \sum_{q=1}^{N} a_{qjl_j} v_{ql_j}(p) = 0 \tag{11.8}$$

with

$$a_{qjl_j} = \int_0^\infty \frac{u^2 \mathrm{d}u}{(2\pi)^3} v^*_{ql_j}(u)\tilde{\varphi}_{jl_j}(u). \tag{11.9}$$

We see that a given angular momentum l defines a group of separable potentials. For example, the states $j = 1s, 2s, 3s, \ldots$ enter the group $l = 0$, $j = 2p, 3p, 4p, \ldots$ enter the group $l = 1$, etc. The functions $g_j(\mathbf{p}) = \left(\varepsilon_j - \frac{p^2}{2}\right)\tilde{\varphi}_j(\mathbf{p})$ are called the vertex functions. They define the components of the separable potentials.

We denote $v_n(\mathbf{p}) = v_{nl}(p)Y_{l,m}(\theta_p, \phi_p)$ and redefine in (11.7) $\sum_{q=1}^{N}\sum_{l=0}^{L}\sum_{m=-l}^{l}$ by $\sum_{n=1}$ (do not confuse n with the principal quantum number). Equation (11.7) can be written in matrix form as follows

$$\mathbf{G} + \mathbf{AV} = 0,$$

or

$$\mathbf{V} = -\mathbf{A}^{-1}\mathbf{G},$$

where the components of $\mathbf{V}$ are the unknown $v_n(\mathbf{p})$. For the matrix $\mathbf{A}$ we obtain a matrix equation from (11.8)

$$\Gamma = \mathbf{AA}^T, \tag{11.10}$$

where $\mathbf{A}^T$ denotes the transpose matrix of $\mathbf{A}$. The elements of the matrix Γ are known and given by:

$$\Gamma_{ij} = \int \frac{\mathrm{d}^3u}{(2\pi)^3}\tilde{\varphi}^*_i(\mathbf{u})\left(\frac{1}{2}u^2 - \varepsilon_j\right)\tilde{\varphi}_j(\mathbf{u}).$$

Equation (11.10) can have an infinite number of solutions for the elements of matrix $\mathbf{A}$ within any ℓ-block, if the number of states we take into account in each block is more than one. If we choose matrix $\mathbf{A}$ to be triangular, we obtain a unique solution for the components of the separable potentials. For example, let us consider $\ell = 0$ and write from (11.8)

$$\left(\varepsilon_{1s} - \frac{p^2}{2}\right)\tilde{\varphi}_{1s}(p) + a_{11}v_{1s}(p) = 0,$$

$$\left(\varepsilon_{2s} - \frac{p^2}{2}\right)\tilde{\varphi}_{2s}(p) + a_{21}v_{1s}(p) + a_{22}v_{2s}(p) = 0.$$

This system of linear equations allows to determine components of the separable potential, and the integrals (11.9) allow to determine all unique positive coefficients

a_{11}, a_{21}, a_{22}. We have tested our approach with several cases of symmetric matrix **A**, but have not found any significant difference.

The continuum states for a separable potential follow from the definition of the potential. They are important for the calculation of the ionisation yield and the electron energy spectrum. We get from (11.7)

$$\left(\frac{k^2}{2}-\frac{p^2}{2}\right)\tilde{\varphi}^-(\mathbf{k},\mathbf{p})+\sum_{n=1}^{N}C_n(\mathbf{k})v_n(\mathbf{p})=0,\quad C_n(\mathbf{k})=\int\frac{d^3u}{(2\pi)^3}v_n^*(\mathbf{u})\tilde{\varphi}^-(\mathbf{k},\mathbf{u}). \tag{11.11}$$

and

$$\tilde{\varphi}^-(\mathbf{k},\mathbf{p})=(2\pi)^3\delta(\mathbf{k}-\mathbf{p})-\frac{2}{k^2-p^2-i0}\sum_{j=1}^{N}C_n(\mathbf{k})v_n(\mathbf{p}) \tag{11.12}$$

For the coefficients $C_n(\mathbf{k})$, we obtain the linear system of algebraic equations

$$C_n(\mathbf{k})[1+\lambda_{nn}]+\sum_{j\neq n}^{N}\lambda_{nj}C_j(\mathbf{k})=v_n^*(\mathbf{k}),$$

with

$$\lambda_{nj}=2\int\frac{d^3u}{(2\pi)^3}\frac{v_n^*(\mathbf{u})v_j(\mathbf{u})}{k^2-p^2-i0}.$$

For atomic hydrogen all the integrals can be evaluated analytically.

11.2.3 TDSE with Separable Potentials

Let us go back to the TDSE (11.4) in the momentum space and define the action

$$S(\mathbf{p},t)=\frac{1}{2}\int_0^t\left(\mathbf{p}+\frac{1}{c}A(\xi)\mathbf{e}\right)^2 d\xi.$$

We also define the functions

$$F_n(t)=\int\frac{d^3u}{(2\pi)^3}v_n^*(\mathbf{u})\tilde{\Phi}_V(\mathbf{u},t), \tag{11.13}$$

and write the solution of (11.4) as

$$\tilde{\Phi}_V(\mathbf{p},t) = e^{-\mathrm{i}S(\mathbf{p},t)}\left[\tilde{\varphi}_0(p) + \mathrm{i}\sum_{n=1}^{N} v_n(\mathbf{p}) \int_0^t \mathrm{d}\xi\, F_n(\xi) e^{\mathrm{i}S(\mathbf{p},\xi)}\right]. \tag{11.14}$$

By inserting (11.14) into (11.13), we obtain a system of coupled Volterra equations for the functions $F_n(t)$. We can write this system in matrix form as follows

$$\mathbf{F}(t) = \mathbf{F}_0(t) + \int_0^t \mathbf{K}(t,\xi)\mathbf{F}(\xi)\mathrm{d}\xi.$$

The analytical expressions for the free term $\mathbf{F}_0(t)$ and the kernel $\mathbf{K}(t,\xi)$ can be found in [8].

Let us discuss (11.14). If we turn off the external field, then, from (11.4), we obtain

$$\tilde{\Phi}_V(\mathbf{p},t) = \tilde{\varphi}_0(p)e^{-\mathrm{i}\varepsilon_0 t}. \tag{11.15}$$

It is thanks to the integral term in (11.14) that $\tilde{\Phi}_V(\mathbf{p},t)$ tends to the hydrogen ground state when the field is zero. It means that we cannot take the term $e^{-\mathrm{i}S(\mathbf{p},t)}\tilde{\varphi}_0(p)$ as a zero order approximation and build an iterative series out of it. Indeed, this free wavepacket in (11.14) disperses in the configuration space even in the absence of the field, which contradicts the meaning of a stationary state. This observation is discussed further in the context of SFA in [22].

If the separable potentials have been generated by imposing that they support N bound states, the ionisation yield $P(t)$ can be calculated as

$$P(t) = 1 - \sum_{n=1}^{N} |\langle\tilde{\varphi}_n|\tilde{\Phi}(t)\rangle|^2. \tag{11.16}$$

11.2.4 Gauge Invariance and Separable Potentials

We have noted above that the TDSE with non-local separable potentials is not gauge invariant. If we put $\tilde{\Phi}_V(\mathbf{p},t) = \tilde{\Phi}_L(\mathbf{p} + \frac{1}{c}A(t)\mathbf{e}, t)$ into (11.4), which we reproduce here for clarity

$$\left[\mathrm{i}\frac{\partial}{\partial t} - \frac{1}{2}(\mathbf{p} + \frac{1}{c}A(t)\mathbf{e})^2\right]\tilde{\Phi}_V(\mathbf{p},t) + \sum_{n=1} v_n(\mathbf{p}) \int \frac{\mathrm{d}^3u}{(2\pi)^3} v_n^*(\mathbf{u})\tilde{\Phi}_V(\mathbf{u},t) = 0, \tag{11.17}$$

we get

$$\left[\mathrm{i}\frac{\partial}{\partial t} - \frac{1}{2}p^2 - \mathrm{i}E(t)(\mathbf{e}\cdot\nabla_p)\right]\tilde{\Phi}_L(\mathbf{p},t) + \sum_{n=1} v_n(\mathbf{p} - \frac{1}{c}A(t)\mathbf{e})\times \int \frac{\mathrm{d}^3u}{(2\pi)^3} v_n^*(\mathbf{u} - \frac{1}{c}A(t)\mathbf{e})\tilde{\Phi}_L(\mathbf{u},t) = 0. \tag{11.18}$$

There are two equivalent ways to obtain (11.18) from (11.5):

$$\frac{4\pi Z}{|\mathbf{p}-\mathbf{u}|^2} \to \sum_{n=1} v_n(\mathbf{p})v_n^*(\mathbf{u}),$$

or

$$\frac{4\pi Z}{|\mathbf{p}-\mathbf{u}|^2} \to \sum_{n=1} v_n(\mathbf{p} - \frac{1}{c}A(t)\mathbf{e})v_n^*(\mathbf{u} - \frac{1}{c}A(t)\mathbf{e}).$$

A question arises: how to generate the time dependent separable potential components v_n in terms of the (time independent) bound state wave functions? To answer this question, we consider the following TDSE

$$\left[\mathrm{i}\frac{\partial}{\partial t} - \frac{1}{2}(\mathbf{p} - \frac{1}{c}A(t)\mathbf{e})^2 - \mathrm{i}E(t)(\mathbf{e}\cdot\nabla_p)\right]\tilde{\varphi}_{jL}(\mathbf{p},t) + \int \frac{\mathrm{d}^3u}{(2\pi)^3}\frac{4\pi Z}{|\mathbf{p}-\mathbf{u}|^2}\tilde{\varphi}_{jL}(\mathbf{u},t) = 0. \tag{11.19}$$

This equation has the same Coulomb spectrum when the external field is zero. It means that in the configuration space, the solution $\tilde{\varphi}_{jL}(\mathbf{r},t)$ is a Coulomb function, multiplied by the phase factor, present in Göppert-Mayer transformation (11.3), while in momentum space we have

$$\tilde{\varphi}_{jL}(\mathbf{p},t) = \tilde{\varphi}_{jV}(\mathbf{p} - \frac{1}{c}A(t)\mathbf{e})e^{-\mathrm{i}\varepsilon_j t}. \tag{11.20}$$

To the best of our knowledge, the first one who noticed that was Faisal in [26]. With time dependent separable potentials (11.19) writes

$$\left[\mathrm{i}\frac{\partial}{\partial t} - \frac{1}{2}(\mathbf{p} - \frac{1}{c}A(t)\mathbf{e})^2 - \mathrm{i}E(t)(\mathbf{e}\cdot\nabla_p)\right]\tilde{\varphi}_{jL}(\mathbf{p},t) + \sum_{n=1} v_n(\mathbf{p} - \frac{1}{c}A(t)\mathbf{e}) \int \frac{\mathrm{d}^3u}{(2\pi)^3} v_n^*(\mathbf{u} - \frac{1}{c}A(t)\mathbf{e})\tilde{\varphi}_{jL}(\mathbf{u},t) = 0. \tag{11.21}$$

By substituting $\tilde{\varphi}_{jL}(\mathbf{u}, t)$ by its expression (11.20) we obtain, after changing variables (compare with (11.7))

$$\left[\varepsilon_j - \frac{1}{2}p^2\right]\tilde{\varphi}_{jV}(\mathbf{p}, t) + \sum_{n=1} v_n(\mathbf{p}) \int \frac{d^3u}{(2\pi)^3} v_n^*(\mathbf{u})\tilde{\varphi}_{jV}(\mathbf{u}, t) = 0. \qquad (11.22)$$

Thus, if one has the TDSE (11.17) in the V-gauge, using the time-independent separable potentials, then (11.18) is its gauge partner, but with time dependent separable potentials. It is a family of gauge invariant solutions.

Equation (11.18) can be written down, however, with time independent separable potentials, defined by the same spectral functions. Equation (11.17) will be defined by time-dependent components with $\mathbf{p} + \frac{1}{c}A(t)\mathbf{e}$ as an argument. We will have another family of gauge invariant solutions as the operators in brackets of (11.17) and (11.18) differ. This is a price to pay for using separable potentials.

11.2.5 Results

From the numerical point of view, the calculations both for atoms and molecules are inexpensive. It can be seen from the form of the $F_n(t)$ function in (11.13) that the spatial dependence is treated analytically. The set of Volterra integral equations of second kind in time for F functions are solved using a block-by-block method (a grid in time), using the GPUs to evaluate the kernels $K(t, \xi)$ for different time points before propagation.

In Figs. 11.1 and 11.2 we compare the predictions of the SPAM model with the experiment for a low frequency pulse, and with TDSE calculations for a high frequency pulse correspondingly (more results are given in [5–8]). For reference, the potential in the case 1s and 2p states included is given by

$$\frac{4\pi}{|\mathbf{p}-\mathbf{u}|^2} \approx \frac{16\pi}{(p^2+1)(u^2+1)} + \frac{32\pi^2}{3}\frac{(\mathbf{p}\cdot\mathbf{u})}{(p^2+1/4)^2(u^2+1/4)^2}$$

In the experiment, presented in [15], the authors consider a hydrogen atom, interacting with a laser pulse of frequency of 0.0723 a.u. (630 nm wave length), 10 cycles pulse duration and of peak intensity of 6.5×10^{13} W/cm^2. In Fig. 11.1 a comparison of this experimental spectrum with a SPAM model calculation, with 1s and 2p states included, is presented (the experimental spectrum is normalised to the theoretical one). One can notice an agreement for the peak positions for higher electron energies ($>$ 1 a.u.). At lower energies we do not observe a good agreement for two reasons:

- the slow emitted electron is described by a plane wave which neglects the influence of the binding potential;
- the ac-Stark shift is not properly included in our SPAM model since it involves only two levels (1s and 2p).

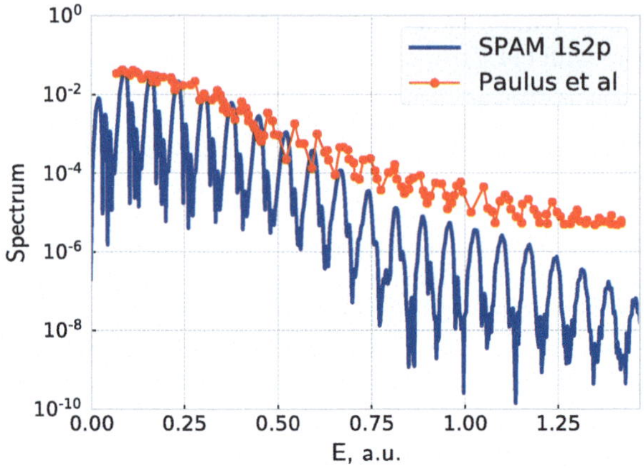

Fig. 11.1 Electron energy spectrum resulting from the interaction of atomic hydrogen with a laser pulse of frequency 0.0723 a.u. (630 nm), 10 cycles duration and with peak intensity 6.5×10^{13} W/cm^2. The red curve is the experimental data from [15], blue curve is the results of calculation with a SPAM model with 1s and 2p states included. Experimental data is normalised to the maximum of the theoretical spectrum. Theoretical vector potential is described by a sine squared envelope

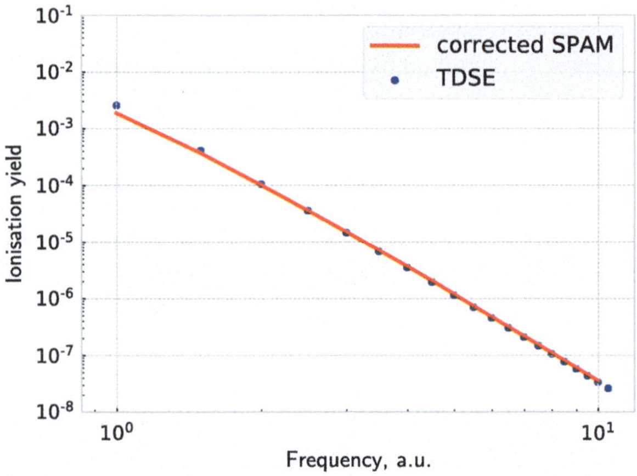

Fig. 11.2 Dependence of the ionisation yield of the H atom on the laser pulse frequency for a sine squared pulse of 2 cycle duration and 10^{14} W/cm^2 peak intensity. The solid line has been obtained by using our SPAM method which includes only 1s state and the dots show the results of the full TDSE calculation. The perturbation theory calculation in the same regime is given in [27]

In Fig. 11.2 the SPAM, containing only 1s state, is compared with the TDSE calculation results. We consider a laser pulse of 2 cycles full duration and 10^{14} W/cm^2 peak intensity. To obtain a better coincidence of the results, we divided the prediction of SPAM by 2 and obtained the so-called corrected SPAM. This correction is due to the lack of interaction in the final state of the system with the laser field, and is obtained analytically for high frequencies in [27].

11.3 Many-Electron Systems

11.3.1 SAE Approximation

The influence of the electron-electron correlations on the dynamics of the processes of excitation and scattering is actively studied in helium atom, hydrogen anion and recently in H_2 molecules [28]. All other atoms and molecules are described with a treatment where electrons are described independently and moving in the mean field of the nuclei and of the all other electrons (see e.g. Hartree-Fock method and, more recently, the multiconfiguration time-dependent Hartree-Fock method [29–31] and the configuration interaction singles method [32, 33]). The best example of such a treatment is the system of non-linear integrodifferential equations of Hartree-Fock.

This system of integrodifferential equations is very often reformulated in terms of one electron local potentials with Coulomb tail. In this case we obtain a TDSE which is particularly attractive to treat single ionisation of a molecule in a pulsed field. This is essentially the SAE approximation, when this "active" electron under the external perturbation, is leaving the atom, and all other electrons are frozen. This approximation is of a particular interest for separable potential application when the separable potentials are built in terms of vertex functions of a single electron. Usually these separable potentials are generated numerically when the one-particle wave function is taken from tables or generated with public codes.

When considering an external electromagnetic field as a perturbation that leads to the emission of an electron from an outer atomic shell, one has to use more complicated separable potentials that account for the excitation of the active electron. It allows one to account for different processes which is more adequate in this case. To find out what can be described with our SPAM model and in which frequency and intensity domains, we consider below a rather complicated process of single ionisation in a laser field, for the hydrogen anion, hydrogen molecule and water molecule.

11.3.2 Hydrogen Anion

A full solution of a TDSE for H^- is feasible at a current level of computer resources. It has two electrons like a helium atom, for which the exact dipole approximation calculations exist since the end of the last century [37–39]. For the hydrogen anion several independent calculations have been made too [35, 39–41].

From the point of view of the applicability of the SPAM model, the hydrogen anion has a major advantage: the anion becomes neutral once one of the electrons is detached, implying that the SPAM model and the short range potential approximation should work beautifully, as the potential does not have a Coulomb tail [42].

The ground state of the anion $1s^2$ is incredibly weakly bound, having binding energy of -0.0277 a.u. only [43].

The experiments on the photodetachment in a strong laser field are quite difficult to carry out, but a few exist [44–48]. The main difficulty is the depletion of negative ions by a low order detachment process at the fore front of the pulse [49, 50]. The intensity there is low, but sufficient for a many photon detachment process. The first observation of such many photon detachment, analogous to above threshold ionisation, was reported in the F^-, Cl^-, Au^- ions [44–47]. Their outer electron binding energy is much higher than the one of H^-, so it was simpler experimentally.

The SPAM model calculations for H^- are very similar to those for atomic hydrogen with only 1s state. However, the ground state wave function, as given by Yamaguchi [51], is slightly different:

$$\tilde{\varphi}(\mathbf{r}) = \frac{\sqrt{2\kappa\beta(\kappa+\beta)}}{\beta-\kappa}\frac{e^{-\kappa r}-e^{-\beta r}}{r}Y_{00}(\theta,\phi), \quad \int d^3r\, |\tilde{\varphi}(\mathbf{r})|^2 = 1, \tag{11.23}$$

with

$$\kappa = 0.235, \quad \varepsilon = -\frac{\kappa^2}{2} = -0.0276, \quad \beta = 0.913.$$

Here κ and β are fitting parameters. On the basis of this wave function we generate the corresponding separable potential in the same way, like we do it for atomic hydrogen (see (11.8) and after).

The comparison of the ionisation yield dependence on the intensity for different calculations is presented in the Fig. 11.3. It has to be mentioned, that the SPAM result in this case is not multiplied by any factor. The prediction of the SPAM model lies quite close to the results obtained by Bachau et al. [34], Scrinzi et al. [35] that take accurately into account the electron-electron correlations.

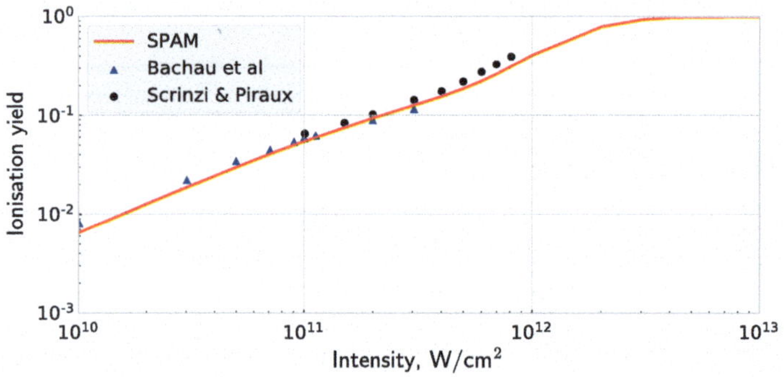

Fig. 11.3 Comparison of the SPAM ionisation yield intensity dependence with more sophisticated calculations from [34, 35] that take electron-electron interaction into account. Hydrogen anion interacts with a sine squared laser pulse of 8 cycles duration and 0.03 a.u. (1520 nm) photon energy. SPAM is presented by a red full line; calculation from [34]—blue triangles; calculation from [35]—black circles

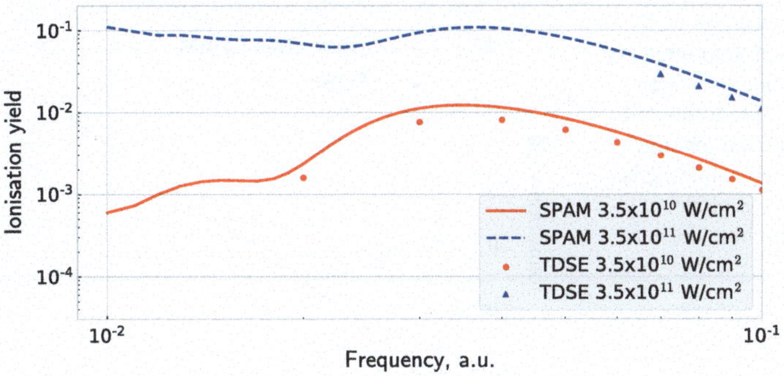

Fig. 11.4 Ionisation yield of H^- versus external field frequency. The duration of the pulse is fixed to 2 cycles, the peak intensity is 3.5×10^{10} W/cm^2 for the full red line (SPAM) and circles (full TDSE [36]) and 3.5×10^{11} W/cm^2 for the dashed blue line (SPAM) and triangles (full TDSE [36])

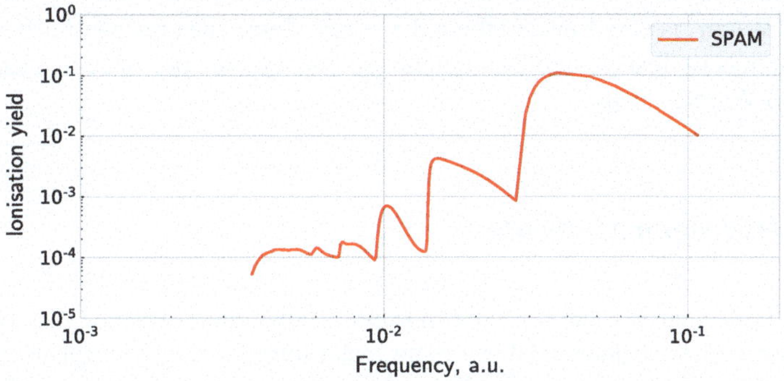

Fig. 11.5 Ionisation yield of the SPAM model of H^- versus external field frequency. The duration of the pulse is fixed to 64 cycles and the peak intensity is 3.5×10^{10} W/cm^2. The cusps result from the closing of different channels

The good agreement of the SPAM prediction with other approaches in the case of the negative ion can be explained by the absence of the Coulomb interaction in the final state. As shown analytically in [8], the difference between the SPAM prediction for atomic hydrogen with respect to the full TDSE calculation is mainly due to the short range of the approximate model potential. In the case of hydrogen anion this approximation is much closer to the real system, leading to a much better agreement.

In Fig. 11.4, we consider the ionisation yield as a function of the frequency for two intensities 3.5×10^{10} W/cm^2 and 3.5×10^{11} W/cm^2 and a total pulse duration of 2 optical cycles. The agreement with the full calculation is slightly worse, although the shape of the curves is the same. The differences are attributed to the fact that the Yamaguchi potential does not describe the electrons on the same footing as it should be.

In Fig. 11.5, the ionisation yield for H^- is plotted versus the external field frequency for a very long pulse of 64 optical cycles full duration. Since the intensity

Fig. 11.6 Ionisation yield versus external field intensity for H_2 molecule, interacting with a pulse of 6 cycles full duration and 5 a.u. photon energy. The red solid line is corrected SPAM model prediction and blue dots are TDSE [53]

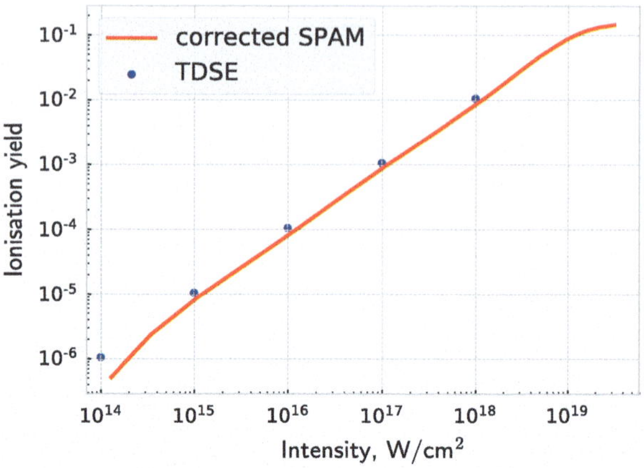

3.5×10^{10} W/cm^2 is very low and the pulse is very long, the closing of different ionisation channels is clearly visible and manifests itself by the presence of a succession of so-called Wigner cusps.[2]

11.3.3 Hydrogen Molecule

Molecular hydrogen is a very simple diatomic molecule. However, it is only very recently that physicists suceeded to obtain exact numerical results for its evolution in a laser field, taking into account all the electronic and nuclear degrees of freedom [28, 54].

In Figs. 11.6 and 11.7 we compare the predictions for the ionisation yield of the SPAM model (with a correction, discussed in detail in [27]) with the full TDSE calculation [53]. In Fig. 11.6 the laser field frequency is fixed to 5 a.u. and the total pulse duration is 6 optical cycles. The single ionisation probability, as a function of the laser pulse peak intensity shows good agreement, despite the fact that in some cases SAE fails to describe the H_2 molecule [55].

In Fig. 11.7 the laser pulse peak intensity is fixed to 4×10^{14} W/cm^2 and the total pulse duration is 2 optical cycles. Again, the single ionisation probability, as a function of the laser pulse frequency shows relatively good agreement. A strong disagreement for lower frequencies around the ionisation threshold -0.566 a.u. is attributed to the strong influence of the binding potential on the slow electrons, and the important role played by the intermediate states in this region (there are no intermediate states in this SPAM model).

[2]Cusp is an abrupt change of the photodetachment cross section which violates the Wigner's threshold law [52].

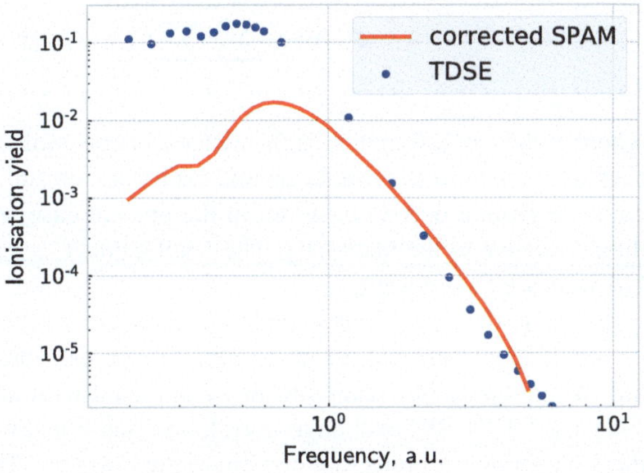

Fig. 11.7 Ionisation yield versus external field frequency for H_2 molecule interacting with a pulse of 2 cycles duration and 4×10^{14} W/cm^2 peak intensity. The red solid line is the corrected SPAM model prediction and blue dots are TDSE [53]

11.3.4 Water Molecule

In this contribution we also consider the $1b_1$ orbital (the HOMO—highest occupied molecular orbital) of water molecule. As the water molecule has 5 occupied orbitals, why do we choose the $1b_1$ orbital?

HOMO $1b_1$ is essentially a 2p atomic orbital of the oxygen atom with very little influence of each hydrogen atom. Ionization of the water HOMO orbital leaves the geometry of the molecule unchanged, thus allowing us to apply the fixed nuclei approximation as the vibrational excitation is low. Farrell et al. showed that by contrast to the ionization of the HOMO, the single ionization of the second least bound orbital $3a_1$ (HOMO-1) triggers a fast nuclear dynamics of the molecular ion and strongly excites the bending mode at photon energies around 0.54 a.u. [57]. In fact, the period of the fastest oscillation in the water molecule, namely the asymmetrical stretching of the OH bonds, is 8.9 fs which is much longer than the pulse durations we consider here. In other words, we can assume that the molecule is frozen during its interaction with the pulse.

To generalize the separable model to the case of the water molecule, we first generate the spatial part $\Phi_0(\mathbf{r})$ of the HOMO $1b_1$ in the configuration space. The molecule lies in the yz plane, with the z axis (and the polarisation vector) passing through the oxygen atom, and the hydrogens are equally distanced from this axis. The spatial part of the molecular orbital is obtained by geometry optimization with the GAMESS(US) program in the Hartree-Fock approximation [58]. Instead of Hartree-Fock one could use DFT, it would not change much, as long as one uses experimental orbital energies instead of the ones generated by GAMESS(US) as they are strongly influenced by the method while the wave functions are not.

The general expression of a molecular orbital α, denoted by $\Phi_\alpha(\mathbf{r})$, is:

$$\Phi_\alpha(\mathbf{r}) = \sum_{j=1}^{3} \sum_{\gamma_j} C_{\gamma_j,\alpha}\, G_{\gamma_j}(\mathbf{r} - \mathbf{R}_j)$$

where index j designates each nucleus in the molecule. The second summation runs over the atomic orbitals around each nucleus and G_{γ_j} is a so-called contracted gaussian from a 6–31G basis set in the present case. $\mathbf{r} - \mathbf{R}_j$ is the electronic coordinate relative to the nucleus j. The coefficients $C_{\gamma_j,\alpha}$ are the ones generated by the GAMESS(US) program.

Since, to the best of our knowledge, there is no experiment performed for a water molecule in a laser field for frequencies above the water molecule ionisation threshold, we compare our results to those obtained with another SAE calculation [56]. They use DFT LB94 exchange-correlation functional with a coulomb tail to generate the Kohn-Sham orbitals that constitute the basis set. The Hamiltonian describes the motion of the active electron in the potential formed by the nuclei and the remaining frozen electrons. In the SAE approximation this leads to the frozen Hartree-Fock, or static-exchange Hamiltonian.

Since the full final momentum wave packet is available in the SPAM model, any information about the system can be easily extracted. The absence of the intermediate states means however that regimes where these states are important, like low frequency ionisation, cannot be treated accurately. Being a SAE approach, one neglects all the dynamical interactions between the particles. Finally, the Born Oppenheimer approximation neglects all the processes related to the motion of the nuclei. Nevertheless, the SPAM model allows one to make predictions for any complex system, where the aforementioned approximations are adequate, in the single photon and ultrashort pulse regimes. The SPAM model is very scalable, so the limits on the size of the system are given by the hardware resources only. All the calculations have been performed in the 6-31G basis set. We are not aware of any significant discrepancies in our calculations that could be attributed to the incompleteness of this basis.

In order to have some idea about how accurate is the prediction of the SPAM model for high frequencies, we ran the SAE-TDSE code from [56] in the 1-photon ionisation regime and compared the ionisation yield prediction of these two SAE models (see Fig. 11.8). Laser pulse total duration was fixed to 2 optical cycles and peak intensity was fixed to 4×10^{14} W/cm^2. A significant difference between these models is the fact that for SAE-TDSE the full Coulomb potential has been taken into account in generating the one-electron orbital basis, thus we suggest to correct the SPAM result with the same factor as for atomic hydrogen. In fact, we have perfect agreement between these models for high frequencies, and poor agreement for the photon energies near the ionisation threshold. This can be explained by the fact that SAE-TDSE uses 6000 Kohn-Sham orbitals to propagate the wave function and thus supports some intermediate configurations, while in SPAM model we don't have any intermediate state at all.

We can see in Fig. 11.9 that the corrected SPAM coincides with SAE-TDSE in a wide intensity range as well. This agreement of the corrected SPAM model and SAE-TDSE approach in a wide intensity and frequency range indicates that the SPAM

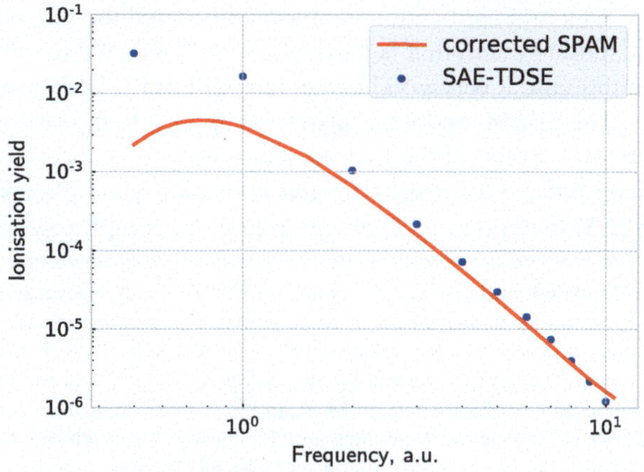

Fig. 11.8 The molecule lies in *yz* plane. Dependence of the ionisation yield of two water molecule models on the photon energy for a sine squared pulse of 2 cycles full duration and 4×10^{14} W/cm^2 peak intensity. The solid line has been obtained by using our corrected SPAM method, the dots show the results of SAE-TDSE calculation [56]

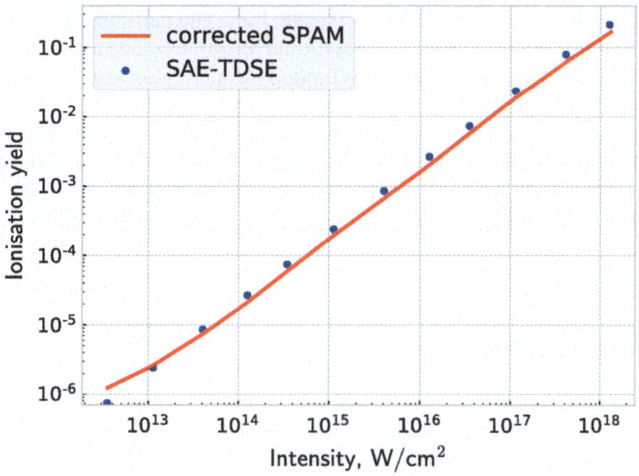

Fig. 11.9 The molecule lies in *yz* plane. Dependence of the ionisation yield of two water molecule models on the laser pulse peak intensity for a sine squared pulse of 6 cycles full duration and 5 a.u. photon energy. The solid line has been obtained by using our corrected SPAM method, the dots show the results of the SAE-TDSE calculation [56]

correction factor does not depend on intensity or on frequency in the high frequency regime.

11.4 Summary and Conclusions

We have developed the SPAM model in the case of the interaction of atomic hydrogen with a laser pulse. In particular, we have analysed in depth the domain of validity of this model and calculated electron energy spectra which have been compared to the experimental one and to those obtained by solving the TDSE numerically.

Furthermore, we have extended SPAM method to the treatment, within the SAE approximation, of the interaction of a complex quantum systems with high frequency ultrashort laser pulses.

As a first application we have considered the photodetachment of H^- which is a quantum system that is particularly well adapted to our SPAM approach. We obtain in this case a very good agreement with full TDSE results.

The SPAM model has also been applied to the case of the single ionisation of the HOMO orbital of the hydrogen and water molecules by a high frequency ultrashort laser pulse. Our results for the ionisation yield clearly show the pertinence of the SPAM method in its application to more complex systems.

Acknowledgements We are grateful to Professor A. Saenz and his group for running their code for molecular hydrogen for us to be able to compare our results with theirs. A. G. is "aspirant au Fonds de la Recherche Scientifique (F. R. S.-FNRS)". Yu.P. thanks the Université catholique de Louvain (UCL) for financially supporting several stays at the Institute of Condensed Matter and Nanosciences of the UCL. F. M. F. and P. F. O'M. gratefully acknowledge the European network COST (Cooperation in Science and Technology) through the Action CM1204 "XUV/X-ray light and fast ions for ultrafast chemistry" (XLIC) for financing several short term scientific missions at UCL. P.D. and A.G. acknowledge COST XLIC and F. R.S-FNRS for financing two short term scientific missions (STSM) in Trieste, Italy, and participation in COST XLIC meetings. The present research benefited from computational resources made available on the Tier-1 supercomputer of the Federation Wallonie-Bruxelles funded by the Region Wallonne under the Grant No. 1117545 as well as on the supercomputer Lomonosov from Moscow State University and on the supercomputing facilities of the UCL and the Consortium des Equipements de Calcul Intensif (CECI) en Federation Wallonie-Bruxelles funded by the F.R.S.-FNRS under the convention 2.5020.11. Y.P. is grateful to the Russian Foundation for Basic Research (RFBR) for financial support under the grant No. 16-02-00049-a. O.Ch. acknowledges support from the Hulubei-Meshcheryakov program JINR-Romania.

References

1. E.W. Schmid, H. Ziegelmann, *The Quantum Mechanical Three- Body Problem: Vieweg Tracts in Pure and Applied Physics*, vol. 2 (Elsevier, 2017)
2. N. Zh. Takibayev, Class of model problems in three-body quantum mechanics that admit exact solutions, Phys. At. Nucl. **71**(3), 460–468 (2008)
3. N. Zh. Takibaev, Solution of electron equation for the ion system H_2^+ (in Russian). Izv. Nats. Akad. Nauk, Ser. Fiz. Mat. **2**, 80–85 (2008)
4. S. Giraud et al., Strong field atomic ionization dynamics: role of the Coulomb potential studied by means of a model. Proc. SPIE **6165**, 61650D (2006)
5. H. M. Tetchou Nganso et al. On the role of the Coulomb potential in strong field atomic ionization dynamics. J. Electron Spectrosc. Relat. Phenom. **161**(1–3), 178–181 (2007), https://doi.org/10.1016/j.elspec.2006.10.006
6. H.M. Tetchou Nganso et al. Ionization of atoms by strong infrared fields: solution of the time-dependent Schrödinger equation in momentum space for a model based on separable potentials. Phys. Rev. A **83**(1), 013401 (2011)
7. H.M. Tetchou Nganso et al., Interaction of a model atom exposed to strong laser pulses: role of the Coulomb potential. Phys. Rev. A **87**(1), 013420 (2013)
8. A. Galstyan et al., Modelling laser-atom interactions in the strong field regime. Eur. Phys. J. D **71**(4), 97 (2017)
9. H. Rottke, K.H. Welge, Photoionization of the hydrogen atom near the ionization limit in strong electric fields. Phys. Rev. A **33**(1), 301–311 (1986)
10. H. Rottke et al., Multiphoton ionization of atomic hydrogen in intense subpicosecond laser pulses. Phys. Rev. Lett. **64**(4), 404–407 (1990)

11. M. Dörr et al., The energy spectrum of photoelectrons produced by multiphoton ionization of atomic hydrogen. J. Phys.B Atomic Mol. Opt. Phys. **25**(12), L275–L280 (1992)
12. Y. Gontier et al., Five-photon ionization of atomic hydrogen at wavelengths around the threshold for four-photon ionization. Phys. Rev. A **46**(9), 5594–5599 (1992)
13. H. Rottke et al., Atomic hydrogen in intense short laser pulses: a new series of photoelectron peaks from above-threshold ionization. J. Phys. B: Atomic, Mol. Opt. Phys. **26**(2), L15–L22 (1993)
14. H. Rottke et al., Atomic hydrogen in a strong optical radiation field. Phys. Rev. A **49**(6), 4837–4851 (1994)
15. G.G. Paulus et al., High-order above-threshold ionization of atomic hydrogen using intense, ultrashort laser pulses. J. Phys. B At. Mol. Opt. Phys. **29**(7), L249–L256 (1996)
16. M.G. Pullen et al., Experimental ionization of atomic hydrogen with fewcycle pulses. Opt. Lett. **36**(18), 3660–3662 (2011). https://doi.org/10.1364/OL.36.003660
17. E. Huens et al., Numerical studies of the dynamics of multiphoton processes with arbitrary field polarization: methodological considerations. Phys. Rev. A **55**(3), 2132–2143 (1997)
18. B. Piraux et al., Excitation of Rydberg wave packets in the tunneling regime. Phys. Rev. A **96**(4), 043403 (2017)
19. T.C. Rensink et al., Model for atomic dielectric response in strong, timedependent laser fields. Phys. Rev. A **89**(3), 033418 (2014)
20. T.C. Rensink, T.M. Antonsen, Strong-field ionization and gauge dependence of nonlocal potentials. Phys. Rev. A **94**(6), 063407 (2016), https://doi.org/10.1103/PhysRevA.94.063407
21. S.V. Popruzhenko, Keldysh theory of strong field ionization: history, applications, difficulties and perspectives.J. Phys. B Atomic, Mol. Opt. Phys. **47**(20), 204001 (2014)
22. A. Galstyan et al., Reformulation of the strong-field approximation for lightmatter interactions. Phys. Rev. A **93**(2), 023422 (2016), https://doi.org/10.1103/PhysRevA.93.023422
23. Yu. Popov et al., Strong field approximation within a Faddeev-like formalism for laser-matter interactions. Eur. Phys. J. D **71**(4), 93 (2017), https://doi.org/10.1140/epjd/e2017-70708-7
24. R. Szmytkowski, Alternative approach to the solution of the momentum-space Schroedinger equation for bound states of the N-dimensional problem. Ann. Phys. **524**(6–7), 345–352 (2012)
25. H.S. Cohl, On a generalization of the generating function for Gegenbauer polynomials. Integral Transforms Spec. Funct. **24**(10), 807–816 (2013), https://doi.org/10.1080/10652469.2012.761613
26. F.H.M. Faisal, Gauge-equivalent intense-field approximations in velocity and length gauges to all orders. Phys. Rev. A **75**(6), 063412 (2007)
27. A. Galstyan et al. Ionisation of H2O by a strong ultrashort XUV pulse: a model within the single active electron approximation. Chem. Phys. **504**, 22–30 (2018), https://doi.org/10.1016/j.chemphys.2018.02.014
28. A. Palacios, H. Bachau, F. Martin, Excitation and ionization of molecular hydrogen by ultrashort vuv laser pulses. Phys. Rev. A **75**(1), 013408 (2007)
29. J. Zanghellini et al., An MCTDHF approach to multielectron dynamics in laser fields. Laser Phys. **13**(8), 1064–1068 (2003)
30. T. Kato, H. Kono, Time-dependent multiconfiguration theory for electronic dynamics of molecules in an intense laser field. Chem. Phys. Lett. **392**(4–6), 533–540 (2004)
31. K.L. Ishikawa, T. Sato, A review on ab initio approaches for multielectron dynamics. IEEE J. Sel. Top. Quantum Electron **21**(5), 1–16 (2015)
32. N. Rohringer, A. Gordon, R. Santra, Configuration-interactionbased time-dependent orbital approach for ab initio treatment of electronic dynamics in a strong optical laser field. Phys. Rev. A **74**(4), 043420 (2006)
33. S. Pabst, R. Santra, Strong-field many-body physics and the giant enhancement in the high-harmonic spectrum of xenon. Phys. Rev. Lett. **111**(23) 233005 (2013)
34. H. Bachau et al., Applications of B-splines in atomic and molecular physics. Rep. Prog. Phys. **64**(12), 1815–1944 (2001)
35. Armin Scrinzi and Bernard Piraux, Two-electron atoms in short intense laser pulses. Phys. Rev. A **58**(2), 1310–1321 (1998)

36. G. Lagmago Kamta et al., Ionization of H^- by a strong ultrashort laser pulse. J. Phys. B Atomic Mol. Opt. Phys. **34**(5), 857–868 (2001)
37. J. Purvis et al., Multiphoton ionization of H^- and He in intense laser fields. Phys. Rev. Lett. **71**(24), 3943–3946 (1993)
38. E.S. Smyth, J.S. Parker, K.T. Taylor, Numerical integration of the time-dependent Schrödinger equation for laser-driven helium. Comput. Phys. Commun. **114**(1–3), 1–14 (1998), https://doi.org/10.1016/S0010-4655(98)00083-6
39. E. Foumouo et al., Theory of multiphoton single and double ionization of two-electron atomic systems driven by short-wavelength electric fields: an ab initio treatment Phys. Rev. A **74**(6), 063409 (2006)
40. L.A.A. Nikolopoulos, P. Lambropoulos, Time-Dependent Nonperturbative Theory of H^- in a Strong Laser Field. Phys. Rev. Lett. **82**(19), 3771–3774 (1999)
41. H.-C. Shao, F. Robicheaux , Photodetachment of H^- from intense, short, high-frequency pulses. Phys. Rev. A **93**(5), 053414 (2016)
42. G.W.F. Drake, Second bound state for the hydrogen negative ion. Phys. Rev. Lett. **24**(4), 126–127 (1970)
43. M.A. Kornberg, Multiphoton ionization of H^- at 160 nm: a study on the production of protons. Europhys. Lett. **52**(2), 130–136 (2000)
44. C. Blondel et al., Excess-photon absorption in a negative ion. J. Phys. B: At. Mol. Opt. Phys. **24**(16), 3575–3588 (1991)
45. M.D. Davidson, H.G. Muller, H.B. van Linden, van den Heuvell, Experimental observation of excess-photon detachment of negative ions. Phys. Rev. Lett. **67**(13), 1712–1715 (1991)
46. M.D. Davidson et al., Longer wavelengths require lower intensity in multiphoton detachment of negative ions. Phys. Rev. Lett. **69**(24), 3459–3462 (1992)
47. H. Stapelfeldt et al., Excess-photon detachment in the negative gold ion. Phys. Rev. Lett. **67**(13), 1731–1734 (1991)
48. C.R. Quick et al., Photodetachment of the H- ion. Nucl. Instrum. Methods Phys. Res. Sect. B Beam Interact. Mater. Atoms **56** 205–210 (1991), https://doi.org/10.1016/0168-583X(91)96007-8
49. R. Reichle, H. Helm, I.Y. Kiyan, Photodetachment of H^- in a strong infrared laser field. Phys. Rev. Lett. **87**(24), 243001 (2001)
50. R. Reichle, H. Helm, I.Y. Kiyan, Detailed comparison of theory and experiment of strong-field photodetachment of the negative hydrogen ion. Phys. Rev. A **68**(6), 063404 (2003)
51. Y. Yamaguchi, Two-nucleon problem when the potential is nonlocal but separable I. Phys. Rev. **95**(6), 1628–1634 (1954)
52. J. Slater et al., Alkali negative ions. III. Multichannel photodetachment study of Cs^- and K^-. Phys. Rev. A **17**(1), 201–213 (1978)
53. A. Saenz and his group. *Private Communication*, (2017)
54. Y. Vanne, A. Saenz, Numerical treatment of diatomic two electron molecules using a B-spline based CI method. J. Phys. B: At. Mol. Opt. Phys. **37**(20), 4101–4118 (2004)
55. M. Awasthi, A. Saenz, Breakdown of the single-activeelectron approximation for one-photon ionization of the B1Su + state of H2 exposed to intense laser fields. Phys. Rev A. **81**(6), 063406 (2010)
56. S. Petretti et al., Water molecules in ultrashort intense laser fields. Chem. Phys. **414**, 45–52 (2013), https://doi.org/10.1016/j.chemphys.2012.01.011
57. J.P. Farrell et al., Strong field ionization to multiple electronic states in water. Phys. Rev. Lett. **107**(8), 083001 (2011)
58. M.W. Schmidt et al., General atomic and molecular electronic structure system. J. Comput. Chem. **14**(11), 1347–1363 (1993), https://doi.org/10.1002/jcc.540141112

Chapter 12
Effects of Hyperfine Interaction in Atomic Photoionization

Elena V. Gryzlova and Alexei N. Grum-Grzhimailo

Abstract The chapter is devoted to theoretical consideration of the role of hyperfine interaction in atomic photoionization. The interaction between nuclear spin and an atomic shell considered to give a small correction in a variety of phenomena may have significant effects on polarization and correlation parameters of a process. We outline a theoretical approach based on statistical tensors and density matrices for determination of photoelectron angular distribution in two-colour atomic ionization accounting for their evolution caused by nuclear spin. As a practical example we consider double resonant ionization of Xe through excitation via discrete and Rydberg autoionizing states, as in the first isotope-selective experiment in VUV domain (O'Keeffe et al in Phys Rev Lett 111:243002(1)–243002(5), 2013 [44]). Variations in the angular pattern for the different isotopes and light polarization are shown. The possibility to determine the hyperfine constant is discussed.

12.1 Introduction

One of the commonest and widely used technique in quantum mechanics, starting from Born-Oppenheimer approximation in molecular physics [9] up to single-particle shell model [23] in nuclear physics is the separation of degrees of freedom. In atomic physics scientists commonly separate nuclear and atomic-shell parameters by treating their interaction as a hyperfine correction.

Even though the hyperfine interaction is widely considered to have only a small effect or correction, it is crucially important for understanding of the interplay between atomic and nuclear degrees of freedom as well as for many applications. The transition between hyperfine structure (HFS) levels of the ground state of a hydrogen atom produces famous 21 cm radiation (Fig. 12.1) which is used for imaging

E. V. Gryzlova (✉) · A. N. Grum-Grzhimailo
Skobeltsyn Institute of Nuclear Physics, Lomonosov Moscow State University, Moscow 119991, Russia
e-mail: gryzlova@gmail.com

A. N. Grum-Grzhimailo
e-mail: grum@sinp.msu.ru

K. Yamanouchi et al. (eds.), *Progress in Photon Science*, Springer Series in Chemical Physics 119, https://doi.org/10.1007/978-3-030-05974-3_12

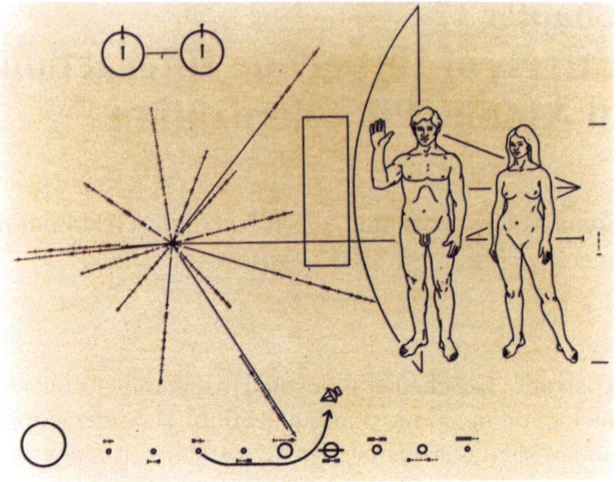

Fig. 12.1 The Pioneer plaques: a gold-anodized aluminium plaques which were placed on board the 1972 Pioneer 10 and 1973 Pioneer 11 spacecraft, featuring a pictorial message, to be intercepted by extraterrestrial life. The scaling is based on the hyperfine transition of a hydrogen atom whose wavelength is 21 cm

galaxies [10, 31, 39] and for serving as a metric to distance determination. Pumping of HFS levels is of importance in generation of coherent radiation starting from the first maser [47] to X-ray lasers [48]. The hyperfine interaction affects polarization of spectral lines used in plasma diagnostics [30]; it may change a forbidden transition to an allowed one [51, 55], or change the transition type [14], sometimes drastically decreasing life-time of an excited state [7, 52]. The atomic clocks are based on transitions between the HFS levels since the early beginning [16] and up to now [11, 21, 58]. Investigations of hyperfine interaction are actively developing because of possible implementation to quantum entanglement and quantum computations.

If an atomic nucleus possesses a nonzero spin, I, the interaction of the total angular momentum of the electronic shell, J, with the magnetic field created by the atomic nucleus leads to the hyperfine splitting. The HFS level is described by the state vector $|\,(JI)FM_F\,\rangle$, where $F,\ M_F$ denote the total angular momentum of atom and its projection, respectively. Each F-sublevel evolves with a frequency corresponding to complex quasienergy $E_F - i\Gamma_F/2$, where the real part corresponds to the energy and the imaginary part describes the width of the HFS level. The density matrix of the electron-shell momentum J, $\rho_{JM,J'M'} \equiv \langle JM \mid \rho \mid J'M'\rangle$, is defined according to the standard prescriptions by a trace of the density matrix of the total atomic angular momentum, $\rho_{FM_F,F'M_{f'}} \equiv \langle (JI)FM_F \mid \rho \mid (JI)F'M_{F'}\rangle$, over the quantum numbers of the unobserved subsystem, i.e., the atomic nucleus. We consider here an isotope with a nuclear spin $\boldsymbol{I}$. The decoupling of $\boldsymbol{F}$ and $\boldsymbol{J}$ leads to

$$\rho_{JM,J'M'} = \sum_{FF'M_FM'_FM_I} (JM_J, IM_I \mid FM_F)\,(J'M'_J, IM_I \mid F'M'_F)\,\rho_{FM_F,F'M'_F}\,, \quad (12.1)$$

where the Clebsch-Gordan coefficients are introduced. The diagonal density matrix elements evolve as $\rho_{FM_F,FM_F} \sim \exp(-\Gamma_F t)$, decaying with time, while the non-diagonal elements evolve as $\rho_{FM_F,F'M'_F} \sim \exp\left(i\omega_{FF'}t - \frac{1}{2}(\Gamma_F + \Gamma'_F)t\right)$, decaying and oscillating with time. Then each element of the density matrix of the electronic

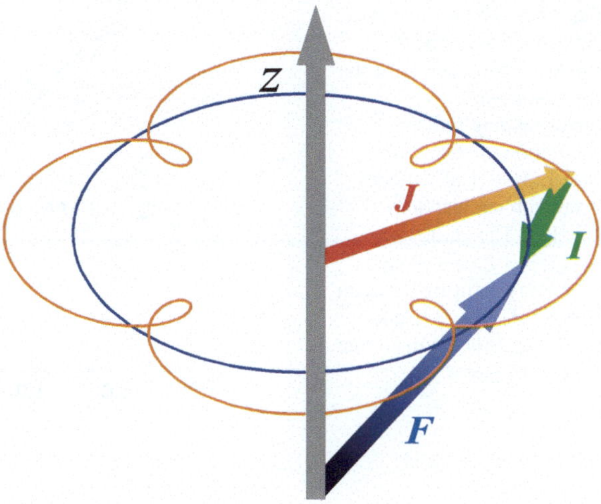

Fig. 12.2 Classical interpretation of hyperfine interaction. Projection M_F of the total angular momentum is a well-defined quantum number, presented as the cone or rotation of $\boldsymbol{F}$ around the quantization axis (gray arrow). At any moment $\boldsymbol{F}=\boldsymbol{J}+\boldsymbol{I}$, but neither M_J or M_I is well-defined and measurements at different times give different values for M_J and M_I. (See text for notations)

shell (12.1) oscillates with the number of frequencies, $\omega_{FF'} = E_F - E_{F'}$ defined by the energy splittings[1] of the HFS levels.

When HFS levels of an atom are coherently excited, a beating in electronic shell arises as result of the precession of the total angular momentum of the electronic shell in magnetic field created by the atomic nucleus (Fig. 12.2). A typical evolution of the density matrix elements of the electronic shell is shown in Fig. 12.3. The effect is similar to the zero-field beating observed due to coherent excitation of different magnetic sublevels split by an external magnetic field (Zeeman effect), and has been known since the middle of 1960s. Alexandrov and Kulyasov [1] and Hadeishi and Nierenberg [28] observed modulation of the fluorescence from atoms excited by electron impact. Alexandrov and his colleagues also observed modification of spectral profiles of the fluorescence lines after the excitation using optical laser [2, 3]. Resolving the time modulation of the fluorescence after photoexcitation was a much more demanding experimental task and the first experiment was performed later by Haroch, Paisne and Shawlow [29], two of them becoming later the Nobel laureates (Shawlow in 1981; Haroch in 2012). Investigations of the quantum beats were further stimulated by the development of time- and angular-resolved photoelectron spectroscopy after the measurements of the photoelectron angular distributions in two-photon resonant ionization became possible. In a series of papers by Berry with coauthors, a general approach for description of the quantum beats was formulated in terms of statistical tensor formalism and applied to lithium and alkaline earth metals [12, 40]. The applications finally became so numerous and various that the new types of spectroscopy were developed into a broad research field named "quantum beat spectroscopy" [27, 53].

The relative population evolution of magnetic substates $\mid J M_J \rangle$ is the evolution of the polarization of the atomic state with the total angular momentum of the electronic

[1] Atomic units are used hereafter $\hbar = e = m_e = 1$, unless otherwise specified.

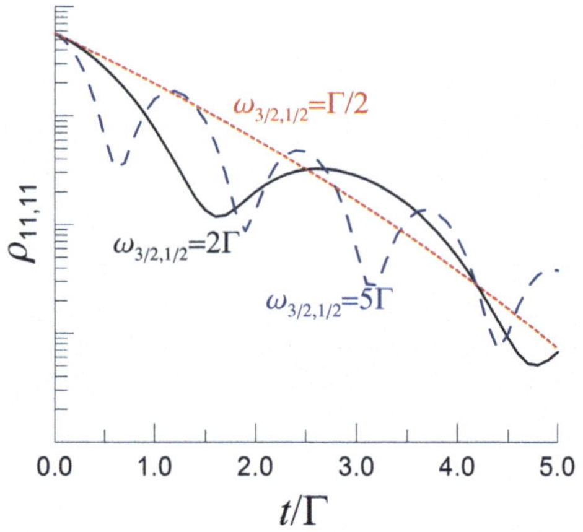

Fig. 12.3 The evolution of (unnormalized) density matrix element $\rho_{11,11}$ of the electronic shell, corresponding to the population of state with $J = 1, M = 1$ of a model atom with nuclear spin $I = 1/2$. The value $\omega_{3/2,1/2} = E_{3/2} - E_{1/2}$ is the energy splitting between HFS levels $F = 3/2$ and $F' = 1/2$. It is assumed that $\Gamma_{3/2} = \Gamma_{1/2} = \Gamma$

shell $\boldsymbol{J}$. To characterize this polarization and to obtain observable quantities, such as photoelectron angular distributions, the density matrix and statistical tensor (state multipole) technique is applied [6, 8]. The statistical tensor of the angular momentum J (which can take different values) is related to the density matrix of this momentum as

$$\rho_{kq}(J, J') = \sum_{MM'} (-1)^{J'-M'} \big(JM, J' - M' \,|\, kq\big)\, \rho_{JM,J'M'} \,, \tag{12.2}$$

where $k = |J - J'|, |J - J'| + 1, ..., J + J' - 1, J + J'$ and $-k \le q \le k$. For isotropic system only zero-rank tensor (12.2) ($k = q = 0$) is non-vanishing. The angular momentum is polarized when at least one of the statistical tensors with $k \neq 0$ is nonzero.

Polarization of an atomic state can be produced by different means. One of the methods is optical excitation, for example, by laser. The advantage of this method is that polarization of optical and IR lasers is easy to control. Excitation of an atom by light with definite polarization produces aligned or oriented state: in the dipole approximation linearly polarized radiation excites an atom from a state with $J = 0$ to the magnetic substate with $J = 1, M = 0$ (quantization axis along the electric vector of the radiation) and right(left) polarized radiation excites the same state to the substate $J = 1, M = +1$ ($J = 1, M = -1$) (quantization axis along the radiation beam).

Polarization of an atomic state may manifest itself, sometimes crucially, in different ways, for example, in the complicated angle dependence of electron emission and fluorescence, generated by the higher rank ($k > 2$) statistical tensors (12.2) [43, 56], or in the modification of ionization probability and creation of inverse population [57]. The quantum beats decrease the effect of the polarization (12.2) as time goes on and can make the polarization disappear, on which a good overview can be found in [18]. An interesting graphical approach was applied in [33, 35] to analysis of the hyperfine quantum beats. It should be mentioned that depolarization occurs due to the collisions in the reaction volume and the radiation trapping in addition to the HFS-caused depolarization. The former will not be discussed here: although their influence may be strong [15, 37, 49], it will not be the case in the experiments discussed below [44, 46].

Evolution of atomic polarization due to HFS can be determined in experiments with two conjugate polarizations, for instance, in experiment on circular magnetic dichroism [13, 20, 45]. The hyperfine interactions play a crucial role in realization of different schemes of optical pumping [19, 36, 42]. The development of synchrotron and free-electron laser facilities, producing brilliant XUV radiation, allows us to change typical targets in the experimental studies on the effects of the hyperfine interaction on the atomic processes from alkali and alkali earth elements to noble gases. Experimental techniques are progressing, and recently, even the direct observation of the quantum beats in this energy domain became feasible [22].

Natural mixtures of isotopes for neon and argon are dominated by the isotopes with zero nuclear spin, which do not generate the hyperfine splitting. Xenon containing nine stable isotopes with the nuclear spins $I = 0,\ 1/2,\ 3/2$ with the corresponding abundance ratios 53 : 26 : 21 is an appropriate target for the studies described below. Conventional experiments on measurements of the photoelectron angular distributions in resonant ionization were performed with the Xe isotopes with the natural abundance ratios [4, 38, 54]. The first measurements of photoelectron angular distributions for the selected isotopes [44] were performed at the French synchrotron SOLEIL. The first VUV photon excited $\mathrm{Xe}^*(5p^5\,{}^2P_{3/2})5d[3/2]_1$ state,[2] and then, a photon from the optical laser excited it into the $\mathrm{Xe}^{**}(5p^5\,{}^2P_{1/2})4f[5/2]_2$ Rydberg autoionizing state, which subsequently decays with emission of an electron. The key point is that these emitted electrons were detected in coincidence with a mass-selected ion from which its spin ($I = 0$, $I = 1/2$, or $I = 3/2$) can be specified. The experiment was done using the photon beams with various polarizations. The analysis of the experiment was given in [26]. Later, a similar joint experimental and theoretical investigation was performed using Kr [46] for the sequence of discrete states $[{}^2P_{3/2}]4d[1/2]_1$, $[{}^2P_{3/2}]5d[1/2]_1$, $[{}^2P_{1/2}]4d[3/2]_1$, $[{}^2P_{3/2}]5d[3/2]_1$, $[{}^2P_{1/2}]6s[1/2]_1$ with the natural isotope mixture containing 11.5% of the isotope with $I = 9/2$.

12.2 Formal Description

We are going to present here detailed description of our method on a practical example of two-photon resonant ionization of xenon. The method has also been applied to krypton [46] and can be applied to any initially unpolarized atoms or ions. The

[2] In the jK-coupling scheme, the $nl[K]_J$ indicates, for the Xe atom, that the total angular momentum j of the $5p_j^5$ core is first coupled to the orbital momentum of the excited electron $\boldsymbol{\ell}$, $\boldsymbol{j} + \boldsymbol{\ell} = \boldsymbol{K}$, with subsequent coupling of spin of this electron, $\boldsymbol{K} + \boldsymbol{s} = \boldsymbol{J}$.

theoretical approach is based on the formalism of the statistical tensors and the polarization density matrix [6, 8].

In the first step, the atom is excited by a photon with ω_1

$$\omega_1 + \mathrm{Xe}\,(5p^6) \to \mathrm{Xe}^*(5p^{5\,2}P_{3/2})5d[3/2]_1\,. \tag{12.3}$$

Initially unpolarized atoms with the total angular momentum J_0 (here $J_0 = 0$) become polarized after excitation into the intermediate state Xe* with the electron-shell angular momentum J_i (here $J_i = 1$). The excited atom is further excited by the photon ω_2, followed by the autoionizing decay and emission of a photoelectron,

$$\begin{aligned}\omega_2 + \mathrm{Xe}^*(5p^{5\,2}P_{3/2})5d[3/2]_1 &\to \mathrm{Xe}^{**}(5p^{5\,2}P_{1/2})4f[5/2]_2\,,\\ &\to \mathrm{Xe}^+\,(5p^{5\,2}P_{3/2}) + e^-(\ell j)\,.\end{aligned} \tag{12.4}$$

We emphasize that the method is applicable either with or without autoionizing states at the second step, but it is important that the final state is described in the representation $|\,(L_f S_f)J_f;\, \ell j : J\,\rangle$, where L_f, S_f, J_f are orbital, spin and total angular momenta of the residual ion, respectively, and ℓ and j are orbital and total angular momenta of the photoelectron.

The scheme of the process is depicted in Fig. 12.4. It should be noted that, while the discrete intermediate state can have either non-generate or degenerate HFS levels, the HFS of the final continuum state can be considered to be degenerate.

Such a scheme in which HFS effects take place only in the intermediate stage for the instantaneous and coherent excitation with long observation times allows us to represent its polarization as a product[3] of the resultant polarization of the electronic shell,

$$\rho_{k_1q_1}(J_i;\, J_i') = \hat{J}_0^{-1}(-1)^{J_0+J_i+k_1+1}\,\hat{J}_i\hat{J}_i'\,\rho^{(\gamma)}_{k_1q_1}\begin{Bmatrix} 1 & J_i & J_0 \\ J_i' & 1 & k_1 \end{Bmatrix} D_{J_0,J_i}\,D^*_{J_0,J_i'} \tag{12.5}$$

and the depolarization factor originating from the nuclear spin precession

$$\begin{aligned} h_k(I) = \mathscr{H}\frac{1}{\hat{I}^2}\sum_{FF'}\hat{F}^2\hat{F}'^2\begin{Bmatrix} F & F' & k \\ J_i' & J_i & I \end{Bmatrix}^2 \\ \times \int_{-\infty}^{\infty}\int_{-\infty}^{t_2} E_1(t_1)E_2(t_2)\,\exp[-i\omega_{FF'}(t_2-t_1) - \Gamma(t_2-t_1)]\,dt_1dt_2\,, \end{aligned} \tag{12.6}$$

[3]The easiest way to obtain this is to consider the system in the initial state as a product of two subspaces, the electronic shell momentum and the nuclear spin $\hat{J} \otimes \hat{I}$, and then, couple them to the total angular momentum $\hat{F}$ only after the excitation.

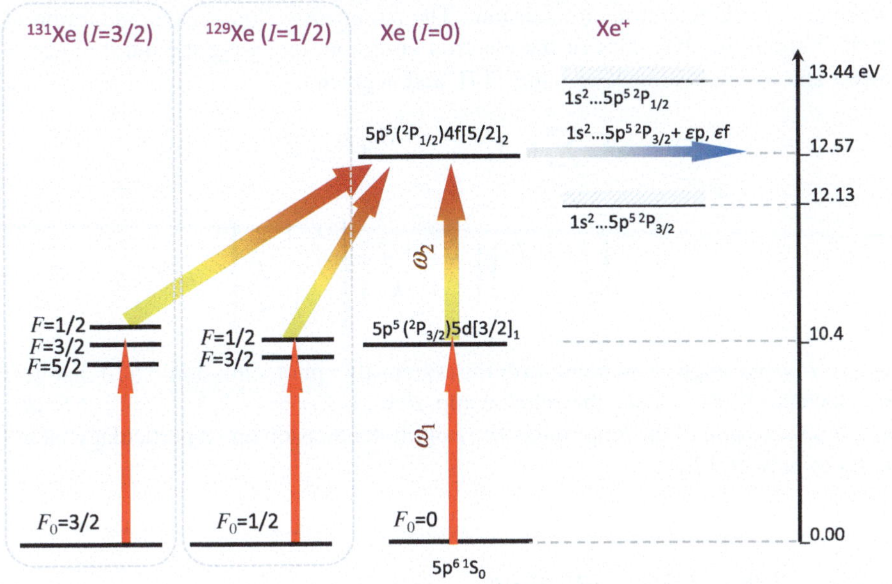

Fig. 12.4 Scheme of two-colour resonant ionization of xenon isotopes with different nuclear spin in the region of the Xe**$(5p^{5\,2}P_{1/2})4f[5/2]_2$ Rydberg autoionizing state

where the sum is taken over HFS levels $F,\ F'$; $E_1(t)$ and $E_2(t)$ are envelopes of the first (ω_1) and second (ω_2) laser fields, and the width Γ is supposed to be common for all HFS levels. In the reduced dipole matrix element, $D_{J_0 J_i} = \langle \alpha_i J_i \,||\, \hat{D} \,||\, \alpha_0 J_0 \rangle$, the initial (excited) state is characterized by the total angular momentum of the electronic shell J_0 (J_i) and a set of other quantum numbers α_0 (α_i); $\rho^{(\gamma)}_{k_1 q_1}$ is a statistical tensor of the photon (see Sect. 12.2.2). We have introduced the abbreviation $\hat{a} = \sqrt{2a+1}$ and the standard notations for nj-symbols. The depolarization factor (12.6), which includes also the depolarization induced by collisions ($\mathscr{H} \approx 1$), depends on the nuclear spin, atomic angular momentum, energy, time and splitting (see Sect. 12.2.3 for further discussion).

The photoelectron angular distribution in the two-step resonant ionization[4] can be cast into the form [5].

$$\frac{d\sigma}{d\Omega} = \frac{\pi\omega}{c} \sum_{\substack{kqk_1q_1\\k_2q_2}} h_{k_1}(I)\, B(k_1, k, k_2)\, \rho^{(\gamma)}_{k_2 q_2}\, \rho_{k_1 q_1}(J_i, J_i') \times (k_1 q_1, k_2 q_2 \,|\, kq)\, \frac{1}{\sqrt{4\pi}\,\hat{k}}\, Y_{kq}(\vartheta, \varphi)\,, \tag{12.7}$$

[4]The statistical tensor (12.5) has a dimension of probability per second, while the dimension of the two-photon cross section (12.7) is cm^2/s.

where c^{-1} is the fine-structure constant. The parameters $B(k_1, k, k_2)$ contain information about the dynamics of the electron emission, including the interference of direct and resonant ionization paths [17], and is given by:

$$B(k_1, k, k_2) = \hat{k}_1 \hat{k}_2 \sum_{\substack{\ell\ell' jj' \\ JJ'}} (-1)^{J_f+J+k-1/2} \hat{\ell}\hat{\ell}' \hat{j}\hat{j}' \hat{J}\hat{J}' (\ell 0, \ell' 0 \,|\, k0)$$

$$\times \begin{Bmatrix} k & J & J' \\ J_f & L' & L \end{Bmatrix} \begin{Bmatrix} j & \ell & 1/2 \\ \ell' & j' & k \end{Bmatrix} \begin{Bmatrix} J_i & 1 & J \\ J_i' & 1 & J' \\ k_1 & k_2 & k \end{Bmatrix} \mathscr{D}_{\ell jJ} \mathscr{D}^*_{\ell' j' J'} . \quad (12.8)$$

where J is total angular momentum of the system after photoemission. The dynamical parameters (12.8)[5] satisfy the relationship $B(k_1, k, k_2) = (-1)^{J_i - J_i' + k_1 + k_2 + k} B^*(k_1, k, k_2)$. Discussion of the amplitudes $\mathscr{D}_{\ell jJ}$ of photoemission into a particular channel is given in Sect. 12.2.1.

12.2.1 Role of Autoionization

Assume that discrete and continuum configurations of atom are mixed by interaction V, which is usually the Coulomb interaction. The decay of the autoionizing state with total angular momentum of the electronic shell J_a and other quantum numbers α_a is then described by the reduced matrix elements ('decay amplitudes') $V_{\ell jJ} = \langle \alpha_f J_f \ell j J \,||\, \hat{V} \,||\, \alpha_a J_a \rangle$. The full autoionization width is given by $\Gamma_a = 2\pi \sum_{\ell jJ} |V_{\ell jJ}|^2$, provided only one final ionic state $\alpha_f J_f$ is possible. The interference between the direct and the resonant ionization paths in the region of an isolated autoionizing state may be taken into account by presenting the dipole matrix elements in (12.8) in the form [5, 32],

$$\mathscr{D}_{\ell jJ} = e^{i\delta_{\ell jJ}} \left(\frac{\sum_{\ell' j' J'} D_{\ell' j' J'} V_{\ell' j' J'}}{\sum_{\ell' j' J'} |V_{\ell' j' J'}|^2} V_{\ell jJ} \frac{q - i}{\epsilon + i} + D_{\ell jJ} \right) , \quad (12.9)$$

where q is the Fano profile index [17], $\epsilon = 2(E - E_a)/\Gamma_a$ is the detuning from the resonance position E_a, $\delta_{\ell jJ}$ is the scattering phase in the corresponding ionization channel, and the reduced amplitudes of the direct ionization is denoted as $D_{\ell jJ} = \langle \alpha_f J_f \ell j J \,||\, \hat{D} \,||\, \alpha_i J_i \rangle$ (see (12.4) and (12.5) of [25]). The amplitude (12.9) turns into the reduced dipole direct ionization amplitude far from the resonance ($\epsilon \to \pm\infty$): $\mathscr{D}_{\ell jJ} \to e^{i\delta_{\ell jJ}} D_{\ell jJ}$.

[5]The first $6j$-symbol was missed in (6) of [26].

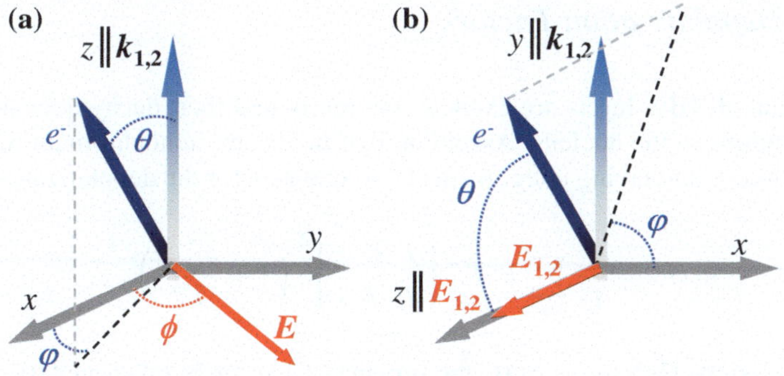

Fig. 12.5 The coordinate system for arbitrary polarization of two collinearly propagating field (**a**) and for particular case when both fields are linearly polarized in the same direction (**b**). Dashed black lines mark the projection of the direction of electron emission on the xy-plane

12.2.2 Statistical Tensors of Photon

We choose z-axis along the radiation beams $\boldsymbol{k}_{1,2}$, see Fig. 12.5a, the direction of the x-axis will be specified in the Sect. 12.3.1 (whether along synchrotron or laser polarization vector, $\boldsymbol{E}_1$ or $\boldsymbol{E}_2$, respectively). In this coordinate system a general form of the photon statistical tensors in the dipole approximation,[6] expressed via the Stocks parameters $p_1,\ p_2,\ p_3$ was presented, for example, in [6]:

$$\rho_{00}^{(\gamma)} = 1/\sqrt{3}\,,\quad \rho_{10}^{(\gamma)} = p_3/\sqrt{2}\,, \tag{12.10}$$

$$\rho_{20}^{(\gamma)} = 1/\sqrt{6}\,,\quad \rho_{2\pm2}^{(\gamma)} = -(p_1 \mp i p_2)/2\,. \tag{12.11}$$

For the radiation linearly polarized in the direction determined by the angle ϕ (see Fig. 12.5a), its statistical tensors with nonzero projection take the form $\rho_{2\pm2}^{(\gamma)} = -\exp(-2i\phi)/2$. There is one geometrically specified case when both fields are linearly polarized in the same direction. For this case it is convenient to choose the z-axis along the polarization vectors (see Fig. 12.5b). Then, only two statistical tensors are nonvanishing:

$$\rho_{00}^{(\gamma)} = 1/\sqrt{3}\,,\quad \rho_{20}^{(\gamma)} = -\sqrt{2/3}\,. \tag{12.12}$$

[6]Statistical tensors of photon in the dipole approximation are defined according to (12.2) for a particle with the angular momentum $J = 1$ having zero projection on the direction of the propagation.

12.2.3 Depolarization Factors

Suppose that all HFS levels are excited coherently and their decay does not affect internal motions in the nucleus. Assuming that fields are monochromatic with constant amplitudes and taking integrals in (12.6) one gets for the depolarization factor

$$h_k(I) = \frac{1}{\hat{I}^2} \sum_{FF'} \hat{F}^2 \hat{F'}^2 \begin{Bmatrix} F & F' & k \\ J_i' & J_i & I \end{Bmatrix}^2 \left(1 + \frac{\omega_{FF'}^2}{\Gamma^2}\right)^{-1} . \qquad (12.13)$$

For the degenerate HFS ($\omega_{FF'} = 0$), the sum in (12.13) can be taken analytically with the result $h_k(I) = 1$. Thus, there is no depolarization for overlapping HFS levels. In the opposite limit, when the HFS levels are well separated ($\omega_{FF'} \gg \Gamma$) only diagonal terms with $F = F'$ survive.

The main term of hyperfine energy correction may be expressed as $A\,(F(F+1) -I(I+1) - J(J+1))\,/2$ [34], where A is the HFS coupling constant for the level under consideration. For our case $J_i = 1$ and nuclear spin $I = 1/2,\ 3/2$ the depolarization factors are

$$h_1(1/2) = \frac{7}{9} + \frac{2}{9} \cdot \frac{1}{1+2.25\alpha} \to 7/9\,, \qquad (12.14)$$

$$h_2(1/2) = \frac{1}{3} + \frac{2}{3} \cdot \frac{1}{1+2.25\alpha} \to 1/3\,,$$

$$h_1(3/2) = \frac{19}{45} + \frac{5}{18} \cdot \frac{1}{1+2.25\alpha} + \frac{3}{10} \cdot \frac{1}{1+6.25\alpha} \to \frac{19}{45}\,, \qquad (12.15)$$

$$h_2(3/2) = \frac{37}{150} + \frac{1}{30} \cdot \frac{1}{1+2.25\alpha} + \frac{3}{10} \cdot \frac{1}{1+16\alpha} + \frac{21}{50} \cdot \frac{1}{1+6.25\alpha} \to \frac{37}{150}\,.$$

We introduced the parameter $\alpha = (A/\Gamma)^2$. Limits in (12.14), (12.15) are indicated for the case when the energy splitting between the HFS levels are much larger than the width Γ ($\alpha \gg 1$). It is worth noting that these limits give lowest possible values for the depolarization produced by the HFS. Partial overlapping of the HFS levels reduces depolarization and increases the factors $h_k(I)$. For a larger nuclear spin the limiting values of $h_k(I)$ are lower and the depolarization is more efficient.

12.3 Observable Quantities

The experiments were performed at the vacuum ultraviolet variable polarization beam line DESIRS [41], the French synchrotron source, SOLEIL, whose scientific scope includes the study of photoionization dynamics and circular dichroism. For more experimental details see [26, 44], here we concentrate on theoretical analysis of the experimental results. As mentioned above, the natural mixture of xenon contains

Table 12.1 Natural abundance (%) of the xenon isotopes and their nuclear spin I

	^{124}Xe	^{126}Xe	^{128}Xe	^{129}Xe	^{130}Xe	^{131}Xe	^{132}Xe	^{134}Xe	^{136}Xe
I	0	0	0	1/2	0	3/2	0	0	0
	0.00095	0.00089	0.0191	0.264	0.0407	0.212	0.269	0.104	0.088

two valuable species of stable isotopes with non-zero spins, ^{129}Xe ($I = 1/2$) and ^{131}Xe ($I = 3/2$) (see Table 12.1).

Photoelectron angular distributions were measured in coincidence with a mass (nuclear isotope) selected ion in the following five different combinations of the polarizations of two collinearly propagating light beams: (i) two linearly polarized light beams having the same polarization direction, (ii) two linearly polarized beams whose polarization directions are perpendicular to each other, (iii) two left-handed circularly polarized light beams, (iv) two right-handed circularly polarized light beams, and (v) a left-handed circularly polarized light beam and a right-handed circularly polarized light beam. In addition, two dichroisms integrated over the angles of the electron emission, that is, integrated circular magnetic dichroism (CMD) and integrated linear dichroism (LD) were measured.

12.3.1 Linear and Circular Dichroism and Determination of Hyperfine Constant

When both of the fields are linearly polarized, we take $z \parallel \boldsymbol{E}_1$ and ϕ is the angle between $\boldsymbol{E}_1$ and $\boldsymbol{E}_2$. For the angle-integrated photoionization cross section one can obtain from (12.7)–(12.8)

$$\sigma^{LL} = \frac{\pi\omega}{c}\left(\left[1 + h_2\frac{1+3\cos 2\phi}{20}\right]\sum_{\ell j}|\mathscr{D}_{\ell j J=2}|^2 + \left[1 - h_2\frac{1+3\cos 2\phi}{4}\right]\sum_{\ell j}|\mathscr{D}_{\ell j J=1}|^2 + \right.$$
$$\left. + \left[1 + h_2\frac{1+3\cos 2\phi}{2}\right]\sum_{\ell j}|\mathscr{D}_{\ell j J=0}|^2\right). \tag{12.16}$$

Linear dichroism is the difference in the cross-sections measured for parallel ($\phi = 0$) and perpendicular ($\phi = 90°$) polarization directions. One obtains from (12.16):

$$\mathrm{LD} = \frac{\sigma^{\parallel} - \sigma^{\perp}}{\sigma^{\parallel} + \sigma^{\perp}} \tag{12.17}$$
$$= 3h_2\frac{\sum_{\ell j}\left(\frac{1}{20}|\mathscr{D}_{\ell j J=2}|^2 - \frac{1}{4}|\mathscr{D}_{\ell j J=1}|^2 + \frac{1}{2}|\mathscr{D}_{\ell j J=0}|^2\right)}{\sum_{\ell j}\left((1+\frac{h_2}{20})|\mathscr{D}_{\ell j J=2}|^2 + (1-\frac{h_2}{4})|\mathscr{D}_{\ell j J=1}|^2 + (1+\frac{h_2}{2})|\mathscr{D}_{\ell j J=0}|^2\right)}.$$

When both fields are circularly polarized with equal (++) or opposite (+−) helicities, the angle-integrated cross section is

$$\sigma^{+\pm} = \frac{\pi\omega}{c}\left(\left[1 \pm \frac{3}{4}h_1 + \frac{h_2}{20}\right]\sum_{\ell j}|\mathscr{D}_{\ell j J=2}|^2 + \left[1 \mp \frac{3}{4}h_1 - \frac{h_2}{4}\right]\sum_{\ell j}|\mathscr{D}_{\ell j J=1}|^2 + \right.$$
$$\left. + \left[1 \mp \frac{3}{2}h_1 + \frac{h_2}{2}\right]\sum_{\ell j}|\mathscr{D}_{\ell j J=0}|^2\right) . \tag{12.18}$$

The circular magnetic dichroism follows from (12.18):

$$\mathrm{CMD} = \frac{\sigma^{++} - \sigma^{+-}}{\sigma^{++} + \sigma^{+-}} \tag{12.19}$$
$$= 3h_1 \frac{\sum_{\ell j}(|\mathscr{D}_{\ell j J=2}|^2 + |\mathscr{D}_{\ell j J=1}|^2 + 2|\mathscr{D}_{\ell j J=0}|^2)}{4\sum_{\ell j}\left((1+\frac{h_2}{20})|\mathscr{D}_{\ell j J=2}|^2 + (1-\frac{h_2}{4})|\mathscr{D}_{\ell j J=1}|^2 + (1+\frac{h_2}{2})|\mathscr{D}_{\ell j J=0}|^2\right)} .$$

Equations (12.17) and (12.19) allow to set limits for both of the two types of dichroism. As a rule, autoionization significantly amplifies the channels with $J = J_a$. In our case of resonant ionization in the vicinity of the autoionizing state $\mathrm{Xe}^{**}(5p^{5\,2}P_{1/2})4f[5/2]_2$, channels with $J = 2$ should be enhanced. Neglecting channels with $J = 0, 1$, we obtain from (12.17) and (12.19),

$$\mathrm{LD} = \frac{3\,h_2}{20 + h_2} , \tag{12.20}$$

$$\mathrm{CMD} = \frac{15\,h_1}{20 + h_2} . \tag{12.21}$$

Measured values of dichroism are expected to be lower than defined by (12.20), (12.21), because resonant approximation is not well satisfied on tails of the resonance profile. Nevertheless, the resonance $\mathrm{Xe}^{**}(5p^5\,(^2P_{1/2})4f[5/2]_2$ chosen here is quite strong with q-index from 4 to 10 in different models [26, 50]. Thus we believe that comparison of the measured dichroism with the limit values is useful. It is seen from Table 12.2 that the agreement for CMD is surprisingly good for isotopes with $I =$ 0, 1/2. As expected from (12.20), (12.21), and (12.14), (12.15), the experimental values of CMD is much larger than of LD, therefore, meaningful conclusions may be drawn only from the CMD values.

The accurate consideration of the weak channels with $J = 0, 1$ performed in multiconfiguration Hartree-Fock approximation [26] gives only a little lower values than in the resonance model. The agreement of the measured dichroism with the algebraic expectation for ^{129}Xe ($I = 1/2$) isotope means that the HFS levels for this isotope are well separated and their depolarization are the strongest.

Surprisingly, the measured dichroism for ^{131}Xe ($I = 3/2$) isotope is noticeably above the expected value. This means that the HFS levels overlap, decreasing the

Table 12.2 Measured and calculated linear and circular dichroism. Numbers in parenthesis give experimental uncertainties

Experiment				Theory		
I	0	1/2	3/2	0	1/2	3/2
LD	0.11 (2)	0.03 (3)	0.04 (3)	$\frac{1}{7} \approx 0.14$	$\frac{3}{61} \approx 0.05$	$\frac{111}{3037} \approx 0.037$
CMD	0.67 (2)	0.54 (4)	0.43 (4)	$\frac{5}{7} \approx 0.71$	$\frac{35}{61} \approx 0.57$	$\frac{950}{3037} \approx 0.31$

effect of depolarization. Based on this observation, we proposed a method to determine the hyperfine structure constant A for partly overlapping HFS levels, which cannot be resolved spectroscopically. By substituting (12.15) into (12.21) and by using the measured CMD, one can obtain a simple equation for α. Our estimate gives $A/\Gamma = 0.85 \pm 0.14$ and the depolarization factors, $h_1(3/2) = 0.58 \pm 0.04$ and $h_2(3/2) = 0.35 \pm 0.04$.

12.3.2 Photoelectron Angular Distribution

In this section we consider three possible geometries: both fields are linearly polarized in the same direction, both fields are circularly polarized with equal helicities and with opposite helicities. The last two cases imply collinearly propagating fields. For these polarization sets the photoelectron angular distribution is axially symmetric with respect to the polarization vector or propagation direction and, in each case, can be determined by two corresponding dimensionless angular anisotropy parameters, β_2^ν and β_4^ν as:

$$\frac{d\sigma}{d\Omega} = \frac{\sigma^\nu}{4\pi}\left(1 + \beta_2^\nu P_2(\cos\theta) + \beta_4^\nu P_4(\cos\theta)\right) , \qquad (12.22)$$

index $^\nu = lin,\ ++,\ +-$ indicates cases when both of the fields are either linearly polarized in one direction, or circularly polarized with equal or opposite helicities; $P_n(x)$ is the Legendre polynomial and θ measured from the symmetry axis. The second-rank polynomial produces an 'eight'-like form of the photoelectron angular distribution, while the fourth-rank polynomial is responsible for its 'butterfly'-like component. The depolarization leads to the reduction of the value of β_4.

For convenience, we introduce modified dynamical parameters instead of (12.8) as follows: $\mathscr{B}_{000} = \frac{1}{3}B(0,0,0)$, $\mathscr{B}_{202} = \frac{\sqrt{5}}{6}B(2,0,2)$, $\mathscr{B}_{022} = \frac{\sqrt{10}}{12}B(0,2,2)$, $\mathscr{B}_{220} = \frac{\sqrt{10}}{12}B(2,2,0)$, $\mathscr{B}_{222} = \frac{5\sqrt{14}}{42}B(2,2,2)$, $\mathscr{B}_{242} = \frac{\sqrt{70}}{672}B(2,4,2)$, $\mathscr{B}_{122} = \frac{\sqrt{30}}{12}B(1,2,2)$, $\mathscr{B}_{221} = \frac{\sqrt{30}}{12}B(2,2,1)$, $\mathscr{B}_{101} = \frac{\sqrt{3}}{2}B(1,0,1)$, $\mathscr{B}_{121} = \frac{\sqrt{6}}{2}B(1,2,1)$. Then, the angular anisotropy parameters can be presented in the following forms:

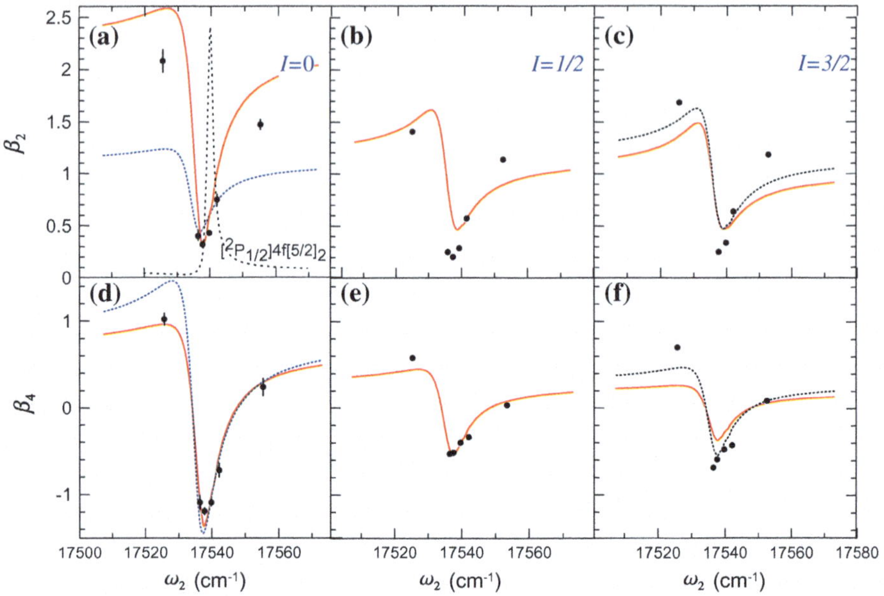

Fig. 12.6 Angular anisotropy parameters as function of second (ionizing) laser energy measured in coincidence with nuclear isotopes of the residual ion. Both fields are linearly polarized in the same direction. Multiconfiguration Hartree-Fock results (red solid) are obtained by assuming that HFS states for the isotopes with nonzero spin $I = 1/2, 3/2$ are well isolated ($\alpha \gg 1$). Blue dashed curves in the panels $I = 3/2$ present results with α obtained in Sect. 12.3.1. The dashed curves in panels (**a**) and (**b**) show the results obtained when only the resonant channels with $J = 2$ are included. The profile of the $Xe^{**}(5p^{5\ 2}P_{1/2})4f[5/2]_2$ resonance in the angle-integrated cross section σ is shown in panel (**a**) for comparison (black dotted line)

$$
\begin{aligned}
\beta_2^{lin} &= -4\,\frac{h_2(\mathscr{B}_{220}+\mathscr{B}_{022})+\mathscr{B}_{222}}{\mathscr{B}_{000}+4h_2\mathscr{B}_{202}}\,,\\
\beta_4^{lin} &= \frac{192\,h_2\mathscr{B}_{242}}{\mathscr{B}_{000}+4h_2\mathscr{B}_{202}}\,,
\end{aligned}
\tag{12.23}
$$

$$
\begin{aligned}
\beta_2^{+\pm} &= \frac{2\mathscr{B}_{022}+h_2(2\mathscr{B}_{220}-\mathscr{B}_{222})\pm h_1\mathscr{B}_{121}}{\mathscr{B}_{000}\mp h_1\mathscr{B}_{101}+h_2\mathscr{B}_{202}}\,,\\
\beta_4^{+\pm} &= \frac{48\,h_2\mathscr{B}_{242}}{\mathscr{B}_{000}\mp h_1\mathscr{B}_{101}+h_2\mathscr{B}_{202}}\,.
\end{aligned}
\tag{12.24}
$$

Figures 12.6 and 12.7 present the asymmetry parameters (12.23) and (12.24) in the region around the $Xe^{**}(5p^{5\ 2}P_{1/2})4f[5/2]_2$ resonance for isotopes with different nuclear spins. The parameter β_4^{ν} would vanish for an unpolarized intermediate Xe^* $(5p^{5\ 2}P_{3/2})5d[3/2]_1$ state. The high values reached by β_4^{ν} indicate high anisotropy of this state.

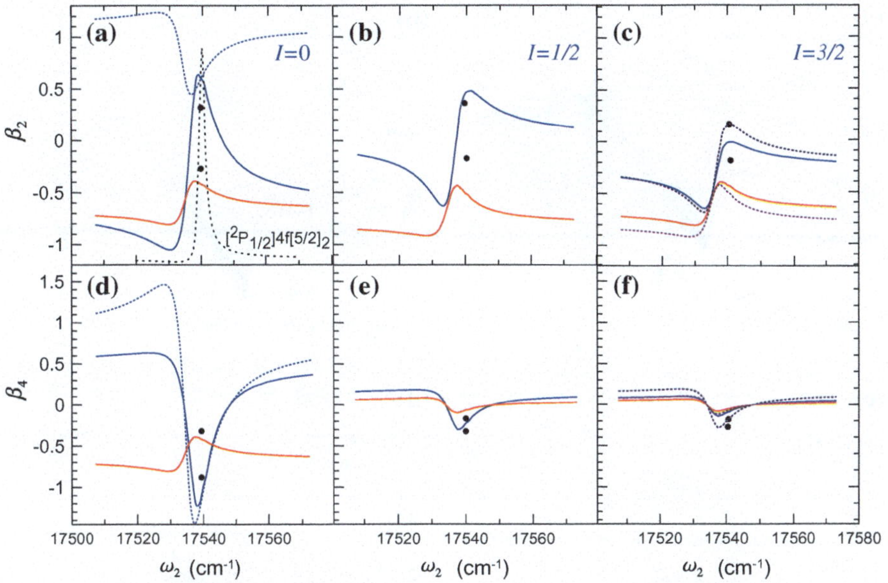

Fig. 12.7 The same as in Fig. 12.6, for circularly polarized fields. Red curves mark equal helicities (++), blue curves mark opposite helicities (+−)

In the panels (a) and (d) of Figs. 12.6 and 12.7 we present angular anisotropy parameters calculated in the complete model [26] and in the model when only the resonant channels with $J = 2$ ("resonant model") are included, with $h_1 = h_2 = 1$. In the panels (a) and (d) of the same figures, the curves are closer in the vicinity of the resonance, but deviate from each other significantly at tails of the resonance. For opposite helicities, β_2 in the resonant and the complete model even inverts the shape (blue dashed and solid curves in Fig. 12.7). For equal helicities the weak channels with $J = 0, 1$ are not excited for the isotope with $I = 0$ (i.e. without the depolarization) due to the selection rules for the magnetic quantum number. Therefore only $\beta_{2,4}^{+-}$ are changed by inclusion of the weak channels and their influence is crucial (panels (a) and (d) of Figs. 12.6 and 12.7). The resonance behavior of the β_2 parameter changes completely from the window-type to the resonance-type.

The calculation of the electronic structure does not differ for the different isotopes and the results for the anisotropy parameters for the isotopes should be automatically consistent once the depolarization factors due to the coupling between the electronic and nuclear angular momenta are properly included (see Sects. 12.2.3, 12.3.1). As can be seen in the panels (c)–(f) of Figs. 12.6 and 12.7, one do may use the same atomic spectroscopic model for all the isotopes.

For the ^{131}Xe ($I = 3/2$) isotope, the depolarization coefficients found above from the analysis of the CMD data (see Sect. 12.3.1), lead to slightly better agreement between theory and experiment than when the assumption of the separated HFS levels (panels (e) and (f) of Figs. 12.6 and 12.7) is made. Note that the profiles for

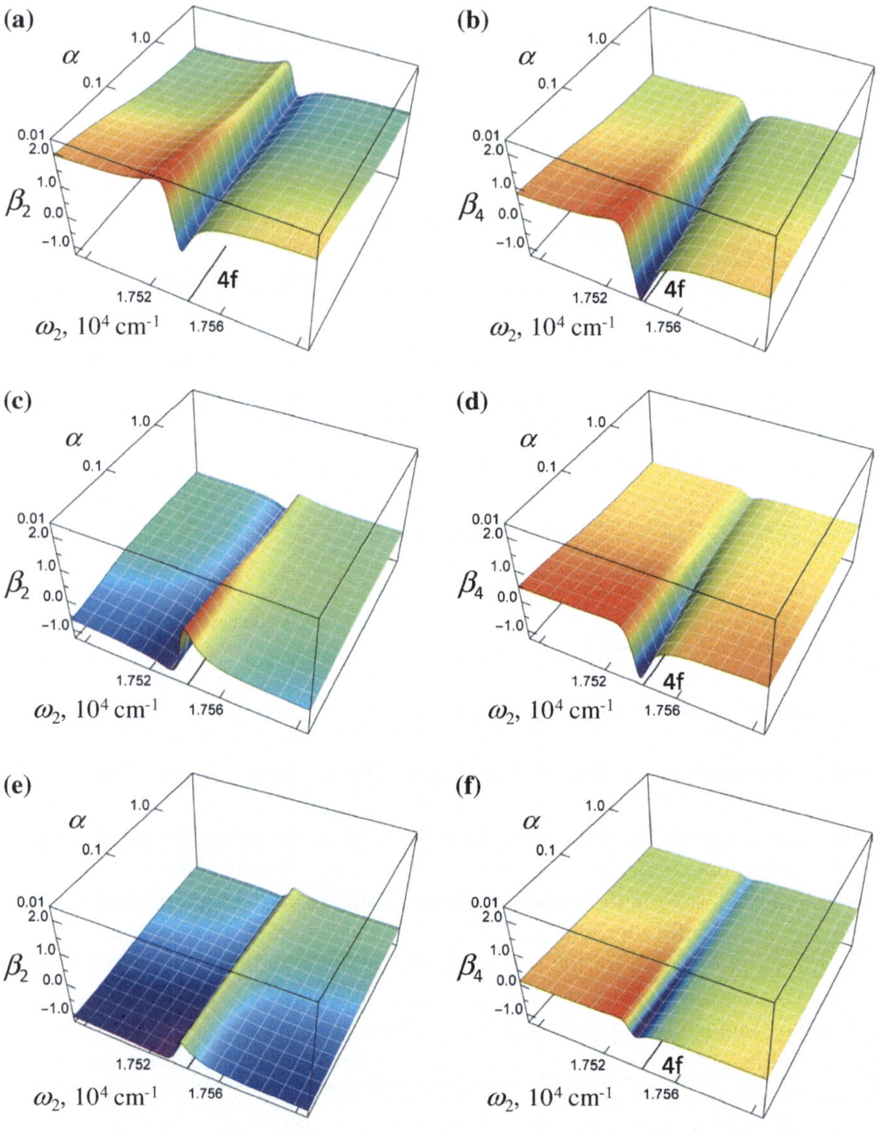

Fig. 12.8 Calculations for the isotope ^{131}Xe ($I = 3/2$): Angular anisotropy parameters as function of second (ionizing) laser energy and the depolarization factor α when both fields are linearly polarized in the same direction (**a**, **b**); one field is right and other field is left polarized (**c**, **d**) and both fields are right polarized (**e**, **f**). The position of the Xe$^{**}(5p^{5\ 2}P_{1/2})4f[5/2]_2$ state is indicated on the ω_2 axis by a long tick, marked as 4f

$\beta_{2,4}^{\nu}$ for all isotopes are equally broadened and shifted with respect to the profile of the resonance in the angle-integrated cross section in accordance with the scaling theorem [24].

To further illustrate the sensitivity of the asymmetry parameters to the hyperfine interaction, Fig. 12.8 presents the resonance profiles of β_2^{ν} and β_4^{ν} as function of laser energy and the parameter α for the isotope with $I = 3/2$. As we have mentioned in Sect. 12.2.3, even for infinite α the HFS depolarization is restricted, therefore, β_k does not change with further increasing α. From the limit value of $h_2(3/2)$ (see (12.15)) one may expect that β_4^{ν} decreases 4 times because of depolarization (Fig. 12.8b, d, f), but does not vanish. Therefore only collisional depolarization may completely demolish β_4. Besides decreasing of high-rank anisotropy parameters, the depolarization also opens some forbidden channels. In the case under consideration, these are channels $Xe^+(5p^{5\,2}P_{3/2}\varepsilon\ell : J = 1)$ for linearly polarized fields and $Xe^+(5p^{5\,2}P_{3/2}\varepsilon\ell : J = 0, 1)$ for circularly polarized light beams with equal helicity. The values of β_2^{ν} at the tails of the resonance are determined by a complicated interplay of the contributing channels (Fig. 12.8a, c, e).

12.4 Future Directions

Recent rapid progress in generation of VUV and X-Ray light pulses as well as their detection introduced an evolution in experimental studies on the effects of the hyperfine interactions from alkali and alkali earth atoms to noble gases. Even the investigation of positively charged ions became possible through sequential ionization (see chapter [25] in this book). The free-electron laser facilities, which are currently in operation, as well as those to be in operation in the near future can generate pulses with good longitudinal coherence, high brightness and short duration, with which we can investigate a variety of phenomena strongly affected by HFS by time- and angular-resolved photoelectron spectroscopy in the high-frequency domain. For example, by controlling and monitoring properties of the high-frequency radiation, we can investigate the effect of hyperfine interactions in the decay process of inner-shell vacancies.

Acknowledgements The authors highly appreciate the team of experimentalists, D. Cubaynes, G. A. Garcia, M. Meyer, L. Nahon and P. O'Keeffe, whose inspiration measurements initiated our theoretical study. We are grateful to kind hospitality of the French synchrotron SOLEIL and the European XFEL. EVG acknowledges financial support from the Basis foundation via the Junior Leader program.

References

1. E.B. Aleksandrov, V.N. Kulyasov, Sov. Phys. JETP **28**, 396–400 (1969)
2. E.B. Aleksandrov, Sov. Phys. JETP **29**, 846–848 (1969)
3. E.B. Aleksandrov, V.P. Kozlov, V.N. Kulyasov, Sov. Phys. JETP **39**, 620–626 (1974)
4. S. Aloïse, P. O'Keeffe, D. Cubaynes, M. Meyer, A.N. Grum-Grzhimailo, Phys. Rev. Lett. **94**, 223002(1)–223002(4) (2005)
5. S. Baier, A.N. Grum-Grzhimailo, N.M. Kabachnik, J. Phys. B **27**, 3363–3388 (1994)
6. V.V. Balashov, A.N. Grum-Grzhimailo, N.M. Kabachnik, *Polarization and Correlation Phenomena in Atomic Collisions. A Practical Theory Course* (Kluwer Plenum, New York, 2000)
7. B.B. Birkett, J.-P. Briand, P. Charles et al., Phys. Rev. A **47**, R2454–R2457 (1993)
8. K. Blum, *Density Matrix Theory and Applications*, 2nd edn (Plenum, New York, 1996)
9. M. Born, J.R. Oppenheimer, Annalen der Physik **389**, 457–484 (1927)
10. T. Brage, P.G. Judge, C.R. Proffitt, Phys. Rev. Lett. **89**, 281101(1)–281101(4) (2002)
11. A.W. Carr, M. Saffman, Phys. Rev. Lett. **117**, 150801(1)–150801(6) (2016)
12. R.-L. Chien, O.C. Mullins, R.S. Berry, Phys. Rev. A **28**, 2678–2084 (1983)
13. L.E. Cuéllar, C.S. Feigerle, H.S. Carman Jr., R.N. Compton, Phys. Rev. A **45**, 6437–6440 (1991)
14. B. Denne, S. Huldt, J. Pihl, R. Hallin, Phys. Scr. **22**, 45–48 (1980)
15. J.A. Duncanson Jr., M.P. Strand, A. Lindgård, R.S. Berry, Phys. Rev. Lett. **37**, 987–990 (1976)
16. L. Essen, J.V.L. Parry, Nature **176**, 280–282 (1955)
17. U. Fano, Phys. Rev. **124**, 1866–1978 (1961)
18. U. Fano, J.H. Macek, Phys. Rev. **45**, 553–573 (1973)
19. P.M. Farrell, W.R. MacGillivray, M.C. Standage, Phys. Rev. A **44**, 1828–1835 (1991)
20. C.S. Feigerle, R.N. Compton, L.E. Cuéllar, N.A. Cherepkov, L.V. Chernysheva, Phys. Rev. A **53**, 4183–4189 (1996)
21. V.V. Flambaum, V.A. Dzuba, A. Derevianko, Phys. Rev. Lett. **101**, 220801(1)–220801(4) (2008)
22. R. Forbes, V. Makhija, A. Stolow, I. Wilkinson, P. Hocketty, R. Lausten, Phys. Rev. A **97**, 063417(1)–063417(11) (2018)
23. M. Göppert-Mayer, Phys. Rev. **78**, 16–21 (1950)
24. A.N. Grum-Grzhimailo, S. Fritzsche, P. O'Keeffe, M. Meyer, J. Phys. B **38**, 2545–2553 (2005)
25. A.N. Grum-Grzhimailo, E.V. Gryzlova, Chapter *New Trends in Complete Experiment on Atomic Photoionization* (in this book)
26. E.V. Gryzlova, P. O'Keeffe, D. Cubaynes, G.A. Garcia, L. Nahon, A.N. Grum-Grzhimailo, M. Meyer, New J. Phys. **17**, 043054(1)–043054(15) (2015)
27. E. Hack, J.R. Huber, Int. Rev. Phys. Chem. **10**, 287–317 (1991)
28. T. Hadeishi, W.A. Nierenberg, Phys. Rev. Lett. **14**, 891–892 (1965)
29. S. Haroche, J.A. Paisner, A.L. Schawlow, Phys. Rev. Lett. **30**, 948–951 (1973)
30. J.R. Henderson, P. Beiersdorfer, C.L. Bennett et al., Phys. Rev. Lett. **65**, 705–708 (1990)
31. D. Huff, W.V. Houston, Phys. Rev. **36**, 842–846 (1930)
32. N.M. Kabachnik, I.P. Sazhina, J. Phys. B **9**, 1681–1697 (1976)
33. A. Kupliauskiene, N. Rakštikas, V. Tutlys, J. Phys. B **34**, 1783–1803 (2001)
34. L.D. Landau, E.M. Lifshitz, *Quantum Mechanics: Non-Relativistic Theory, 3rd edn* (Pergamon, London, 1977)
35. R. Luypaert, J. Van Craen, J. Phys. B **10**, 3627–3636 (1977)
36. J.J. McClelland, M.H. Kelley, Phys. Rev. A **31**, 3704–3710 (1985)
37. M. Meyer, B. Müller, A. Nunnemann, Th. Prescher, E.V. Raven, M. Richter, M. Schmidt, B. Sonntag, P. Zimmermann, Phys. Rev. Lett. **59**, 2963–2966 (1987)
38. M. Meyer, M. Gisselbrecht, A. Marquette, C. Delisle, M. Larzillière, I.D. Petrov, N.V. Demekhina, V.L. Sukhorukov, J. Phys. B **38**, 285–295 (2005)
39. S. Mrozowski, Phys. Rev. **57**, 207–211 (1940)
40. O.C. Mullins, R.-L. Chien, J.E. Hunter III, J.S. Keller, R.S. Berry, Phys. Rev. A **31**, 321–328 (1985)

41. L. Nahon, N. de Oliveira, G.A. Garcia, J.-F. Gil, B. Pilette, O. Marcouillé, B. Lagarde, F. Polack, J. Synchrotron Radiat. **19**, 508–520 (2012)
42. D.E. Nikonov, U.W. Rathe, M.O. Scully, S.-Y. Zhu, E.S. Fry, X. Li, G.G. Padmabandu, M. Fleischhauer, Quantum Opt. **6**, 245–260 (1994)
43. P. O'Keeffe, S. Aloïse, S. Fritzsche, B. Lohmann, U. Kleiman, M. Meyer, A.N. Grum-Grzhimailo, Phys. Rev. A **70**, 012705(1)–012705(14) (2004)
44. P. O'Keeffe, E.V. Gryzlova, D. Cubaynes, G.A. Garcia, L. Nahon, A.N. Grum-Grzhimailo, M. Meyer, Phys. Rev. Lett. **111**, 243002(1)–243002(5) (2013)
45. K.L. Reid, S.P. Duxon, M. Towrie, Chem. Phys. Lett. **228**, 351–356 (1994)
46. N. Saquet, D.M.P. Holland, S.T. Pratt, D. Cubaynes, X. Tang, G.A. Garcia, L. Nahon, K.L. Reid, Phys. Rev. A **93**, 033419(1)–033419(11) (2016)
47. A.L. Schawlow, C.H. Townes, Phys. Rev. **112**, 1940–1949 (1958)
48. J.H. Scofield, J. Nilsen, Phys. Rev. A **49**, 2381–2388 (1994)
49. A. Siegel, J. Ganz, W. Bußert, H. Hotop, J. Phys. B **16**, 2945–2959 (1983)
50. V. Sukhorukov, I. Petrov, M. Schäfer, F. Merkt, M.-W. Ruf, H. Hotop, J. Phys. B **45**, 092001(1)–092001(43) (2012)
51. J.R. Tolsma, D.J. Haxton, C.H. Greene, R. Yamazaki, D.S. Elliott, Phys. Rev. A **80**, 033401(1)–033401(9) (2009)
52. E. Träbert, P. Beiersdorfer, G.V. Brown, Phys. Rev. Lett. **98**, 263001(1)–263001(4) (2007)
53. W.A. van Wijngaarden, J. Sagle, J. Phys. B **24**, 897–903 (1991)
54. F. Wuilleumier, M. Meyer, J. Phys. B **39**, R425–R477 (2006)
55. R.P. Wood, C.H. Greene, D. Armstrong, Phys. Rev. A **47**, 229–235 (1993)
56. Z.W. Wu, A. Surzhykov, S. Fritzsche, Phys. Rev. A **89**, 022513(1)–022513(7) (2014)
57. H. Xu, E. Lötstedt, T. Ando, A. Iwasaki, K. Yamanouchi, Phys. Rev. A **96**, (R)041401(1)–041401(6) (2017)
58. J. von Zanthier, T. Becker, M. Eichenseer et al., Opt. Lett. **25**, 1729–1731 (2000)

Chapter 13
New Trends in 'Complete' Experiment on Atomic Photoionization

Alexei N. Grum-Grzhimailo and Elena V. Gryzlova

Abstract Complete experiments played a crucial role in the progress of the physics of elementary atomic processes over decades, providing most detailed information about the dynamics of the reactions. The overview of recent developments on complete experiments in atomic photoionization is presented. The advent of free-electron lasers operating in the XUV significantly increased the potential of these experiments. Unprecedented intensity of the XUV radiation concentrated in femtosecond pulses with variable energy and polarization opens the door to new type of complete experiments, which are now becoming realistic: on the two-colour above threshold ionization, on ionization by the fundamental frequency in the XUV and its second harmonic, on photoionization of positive ions, on non-dipole contribution in photoionization.

13.1 Introduction

A concept of complete experiment was formulated at the end of 1950s in a series of papers on nuclear reactions, that is, scattering of nucleons by nucleons [64], extended to the case of arbitrary spins of the colliding particles [7], inelastic processes [63], the β-decay [70]. The quintessence of the complete experiment was formulated by Yakov Smorodinskii as [70].

> ... how many experiments are required for the reconstruction of the scattering matrix for scattering of nucleons by nucleons (the "complete experiment").

The concept of complete (originally 'perfect') experiment in atomic physics was first formulated 10 year later, up to our knowledge independently, by Benjamin Bederson, who put a similar question [5]

A. N. Grum-Grzhimailo (✉) · E. V. Gryzlova
Skobeltsyn Institute of Nuclear Physics, Lomonosov Moscow State University, Leninskie Gory, 19991 Moscow, Russia
e-mail: grum@sinp.msu.ru

E. V. Gryzlova
e-mail: gryzlova@gmail.com

K. Yamanouchi et al. (eds.), *Progress in Photon Science*, Springer Series in Chemical Physics 119, https://doi.org/10.1007/978-3-030-05974-3_13

How many independent differential scattering experiments are required in order to fully determine all the parameters which are calculable from potential scattering theory?

This question was first related to electron-atom scattering [50]. Later the definition of complete experiment transformed into an equivalent, more transparent and universal, as an experiment or set of experiments which allow to determine all the quantum mechanical amplitudes related to the process. This implies measurements of both, absolute values and relative phases of the amplitudes. From this information any observable can be predicted and the most comprehensive test of theory can be provided up to the degree allowed by quantum mechanics. Later the concept of complete experiment was extended from scattering processes to half-scattering processes, such as atomic photoionization [11, 19, 41, 42, 48], molecular photoionization [12, 23, 65], and atomic Auger decay [27, 44, 46, 76], and generally played a crucial role in the progress of the physics of elementary atomic processes over decades [1, 4, 51].

Assume that a detector (or a set of detectors) is tuned to a state described by the density matrix ε. According to quantum mechanics, the probability for the detector to register an event is $W = \mathrm{Tr}\,\rho\varepsilon$, where ρ is the density matrix of the final state after the reaction. Introducing transition operator T, expressing ρ in terms of the initial density matrix (ρ_0) before the reaction, $\rho = T\rho_0 T^+$, and choosing convenient basis sets, $|\,\xi_i\,\rangle$ and $|\,\chi_f\,\rangle$, for the initial and final states of the system, respectively, we obtain the measured intensity as

$$\begin{aligned} W &= \sum_{ff'} \langle \chi_f \,|\, T\rho_0 T^+ \,|\, \chi_{f'} \rangle \, \langle \chi_{f'} \,|\, \varepsilon \,|\, \chi_f \rangle \\ &= \sum_{ff'ii'} \langle \xi_i \,|\, \rho_0 \,|\, \xi_{i'} \rangle \, \langle \chi_{f'} \,|\, \varepsilon \,|\, \chi_f \rangle \, T_{i\to f} \, T^*_{i'\to f'} \,, \end{aligned} \tag{13.1}$$

where the transition amplitudes $T_{i\to f} \equiv \langle \chi_f \,|\, T \,|\, \xi_i \rangle$ have been introduced. The coefficients in the bilinear combination of the amplitudes $T_{i\to f}$ are to be determined within a certain theoretical framework. Tuning the detector to a different state ε', another bilinear combination of the same amplitudes may be measured and so on. To perform the complete experiment means to choose the detector states $\varepsilon, \varepsilon', ...$ and/or to change the initial state ρ_0 in order to find all the independent amplitudes $T_{i\to f}$, relevant to the process under study, from the set of equations provided by the measurements of W, W', ... The concept of complete experiment at the present stage implies 'stationary' amplitudes $T_{i\to f}$, i.e. the infinite time limit, when the process is finished.

Having the full set of the amplitudes $T_{i\to f}$ any observable Q in the final state can be found according to the quantum mechanical prescription.

$$\langle Q \rangle = \mathrm{Tr}\,(\rho Q) = \sum_{ff'ii'} \langle \xi_i \,|\, \rho_0 \,|\, \xi_{i'} \rangle \langle \chi_{f'} \,|\, Q \,|\, \chi_f \rangle T_{i\to f} T^*_{i'\to f'} \,. \tag{13.2}$$

In the case when Q describes a subsystem, the corresponding reduced density matrix ρ should be taken.

It is clear from the above formulation that an experiment can be complete only within a certain theoretical framework, because only a certain model, or a class of models, allow to count the number of amplitudes. It follows that recipes for complete experiments and, hence, justifying them are model dependent, as well as analyzing their results and extracting the amplitudes from the measurements. This shows the exceptional role of theory in the problem of complete experiment.

Actually, the complete experiment is not necessarily one experiment, but rather a set of experiments, which could be performed at different time in different labs, measuring different complementary quantities, in total giving enough data for extracting the amplitudes. In this sense, any data on differential characteristics of the reaction may be considered as a step towards the complete experiment.

The significance of the complete experiment on photoionization was eloquently expressed by Katherine Reid [66]:

> In experiments aimed at determining information about photoionization dynamics, the 'Holy Grail' is the complete experiment.

A comprehensive review of the complete experiments on atomic photoionization was published in 2013 [51]. The purpose of the present overview is to cover some recent developments in the field.

The structure of the review is as follows. In Sect. 13.2 we remind some traditional issues of the complete experiment on photoionization; Sect. 13.3 discusses generalization to the two-photon ionization; Sect. 13.4 presents a photoionization by bichromatic light $\omega + 2\omega$ as a method for the complete experiment, which incorporates final states with opposite parities; Sect. 13.5 discusses complete experiments with positively charged ions, which became now feasible with free-electron lasers; Sect. 13.6 generalizes the complete experiment beyond the dipole approximation; Sect. 13.7 discusses additional information, electron density and electron current, which can be found from the results of complete experiment; Sect. 13.8 contains some concluding remarks.

13.2 Complete Experiment on Photoionization: A Conventional Approach

Consider atomic photoionization

$$h\nu + A(\alpha_i J_i) \to A^+(\alpha_f J_f) + e^-_{ph}\,, \tag{13.3}$$

where both the initial atomic state and the final ionic state are characterized by their total electronic angular momenta, J_i and J_f, respectively. All other quantum numbers that are necessary for specifying the states are denoted by α_i and α_f. The natural basis set for the initial atomic state is $|\,\xi_i\,\rangle = |\,\alpha_i J_i M_i\,\rangle$, where M_i is the projection of the angular momentum J_i. The convenient basis set $|\, \chi_f \,\rangle$ for the final state depends on the transition operator. A standard method of theoretical description of atomic photoionization uses multipole expansion of the vector potential in combination with the perturbation theory with respect to interaction of the electromagnetic field with the atom. It is implied that the radiation pulse contains many optical cycles and the vector potential corresponds to the plane electromagnetic wave. Leaving the first term in the multipole expansion leads to so-called 'dipole' approximation with the transition operator $D_\lambda = e\sum_i (r_\lambda)_i$, where the summation is taken over all electrons in the target; $\lambda = 0, \pm 1$ is a spherical component of a vector and $\mathbf{r}$ is the radius vector of the electron. In this case the basis set $|\, \chi_f \,\rangle = |\, (\alpha_f J_f, E\ell j)\, JM \,\rangle$ is convenient in order to use directly the conservation of the total angular momentum and parity. Here, E, ℓ and j are the energy, orbital and total angular momenta of the photoelectron; J is the total angular momentum of the final state in (13.3), $\mathbf{J} = \mathbf{J}_f + \mathbf{j}$, and M is its projection. Due to the Wigner-Eckart theorem,

$$T_{i\to f} \sim \langle (\alpha_f J_f, E\ell j) JM \,|\, D_\lambda \,|\, \alpha_i J_i M_i \rangle = (2J+1)^{-\frac{1}{2}} (J_i M_i, 1\lambda \,|\, JM)\, D_E(J_f \ell j, J), \tag{13.4}$$

the dependence of the amplitudes on the magnetic quantum numbers is explicitly given by the known Clebsch-Gordan coefficient. Thus, measurement of the partial wave reduced amplitudes

$$D_E(J_f \ell j, J) \equiv \langle (\alpha_f J_f, E\ell j)\, J \,||\, D \,||\, \alpha_i J_i \rangle \equiv d^E_{J_f \ell j J} \exp \mathrm{i}\, \delta^E_{J_f \ell j J}\,, \tag{13.5}$$

which contain the dynamics of the atomic photoionization, is the goal of the complete photoionization experiment within the dipole approximation. In (13.5), $d^E_{J_f \ell j J}$ is the (non-negative) absolute value of the amplitude and $\delta^E_{J_f \ell j J}$ is its phase. Expressions of the type (13.1) for observable quantities in terms of the photoionization amplitudes (13.5) can be obtained by standard algebra of the angular momentum. Many of such expressions are presented, for example, in [2, 53].

Further simplification can be done by neglecting relativistic interactions, such as spin-orbit interaction. Then the electron wavefunctions are independent of j, while spin and orbital momenta are individually conserved. The amplitudes (13.5) then reduce to

$$D_E(L_f \ell, L) \equiv \langle (\alpha_f L_f, E\ell)\, L \,||\, D \,||\, \alpha_i L_i \rangle \equiv d^E_{L_f \ell L} \exp \mathrm{i}\, \delta^E_{L_f \ell L} \tag{13.6}$$

and their total number essentially reduces too. In fact the amplitudes (13.6) can still depend on S_f and S via electron exchange and genealogy of the many-electron atomic and ionic states (genealogy coefficients). Finally, assuming independence of the electron wave functions of the global quantum numbers L (and S_f, S) results in not more than two amplitudes. In the cases of single-electron transitions, the amplitudes can be written as (we omit the photoelectron energy E for brevity).

$$D_\ell \equiv \langle E\ell \,||\, r \,||\, \ell_i \rangle \equiv d_\ell \exp \mathrm{i}\, \delta_\ell\,, \qquad \ell = \ell_i \pm 1\,. \tag{13.7}$$

The latter approximation characterizes the so-called Cooper-Zare model [15, 16] and has a rather large domain of applicability. Note that even with the noticeable fine-structure splitting, one can still in many cases neglect the dependence of the wavefunctions on the total angular momentum, because the energy splitting is much more sensitive to spin-orbit interaction than the wave-functions. Intermediate situations are possible and often realized. For example, the amplitudes may be considered as depending only on ℓ and j, $D_{\ell j}$. Furthermore, here the phases $\delta_{\ell j}$ may be sometimes considered as independent of j, in contrast to the absolute value $d_{\ell j}$ and visa versa. Note that the phases of the amplitudes δ_ℓ, $\delta_{\ell j}$, etc. are directly related to the scattering phase of electron in the field of the residual ion and absolute squares d_ℓ, $d_{\ell j}$, etc. define the partial ionization cross sections in the corresponding channel. Therefore, the phase and the absolute value of the amplitude have a clear physical meaning. The general absolute normalization of the photoionization cross sections is a separate procedure. This is a reason why nowadays most often the notion 'complete experiment' is used with respect to extraction only dimensionless parameters: absolute ratio and relative phases of the amplitudes.

Even in the simplest nontrivial case of two amplitudes, the complete information on photoionization cannot be obtained, in general, by measuring only the photoelectron angular distributions (PADs) and photoionization cross section, but additional observables are required [3, 52]. For example, in the dipole approximation for linearly polarized light in the case of isotropic (unpolarized) target the PAD is of the form

$$\frac{\mathrm{d}^2\sigma}{\mathrm{d}\Omega \mathrm{d}E} = \frac{\sigma(E)}{4\pi}\left(1 + \beta_2(E) P_2(\cos\vartheta)\right) , \tag{13.8}$$

where $P_n(x)$ is the n-th Legendre polynomial, ϑ is the angle between the emission direction of the photoelectron and the polarization direction of the photon beam. The anisotropy (asymmetry) parameter β_2 (hereafter we omit the argument E for brevity) is expressed in terms of the photoionization amplitudes. It may be shown that the PAD for arbitrary polarized photon beam is expressed in terms of the same dimensionless parameter β_2. For example, (13.8) holds for the circularly polarized light with the substitution $\beta_2 \to -\frac{1}{2}\beta_2$ provided the angle ϑ is counted from the direction of the photon beam. Nevertheless, the complete experiment was achieved by combining angle- and spin-resolved photoelectron spectroscopy [40, 68], angle-resolved photoelectron spectroscopy together with secondary electron spectroscopy [39, 54, 71], together with secondary fluorescence polarimetry [6, 59], in photoelectron-Auger electron [8, 47, 67] and photoelectron-fluorescence [73] coincidences.

If the initial state of an open-shell atom is polarized (i.e. the population of the magnetic substates is not uniform), the measurement of the PAD allows one to perform the complete experiment even without information on other observables, provided the target polarisation can be accurately controlled [49]. More complicated PADs and controlling the polarization of both, the target and the photons, give more independent dynamical parameters (combinations of amplitudes), than for unpolarized atom. This property was exploited to perform the complete photoionization experiments with atoms polarized by inhomogeneous magnetic field [61], by laser optical pumping [25], and by single photon absorption [37, 60]. For example, the PAD in the case of initially aligned P-states[1] (i.e. states with $\mathcal{A}_{20} \neq 0$, $\mathcal{A}_{10} = 0$) with the axis of symmetry along the linear polarization of the photons is of the form

$$\frac{\mathrm{d}^2\sigma}{\mathrm{d}\Omega \mathrm{d}E} = \frac{\sigma}{4\pi}\left(1 + \beta_2 P_2(\cos\vartheta) + \beta_4 P_4(\cos\vartheta)\right) , \tag{13.9}$$

containing one more parameter, β_4, in comparison with (13.8).

13.3 Two-Colour Above-Threshold Ionization

So far it was assumed that only one photon participates in the reaction (see (13.3)). The concept of complete experiment for the two-photon and, generally, multiphoton ionization can be formulated in a similar way. The main formal difference would be another tensor srtructure of the transition operator and, hence, other selection rules and list of photoionization channels. However, the complexity of the multiphoton case can come from the fact that the field is too strong to consider absorption of fixed number of photons and use the lowest possible order of the perturbation theory. Then the concept of complete experiment on multiphoton ionization needs reformulating, using another language. Here we stay within the framework of the lowest possible order of the perturbation theory.

In the simplest case of two-photon ionization the partial wave second-order photoionization amplitude in the dipole approximation is expressed in terms of the second-order reduced matrix elements

$$S_n(J_f \ell j, J; \omega_1, \omega_2) \equiv \langle (\alpha_f J_f, \ell j) J \,||\, D G_n D \,||\, \alpha_i J_i \rangle , \tag{13.10}$$

where notations for the quantum numbers are similar to (13.5), n symbolizes symmetry of intermediate virtual states, G is the corresponding part of the atomic Green's function and $\omega_{1,2}$ indicate that the amplitude depends on the frequencies of both light sources. The number of independent two-photon amplitudes S_n is determined by two sets of individual selection rules for two dipole momentum operators D in (13.10). The generally larger number of the independent two-photon amplitudes in comparison with the one-photon ionization is due to larger possible interval of the angular momentum of ionization channels and various possible symmetries of the intermediate states. From another side, the presence of the two photons, gives more flexibility in using polarization properties of the radiation for the needs of complete experiment.

---

[1]The alignment parameter of the P-state is defined as $\mathcal{A}_{20} = 2^{-\frac{1}{2}} N^{-1} (N_{+1} + N_{-1} - 2N_0)$, where N_m is the number of atoms in the magnetic substate m of the P-state and $N = N_{+1} + N_0 + N_{-1}$ is the total number of atoms in the P-state. The orientation parameter for the P-state is defined as $\mathcal{A}_{10} = \sqrt{\frac{3}{2}}\, N^{-1} (N_{+1} - N_{-1})$.

The complete experiments on the direct two-photon ionization has been realized in Rb atom with elliptically polarized optical laser for the case when the photon energy is below the ionization threshold [74]. It was not possible to analyse the five relative amplitudes, which occur in the dipole approximation ($s_{1/2} \rightarrow p_{1/2} \rightarrow s_{1/2}, d_{3/2}$; $s_{1/2} \rightarrow p_{3/2} \rightarrow s_{1/2}, d_{3/2}, d_{5/2}$), but the PADs produced by the elliptically polarized light were enough to measure the two relative cross sections for ionization into $s_{1/2}$, $d_{3/2}$ and $d_{5/2}$ channels and a phase difference between s and d amplitudes, neglecting the phase difference between $d_{3/2}$ and $d_{5/2}$ amplitudes. The results of this complete experiment were only partly in accordance with theory [13, 17]. The authors of [74] demonstrate advantages of the elliptically polarized light for complete experiment on the two-photon ionization.

It was mentioned in 2013 [51] that the complete experiments on two-colour two-photon non-resonant ionization have not yet been performed, as well as complete experiments for above threshold photoionization, when the energy of one of the photons is enough to ionize the atom. Such an experiment now has been performed. The new brilliant XUV sources can produce high density of the photoelectrons in ionization from the ground state providing a noticeable probability for the atom to further absorb photons from an optical laser, producing sidebands in the photoelectron spectra. If the optical laser field is not strong and produces only one pair of sidebands, the photoelectrons in the first sideband may be considered as a result of the two-photon two-colour above-threshold ionization (ATI), which can be treated in the second-order perturbation theory. The complete experiment on the two-colour ATI have been realized in helium atom [55] at free electron laser (FEL) FERMI. The scheme of the transitions is shown in Fig. 13.1. By combining different polarizations

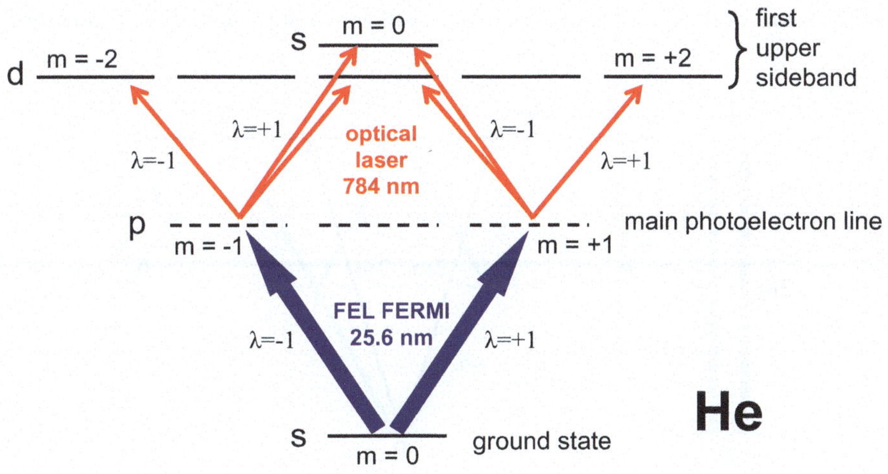

Fig. 13.1 Scheme of transitions between magnetic substates in two-colour ATI of atomic helium by circularly polarized photon beams [55]. Only the first upper sideband is shown. The light chirality λ is indicated

of the two lasers, FEL and optical, and measuring the PADs of the upper and lower sidebands, the absolute ratio of transition amplitudes into Es and Ed continuum channels, $|D_s|/|D_d|$, see (13.7), for the ATI from the ground state was determined, as well as the relative phase between the two amplitudes. The extracted values are in accordance with the numerical calculations in the second-order perturbation theory [55]. The contribution into the ATI from s and d channels was found almost equal for the sideband, in accordance with earlier observations [56], and deviating strongly from the propensity rules predicting that, upon absorption of one photon, transitions related to an increase in the angular momentum should be favored [20]. This was discussed in more detail in [56].

13.4 Ionization by Coherent First and Second Harmonics

In 'conventional' complete experiment on photoionization within the dipole approximation, the amplitudes correspond to ionization into channels with one parity. In the simplest case of ionization of the s electron, as in He, only one p-wave is possible due to the selection rule for absorption of one photon. When ionizing p-electron, for example, from the outer shell of the noble gas atom, the dipole transition leads to s- and d-photoelectron waves with the two (in the Cooper-Zare model) corresponding amplitudes. In the two-photon ionization also the channels with fixed (opposite to the one-photon ionization) parity are accessible.

However, there is an interesting possibility to involve ionization channels of both parities into the complete experiment. The scheme of the process is shown in Fig. 13.2a and imply two-photon ionization by fundamental frequency (ω) and one-photon ionization by its second harmonic (2ω), which are longitudinally coherent.

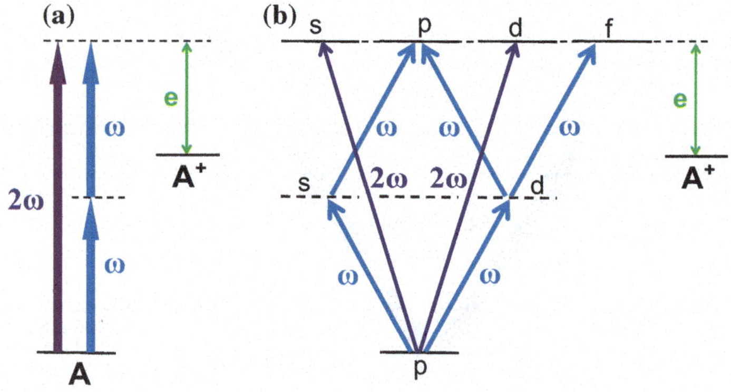

Fig. 13.2 Scheme of simultaneous atomic ionization by the fundamental (ω) and its second harmonic (2ω): the general scheme (**a**); five different paths for promoting p-electron to continuum in the $\omega + 2\omega$ ionization (**b**)

The latter is necessary to generate interference between the amplitudes of the one-photon and the two-photon ionization. Note that this interference term contributes to the PAD and vanishes after integrating over the angle of the electron emission. In the above examples, ionization of He leads to the p-wave in one photon ionization (2ω) and to s- and d-waves in two-photon ionization (ω), giving three amplitudes: two-photon amplitudes and one-photon amplitude. The $\omega + 2\omega$ ionization from the p-shell, for example, of the noble gas atom leads to the final channels with p- and f-waves (ω) and s- and d-waves (2ω), giving three two-photon amplitudes of the type (13.10) and two one-photon amplitudes, respectively (see Fig. 13.2b). With the increased number of the amplitudes in comparison with 'pure' ω or 'pure' 2ω ionization, the complete experiment seems becoming more difficult, but note that the number of the independent measurable parameters also increases: the PADs contain more anisotropy coefficients, while changing mutual polarization of the fundamental and its second harmonic gives even more parameters. Of course, individual measurements for ω and 2ω ionization is another source of independent information on the photoionization amplitudes. An important step to the $\omega + 2\omega$ type of complete experiment has been done in [75] in the optical range, based on the theoretical prescriptions [58]: the relative phases $\delta_p - \delta_d$ and $\delta_p - \delta_s$ were determined in the ($\omega + 2\omega$) ionization from the ground state of rubidium.

With the advent of FEL FERMI and having in mind other future seeded FELs, generating pulses with the longitudinal coherence [24, 62], the $\omega + 2\omega$ complete experiments may be performed in the XUV and first attempts are under way [72]. These experiments are closely related to the coherent control of PADs in the XUV, currently under study theoretically and experimentally [18, 31, 35, 62].

13.5 Complete Experiments on Photoionization of Ions

As mentioned above, measurements of PADs from unpolarized atoms are not enough to extract all complex photoionization amplitudes even within the dipole approximation. To overcome this obstacle different methods of producing polarized targets have been applied, usually involving additional field(s). Up to very recently all the complete experiments on photoionization have been performed with neutral targets. However, atomic photoionization leads to a polarized residual ion [10, 21, 45], a property that can be exploited for the complete photoionization experiment with positively charged ions. This general idea now can be realized with the FELs. Due to the exceptionally high intensity delivered by these light sources, ionic samples can be created and probed by sequential photoabsorption within the very same femtosecond FEL pulse [9, 57].

The simplest of such a process is the sequential two-photon double ionization (Fig. 13.3). The PADs of the first- and the second-step electrons, e_1 and e_2, respectively, have been many times observed and theoretically analysed for the noble gas atoms (see [9, 30, 34] and references therein). The important feature of the process is that the intermediate ion, A^+, is either aligned or oriented depending on

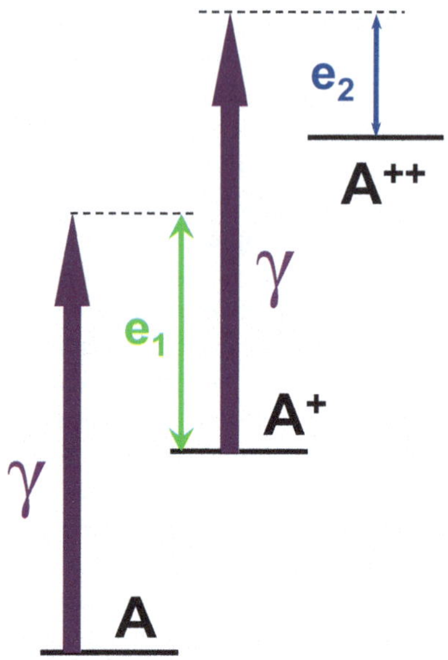

Fig. 13.3 General scheme of atomic sequential two-photon double ionization

polarization of the VUV pulse. Accordingly, the stepwise mechanism is used in the theoretical description of the sequential two-photon double ionization with creation of a polarized intermediate single-charged ion. Then the PAD of the second electron is described by (13.9), where β_2 and β_4 are not only functions of the second-step ionization amplitudes, but also functions of $\mathcal{A}_{10}$ and/or $\mathcal{A}_{20}$ of the intermediate ion. The latter parameters, in turn, are functions of the first-step ionization amplitudes. It can be shown that by measurements of anisotropy parameters β_2, β_4 of the second electron with circularly and linearly[2] polarized FEL it is possible, within the Cooper-Zare model, not only to perform the complete experiment on the photoionization of ion, but also to extract one more independent parameter for the photoionization of the neutral atom and thus to contribute to the complete experiment on photoionization of neutral atom. The situation is even better: taking into account that within the Cooper-Zare model, similar amplitudes describe the second-step ionization to different multiplet levels (e.g. to $A^{2+}(np^4\,{}^3P,\ {}^1D,\ {}^1S)$ for the noble gases), which are distinguished experimentally by the photoelectron energy, information on the amplitudes is even redundant and one can select most reliable data and perform a cross checking.

We theoretically simulated such a complete experiment on photoionization of the ion Ar^+ [29], taking calculated values of the anisotropy parameters, giving them 10% 'experimental inaccuracy' and solving the reverse problem of extracting the amplitudes from the 'measured' parameters. Some of the results are shown in Fig. 13.4,

[2]In contrast to one-photon ionization, here β_2 for circularly and linearly polarized light are independent.

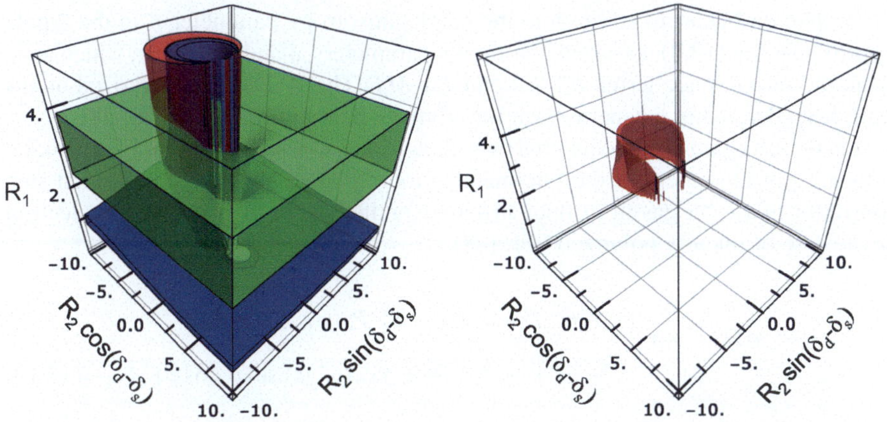

Fig. 13.4 Results of simulated complete experiment on photoionization of Ar^+ at the photon energy of 1 keV. Left panel: Restrictions imposed by imagined measurements of dipole parameters β_2 (red) and β_4 (green) in sequential two-photon double ionization by linearly polarized light and β_2 (blue) by circularly polarized light. Right panel: Corresponding restricted space of the absolute amplitude ratios for the second ionization step, $R_2 = (d_d/d_s)_2$, and for the first ionization step, $R_1 = (d_d/d_s)_1$. We suppose that each of the PAD parameters possesses a theoretical value with an error-bar of 10%. The residual ion is in the $Ar^{2+}(3p^4\,{}^3P)$ state

which demonstrates how the allowed space for the amplitudes is restricted by including data from different measured quantities.

13.6 Beyond the Dipole Approximation

While complete photoionization experiments have been considered within the dipole approximation, non–dipole effects should be taken into account in future research within two respects. First, specific results may be modified by unwanted non-dipole parts of the photoionization amplitudes (see reviews [36, 43]), and, second, in principle one can put a task to extract the non–dipole parts of the photoionization amplitudes.

Within the first order non–dipole corrections, i.e. taking into account electric quadrupole (E2) and magnetic dipole (M1) ionization amplitudes the PADs from unpolarized atoms, produced by linearly polarized light is usually parameterized by [14, 69].

$$\frac{d^2\sigma}{d\Omega dE} = \frac{\sigma}{4\pi}\Big(1 + \beta P_2(\cos\vartheta) + \big(\delta + \gamma\cos^2\vartheta\big)\sin\vartheta\cos\varphi\Big)\,, \tag{13.11}$$

where the angle ϑ is counted relative to the polarization of the incoming radiation, similar to (13.8), (13.9), and φ is the azimuthal angle with respect to the direction of the photon beam. In addition to the single anisotropy parameter β in the dipole approximation, (13.8), two new non-dipole parameters, δ and γ, appear, which arise from the interference terms $E1$-$E2$ and $E1$-$M1$. The PADs from polarized atoms have been derived in a general form for arbitrary field multipoles in [26, 28].

Let us turn again to the sequential two-photon double ionization (Fig. 13.3). Taking into account alignment of the intermediate ionic P-state generated at the first-step ionization of a noble gas atom, one can derive for the PAD of the second photoelectron in the case of linearly polarized FEL [28].

$$\frac{d\sigma}{d\Omega} = \frac{\sigma}{4\pi}\Big(1 + \beta_2^L\, P_2(\cos\vartheta) + \beta_4^L\, P_4(\cos\vartheta) + (\delta + \gamma_2\cos^2\vartheta + \gamma_4\cos^4\vartheta)\sin\vartheta\cos\varphi\Big)\,, \tag{13.12}$$

which generalizes (13.9). The superscript 'L' denotes linear polarization of the FEL. The first three terms here correspond to the pure dipole contribution; the last three terms give the first-order non-dipole corrections and include $E1$-$E2$ and $E1$-$M1$ interference. Due to specific selection rules, the M1 photoionization amplitudes may be neglected in the vast majority of cases. The last term with γ_4 is a non-dipole correction specific for the two-photon ionization. The PAD for the circularly polarized FEL takes the form [33]

$$\frac{d^2\sigma}{d\Omega dE} = \frac{\sigma}{4\pi}\left(1 + \sum_{n=1}^{5}\beta_n^C\, P_n(\cos\vartheta)\right)\,, \tag{13.13}$$

where the angle ϑ is counted from the direction of the photon beam and superscript 'C' denotes circular polarization of the FEL. In (13.13), β_2^C and β_4^C represent the pure dipole contribution, while β_1^C, β_3^C β_5^C represent the first-order non-dipole contribution. Some of the 10 parameters in (13.12) and (13.13) may be not independent. For example, within the Cooper-Zare model (were also the M1 photoionization amplitudes identically vanish), the 10 independent anisotropy parameters characterizing the PADs in the second-step ionization are expressed in terms of only four (complex) amplitudes: two dipole d_s, d_d and two quadrupole q_p, q_f. Additionally, usually unknown absolute ratio of the two dipole amplitudes of the first-step ionization are also to be determined from the PADs (13.12) and (13.13).

Thus, the PADs in the second ionization step for linearly and circularly polarized field are enough to extract three ratios of the complex second-step amplitudes and the absolute ratio of the first-step dipole amplitudes, i.e. enabling the complete experiment on the second-step photoionization. The latter absolute ratio, combined with the PADs of the first ionization step, as mentioned above, completes also the experiment on the first-step photoionization.

The PADs in the second ionization step have been measured many times with FEL for noble gas atoms [9] and the non-dipole effects in these PADs have been reported recently in argon [38]. This opens a way, as recently proposed [29], to generalize a complete experiment on photoionization for accounting the quadrupole amplitude

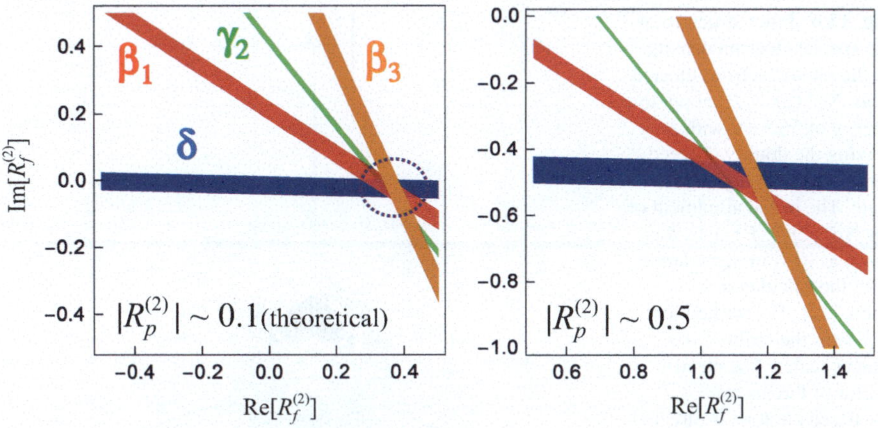

Fig. 13.5 Theoretical simulation of complete experiment on photoionization of $Ar^+(3p^{5\ 2}P)$ with accounting for the quadrupole ionization amplitude. The photon energy of light, generating the sequential two-photon double ionization is 1 keV. Left panel: Restrictions imposed on $R_f^{(2)} = q_f/d_s$ by imagined measurements of non-dipole parameters in sequential two-photon double ionization (see (13.12), (13.13)). δ (blue), γ_2 (green) by linearly polarized light and β_1 (red), β_3 (orange) by circularly polarized light, provided $R_p^{(2)} = q_p/d_s$ is close to theoretical expectation. Right panel demonstrates the loosing of consistency when $R_p^{(2)}$ is assumed 5 times larger than the theoretical expectation. The value of $R^{(1)} = d_d/d_s$ is taken 2.8 in accordance with calculations. The superscripts '(1)', '(2)' indicate the first and the second ionization steps, respectively

by the measurement of PADs in sequential two-photon double ionization. Such a complete experiment with extracting both dipole and non-dipole photoionization amplitudes has not been yet realized. We theoretically simulated such an experiment for the sequential two-photon double ionization of argon within the Cooper-Zare model, similar to the case, described in Sect. 13.5. Some of the results of the simulation are demonstrated in Fig. 13.5. It is seen how deviation of one of the absolute amplitude ratio from correct value generates an inconsistency: the allowed amplitude space vanishes, within prescribed 10% accuracy of the imaged measurements.

13.7 Spacial Electron Density

Besides observable quantities which are usually treated in connection with the complete experiment, such as angular distribution and polarization of the reaction products, including secondary products, different kind of vector correlations, partial cross sections, it is possible to find other physical values, characterizing the final system or its subsystem using (13.2). One can find, for example, such formally observable physical quantities as probability density of finding photoelectron at the coordinate $\mathbf{r}_{ph}$ ('photoelectron charge cloud'), or electron density of the residual ion, or elec-

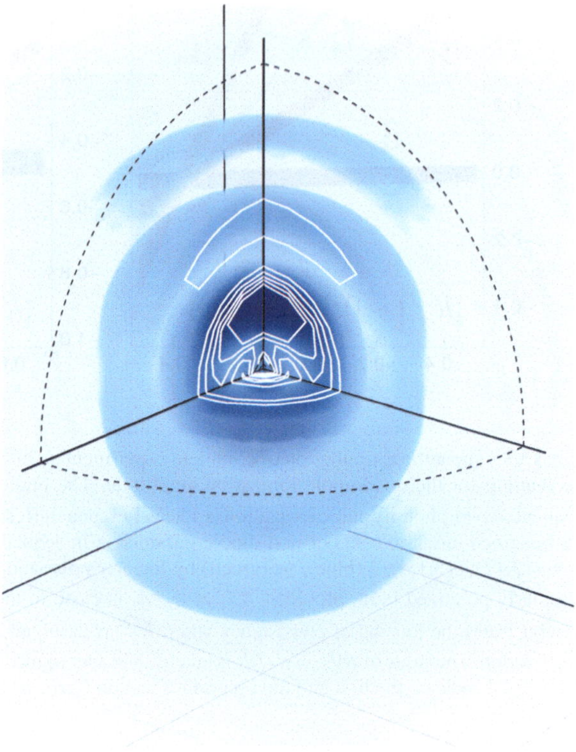

Fig. 13.6 Inner structure of the spacial electron density of the photoelectrons ejected from $Ne^+(2p^5\,{}^2P)$ at photon energy of 56.5 eV with leaving the doubly charged ion in the $Ne^{2+}(2p^4\,{}^3P)$ state. The initial alignment of the $Ne^+(2p^5\,{}^2P)$ state was taken as -0.1, in accordance with the calculated value [22]. The dashed line indicates the radius 5 a.u. (linear scale). The electric vector of the linearly polarized photons is directed along the vertical axis

tric currents within the electron clouds. But to do this, we have to express the basis functions $|\,\chi_f\,\rangle$ in the coordinate representation.

Consider, for example, the photoelectron cloud. The starting point is (13.2) $\langle n_{ph}\rangle = \mathrm{Tr}\,\rho_{ph}n_{ph}$, where ρ_{ph} is the density matrix of the photoelectron, which is calculated as a trace of the density matrix of the system 'residual ion + photoelectron', and n_{ph} is operator of the photoelectron density, $n_{ph}(\mathbf{r}) = \delta(\mathbf{r} - \mathbf{r}_{ph})$ in the coordinate representation; $\mathbf{r}_{ph}$ is the coordinate of the photoelectron. For fixed energy of the photoelectron E and fixed fine structure state of the residual ion $\alpha_f J_f$:

$$\begin{aligned}\mathrm{Tr}\,\rho_{ph}n_{ph} &= \sum_{\substack{\ell jm\\ \ell' j'm'}} \langle \ell jm\,|\,\rho_{ph}\,|\,\ell' j'm'\rangle\langle \ell' j'm'\,|\,n_{ph}\,|\,\ell jm\rangle \qquad (13.14)\\ &= \sum_{\substack{\ell jm\\ \ell' j'm'}} \left(\sum_{M_f} \langle \alpha_f J_f M_f, \ell jm\,|\,\rho\,|\,\alpha_f J_f M_f, \ell' j'm'\rangle\right) \langle \ell' j'm'\,|\,\delta(\mathbf{r} - \mathbf{r}_{ph})\,|\,\ell jm\rangle.\end{aligned}$$

After standard algebra [2] one finally obtains for the photoelectron cloud $W(\mathbf{r}_{ph})$ equations (12.7) and (12.8) of [32] with the trivial substitution $Y_{kq}(\vartheta, \varphi) \;\rightarrow\; R_{E\ell}(r_{ph})\,R^*_{E\ell'}(r_{ph})\,Y_{kq}(\theta_{ph}, \phi_{ph})$. Here $R_{E\ell}(r_{ph})$ is the radial function of the photoelectron partial wave.

As an example, for a particular case of the second-step photoelectron in sequential two-photon double ionization by linearly polarized light of the p-subshell of a noble gas atom (see Fig. 13.3), the photoelectron cloud is described by the general formula

$$W(\mathbf{r}_{ph}) = b_0(r_{ph}) + b_2(r_{ph})\, P_2(\cos\theta_{ph}) + b_4(r_{ph})\, P_4(\cos\theta_{ph})\,, \qquad (13.15)$$

where θ_{ph} is counted from the direction of the electric field, $b_k(r_{ph})$, $(k = 0,\, 2,\, 4)$ are known functions of the dipole matrix elements (13.5) and the radial functions $R_{E\ell}(r_{ph})$. The cloud depends on the latter functions, which in turn depend on the theoretical model. Obviously, these functions should be calculated in the model, which was used in the theoretical description of the complete experiment. But there is still a room for choosing these functions. For example, a variety of electron wavefunctions satisfy the Cooper-Zare model. Figure 13.6 presents an example of the electron cloud, which could be obtained from the results of the complete photoionization experiment on neon. The frozen-core Hartree-Fock radial functions $R_{Es}(r_{ph})$ and $R_{Ed}(r_{ph})$ of the photoelectron were substituted into (13.15).

13.8 Conclusion

Starting from the end of 1970s , when the concept of a complete photoionization experiment began to be fitted, the field have been grown vigorously, using to advantage the third-generation synchrotron radiation sources. In 2005 the first free-electron laser FLASH started operation in the XUV. With the creation of the XFELs X-ray scientists associate the possibility of progress in a number of very important scientific and practical areas; new avenues are opening in physics of interaction of radiation with matter. Unprecedented intensity and femtosecond time duration of the XUV pulses allow to investigate new phenomena in condensed matter, nanostructures, plasma. To understand phenomena in complex systems, it is necessary to understand how radiation interacts with simple systems. The complete experiments on photoionization, as a method of detailed study of the elementary photoprocesses with small quantum systems, can significantly contribute into this understanding. We hope that our overview shows that the field of complete photoionization experiments has received a push for further development.

Acknowledgements We like to acknowledge many colleagues, experimentalists and theoreticians, for fruitful discussions on the topic of the overview and collaboration during last few years, especially K. Bartschat, S. M. Burkov, P. Carpeggiani, N. Douguet, G. Hartmann, M. Ilchen, N. M. Kabachnik, A. K. Kazansky, T. Mazza, M. Meyer, R. Moshammer, K. C. Prince, A. Rudenko, G. Sansone, E. I. Staroselskaya, S. I. Strakhova, K. Ueda, J. Viefhaus, and many others. ANG gratefully acknowledges kind hospitality of the European XFEL. EVG acknowledges financial support from the Basis foundation via the "Junior Leader" program.

References

1. N. Andersen, K. Bartschat, *Polarization, Alignment, and Orientation in Atomic Collisions* (Springer, 2017)
2. V.V. Balashov, A.N. Grum-Grzhimailo, N.M. Kabachnik, *Polarization and Correlation Phenomena in Atomic Collisions. A Practical Theory Course* (Kluwer Academic/Plenum Publishers, 2000)
3. U. Becker, Complete photoionisation experiments. J. Electr. Spectrosc. Rel. Phenom. **96**, 105–115 (1998)
4. U. Becker, A. Crowe (eds.), *Complete Scattering Experiments* (Kluwer Academic/Plenum Publishers, 2001)
5. B. Bederson, The 'Perfect' scattering experiment. I. Comm. Atom. Mol. Phys. **1**, 41–45 (1969)
6. H.-J. Beyer, J.B. West, K.J. Ross, K. Ueda, N.M. Kabachnik, H. Hamdy, H. Kleinpoppen, A new approach to the complete photoionization experiment, by means of a coincidence measurement between autoionized electrons and polarized fluorescent photons, in the region of the 3p3d resonance in calcium. J. Phys. B: At. Mol. Opt. Phys. **28**, L47–L52 (1995)
7. S.M. Bilen'kii, L.I. Lapidus, L.D. Puzikov, R.M. Ryndin, On finding the matrix of the $a + a \rightarrow b + b'$ reaction. J. Exptl. Theoret. Phys. (U.S.S.R.) **35**, 959–961 (1958), Soviet Phys. JETP **8**, 669–670 (1959)
8. P. Bolognesi, A. De Fanis, M. Coreno, L. Avaldi, Complete characterization of the Ar $2p_{3/2}$ photoionization via Auger-electron photoelectron coincidence experiments. Phys. Rev. A **70**, 022701(1)–022701(5) (2004)
9. M. Braune, G. Hartmann, M. Ilchen, A. Knie, T. Lischke, A. Reinköster, A. Meissner, S. Deinert, L. Glaser, O. Al-Dossary, A. Ehresmann, A.S. Kheifets, J. Viefhaus, Electron angular distributions of noble gases in sequential two-photon double ionization. J. Mod. Optics **63**, 324–333 (2016)
10. C.D. Caldwell, R.N. Zare, Alignment of Cd atoms by photoionization. Phys. Rev. A **16**, 255–262 (1977)
11. N.A. Cherepkov, Spin polarization of photoelectrons ejected from unpolarized atoms. J. Phys. B: At. Mol. Phys. **12**, 1279–1296 (1979)
12. N.A. Cherepkov, G. Raseev, J. Adachi, Y. Hikosaka, K. Ito, S. Motoki, M. Sano, K. Soejima, A. Yagishita, K-shell photoionization of CO: II. Determination of dipole matrix elements and phase differences. J. Phys. B: At. Mol. Opt. Phys. **33**, 4213–4236 (2000)
13. J. Colgan, M.S. Pindzola, Fine structure continuum cross section ratios in the two-photon ionization of rubidium using elliptically polarized light. Phys. Rev. Lett. **86**, 1998–2001 (2001)
14. J.W. Cooper, Photoelectron-angular-distribution parameters for rare-gas atoms. Phys. Rev. A **47**, 1841–1851 (1993)
15. J.W. Cooper, R.N. Zare, Angular distribution of photoelectrons. J. Chem. Phys. **48**, 942–943 (1968)
16. J.W. Cooper, R.N. Zare: In S. Geltman, K.T. Mahanthappa, W.E. Britten (eds.), *Lectures in Theoretical Physics*, vol. XI-C (Gordon and Breach, New York, 1969), p. 317
17. A. Dodhy, R.N. Compton, J.A.D. Stockdale, Photoelectron angular distributions for near-threshold two-photon ionization of cesium and rubidium atoms. Phys. Rev. Lett. **54**, 422–425 (1985)
18. N. Douguet, A.N. Grum-Grzhimailo, E.V. Gryzlova, E.I. Staroselskaya, J. Venzke, K. Bartschat, Photoelectron angular distributions in bichromatic atomic ionization induced by circularly polarized VUV femtosecond pulses. Phys. Rev. A **93**, 033402(1)–033402(10) (2016)
19. J.A. Duncanson Jr., M.P. Strand, A. Lindgard, R.S. Berry, Angular distribution of electrons from resonant two-photon ionization of sodium. Phys. Rev. Lett. **37**, 987–990 (1976)
20. U. Fano, Propensity rules: an analytical approach. Phys. Rev. A **32**, 617–618 (1985)
21. S. Flügge, W. Mehlhorn, V. Schmidt, Angular distribution of Auger electrons following photoionization. Phys. Rev. Lett. **29**, 7–9 (1972)

22. S. Fritzsche, A.N. Grum-Grzhimailo, E.V. Gryzlova, N.M. Kabachnik, Angular distributions and angular correlations in sequential two-photon double ionization of atoms. J. Phys. B: At. Mol. Opt. Phys. **41**, 165601(1)–165601(12) (2008)
23. O. Geßner, Y. Hikosaka, B. Zimmermann, A. Hempelmann, R.R. Lucchese, J.H.D. Eland, P.-M. Guyon, U. Becker, $4\sigma^{-1}$ Inner valence photoionization dynamics of NO derived from photoelectron-photoion angular correlations. Phys. Rev. Lett. **88**, 193002(1)–193002(4) (2002)
24. L. Giannessi, E. Allaria, K.C. Prince, C. Callegari, G. Sansone, K. Ueda, T. Morishita, C.N. Liu, A.N. Grum-Grzhimailo, E.V. Gryzlova, N. Douguet, K. Bartschat, Coherent control schemes for the photoionization of neon and helium in the Extreme Ultraviolet spectral region. Sci. Rep. **8**, 7774(1)–7774(12) (2018)
25. K. Godehusen, P. Zimmermann, A. Verweyen, A. von dem Borne, P.H. Wernet, B. Sonntag, A complete photoionization experiment with polarized atoms using magnetic dichroism and phase tilt measurements. Phys. Rev. A **58**, R3371–R3374 (1998)
26. A.N. Grum-Grzhimailo, Non-dipole effects in magnetic dichroism in atomic photoionization. J. Phys. B At. Mol. Opt. Phys. **34**, L359–L365 (2001)
27. A.N. Grum-Grzhimailo, A. Dorn, W. Mehlhorn, On complete experiments for Auger decay. Comm. Atom. Mol. Phys. Comm. Mod. Phys. D **1**, 29–39 (1999)
28. A.N. Grum-Grzhimailo, E.V. Gryzlova, M. Meyer, Non-dipole effects in the angular distribution of photoelectrons in sequential two-photon atomic double ionization. J. Phys. B At. Mol. Opt. Phys. **45**, 215602(1)–215602(9) (2012)
29. A.N. Grum-Grzhimailo, E.V. Gryzlova, M. Meyer, Towards complete photoionization experiments beyond the dipole approximation, in *Book of abstracts, IWP-RIXS-2017 International Workshop on Photoionization & Resonant Inelastic X-ray Scattering* (Aussois, France, 2017), p. 26
30. A.N. Grum-Grzhimailo, E.V. Gryzlova, S. Fritzsche, N.M. Kabachnik, Photoelectron angular distributions and correlations in sequential double and triple atomic ionization by free electron lasers. J. Mod. Optics **63**, 334–357 (2016)
31. A.N. Grum-Grzhimailo, E.V. Gryzlova, E.I. Staroselskaya, J. Venzke, K. Bartschat, Interfering one-photon and two-photon ionization by femtosecond VUV pulses in the region of an intermediate resonance. Phys. Rev. A **91**, 063418(1)–063418(9) (2015)
32. E.V. Gryzlova, A.N. Grum-Grzhimailo, Chapter *Effects of hyperfine interaction in atomic photoionization*, (in this book)
33. E.V. Gryzlova, A.N. Grum-Grzhimailo, E.I. Kuzmina, S.I. Strakhova, Sequential two-photon double ionization of noble gases by circularly polarized XUV radiation. J. Phys. B: At. Mol. Opt. Phys. **47**, 195601(1)–195601(11) (2014)
34. E.V. Gryzlova, A.N. Grum-Grzhimailo, E.I. Staroselskaya, S.I. Strakhova, Similarity between the angular distributions of the first- and second-step electrons in sequential two-photon atomic double ionization. J. Electr. Spectrosc. Rel. Phenom. **204**, 277–283 (2015)
35. E.V. Gryzlova, A.N. Grum-Grzhimailo, E.I. Staroselskaya, N. Douguet, K. Bartschat, Quantum coherent control of the photoelectron angular distribution in bichromatic-field ionization of atomic neon. Phys. Rev. A **97**, 013420(1)–013420(9) (2018)
36. R. Guillemin, O. Hemmers, D.W. Lindle, S.T. Manson, Experimental investigation of nondipole effects in photoemission at the advanced light source. Rad. Phys. Chem. **75**, 2258–2274 (2006)
37. L.H. Haber, B. Doughty, S.R. Leone, Continuum phase shifts and partial cross sections for photoionization from excited states of atomic helium measured by high-order harmonic optical pump-probe velocity map imaging. Phys. Rev. A **79**, 031401(R1)–031401(R4) (2009)
38. G. Hartmann, M. Ilchen, A. Achner, A. Beckmann, C. Callegari, R. Cucini, A. de Fanis, E. Ferrari, P. Finetti, L. Glaser, A.N. Grum-Grzhimailo, E.V. Gryzlova, J. Buck, A. Knie, A. Lindahl, T. Mazza, M. Meyer, E. Roussel, F. Scholz, I. Shevchuk, J. Seltmann, J. Viefhaus, P. Walter, M. Zagrando, Observation of strong non-dipole effects in sequential multi-photon ionization using VUV FEL radiation, in *International Conference on Many Particle Spectroscopy of Atoms, Molecules, Clusters and Surfaces (MPS-2016), Book of Abstracts* (Moscow, Russia 23–26 Aug 2016), p. 19

39. A. Hausmann, B. Kämmerling, H. Kossmann, V. Schmidt, New approach for a perfect experiment: 2p photoionization in atomic magnesium. Phys. Rev. Lett. **61**, 2669–2672 (1988)
40. C. Heckenkamp, F. Schäfers, G. Schönhense, U. Heinzmann, Experimental characterization of the Xe 5p photoionization by angle-and spin-resolved photoelectron spectroscopy. Z. Phys. D **2**, 257–274 (1986)
41. U. Heinzmann, Experimental determination of the phase differences of continuum wavefunctions describing the photoionisation process of xenon atoms: I. Measurements of the spin polarisations of photoelectrons and their comparison with theoretical results. J. Phys. B: At. Mol. Phys. **13**, 4353–4366 (1980)
42. U. Heinzmann, Experimental determination of the phase differences of continuum wavefunctions describing the photoionisation process of xenon atoms: II. Evaluation of the matrix elements and the phase differences and their comparison with data in the discrete spectral range in application of the multichannel quantum defect theory. J. Phys. B: At. Mol. Phys. **13**, 4367–4381 (1980)
43. O. Hemmers, R. Guillemin, D.W. Lindle, Nondipole effects in soft X-ray photoemission. Rad. Phys. Chem. **70**, 123–147 (2004)
44. U. Hergenhahn, G. Snell, M. Drescher, B. Schmidtke, N. Müller, U. Heinzmann, M. Wiedenhöft, U. Becker, Dynamically induced spin polarization of resonant Auger electrons. Phys. Rev. Lett. **82**, 5020–5023 (1999)
45. V.L. Jacobs, Theory of atomic photoionization measurements. J. Phys. B: At. Mol. Phys. **5**, 2257–2271 (1972)
46. N.M. Kabachnik, I.P. Sazhina, On the problem of a complete experimental characterisation of Auger decay. J. Phys. B: At. Mol. Opt. Phys. **23**, L353–L357 (1990)
47. B. Kämmerling, V. Schmidt, Complete fragmentation pattern for two-step double photoionization in xenon. Phys. Rev. Lett. **67**, 1848–1851 (1991)
48. J. Kessler, The "Perfect" photoionization experiment. Comm. Atom. Mol. Phys. **10**, 47–55 (1981)
49. H. Klar, H. Kleinpoppen, Angular distribution of photoelectrons from polarised atoms exposed to polarised radiation. J. Phys. B: At. Mol. Opt. Phys. **15**, 933–950 (1982)
50. H. Kleinpoppen, Analysis of scattering amplitudes in polarized-electron-atom collisions I. Elastic scattering on one-electron atoms and the excitation process $^2S_{1/2} \rightarrow ^2P_{1/2,3/2}$. Phys. Rev. A **3**, 2015–2027 (1971)
51. H. Kleinpoppen, B. Lohmann, A.N. Grum-Grzhimailo, *Perfect/Complete Scattering Experiments. Probing Quantum Mechanics on Atomic and Molecular Collisions and Coincidences* (Springer, 2013)
52. K.J. Kollath, Theory for laser photoionisation of excited atoms: $n^2P_{1/2,3/2}$ states of Cs. J. Phys. B: Atom. Mol. Phys. **13**, 2901–2919 (1980)
53. A. Kupliauskienė, N. Rakštikas, V. Tutlis, Polarization studies in the photoionization of atoms using a graphical technique. J. Phys. B: At. Mol. Opt. Phys. **34**, 1783–1803 (2001)
54. H. Lörch, J.M. Bizau, N. Scherer, S. Diehl, D. Cubaynes, O. Zerouni, F.J. Wuilleumier, V. Schmidt, W.R. Johnson, Complete description of 3p photoionization in calcium. J. Phys. B: At. Mol. Opt. Phys. **32**, 2215–2226 (1999)
55. T. Mazza, M. Ilchen, A.J. Rafipoor, C. Callegari, P. Finetti, O. Plekan, K.C. Prince, R. Richter, A. Demidovich, C. Grazioli, L. Avaldi, P. Bolognesi, M. Coreno, P. O'Keeffe, M. Di Fraia, M. Devetta, Y. Ovcharenko, V. Lyamayev, S. Düsterer, K. Ueda, J.T. Costello, E.V. Gryzlova, S.T. Strakhova, A.N. Grum-Grzhimailo, A.V. Bozhevolnov, A.K. Kazansky, N.M. Kabachnik, M. Meyer, Angular distribution and circular dichroism in the two-colour XUV+NIR above-threshold ionization of helium. J. Mod. Optics **63**, 367–382 (2016)
56. M. Meyer, D. Cubaynes, D. Glijer, J. Dardis, P. Hayden, P. Hough, V. Richardson, E.T. Kennedy, J.T. Costello, P. Radcliffe, S. Düsterer, A. Azima, W.B. Li, H. Redlin, J. Feldhaus, R. Taïeb, A. Maquet, A.N. Grum-Grzhimailo, E.V. Gryzlova, S.I. Strakhova, Polarization control in two-color above-threshold ionization of atomic helium. Phys. Rev. Lett. **101**, 193002(1)–193002(4) (2008)

57. R. Moshammer, Y.H. Jiang, L. Foucar, A. Rudenko, Th. Ergler, C.D. Schröter, S. Lüdemann, K. Zrost, D. Fischer, J. Titze, T. Jahnke, M. Schöffler, T. Weber, R. Dörner, T.J.M. Zouros, A. Dorn, T. Ferger, K.U. Kühnel, S. Düsterer, R. Treusch, P. Radcliffe, E. Plönjes, J. Ullrich, Few-photon multiple ionization of Ne and Ar by strong free-electron-laser pulses. Phys. Rev. Lett. **98**, 203001(1)–203001(4) (2007)
58. T. Nakajima, Possibility of direct determination of the quantum phase of continua utilizing the phase of lasers. Phys. Rev. A **61**, 041403(R1)–041403(R4) (2000)
59. P. O'Keeffe, S. Aloïse, S. Fritzsche, B. Lohmann, U. Kleiman, M. Meyer, A.N. Grum-Grzhimailo, Resonant Auger decay of $Xe^*4d_{5/2}^{-1}6p$: A contribution to the complete experiment from fluorescence polarization studies. Phys. Rev. A **70**, 012705(1)–012705(14) (2004)
60. P. O'Keeffe, P. Bolognesi, A. Mihelič, A. Moise, R. Richter, G. Cautero, L. Stebel, R. Sergo, L. Pravica, E. Ovcharenko, P. Decleva, L. Avaldi, Photoelectron angular distributions from polarized Ne* atoms near threshold. Phys. Rev. A **82**, 052522(1)–052522(11) (2010)
61. O. Plotzke, G. Prümper, B. Zimmermann, U. Becker, H. Kleinpoppen, Magnetic dichroism in the angular distribution of atomic oxygen 2p photoelectrons. Phys. Rev. Lett. **77**, 2642–2645 (1996)
62. K.C. Prince, E. Allaria, C. Callegari, R. Cucini, G. De Ninno, S. Di Mitri, B. Diviacco, E. Ferrari, P. Finetti, D. Gauthier, L. Giannessi, N. Mahne, G. Penco, O. Plekan, L. Raimondi, P. Rebernik, E. Roussel, C. Svetina, M. Trov, M. Zangrando, M. Negro, P. Carpeggiani, M. Reduzzi, G. Sansone, A.N. Grum-Grzhimailo, E.V. Gryzlova, S.I. Strakhova, K. Bartschat, N. Douguet, J. Venzke, D. Iablonskyi, Y. Kumagai, T. Takanashi, K. Ueda, A. Fischer, M. Coreno, F. Stienkemeier, Y. Ovcharenko, T. Mazza, M. Meyer, Coherent control with a short-wavelength free-electron laser. Nat. Photonics **10**, 176–179 (2016)
63. L.D. Puzikov, Scattering of particles of arbitrary spin. J. Exptl. Theoret. Phys. (U.S.S.R.) **34**, 947–952 (1958), Soviet Phys. JETP **7**, 655–658 (1958)
64. L. Puzikov, R. Ryndin, Ia. Smorodinskii, Reconstruction of the scattering matrix of a two-nucleon system. J. Exptl. Theoret. Phys. (U.S.S.R.) **32**, 592–600 (1957), Soviet Phys. JETP **5**, 489–495 (1957)
65. K.L. Reid, D.H. Leahy, R.N. Zare, Complete description of molecular photoionization from circular dichroism of rotationally resolved photoelectron angular distributions. Phys. Rev. Lett. **68**, 3527–3530 (1992)
66. K.L. Reid, Photoelectron angular distributions. Annu. Rev. Phys. Chem. **54**, 397–424 (2003)
67. S.J. Schaphorst, Q. Qian, B. Krässig, P. van Kampen, N. Scherer, V. Schmidt, Matrix elements for $4d_{5/2}$ photoionization in xenon derived from coincidence electron spectrometry. J. Phys. B: At. Mol. Opt. Phys. **30**, 4003–4017 (1997)
68. G. Schönhense, U. Heinzmann, Evidence of strong international coupling in Hg 5d photoionization by "experimental" transition matrix elements. Phys. Rev. A **29**, 987–990 (1984)
69. P.S. Shaw, U. Arp, S.H. Southworth, Measuring nondipolar asymmetries of photoelectron angular distributions. Phys. Rev. A **54**, 1463–1472 (1996)
70. Y.A. Smorodinsky, The complete experiment in beta-decay. Soviet Physics JETP **9**, 1142–1143 (1959), J. Exptl. Theoret. Phys. (U.S.S.R.) **36**, 1606–1608 (1959)
71. G. Snell, B. Langer, M. Drescher, N. Müller, B. Zimmermann, U. Hergenhahn, J. Viefhaus, U. Heinzmann, U. Becker, Complete description of the Xe 4d photoionization by spin-resolved photoelectron and Auger spectroscopy. Phys. Rev. Lett. **82**, 2480–2483 (1999)
72. K. Ueda, Catching and controlling electrons in action with fully coherent FEL. Invited talk at the conference WAVEFRONT: New Frontiers and Advanced Applications of 4th generation light sources to Atomic, Molecular, Optical, and Cluster Science. ICPT, Trieste, Italy/30 November—1 December 2016
73. K. Ueda, J. West, K.J. Ross, H.J. Beyer, N.M. Kabachnik, A study of the 4p-excited autoionization resonances in Sr by measurement of the angular correlation between the photoejected electrons and polarized fluorescent photons. J. Phys. B: At. Mol. Opt. Phys. **31**, 4801–4812 (1998)
74. Z.M. Wang, D.S. Elliott, Complete measurements of two-photon ionization of atomic rubidium using elliptically polarized light. Phys. Rev. A **62**, 053404(1)–053404(14) (2000)

75. Z.-M. Wang, D.S. Elliott, Determination of the phase difference between even and odd continuum wave functions in atoms through quantum interference measurements. Phys. Rev. Lett. **87**, 173001(1)–173001(4) (2001)
76. J.B. West, K.J. Ross, K. Ueda, H.J. Beyer, Angular correlation measurement between the photo-excited autoionized electron and subsequent polarized fluorescent photon at an autoionization resonance of Sr. J. Phys. B: At. Mol. Opt. Phys. **31**, L647–L654 (1998)

Chapter 14
Theoretical Aspects of Laser-Assisted $(e, 2e)$ Collisions in Atoms

Konstantin A. Kouzakov

Abstract An overview of theoretical approaches to ionization of atomic systems by electron impact in the presence of laser radiation is given. Basic approximations for calculating multiphoton $(e, 2e)$ transition amplitudes are discussed, with special emphasis on the first Born approximation in the projectile-target interaction. Various methods for the treatment of the dressing of initial and final (ionized) atomic-target states by a laser field are brought into focus, ranging respectively from the Floquet theory to the time-dependent perturbation theory and two-level approximation and from the Coulomb-Volkov models to the Sturmian-Floquet approach.

14.1 Introduction

The advance in laser technologies stimulates laser applications in various fields of atomic physics [1–3]. One of such promising developments is laser-assisted elastic and inelastic electron-atom scattering [4], which also includes an ionization channel, namely, the $(e, 2e)$ ionization of atoms in the presence of laser radiation. The latter was measured for the first time by Höhr et al. [5, 6] in the case of a helium atomic target and a Nd:YAG laser beam ($\lambda = 1064$ nm) with intensity $I = 4 \times 10^{12}$ W/cm^2. While the pioneering experiment was performed not so long ago, theoretical investigations of laser-assisted $(e, 2e)$ processes on atoms began much earlier, dating back to the 1970s (see, for instance, [7]). A number of results concerning the dependence of the multiphoton $(e, 2e)$ cross sections on laser-field parameters, such as polarization, frequency and intensity, have been obtained (see [4] for a review of some earlier works and also more recent articles [8–18]). The main theoretical findings can be briefly summarized as follows: (i) the cross sections are seriously modified even by the presence of low-intensity laser radiation and (ii) they strongly depend on the dressing of the atomic target states. The aforementioned experiment of Höhr et al. [5, 6] confirmed the existence of distinct differences in the $(e, 2e)$ differential cross sections

K. A. Kouzakov (✉)
Faculty of Physics, Lomonosov Moscow State University, Moscow 119991, Russia
e-mail: kouzakov@srd.sinp.msu.ru; kouzakov@gmail.com

K. Yamanouchi et al. (eds.), *Progress in Photon Science*, Springer Series in Chemical Physics 119, https://doi.org/10.1007/978-3-030-05974-3_14

between laser-on and laser-off conditions. This indicates the importance of further developing the theory of laser-assisted $(e, 2e)$ collisions for future experiments.

The present contribution aims at outlining a general formalism along with the main methods and approximations, which are typically employed in the theoretical treatment of laser-assisted $(e, 2e)$ collisions in atoms. Atomic units (a.u., $\hbar = e = m_e$) are used throughout unless otherwise specified.

14.2 General Formulation

We consider the process

$$e^- + \mathrm{A} + \ell\omega \rightarrow 2e^- + \mathrm{A}^+, \tag{14.1}$$

where, in the presence of a laser field, an electron impinges on an atomic target A and induces an ionizing collision, in which a net number ℓ of photons with frequency ω is exchanged between the colliding system and the field. As a result, an outgoing electron pair emerges which is formed by the scattered and ejected electrons. In what follows, the incident, scattered, and ejected electron energies and momenta are specified respectively by $(E_0,\mathbf{p}_0)$, $(E_s, \mathbf{p}_s)$, and $(E_e,\mathbf{p}_e)$.

The laser field is assumed to switch on and off adiabatically at $t \rightarrow -\infty$ and $t \rightarrow +\infty$, respectively. More specifically, the turn on and off time δT of the laser pulse as well as the laser pulse duration T are very long on a time scale typical for the target and much longer than the $(e, 2e)$ collision duration. We consider the case of a monochromatic elliptically polarized laser wave with a wave vector $\mathbf{k}$ $(k = \omega/c)$. The frequency ω and intensity I of the wave are such that the laser electric-field amplitude F_0 is much less than the typical intra-atomic field F_A and the Keldysh parameter [19] is $\gamma = \omega F_\mathrm{A}/(2E_I F_0) \gg 1$, where E_I is the atomic ionization energy. This means that the ionization due to the laser electric field occurs in the perturbative regime via multiphoton transitions and hence it does not produce any appreciable effect compared to that due to the electron-atom collision. Without loss of generality we suppose that the z axis is directed along $\mathbf{k}$. A typical situation is when the laser wavelength $\lambda = 2\pi/k$ is much greater than the spatial extent both of the target and of the region where the electron-electron collision takes place. This validates the use of the dipole approximation for the electric component of the laser field:

$$\mathbf{F}(t) = F_x\mathbf{e}_x \cos\omega t + F_y\mathbf{e}_y \sin\omega t, \tag{14.2}$$

where $F_x > 0$ and $F_y > 0$ $(F_y < 0)$ for right (left) polarization. The vector potential corresponding to (14.2) is

$$\mathbf{A}(t) = A_x\mathbf{e}_x \sin\omega t + A_y\mathbf{e}_y \cos\omega t, \tag{14.3}$$

with $A_x = -cF_x/\omega$, $A_y = cF_y/\omega$. Note that the case of linear polarization derives from (14.2) upon setting $F_x = F_0$ and $F_y = 0$, while that of circular polarization amounts to $F_x = |F_y| = F_0/\sqrt{2}$.

14.2.1 S Matrix

The rate of the discussed laser-assisted $(e, 2e)$ reaction is governed by the matrix element of the scattering operator, which is usually called the S operator. Using the Furry representation [20], in which the effect of an external field is included in the asymptotic Hamiltonians of the initial and final channels of the reaction, the S matrix can be presented as

$$S_{(e,2e)} = -i \int_{-\infty}^{\infty} \mathrm{d}t \left\langle \Psi_f^{(-)}(\mathbf{p}_s, \mathbf{p}_e; t) | V_{eA} | \chi_{\mathbf{p}_0}(t) \Phi_i(t) \right\rangle, \tag{14.4}$$

where V_{eA} is the projectile-atom potential, $\Phi_i(t)$ is the laser-dressed initial atomic state, and $\Psi_f^{(-)}(\mathbf{p}_s, \mathbf{p}_e; t)$ is the final (time-reversed) scattering state of the colliding system in the presence of the laser field. The incident electron state $\chi_{\mathbf{p}_0}(t)$ is given by the Gordon-Volkov function, which solves the following Schrödinger equation:

$$i \frac{\partial}{\partial t} \chi_{\mathbf{p}_0}(\mathbf{r}, t) = \frac{1}{2} \left[\hat{\mathbf{p}} + \frac{1}{c} \mathbf{A}(t) \right]^2 \chi_{\mathbf{p}_0}(\mathbf{r}, t). \tag{14.5}$$

For the vector potential given by (14.3) one has (see, for instance, [2])

$$\chi_{\mathbf{p}_0}(\mathbf{r}, t) = \exp \left\{ i \left[\mathbf{p}_0 \mathbf{r} - \alpha_{\mathbf{p}_0} \sin(\omega t + \delta_{\mathbf{p}_0}) - E_0 t - \zeta(t) \right] \right\}, \tag{14.6}$$

where $E_0 = p_0^2/2$ and

$$\alpha_{\mathbf{p}_0} = \frac{\sqrt{F_x^2 p_{0,x}^2 + F_y^2 p_{0,y}^2}}{\omega^2}, \qquad \delta_{\mathbf{p}_0} = \arcsin \left(\frac{F_x p_{0,x}}{\sqrt{F_x^2 p_{0,x}^2 + F_y^2 p_{0,y}^2}} \right),$$

$$\zeta(t) = \frac{1}{2c^2} \int_{-\infty}^{t} A^2(t') \mathrm{d}t'.$$

By definition, the S matrix remains invariant under unitary transformations. Therefore, (14.4) is gauge-invariant, since the gauge transformation of the vector and scalar potentials of the laser field is equivalent to the unitary transformation in quantum mechanics (see [21] for detail).

14.2.2 Cross Sections

It can be shown that, after integrating over time in (14.4), the S matrix has the general form

$$S_{(e,2e)} = -2\pi i \sum_{\ell=-\infty}^{\infty} T_{fi}^{(\ell)} \delta(E_0 + \mathscr{E}_i - \mathscr{E}_f - E_s - E_e - U_p + \ell\omega), \qquad (14.7)$$

where $\mathscr{E}_i$ and $\mathscr{E}_f$ are the quasienergies (see below) of the laser-dressed initial atomic and final ionic states, respectively, $U_p = F_0^2/4\omega^2$ is a ponderomotive potential, and $T_{fi}^{(\ell)}$ are the ℓ-photon transition amplitudes. For the fully differential cross section (FDCS), which provides the most detailed information about the scattering process, one thus has

$$\frac{\mathrm{d}^4\sigma}{\mathrm{d}E_s\mathrm{d}E_e\mathrm{d}\Omega_s\mathrm{d}\Omega_e} = \sum_{\ell=-\infty}^{\infty} \frac{\mathrm{d}^3\sigma^{(\ell)}}{\mathrm{d}E_e\mathrm{d}\Omega_s\mathrm{d}\Omega_e} \delta(E_0 + \mathscr{E}_i - \mathscr{E}_f - E_s - E_e - U_p + \ell\omega), \qquad (14.8)$$

where the ℓ-photon triple differential cross section (TDCS) is

$$\frac{\mathrm{d}^3\sigma^{(\ell)}}{\mathrm{d}E_e\mathrm{d}\Omega_s\mathrm{d}\Omega_e} = \frac{p_s p_e}{(2\pi)^5 p_0} |T_{fi}^{(\ell)}|^2. \qquad (14.9)$$

14.3 Theoretical Methods and Approximations

For calculating the S matrix and cross sections one has to know the two states: (i) the initial laser-dressed atomic state $\Phi_i(t)$ and (ii) the final scattering state $\Psi_f^{(-)}(\mathbf{p}_s, \mathbf{p}_e; t)$ of the colliding system in the presence of the laser field. The first is the solution of the time-dependent Schrödinger equation (TDSE) for an atom in the laser field, and the second solves the TDSE for the interacting projectile-target system in the laser field and obeys the proper asymptotic behavior (when $t \to \infty$ and relative positions of the final ion and two outgoing electrons tend to infinity). Since both Hamiltonians are periodic in time, the TDSE can be solved employing the Floquet theory [22].

14.3.1 Initial Laser-Dressed Atomic State

In the Floquet theory, one seeks the solution to the TDSE for the initial state,

$$i\frac{\partial}{\partial t}|\Phi_i(t)\rangle = [H_\mathrm{A} + H_\mathrm{int}(t)]|\Phi_i(t)\rangle, \qquad (14.10)$$

where H_A is the field-free atomic Hamiltonian and $H_{int}(t)$ is the atom-field interaction Hamiltonian, in the form of the following (Floquet-Fourier) expansion:

$$|\Phi_i(t)\rangle = e^{-i\mathcal{E}_i t} \sum_{n=-\infty}^{\infty} e^{-in\omega t} |\Phi_i^{(n)}(\mathcal{E}_i)\rangle. \tag{14.11}$$

The Floquet-Fourier components $|\Phi_i^{(n)}(\mathcal{E}_i)\rangle$ satisfy the system of coupled time-independent equations

$$(H_A - n\omega - \mathcal{E}_i)|\Phi_i^{(n)}(\mathcal{E}_i)\rangle + \sum_{k=-\infty}^{\infty} (H_{int})_{n-k}|\Phi_i^{(k)}(\mathcal{E}_i)\rangle = 0, \qquad n = 0, \pm 1, \pm 2, \ldots \tag{14.12}$$

Here $(H_{int})_{n-k}$ are the components of the Fourier expansion

$$H_{int}(t) = \sum_{n=-\infty}^{\infty} e^{-in\omega t} (H_{int})_n. \tag{14.13}$$

It can be seen that the solution of the system (14.12) does not define the quasienergy $\mathcal{E}_i$ uniquely, since the latter can be changed to $\mathcal{E}_i + m\omega$, where m is an arbitrary integer. The customary way of defining the quasienergy consists in using the boundary condition at $t \to -\infty$,

$$|\Phi_i(t \to -\infty)\rangle \to e^{-iE_A^{(i)}t}|\Phi_A^{(i)}\rangle, \tag{14.14}$$

where $E_A^{(i)}$ is the energy of the field-free atomic state $|\Phi_A^{(i)}\rangle$, and requiring that only the $n = 0$ component of the expansion (14.11) remains nonvanishing in the limit $t \to -\infty$.

14.3.1.1 Time-Dependent Perturbation Theory

If the laser field is nonresonant with atomic transitions, the laser-atom interaction $H_{int}(t)$ appears to be weak due to the imposed condition $F_0 \ll F_A$ and, hence, can be treated as a perturbation. It is efficient to develop the time-dependent perturbation theory in the length gauge (L-gauge), where

$$H_{int}^{L}(t) = \mathbf{F}(t) \cdot \mathbf{R}, \qquad \mathbf{R} = \sum_{k=1}^{Z} \mathbf{r}_k, \tag{14.15}$$

with Z being the number of atomic electrons (or the nuclear charge) and $\mathbf{r}_k$ being their positions. The choice of the L-gauge is based on the observation that over the atomic region the electron-laser interaction (14.15) is much weaker than the intra-atomic potential experienced by electrons. Developing the time-dependent perturbation theory for

$$|\Phi_i^{\mathrm{L}}(t)\rangle = \exp\left(\frac{i}{c}\mathbf{A}(t)\cdot\mathbf{R}\right)|\Phi_i(t)\rangle \tag{14.16}$$

to first order, one obtains

$$|\Phi_i^{\mathrm{L}}(t)\rangle = \mathrm{e}^{-iE_{\mathrm{A}}^{(i)}t}\left[|\Phi_{\mathrm{A}}^{(i)}\rangle - \frac{1}{2}\sum_{j\neq 0}\left(\mathrm{e}^{i\omega t}\frac{M_{ji}^{+}}{\omega_{ji}+\omega} + \mathrm{e}^{-i\omega t}\frac{M_{ji}^{-}}{\omega_{ji}-\omega}\right)|\Phi_{\mathrm{A}}^{(j)}\rangle\right], \tag{14.17}$$

where $\omega_{ji} = E_{\mathrm{A}}^{(j)} - E_{\mathrm{A}}^{(i)}$ are the field-free atomic transition energies, and

$$M_{ji}^{\pm} = \left\langle\Phi_{\mathrm{A}}^{(j)}|F_x R_x \mp i F_y R_y|\Phi_{\mathrm{A}}^{(i)}\right\rangle$$

are the dipole transition matrix elements.

When $\omega \ll |\omega_{ji}|$ for all $j \neq i$ (the low-frequency regime), one can readily perform the n summation in (14.17) using the low-frequency ($\omega_{ji} \pm \omega \approx \omega_{ji}$) and closure ($\omega_{ji} \approx \omega_{\mathrm{cl}}$) approximations. This yields

$$\Phi_i^{\mathrm{L}}(X,t) = \mathrm{e}^{-iE_{\mathrm{A}}^{(i)}t}\left[1 - \frac{1}{\omega_{\mathrm{cl}}}\left(F_x R_x \cos\omega t + F_y R_y \sin\omega t\right)\right]\Phi_{\mathrm{A}}^{(i)}(X), \tag{14.18}$$

where $X = \{\mathbf{r}_1, \mathbf{r}_2, \ldots \mathbf{r}_Z\}$.

14.3.1.2 Two-Level Approximation

When the laser frequency ω is close to or coincides with a particular atomic transition energy ω_{ji}, the perturbation theory (14.17) is not applicable anymore. A more appropriate theoretical approach in such a case consists in using the two-level approximation[1]

$$|\Phi_i^{\mathrm{L}}(t)\rangle = \mathrm{e}^{-iE_{\mathrm{A}}^{(i)}t}a_i(t)|\Phi_{\mathrm{A}}^{(i)}\rangle + \mathrm{e}^{-iE_{\mathrm{A}}^{(j)}t}a_j(t)|\Phi_{\mathrm{A}}^{(j)}\rangle, \tag{14.19}$$

where the coefficients $a_i(t)$ and $a_j(t)$ are determined by solving the TDSE (14.10) in the L-gauge with ansatz (14.19). The latter can be done using the so-called rotating wave approximation (RWA) [1] which neglects the fast oscillating terms $\propto \mathrm{e}^{\pm i(\omega_{ji}+\omega)t}$ and $\propto \mathrm{e}^{\pm i\omega t}$ in comparison with the slow oscillating terms $\propto \mathrm{e}^{\pm i(\omega_{ji}-\omega)t}$. This procedure yields

[1]Here the field-free atomic states are assumed to be nondegenerate. Generalization to the case of degenerate states is straightforward (see, for instance, [13]).

$$a_i^{\pm}(t) = \sqrt{\frac{|\Delta| + \Omega}{2\Omega}} \exp\left[-\frac{i}{2}(\Delta \mp \Omega)t\right],$$

$$a_j^{\pm}(t) = \mp \frac{M_{ji}^{-}}{\sqrt{2\Omega(|\Delta| + \Omega)}} \exp\left[\frac{i}{2}(\Delta \pm \Omega)t\right]. \quad (14.20)$$

Here

$$\Delta = \omega_{ji} - \omega, \qquad \Omega = \sqrt{\Delta^2 + \chi_{ji}^2} \quad (14.21)$$

are the resonance detuning and the generalized Rabi frequency, respectively, while

$$\chi_{ji} = |M_{ji}^{-}|$$

stands for the Rabi frequency. Note that according to (14.20) the following relations hold true:

$$|a_i^{\pm}(t)|^2 + |a_j^{\pm}(t)|^2 = 1, \qquad |a_i^{\pm}(t)|^2 \geq \frac{1}{2}, \qquad |a_j^{\pm}(t)|^2 \leq \frac{1}{2}. \quad (14.22)$$

The laser-dressed target state is thus given by

$$\Phi_i^{L\pm}(X,t) = \exp\left\{-i\left[E_A^{(i)}t + \frac{1}{2}(\Delta \mp \Omega)t\right]\right\}$$
$$\times \sqrt{\frac{|\Delta| + \Omega}{2\Omega}} \left[\Phi_A^{(i)}(X) \mp e^{-i\omega t} \frac{M_{ji}^{-}}{|\Delta| + \Omega} \Phi_A^{(j)}(X)\right]. \quad (14.23)$$

The target wave function evolves into $\Phi_i^{L+}(X,t)$ or $\Phi_i^{L-}(X,t)$ according to whether $\Delta \geq 0$ or $\Delta < 0$.

14.3.2 Final Laser-Dressed Scattering State

In general, finding the final state $\Psi_f^{(-)}(\mathbf{p}_s, \mathbf{p}_e; t)$ is a more difficult task than in the $\Phi_i(t)$ case. Apart from dealing with a system that has an additional interacting electron (i.e., the projectile electron), one faces the three-body scattering problem involving Coulomb-tail potentials in the presence of a laser field. Since the solution is not known already in the field-free case [23], one has to resort to approximate treatments.

14.3.2.1 Asymptotic Behavior

Due to specifics of scattering on long-range potentials, such as Coulomb-tail potentials, a nontrivial issue which arises in the case of the presence of a laser field consists in imposing a proper asymptotic condition on the solution of the TDSE for $\Psi_f^{(-)}(\mathbf{p}_s, \mathbf{p}_e; t)$ when r_s, r_e, and $r_{se} = |\mathbf{r}_s - \mathbf{r}_e| \to \infty$. This issue is convenient to address in the accelerated, or Kramers-Henneberger (KH), frame [24]:

$$\Psi_f^{\mathrm{KH}(-)}(\mathbf{p}_s, \mathbf{p}_e; X_0, t) = \exp\left[\mathbf{a}(t)\cdot\sum_{k=0}^{Z+1}\nabla_k + i(Z+1)\zeta(t)\right]\Psi_f^{(-)}(\mathbf{p}_s, \mathbf{p}_e; X_0, t), \tag{14.24}$$

where $X_0 = \{\mathbf{r}_0, X\}$, and

$$\mathbf{a}(t) = \frac{1}{c}\int_{-\infty}^{t} \mathrm{d}t'\,\mathbf{A}(t') = a_x\mathbf{e}_x\cos\omega t + a_y\mathbf{e}_y\sin\omega t \tag{14.25}$$

is the displacement vector of a classical electron in the laser field, with $a_{x(y)} = F_{x(y)}/\omega^2$. When working within the KH frame, a laser field is effectively absent, and the scattered and ejected electrons move in a Coulomb-tail potential of an ion A^+ that oscillates in time (the oscillations are equivalent to those of a classical free electron in a laser field in the laboratory frame). At large distances from the ion, the role of its oscillating motion vanishes: the outgoing electrons experience therefore a usual, time-independent, Coulomb-tail force and, accordingly, the leading asymptotic behavior of the wave function (14.24) is

$$\left\langle \mathbf{r}_s, \mathbf{r}_e|\Psi_f^{\mathrm{KH}(-)}(\mathbf{p}_s, \mathbf{p}_e; t)\right\rangle \xrightarrow[r_s, r_e, r_{se}\to\infty]{} \mathrm{e}^{i(\mathbf{p}_s\cdot\mathbf{r}_s - E_s t)}\mathrm{e}^{i(\mathbf{p}_e\cdot\mathbf{r}_e - E_e t)}C_{\mathbf{p}_s}^{(-)}(\eta_s, \mathbf{r}_s)$$
$$\times C_{\mathbf{p}_e}^{(-)}(\eta_e, \mathbf{r}_e)C_{\mathbf{p}_{se}}^{(-)}(\eta_{se}, \mathbf{r}_{se})|\tilde{\Phi}_f^{\mathrm{KH}}(t)\rangle. \tag{14.26}$$

Here the Coulomb-distortion factors are given by

$$C_{\mathbf{p}}^{(-)}(\eta, \mathbf{r}) = \exp\left[-i\eta\ln(pr + \mathbf{p}\cdot\mathbf{r})\right],$$

with the Sommerfeld parameters $\eta_{s(e)} = -1/p_{s(e)}$ and $\eta_{se} = 1/p_{se}$ ($p_{se} = |\mathbf{p}_s - \mathbf{p}_e|$), and $\tilde{\Phi}_f^{\mathrm{KH}}(t)$ is the laser-dressed ionic state satisfying the boundary condition

$$|\tilde{\Phi}_f^{\mathrm{KH}}(t\to\infty)\rangle \to \mathrm{e}^{-iE_{\mathrm{A}^+}^{(f)}t}|\tilde{\Phi}_{\mathrm{A}^+}^{(f)}\rangle, \tag{14.27}$$

where $E_{\mathrm{A}^+}^{(f)}$ and $|\tilde{\Phi}_{\mathrm{A}^+}^{(f)}\rangle$ are the field-free energy and state of the final ion.

14.3.2.2 3C-Volkov Wave Function

The approximate final-state wave function that accounts for the proper asymptotic behavior (14.26) can be formulated on the basis of the 3C model [25], which proved to be useful in the theoretical treatment of field-free $(e, 2e)$ collisions in atoms, and Coulomb-Volkov approximation [26]. In the KH frame, it reads [27]

$$\left\langle \mathbf{r}_s, \mathbf{r}_e | \Psi_f^{\mathrm{KH}(-)}(\mathbf{p}_s, \mathbf{p}_e; t) \right\rangle = \mathrm{e}^{-i(E_s+E_e)t} \mathrm{e}^{-i(\mathbf{p}_s+\mathbf{p}_e)\cdot\mathbf{a}(t)} \mathrm{e}^{-i\mathbf{p}_{se}\cdot\mathbf{r}_{se}} \psi_{\mathbf{p}_s}^{c(-)}(\eta_s, \mathbf{r}_s + \mathbf{a}(t)) \times \psi_{\mathbf{p}_e}^{c(-)}(\eta_e, \mathbf{r}_e + \mathbf{a}(t)) \psi_{\mathbf{p}_{se}}^{c(-)}(\eta_{se}, \mathbf{r}_{se}) |\tilde{\Phi}_f^{\mathrm{KH}}(t)\rangle, \quad (14.28)$$

where $\psi_{\mathbf{p}}^{c(-)}(\eta, \mathbf{r})$ stands for a stationary Coulomb wave function with incoming spherical wave behavior (see, for instance [28]):

$$\psi_{\mathbf{p}}^{c(-)}(\eta, \mathbf{r}) = \mathrm{e}^{-\frac{1}{2}\pi\eta} \Gamma(1 - i\eta) \mathrm{e}^{i\mathbf{p}\cdot\mathbf{r}} {}_1F_1(i\eta, 1; -i(pr + \mathbf{pr})), \quad (14.29)$$

where ${}_1F_1$ is the confluent hypergeometric function. In the field-free case, (14.28) reduces to the field-free 3C model [25]. It should be also noted that the laser-dressed final ionic state in (14.28) can be calculated using the methods and approximations outlined above in regard to the laser-dressed initial atomic state.

14.4 First Born Approximation

One of the most frequently used approaches in the theory of $(e, 2e)$ collisions is the first Born approximation (FBA), which treats the projectile-atom interaction V_{eA} in the S matrix (14.4) only to first order. It is supposed to be generally applicable if both the incident and scattered electrons are fast. In the FBA, the laser-dressed final scattering state of the colliding system is approximated as

$$|\Psi_f^{(-)}(\mathbf{p}_s, \mathbf{p}_e; t)\rangle = |\chi_{\mathbf{p}_s}(t) \Phi_f^{(-)}(\mathbf{p}_e; t)\rangle, \quad (14.30)$$

where $|\Phi_f^{(-)}(\mathbf{p}_e; t)\rangle$ is the laser-dressed final atomic state with one (ejected) electron in continuum having the asymptotic momentum $\mathbf{p}_e$. Note that the final state (14.30) can be derived from the 3C-Volkov model (14.28) upon setting $\eta_s = \eta_{se} = 0$.

Using (14.30) and the explicit form of the Gordon-Volkov functions (14.6), one obtains for the S matrix the following expression:

$$S_{(e,2e)}^{\mathrm{FBA}} = -i \frac{4\pi}{Q^2} \int_{-\infty}^{\infty} \mathrm{d}t \, \mathrm{e}^{-i[\Delta E t + \alpha_{\mathbf{Q}} \sin(\omega t + \delta_{\mathbf{Q}})]} \left\langle \Phi_f^{(-)}(\mathbf{p}_e; t) \right| \sum_{k=1}^{Z} \mathrm{e}^{i\mathbf{Q}\cdot\mathbf{r}_k} - Z \left| \Phi_i(t) \right\rangle, \quad (14.31)$$

where $\Delta E = E_0 - E_s$ and $\mathbf{Q} = \mathbf{p}_0 - \mathbf{p}_s$ are the energy and momentum transfers. Provided the momentum-transfer value is small, the contributions due to exchange between the projectile and atomic electrons are typically omitted in (14.31) due to large p_0 and p_s values (on the atomic scale) in the case of fast incident and scattered electrons.

If neglecting the dressing of the initial and final atomic states by the laser field, one can readily perform the time integration in (14.31) using the following formula [29]:

$$e^{iz\cos\xi} = \sum_{\ell=-\infty}^{\infty} J_\ell(z)\, e^{i\ell z},$$

where J_ℓ are the Bessel functions of integer order. As a result, one obtains the ℓ-photon TDCS (14.9) in the form [7]

$$\frac{d^3\sigma_{\mathrm{FBA}}^{(\ell)}}{dE_e d\Omega_s d\Omega_e} = |J_\ell(\alpha_{\mathbf{Q}})|^2 \frac{d^3\sigma_{\mathrm{FBA}}}{dE_e d\Omega_s d\Omega_e}, \tag{14.32}$$

where $d^3\sigma_{\mathrm{FBA}}$ is the field-free TDCS in the FBA approach, where, however, the energy balance is

$$E_0 + E_{\mathrm{A}}^{(i)} + \ell\omega = E_{\mathrm{A}^+}^{(f)} + E_s + E_e,$$

i.e., it takes into account the transfer of ℓ photons in the laser-assisted $(e, 2e)$ collision. From the properties of the Bessel functions it follows that the cross section (14.32) turns to zero for all $\ell \neq 0$ if $\alpha_{\mathbf{Q}} = 0$ or, in other words, if the momentum transfer is perpendicular to the laser polarization.

14.4.1 Laser-Dressed Final Atomic State

The asymptotic behavior of the state $\Phi_f^{(-)}(\mathbf{p}_e; t)$ is also convenient to formulate in the KH frame:

$$\langle \mathbf{r}_e | \Phi_f^{\mathrm{KH}(-)}(\mathbf{p}_e; t)\rangle \xrightarrow[r_e\to\infty]{} e^{i(\mathbf{p}_e\cdot\mathbf{r}_e - E_e t)} C_{\mathbf{p}_e}^{(-)}(\eta_e, \mathbf{r}_e) |\tilde{\Phi}_f^{\mathrm{KH}}(t)\rangle. \tag{14.33}$$

Below the approaches are described, in which the laser-dressed final atomic state is constructed to obey the proper asymptotic behavior.

14.4.1.1 Coulomb-Volkov Approximation

The condition (14.33) can be fulfilled by neglecting correlations between the ejected and ionic electrons and using the Coulomb-Volkov wave function [26] for the laser-dressed ejected electron state, namely,

$$|\Phi_f^{\mathrm{KH}(-)}(\mathbf{p}_e;t)\rangle = |\psi_{\mathrm{CV}}^{\mathrm{KH}(-)}(\mathbf{p}_e,\eta_e;t)\tilde{\Phi}_f^{\mathrm{KH}}(t)\rangle, \tag{14.34}$$

where the Coulomb-Volkov function in the KH frame is given by

$$\psi_{\mathrm{CV}}^{\mathrm{KH}(-)}(\mathbf{p}_e,\eta_e;\mathbf{r}_e,t) = \mathrm{e}^{-i[E_e t+\mathbf{p}_e\cdot\mathbf{a}(t)]}\psi_{\mathbf{p}_e}^{c(-)}(\eta_e,\mathbf{r}_e+\mathbf{a}(t)). \tag{14.35}$$

If now using the approximation (14.34) in the FBA S-matrix (14.31) and neglecting the dressing of the initial atomic $\Phi_i(t)$ and final ionic $\tilde{\Phi}_f(t)$ states by the laser field, it is also possible to relate the multiphoton TDCS to the field-free FBA cross section. In particular, in the case of a circular polarized laser beam or when

$$|\alpha_{\mathbf{Q}}| \gg \left|\frac{F_x^2 - F_y^2}{8\omega^3}\right|, \tag{14.36}$$

one derives [30]

$$\frac{\mathrm{d}^3\sigma_{\mathrm{FBA}}^{(\ell)}}{\mathrm{d}E_e\mathrm{d}\Omega_s\mathrm{d}\Omega_e} = |J_\ell(\alpha_{\mathbf{q}_{\mathrm{ion}}})|^2\frac{\mathrm{d}^3\tilde{\sigma}_{\mathrm{FBA}}}{\mathrm{d}E_e\mathrm{d}\Omega_s\mathrm{d}\Omega_e}, \tag{14.37}$$

where the field-free FBA cross section $\mathrm{d}^3\tilde{\sigma}_{\mathrm{FBA}}$ is calculated with the model field-free final atomic state

$$|\Phi_{\mathrm{A}}^{(f)}\rangle = |\psi_{\mathbf{p}_e}^{c(-)}\tilde{\Phi}_{\mathrm{A}^+}^{(f)}\rangle$$

and for the energy balance

$$E_0 + E_{\mathrm{A}}^{(i)} + \ell\omega = E_{\mathrm{A}^+}^{(f)} + E_s + E_e + U_p,$$

and $\mathbf{q}_{\mathrm{ion}} = \mathbf{Q} - \mathbf{p}_e$ is the recoil-ion momentum. The cross section (14.37) has a similar structure as that given by (14.32). However, it turns to zero for all $\ell \neq 0$ if the laser polarization is perpendicular to the recoil-ion momentum $\mathbf{q}_{\mathrm{ion}}$ rather than the momentum transfer $\mathbf{Q}$.

The approximation (14.35) can be further improved by taking into account the role of all unperturbed electron states in the dressing of the ejected-electron wave function by the laser field. This can be done within the first-order time-dependent perturbation theory as was initially proposed by Joachain et al. [31]. For atomic hydrogen and a linearly polarized laser beam such a modification of the Coulomb-Volkov wave function in the laboratory frame reads

$$\psi_{\mathrm{MCV}}^{(-)}(\mathbf{p}_e,\eta_e;\mathbf{r}_e,t) = \mathrm{e}^{-i[E_e t+\mathbf{a}(t)\cdot\mathbf{r}_e+\alpha_{\mathbf{p}_e}\sin\omega t]}\Bigg[(1+i\alpha_{\mathbf{p}_e}\sin\omega t)\psi_{\mathbf{p}_e}^{c(-)}(\eta_e,\mathbf{r}_e) \\ -i\frac{F_0}{2}\sum_j\left(\frac{\mathrm{e}^{i\omega t}}{\omega_{j\mathbf{k}_e}+\omega} - \frac{\mathrm{e}^{-i\omega t}}{\omega_{j\mathbf{k}_e}-\omega}\right)M_{j\mathbf{k}_e}\psi_j(\mathbf{r}_e)\rangle\Bigg], \tag{14.38}$$

where

$$M_{j\mathbf{k}_e} = -\langle\psi_j|x|\psi_{\mathbf{p}_e}^{c(-)}(\eta_e)\rangle$$

are the dipole matrix elements, with ψ_j being the unperturbed hydrogen wave functions.

14.4.1.2 Sturmian-Floquet Approach

An alternative, nonperturbative method for calculating the laser-dressed ejected-electron state consists in using the Hermitian Floquet theory and employing the basis set of Sturmian functions, which proved to be efficient in the treatment of the laser-assisted scattering processes [3, 33–36]. The dynamics of the ejected electron in the KH frame is governed by the following TDSE:

$$i\frac{\partial}{\partial t}\psi_{\mathbf{p}_e}^{\mathrm{KH}(-)}(\mathbf{r}_e, t) = \left(-\frac{1}{2}\Delta + V[\mathbf{r}_e + \mathbf{a}(t)]\right)\psi_{\mathbf{p}_e}^{\mathrm{KH}(-)}(\mathbf{r}_e, t), \tag{14.39}$$

where $V[\mathbf{r}_e + \mathbf{a}(t)]$ is a space-translated electron-ion potential, which has the Coulomb tail

$$V[\mathbf{r}_e + \mathbf{a}(t)] \xrightarrow[r_e\to\infty]{} -\frac{1}{r_e}.$$

Using the Floquet-Fourier expansion

$$\psi_{\mathbf{p}_e}^{\mathrm{KH}(-)}(\mathbf{r}_e, t) = \mathrm{e}^{-iE_e t}\sum_{n=-\infty}^{\infty}\mathrm{e}^{-in\omega t}\mathscr{F}_n^{\mathrm{KH}(-)}(\mathbf{p}_e, \mathbf{r}_e), \tag{14.40}$$

one arrives at an infinite set of coupled time-independent equations:

$$\begin{aligned}&\left(H_n + \widetilde{V}_0(a_x, a_y; \mathbf{r}_e) - E_e\right)\mathscr{F}_n^{\mathrm{KH}(-)}(\mathbf{p}_e, \mathbf{r}_e)\\ &+\sum_{\nu\neq n}V_{n-\nu}(a_x, a_y; \mathbf{r}_e)\mathscr{F}_\nu^{\mathrm{KH}(-)}(\mathbf{p}_e, \mathbf{r}_e) = 0, \qquad n = 0, \pm1, \pm2, \ldots\end{aligned} \tag{14.41}$$

Here

$$H_n = H_\mathrm{C} - n\omega, \qquad H_\mathrm{C} = -\frac{1}{2}\Delta - \frac{1}{r_e}, \qquad \widetilde{V}_0(a_x, a_y; \mathbf{r}_e) = V_0(a_x, a_y; \mathbf{r}_e) + \frac{1}{r_e}, \tag{14.42}$$

and V_n are the Fourier components of $V[\mathbf{r}_e + \mathbf{a}(t)]$,

$$V_n(a_x, a_y; \mathbf{r}_e) = \frac{1}{T}\int_0^T \mathrm{d}t\,\exp(in\omega t)\,V\left[\mathbf{r}_e + \mathbf{a}(t)\right], \tag{14.43}$$

with $T = 2\pi/\omega$ being the optical cycle.

The solutions $\mathscr{F}_n^{\mathrm{KH}(-)}$ must satisfy the incoming boundary conditions in the form [32]

$$\mathscr{F}_n^{\mathrm{KH}(-)}(\mathbf{p}_e, \mathbf{r}_e) \xrightarrow[r_e\to\infty]{} \delta_{n0} \exp\left[i\mathbf{p}_e \cdot \mathbf{r}_e - i\eta_e \ln(p_e r_e + \mathbf{p}_e \cdot \mathbf{r}_e)\right] + f_n^{(-)}(p_e, \hat{\mathbf{r}}_e) \frac{\exp\left[-i p_e^{(n)} r_e + i\eta_e^{(n)} \ln(2p_e^{(n)} r_e)\right]}{r_e}, \quad (14.44)$$

where $\eta_e^{(n)} = -1/p_e^{(n)}$, and

$$p_e^{(n)} = \begin{cases} \sqrt{2(E_e + n\omega)}, & E_e + n\omega \geq 0, \\ \pm i\sqrt{-2(E_e + n\omega)}, & E_e + n\omega < 0. \end{cases} \quad (14.45)$$

The Coulomb-specific asymptotic behavior (14.44) can be taken into account by recasting the system of Floquet equations into the Lippman–Schwinger form

$$\mathscr{F}_n^{\mathrm{KH}(-)}(\mathbf{p}_e, \mathbf{r}_e) = \delta_{n0}\psi_{\mathbf{p}_e}^{c(-)}(\eta_e, \mathbf{r}_e) + \sum_\nu \int d\mathbf{r}_e' G_c^{(-)}(p_e^{(n)}; \mathbf{r}_e, \mathbf{r}_e') \mathscr{V}_{n-\nu}(a_x, a_y; \mathbf{r}_e') \mathscr{F}_\nu^{\mathrm{KH}(-)}(\mathbf{p}_e, \mathbf{r}_e'),$$
$$n = 0, \pm 1, \pm 2, \ldots \quad (14.46)$$

Here the notation $\mathscr{V}_n = \delta_{n0}\tilde{V}_0 + (1 - \delta_{n0})V_n$ is introduced, and $G_c^{(-)}$ is the advanced Coulomb Green's function, which satisfies the equation

$$\left(\frac{p_e^{(n)2}}{2} + \frac{1}{2}\Delta + \frac{1}{r_e}\right) G_c^{(-)}\left(p_e^{(n)}; \mathbf{r}_e, \mathbf{r}_e'\right) = \delta(\mathbf{r}_e - \mathbf{r}_e'). \quad (14.47)$$

The solution of the system of the Lippmann–Schwinger–Floquet equations (14.46) can be sought using the expansion of the Floquet-Fourier components in terms of Sturmian functions. Recently, a new efficient method has been proposed [37] based on employing the basis set of quasi-Sturmian functions [38] in parabolic coordinates. A marked advantage of the quasi-Sturmians is that they possess an appropriate incoming Coulomb asymptotic behavior, thus providing the proper asymptotic form (14.44) of the solution.

14.4.2 Laser-Assisted Electron Momentum Spectroscopy

The $(e, 2e)$ reactions involving large momentum transfer under kinematical conditions close to a free electron-electron collision are usually referred to as electron momentum spectroscopy (EMS) [39, 40]. EMS is a well-known method for exploring the electronic structure of various systems ranging from atoms and molecules to clusters and solids.

The theoretical formulation of EMS in the presence of laser radiation was given in [13]. According to this formulation, the S matrix of the laser-assisted $(e, 2e)$ EMS process can be presented as

$$S^{\mathrm{EMS}}_{(e,2e)} = -i\,\frac{4\pi}{Q^2}\int\limits_{-\infty}^{\infty} \mathrm{d}t\,\exp\left[-i\left(\varepsilon - \frac{q^2}{2}\right)t\right]\langle\chi_{\mathbf{q}}(t)|\varphi_{fi}(t)\rangle, \tag{14.48}$$

where $\varepsilon = E_s + E_e - E_0$, $\mathbf{q} = -\mathbf{q}_{\mathrm{ion}}$, and

$$|\varphi_{fi}(t)\rangle = \langle\tilde{\Phi}_f(t)|\Phi_i(t)\rangle \tag{14.49}$$

is the laser-dressed Kohn-Sham orbital. The expression (14.48) follows from the FBA result (14.31) using the binary-encounter approximation, which accounts only for the interaction between the colliding electrons in the projectile-atom potential $V_{e\mathrm{A}}$, and the laser-dressed final atomic state in the form

$$|\Phi^{(-)}_f(\mathbf{p}_e; t)\rangle = |\chi_{\mathbf{p}_e}(t)\tilde{\Phi}_f(t)\rangle. \tag{14.50}$$

The time integration in (14.48) can be readily performed using the Floquet expansion of the laser-dressed Kohn-Sham orbital

$$|\varphi_{fi}(t)\rangle = \mathrm{e}^{-i\mathscr{E}_{fi}t}\sum_{n=-\infty}^{\infty}\mathrm{e}^{-in\omega t}|\varphi^{(n)}_{fi}(\mathscr{E}_{fi})\rangle, \tag{14.51}$$

where $\mathscr{E}_{fi}$ is the Kohn-Sham quasienergy.

If the laser-dressing effect in (14.51) is negligible, then

$$|\varphi_{fi}(t)\rangle = \mathrm{e}^{-iE_{fi}t}|\varphi_{fi}\rangle, \tag{14.52}$$

where $E_{fi} = E^{(i)}_{\mathrm{A}} - E^{(f)}_{\mathrm{A}+}$ and φ_{fi} are the field-free (unperturbed) Kohn-Sham energy and orbital. In such a case, provided the laser beam is circularly polarized or the inequality (14.36) holds, one derives the ℓ-photon TDCS as (cf. 14.37)

$$\frac{\mathrm{d}^3\sigma^{(\ell)}_{\mathrm{EMS}}}{\mathrm{d}E_e\mathrm{d}\Omega_s\mathrm{d}\Omega_e} = |J_\ell(\alpha_{\mathbf{q}})|^2\frac{\mathrm{d}^3\sigma_{\mathrm{EMS}}}{\mathrm{d}E_e\mathrm{d}\Omega_s\mathrm{d}\Omega_e}, \tag{14.53}$$

where $\mathrm{d}^3\sigma_{\mathrm{EMS}}$ is the field-free EMS cross section[2] [39]

[2]The exchange between the colliding electrons is taken into account, for in the EMS kinematics both outgoing electrons (scattered and ejected) are fast.

$$\frac{d^3\sigma_{\text{EMS}}}{dE_e d\Omega_s d\Omega_e} = \frac{p_s p_e}{2\pi^3 p_0}\left(\frac{1}{|\mathbf{p}_0 - \mathbf{p}_s|^4} + \frac{1}{|\mathbf{p}_0 - \mathbf{p}_e|^4} - \frac{1}{|\mathbf{p}_0 - \mathbf{p}_s|^2|\mathbf{p}_0 - \mathbf{p}_e|^2}\right)|\varphi_{fi}(\mathbf{q})|^2 \tag{14.54}$$

which is calculated for the energy balance

$$E_{if} + \ell\omega = \varepsilon + U_p.$$

If the laser-dressing effect in (14.51) is substantial, one can minimize the role of the interaction of the fast ingoing and outgoing electrons with the laser field by considering such laser-field orientations that $\alpha_{\mathbf{q}} = 0$ [15]. Assuming again a circularly polarized laser beam or the validity of (14.36), the resultant ℓ-photon TDCS is given by

$$\frac{d^3\sigma^{(\ell)}_{\text{EMS}}}{dE_e d\Omega_s d\Omega_e} = \frac{p_s p_e}{2\pi^3 p_0}\left(\frac{1}{|\mathbf{p}_0 - \mathbf{p}_s|^4} + \frac{1}{|\mathbf{p}_0 - \mathbf{p}_e|^4} - \frac{1}{|\mathbf{p}_0 - \mathbf{p}_s|^2|\mathbf{p}_0 - \mathbf{p}_e|^2}\right)|\varphi^{(\ell)}_{fi}(\mathbf{q})|^2, \tag{14.55}$$

and the energy balance is

$$\mathscr{E}_{if} + \ell\omega = \varepsilon + U_p.$$

Thus, the cross section (14.55) contains the direct information about the ℓth Floquet-Fourier component of the laser-dressed Kohn-Sham orbital in momentum space.

14.5 Concluding Remarks

In this work an account of the basic theoretical methods and approximations in the field of laser-assisted $(e, 2e)$ collisions in atoms has been given. At the same time, a number of issues related to the theory of laser-assisted $(e, 2e)$ processes inevitably remained beyond the scope of the present contribution. In particular, the high-order Born approximations, for example, such as the second Born approximation, has not been discussed here. The reason is that no theoretical study of the Born series in the laser-assisted $(e, 2e)$ case has been carried out so far, except for the second-Born calculations performed in [14, 17]. Owing to the long-range Coulomb-tail potentials involved in the $(e, 2e)$ scattering processes, the higher Born approximations are known to diverge in the field-free case. There are theoretical methods allowing to cope with these divergences (see [41] and references therein), but they are not directly applicable to the laser-assisted case. It should be noted that the authors of the works [14, 17] left the problem of divergences unaddressed, and therefore it still awaits a rigorous theoretical analysis.

Some comments should be made about testing the presented theoretical approaches. Currently the laser-assisted $(e, 2e)$ measurements are lacking: only one experimental study [5] has been conducted so far. Notable discrepancies were found [6] between the experimental data and the FBA calculations using the initial laser-dressed atomic wave function in the form (14.18) and the Coulomb-

Volkov model (14.34) for the final laser-dressed atomic state. The disagreement was attributed in [6] mainly to the deficiencies of the Coulomb-Volkov approximation, thus suggesting that more advanced non-perturbative treatments (for example, such as the R-matrix-Floquet theory [42]) are needed. One of novel advanced non-perturbative approaches, namely, the quasi-Sturmian-Floquet approach [37], has been outlined above. It might be expected that the further progress in laser-assisted $(e, 2e)$ experimental studies, including the improvement of energy and momentum resolutions and investigation of the atomic targets other than helium, will provide more stringent tests for the current theoretical understanding.

Acknowledgements I am grateful to Yuri Popov and Sergey Zaytsev for useful discussions. I cordially thank Kaoru Yamanouchi for kindly inviting me to contribute to this volume.

References

1. M.H. Mittleman, *Introduction to the Theory of Laser-Atom Interaction* (Plenum, New York, 1993)
2. M.V. Fedorov, *Atomic and Free Electrons in a Strong Light Field* (World Scientific, Singapore, 1997)
3. C.J. Joachain, N.J. Kylstra, R.M. Potvliege, *Atoms in Intense Laser Fields* (Cambridge University Press, Cambridge, 2011)
4. F. Ehlotzky, Phys. Rep. **345**, 175 (2001)
5. C. Höhr, A. Dorn, B. Najjari, D. Fischer, C.D. Schroter, J. Ullrich, Phys. Rev. Lett. **94**, 153201 (2005)
6. C. Höhr, A. Dorn, B. Najjari, D. Fischer, C.D. Schroter, J. Ullrich, J. Electron Spectrosc. Relat. Phenom. **161**, 172 (2007)
7. M. Mohan, P. Chand, Phys. Lett. A **65**, 399 (1978)
8. A. Makhoute, D. Khalil, A. Maquet, C.J. Joachain, R. Taeb, J. Phys. B **32**, 3255 (1999)
9. S.-M. Li, J. Berakdar, S.-T. Zhang, J. Chen, J. Phys. B **38**, 1291 (2005)
10. A. Chattopadhyay, C. Sinha, Phys. Rev. A **72**, 053406 (2005)
11. S. Ghosh Deb, S. Roy, C. Sinha, Eur. Phys. J. D **55**, 591 (2009)
12. S. Ghosh Deb, C. Sinha, Eur. Phys. J. D **60**, 287 (2010)
13. K.A. Kouzakov, Y.V. Popov, M. Takahashi, Phys. Rev. A **82**, 023410 (2010)
14. M.-Y. Zheng, S.-M. Li, Phys. Rev. A **82**, 023414 (2010)
15. A.A. Bulychev, K.A. Kouzakov, Y.V. Popov, Phys. Lett. A **376**, 484 (2012)
16. A.A. Bulychev, K.A. Kouzakov, Eur. Phys. J. D **68**, 354 (2014)
17. I. Ajana, A. Makhoute, D. Khalil, A. Dubois, J. Phys. B **47**, 175001 (2014)
18. Y.V. Popov, K.A. Kouzakov, A.A. Bulychev, S.I. Vinitsky, Theor. Math. Phys. **186**, 93 (2016)
19. L.V. Keldysh, Sov. Phys. JETP **20**, 1307 (1965)
20. W.H. Furry, Phys. Rev. **81**, 115 (1951)
21. D.H. Kobe, A.L. Smirl, Am. J. Phys. **46**, 624 (1978)
22. G. Floquet, Ann. Ec. Norm. **13**, 47 (1883)
23. L.D. Faddeev, S.P. Merkuriev, *Quantum Scattering Theory for Several Particle Systems* (Kluwer Academic Publishers, Dordrecht, 1993)
24. W.C. Henneberger, Phys. Rev. Lett. **21**, 838 (1968)
25. M. Brauner, J. Briggs, H. Klar, J. Phys. B At. Mol. Opt. Phys. **22**, 2265 (1989)
26. M. Jain, N. Tzoar, Phys. Rev. A **18**, 538 (1978)
27. K.A. Kouzakov, Y.V. Popov, M. Takahashi, J. Phys. Conf. Ser. **288**, 012009 (2011)
28. R.G. Newton, *Scattering Theory of Waves and Particles* (McGraw-Hill, New York, 1966)

29. M. Abramowitz, I.A. Stegun (eds.), *Handbook of Mathematical Functions with Formulas, Graphs, and Mathematical Tables* (Dover, New York, 1972)
30. P. Cavaliere, G. Ferrante, C. Leone, J. Phys. B **13**, 4495 (1980)
31. C.J. Joachain, P. Francken, A. Maquet, P. Martin, V. Véniard, Phys. Rev. Lett. **61**, 165 (1988)
32. M. Gavrila, J.Z. Kamiński, Phys. Rev. Lett. **52**, 613 (1984)
33. A. Maquet, Phys. Rev. A **15**, 1088 (1977)
34. R.M. Potvliege, R. Shakeshaft, Phys. Rev. A **38**, 1098 (1988)
35. R.M. Potvliege, R. Shakeshaft, Phys. Rev. A **40**, 3061 (1989)
36. M. Dörr, R.M. Potvliege, R. Shakeshaft, Phys. Rev. A **41**, 558 (1990)
37. A.S. Zaytsev, S.A. Zaytsev, L.A. Ancarani, K.A. Kouzakov, Phys. Rev. A **97**, 043417 (2018)
38. J.A. Del Punta, M.J. Ambrosio, G. Gasaneo, S.A. Zaytsev, L.U. Ancarani, J. Math. Phys. **55**, 052101 (2014)
39. E. Weigold, I.E. McCarthy, *Electron Momentum Spectroscopy* (Kluwer Academic/Plenum, New York, 1999)
40. V.G. Neudatchin, Y.V. Popov, Y.F. Smirnov, Phys. Usp. **42**, 1017 (1999)
41. V.L. Shablov, P.S. Vinitsky, Y.V. Popov, O. Chuluunbaatar, K.A. Kouzakov, Phys. Part. Nucl. **41**, 335 (2010)
42. P.G. Burke, *R-Matrix Theory of Atomic Collisions* (Springer, Berlin, 2011)

Chapter 15
Interference Femtosecond Linear and Nonlinear Comb-Spectroscopy in Octave Expanded Range

Sergey Pulkin, Eugene Borisov, Alexey Kalinichev, Irina Korshok, Dmitrii Venediktov, Nickolai Pulkin, Vladimir Venediktov, Mikhail Balabas, Vladislav Shoev, Svetlana Uvarova, Denis Ivanov, Tatiana Ivanova and Svetlana Savel'eva

Abstract We proposed linear and nonlinear interference comb-spectroscopy in order to study the interaction of a train of femtosecond pulses with an atomic medium experimentally and theoretically. Numerical calculations of the interaction of a train of short femtosecond pulses with atomic vapor has been made for comparison with experimental results. It is shown that interference of atomic states produces unidirectional coherent radiation on an adjacent transition (blue radiation at 420 nm) when input near IR radiation (780 nm) is focused into the cell filled with Rb atoms.

S. Pulkin (✉) · I. Korshok · D. Venediktov · V. Venediktov · M. Balabas · V. Shoev · S. Uvarova · D. Ivanov · T. Ivanova · S. Savel'eva
Department of General Physics 1, Saint-Petersburg State University, St. Petersburg, Russia
e-mail: spulkin@mail.ru

V. Venediktov
e-mail: vlad.venediktov@mail.ru

E. Borisov · A. Kalinichev
Research Centre "Optical and Laser Methods of Researches of Matter", Saint-Petersburg State University, St. Petersburg, Russia
e-mail: enbor@bk.ru

N. Pulkin
Department of Laser Technologies and Systems, ITMO University, St. Petersburg, Russia
e-mail: kpulkin@mail.ru

V. Venediktov
Laser Measurement and Navigation Systems Department, Electrotechnical University "LETI", St. Petersburg, Russia

K. Yamanouchi et al. (eds.), *Progress in Photon Science*, Springer Series in Chemical Physics 119, https://doi.org/10.1007/978-3-030-05974-3_15

15.1 Introduction

In this chapter, we investigate the interaction of femtosecond pulses with an atomic medium. We propose to use radiation of comb-generator of femtosecond laser as a bright source of "white" light for interference spectroscopy of atoms and molecules. The generation of octave expanded spectrum is possible now by using the radiation of the comb generator with micro-structured fiber to octave expanded spectral range from 400 to 1100 nm. It allows us to revitalize the famous "Rozhdestvensky hook" method [1] by which anomalous dispersion can be measured. Such wide coherent radiation gives us opportunity to measure the atomic oscillator strengths simultaneously for all the transitions in the wide spectral range. The interferometric technique with increasing sensitivity [2] allows us to make measurements of oscillator strengths for weak transitions ($f < 0.01$). The atomic and molecular dynamics can be studied with femtosecond time resolution for different scientific areas such as laser physics, plasma physics, medicine and biology. Now the comb-spectroscopy is a promising method developing very rapidly [3]. Because of the high brightness and high spectral and temporal resolutions, the comb spectroscopy has been widely applied to linear molecular spectroscopy in IR-range [3]. The nonlinear phenomena of atoms and molecules interacting with light can also be investigated thanks to the high laser-field intensity achieved when the radiation of a comb generator is focused [1, 4]. The purpose of this paper to show of the first results of interference comb-spectroscopy in the field of linear and nonlinear spectroscopy.

In the case of the comb generator of fs-laser excitation, the accumulation of signals during the acceptable exposition time, for example, 10^{-4} s, 10^4 interferograms are detected by the CCD camera. Thus, during the exposure time the phase of optical radiation does not change, i.e. there is no movement of mirrors in the Michelson interferometer more than on a half of wavelength. It allowed the authors [1] to develop a method of interference comb-spectroscopy in which interferential hooks arise on the slit placed on the entrance focal plane of the spectrometer. In one of the arms in the interferometer, a cell filled with an atomic medium is placed. The Rozhdestvensky hooks are shown in Fig. 15.1 [1] whose ordinate represents wavelength. In the interference fringes projected on the exit plane of the spectrograph, "hooks" arise near the resonance doublet of Rb atoms. Because the square of the distance in wavelength between two tops within a hook is proportional to the oscillator strength, and the ratio of the oscillator strengths for the first resonance doublet of Rb atoms in Fig. 15.1 can be given by $(d_1/d_2)^2$, which the ratio of the squared distances between the tops of the respective components in the doublet. In Fig. 15.1a, the bandwidth of the spectral range is about 7 nm, and in Fig. 15.1b, the bandwidth of the spectral range is expanded by a micro-structured fiber to about 700 nm, which covers one octave bandwidth.

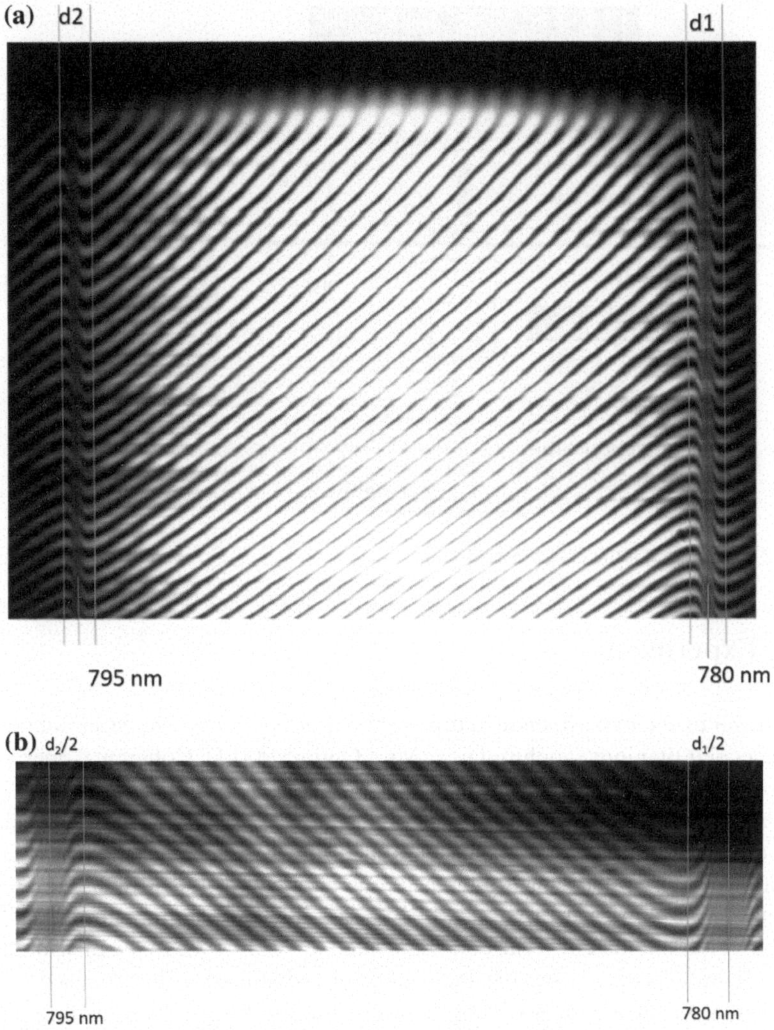

Fig. 15.1 **a** The interference hook near the resonance doublet lines of Rb atoms for the spectral bandwidth of the laser pulses of about 7 nm in the comb spectrum. **b** The interference hook recorded when the spectral bandwidth is expanded by a microstructered fiber to cover the octave range of about 700 nm in the comb spectrum

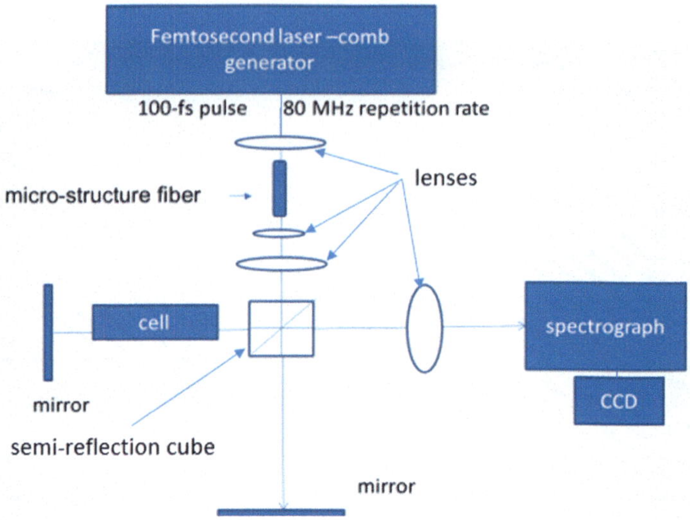

Fig. 15.2 Schematic of the experimental setup

15.2 Experiment

The scheme of the experimental setup is shown in Fig. 15.2. The laser source was a femtosecond titanium sapphire laser (Mira Optima 900-D, Coherent), operated in a comb-generator mode. A sequence of pulses with the pulse duration of 120 fs, the pulse energy of 10 nJ and the repetition rate of 76 MHz was generated. For pumping Mira Optima 900-D, a diode-pumped solid-state laser (Verdi V10, Coherent) was used. The sample medium was prepared in a sealed cell with a mixture of isotopes of rubidium atoms (Rb^{85} and Rb^{87}). The density of atoms in this cell was about 10^{12}–10^{13} atoms/cm^3. The cell was heated to 90–130 °C. In the measurements of linear comb-spectroscopy, the incident laser radiation was not focused. A narrow coherent beam whose wavelength is tuned to the resonance transition of $\lambda =$ 780 nm was generated when incident laser beam is focused inside the cell. Unidirectional coherent radiation was generated from the waist of the laser beam focus whose diameter and length are 100 μm and 0.5 cm, respectively. The divergence of the generated beam is 2–3 mrad. Figure 15.3a illustrates the one octave expanded spectrum in the wide wavelength range of 400–1,100 nm. For the demonstration of the interferometric measurements in the wide spectral range, we placed an iodine cell in the interferometer in place of the Rb cell. Figure 15.3b shows an absorption spectrogram for the $B^3\Pi_u^+ - X^1\Sigma_g^+$ $(v,0)$ $(v = 0$–$5)$ bands of I_2 in the wavelength range around 540 nm and demonstrates that the absorption spectrum and the dispersion can be recorded simultaneously by interference comb spectroscopy with the wide spectral bandwidth of 700 nm, enabling us to use the hook method in one octave spectral range.

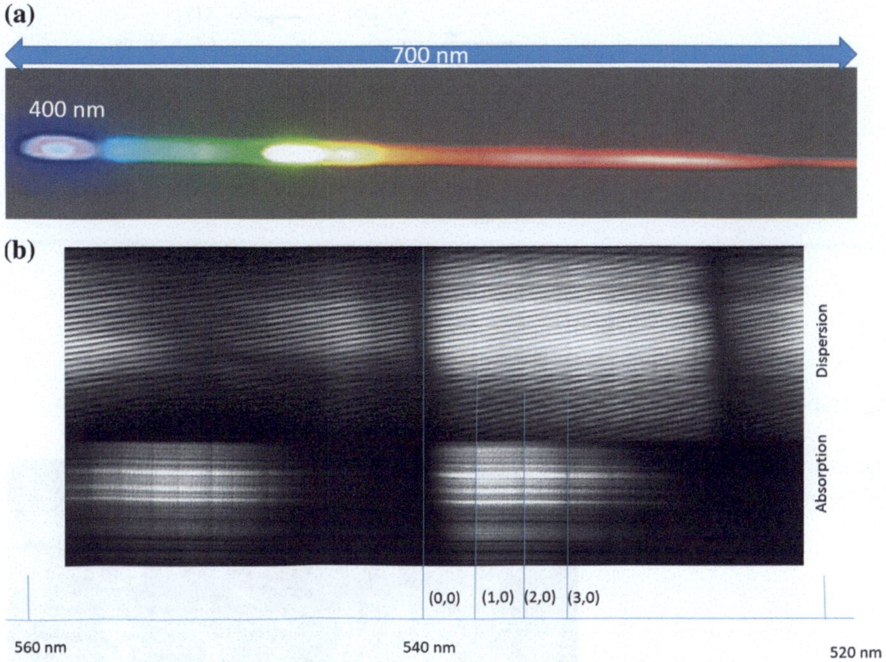

Fig. 15.3 **a** The supercontinuum spectrum in the wavelength region between 400 nm and 1.1 μm. **b** The spectrogram of the photoabsorption of iodine I_2 exhibiting the $B^3\Pi_u^+ - X^1\Sigma_g^+(v,0)$ ($v =$ 0–5) bands. Interference fringes intersect these absorption lines. In the wavelength region longer than 545 nm, the spectrum becomes more complicated because of the appearance of the hot band transitions from the vibrationally excited states ($\nu = 1$ and $\nu = 2$) in the ground electronic state

Figure 15.4 shows a diagram of the investigated levels of Rb atoms. By the irradiation of the femtosecond laser pulses, the $5D_{5/2}$ level can be populated through the cascade transition of $5S_{1/2} \rightarrow 5P_{3/2} \rightarrow 5D_{5/2}$ and through the two photon transition of $5S_{1/2} \rightarrow 5D_{5/2}$ [5]. It is known from [5] that the $5D_{5/2} \rightarrow 6P_{3/2}$ transition at 5.2 μm can be superradiant. As explained later, we introduce a coupling field for the $5D_{5/2} \rightarrow 6P_{3/2}$ transition so that superradiation of coherent light at 5.2 μm arises, as demonstrated in [5], through the population inversion realized during the period of the laser irradiation. Due to the interference matrix element between $5S_{1/2}$ and $5D_{5/2}$ and the coupling field resonance at the $5D_{5/2} \rightarrow 6P_{3/2}$ transition, coherent narrow-beam radiation at 420 nm arises as the $6P_{3/2} \rightarrow 5S_{1/2}$ by the nonlinear interference effect [6] in the four-level system of Rb.

Figure 15.5 shows a photograph of the narrow-beam coherent radiation emanating from the cell filled with a Rb vapor. The cell is placed on the one arm of the Michelson interferometer. The laser beam was focused on the different positions on the cell axes. In Fig. 15.5, the focus is placed at the central position of the cell. A similar phenomenon of blue radiation was discovered by Scully's group in very different conditions [5]. In addition, coherent IR radiation along the laser beam axis with the

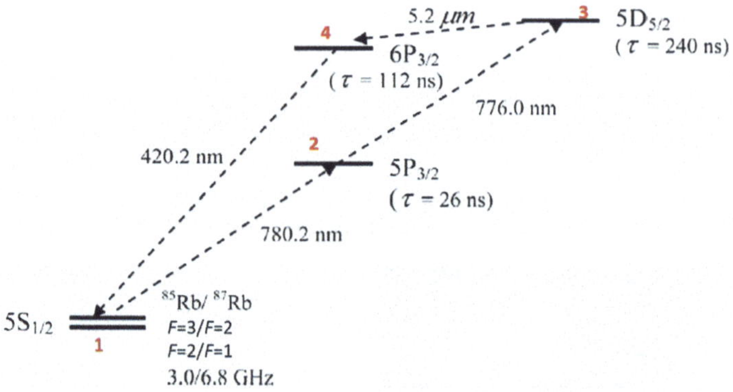

Fig. 15.4 Energy level diagram and the optical transitions of Rb

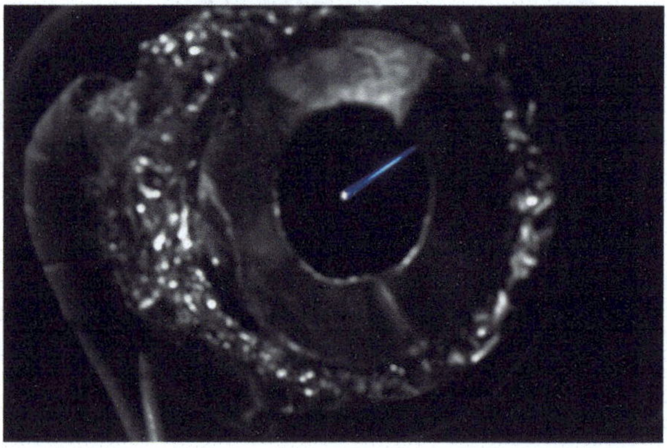

Fig. 15.5 Radiation from the cell with a Rb vapor

small divergence angle was observed at the $5P_{3/2} \rightarrow 5S_{1/2}$ (780 nm) transition and at the $5D_{5/2} \rightarrow 5P_{3/2}$ transition (795 nm) [1].

15.3 Results and Discussion

It is demonstrated in Fig. 15.6 that the radiation power spectrum obtained from a femtosecond laser comb-generator at the $5S_{1/2} \rightarrow 5P_{1/2}$ transition (λ = 795 nm) when the laser pulses are focused. The transition $5S_{1/2} \rightarrow 5P_{1/2}$ is more available for researching as transition in pure 2-level atoms comparable $5S_{1/2} \rightarrow 5P_{3/2}$ (λ = 780 nm). The band width of the generated narrow-band coherent radiation is 7.6 nm, which corresponds to the input pulse duration of 120 fs. The inset shows the temperatures of the Rb vapor in the cell. The temperature 59 °C corresponds to the concentration of atoms in the cell of $N \sim 10^{13}$ cm^{-3}, with the absorption coefficient of $k_0 \sim 30$ cm^{-1}. With the cell length of $l = 10$ cm, the optical density becomes $k_0 l \sim$

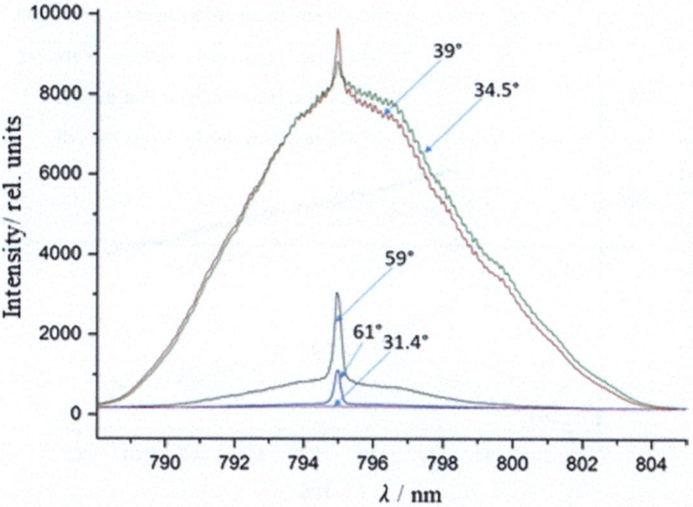

Fig. 15.6 Radiation power spectra of the $5S_{1/2}\rightarrow 5P_{1/2}$ transition of Rb atoms ($\lambda = 795$ nm) at the five different cell temperature in °C. As the temperature increases, the radiation intensity decreases because of the photoabsorption by Rb atoms within the cell. At the lowest temperature ($t = 31.4°$), the radiation intensity is negligibly small because the density of Rb atoms is too low

300. Thus, almost all the incident radiation before focus is absorbed and the coherent narrow-beam radiation almost disappears as shown by the black curve (59 °C) in Fig. 15.6. At high temperatures (59 and 61 °C), we can identify a strong narrow peak on a weak broad profile in the power density curves. As explained in the next section, the low intensity of the broad feature can be ascribed to the fact that the intensity of the incident laser beam was not sufficiently strong. At low temperatures (30–35 °C), the broad spectra presented by the red and green curves, corresponding to narrowly directed coherent radiation ($N \sim 10^{10}$ cm^{-3}), are observed. In this case, the linear absorption is weak because $k_0 l \sim 0.3$ is small, and the intensity of focused radiation is strong enough so that the broad part of the curve appears. For the confirmation, we placed the focus at the entrance, in the center, and at the exit of the cell.

Figure 15.7 shows the time dependence of the radiation intensity of the $5^2S_{1/2}$–$6^2P_{1/2}$ transition at $\lambda = 422$ nm and that of the $5^2S_{1/2}$–$6^2P_{3/2}$ transition at $\lambda = 420$ nm at the different focal position in the cell recorded by keeping the same atomic concentrations ($N \sim 10^{13}$ cm^{-3}).

In Fig. 15.7, the curve 1 shows the presence of two decay regions: a fast decay corresponding to a short excitation pulse, and a slow decay corresponding to spontaneous decay from the $6P_{3/2}$ level. The lifetime calculated for the slow decay is $\tau = 112 \pm 5$ ns. This decay time coincides with the lifetime of the $6P_{3/2}$ state [7]. The rapid decay is associated with a nonlinear interference effect, showing the existence of two-photon $6P_{3/2} \rightarrow 5P_{3/2}$ transition and/or the two photon $5D_{5/2} \rightarrow 5S_{1/2}$ transition. In this case, the focus is placed at the entrance of the cell, and consequently, the

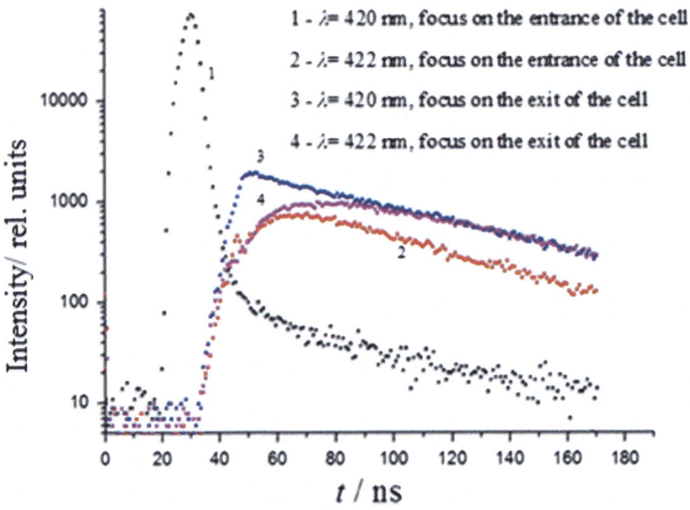

Fig. 15.7 Time dependences of the radiation intensity of the $5^2S_{1/2}$–$6^2P_{1/2}$ transition at $\lambda = 422$ nm (the curves 2 and 4) and that of the $5^2S_{1/2}$–$6^2P_{3/2}$ transition at $\lambda = 420$ nm (the curves 1 and 3)

absorption is smaller than that when the focus is placed at the center or at the exit. The curve 3 was obtained by placing the focus on the exit of the cell.

The remaining curves 2 and 4 show the decay of the $6^2P_{1/2}$ level without rapid part of decay, this is due to the less efficient excitation to the $5^2D_{5/2}$ level, compared with the more efficient excitation through the two-photon $5S_{1/2} \rightarrow 5D_{5/2}$ transitions by a laser pulse with the duration of $\tau_{\text{pulse}} \approx 120$ fs whose spectral pulse width is $\Delta\omega \approx 7.6$ nm. The transitions 1–2 and 2–3 are almost overlapped and the two-photon excitation $5^2D_{5/2}$ is very effective. Because the bandwidth of the laser pulses covers both of the transitions, two-photon excitation occurs efficiently.

When the focus is placed at the exit of the cell, a rapid decay is not identified. This is because the radiation is absorbed in the layer of the atomic vapor before being focused near the end of the cell, and the light intensity is not strong enough so that a nonlinear interference effect can appear. When the focus is placed at the entrance of the cell, the radiation at the $6P_{3/2} \rightarrow 5S_{1/2}$ transition, which passes through the cell with almost no absorption losses, is observed. At the same time, the intensity of the incident radiation is sufficiently large so that a nonlinear interference effect can appear. The radiation at the wavelength of 420 nm is produced at the waist of a Gaussian beam. The diameter of the waist is 100 μm and the length of the waist is 7 mm. When the focus is placed at the outlet of the cell with the length of l = 10 cm, unfocused radiation is absorbed before the focus, and the intensity is not strong enough for inducing coherent unidirectional radiation.

It is noteworthy that the radiation at the wavelengths, corresponding to the $5S_{1/2} \rightarrow 5P_{3/2}$ (780 nm), $5P_{3/2} \rightarrow 5D_{5/2}$ (776 nm), and $6P_{3/2} \rightarrow 5S_{1/2}$ (420 nm) transitions, has a near-diffraction divergence of the order of 2.0–3.0 mrad. Figure 15.8 shows

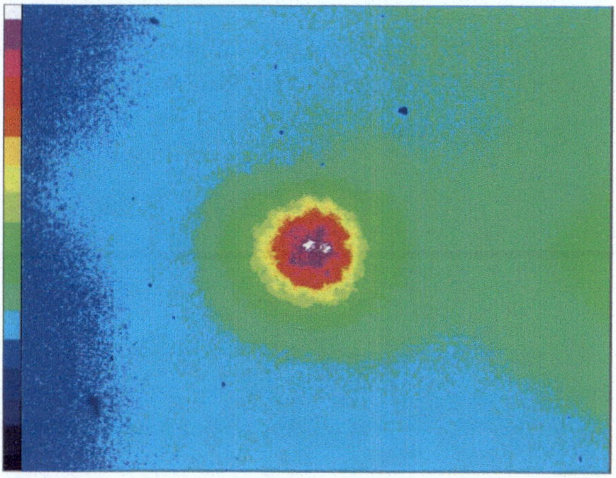

Fig. 15.8 Divergence of the narrow unidirectional radiation at the wavelength of 420 nm in linear color scale. The diameter of the yellow ring is 10 mm on the screen and the distance between the focus and the screen is about 3 m

the small divergence of the radiation at the wavelength of 420 nm, represented by the small diameter (10 mm) of the yellow ring, representing also the beam shape is Gaussian. The color scale in Fig. 15.8 is linear, and the distance between the focus and the screen is about 3 m.

When the focus is placed at the entrance or in the center of the cell, the narrow-beam coherent radiation, which is generated from the waist of laser beam, is absorbed only slightly, resulting in the small divergence.

The coherent narrow-beam radiation, which is generated from the waist of laser beam, when the focus is placed at the entrance or in the center of the cell, is almost not absorbed. This fact connected with small divergence, when self-absorption is small. In the case of reabsorption, the incident radiation spreads on 4π radians and absorption is large. For unidirectional radiation, absorption is small because radiation spread on the one spectral mode in small angle. The direct evidence that unidirectional coherent radiation spread through atomic media without reabsorption is the equality of the effective lifetime calculated from experimental decay curves (Fig. 15.7, slow decay parts) to the radiative lifetime.

In order to interpret the experimental results, the system of equations for the density matrix in the rotating-wave approximation was solved numerically. Equations are written for a two-level atomic medium for the $5S_{1/2}$ and $5P_{3/2}$ levels of Rb and for a four-level atomic medium for the $5S_{1/2}$, $5P_{3/2}$, $5D_{5/2}$, and $6P_{3/2}$ levels of Rb. The equations were derived from the quantum Liouville equation, which is equivalent to the von Neumann equation (15.1),

$$i\hbar\left(\frac{\partial\rho}{\partial t}\right) = \left[\hat{H}, \hat{\rho}\right], \tag{15.1}$$

where $\hat{H}$ is the total energy operator of the system and $\hat{\rho}$ is the density matrix operator. In the following equations, the numbers, 1, 2, 3, and 4, appearing as subscripts represent $5S_{1/2}$, $5P_{3/2}$, $5D_{5/2}$ and $6P_{3/2}$, respectively.

Two-level system:

$$\frac{d\rho_{11}}{dt} = \left(2V'_{12}\rho''_{12} - 2V''_{12}\rho'_{12}\right) + g_{21}\rho_{22} \tag{15.2}$$

$$\begin{aligned}\frac{d\tilde{\rho}'_{12}}{dt} + \frac{id\tilde{\rho}''_{12}}{dt} &= -i\Delta_{12}\tilde{\rho}'_{12} - i\Delta_{12}\tilde{\rho}''_{12} + i\tilde{V}'_{12}\tilde{\rho}_{11} - V^{\sim\prime\prime}_{12}\tilde{\rho}_{11} - i\tilde{V}'_{12}\tilde{\rho}_{22} \\ &\quad - \tilde{V}''_{12}\tilde{\rho}_{22} + 0.5\tilde{\rho}'_{12}g_{21} + 0.5\tilde{\rho}''_{12}g_{21}\end{aligned} \tag{15.3}$$

$$\begin{aligned}\frac{d\tilde{\rho}'_{12}}{dt} - \frac{id\tilde{\rho}''_{12}}{dt} &= -i\Delta_{21}\tilde{\rho}'_{21} - i\Delta_{21}\tilde{\rho}''_{21} - i\tilde{V}'_{12}\rho^{\sim}_{11} + \tilde{V}''_{12}\tilde{\rho}_{11} - i\tilde{V}'_{12}\tilde{\rho}_{22} \\ &\quad - \tilde{V}''_{12}\tilde{\rho}_{22} + 0.5\tilde{\rho}'_{12}g_{21} + 0.5\tilde{\rho}''_{12}g_{21}\end{aligned} \tag{15.4}$$

$$\frac{d\rho_{22}}{dt} = -\left(2V'_{12}\rho''_{12} - 2V''_{12}\rho'_{12}\right) - g_{21}\rho_{22} \tag{15.5}$$

Four-level system:
Off-diagonal elements:

$$\begin{aligned}\frac{d\tilde{\rho}'_{12}}{dt} + \frac{id\tilde{\rho}''_{12}}{dt} &= -i\Delta_{12}\tilde{\rho}'_{12} - \Delta_{12}\tilde{\rho}''_{12} - i\tilde{V}'_{32}\tilde{\rho}'_{13} + \tilde{V}''_{32}\tilde{\rho}'_{13} - \tilde{V}'_{32}\tilde{\rho}''_{13} + i\tilde{V}''_{32}\tilde{\rho}''_{13} \\ &\quad + i\tilde{V}'_{12}\tilde{\rho}_{11} - \tilde{V}''_{12}\tilde{\rho}_{11} - i\tilde{V}'_{12}\tilde{\rho}_{22} - \tilde{V}''_{12}\tilde{\rho}_{22} - i\tilde{V}'_{14}\tilde{\rho}'_{24} - \tilde{V}''_{14}\tilde{\rho}'_{24} \\ &\quad - \tilde{V}'_{14}\tilde{\rho}''_{24} + i\tilde{V}''_{14}\tilde{\rho}''_{24} - 0.5g_{21}\tilde{\rho}'_{12} - i0.5g_{21}\tilde{\rho}''_{12}\end{aligned} \tag{15.6}$$

$$\begin{aligned}\frac{d\tilde{\rho}'_{13}}{dt} + \frac{id\tilde{\rho}''_{13}}{dt} &= -i\Delta_{13}\tilde{\rho}'_{13} - \Delta_{13}\tilde{\rho}''_{13} - i\tilde{V}'_{23}\rho^{\sim\prime}_{12} - \tilde{V}''_{23}\rho^{\sim\prime}_{12} + \tilde{V}'_{23}\rho^{\sim\prime\prime}_{12} - i\tilde{V}''_{23}\tilde{\rho}''_{12} \\ &\quad + i\tilde{V}'_{12}\tilde{\rho}'_{23} + \tilde{V}''_{12}\tilde{\rho}'_{23} - \tilde{V}'_{12}\tilde{\rho}'_{23} + i\tilde{V}''_{12}\tilde{\rho}''_{23} - i\tilde{V}'_{43}\tilde{\rho}'_{14} - \tilde{V}''_{43}\tilde{\rho}'_{14} \\ &\quad + \tilde{V}'_{43}\tilde{\rho}''_{14} - i\tilde{V}''_{43}\tilde{\rho}''_{14} + i\tilde{V}'_{14}\tilde{\rho}'_{43} + \tilde{V}''_{14}\tilde{\rho}'_{43} - \tilde{V}'_{14}\tilde{\rho}''_{43} + i\tilde{V}''_{14}\tilde{\rho}''_{43} \\ &\quad + 0.5g_{31}\tilde{\rho}'_{13} - i0.5g_{31}\tilde{\rho}''_{13}\end{aligned} \tag{15.7}$$

$$\begin{aligned}\frac{d\tilde{\rho}'_{14}}{dt} + \frac{id\tilde{\rho}''_{14}}{dt} &= -i\Delta_{14}\tilde{\rho}'_{14} - \Delta_{14}\tilde{\rho}''_{14} - i\tilde{V}'_{34}\tilde{\rho}'_{13} + \tilde{V}''_{34}\tilde{\rho}'_{13} + \tilde{V}'_{34}\tilde{\rho}''_{13} + i\tilde{V}''_{34}\tilde{\rho}''_{13} \\ &\quad - i\tilde{V}'_{14}\tilde{\rho}_{11} + \tilde{V}''_{14}\tilde{\rho}_{11} + i\tilde{V}'_{14}\tilde{\rho}_{44} - \tilde{V}''_{14}\tilde{\rho}_{44} - i\tilde{V}'_{12}\tilde{\rho}'_{24} - \tilde{V}''_{12}\tilde{\rho}'_{24} \\ &\quad - \tilde{V}'_{12}\tilde{\rho}''_{24} - i\tilde{V}''_{12}\tilde{\rho}''_{24} - 0.5g_{41}\tilde{\rho}'_{14} - i0.5g_{41}\tilde{\rho}''_{14}\end{aligned} \tag{15.8}$$

$$\begin{aligned}\frac{d\tilde{\rho}'_{12}}{dt} - \frac{id\tilde{\rho}''_{12}}{dt} &= -i\Delta_{21}\tilde{\rho}'_{21} - \Delta_{21}\tilde{\rho}''_{21} - i\tilde{V}'_{23}\tilde{\rho}'_{13} + \tilde{V}''_{23}\tilde{\rho}''_{13} - \tilde{V}'_{23}\tilde{\rho}''_{13} + i\tilde{V}''_{23}\tilde{\rho}''_{13} \\ &\quad - i\tilde{V}'_{12}\tilde{\rho}_{11} + \tilde{V}''_{12}\tilde{\rho}_{11} - i\tilde{V}'_{12}\tilde{\rho}_{22} - \tilde{V}''_{12}\rho_{22} + i\tilde{V}'_{14}\tilde{\rho}'_{14} - \tilde{V}''_{14}\tilde{\rho}'_{14} \\ &\quad + \tilde{V}'_{14}\tilde{\rho}''_{14} + i\tilde{V}''_{14}\tilde{\rho}''_{14} - \tilde{V}'_{14}\tilde{\rho}'_{24} + i\tilde{V}''_{14}\tilde{\rho}'_{24} - i\tilde{V}'_{14}\tilde{\rho}''_{24} - \tilde{V}''_{14}\tilde{\rho}''_{24} \\ &\quad - 0.5g_{21}\tilde{\rho}'_{12} + i0.5g_{21}\tilde{\rho}''_{12}\end{aligned} \tag{15.9}$$

$$\begin{aligned}\frac{\mathrm{d}\tilde{\rho}'_{23}}{\mathrm{d}t}+\frac{i\mathrm{d}\tilde{\rho}''_{23}}{\mathrm{d}t}&=-i\Delta_{23}\tilde{\rho}'_{23}-\Delta_{23}\tilde{\rho}''_{23}-i\tilde{V}'_{23}\tilde{\rho}_{22}-\tilde{V}''_{23}\rho_{22}-i\tilde{V}'_{23}\rho_{33}-\tilde{V}''_{23}\tilde{\rho}_{33}\\&\quad+i\tilde{V}'_{12}\tilde{\rho}'_{13}+\tilde{V}''_{12}\tilde{\rho}'_{13}-\tilde{V}'_{12}\tilde{\rho}''_{13}+i\tilde{V}''_{12}\tilde{\rho}''_{13}-i\tilde{V}'_{23}\tilde{\rho}'_{13}-\tilde{V}''_{23}\tilde{\rho}'_{13}\\&\quad+\tilde{V}'_{23}\tilde{\rho}''_{13}-i\tilde{V}''_{23}\tilde{\rho}''_{13}-0.5g_{32}\tilde{\rho}'_{23}-i0.5g_{32}\tilde{\rho}''_{23}\end{aligned}\tag{15.10}$$

$$\begin{aligned}\frac{\mathrm{d}\tilde{\rho}'_{24}}{\mathrm{d}t}+\frac{i\mathrm{d}\tilde{\rho}''_{24}}{\mathrm{d}t}&=-i\Delta_{24}\tilde{\rho}'_{24}-\Delta_{24}\tilde{\rho}''_{24}-i\tilde{V}'_{14}\tilde{\rho}'_{21}+\tilde{V}''_{14}\tilde{\rho}'_{21}+\tilde{V}'_{14}\tilde{\rho}''_{21}+\tilde{V}''_{14}\tilde{\rho}''_{21}\\&\quad+i\tilde{V}'_{21}\tilde{\rho}'_{14}-\tilde{V}''_{21}\tilde{\rho}'_{14}-\tilde{V}'_{21}\tilde{\rho}''_{14}+i\tilde{V}''_{21}\tilde{\rho}''_{14}+i\tilde{V}'_{34}\tilde{\rho}'_{13}-\tilde{V}''_{34}\tilde{\rho}'_{13}\\&\quad-\tilde{V}'_{34}\tilde{\rho}''_{13}-i\tilde{V}''_{34}\tilde{\rho}''_{13}-0.5(g_{41}+g_{21})\tilde{\rho}'_{24}-i0.5(g_{41}+g_{21})\tilde{\rho}''_{24}\end{aligned}\tag{15.11}$$

$$\begin{aligned}\frac{\mathrm{d}\tilde{\rho}'_{13}}{\mathrm{d}t}-\frac{i\mathrm{d}\tilde{\rho}''_{13}}{\mathrm{d}t}&=-i\Delta_{31}\tilde{\rho}'_{31}-\Delta_{31}\tilde{\rho}''_{31}-i\tilde{V}'_{12}\tilde{\rho}'_{23}-\tilde{V}''_{12}\tilde{\rho}'_{23}-\tilde{V}'_{12}\tilde{\rho}''_{23}-i\tilde{V}''_{12}\tilde{\rho}''_{23}\\&\quad+i\tilde{V}'_{23}\tilde{\rho}'_{12}+\tilde{V}''_{23}\tilde{\rho}'_{12}-\tilde{V}'_{23}\tilde{\rho}''_{12}+i\tilde{V}''_{23}\tilde{\rho}''_{12}-i\tilde{V}'_{14}\tilde{\rho}'_{34}-\tilde{V}''_{14}\tilde{\rho}'_{34}\\&\quad+\tilde{V}'_{14}\tilde{\rho}''_{34}+i\tilde{V}''_{14}\tilde{\rho}''_{34}+i\tilde{V}'_{34}\tilde{\rho}'_{14}-i\tilde{V}''_{34}\tilde{\rho}'_{14}-\tilde{V}'_{34}\tilde{\rho}''_{14}+i\tilde{V}''_{34}\tilde{\rho}''_{14}\\&\quad-0.5(g_{32}+g_{34})\tilde{\rho}'_{13}+i0.5(g_{32}+g_{34})\tilde{\rho}''_{13}\end{aligned}\tag{15.12}$$

$$\begin{aligned}\frac{\mathrm{d}\tilde{\rho}'_{23}}{\mathrm{d}t}-\frac{i\mathrm{d}\tilde{\rho}''_{23}}{\mathrm{d}t}&=-i\Delta_{32}\tilde{\rho}'_{32}-\Delta_{32}\tilde{\rho}''_{32}-i\tilde{V}'_{12}\tilde{\rho}'_{13}+\tilde{V}''_{12}\tilde{\rho}'_{13}-\tilde{V}'_{12}\tilde{\rho}''_{13}-i\tilde{V}''_{12}\tilde{\rho}''_{13}\\&\quad+i\tilde{V}'_{23}\tilde{\rho}_{22}+\tilde{V}''_{23}\tilde{\rho}_{22}-i\tilde{V}'_{23}\tilde{\rho}_{33}-\tilde{V}''_{23}\tilde{\rho}_{33}+i\tilde{V}'_{34}\tilde{\rho}'_{24}-\tilde{V}''_{34}\tilde{\rho}'_{24}\\&\quad+\tilde{V}'_{34}\tilde{\rho}''_{24}+i\tilde{V}''_{34}\tilde{\rho}''_{24}-0.5(g_{32}+g_{34}+g_{41})\tilde{\rho}'_{23}\\&\quad+i0.5(g_{32}+g_{34})\tilde{\rho}''_{23}\end{aligned}\tag{15.13}$$

$$\begin{aligned}\frac{\mathrm{d}\tilde{\rho}'_{34}}{\mathrm{d}t}+\frac{i\mathrm{d}\tilde{\rho}''_{34}}{\mathrm{d}t}&=-i\Delta_{34}\tilde{\rho}'_{34}-\Delta_{34}\tilde{\rho}''_{34}-i\tilde{V}'_{14}\tilde{\rho}'_{13}+\tilde{V}''_{14}\tilde{\rho}'_{13}-\tilde{V}'_{14}\tilde{\rho}''_{13}-i\tilde{V}''_{14}\tilde{\rho}''_{13}\\&\quad+i\tilde{V}'_{13}\tilde{\rho}'_{14}+\tilde{V}''_{13}v'_{14}-\tilde{V}'_{13}i\tilde{\rho}''_{14}+i\tilde{V}''_{13}\tilde{\rho}''_{14}-i\tilde{V}'_{34}\tilde{\rho}_{33}+\tilde{V}''_{34}\tilde{\rho}_{33}\\&\quad+i\tilde{V}'_{34}\tilde{\rho}_{44}-\tilde{V}''_{34}\rho_{44}+i\tilde{V}'_{23}\tilde{\rho}'_{24}-\tilde{V}''_{23}\tilde{\rho}'_{24}+\tilde{V}'_{23}\tilde{\rho}''_{24}-i\tilde{V}''_{23}\tilde{\rho}''_{24}\\&\quad-0.5(g_{32}+g_{34}+g_{41})\tilde{\rho}'_{34}+i0.5(g_{32}+g_{34}+g_{41})\tilde{\rho}''_{34}\end{aligned}\tag{15.14}$$

$$\begin{aligned}\frac{\mathrm{d}\tilde{\rho}'_{14}}{\mathrm{d}t}-\frac{i\mathrm{d}\tilde{\rho}''_{14}}{\mathrm{d}t}&=-i\Delta_{41}\tilde{\rho}'_{41}-\Delta_{41}\tilde{\rho}''_{41}-i\tilde{V}'_{12}\tilde{\rho}'_{24}+\tilde{V}''_{12}\tilde{\rho}'_{24}-\tilde{V}'_{12}\tilde{\rho}''_{24}+i\tilde{V}''_{12}\tilde{\rho}''_{24}\\&\quad-i\tilde{V}'_{34}\tilde{\rho}'_{13}-\tilde{V}''_{34}\tilde{\rho}'_{13}-v'_{34}\tilde{\rho}''_{13}-\tilde{V}''_{34}\tilde{\rho}''_{13}+i\tilde{V}'_{14}\tilde{\rho}_{11}+\tilde{V}''_{14}\tilde{\rho}_{11}\\&\quad-i\tilde{V}'_{14}\tilde{\rho}_{44}-\tilde{V}''_{14}\tilde{\rho}_{44}+0.5g_{41}\tilde{\rho}'_{41}-i0.5g_{41}\tilde{\rho}''_{41}\end{aligned}\tag{15.15}$$

$$\begin{aligned}\frac{\mathrm{d}\tilde{\rho}'_{24}}{\mathrm{d}t}-\frac{i\mathrm{d}\tilde{\rho}''_{24}}{\mathrm{d}t}&=-i\Delta_{42}\tilde{\rho}'_{42}-\Delta_{42}\tilde{\rho}''_{42}-i\tilde{V}'_{12}\tilde{\rho}'_{14}+\tilde{V}''_{12}\tilde{\rho}'_{14}-\tilde{V}'_{12}\tilde{\rho}''_{14}-i\tilde{V}''_{12}\tilde{\rho}''_{14}\\&\quad+i\tilde{V}'_{14}\tilde{\rho}'_{12}+\tilde{V}''_{14}\tilde{\rho}'_{12}+\tilde{V}'_{14}\tilde{\rho}''_{12}-i\tilde{V}''_{14}\tilde{\rho}''_{12}-i\tilde{V}'_{23}\tilde{\rho}'_{34}-\tilde{V}''_{23}\tilde{\rho}'_{34}\\&\quad-\tilde{V}'_{23}\tilde{\rho}''_{34}-i\tilde{V}''_{23}\tilde{\rho}''_{34}-i\tilde{V}'_{34}\tilde{\rho}'_{23}-\tilde{V}''_{34}\tilde{\rho}'_{23}-\tilde{V}'_{34}\tilde{\rho}''_{23}+i\tilde{V}''_{34}\tilde{\rho}''_{23}\\&\quad+0.5(g_{41}+g_{21})\tilde{\rho}'_{42}+0.5(g_{41}+g_{21})\tilde{\rho}''_{42}\end{aligned}\tag{15.16}$$

$$\frac{d\tilde{\rho}'_{34}}{dt} - \frac{i d\tilde{\rho}''_{34}}{dt} = -i\Delta_{43}\tilde{\rho}'_{43} - \Delta_{43}\tilde{\rho}''_{43} + i\tilde{V}'_{14}\tilde{\rho}'_{14} + \tilde{V}''_{14}\tilde{\rho}'_{14} + \tilde{V}'_{14}\tilde{\rho}''_{14} + i\tilde{V}''_{14}\tilde{\rho}''_{14}$$
$$- i\tilde{V}'_{23}\tilde{\rho}'_{24} + \tilde{V}''_{23}\tilde{\rho}'_{24} - \tilde{V}'_{23}\tilde{\rho}''_{24} - i\tilde{V}''_{23}\tilde{\rho}''_{24} + i\tilde{V}'_{34}\tilde{\rho}_{33} + \tilde{V}''_{34}\tilde{\rho}_{33}$$
$$- i\tilde{V}'_{34}\tilde{\rho}_{44} - \tilde{V}''_{34}\tilde{\rho}_{44} + 0.5(g_{34} + g_{41} + g_{32})\tilde{\rho}'_{43}$$
$$- i0.5(g_{34} + g_{41} + g_{32})\tilde{\rho}''_{43} \tag{15.17}$$

The number of equations for off-diagonal elements can be reduced because $\rho_{ij} = \rho^*_{ij}$.

Now we give the expressions for the diagonal elements

$$\frac{d\rho_{11}}{dt} = \left(2V'_{12}\rho''_{12} - 2V''_{12}\rho'_{12}\right) + \left(2V'_{14}\rho''_{14} - 2V''_{14}\rho'_{14}\right) + g_{21}\rho_{22} + g_{41}\rho_{44} \tag{15.18}$$

$$\frac{d\rho_{22}}{dt} = -\left(2V'_{12}\rho''_{12} - 2V''_{12}\rho'_{12}\right) - \left(2V'_{14}\rho''_{14} - 2V''_{14}\rho'_{14}\right) - g_{21}\rho_{22} + g_{32}\rho_{33} \tag{15.19}$$

$$\frac{d\rho_{33}}{dt} = \left(2V'_{23}\rho''_{23} - 2V''_{23}\rho'_{23}\right) + \left(2V'_{34}\rho''_{34} - 2V''_{34}\rho'_{34}\right) - g_{32}\rho_{33} - g_{34}\rho_{33} \tag{15.20}$$

$$\frac{d\rho_{44}}{dt} = \left(2V'_{14}\rho''_{14} - 2V''_{14}\rho'_{14}\right) + \left(2V'_{34}\rho''_{34} - 2V''_{34}\rho'_{34}\right) + g_{34}\rho_{33} - g_{41}\rho_{44} \tag{15.21}$$

where g_{ij} represents the decay constant, $\tilde{\rho}'_{ij}$ the real part of the density matrix element, $\tilde{\rho}''_{ij}$ the imaginary part of the density matrix element, V'_{ij} the real part of the energy of interaction of atoms with the field, $\tilde{V}''_{ij}$ the imaginary part of the energy of interaction of atoms with the field, $\Delta_{ij} = \omega_L - \omega_{ij}$ detuning, ω_L laser frequency, ω_{ij} transition frequency and $\hbar = 1$. For diagonal elements, $g_{ij} = \frac{1}{T_{1j}}$, where T_{1j} is the longitude relaxation time. For off-diagonal elements, $g_{ij} = \frac{1}{T_{2ij}}$, where T_{ij} is the time of transverse relaxation time.

The numerical calculation is carried out in the density matrix formalism using the equations of the density matrix with $g_{21} = 1$, $\Delta = 0$, and the laser pulse width of $\tau = 120$ fs. The detuning (Δ) we put equal 0 for simplicity.

In order to obtain the intensity spectra of the off-diagonal elements, a Fourier transform of the temporal variations of density matrix elements was performed.

The input field for a single pulse case is assumed to have the following form:

$$V_{ij}(t) = V^0_{ij} \cdot \exp\left(-\left(\left(\frac{t - t_1}{\tau_{\text{pulse}}}\right)^2\right)\right), \tag{15.22}$$

where $V^0_{ij} = d_{ij}E^0_{ij}$ represents the energy of a two-level atom interacting with the laser field at the *i*-*j* transition, E^0_{ij} is the amplitude of an electromagnetic field, t_1 is the delay of the beginning of the pulse in time with respect to the origin.

The input field in the case of comb will have the following form:

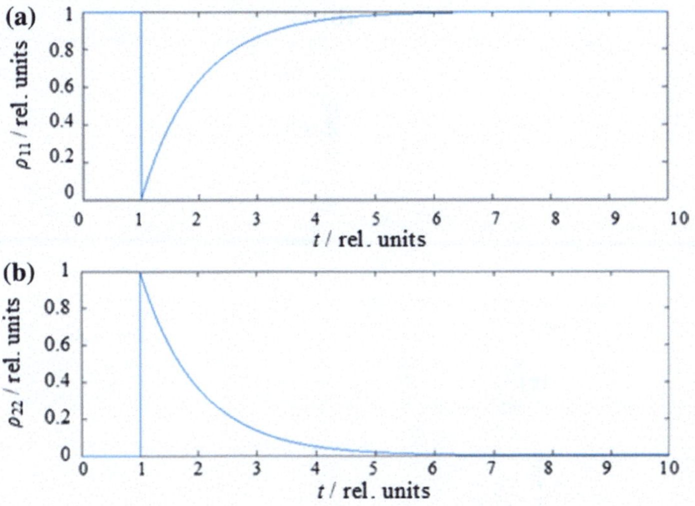

Fig. 15.9 The time dependence of **a** ρ_{11} and that of **b** ρ_{22}. The time represented in the relative units in the abscissa is scaled by $1/\gamma_{21} \approx 30$ ns

$$V_{ij}(t) = V_{ij}^{0} \cdot \sum \exp\left(-\left(\left(\frac{t - t_1 - Nt_2}{\tau_{\text{pulse}}}\right)^2\right)\right), \tag{15.23}$$

where N is the number of pulses in the comb and t_2 is the pulse repetition period. The parameters adopted in the calculations are set as follows:

$$\begin{gathered}
\tau_{\text{pulse}} = 3.8 \cdot 10^{-3}, t_1 = 1, t_2 = 1, \\
g_{21} = 1, g_{41} = 0.3 \cdot g_{21}, \\
\Gamma_{21} = 0.5 \cdot g_{21}, \Gamma_{32} = 0.5 \cdot (g_{32} + g_{21}), \Gamma_{41} = 0.5 \cdot g_{41}, \\
\Gamma_{34} = 0.5 \cdot (g_{34} + g_{32} + g_{41}), \\
N = 1000, \\
V_{12}^{0} = 233, V_{23} = 0.3 \cdot V_{12}, V_{14} = 0.1 \cdot V_{12}, V_{34} = 0.1 \cdot V_{12}
\end{gathered}$$

These parameters correspond to normalized parameters for Rb atoms [5], and are in relative units.

Figure 15.9a, b show the temporal behavior of the diagonal elements of the density matrix, corresponding to the populations of the respective levels, for the single pulse case. In these figures, $t = 1$ in the relative units correspond to $1/\gamma_{21} \approx 30$ ns.

Figure 15.10 shows the temporal behavior of the off-diagonal element representing the polarization and Fig. 15.11 shows the power spectra obtained after the Fourier transform. In both figures, the three cases of the pulse area are shown. The pulse areas are π (a), 0.75 π (b), and 0.3 π (c). Figures 15.10c and 15.11c are qualitatively consistent with the experimental results shown in Fig. 15.7 and Fig. 15.6, respectively.

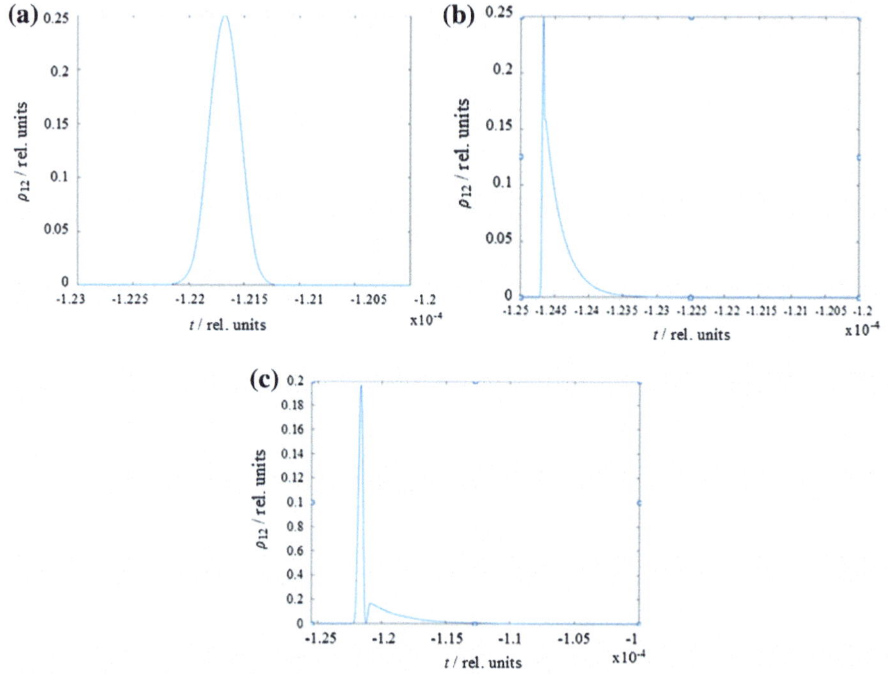

Fig. 15.10 The time dependence of ρ_{12}: **a** π-pulse, **b** 0.75 π-pulse and **c** 0.3 π-pulse. The time represented in the relative units in the abscissa is scaled by $1/\gamma_{21} \approx 30$ ns

For more quantitative discussion, we need to solve the Maxwell-Bloch equations, that is, the Maxwell wave equations together with the density matrix equations [6].

It can be concluded that when the pulse area is $n\pi$ radians (n is an integer), coherent π-pulses are generated from the medium. In other cases, when the area is smaller or larger than π, the upper level decays spontaneously because the population of the upper level is not equal to 0.

The narrowly directed coherent radiation of the $5S_{1/2} \rightarrow 5P_{3/2}$ transition at $\lambda = 780$ nm is associated with the creation of the inversion $\rho_{22}(t) > \rho_{11}(t)$. In this case, the radiation arises in a narrow region at the focal waist of a Gaussian beam. The narrowly directed coherent radiation of the $5P_{3/2} \rightarrow 5D_{5/2}$ transition at $\lambda = 776$ nm is also associated with the inversion of the populations between ρ_{22} and ρ_{11} at around $t = 1.0$ as shown in Fig. 15.12. This is because the rate of the $5S_{1/2} \rightarrow 5P_{3/2}$ transition (0.3×10^8 s^{-1}) is larger than that of the $5P_{3/2} \rightarrow 5D_{5/2}$ transition (0.4×10^7 s^{-1}), i.e., the $5D_{5/2}$ level decays more slowly than the $5P_{3/2}$ level.

The narrowly directed coherent radiation of the $6P_{3/2} \rightarrow 5S_{1/2}$ transition at $\lambda = 420$ nm is of particular interest. An analysis of the equation for the non-diagonal matrix element ρ_{41}, which was introduced above as (15.15),

$$\frac{d\tilde{\rho}'_{14}}{dt} - \frac{i d\tilde{\rho}''_{14}}{dt} = -i\Delta_{41}\tilde{\rho}'_{41} - i\Delta_{41}\tilde{\rho}''_{41} - i\tilde{V}'_{12}\tilde{\rho}'_{24} + \tilde{V}''_{12}\tilde{\rho}'_{24} - \tilde{V}'_{12}\tilde{\rho}''_{24}$$

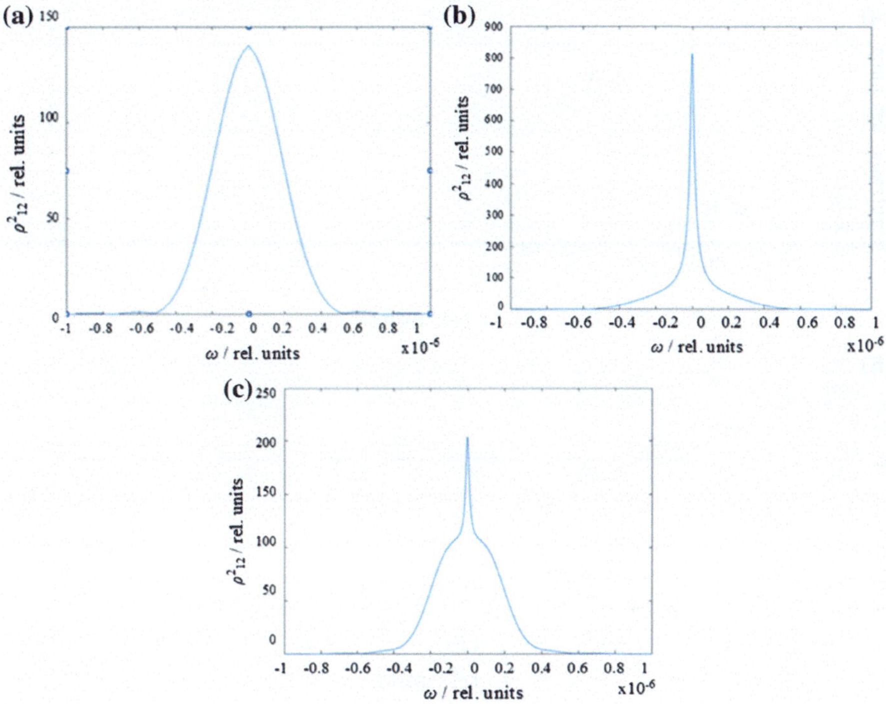

Fig. 15.11 Power spectra of the radiation at the $5S_{1/2} \rightarrow 5P_{3/2}$ transition: **a** π-pulse, **b** 0.75 π-pulse and **c** 0.3 π-pulse

$$+ \boldsymbol{i\tilde{V}''_{12}\tilde{\rho}''_{24}} - \boldsymbol{i\tilde{V}'_{34}\tilde{\rho}'_{13}} - \tilde{V}''_{34}\tilde{\rho}'_{13} - \tilde{V}'_{34}\tilde{\rho}''_{13} - \tilde{V}''_{34}\tilde{\rho}''_{13} + i\tilde{V}'_{14}\tilde{\rho}_{11} + \tilde{V}''_{14}\tilde{\rho}_{11}$$
$$- i\tilde{V}'_{14}\tilde{\rho}_{44} - \tilde{V}''_{14}\tilde{\rho}_{44} + 0.5g_{41}\tilde{\rho}'_{41} - i0.5g_{41}\tilde{\rho}''_{41}, \quad (15.24)$$

shows that coherent radiation of this transition can be generated through a nonlinear interference effect originating from the interference terms shown in bold face above. When we set $\Delta = 0$ and $\tilde{V}'_{ij} = 0$, for simplicity, in (15.24), we obtain

$$\frac{d\tilde{\rho}''_{14}}{dt} = -\tilde{V}''_{12}\tilde{\rho}'_{24} - \tilde{V}'_{34}\tilde{\rho}'_{13} + \tilde{V}'_{14}(\tilde{\rho}_{11} - \tilde{\rho}_{44}) \quad (15.25)$$

from the imaginary part of (15.24). Equation (15.25) shows that the amplification without inversion (AWI) can take place by the interference matrix element $\tilde{\rho}'_{24}$ and the generation of narrow coherent unidirectional radiation can take place by the interference matrix element $\tilde{\rho}'_{13}$. When the sum of these two terms becomes larger than the last term in (15.25), the narrow unidirectional coherent radiation is generated at the transition $5S_{1/2} \rightarrow 5P_{3/2}$ ($\lambda = 420$ nm).

Therefore, in the presence of the coupling radiation for the $5D_{5/2} \rightarrow 6P_{3/2}$ transition at $\lambda = 5.2$ μm [1], the radiation of the $6P_{3/2} \rightarrow 5S_{1/2}$ transition at $\lambda = 420$ nm becomes possible through the two photon $5S_{1/2} \rightarrow 5D_{5/2}$ transition [5]. This coupling radiation

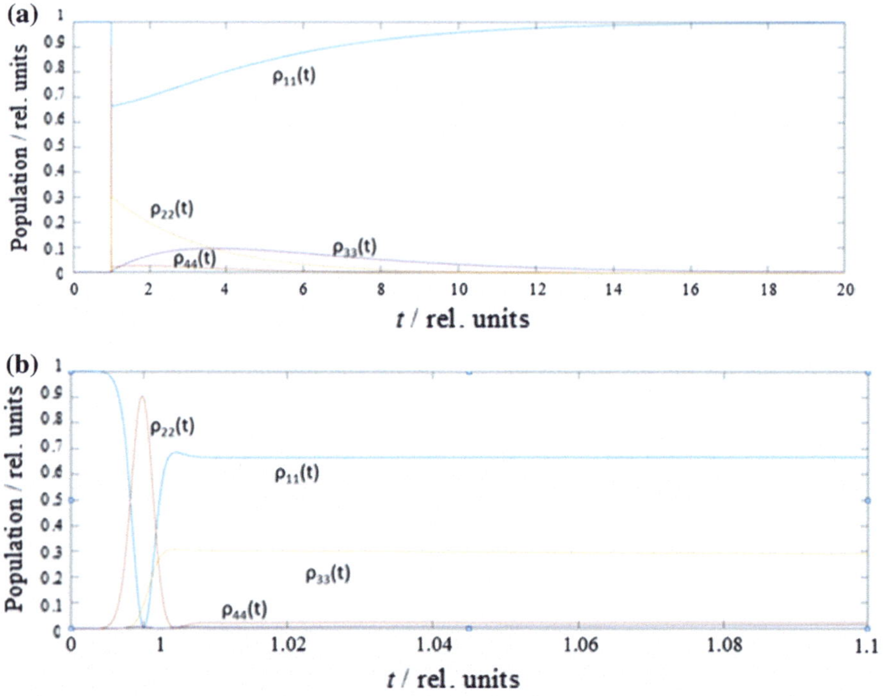

Fig. 15.12 Temporal dependences of the populations in the two different time scales. The time represented in the relative units in the abscissa is scaled by $1/\gamma_{21} \approx 30$ ns.

exists because superradiance [5] or coherent radiation originating from the $5D_{5/2} \rightarrow 6P_{3/2}$ transition at $\lambda = 5.2$ μm [5].

The second mechanism for creating the coherent radiation is based on the presence of a weak field at the transition frequency of the $6P_{3/2} \rightarrow 5S_{1/2}$ transition. In this mechanism, the amplification without population inversion becomes possible through the interference matrix element ρ_{24} on the forbidden $6P_{3/2} \rightarrow 5P_{3/2}$ transition. In this case, a two-photon $5S_{1/2} \rightarrow 6P_{3/2}$ transition becomes possible through the $5S_{1/2} \rightarrow 5P_{3/2}$ and $6P_{3/2} \rightarrow 5S_{1/2}$ transitions.

So far we have been discussing the case of the one pulse incident field (15.22), but the mechanism of the generation of the unidirectional coherent radiation induced by the comb field (15.23) is basically the same as that of the one pulse incident field case as long as the repetition rate of comb components are larger than the decay rates of the populations and non-diagonal (polarization) elements. These conditions are fulfilled in our experiments using of Rb atoms.

Figures 15.13 and 15.14 are the temporal behaviors of the polarization of the radiation and the power spectrum of the radiation obtained by the comb field at the 780 nm radiation for the different pulse areas (1π and 0.7π). These results are consistent with those obtained by the one-pulse excitation, showing that the unidirectional

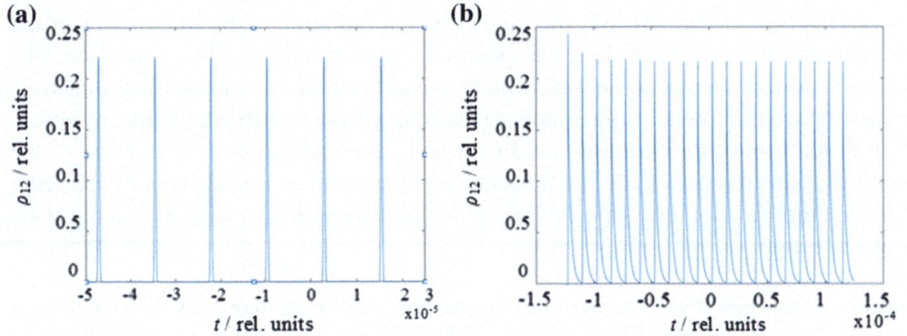

Fig. 15.13 Time dependences of non-diagonal element for **a** a π-pulse and **b** a 0.7 π-pulse. The time represented in the relative units in the abscissa is scaled by $1/\gamma_{21} \approx 30$ ns

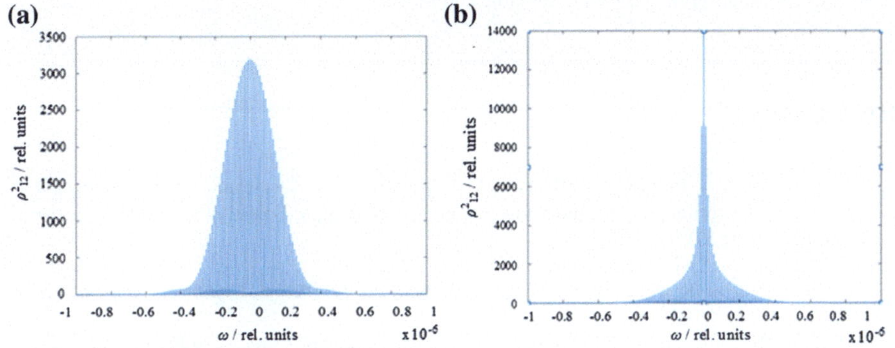

Fig. 15.14 **a** Power spectrum of radiation for a π-pulse and **b** that for a 0.7 π-pulse. The time represented in the relative units in the abscissa is scaled by $1/\gamma_{21} \approx 30$ ns.

coherent radiation on 420 nm originates from the nonlinear interference effect in both one-pulse excitation and comb-field excitation cases.

15.4 Summary

In this chapter we have presented the results of our recent experimental studies on linear and nonlinear comb-spectroscopy in the octave expanded spectral range. The oscillator strengths for the atomic transitions and molecular transitions in the wide spectral range covering the entire visible wavelength range can be measured by the Rozhdestvensky hook method. The accumulation of as many as 20,000–30,000 interference patterns during the long exposure time become possible because the phase differences among the different patterns is kept during the measurements. The different nonlinear effects in the tightly focused radiation from the comb generator of the femtosecond laser pulses appear in the atomic polarization, for example, as the production of the narrowly directed coherent radiation at the frequencies of the pump radiation in the red side of the spectrum and in the adjacent blue transition. The strong unidirectional coherent radiation on the resonant transition and on the cascade transition are interpreted as phenomena originating from pulsed population inversion. The generation of the blue-side unidirectional coherent radiation is ascribed to the nonlinear interference effect. It is possible to use an octave-expanded comb-spectrum for the amplification and gain in "lasing without inversion or mirrors" in the wide spectral range.

Acknowledgements The experiments results were performed in the Research Center for "Optical and laser methods of substance research" at the St. Petersburg State University. The financial support provided by RFBR grant N 18-02-01095 is gratefully acknowledged. V. Yu. Venediktov fulfills the State Contract of the Russian Federal Ministry of Education and Science, "Organization of Scientific Researches." The authors thank Prof. Kaoru Yamanouchi (The University of Tokyo) for his careful reading of the manuscript and for his valuable comments.

References

1. S.V. Uvarova, S.A. Pulkin, Y.N. Borisov, M.V. Balabas, S.V. Saveleva, A.A. Kalinichev, D.V. Venediktov, Experimental nonlinear interference spectroscopy. Izvestiya Rossiiskoi Akademii Nauk **81**, 1466 (2017)
2. I.S. Zeilikovich, A.M. Lyalikov, Sov. Phys. Usp. **34**, 74 (1991)
3. M. Yan, P.L. Luo, K. Iwakuni, G. Millot, T.W. Hänsch, N. Picqué, Light Sci. Appl. **6**, 17076 (2017)
4. S. Pulkin, High-resolution femtosecond comb spectroscopy, in *Progress in Photon Science*, ed. K. Yamanouchi. Springer Series in Chemical Physics, vol. 115 (Springer, 2017), pp. 161–168
5. G. Papadimitriou, D. Pentaris, T. Efthimiopoulos, A. Lyras, J. Phys. B: At. Mol. Opt. Phys. **50**, 125401 (2017)
6. A.A. Radtsig, B.M. Smirnov, Parameters of atoms and atomic ions. Energo-atomizdat 344 (1986)
7. G.O. Ariunbold, M.M. Kash, V.A. Sautenkov, H. Li, Y.V. Rostovtsev, G.R. Welch, M.O. Scully, Phys. Rev. A **82**, 043421 (2010)

Part IV
Plasma and Relativistic Phenomena

Chapter 16
Terahertz and X-Ray Emission from Clustered Plasma and Dynamics of the Cluster Formation in the Expanding Jet

A. V. Balakin, M. S. Dzhidzhoev, V. M. Gordienko, I. A. Zhvaniya, I. E. Ivanov, N. A. Kuzechkin, P. M. Solyankin and A. P. Shkurinov

Abstract In this chapter we introduce our recent studies on terahertz (THz) and X-ray emission generated in an Ar gas-cluster jet under irradiation with high-intense ultrashort laser pulses. We carried out a numerical simulation of cluster formation

---

A. V. Balakin · M. S. Dzhidzhoev · V. M. Gordienko · I. A. Zhvaniya · I. E. Ivanov · N. A. Kuzechkin · P. M. Solyankin · A. P. Shkurinov (✉)
Faculty of Physics & International Laser Center, Lomonosov Moscow State University, Moscow 119991, Russia
e-mail: ashkurinov@physics.msu.ru

A. V. Balakin
e-mail: a.v.balakin@physics.msu.ru

M. S. Dzhidzhoev
e-mail: djidjoevms@mail.ru

V. M. Gordienko
e-mail: gord@phys.msu.ru

I. A. Zhvaniya
e-mail: zhvania@mail.ru

I. E. Ivanov
e-mail: ivanovmai@gmail.com

A. V. Balakin · N. A. Kuzechkin · P. M. Solyankin · A. P. Shkurinov
Institute on Laser and Information Technologies, Branch of the Federal Scientific Research Centre "Crystallography and Photonics" of Russian Academy of Sciences, Svyatoozerskaya 1, 140700 Shatura, Moscow, Russia
e-mail: lents125@mail.ru

P. M. Solyankin
e-mail: solyankin@physics.msu.ru

A. P. Shkurinov
The National University of Science and Technology MISiS, Moscow 119049, Russia

K. Yamanouchi et al. (eds.), *Progress in Photon Science*, Springer Series in Chemical Physics 119, https://doi.org/10.1007/978-3-030-05974-3_16

processes in a supersonic jet produced under adiabatic extension of gaseous Ar into vacuum, which shows that the concentration ratios among non-clusterized Ar monomers, small-size Ar clusters and large-size Ar clusters significantly vary along the jet. The chapter presents experimental results on THz and X-ray emission generated by irradiation of the jet with intense femtosecond laser pulses at various downstream distances measured from the nozzle outlet along the axis of symmetry of the nozzle. It is shown that THz and X-ray emission from the jet is a useful tool for the study of clustering dynamics in the course of free expansion of gas through a nozzle into vacuum.

16.1 Introduction

Over the past years the interaction of intense femtosecond laser pulses with cluster targets has been an attractive research subject because atomic and molecular clusters can couple with intense femtosecond laser pulses very efficiently and induce a range of nonlinear effects [1–8]. It is also known that a cluster beam can absorb intense ultrashort laser pulses almost completely (up to 95%) because of linear (Mie) and nonlinear resonance interactions [9, 10]. That is why, it has been regarded as a promising medium for generation of intense coherent electromagnetic pulses with sub-picosecond duration in a wide spectral range from X-ray up to THz.

The efficient emission of powerful X-ray pulses is one of the most attractive phenomena that can be observed when clusters interact with intense femtosecond laser pulses [6, 7]. When clusters are exposed to an intense femtosecond laser pulse, constituent atoms are multiply ionized, and the clusters turn into plasma with solid-state density. Because of the high local density of atoms in a cluster, which exceeds typical gas density by three orders of magnitude, part of the ions can be excited by the collision with electrons in the plasma, resulting in the emission of X-ray quanta. In a gaseous medium, the probability of electronic excitation of ions may not proceed efficiently because of the relatively low electron density. It should also be noted that the density of ions in the cluster is in general higher than that in an ionized gaseous medium. X-ray radiation from cluster plasma is comparable with X-ray yield from a solid-state target. Since the generated X-ray intensity can be dramatically increased by the clusters, their presence can be also probed by measuring the intensity of the X-ray emission.

On the other hand, it was recently found that THz radiation can be effectively generated through the interaction of a cluster medium with intense femtosecond laser pulses. Indeed, through a number of pioneering experiments, it has been revealed that THz radiation can be generated efficiently using a cluster target. It was reported [5, 11, 12] that intense THz pulses were generated when an Ar cluster beam was irradiated with femtosecond laser pulses. Nagashima et al. [5] and Jahangiri et al. [11] showed that THz pulse generation is enhanced by more than two orders of magnitude when Ar clusters are formed in an Ar cluster jet. Jahangiri et al. [11, 12] reported that the spatial distribution of emitted THz radiation from a cluster plasma exhibited a four-lobed structure and that the dependence of the intensity of THz pulses on the laser pulse energy exhibited nonlinear growth and its quadratic increase did not show any saturation at least up to 70 mJ/pulse, which was the maximum pulse energy they were able to achieve, corresponding to the light field intensity of ~10^{17} W/cm^2. Jahangiri et al. [12] suggested that this THz radiation originates from a temporal variation of electrical quadrupoles produced by the charge separation in cluster plasma induced by a ponderomotive force of the laser pulses.

In order to understand better the processes occurring in the cluster plasma created by the irradiation of cluster jet with intense femtosecond laser pulses and to optimize the plasma formation, we need to know in detail the properties that characterize the jet such as the cluster size distribution and concentration of the clusters, the degree of condensation and the average atomic density along the beam propagation direction. This is due to the fact, that the jet formed by the adiabatic expansion of atoms (or molecules) of gas through a supersonic nozzle into vacuum, contains clusters of different sizes, as well as non-clustered atoms (or molecules) of the gas. Thus, the resulting jet is an object with significant spatial dispersion, and in this chapter, we will use the term "gas-cluster jet" when there is a need to emphasize its inherent feature. Normally, for the estimation of the mean size of the clusters produced via adiabatic expansion of gases into vacuum, a semi-empirical approach developed by Hagena [13, 14], is widely used. According to Hagena the mean cluster size $\langle N \rangle$, where N represents the number of atoms in a cluster, can be estimated using the geometric dimensions of the nozzle, empirical gas constant, the stagnation pressure and the temperature of the gas in high-pressure camera. It is true that we can use the Hagena formula only for the estimation of $\langle N \rangle$ and we can change $\langle N \rangle$ by adjusting the stagnation pressure and the temperature of the gas flowing through the nozzle by taking advantage of the Hagena formula.

Later on, numerical simulations [15, 16] were performed for the cluster jet formation through the gas expansion from a nozzle. Moreover, a variety of experimental techniques for measuring the cluster size, such as Rayleigh scattering [15, 17, 18], optical interferometry [15–17], Raman scattering [18], and atomic beam scattering [19] were developed. It was shown that the mean cluster size and the concentration of clusters can vary significantly and non-uniformly depending on the distance from the jet axis along the radial direction [15]. This inhomogeneous distribution of clusters is certainly an important factor when we investigate the interaction of intense laser pulses with a cluster jet. For example, it was found in [20] that the position of the laser beam focus in the direction perpendicular to the symmetry axis of the nozzle is a sensitive control parameter for the X-ray generation from the laser induced plasma produced from a cluster jet. Therefore, it is expected that the distance of the laser beam focus measured from the nozzle edge along the symmetry axis of the nozzle can also be a valuable parameter to control the X-ray yield, but this has not been explored yet.

In general, the clusters' formation process in the gas jet is quite complex and is of probabilistic nature. Therefore, numerical simulations of the cluster formation processes in the free jet expansion from a nozzle could help us understand their characteristic properties. In this chapter, we introduce the results of the numerical simulation of the formation of argon atom clusters when Ar gas passes through a supersonic conical nozzle and expands into vacuum until the jet reaches the distance of 60 mm measured from the nozzle throat along the symmetry axis of the nozzle, and demonstrates that the ratio between Ar monomers, small and large clusters fractions can change dramatically both along the jet propagation direction, and across the jet in the radial direction when the distance from the nozzle throat increases. We believe that we can take advantage of the results of this numerical simulation to optimize the yields of X-ray and THz radiation by focusing intense femtosecond laser pulses in the Ar gas-cluster jet.

16.2 Numerical Simulation of Clustering Process

The numerical simulation of condensation of Ar atoms in the supersonic conical nozzle and in the jet expanding through the nozzle was carried out taking into account the measurements of a nozzle which were used in our further experiments, its cross-section is depicted in Fig. 16.1a. The simulation was performed in two separate stages. A steady-state flow of a viscous gas in an axis-symmetric conical nozzle and in the downstream of the jet was calculated at the first stage. We solved numerically the system of two-dimensional axially symmetric non-stationary Navier-Stokes equations written in a divergent form accompanied by the ideal gas law and the boundary conditions on the edges of the computational domain. The terms responsible for convective transfer were approximated in this system by a modified Godunov scheme of high-order of accuracy [21]. The terms of the equations describing viscous diffusive transfer were approximated with the control volume scheme. The time evolution was performed by using the third-order Runge-Kutta method. It should be noted that the numerical calculation does not describe a non-stationary dynamical transition into the steady flow just after the pulsed gas beam is generated, but describes the flow of the gas beam after it becomes the steady flow.

Since the degree of expansion of the gas in the nozzle and that in the jet is sufficiently high, the gas temperature, which has a room value at the nozzle throat, rapidly drops at the nozzle output to below the critical temperature at which the clustering starts. Therefore, the temperature dependence of a viscous transfer coefficient μ was described by the modified Sutherland formula [22] as

$$\mu = \begin{cases} \mu(T_c)\left(\frac{T}{T_c}\right)^a, & T < T_c \\ \mu(T_c)\left(\frac{T}{T_c}\right)^{3/2}\frac{T_c+S}{T+S}, & T \geq T_c \end{cases} \quad (16.1)$$

where T_c represents the critical temperature of the gas, S—Sutherland constant and a—power factor. For an Ar gas, $T_c = 150$ K, $S = 128.35$ K, $a = 0.945$.

The effect of sparseness that might occur near the solid surfaces was accounted for by the boundary conditions for the velocity slip and the temperature jump at the wall of the first order [23] as

$$u_{\text{slip}} = u_s - u_{\text{wall}} = \frac{2-\sigma}{\sigma}\lambda\frac{\partial u_s}{\partial n} + \frac{3}{4}\frac{\mu}{\rho T_G}\frac{\partial T_{\text{wall}}}{\partial s} \quad (16.2)$$

$$T_{\text{jump}} = T_G - T_{\text{wall}} = \frac{2-\sigma_T}{\sigma_T}\frac{2\gamma}{\gamma+1}\frac{\lambda}{\text{Pr}}\frac{\partial T_G}{\partial n}, \quad (16.3)$$

where σ is the tangential-momentum accommodation coefficient, σ_T is the thermal accommodation coefficient (for example, $\sigma = 1$ in the case of diffuse reflection from the wall with the zero tangential velocity and $\sigma = 0$ in the case of specular reflection), $\frac{\partial}{\partial n}$ and $\frac{\partial}{\partial s}$ denote the spatial derivatives in the normal and tangential directions relative to the wall, u_s is the velocity of the gas along the wall, u_{wall} is the wall velocity along the tangential direction, T_G is the temperature of the gas, T_{wall} is the wall temperature, which was fixed to be $T_{\text{wall}} = 297$ K, $\lambda = \frac{16\mu}{5\rho\sqrt{2\pi RT_G}}$ is the mean-free-path of atoms between collisions, ρ is the density of the gas, R is the specific gas constant, γ is the ratio of specific heats, and Pr is the non-dimensional Prandtl number.

At the second stage, we carried out a numerical simulation of a non-equilibrium clustering process by taking into account the previously obtained density distributions along the stream lines. A quasi-chemical model [24, 25] representing the association of a monomer to a cluster and dissociation of a monomer from a cluster was used to describe the clustering of Ar, i.e.,

$$M_1 + M_i \rightleftarrows M_{i+1}, \quad \text{i} = 1, 2, \ldots \quad (16.4)$$

$$M_1 + M_i \rightleftarrows M_1 + M_{i-1} + M_1, \quad \text{i} = 2, 3, \ldots \quad (16.5)$$

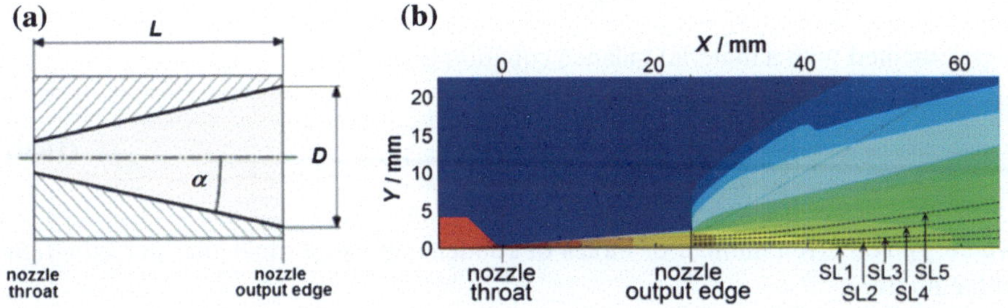

Fig. 16.1 **a** Cross-section of the conical nozzle used in the experiments: L = 24.7 mm, D = 4.7 mm, $\alpha \approx 4.6°$; **b** distribution of Ar atom density inside and outside of the conical nozzle ($Y = 0$ corresponds to the axis of symmetry of the nozzle). The magnitude of the density is represented by a color scale: ~5 kg/m^3 (red), ~0.3 kg/m^3 (yellow), ~10^{-2} kg/m^3 (green), ~3×10^{-4} kg/m^3 (blue), and ~10^{-5} kg/m^3 (dark blue)

where M_I represents a monomer and M_i represents the i-mer, which is a cluster consisting of i atoms.

An Ar vapor is considered as a mixture of perfect gases, each of which consists of i-mers. The liquid-drop model [24, 25] based on an expression for the Gibbs potential [24–26] is used for the description of the thermodynamic properties of the clusters. The Gibbs potential is given by

$$G(\rho, T, \gamma) = \sum_{i=1}^{\infty} \gamma_i G_i(p_i, T), \tag{16.6}$$

where $G_i(p_i, T) = RT\ \ln(p_i/p_0) + G_i^0(T)$, p and T represent respectively the pressure and the temperature of the mixture and $p_0 = 101325$ Pa is the standard pressure. The partial pressure p_i is expressed using the molar fraction of the i-mer x_i as $p_i = px_i$, and γ_i represents the molar mass concentration of i-mer. The standard molar Gibbs potential, $G_i^0(T)$, in (16.3) is given by

$$G_i^0(T) = iG_L^0(T) + \sigma_i(T)(36\pi)^{1/3} N_A(m_1 / \rho_L(T))^{2/3} i^{2/3}, \tag{16.7}$$

where $G_L^0(T)$ is the standard molar Gibbs potential for the liquid phase [26], $\sigma_i(T)$ is the surface tension of i-mer, $\rho_L(T)$ is the density of the liquid, N_A is the Avogadro's constant, m_I is the mass of a monomer atom (or molecule). The standard molar Gibbs potential of $G_1^0(T)$ ($i = 1$) represents the standard molar Gibbs potential for the gaseous phase [26].

In accordance with (16.4) and (16.5), the temporal evolution of the concentrations γ_i could be described by an infinite system of the ordinary differential equations [24] as

$$\frac{d\gamma_{i+1}}{dt} = I_i - I_{i+1}, \tag{16.8}$$

supplemented with a material balance equation given by

$$\sum_{i=1}^{\infty} i\gamma_i = \text{const} = \gamma_0, \tag{16.9}$$

where γ_0 is a given number of moles of condensing vapor molecules per kilogram in the mixture.

According to [25, 26], the specific stream of i-mer, I_i, is given by

$$I_i = \nu_{\Sigma i}\, \varepsilon_i \left(\frac{\gamma_i}{\varepsilon_i} - \frac{\gamma_{i+1}}{\varepsilon_{i+1}} \right), \tag{16.10}$$

with

$$\varepsilon_i = \exp\left[-\left(\frac{G_{i+1}(p,T) - G_i(p,T)}{RT}\right)\right], \tag{16.11}$$

and

$$\nu_{\Sigma i} = \sum_{r=1}^{2} \nu_i^{(r)}, \tag{16.12}$$

where $\nu_i^{(r)}$ represents the frequency of the addition of monomers to the i-mer [21, 23].

As the concentrations of i-mer and $(i+1)$-mer could differ by several orders of magnitude, we modified the system (16.5) according to [25] and also used finite-dimensional systems of N-size, obtained by truncating the original infinite system in the numerical simulation. In order to solve simultaneously the "rigid" system of the ordinary differential equations describing the kinetics of the condensation and the system of differential-algebraic equations, in which the laws of conservation of mass, momentum and energy of the mixture and the model of thermodynamics are included, we adopted an iterative computational algorithm that we developed especially for solving this problem. At each iteration step, we used the sweep method for solving the linear system with three-diagonal matrices, which were obtained by approximating the system of condensation kinetics equations according to the implicit scheme. At the same time, we solved the system of nonlinear algebraic equations, obtained from the corresponding approximations of the conservation laws and thermodynamic relations, in order to determine with high accuracy the monomer concentration and the pressure, temperature and velocity of the mixture [25].

Some of the results obtained by the numerical calculations of the cluster formation process described above are shown in Figs. 16.1b and 16.2. In the calculations, the half opening angle of the conical nozzle was set to be 4.6°, the output diameter was set to be 4.7 mm, and the nozzle length was set to be 24.7 mm, which are the same as those of the conical nozzle used in the experiment and depicted in Fig. 16.1a. The stagnation pressure of Ar gas was set to be 2 MPa. Figure 16.1b shows the spatial distribution of Ar atom density inside and outside of the conical nozzle.

The direction along the symmetry axis of the nozzle is set to be the X axis ($Y = 0$, see Fig. 16.1b), which is the same as the jet propagation direction. The nozzle throat position is set to be at the origin of the X axis, that is at $X = 0$ mm, and the nozzle output edge is placed at $X = 24.7$ mm. The radial direction, which is perpendicular to the symmetry axis of the nozzle, is denoted as the Y axis. The five stream lines, SL1-SL5, are indicated by dashed lines at Fig. 16.1b. SL1 represents the symmetry axis of the nozzle and SL2, SL3, SL4 and SL5 are the stream lines passing the spatial position at $Y = 0.4$ mm, 0.9 mm, 1.3 mm, and 1.7 mm, respectively, at $X = 24.7$ mm.

The distributions of the mean cluster size $<N>$ and mean cluster concentration $<n_{\text{cl}}>$ along the respective five stream lines as a function of the distance from the nozzle throat are shown at Fig. 16.2a and 16.2b correspondingly. The four vertical lines, CS1-CS4, show the positions of the four cross-sections located at $X =$

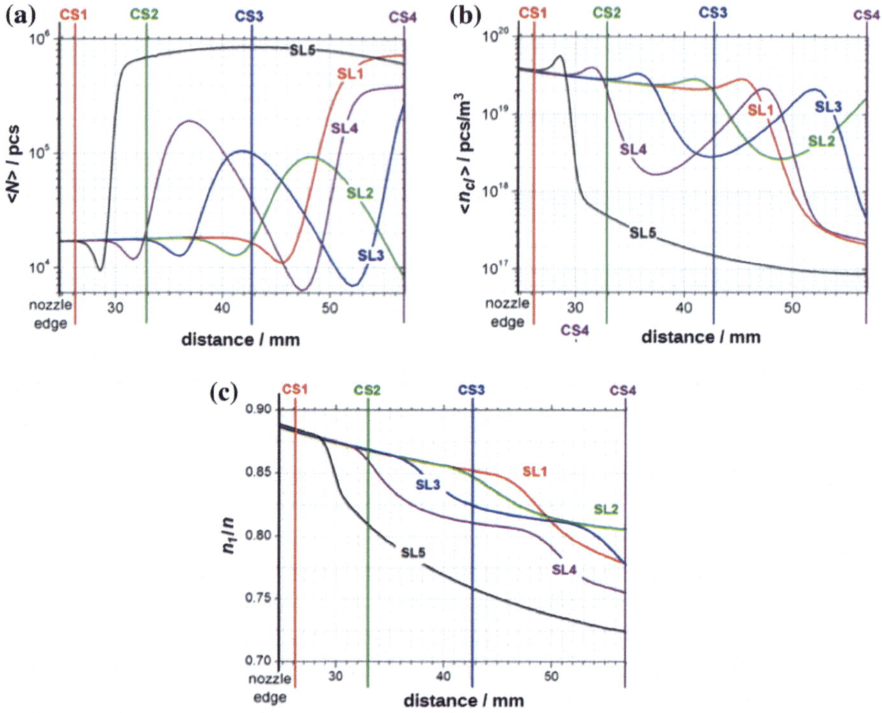

Fig. 16.2 The distribution of **a** mean cluster size $<N>$, **b** mean cluster concentration $<n_{cl}>$, and **c** ratio of monomer concentration n_1 to the total atomic concentration n along five stream lines as a function of the distance from the nozzle throat. The nozzle throat is located at X = 0 mm, the nozzle output edge is located at X = 24.7 mm (for reference see Fig. 16.1). The units of "pcs" represents the number of Ar atoms in a cluster

26.2 mm, 32.9 mm, 42.7 mm and 56.9 mm away from the nozzle throat, respectively. Accordingly, they are located downstream at 1.5 mm, 8.2 mm, 18 mm and 32.2 mm distances measured from the nozzle output edge, respectively. It is clearly seen that spatial properties of the cluster jet possesses significant dispersion in both directions both along the jet downstream and across to the propagation direction. The mean cluster size remains unchanged until SC1 cross-section and then decreases starting from the periphery of the jet, while the paraxial area of the jet still contains "frozen" size clusters up to CS2. After leaving the nozzle the mean cluster size first decreases and next significantly increases along all stream lines. Opposite to that, the mean cluster concentration first increases and then drastically decreases. This means that small-size clusters born in the vicinity of the nozzle output edge first collapse starting from the periphery toward the core of the jet and their concentration correspondingly increases, then a fast association process begins—and small clusters merge and large-size clusters appear, but their concentration consequentially drops. Also, one may note an oscillation behavior of the mean cluster size as the distance from the nozzle output edge increases along SL2, SL3, SL4.

The relative proportion of monomers in the total Ar gas atoms in the jet along five stream lines as a function of the distance from the nozzle throat is depicted at Fig. 16.2c. It can be seen that the percentage of monomers in the jet decreases when moving along the axis X from CS1 to CS4. But there are still more than 70% of monomers in SL5 even in CS4. Thus, the resulting gas-cluster jet contains not only different size clusters but plenty of non-clustered Ar atoms as well.

For better perception of the results of the numerical calculations presented at Fig. 16.2 we plotted the radial distributions of the mean cluster size and the mean cluster concentration in the cross-sections CS1, CS2, CS3 and CS4, which are depicted at Fig. 16.3a, 16.3b, 16.3c, and 16.3d, respectively. Here we denoted the radial abscissa as r to emphasize the radial symmetry of the cluster jet, and r abscissa has the same meaning as Y abscissa in Fig. 16.1b. It is clearly seen that distributions of the mean cluster size and the mean cluster concentration close to the nozzle output edge in the CS1 are homogeneous, the jet consists of small-size clusters containing around 1.8×10^4 Ar atoms each. Next three Figures: 16.3b, 16.3c, and 16.3d demonstrate the dynamics of the spatial structure of the jet along the stream.

Thus, the results of the numerical simulations presented in Figs. 16.1, 16.2 and 16.3 show that for the correct description of interaction of intense femtosecond laser pulses with the gas-cluster target one should take into account that the spatial structure of the gas-cluster jet expanding through the conical nozzle possesses significant dispersion and drastically changes both along the jet downstream and across to the

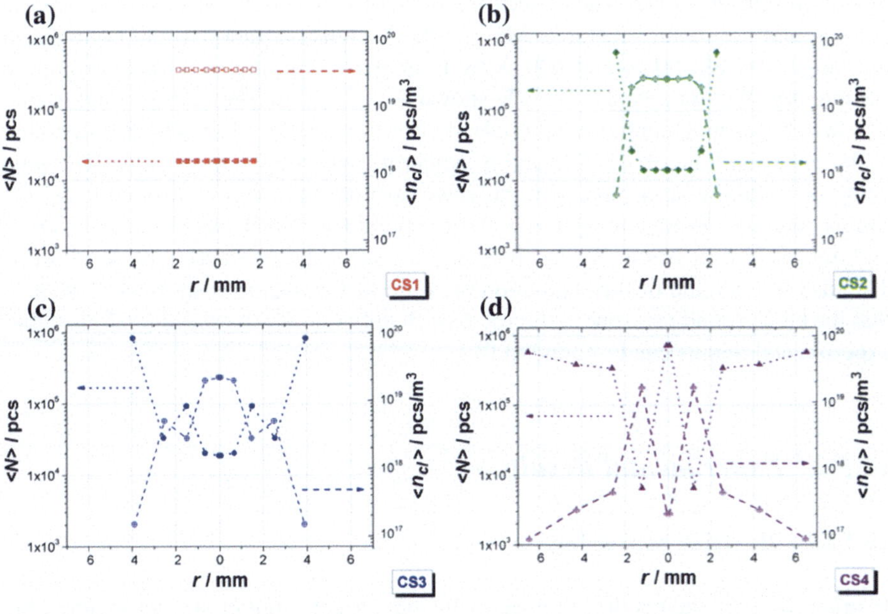

Fig. 16.3 The radial distributions of the mean cluster size and the mean concentration of clusters in the cross-sections CS1–CS4

propagation direction. So, the distance from the nozzle throat is an important parameter which defines the properties of the gas-cluster jet.

16.3 Experimental Section

16.3.1 Experimental Setup

In our experiments for the generation of THz radiation and X-ray in the cluster jet, we modified slightly the experimental setup that was described in detail in our previous report [8] so that we can irradiate the jet with a focused laser pulse at different downstream distances measured from the nozzle output edge. The focal point was located at the symmetry axis of the nozzle and could be positioned discretely along the X axis in the range between 1.5 and 32.3 mm downstream of the nozzle output edge at the locations of CS1–CS4.

Here we would like to list just the main parameters of the experimental setup. It is composed of three main modules: (i) a near-IR ultrashort laser light source generating ultrashort high-energy optical pulses, (ii) a cluster production module, and (iii) a detection system of THz and X-ray radiation. As the laser light source, we use a CPA laser system based on a femtosecond Ti:sapphire laser with a multipass amplifier, which generates pulses whose energy can be raised to 30 mJ/pulse at the repetition rate of 10 Hz. The central wavelength is 810 nm, the beam diameter is 1.5 cm, and the quality factor is $M^2 = 1.6$. The pulse duration τ can be tuned in the range between 50 fs and 600 fs by laser pulse chirping in a vacuum grating compressor. We use a conical nozzle shown at Fig. 16.1a. The nozzle is connected to a high-pressure chamber with a pulsed electromagnetic valve, which is operated at the repetition rate of 1.25 Hz and is synchronized with the laser pulse. The time delay between the laser pulse and the valve opening time was controlled using the timing module in order to optimize the temporal overlap between the laser pulse and the gas-cluster beam. We use pure argon gas for the production of Ar clusters. The maximum value of the stagnation pressure of the nozzle is set to be 2 MPa so that the background pressure in the vacuum chamber does not exceed 0.67 Pa. Other experimental details can be found in [8].

16.3.2 Experimental Results

16.3.2.1 Rayleigh Scattering

In order to verify the results obtained by the numerical simulation, we measured the Rayleigh scattering signals from the gas-cluster jet. For these Rayleigh scattering measurements, we use a continuous semiconductor laser equipped with a spatial

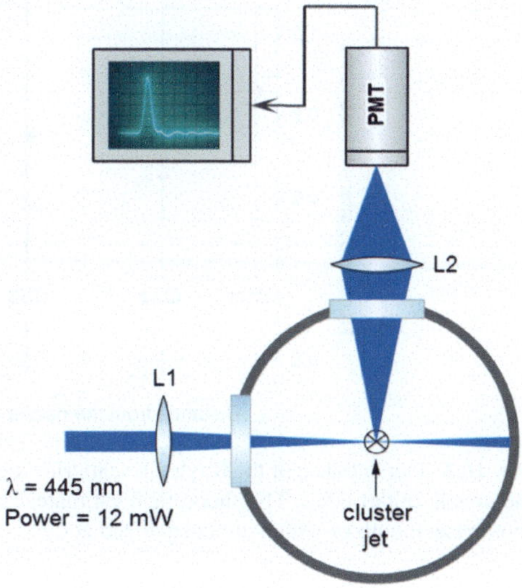

Fig. 16.4 The scheme of the experimental setup for the measurement of the Rayleigh scattering of laser radiation by a gas-cluster jet

filter for the TEM_{00} mode as a light source. The wavelength of the laser is 445 nm, the power is 12 mW, and the beam diameter is 1 cm. The laser light was focused on the gas-cluster jet by the L1 lens ($f = 20$ cm, 3 cmϕ), and the scattered radiation emitted in the direction perpendicular to the gas-cluster beam axis was collected through the quartz window of the vacuum chamber as shown in Fig. 16.4.

The scattered radiation was collected and focused by a lens L1 ($f = 7$ cm, 5 cmϕ) on the photomultiplier tube (PMT) detector. From the measurement of the intensity of the scattered radiation, we can estimate the average number of particles in a cluster and the volume-integrated concentration of clusters. The differential cross-section for the Rayleigh scattering by a dielectric spherical particle is given by

$$\sigma_d(\theta) = \frac{\omega^4}{c^4}\left|\frac{3}{4\pi}\frac{\varepsilon_r - 1}{\varepsilon_r + 2}\right|^2 \left(\frac{N\,m_a}{\rho}\right)^2 \sin^2(\theta), \tag{16.13}$$

where N is the number of atoms in a cluster. The intensity of the radiation scattered by the gas-cluster jet is proportional to the $\langle N^2\rangle \cdot \langle n_{cl}\rangle$. The squares in Fig. 16.5 represent the calculated values of the Rayleigh scattering intensities, $\langle N^2\rangle \cdot \langle n_{cl}\rangle$, as a function of the downstream distance measured from the nozzle outlet edge. The rhombuses in Fig. 16.4 represent the observed Rayleigh scattering signal intensities. The experimental and theoretical Rayleigh scattering intensities are normalized at the distance of CS1. It can be said that the theoretical estimates reproduce the overall trend of the experimental data, even though they exhibit certain deviations from the experimental data beyond their uncertainties in the range between 2.5 and 12%.

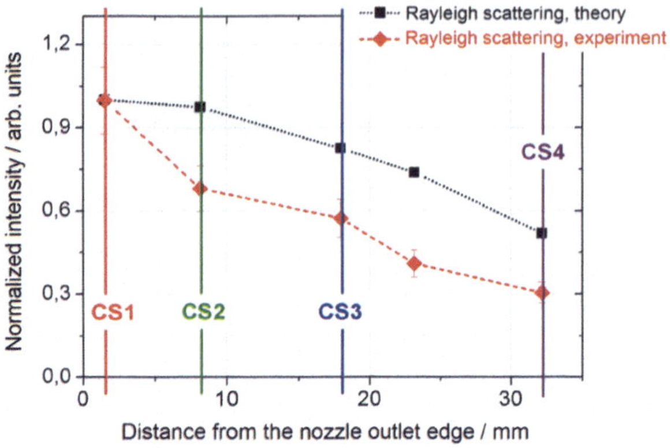

Fig. 16.5 Dependence of the Rayleigh scattering signal intensity on the downstream distance from the nozzle outlet edge. The theoretical estimates (black squares) and the experimental data (red rhombuses), both of which are normalized at CS1

16.3.2.2 Terahertz and X-Ray Emission

In our experiments THz emission generated under the single-color excitation was observed only at the angle of 30° relative to the direction of laser beam propagation. No THz signal from the Ar gas-cluster beam was registered in the forward direction. Figure 16.6 shows the dependences of the intensities of THz and X-ray pulses simultaneously generated from the Ar gas-cluster beam as a function of the laser pulse duration for both positively and negatively chirped laser pulses recorded at the three cross-sections, CS1, CS3 and CS4. The pulse duration of around 55 fs corresponds to a Fourier-limited pulse and is indicated by the central vertical dotted line in Fig. 16.6. As we mentioned above, there are mainly small-size clusters at the cross-section CS1 with the average number of atoms $\langle N \rangle \approx 1.7 \times 10^4$, and cluster size distributions in cross-sections CS3 and CS4 are non-uniform: the value of $\langle N \rangle$ varies from 1.6×10^4 to 8.4×10^5 in CS3 and from 8.4×10^3 to 6.0×10^5 in CS4.

As can be seen in Fig. 16.6a, the X-ray intensity takes almost the maximum value at around the minimum pulse duration and decreases as the pulse duration increases, and the maximum intensity becomes lower steadily as the distance between the excitation area and the nozzle output edge increases from CS1 to CS4. On the contrary, as can be seen in Fig. 16.6b, the THz intensity takes the lowest value at the minimum pulse duration and becomes larger as the pulse duration increases up to around 250 fs, and the THz intensities are almost the same at CS1 and CS3, and decrease largely at CS4.

The energy spectrum of the X-ray emission generated from the Ar gas-cluster beam by the single-color excitation is shown in Fig. 16.7. The presence of the gap in the energy spectrum between 3.5 and 5 keV is ascribed to the measurement procedure. The X-ray photons in the low-photon energy region below 3.5 keV where the K-line emission of Ar appears and those in the higher photon energy region above 5 keV were measured separately using different sets of filters and diaphragms which were

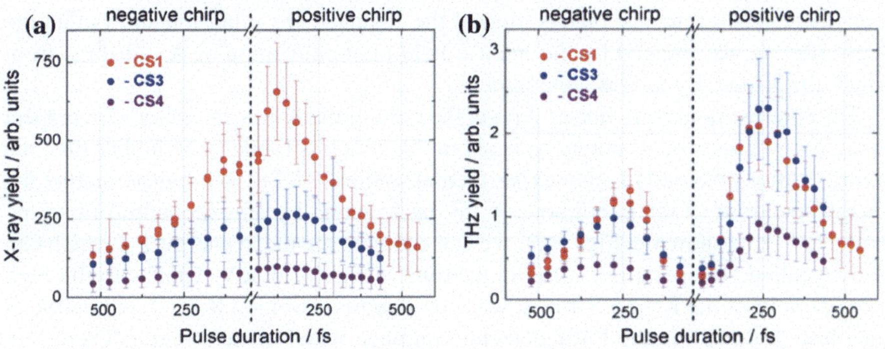

Fig. 16.6 X-Ray (**a**) and THz (**b**) yields from an Ar gas-clusters jet as a function of the temporal duration of the excitation laser pulses recorded at the cross-sections of CS1 (red dots), CS3 (blue dots), and CS4 (purple dots). The vertical dotted line in the center represents the pulse duration for a Fourier limited pulse

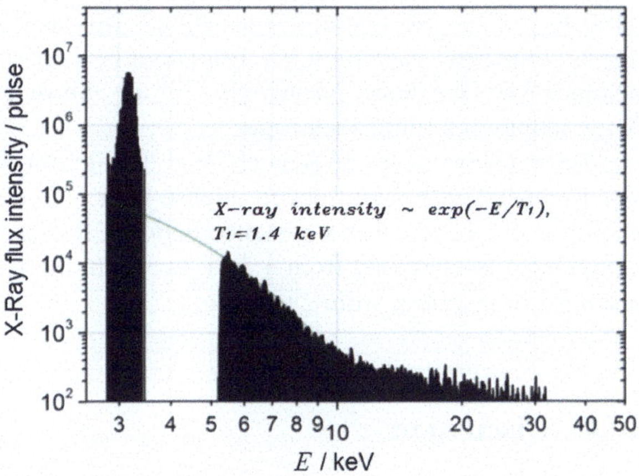

Fig. 16.7 Energy spectrum of the X-ray emission generated when an Ar gas cluster jet was excited at CS1 by focused negatively chirped femtosecond laser pulses (20 mJ/pulse, pulse duration of 250 fs). The backing pressure of Ar was 2 MPa

used to attenuate X-Ray intensity for achieving the single-photon counting mode of the X-ray spectrometer. It should be noted that X-ray emission generated from the Ar gas-cluster beam can be used not only for the characterization of cluster plasma itself but also for the confirmation of the formation of Ar clusters in the jet by the presence of the strong characteristic K-line of Ar in the X-ray spectrum at around 3 keV.

As long as the ponderomotive potential of electrons in the laser field exceeds the ionization potential of the K shell of Ar, the K-line X-ray can be generated efficiently from a non-clustered Ar gas. As the ponderomotive potential of electrons is given by $U_p = 9.3 \times 10^{-14}\, I\ (\mathrm{W/cm^2}) \cdot \lambda^2\ (\mu\mathrm{m}^2)$, the ponderomotive potential in our experiment can be estimated to be $U_p = 1.8$ keV using the laser field intensity of around 3×10^{16} W/cm^2, which is smaller than 3.2 keV, the ionization potential of the K shell of Ar. This means that we could not generate the K-line X-ray using Ar monomers. Therefore, the observation of the K-line X-ray emission shows that the dense plasma was created by the intense-field photoionization of Ar clusters, from which the K-line X-ray was generated.

The contrast ratio of K-line of Ar, i.e. the ratio of the K-line intensity with respect to the X-ray background intensity, is about 90, which is found to be higher than the contrast ratios obtained in other experimental studies [27, 28]. By approximating the energy spectrum of the continuous X-ray background by an exponential function, $\exp(-E/T_1)$, as shown in Fig. 16.7, we can estimate the mean energy T_1, which can also be called "temperature," of hot electrons as $T_1 = 1.4$ keV. Thanks to the high contrast ratio of the K-line X-ray, it was possible to record the X-ray yield online at each laser pulse using the PMT detector equipped with NaI scintillator. Therefore, the measurements of the X-ray emission to be discussed below were performed using the PMT detector for the X-ray registration. The total energy of all the X-ray quanta that reached the scintillator was recorded in each laser pulse. By taking into account the solid angle of the PMT detector and the efficiency of the X-ray generation, we estimated the number of characteristic K-line photons that was detected in each laser pulse to be 10^5. We confirmed that the X-ray yield obtained when the detection angle for the PMT detector is 45° from laser beam propagation direction and the one obtained when the detection angle is 135° are almost the same, which means that the X-ray radiation is almost isotropic.

The snapshots of the plasma channel that is formed when the Ar gas cluster jet was irradiated with femtosecond intense laser pulses are shown in Fig. 16.8. As is shown in this figure, as the distance between the laser beam focus and the nozzle output edge is increased from 1.5 to 32.3 mm, the length of the filament emitting visible light increases from 2 to 8 mm.

16.4 Discussion

It is possible that the dependences of the X-ray and THz emission yields on the laser pulse duration shown in Fig. 16.6 can be attributed to certain properties of the free electrons produced in clusters by intense femtosecond laser pulses as well as to the growth process of the clusters [29].

The maximum laser field intensity in our experiments was $I_{max} = 1.3 \times 10^{17}$ W/cm^2, which was achieved with the shortest laser pulse of about 50 fs. This light field intensity exceeds the Coulombic field within an atom. For a hydrogen atom, the intra-field at the first Bohr orbit with radius $a_B \approx 0.05$ nm is given by a well-known formula,

$$E_0 = \frac{e}{a_B^2} \approx 5 \times 10^9 (\text{V/cm}) \tag{16.14}$$

where e is the elementary charge. Consequently, the characteristic light field intensity, i.e., the characteristic energy flux density I_0 becomes

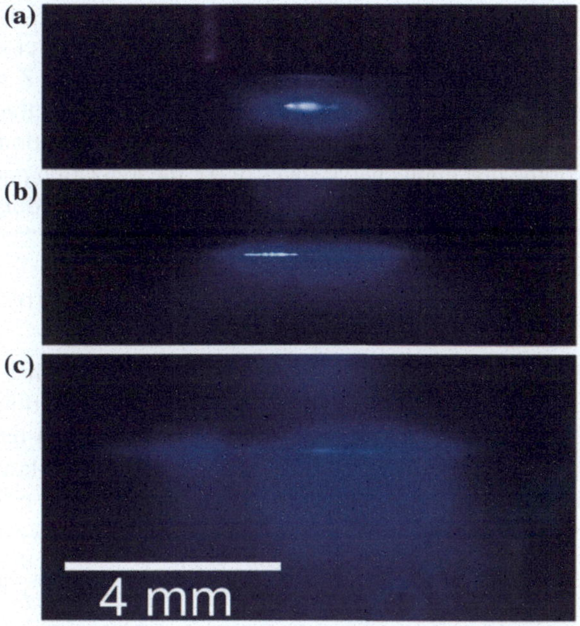

Fig. 16.8 CCD images of the plasma channel generated in the Ar gas cluster jet. The laser beam is focused at **a** CS1, **b** CS3, and **c** CS4

$$I_0 = \frac{cE_0^2}{8\pi} = 3.5 \cdot 10^{16}\left(\mathrm{W/cm^2}\right). \tag{16.15}$$

At this light field intensity, the barrier-suppression ionization becomes possible, and an electron in the 1 s orbital is pulled out from a hydrogen atom by the laser field. According to the existing theoretical description of the barrier-suppression ionization of an atom [29–31], the ionization of an atom into an atomic ion having the positive charge Z occurs when the strength of the external electric field of light exceeds the critical value,

$$E_{cr} = \frac{J_Z^2}{4e^3Z}, \tag{16.16}$$

where J_z represents the ionization potential for producing an atomic ion with the positive charge Z from an atom with the positive charge $Z - 1$. The field ionization produces free electrons with equal probability both in atoms in a cluster and in ordinary gas atoms because there are no free electrons in the cluster before the ionization and the laser field inside the cluster is not screened.

According to [32], the ionization potentials for removing the first 11 electrons in Ar are $J_1 = 15.756$, $J_2 = 27.62$, $J_3 = 40.90$, $J_4 = 59.79$, $J_5 = 75.0$, $J_6 = 91.3$, $J_7 = 123.9$, $J_8 = 143.4$, $J_9 = 422.6$, $J_{10} = 479.0$, and $J_{11} = 539.5$ eV, respectively. It can be seen that J_z increases almost 3 times from J_8 to J_9. Accordingly, the required laser field amplitude increases 7.7 times from $0.87E_0$ to $6.7E_0$, which corresponds to the increase in the laser field intensity about 60 times from $I_8 = 2.64 \times 10^{16}$ W/cm^2 to $I_9 = 1.58 \times 10^{18}$ W/cm^2. Since the laser field intensity in our experiment did not exceed $I_{\max} = 1.3 \times 10^{17}$ W/cm^2, the ion charge Z was considered to be around 8. The chirping of the laser pulse could not change this value because $I_{\max}$ changes only by a factor of 12 in the range of the pulse duration variation in our experiment. For the first eight charge states of Ar, the ionization potential is almost proportional to Z and can be evaluated as $J_Z \sim 16.5Z$ eV. According to [29], the ionization time of these charge states can be estimated as

$$\tau_Z \sim \sqrt{2}e^2 Z\sqrt{m_e}\Big/ J_Z^{3/2} \sim 10^{16} Z^{-1}(\mathrm{s}) \tag{16.17}$$

where *me* is the mass of an electron. This means that the ionization occurs almost immediately during the period less than one oscillation of the laser electric field.

The deeper levels with $Z \geq 9$ could be ionized through the mechanism of the impact ionization. Indeed, any free electron in the laser field with the electric field strength E and the frequency ω gains the energy W of the order of

$$W = \frac{m_e}{2}\left(\frac{eE}{m_e\omega}\right)^2 = 4300\frac{I}{I_0}(\mathrm{eV}), \tag{16.18}$$

which can exceed the ionization energy of the deeper levels. However, the electron impact ionization is known to take a few tens of femtoseconds in the case of atomic clusters and much longer time in the case of monomers. A cluster has high local density which is close to solid density and therefore inelastic electron-ion collisions occur quite often during the duration of the ultrashort laser pulse [33].

It is possible that the interaction of a femtosecond laser pulse with Ar clusters leads to the formation a quasi-neutral plasma inside the cluster core, and the quasi-neutrality persists during the initial stage of the laser pulse propagation. In the course of the interaction with the rising edge of the laser field, the radial distribution of the constituting atomic ions gradually becomes non-uniform with its spherical symmetry is almost being kept, but the distribution of electrons becomes asymmetric through the oscillation of electrons between the poles of the clusters [34]. The intense laser field drives electrons out of the cluster at every half period of the field oscillations, while the ion cloud gradually expands in the radial direction. Associated with the fast losses of electrons, the cluster gains a net electric charge Q. This process is the so-called outer ionization of the cluster [29] and it differs from aforementioned inner ionization when single atom inside the cluster gains a charge Z.

The effect of outer ionization is due to the different field strength in different parts of the cluster. Indeed, the total electric field affecting a single atomic ion within a cluster is composed of an external laser field averaged over a period of the laser field, the static field within the charged cluster and the dynamic field created by the oscillations of the electron cloud in the cluster driven by the laser field. On the contrary, electrons quickly respond to the variation of the electric field of the intense laser light.

In order to evaluate Q, we assume that the charge density is evenly distributed across the cluster so that the static field is radially symmetric and given by

$$E_Q(r) = \frac{Q r}{R^3}, \tag{16.19}$$

where r is the radial distance from the center of the cluster and R is the radius of the cluster. Using the relation between the cluster charge and the laser field strength,

$$Q(R) = 4ER^2, \tag{16.20}$$

we obtain

$$E_Q(R) = 4E \tag{16.21}$$

The sum of the laser field E and cluster field E_Q has anisotropic radial distribution. Indeed, the laser field applied to the poles and cluster field E_Q are collinear to each other and their sum $E + 4E = 5E$, whereas the sum at the equator is $\sqrt{E^2 + (4E)^2} = \sqrt{17}E$, because these fields are orthogonal at the equator. It is possible that the difference between the field amplitude at the poles and the field amplitude at the equator becomes much larger by the non-uniformity of the spatial distribution of electrons. As reported in [35], the polarization-induced electric field at the poles exceeds the laser electric field by as large as one order of magnitude, which may be sufficiently large for the 9th ionization and the further ejection of electrons from Ar^{9+}. However, further calculations are needed to verify this scenario.

We can distinguish two subsystems of electrons in the cluster: the outer one, which oscillates back and forth through the cluster and the inner one, which remains inside the cluster. Energetic oscillations of the electrons from outer subsystem may lead to strong increase of electron temperature inside the cluster, which is called nonlinear resonance heating process [6]. The dependence of the X-ray yield from cluster plasma on laser pulse duration similar to that obtained in our experiments (see Fig. 16.6a), was observed in [6] and explained by the nonlinear resonance heating. The effectivity of this process is maximal at the highest intensity of laser radiation which is achieved with the shortest duration of the laser pulse.

Another process, linear resonant absorption or Mie resonance, is also possible during laser-cluster interaction [2]. After ionization by the laser field, the cluster begins to expand due to hydrodynamic forces and Coulomb pressure forces. Simplified, if we assume that the electron density inside the cluster is homogeneous, the absorption cross-section of the cluster is described by the formula:

$$\sigma_a = \frac{\pi R^3 \nu}{c} \frac{\omega_p^2 / 9}{(\omega - \omega_p / \sqrt{3})^2 + \nu^2 / 4} \tag{16.22}$$

At a certain moment during expansion, when the condition $\omega = \omega_p / \sqrt{3}$ is reached, laser pulse energy is resonantly absorbed by the cluster. Linear resonance absorption leads to increase of the number of electrons in the inner subsystem of the cluster. We assume that the dependence of THz intensity on laser pulse duration presented in Fig. 16.6b can be explained by the effect of linear resonance absorption.

By following the discussion in [12], we assume that the THz radiation produced by the laser radiation is generated by a time-varying axially symmetric linear quadrupole, which is produced by the ponderomotive force of the laser pulse. Indeed, because of the ponderomotive force, the plasma filament produced by the laser pulse expands radially first, and then, shrinks back after the propagation of the laser pulse, resulting in the formation of the time-varying linear quadrupole.

The decrease in the X-ray emission yield as the distance X between the nozzle output edge and the laser focus increases from CS1 to CS4 may be ascribed to the characteristic propagation of the intense laser pulse in an Ar gas-cluster jet. As the distance X increases, the length of the interaction region increases and the density of clusters in the laser excitation volume decreases. Furthermore, because the laser beam can be defocused by the free electrons formed by the ionization of Ar monomers in the Ar gas-cluster jet, the laser radiation could interact with clusters less efficiently.

As shown in [5], the waveform of the pulsed THz radiation comprises two half-periods corresponding to the one-time radial expansion and subsequent compression of the filament. A phenomenological model of the THz radiation developed in [12] takes into account the effect of the finite length L of the filament, which is larger than the wavelength λ of the THz radiation. It was shown in [12] that the transition radiation and Cherenkov radiation could not explain the experimental angular distribution of THz radiation generated from the plasma produced from a gas-cluster jet. A mechanism of low-frequency quadrupole emission from a plasma channel, which is created by femtosecond laser pulse in gaseous medium, was studied theoretically in [36]. The interference of quadrupole THz sources in the plasma channel results in the formation of a specific angular distribution of the THz field, which takes a minimal value along the filament axis and a maximum value at the angle $\theta_{\max}$ between the detector and the filament axis given by

$$\theta_{\max} = 69^\circ \sqrt{L/\lambda} \tag{16.23}$$

In our present experiments, the THz radiation was collected at the deflection angle of 30° from the laser propagation direction. The laser beam is focused at different downstream distances from the nozzle output edge and the length of the filament should vary, so the angular distribution of the THz radiation is expected to change. In accordance with formula (16.23) we expect a decrease of $\theta_{\max}$ when X increases, since the value of L grows (see Fig. 16.8).

16.5 Conclusion

We have carried out numerical simulations of the cluster formation process in a supersonic jet of Ar, and found that the spatial distribution of the cluster concentration represents the clustering dynamics in the jet expansion. We have shown that the distance from the nozzle output edge along the symmetry axis of a cylindrical nozzle is an important parameter describing the properties of the cluster target.

We have shown that the difference in the optimal laser pulse duration for the efficient generation of X-ray and that for the efficient generation of THz radiation can be explained by the different time periods required for the formation of electron subsystems in an Ar cluster. The population of the outer electrons increases quickly during the first 50 fs and they are responsible for the generation of X-ray, which explains the maximum X-ray yield at the minimal pulse duration in our experiments. The number of inner electrons increases only slowly and reaches maximum at the time scale of the order of 100 fs, which can be attributed to the enhancement of THz yield at $\tau \approx 250$ fs in our experiments. At the later stage, the number of the outer electrons begins to rise again, which may decrease the intensity of the THz radiation.

The authors thank I. A. Ozheredov, A. B. Savel'ev for continuous support. This work was supported by the Ministry of Science and Higher Education within the State assignment FSRC "Crystallography and Photonics" RAS. This work was supported by the RFBR under Grants 18-52-16016 and 17-02-01217. This work was carried out with financial support from the Ministry of Education and Science of the Russian Federation in the framework of the Increase Competitiveness Program of NUST "MISiS" (no. K2-2017-003).

References

1. T. Ditmire, T. Donnelly, A.M. Rubenchik, R.W. Falcone, M.D. Perry, Interaction of intense laser pulses with atomic clusters. Phys. Rev. A Gen. Phys. **53**, 3379 (1996)
2. T.D. Donnelly, T. Ditmire, K. Neuman, M.D. Perry, R.W. Falcone, High-order harmonic generation in atom clusters. Phys. Rev. Lett. **76**, 2472 (1996)
3. I. Alexeev, T.M. Antonsen, K.Y. Kim, H.M. Milchberg, Self-focusing of intense laser pulses in a clustered gas. Phys. Rev. Lett. **90**, 103402 (2003)
4. P.K. Tiwari, V.K. Tripathi, Stimulated raman scattering of a laser in a plasma with clusters. Phys. Plasmas **11**, 1674 (2004)
5. T. Nagashima, H. Hirayama, K. Shibuya, M. Hangyo, M. Hashida, S. Tokita, S. Sakabe, Terahertz pulse radiation from argon clusters irradiated with intense femtosecond laser pulses. Opt. Exp. **17**, 8907 (2009)
6. L.M. Chen, F. Liu, W.M. Wang, M. Kando, J.Y. Mao, L. Zhang, J.L. Ma, Y.T. Li, S.V. Bulanov, T. Tajima, Y. Kato, Z.M. Sheng, Z.Y. Wei, J. Zhang, Intense high-contrast femtosecond K-shell x-ray source from laser-driven Ar clusters. Phys. Rev. Lett. **104**, 215004 (2010)
7. V.M. Gordienko, M.S. Dzhidzhoev, I.A. Zhvaniya, V.T. Platonenko, D.N. Trubnikov, D.O. Fedorov, Hard X-ray generation and plasma filament formation under interaction of femtosecond laser pulses with large molecular clusters. Eur. Phys. J. D **67**, 55 (2013)

8. A.V. Balakin, M.S. Dzhidzhoev, V.M. Gordienko, M.N. Esaulkov, I.A. Zhvaniya, K.A. Ivanov, I.A. Kotelnikov, N.A. Kuzechkin, I.A. Ozheredov, V.Y. Panchenko, A.B. Savel'ev, M.B. Smirnov, P.M. Solyankin, A.P. Shkurinov, IEEE Trans. Terahertz Sci. Technol. **7**, 79 (2017)
9. M. Kundu, D. Bauer, Optimizing the ionization and energy absorption of laser-irradiated clusters. Phys. Plasmas **15**, 033303 (2008)
10. T. Ditmire, R.A. Smith, J.W.G. Tisch, M.H.R. Hutchinson, High intensity laser absorption by gases of atomic clusters. Phys. Rev. Lett. **78**, 3121 (1997)
11. F. Jahangiri, M. Hashida, T. Nagashima, S. Tokita, M. Hangyo, S. Sakabe, Intense terahertz emission from atomic cluster plasma produced by intense femtosecond laser pulses. Appl. Phys. Lett. **99**, 261503 (2011)
12. F. Jahangiri, M. Hashida, S. Tokita, T. Nagashima, M. Hangyo, S. Sakabe, Enhancing the energy of terahertz radiation from plasma produced by intense femtosecond laser pulses. Appl. Phys. Lett. **102**, 191106 (2013)
13. O.F. Hagena, W. Obert, Cluster formation in expanding supersonic jets: effect of pressure, temperature, nozzle size, and test gas. J. Chem. Phys. **56**, 1793 (1972)
14. O.F. Hagena, Cluster ion sources. Sci. Instrum. **63**, 2374 (1992)
15. F. Dorchies, F. Blasco, T. Caillaud, J. Stevefelt, C. Stenz, A.S. Boldarev, V.A. Gasilov, Spatial distribution of cluster size and density in supersonic jets as targets for intense laser pulses. Phys. Rev. A **68**, 023201 (2003)
16. F. Blasco, T. Caillaud, F. Dorchies, C. Stenz, J. Stevefelt, A.S. Boldarev, V.A. Gasilov, Characterization of argon cluster jets for laser interaction studies. Nucl. Instr. Meth. Phys. Res. B **205**, 324 (2003)
17. K.Y. Kim, V. Kumarappan, H.M. Milchberg, Measurement of the average size and density of clusters in a gas jet. Appl. Phys. Lett. **83**, 3210 (2003)
18. A. Ramos, J.M. Fernández, G. Tejeda, S. Montero, Quantitative study of cluster growth in free-jet expansions of CO_2 by Rayleigh and Raman scattering. Phys. Rev. A **72**, 053204 (2005)
19. U. Buck, R. Krohne, Cluster size determination from diffractive He atom scattering. J. Chem. Phys. **105**, 5408 (1996)
20. J.-Y. Lin, H. Chu, M. Shen, Y. Xiao, C.-H. Lee, S. Chen, J. Wang, Opt. Commun. **231**, 375 (2004)
21. G.S. Glushko, I.E. Ivanov, I.A. Kryukov, Math. Model. Comput. Simul. **2**, 407 (2010)
22. D.Y. Dubov, *Proceedings of All-Russian Conference Contemporary Problems of Rarefied Gas Dynamics* (2013)
23. S. Roy, R. Raju, H.F. Chuang, B.A. Cruden, M. Meyyappan, J. Appl. Phys. **93**, 4870 (2003)
24. Y.A. Ryzhov, U.G. Pirumov, V.N. Gorbunov, *Nonequilibrium condensation in high-speed gas flows* (Gordon and Breach Science Publishers, Philadelphia, 1989)
25. V.A. Volkov, A.V. Muslaev, U.G. Pirumov, P.V. Rozovskii, Fluid Dyn. **30**, 399 (1995)
26. L.V. Gurvich, I.V. Veiz, V.A. Medvedev, *Thermodynamic Properties of Individual Substances*, Book 2, vol. 4 (Moscow, Nauka, 1981)
27. L.M. Chen, M. Kando, J. Ma, H. Kotaki, Y. Fukuda, Y. Hayashi, I. Daito, T. Homma, K. Ogura, M. Mori, A.S. Pirozhkov, J. Koga, H. Daido, S.V. Bulanov, T. Kimura, T. Tajima, Y. Kato, Phase-contrast x-ray imaging with intense Ar Kα radiation from femtosecond-laser-driven gas target. Appl. Phys. Lett. **90**, 211501 (2007)
28. L. Zhang, L.-M. Chen, D.-W. Yuan, W.-C. Yan, Z.-H. Wang, C. Liu, Z.-W. Shen, A. Faenov, T. Pikuz, I. Skobelev, V. Gasilov, A. Boldarev, J.-Y. Mao, Y.-T. Li, Q.-L. Dong, X. Lu, J.-L. Ma, W.-M. Wang, Z.-M. Sheng, J. Zhang, Enhanced Kα output of Ar and Kr using size optimized cluster target irradiated by high-contrast laser pulses. Opt. Exp. **19**, 25812 (2011)
29. V. Krainov, M. Smirnov, Cluster beams in the super-intense femtosecond laser pulse. Phys. Rep. **370**, 237 (2002)
30. V.P. Krainov, B.M. Smirnov, M.B. Smirnov, Femtosecond excitation of cluster beams. Phys. Usp. **50**, 907 (2007)
31. N.B. Delone, V.P. Krainov, Tunneling and barrier-suppression ionization of atoms and ions in a laser radiation field. Phys. Usp. **41**, 469 (1998)

32. I.K. Kikoin, *Tables of Physical Quantities (Handbook)* (Atomizdat, Moscow, 1976). (in Russian)
33. B.M. Smirnov, Cluster plasma. Phys. Usp. **43**, 453 (2000)
34. Y. Fukuda, Y. Kishimoto, T. Masaki, K. Yamakawa, Structure and dynamics of cluster plasmas created by ultrashort intense laser fields. Phys. Rev. A **73**, 031201 (2006)
35. C. Jungreuthmayer, M. Geissler, J. Zanghellini, T. Brabec, Microscopic analysis of large-cluster explosion in intense laser fields. Phys. Rev. Lett. **92**, 133401 (2004)
36. N.A. Panov, O.G. Kosareva, V.A. Andreeva, A.B. Savel'ev, D.S. Uryupina, R.V. Volkov, V.A. Makarov, A.P. Shkurinov, JETP. Lett. **93**, 638 (2011)

Chapter 17
Acoustic Signal for Femtosecond Filament Plasma Grating Characterization in Air

Daniil E. Shipilo, Vladislav V. Pankratov, Nikolay A. Panov, Vladimir A. Makarov, Andrei B. Savel'ev and Olga G. Kosareva

Abstract We have designed the physical model and numerical algorithm for the simulations of acoustic wave propagation from the source induced by relaxation of a femtosecond plasma grating formed by two crossing filaments in atmospheric air. The model is based on the wave equation with the wave velocity depending on the transverse temperature variation. The algorithm is based on the fine resolution ($\approx 2\,\mu$m) numerical grid employed for the description of the plasma channel substructures in the course of femtosecond filamentation. We have shown that the femtosecond plasma grating emits the superposition of two acoustic signals after plasma recombination. One acoustic signal is represented by an isotropic cylindrical waveform with the characteristic spatial scale equal to the filament diameter (100–200 μm) while the other has the spatial scale equal to the plasma grating period in the range $20-40\,\mu$m. This second wave propagates in the direction parallel to the axis of plasma grating modulation. Based on the simulations, we suggested the noninvasive method for the plasma grating period and the beam convergence angle detection.

17.1 Introduction

Diagnostics of the high-intensity zone produced by ultrashort laser pulse filamentation [1–3] is a nontrivial task requiring specially designed setup, since almost any material inserted into a high-intensity beam will be destroyed. The example of such invasive diagnostics is insertion of a glass plate into the terawatt peak power 800 nm

---

D. E. Shipilo · V. V. Pankratov · N. A. Panov · V. A. Makarov · A. B. Savel'ev · O. G. Kosareva (✉)
Faculty of Physics and International Laser Center, Lomonosov Moscow State University, 1, bld. 62 Leninskiye gori, Moscow 119991, Russia
e-mail: kosareva@physics.msu.ru

D. E. Shipilo · V. V. Pankratov · N. A. Panov · O. G. Kosareva
Lebedev Physical Institute of the Russian Academy of Sciences, 53 Leninskiy prospect, Moscow 119991, Russia

O. G. Kosareva
Institute of Modern Optics, Nankai University, 300350 Tianjin, China

K. Yamanouchi et al. (eds.), *Progress in Photon Science*, Springer Series in Chemical Physics 119, https://doi.org/10.1007/978-3-030-05974-3_17

beam and identification of the system of multiple rings surrounding the filament by the ablation from the glass surface [4]. Quantitatively, the invasive diagnostics might be performed by reflecting the beam from the white paper screen or a wedge and sending a small part of the overall energy to a CCD for transverse and temporal measurements (see, e.g., [5–7]). Such type of invasive measurements can be performed at a certain propagation position because the measurement destroys the filament. The noninvasive longitudinal diagnostics of femtosecond filament is done by means of ICCD camera and the UV filter in front of it in order to observe the nitrogen fluorescence from aside [8, 9]. With this technique the nitrogen fluorescence signal is resolved in the longitudinal direction of the pulse propagation but integrated over the filament transverse section.

In parallel with filament study by optical methods, photoacoustic methods were employed for filament characterization [10, 11]. The sensitivity of the photoacoustic method is high enough to detect the energy deposition into air in the prefilamentation regime without plasma generation. The amount of energy absorbed into molecular gases of air is relatively low (the peak intensity is below 10^{12} W/cm^2), however the acoustic signal was securely detected already at 1-mJ energy of 80-fs 795-nm initial pulse in [11]. The typical time of acoustic wave formation is $d/c_0 \approx 0.3\,\mu$s, where $d \approx 100\,\mu$m is the filament transverse size and $c_0 \approx 340$ m/s is the speed of sound in air. The microphone used in [11] had the bandwidth of 100 kHz. The corresponding temporal resolution was about 10 μs or worse. With this resolution the identification of filament transverse size was not possible, since the detected acoustic waveform characteristic spatial scale was 6.6 mm exceeding the filament diameter and any fine transverse spatial structure within the plasma channel.

The nicely resolved evolution of acoustic wave was observed by performing imaging interferometry for direct measurement of the evolving refractive index profile. The authors used folded wavefront interferometry [12], which measures the 2D phase shift of the probe beam, from which the refractive index perturbation δn is extracted. According to [12] the measured refractive index variation due to air density perturbation in the filament core is $\delta n \approx 10^{-5}$. Taking into account that the refractive index of air at standard temperature and pressure is $n_0 - 1 \approx 2.7 \times 10^{-4}$ [13], which is due to the standard atmospheric density 2.68×10^{19} cm^{-3} or $\rho_0 = 1.3$ kg/m^3, we can estimate the relative density fluctuation $\delta\rho$ due to the plasma-induced heating. Indeed,

$$\frac{\delta\rho}{\rho_0} = \frac{\delta n}{n_0 - 1} = \frac{10^{-5}}{2.7 \times 10^{-4}} \approx 0.04\,. \tag{17.1}$$

Four-filament regular structures produce the acoustic waveguide on the microsecond time scale and the thermal waveguide on the millisecond time scale [14]. Long-lived refractive index perturbations responsible for these guiding structures are analyzed by means of imaging interferometry.

The high-resolution imaging interferometry technique [12, 14] may be complemented by straightforward measurement of air density variation using a wideband piezoelectric transducer [15, 16]. In the experiment [15] the laser beam from 1-TW Ti:Sapphire laser (805 nm, 55 fs, 10 Hz, beam diameter 7 mm FWHM) was focused by the plane-convex lens with a focal length of 3 m. A single filament was formed if the laser pulse energy was 2.5–3 mJ. To create multiple filaments, the energy of the pulse was increased to 8 mJ. Both optical and acoustic diagnostics were used in each laser shot. The wedge prism was used to direct a small part of the optical radiation after the filament to a CCD camera to visualize the transverse distribution of the beam fluence. The wedge prism was located more than 1 m beyond the focal point of the lens, so it was not damaged by the filament. The acoustic diagnostics included a broadband piezoelectric transducer utilizing a 110-μm-thick polyvinylidene fluoride (PVDF) polymer film (the operating bandwidth is up to 6 MHz) with an operating area 6 mm in diameter. This transducer was placed at a distance of 286 cm from the lens and at a distance of 3–4 mm from the filament. Low acoustic impedance and ability to operate in a wide ultrasonic frequency band are the advantages of PVDF piezoelectric films for detection of optoacoustic signals [17]. The signal from the piezoelectric transducer was amplified by the broadband amplifier with a gain of 12 and recorded using ADC PLX9054 PCI PC card (digitization frequency 500 MHz, 8-bit resolution, Rudnev-Shilyaev Corporation). In a single filament regime acoustic waveform detected by the piezoelectric transducer had one maximum with a characteristic transverse spatial scale of 0.4 μs corresponding to the filament diameter $d \approx 100\,\mu$m. In the multiple filament regime several local maxima were observed with peak-to-peak separation of the order of the filament size $d \approx 100\,\mu$m. So, the acoustic method with the 6 MHz broadband piezoelectric transducer allows one to resolve the multiple filament structure.

The advantage of using this broadband acoustic method [15, 16] is the possibility to perform non-invasive analysis as compared with the transverse beam section registration using a CCD and a wedge. The frequency bandwidth larger than 1 MHz ensures submicrosecond temporal resolution. This high resolution acoustic method requires just one or several detectors and does not require the probe beam or optical scheme adjustment. Indeed, the broadband microphone can be used in outdoor conditions for express analysis of multiple filament structure and energy deposition into the medium. Energy deposition into air in the course of single 800-nm filament regime was studied in [18] and was found to be in the range 1–3% of the initial 800-nm pulse energy of 2–4 mJ. In the multiple filament regime the acoustic method was used to follow the lineic (energy per unit length) energy deposition along the superfilament [19]. For the 800-nm pump pulse of 165 mJ, duration 50 fs and geometrical focusing distance in air of 30 cm, the linear deposited energy per 1 cm was found to be 10 mJ using a sonographic technique. In the regularized superfilamentation regime ensured by merging of 4 filaments produced by the amplitude mask the sonographic method showed the deposition of approximately 0.15 mJ per 1 cm out of 10 mJ of the initial pulse energy [20]. The deposited energy was growing faster than linearly with the initial pulse energy increase [19, 20].

Thus, for the past 4 years the acoustic method in studying single and multiple femtosecond filaments has advanced significantly. It was shown that broadband acoustic detectors can be used to noninvasively analyze both energy deposition and fine structure of the plasma filaments.

In this paper we suggest the method for the noninvasive analysis of the dynamic plasma grating produced by interference of two noncollinearly propagating beams from the same 50 fs, 800 nm pulse with the energy of 1.2 mJ, beam diameter of 8 mm at e^{-2} level and 1 kHz repetition rate [22]. The analysis will be performed numerically in the experimental conditions [22]. According to the suggested method the wideband (more than 6 MHz) piezoelectric transducer is located within the distance of several millimeters from the dynamic plasma grating. We simulate the acoustic wave detected by this wideband transducer. The origin of the acoustic wave is the air density perturbation after the plasma recombination in the grating. The directional diagram and frequency spectrum of the acoustic wave induced by the grating are studied and two characteristic spectral maxima corresponding to the overall plasma width in the filament and the width of one strip of the grating (the grating period) are identified numerically based on the solution of the full wave equation for acoustic signal propagation.

17.2 Numerical Model of Plasma-Induced Acoustic Wave Propagation

We suppose that the medium with the background temperature $\theta_0 = 293$ K has a temperature perturbation $T_0(x, y, z)$ induced by the filament's plasma recombination. We start from the gas dynamics equation system:

$$\begin{cases} \frac{\partial \rho}{\partial t} + \nabla(\rho \mathbf{u}) = 0 \\ \rho \left(\frac{\partial \mathbf{u}}{\partial t} + (\mathbf{u} \cdot \nabla)\, \mathbf{u} \right) = -\nabla p \\ p = p_0 \left(\frac{\rho}{\rho_0} \right)^{\gamma} \end{cases} \tag{17.2}$$

where ρ is the air density, p is the air pressure, $\mathbf{u}$ is the flow velocity vector, $p_0 = \rho_0 R\theta_0/\mu$ is the standard atmospheric pressure, $R = 8.31$ J/(K mol), $\mu = 29$ g/mol is the molar mass of air. The dependence $p(\rho)$ corresponds to the adiabatic law with the adiabatic index $\gamma = 1.4$. After linearization with the density perturbation $\rho' = \rho - \rho_0$ as a small parameter, we obtained the wave equation

$$\Delta\xi - \frac{1}{c^2(x, y, z)} \frac{\partial^2 \xi}{\partial t^2} = 0 \tag{17.3}$$

for the medium excitation function [23]:

$$\xi\,(x, y, z, t) = \left[\rho_0 + \gamma\rho'\,(x, y, z, t)\right] \times \left[\theta_0 + T_0\,(x, y, z)\right] , \tag{17.4}$$

where the velocity is expressed as

$$c^2 = c_0^2 \left(1 + \frac{T_0}{\theta_0} \right) , \tag{17.5}$$

and $c_0 = \sqrt{(\gamma R\theta_0/\mu)}$ is the speed of sound in the unperturbed medium. Please, note that the function ξ is not directly proportional to the air pressure $p \propto \left[\rho_0 + \rho'\right] \times [\theta_0 + T_0]$ since the density perturbation ρ' is multiplied by the adiabatic index coefficient in (17.4).

Thus, after the plasma recombination with the characteristic time of 10 ns [21], we obtain the extended region with the characteristic spatial scales of the plasma channel size heated above the room temperature by $T_{00} \approx 70$ K, which can be estimated from

$$\frac{5}{2} k_B T_{00} = U_i N_e , \tag{17.6}$$

where $U_i = 12.2$ eV is the ionization potential, $N_e \sim 10^{16}\,\text{cm}^{-3}$ is the filament plasma density and k_B is the Boltzmann constant. The longitudinal size of this extended region exceeds its transverse size by 3 orders of magnitude at least (the ratio of 10 cm to 100 μm). Thanks to this large ratio we can neglect the derivatives along the propagation coordinate z and in (17.3) we retain the transverse Laplacian only:

$$\Delta \approx \Delta_\perp = \frac{\partial^2}{\partial x^2} + \frac{\partial^2}{\partial y^2} . \tag{17.7}$$

For our case of the dynamic plasma grating produced by the interference of two noncollinearly propagating beams [22] we assume that the initial temperature perturbation $T_0(x, y)$ reproduces the shape of the plasma channels in the vicinity of the geometrical focus and remains the same for all positions z along the filament. This shape consists of the Gaussian intensity distribution with the characteristic transverse size of the order of the filament diameter $d = 2a_{\text{fil}}$. Moreover, due to the interference of the two beams converging towards each other, the Gaussian intensity distribution is modulated with the period

$$d_{\text{pl}} = \frac{\lambda}{\sin \alpha} \tag{17.8}$$

in the plane perpendicular to the laser pulse propagation direction. Here α is the convergence angle, $\lambda = 800$ nm is the fundamental laser wavelength. In the simulations, we considered the initial temperature distribution

$$T_0(x, y) = T_{00} \exp\left(-\frac{x^2 + y^2}{a_{\text{fil}}^2} \right) \times (1 + \cos[kx \sin \alpha]) , \tag{17.9}$$

where $T_{00} = 70$ K and $k = 2\pi/\lambda$ is a wavenumber. The angle α between the propagation axes of the two noncollinearly converging beams was varied between 0 and 2°, therefore the minimum grating period studied $d_{\text{pl}} \approx 23\,\mu$m.

The grid step in the temporal domain was linked to the spatial grid size through the Courant condition

$$\frac{c_{\mathrm{m}}\,\delta t}{\delta x}+\frac{c_{\mathrm{m}}\,\delta t}{\delta y}\leq 1\,, \tag{17.10}$$

where c_{m} is the maximal velocity (17.5) over the (x, y) domain.

The solution to the wave equation (17.3) with the initial temperature perturbation (17.9) is represented by the waveform $\xi(x, y, t)$. The simulation results presented in Sect. 17.3 show the distributions of $\rho'(x, y)$ at the characteristic time moments. The cross sections along one of the transverse coordinates are provided. The Fourier transform is taken from the waveform at the chosen cross sections for the analysis of the waveform spatial scales.

17.3 Plasma Grating Characterization by Acoustic Waveform and Its Spectrum

The free electrons produced by the two converging beams overlapping in the geometrical focus recombine with the parent ions and disturb the flat background temperature

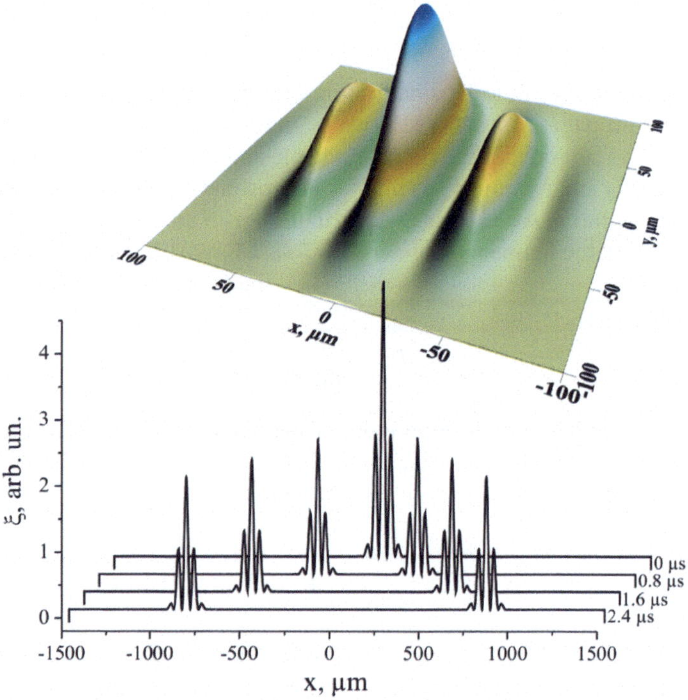

Fig. 17.1 (Upper panel) Temperature profile of the medium $T_0(x, y)$ after recombination of the plasma grating produced by the two 800-nm beams converging at an angle $\alpha = 1°$. (Lower panel) The solution of one-dimensional wave equation for the initial conditions $T_0(x, y = 0)$ from the upper panel

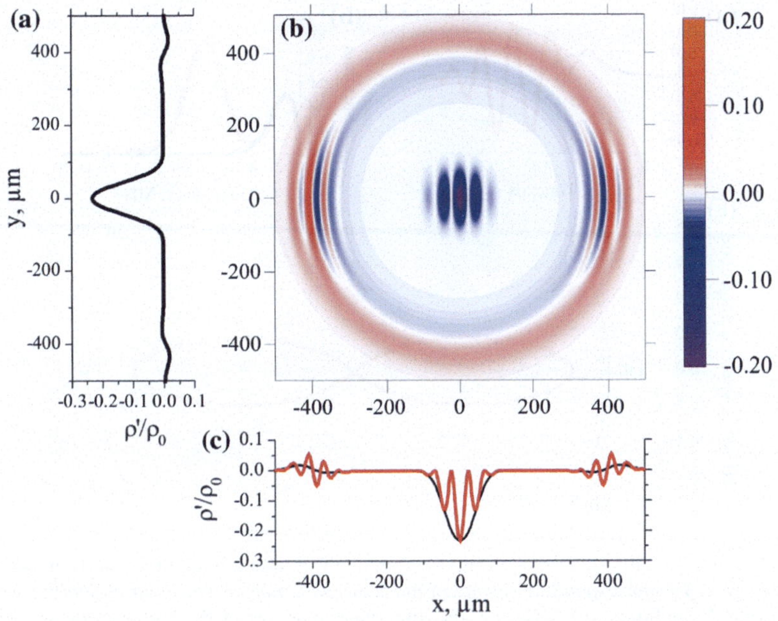

Fig. 17.2 **a**, **c** Air density perturbation $\rho'(x, y)/\rho_0$ cross sections at the time moment $t = 1.2\,\mu s$ after the ionizing laser pulse. The black solid line in panel (**c**) is the same as in panel (**a**) and shows isotropic waveform for the reference. **b** The two-dimensional distribution of the acoustic waveform in the plane perpendicular to the laser pulse propagation direction

profile θ_0. The resulting temperature perturbation $T_0(x, y)$ is given by (17.9) and is shown in the upper panel of Fig. 17.1 at the time moment $t = 0$. The temperature perturbation at the zero time moment is directly proportional to the medium excitation function $\xi(x, y, t = 0)$, since $\rho' = 0$, see (17.4). With the time, the excitation $\xi(x, y, t)$ travels away from the source. By considering one-dimensional excitation $\xi(x, t)$, we obtain motion of the two waveforms in the opposite directions along the x-axis with the velocities of ± 340 m/s (the lower panel of Fig. 17.1).

In the two-dimensional simulations the medium excitation function $\xi(x, y, t)$ is represented by the superposition of the two waves: the isotropic cylindrical one surrounding the overall extended plasma channel and the wave directed along the plasma grating. The medium density perturbation $\rho'(x, y, t)$ is recalculated from the excitation function $\xi(x, y, t)$ according to (17.4) and plotted in the transverse (x, y) plane as the color map (Fig. 17.2b) and the cross sections in the direction perpendicular (Fig. 17.2a) and parallel (Fig. 17.2c) to the plasma grating at the time moment $t = 1.2\,\mu s$ after the initial femtosecond laser pulse producing the filament.

Analysis of the plasma grating can be performed based on the acoustic signals, which can be detected in either parallel (x-axis) or perpendicular (y-axis) direction with respect to the axis of plasma grating modulation (Fig. 17.3a, compare red and black solid curves). In x-direction parallel to the grating orientation the signal wave-

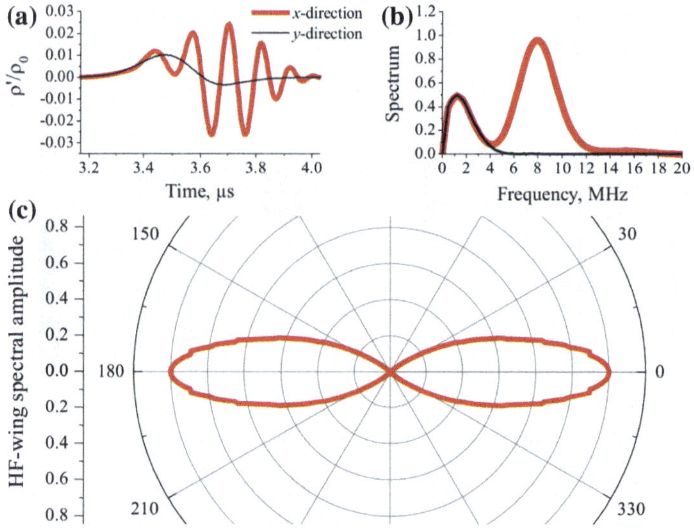

Fig. 17.3 **a** The acoustic signal waveforms detected in the x- and y-axes directions (red and black, respectively). **b** The corresponding spectra of the acoustic signals. **c** The polar plot of the acoustic signal spectral amplitude at 8 MHz, i.e. the directional diagram of the higher-frequency part of acoustic waveform spectrum

form exhibits two characteristic temporal scales: the larger one corresponding to the overall transverse plasma grating size and equal to 0.5 μs, that is 340 m/s × 0.5 μs = 170 μm in the transverse spatial domain, and the smaller one corresponding to the plasma grating period and equal to 0.125 μs or 340 m/s × 0.125 μs = 42.5 μm (red solid curve in Fig. 17.3a). Just a single-cycle waveform is obtained in the y-direction perpendicular to the grating orientation (black solid curve in Fig. 17.3a). The spectral analysis of the two waveforms propagating in the perpendicular directions shows the specific acoustic wave frequencies of 2 MHz and 8 MHz corresponding to the larger and smaller plasma grating spatial scales (Fig. 17.3b, red and black solid curves), respectively. The directional diagram of acoustic signal filtered at 8 MHz shows a factor of 1.4 amplitude decrease within 2 × 15° = 30° relative to the plasma grating parallel axis (the x-axis) (Fig. 17.3c). This means that half of the high-frequency signal energy propagates along the plasma grating and almost zero of this signal energy propagates perpendicular to the plasma grating including the wide sector 30–150°. The low-frequency component propagates uniformly into the cylinder extended along the z-axis, which is the initial femtosecond pulse propagation axis (Fig. 17.2b). Thus, our dynamic plasma grating produced by the two colliding filaments is anisotropic source of acoustic waves and simultaneously the unique source of the directed higher-frequency acoustic wave (Figs. 17.2b and 17.3c).

The transverse size of converging filaments in the vicinity of the geometrical focus (100–200 μm) is much larger than the plasma grating period (20–40 μm). Therefore, the interference pattern of the two beams is similar to that of the two

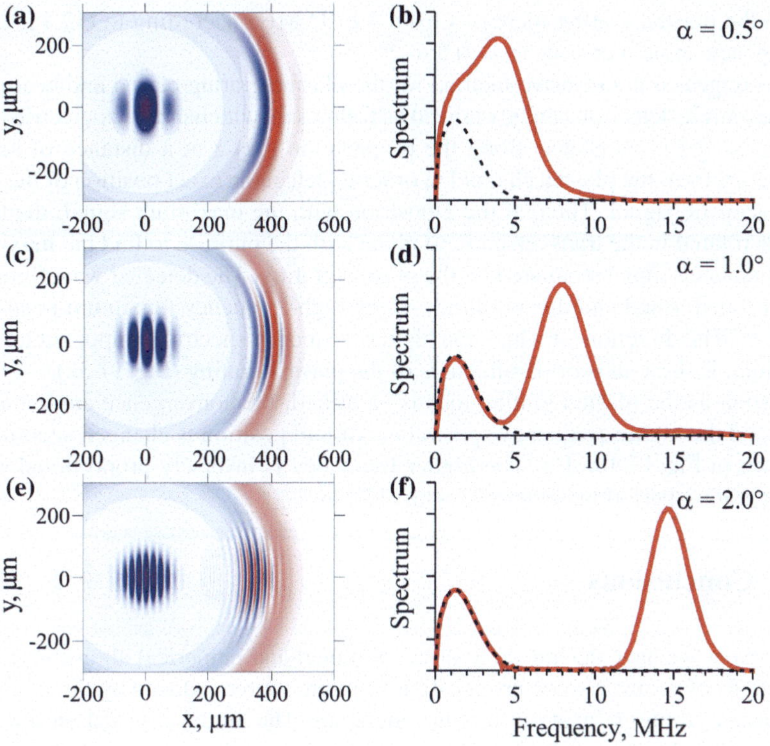

Fig. 17.4 (Left column) The transverse distribution of the air density perturbation $\rho'(x, y)/\rho_0$ at the time moment $t = 1.2\,\mu s$ after plasma has recombined and (right column) the corresponding spectra from the plasma grating formed by filaments converging at **a**, **b** 0.5°, **c**, **d** 1° and **e**, **f** 2°. Black line in (**b**, **d**, **f**) is an isotropic part of the signal, which is the same for all the converging angles α studied

plane waves converging at a small angle of several degrees or less than that. We can introduce the characteristic frequencies for acoustic waveforms:

$$f_{\rm pl} = \frac{c_0}{d_{\rm pl}} = \frac{c_0 \sin\alpha}{\lambda}; \qquad f_{\rm fil} = \frac{c_0}{d} = \frac{c_0}{2a_{\rm fil}} . \tag{17.11}$$

The exact angle of convergence can be reconstructed from the frequency $f_{\rm pl}$ obtained from the waveform spectrum (Fig. 17.4a, c, e). By increasing the convergence angle α we decrease the period of the plasma grating $d_{\rm pl}$ and increase the frequency $f_{\rm pl}$ given by (17.8), (17.11). Because the overall transverse size of converging filaments remains the same, the lower frequency $f_{\rm fil}$ of the isotropic cylindrical acoustic wave corresponding to approximately 170 μm results in the spectral maximum at 2 MHz (Fig. 17.4b, d, f). The higher frequency $f_{\rm pl}$ of the acoustic wave propagating parallel to the plasma grating increases from 4 to 15 MHz according to (17.11) as the convergence angle increases from 0.5 to 2°.

The suggested noninvasive method for the plasma grating period and beam convergence angle detection can be realized as follows. A wideband piezoelectric transducer [15, 16] is translated along the propagation axis z at a distance of several millimeters from the plasma channel in order to detect the exact position of the maximum acoustic signal. Then, at the z-position with the maximum signal, the transducer is rotated in the transverse (x, y) plane with the purpose to find the maximum high-frequency signal irradiated by the plasma grating. The detected waveforms are Fourier transformed and the amplitude of the high-frequency maximum position is identified. The direction, at which the higher-frequency spectral component takes its maximum, is the x-axis or parallel axis of the plasma grating (Fig. 17.3c).

As soon as the plasma grating axis is identified, the convergence angle may be varied and the higher-frequency spectral maximum position is changed accordingly as shown in Fig. 17.4b, d, f. The higher frequency is inversely proportional to the plasma grating period in agreement with (17.11).

17.4 Conclusions

In this paper we have designed the physical model and numerical algorithm for the simulations of acoustic wave propagation from the source induced by the femtosecond plasma grating relaxation in atmospheric air. The model is based on the wave equation with the wave velocity depending on the transverse temperature variation. The numerical grid has the size of $4 \times 4\,\mathrm{mm}^2$ and resolution of about 2 μm. The grid is aimed at the description of the fine plasma channel substructures in the course of femtosecond filamentation.

We have shown that the femtosecond plasma grating emits the superposition of two acoustic signals after plasma recombination. One acoustic signal is represented by an isotropic cylindrical waveform with the characteristic spatial scale equal to the filament diameter in the range 100–200 μm. The other acoustic signal is the wave with the characteristics spatial scale equal to the plasma grating period in the range 20–40 μm. This second wave propagates almost parallel to the plasma grating axis. The directional diagram corresponds to 50% energy of high-frequency wave localized within 30° angle (±15° from the plasma grating parallel axis).

The presence of the two characteristic spatial scales in the acoustic signal from the plasma grating is readily recognized in the acoustics signal spectrum. The spectrum reveals the maxima at two frequencies: the first one at approximately 2 MHz corresponds to ~170 μm of the overall filament diameter, the second one increases from 4 to 15 MHz as the convergence angle increases from 0.5 to 2°.

We suggested the noninvasive method for the plasma grating period and beam convergence angle detection. The method is based on the translation of a wideband (more than 6 MHz) piezoelectric transducer along the pulse propagation axis in order to detect the exact z-position of the maximum acoustic signal. Then, at the z-position with the maximum signal, the transducer is rotated in the (x, y) plane and the plasma grating parallel axis is revealed as the direction at which the higher frequency spectral component takes its maximum value. The value of this higher frequency itself is the inverse period of the plasma grating multiplied by the speed of sound in air.

Acknowledgements This work was partially supported by the Russian Foundation for Basic Research (Grant Nos. 18-52-16020, 18-02-00954, 18-32-01000) and the National key research and development program (2018YFB0504400). D.E.S. acknowledges the program "UMNIK" of Foundation of assistance to development of small forms of enterprises in scientific-technical sphere (FASIE) (11522GU/2017), Scholarship of "Basis" Foundation, Scholarship of RF President SP-2453.2018.2, and SPIE 2018 Optics and Photonics Education Scholarship.

References

1. S.L. Chin, S.A. Hosseini, W. Liu, Q. Luo, F. Theberge, N. Akozbek, A. Becker, V.P. Kandidov, O.G. Kosareva, H. Schroeder, Can. J. Phys. **83**, 863 (2005)
2. A. Couairon, A. Mysyrowicz, Phys. Rep. **441**, 47 (2007)
3. V.P. Kandidov, S.A. Shlenov, O.G. Kosareva, Quant. Electron. **39**, 205 (2009)
4. S.L. Chin, N. Akozbek, A. Proulx, S. Petit, C.M. Bowden, Opt. Comm. **188**, 181–186 (2001)
5. Y. Chen, F. Théberge, O. Kosareva, N. Panov, V.P. Kandidov, S.L. Chin, Opt. Lett. **32**, 3477–3479 (2007)
6. W. Liu, S.A. Hosseini, Q. Luo, B. Ferland, S.L. Chin, O.G. Kosareva, N.A. Panov, V.P. Kandidov, New J. Phys. **6**, 6 (2004)
7. D. Uryupina, M. Kurilova, A. Mazhorova, N. Panov, R. Volkov, S. Gorgutsa, O. Kosareva, A. Savel'ev, S.L. Chin, J. Opt. Soc. Am. B **27**, 667–674 (2010)
8. S.A. Hosseini, Q. Luo, B. Ferland, W. Liu, N. Akozbek, G. Roy, S.L. Chin, Appl. Phys. B **77**, 697–702 (2003)
9. Q. Luo, W. Liu, S.L. Chin, Appl. Phys. B **76**, 337–340 (2003)
10. J. Yu, D. Mondelain, J. Kasparian, E. Salmon, S. Geffroy, C. Favre, V. Boutou, J.-P. Wolf, Appl. Opt. **42**, 7117 (2003)
11. D.V. Kartashov, A.V. Kirsanov, A.M. Kiselev, A.N. Stepanov, N.N. Bochkarev, Y.N. Ponomarev, B.A. Tikhomirov, Opt. Express **14**, 7552 (2006)
12. J.K. Wahlstrand, N. Jhajj, E.W. Rosenthal, S. Zahedpour, H.M. Milchberg, Opt. Lett. **39**, 1290–1293 (2014)
13. K.P. Birch, J. Opt. Soc. Am. A **8**, 647–651 (1991)
14. N. Jhajj, E.W. Rosenthal, R. Birnbaum, J.K. Wahlstrand, H.M. Milchberg, Phys. Rev. X **4**, 011027 (2014)
15. D.S. Uryupina, A.S. Bychkov, D.V. Pushkarev, E.V. Mitina, A.B. Savel'ev, O.G. Kosareva, N.A. Panov, A.A. Karabutov, E.B. Cherepetskaya, Laser Phys. Lett. **13**, 095401 (2016)
16. A.S. Bychkov, E.B. Cherepetskaya, A.A. Karabutov, V.A. Makarov, Laser Phys. Lett. **13**, 085401 (2016)
17. V.G. Andreev, A.A. Karabutov, S.V. Solomatin, E.V. Savateeva, V. Aleinikov, Yu.V. Zhulina, R.D. Fleming, A.A. Oraevsky, Proc. SPIE **3916**, 36 (2000)
18. E.W. Rosenthal, N. Jhajj, I. Larkin, S. Zahedpour, J.K. Wahlstrand, H.M. Milchberg, Opt. Lett. **41**, 3908–3911 (2016)
19. G. Point, E. Thouin, A. Mysyrowicz, A. Houard, Opt. Express **24**, 6271–6282 (2016)
20. D.V. Pushkarev, E.V. Mitina, D.S. Uryupina, R.V. Volkov, N.A. Panov, A.A. Karabutov, O.G. Kosareva, A.B. Savel'ev, JETP Lett. **106**, 561–564 (2017)
21. S. Tzortzakis, B. Prade, M. Franco, A. Mysyrowicz, Opt. Commun. **181**, 123 (2000)
22. S. Xu, Y. Zheng, Y. Liu, W. Liu, Laser Phys. **20**, 1968–1972 (2010)
23. V.V. Pankratov, D.E. Shipilo, M.M. Yandulsky, N.A. Panov, O.G. Kosareva, Proc. SPIE **9990**, 99900N–1 (2016)

Chapter 18
Relativistic Laser Plasma Atto-Physics

A. A. Andreev, Zs. Lecz and S. K. Mishra

Abstract Interaction of an ultrashort and ultraintense laser pulse with overdense/underdense plasmas is considered. Efficient conversion of fundamental laser radiation into sub-femtosecond XUV/X-ray radiation and its significant amplification in laser plasmas are obtained. The results of the simulations were compared with the experimental data and have shown a good coexistence.

18.1 Introduction

The interaction of high intensity laser pulses with matter (including plasma) is usually accompanied by some dynamical processes of a time scales in the range of femto (fs)- or even attoseconds (as). The study of these processes (attophysics) is possible only if the appropriate laser system with high enough (relativistic) intensity is available to initiate them. Recently the researchers direct substantial efforts to the development of coherent power attosecond light pulses because wide ranging applications [1, 2]. Several methods of generation of such pulses have been explored involving high-order harmonic generation (HHG) through the interaction of high-intensity fs laser pulses with gases [3, 4], including low density partly ionized plasma [5] and highly ionized overdense plasmas [6]. Materials start breaking down at relatively low laser intensity and plasma is the only medium that can be used for modern high-power fs lasers operating at relativistic intensities. Such a pulse focused on a solid target, instantly ionizes the surface in its leading edge triggering plasma that subsequently interacts

A. A. Andreev (✉)
Saint Petersburg State University, 199034 Birzhevaya Line 12, St. Petersburg, Russia
e-mail: alexanderandreev72@yahoo.com

A. A. Andreev · Zs. Lecz · S. K. Mishra
ELI-ALPS, Szeged 6720, Hungary

A. A. Andreev
MBI, Max-Born Str. 2a, 12489 Berlin, Germany

S. K. Mishra
Physical Research Laboratory (PRL), Ahmedabad 380009, India

K. Yamanouchi et al. (eds.), *Progress in Photon Science*, Springer Series in Chemical Physics 119, https://doi.org/10.1007/978-3-030-05974-3_18

with the rest of the pulse energy. The specific nature of the interaction depends strongly upon the driving laser properties, the interaction geometry and plasma characteristics. The ordinary laser plasma theory is considering processes with duration comparable or bigger than few fs (laser wave period). Recent experimental technique gives us an opportunity to operate with even smaller (as) time scales this demanding to revise some assumptions of standard models. One important example is the investigation of fast dynamical processes in overdense plasma because in the field of dense plasma physics, in particular plasmas close to solid densities originating from the sudden heating of solid matter, the time scale of all plasma oscillations is set by the plasma frequency $\omega_p = 2\pi/T_p$. The period of these oscillations relative to the laser period is given by $T_p/T_L = (n_c/n_e)^{0.5}$ where the critical density $n_c \sim 10^{21}$ cm^{-3} for laser light of period $T_L \sim 3$ fs and n_e is electron density $\sim 10^{24}$ cm^{-3}. This implies oscillation periods of hundreds attoseconds, and thus time-resolved measurements will require pulses in (as) range. Under appropriate conditions such an interaction can lead to nonlinear specular reflection of light with generation of new frequencies through the surface high-harmonic generation process [7]. This approach is essential since it is important for interdisciplinary applications in the fields of laboratory astrophysics [8], high-energy density physics [9], fast particle generation [10] and has the potential to provide an intense (as) pulses with enough high conversion efficiencies [11, 12].

We start from plasma of high (solid) density because as it was mentioned above harmonic generated from such plasma is the most promising mechanism for future radiation sources based on reflecting plasmas, since the maximum generated frequency grows with laser intensity, and has been predicted to eventually reach multi-keV photon energies and pulse durations down to the zepto-second range for ultra-relativistic laser intensities [13, 14]. Interaction of intense laser pulses with solid surfaces has been intensively studied in the last two decades. Precise manufacturing and high power (~100 TW) laser systems allow physicists to investigate incoherent heating of electrons leading to ion acceleration in the sheath field generated at the target surfaces. If the laser pulse is compressed down to a few 10 s of femtoseconds the regime of coherent heating (acceleration) can be also explored, where the electron nano-bunching results in intense coherent radiation. The electrons move around the laser-plasma boundary at nearly the same trajectory and emit high frequency photons in form of atto-pulses [15]. The ultra-short laser pulses have the advantage in applications where high repetition rate is required. The lower is the energy contained in one pulse the higher the repetition rate can be.

At laser intensities higher than 10^{18} W/cm^2 the electrons acquire relativistic velocity during a quarter laser cycle and get pushed into the plasma by the ponderomotive force. This is illustrated by the first step shown in Fig. 18.1. When this force changes its sign the electrons get accelerated towards the incident pulse and they are pulled also by the charge separation field, because the ions stay still during this short time period (step 2). In this phase the electron momentum is the highest. In the next quarter period the electrons are compressed by the counter acting ponderomotive force and emit coherent synchrotron radiation (step 3) until they are slowed down by the ions. This mechanism of attopulse generation is very general and often referred as

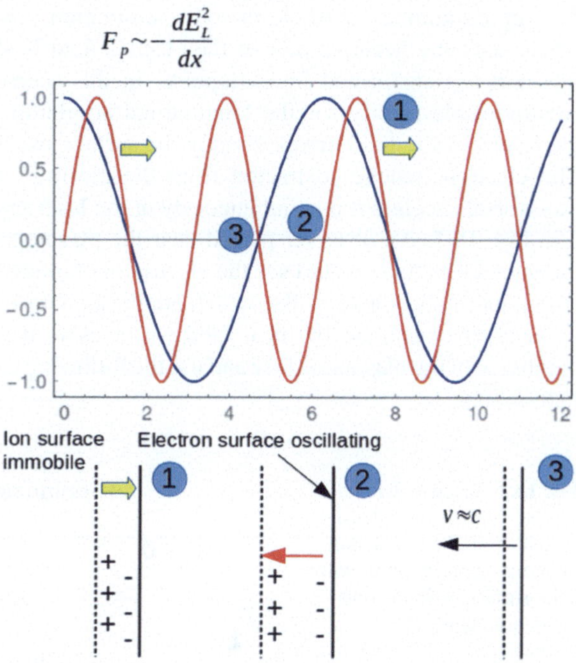

Fig. 18.1 The three major steps in the relativistic oscillating mirror model. In the upper picture the blue line represents the laser field, while the red line is the ponderomotive force

ROM (Relativistic Oscillating Mirror) model [15] at close to normal incidence and it is interpreted as coherent synchrotron emission (CSE) [16] at large incident angles, which is described in the next section.

18.2 Attopulse Generation at Low Repetition Rates

In the case of nowadays high power laser systems, in which a few Joule can be pumped in one pulse leading to peak intensities about several 10^{20} W/cm^2 and repetition rate is limited to maximum 10 Hz. At such high intensities intense attopulses can be generated with broad spectrum spanning up to the 100th harmonics of the fundamental laser wave. For the spectral intensity of reflected wave a universal model has been developed, the so called BGP model [15], where the $I(\omega) \sim \omega^{-8/3}$ scaling was obtained and a cut-off frequency $\omega_c \sim \gamma^3$, where γ is the relativistic Lorentz factor of electrons. It has been quickly realized that this power scaling can be different in the case of oblique incidence [17], where the Brunel electrons escape from the plasma and get accelerated continuously by the laser field along the plasma surface having a long trajectory, not oscillating like in the previous case. The electrons are strongly bunched near the plasma surface and have more synchrotron-like trajectories, therefore this mechanism acquired the name coherent synchrotron emission (CSE), which has been proven to be highly efficient for atto-pulse generation [18].

A representative particle-in-cell simulation is shown in Fig. 18.2, where only the negative part of the electric field is shown by color code because this part is transformed into attopulse. In this scenario the ponderomotive force is responsible mostly for the longitudinal acceleration (along x direction), while the laser electric field drives directly the surface oscillation described in Fig. 18.1. The electron bunches extracted from the plasma experience a longitudinal and transversal acceleration simultaneously at the laser phase corresponding to the step 3 in Fig. 18.1. An observer placed near the plasma surface at some distance from the interaction zone would see the electric field plotted in Fig. 18.3, which contains strong narrow negative peaks, which are the signature of atto-pulses.

In order to analyse the near field we measure the current density along a line parallel with the plasma surface and use the following expression to obtain the electric field in time:

Fig. 18.2 Electric field (purple) and electron density (green) from the interaction of an intense laser pulse with a flat surface at 75 degrees incidence angle. The intensity is $5 \times 10^{19}\ \mathrm{W/cm^2}$, the pulse duration is 30 fs and the plasma density is $50n_{cr}$

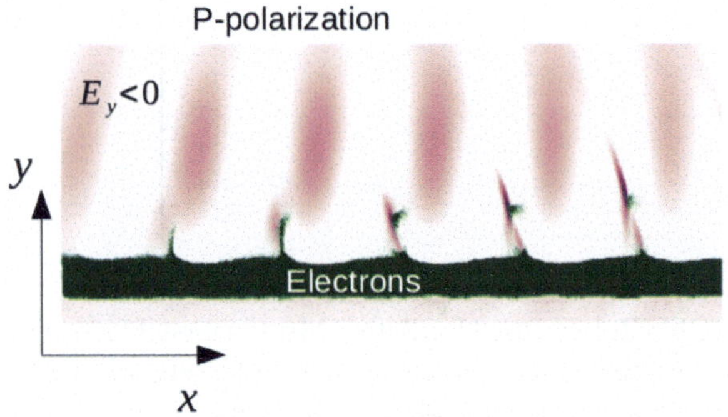

Fig. 18.3 Measured electric field at 1 μm away from the surface in Fig. 18.2

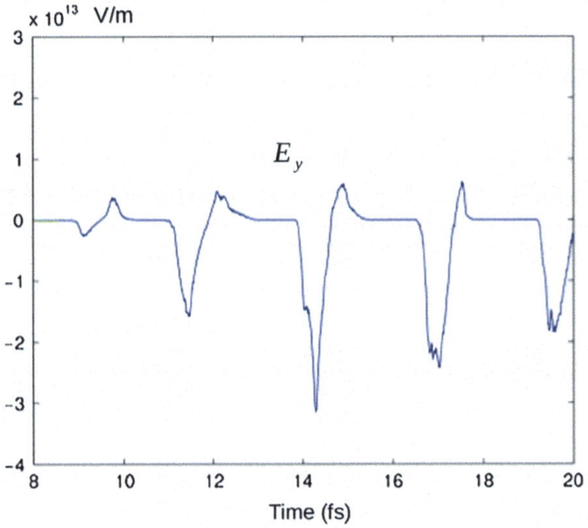

$$E_y(x_0, y_0, t) = \frac{1}{2\varepsilon_0} \int_{j_y} (y_0, x', t')\mathrm{d}t', \tag{18.1}$$

where the concept of retarded time has been used: $t' = t - (x' - x_0)/c$ and (x_0, y_0) are the coordinates of the observer. In Fig. 18.2 one can see that the emitted attopulses have some angular spread and they are emitted at a small angle, comparable to the reflection angle. The angle of emission at each instance of time can be calculated by the following formula:

$$\phi(x', t') = \arctan\left(j_x t' / \int j_y \mathrm{d}t'' \right) \tag{18.2}$$

Using this it is possible to obtain the angular distribution of emission at any time-space coordinate:

$$E_y(x, t, \theta) = \frac{1}{2\varepsilon_0} \int \delta\left(\theta - \phi(x', t')\right) j_y(x', t')\mathrm{d}t', \tag{18.3}$$

where θ is a chosen angular interval (cone angle) which defines the k vector of measurable radiation. By performing a Fourier transformation on $E_y(x, t, \theta)$ one obtains the angle resolved spectral intensity of the emitted radiation and from the phase information one can select the cone angle, where the coherency is the highest. Basically it is necessary to find the angular interval where the variation of $\mathrm{d}\varphi/\mathrm{d}\omega$ is the smallest, where φ is the phase obtained from Fourier transform of the field. By doing so it is possible to obtain the temporal shape of the coherently emitted radiation, which is plotted in Fig. 18.4. It can be seen that the duration of the atto-pulse emitted within the 0.15 rad angle is less than 100 as and its amplitude is 10 times higher than the amplitude of the electric field in the laser pulse.

The modeling of the attopulse emission can be done by directly calculating the electric field starting from the Lienard–Wiechert potential. We use the expression derived in [16]:

$$E_y(t) = C \frac{a_y(t')}{(1 - v_x(t')/c)^2}, \tag{18.4}$$

where C is the constant and $a_y(t) = \exp(-t^2)$ is the approximated transversal acceleration which is a fit to the simulation results. In the extended version of the theory [17] the transversal velocity was approximated by the function $v_y \sim t^n$, where $n = 1$ and 2 were considered, which also results in similar acceleration around $t \approx 0$. In our work we have extended this model and we have shown that higher values of n are also possible. The longitudinal velocity is often approximated by a polinomial function of t around the zero point, which behaves very similarly to the electron velocity around the point where it reaches the maximum value. The velocity $v_x \approx c(1 - \alpha_1 t'^{2\nu})$ is derived in [19] which can be used to obtain the expression of retarded time:

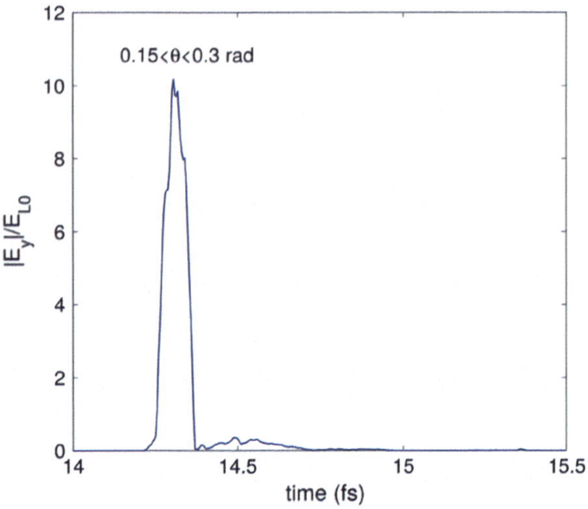

Fig. 18.4 Absolute value of the electric field emitted in a small cone angle, where the coherency is the highest. Obtained from the simulation shown in Figs. 18.2 and 18.3 and using (18.3)

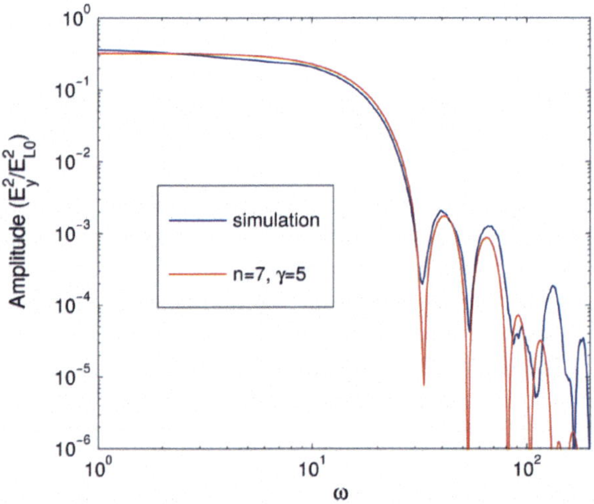

Fig. 18.5 Comparison between the spectrum of the field shown in Fig. 18.4 and the model described in the text [red curve, (18.4, 18.5)]

$$t' = \frac{t2\gamma^2}{1 + 2\alpha_1\gamma^2 t'^{2\nu}/(2n+1)}, \tag{18.5}$$

where $\alpha_1 \approx 6/a_0^3$ and $a_0 = eE_{L0}/cm_e\omega_L$ is the normalized laser field. Inserting (18.5) into (18.4) one obtains the electric field of a single unfiltered attopulse and its Fourier transform is shown in Fig. 18.5, which is in good agreement with the simulation for the given parameters. One can see that the spectrum has a plateau region where the intensity is almost constant and it is followed a sudden drop and exponential decay at higher frequencies. The harmonic order where the spectral intensity drops is approximately $N_{dr} \approx (3/2)a_0^2$, which is close to 32 in our case.

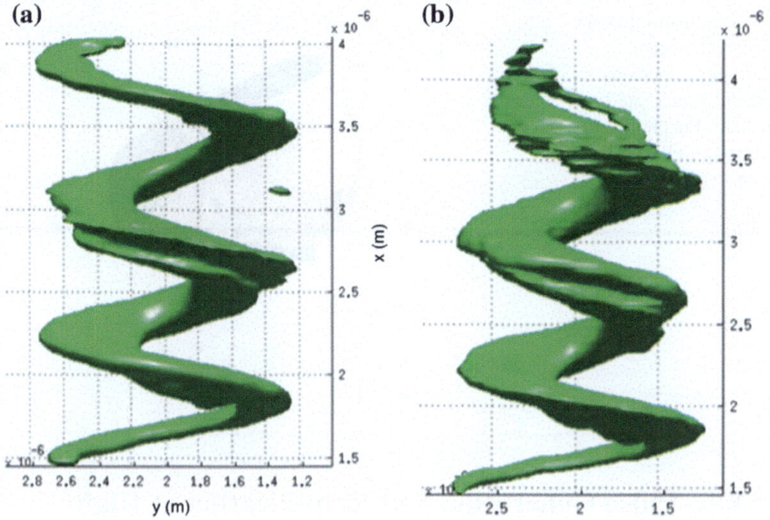

Fig. 18.6 The iso-surface of energy density of electromagnetic radiation produced in the case of cylinder (**a**) and cone (**b**) targets [6]

So far we have used only 2D geometry, but the same atto-pulse generation can be realized in 3D as well and it is more efficient if the interaction of circularly polarized laser pulse with cylindrical symmetric targets is considered. In this case the P-polarized interaction is ensured at each moment of time at one point of a cylinder (or cone) inner surface [20]. The iso-value surface of the generated radiation is shown in Fig. 18.6 for the case of cylinder (a) and cone (b) targets. The radius of the rotation symmetric targets is 0.8 μm and their axes coincide with the laser propagation axis. In the case of cone target the atto-spiral gets focused at the exit side (see Fig. 18.7), where the radius is 30% smaller, thus the intensity can be even higher than that of the incident pulse.

Thus we can conclude in this section that high intensity laser pulses are capable of generating intense atto-pulses via the nano-bunching of electrons near the plasma surface. The process of coherent synchrotron emission results in not so steep spectral intensity scaling and peak amplitudes of attopulses larger than that of the incident pulse. The conversion efficiency from laser to attopulse is about 2–5%, depending on the filtering. The radiated fields presented here are near-fields, the far-field would have different distribution and it should be investigated in further studies. However, the repetition rate of these attopulses is quite low and other means should be found to generate similar pulses with higher repetition rates.

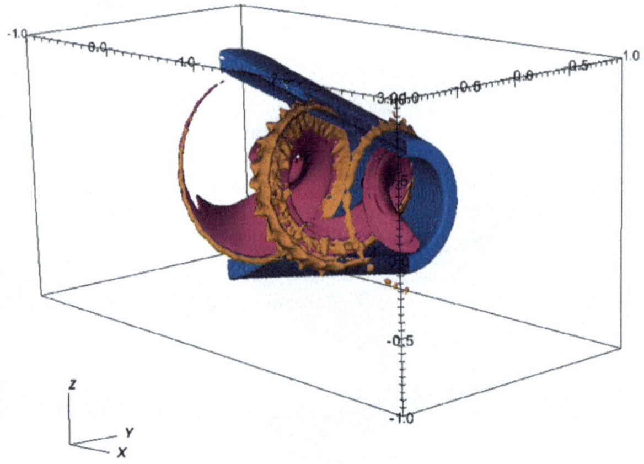

Fig. 18.7 Blue: initial shape of the target. Orange: ultra-relativistic electrons extracted and accelerated by the laser field. Purple: high energy density radiation, i.e. focused atto-spiral

18.3 Atto-Pulse Generation and Amplification at High Repetition Rates

In the previous section high harmonic generation mechanisms up to the 100th harmonic of the laser pulse can be produced with high efficiency resulting in intense attopulses. The drawback if the high intensity pulses is the low repetition rate and the limited contrast ratio, which makes difficult to control the plasma surface before the main pulse arrives. It is possible to produce shorter and less intense laser pulses with high repetition rate (up to 100 kHz), which contain energy on the order of 10 mJ. In this case very high contrast ratio can be also maintained and the remaining challenge is ensuring the new and fresh plasma surfaces after each pulse, which requires rotating or moving target holder. One option for decreasing constrain on the target movement speed is to exploit the possibility of multiple reflection of the short pulse between two plasma surfaces, which was first proposed in [20].

Recently we have investigated the problem of multiple reflections at oblique incidence with the help of 1D (boosted frame) and 2D (lab-frame) simulations [21]. The basic setup is illustrated in Fig. 18.8. The distance between the two foils has to be large enough to ensure non-overlapping laser spots, i.e. after each reflection the laser pulse and its harmonics interacts with fresh surfaces. On the other hand this distance has to be much smaller than the Rayleigh length of the loosely focused pulse in order to avoid significant divergence during consecutive reflections. The intensity of individual harmonics increases after each reflection because the reflected pulse inherently contains the low-order harmonics with the right phase thus the waveform is modified such that in the next interaction the electron bunch motion is more optimal for the described process. We have considered S and P polarization separately and we found that the S-polarized interaction leads to stronger harmonic amplification. After 6–7 reflections the intensity of 10th–30th harmonics can be enhanced by 3 orders of magnitudes. A detailed numerical model is developed in [21], which models each reflection and includes the spectral change caused by the previous reflection.

Fig. 18.8 Schematics of the multiple reflection setup. The two foils are parallel and are moved perpendicular to the interaction plane with 100 cm/s velocity

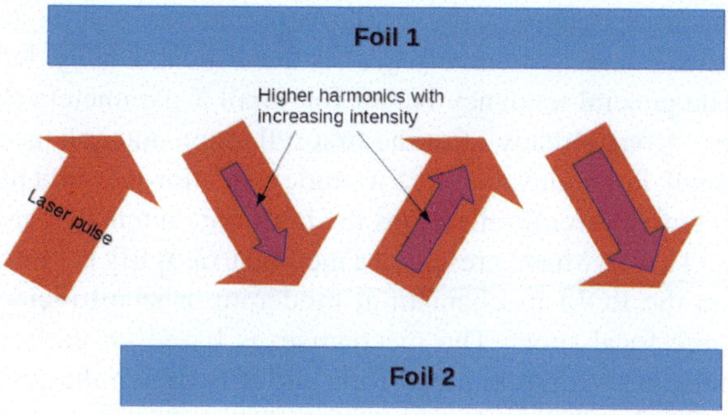

The spectral intensity distribution after the first reflection (obtained from the model) is shown in Fig. 18.9a while the enhancement of harmonics after 8 reflections is shown in Fig. 18.9b. In this idealized model, where laser energy absorption, i.e. electron heating, and dispersion effects are not included the maximum amplification factor can reach the value of 10^5 for the harmonic numbers between 10 and 20. It is interesting to note that this strong amplification happens at low intensity, at higher intensities the enhancement is lower, but it spans over a broader spectral range.

The agreement between the model and 1D simulations is good in [21], although the simulations show one order of magnitude lower amplification because of energy absorption effects. In 2D this difference is even higher, because of the non-uniform transversal intensity of the Gaussian beam, but the amplification factor is higher than 10^3, which is already a good achievement. The change of spectral intensity during consecutive reflections in 2D simulation is shown in Fig. 18.9 for two laser field amplitudes. Due to the large focal spot area the beam divergence does not play a role in the amplitude evolution. There is another important parameter which influences

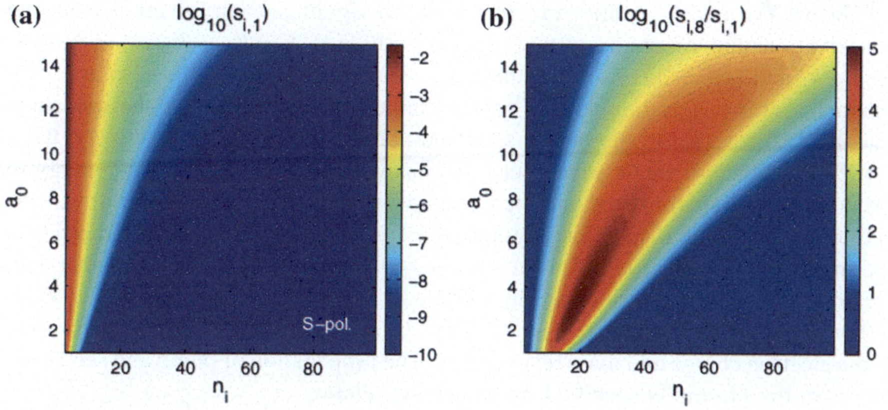

Fig. 18.9 **a** Intensity distribution of harmonics (n_i) for different laser amplitudes. **b** Amplification factor of each harmonics after the 8th reflection

the process, namely the similarity parameter: $S = n_0/(a_0 n_{cr})$, where n_0 is the plasma density and $n_{cr} = \omega_L^2 m_e \varepsilon_0/e^2$ is the critical density defined by the laser wavelength. The general tendency is that for small S parameters (high intensity) the harmonics are strong already after the first reflection and enhancement in further reflections is small. For high values of S a single reflection does not provide high harmonic content, but after several reflections the harmonic intensity can be enhanced significantly.

Thus, we have presented a method to amplify the intensity of atto-pulses generated via the ROM mechanism at moderate or small incident angles and in the case of large focal spots. The mechanism is based on the multiple reflection of the same pulse between two metal foils and it is shown that the amplification saturates after 4–5 reflections for low harmonics ($\omega < \sqrt{n_0/n_{cr}}\omega_L$), but the amplification of higher harmonics continues aver more reflections. This technique can be applied at laser systems with above kHz repetition rate where the laser intensity is slightly above the relativistic threshold. From laser to higher harmonics energy conversion efficiency can reach 1% level. The amplification of relatively low number harmonics by using inhomogeneous plasma layer was considered in [22] with help the developed theory and was confirmed in the experiment.

18.4 Atto-Pulse Amplification in Low Density Plasmas

Another method, which permits to amplify a weak atto-pulse, is connected with nonlinear wave interaction in under-dense plasma. In particularly, it can be Backward Raman Amplification (BRA) scheme, which considers the resonant energy transfer from large energy pump pulse to short Raman down-shifted and counter propagating seed pulse via Langmuir plasma wave [23, 24]. The amplifying medium in this schema is plasma, which can tolerate much larger energy density than any standard grating. We used this schema in the amplifying of a low energy ultra-short pulses.

The BRA resonant condition in order to achieve significant amplification, using the conservation of momentum and energy can be written as: $\boldsymbol{k_0} - \boldsymbol{k_1} = \boldsymbol{k_2}, \omega_0 = \omega_1 + \omega_2$, where $\boldsymbol{k_j}$ and $\omega_j (= 2\pi c/\lambda_j)$ are the wave vector and frequency associated with interacting waves, c is the speed of light, λ_j indicates the wavelength; the subscripts $j = 0,1,2$ correspond to the pump, seed and plasma parameters respectively. BRA is a consequence of three wave decay process where the pump loses its energy to the counter propagating seed and the plasma wave. The Langmuir (plasma) wave is characterized by the dispersion relation $\omega_2^2 = \omega_p^2 + v_{th}^2 k_2^2$, where v_{th} is the electron thermal velocity and ω_p is the plasma frequency corresponding to plasma slab with electron density n_e. Considering the plasma is not too hot (*i.e.* $k_2 v_{th} \ll \omega_p T_e \leq 0.01 m_e c^2 \sim 5$ keV), thus $\omega_2 \approx \omega_p = (4\pi n_e e^2/m_e)^{1/2}$, where e and m_e correspond to the electron charge and mass respectively. The propagation of pump and seed laser pulses in the plasma is specified by dispersion relation $\omega_{0,1}^2 = \omega_p^2 + c^2 k_{0,1}^2$; in this configuration the critical plasma density corresponds to downshifted seed pulse and can be written as $n_{cr} = (m_e \omega_1^2/4\pi e^2)$. The propagation of em waves under resonant BRA condition immediately gives $\lambda_1 \in (\lambda_0, 2\lambda_0)$ i.e., the seed wavelength should be smaller than twice of the pump wavelength and it is reasonable to keep them (seed/ pump) spectrally close as it reduces the required resonant plasma density. The resonant plasma density may be expressed as $n_e = (\pi mc^2/e^2)(\lambda_0^{-1} - \lambda_1^{-1})^2$. This expression indicates that the resonant plasma density for an efficient BRA operation resembles with the order of the solid density in XUV range. It is well understood from the earlier investigations that the BRA process holds efficiently in the under critical plasma regime. To scale and estimate the features of output signal and necessary physical properties of the BRA; we use the analysis [23–27] for the amplification of short pulses, takes account of simple analytical estimates based on slowly varying envelope approximation (*svea*).

In three waves interaction process the excited Langmuir (*L*-) plasma wave procures the fraction (ω_2/ω_0) of the energy from the laser pump. If any damping loss from the *L*-wave is ignored during the energy acquisition from the pump and its transfer to the seed pulse, the sustenance of *L*-wave is limited ideally by the wave breaking phenomenon, occurs when the electron gains quiver velocity larger than the phase velocity of the *L*-wave. Hence, the peak intensity of the pump corresponding to the *L*-wave breaking threshold can be given by $I_{br} = (n_e/n_{cr})^{3/2}(4\omega_1\omega_0/c^2k_2^2)I_M$, where $I_M \approx (n_{cr}m_ec^3/16)$. The maximum achievable duration of the leading spike of the seed thus can be given by

$$\Delta t_1 \sim (\Lambda_0/\gamma_0^2 t) > \Delta t_{1m} \approx (\omega_0/\omega_1)^2(4/\omega_0)[(2\Lambda_0 I_M/3\delta_o I_0)(\omega_1/\omega_0)]^{1/3}.$$

This refers to the largest achievable intensity as

$$I_1 < I_{1m} \approx (w_{1m}/\Delta t_{1m}) \sim (\omega_1/\omega_0)^2(16 I_M\omega_0/\omega_2)[(3\delta_o I_0/2\Lambda_0 I_M)(\omega_0/\omega_1)]^{2/3}.$$

Using the above expressions which represents the optimal seed pulse parameters after its amplification, a parametric space between (λ_0/λ_1) and $(\lambda_{0,\mathrm{nm}}/I_{0,18})$, defining the maximum compression and intensity in the resonant BRA operation has been identified in [27] (see Fig. 18.10; here λ_{nm}, I_{18} and t_{fs} refer wavelength in nm, intensity in 10^{18} W cm^{-2} and time in fs units.

As an illustrative case a contour plot demonstrating this region for $I_{1o}\Delta t_{1o} \sim$ 0.08 J cm^{-2} having leading spike <1 fs, has been displayed in Fig. 18.10a; this seed parameter is consistent with $I_{1o} \sim 10^{14}$ W cm^{-2} and $\Delta t_{1o} \sim 0.08$ fs which has further been used for the numerical calculations. It may readily be seen that the pump with shorter wavelength and large intensity yield larger compression. Similar region has been specified for maximum intensity of the leading spike in Fig. 18.10b; the colored region in figure represents optimum intensity of the leading spike. This concludes that the maximum intensified seed pulse via resonant BRA could be achieved for a moderate plasma density and large intensity pump operating with the shorter wavelengths. These expressions and results (i.e. contour plots) are general in nature and are applicable to arbitrary system parameters for a given seed fluence. The largest seed intensity is achieved in the plasma of lowest possible density (wave-breaking density) i.e. $\omega_2 \sim \omega_1(c^2k_2^2 I_0/4\omega_1\omega_0 I_M)^{1/3}$. Maximizing the above equations over

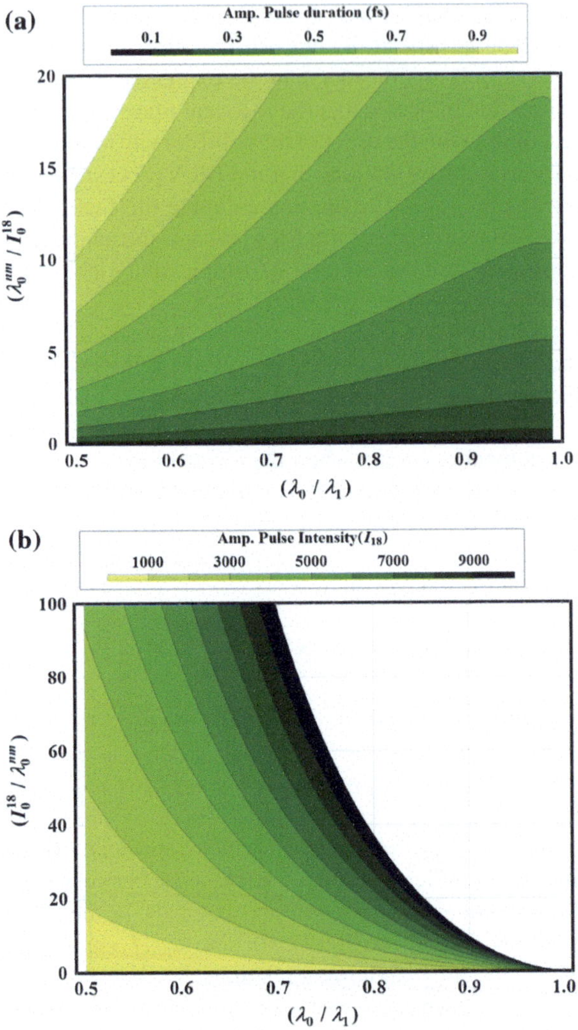

Fig. 18.10 **a** Contour plot describing temporal width of the leading seed spike ($\Delta t_{1,\mathrm{fs}}$) in terms of ($\lambda_{0,\mathrm{nm}}/I_{0,18}$) and ($\lambda_0/\lambda_1$). **b** Contour plot describing optimum intensity of the leading seed spike ($I_{1,18}$) in terms of ($I_{0,18}/\lambda_{0,\mathrm{nm}}$) and ($\lambda_0/\lambda_1$)

the plasma density at wave breaking density, the maximum achievable seed intensity can be obtained as the following: $I_{\max} \approx 16(I_M^2 I_0)^{1/3}(3\delta_o\omega_1/ck_2\Lambda_0)^{2/3}$.

From the above discussion one can conclude that a pump source could be a free electron laser (FEL) [28], which can be produced by using PW lasers, for electron bunch acceleration ($I_0 \sim 10^{21}$ W/cm^2, $\lambda \approx 800$ nm) up to energy ~10 meV. Such electron bunch can be used as electron source for preliminary injection to FEL. For example, FEL output pulse of Fermi light source [28] is specified by spectral range ~60 nm, peak power ~10 GW, pulse length ~50 fs, spot size ~200 μm, and intensity ~10^{13} W/cm^2. A system like [29] can be used to optimize the focal spot and pulse intensity, thus one can say that the anticipated FEL pump pulse can further

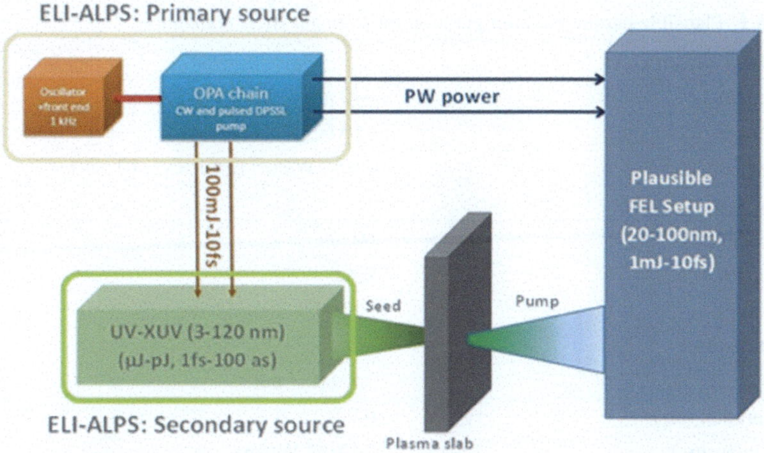

Fig. 18.11 A schematic of the feasible BRA setup, relevant to ALPS infrastructure

be compressed spatially to a ~μm size spot and thus the intensities of the order of ~10^{18} W/cm^2 can be achieved. The schema of BRA setup, consistent with ELI-ALPS parameters is shown in Fig. 18.11.

The thickness of the amplified overdense plasma layer should be of the order of half of the pump pulse duration (~10's μm). Such plasma layers can be generated via illuminating of target surface through the long duration intense laser pulses; for example ~100 μm width plasma of density (~10^{22} cm^{-3}, 100 eV) has experimentally been obtained (see for example [30]). We will use quasi-FEL source and the pump may be specified with intensity $I_0 \approx 10^{17}$ W/cm^2 ($\lambda_0 \approx 40$ nm). The parameters for the ultra-short seed pulse ($\lambda_1 \approx 60$ nm) viz. the pulse length and energy has been taken from above. We take $\tau_1 \leq 1$ fs seed pulse into account to commence further parametric configuration. The plasma density, consistent with the resonant conditions is found to acquire a value ($n_e \leq n_{cr}/4$). The pump intensity is taken below the L-wave breaking threshold intensity. For this case, the seed pulse may acquire the linear growth rate. For the initial seed pulse features consistent with parameter access ELI-ALPS facility, one gets the number of seed pulse exponent: $\Lambda_0 \approx 8$. For the pump laser and plasma parameters one gets $T_e \sim 1.5$ keV and corresponds to plasma wave noise exponentiations Λ_p is slightly higher than Λ_0 and consistent with the obligation to achieve prominent amplification. The maximum intensity achieved by the seed pulse after amplification is $I_{1m} \approx 10^{19}$ W/cm^2. Subsequently, the shortest achievable duration of the output seed pulse is $\delta t_s \leq 0.9$ fs (the pulse may be compressed to 400 as). The seed pulse evaluation leads to maximal possible amplification time $t_M \approx 160$ fs, this certainly limits the maximum duration of pump pulse ($\tau_0 \sim 2t_M$) and hence the plasma width ($l \leq ct_M \approx 48$ μm). The criteria for the length of the plasma can be expressed as $l < c/\nu_{ib} \sim 700$ μm. The sterner one between the two limits should be preferred for plasma scaling. In this case $T_e \ll T_M$ and Landau damping of plasma waves during amplification can be ignored. The preliminary pump/seed laser

Table 18.1 Plausible parametric configuration for resonant BRA realization

Pump laser parameters	
Wave length λ_0	~40 nm
Pulse length τ_0	~10 fs
Peak intensity (I_0)	~10^{17} W/cm^2 (*spot size* ~10 μm)
Seed laser parameters	
wave length λ_1	60 nm
Pulse length τ_1	0.8 fs
Pulse energy ε_1	~10 nJ
Output results	
Plasma density n_e	7.8×10^{22} cm^{-3}
Pulse duration (δt_s)	0.9 fs
Output intensity I_{1m}	7×10^{18} W/cm^2
Amplification factor (I_{1m}/I_1)	7×10^4

pulse parameters exploring the possibility of seed pulse amplification/ compression, evaluated in this section has been summarized in [31] and listed in Table 18.1.

The seed pulse amplification has been verified via the simulations with help of 1D relativistic electromagnetic particle in cell code. In order to simulate BRA, in the scheme we take into account of the interaction between the counter propagating harmonics of a reference fundamental frequency (ω_f). To mimic laser and seed wavelengths we use $\omega_f = 1.57 \cdot 10^{16}\ \mathrm{s}^{-1}$ and plasma parameters are established via the frequency matching between plasma, pump and seed frequencies, thus the pump and seed frequencies can be referred as $\omega_0 = 3\omega_f$ and $\omega_1 = 2\omega_f$. In the case $\omega_2 = \omega_0 - \omega_1$ this should be equal to the fundamental frequency and the resonant plasma density $n_c = 7.82 \times 10^{22}\ \mathrm{cm}^{-3}$. The temporal profile of the pulse as $\sin^2$ is used. The duration of pump and seed pulses are chosen as ~10 and 1 fs respectively. For the simulation $a_0 \sim 0.3(I_0 \sim 10^{17}\ \mathrm{W/cm^2})$ and $a_1 \sim 0.01(I_1 \sim 10^{14}\ \mathrm{W/cm^2})$ corresponding to pump and seed pulses, are used as the normalized laser fields. In the simulations one can see that the peak electric field of the seed pulse is amplified by a large factor (say two orders of magnitude) due to resonant BRA and thus the intensity of the input seed is enhanced by four orders of magnitude. For the same set of data used for these simulations, we have made analytical amplification estimates [27]. The corresponding intensities of the amplified seed pulse for both the simulations and analytical model, are shown in Fig. 18.12.

From the above it is seen that the simulation results are in order of magnitude agreed with the calculations based on the analytical model. It is also seen that the analytical and simulation results are converging for the high intensity pump. For smaller amplitude seed pulses the decrease in the amplification can be attributed to the smaller energy transfer from the pump during amplification and some stretching of the seed pulse duration, which results in decrease of seed peak intensity. For the optimal choice of the system parameters, seed pulse can be intensified by ~four

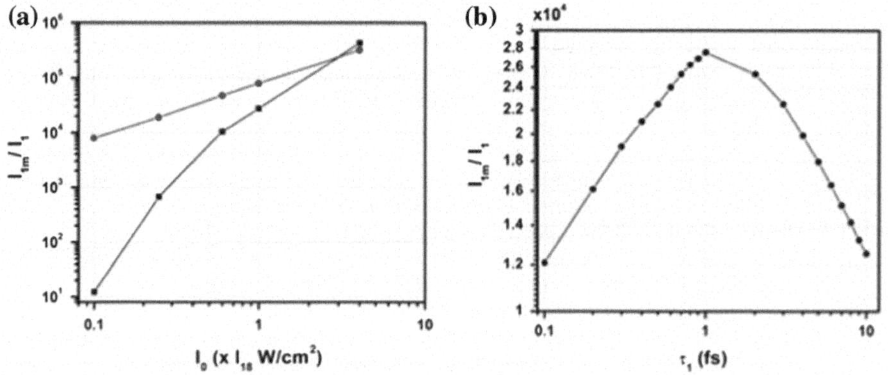

Fig. 18.12 **a** The amplification factor (I_{1m}/I_1) of the amplified seed pulse as a function of pump intensity (I_0); the results refer to pump pulse $\lambda_0 = 40\,\text{nm}$, $\tau_0 = 10\,\text{fs}$, seed with $I_1 \approx 0.1 I_{15}$, $\lambda_1 = 60\,\text{nm}$, $\tau_1 = 1\,\text{fs}$ and the plasma density corresponds to the resonant condition $\lambda_2 = 120\,\text{nm}$ ($\sim n_e \sim n_{cr}/4$). The black and red color marks refer to PIC and analytical results; **b** the amplification factor (I_{1m}/I_1) of the amplified seed pulse as a function of the seed duration (τ_1); the results refer to $\lambda_0 = 40\,\text{nm}$, $I_0 \approx I_{18}$, $I_1 \approx 0.1 I_{15}$, $\lambda_1 = 60\,\text{nm}$, $\tau_0 = 10\,\text{fs}$ and $n_e \sim n_{cr}/4$ (resonant case); the curves refer to PIC results

orders of magnitude, but the pulse compression is not so simple and in this case one need to operate pump at order of magnitude higher intensity. Nonetheless, based on the present analysis, it may be concluded that the resonant BRA operating in XUV regime may efficiently be utilized to amplify and compress the weak ultra-short pulses to Exawatt cm^{-2} and sub fs time scale.

References

1. F. Krausz, M. Ivanov, Attosecond physics. Rev. Mod. Phys. **81**, 163 (2009)
2. U. Teubner, P. Gibbon, High-order harmonics from laser irradiated plasma surfaces. Rev. Mod. Phys. **81**, 445 (2009)
3. S. Chatziathanasiou, S. Kahaly, E. Skantzakis, G. Sansone, R. Lopez-Martens, S. Haessler, K. Varju, G. Tsakiris, D. Charalambidis, P. Tzallas, Generation of attosecond light pulses from gas and solid state media. Photonics **4**, 26 (2017)
4. L. Plaja, R. Torres, A. Zaïr (eds.), *Attosecond Physics*, vol. 177 of Springer Series in Optical Sciences (Springer, Berlin, 2013)
5. R.A. Ganeev, *High-Order Harmonic Generation in Laser Plasma Plumes* (World Scientific, 2013)
6. G. Vampa et al., Linking high harmonics from gases and solids. Nature **522**, 462 (2015)
7. S. Mondal et al., Surface plasma attosecond beamlines. JOSA B **35**, A93 (2018)
8. B.A. Remington, High energy density laboratory astrophysics. Plasma Phys. Controlled Fus. **47**, A191 (2005)
9. R.P. Drake, *High-Energy-Density Physics: Fundamentals, Inertial Fusion, and Experimental Astrophysics* (Springer, 2006)
10. H. Vincenti, S. Monchocé, S. Kahaly, G. Bonnaud, P. Martin, F. Quéré, Optical properties of relativistic plasma mirrors. Nat. Commun. **5**, 3403 (2014)

11. G.D. Tsakiris, K. Eidmann, J. Meyer-ter Vehn, F. Krausz, Route to intense single attosecond pulses. New J. Phys. **8**, 19 (2006)
12. P. Heissler, A. Barna, J.M. Mikhailova, G. Ma, K. Khrennikov, S. Karsch, L. Veisz, I.B. Földes, G.D. Tsakiris, Multi-μJ harmonic emission energy from laser-driven plasma. Appl. Phys. B **118**, 195 (2015)
13. A. Andreev, A.L. Galkin, M.P. Kalashnikov, V.V. Korobkin, M.Y. Romanovski, O.B. Shiryaev, Electrons in relativistically intense laser field: generations of zeptosecond electromagnetic pulses and electron energy spectrum. Quant. Electron. **41**, 729 (2011)
14. G.A. Mourou, T. Tajima, More intense Shorter Pulse. Sci. **331**(7), 41 (2011)
15. T. Baeva, S. Gordienko, A. Pukhov, Theory of high-order harmonic generation in relativistic laser interaction with overdense plasma. Phys. Rev. E **74**, 046404 (2006)
16. D. An der Brugge, A. Pukhov, Phys. Plasmas **17**, 033110 (2010)
17. A. Andreev, K. Platonov, Generation of electron nano-bunches and short wavelength radiation upon reflection of a relativistic intensity laser pulse from a finite size target. Opt. Spectrosc. **114**, 788 (2013)
18. J.M. Mikhailova, M.V. Fedorov, N. Karpowicz, P. Gibbon, V.T. Platonenko, A.M. Zheltikov, F. Krausz, Phys. Rev. Lett. **109**, 245005 (2012)
19. Z. Lecz, A. Andreev, Attospiral generation upon interaction of circularly polarized intense laser pulses with conelike targets. Phys. Rev. E **93**, 013207 (2016)
20. P. Zhang, A.G.R. Thomas, Enhancement of high-order harmonic generation in intense laser interactions with solid density plasma by multiple reflections and harmonic amplification. Appl. Phys. Lett. **106**, 131102 (2015)
21. Z. Lecz, A. Andreev, Enhancement of high harmonic generation by multiple reflection of ultrashort pulses. JOSA B **35**, A51 (2018)
22. J. Braenzel, K. Platonov, L. Ehrentraut, A.A. Andreev, M. Schnürer, Amplification of coherent synchrotron-like high harmonic emission from ultra-thin foils in relativistic light fields. PoP **24**, 080704 (2017)
23. V.M. Malkin, G. Shvets, N.J. Fisch, Phys. Rev. Lett. **82**, 4448 (1999)
24. V.M. Malkin, N.J. Fisch, J.S. Wurtele, Phys. Rev. E **75**, 026404 (2007)
25. R.M.G.M. Trines, F. Fiuza, R. Bingham, R.A. Fonseca, L.O. Silva, R.A. Cairns, P.A. Norreys, Nat. Phys. **7**, 87 (2011)
26. V.M. Malkin, Z. Toroker, N.J. Fisch, Phys. Plasmas **21**, 093112 (2014)
27. S.K. Mishra, A. Andreev, Amplification of ultra-short laser pulses via resonant backward Raman amplification in plasma. Phys. Plasmas **23**, 083108 (2016)
28. http://flash.desy.de/accelerator/
29. G. Vieux et al., New J. Phys. **13**, 063042 (2011)
30. J.D. Sadler, R. Nathvani, P. Oleśkiewicz, L.A. Ceurvorst, N. Ratan, M.F. Kasim, R.M.G.M. Trines, R. Bingham, P.A. Norreys, Compression of X-ray free electron laser pulses to attosecond duration. Sci. Rep. **5**, 16755 (2015)
31. S.K. Mishra, A. Andreev, Scaling for ultrashort pulse amplification in plasma via backward Raman amplification scheme operating in the short wavelength regime. J. Opt. Soc. Am. B **35**, A51 (2018)

Chapter 19
The Efficiency of Coherent Radiation from Relativistic Shocks

Takanobu Amano, Masanori Iwamoto, Yosuke Matsumoto and Masahiro Hoshino

Abstract We discuss a mechanism for intense electromagnetic wave emission at an astrophysical relativistic shock in a magnetized collisionless plasma. At the magnetized shock, the particle reflection by a compressed magnetic field of the shock produces a ring-like distribution in momentum, which gives rise to plasma instabilities. Intense and coherent high-frequency electromagnetic waves will be emitted if the synchrotron maser instability (SMI) is excited, whereas non-propagating magnetic fluctuations will be generated when the Weibel instability (WI) is the dominant mode. The problem is of great astrophysical interest because if intense radiation is emitted, the interaction with the upstream medium induces a large-amplitude electrostatic field (or Wakefield), which may play a role for the acceleration of ultra-high-energy cosmic rays. We review our recent effort to measure the efficiency of the electromagnetic wave emission using fully self-consistent, two-dimensional (2D) particle-in-cell (PIC) simulations for pair plasmas. We found that the emission efficiency in 2D was systematically lower than one dimensional (1D) PIC simulation results. However, the power remains finite even when the WI is active to generate large-amplitude magnetic fluctuations. Astrophysical implications of the present results are briefly discussed.

T. Amano (✉) · M. Iwamoto · M. Hoshino
Department of Earth and Planetary Science, University of Tokyo, 7-3-1, Bunkyo-ku, Hongo, Tokyo 113-0033, Japan
e-mail: amano@eps.s.u-tokyo.ac.jp

M. Iwamoto
e-mail: iwamoto@eps.s.u-tokyo.ac.jp

M. Hoshino
e-mail: hoshino@eps.s.u-tokyo.ac.jp

Y. Matsumoto
Department of Physics, Chiba University, 1-33, Yayoi, Inage-ku, Chiba 263-8522, Japan
e-mail: ymatumot@chiba-u.jp

K. Yamanouchi et al. (eds.), *Progress in Photon Science*, Springer Series in Chemical Physics 119, https://doi.org/10.1007/978-3-030-05974-3_19

19.1 Introduction

Powerful low-frequency electromagnetic radiations are often observed associated with collective space and astrophysical plasma dynamics [1]. An extremely efficient electromagnetic wave emission may result from groups of charged particles moving in phase with each other, just like a single particle that carries a much larger number of charges. In such coherent radiation, the emitted power is proportional to N^2 (where N is the number of charged particles moving in phase) rather than N in incoherent radiation. Therefore, the coherent radiation is much stronger than the incoherent radiation, provided that such coherent charged particle bunches are generated and maintained for a sufficiently long time in natural collisionless plasma environments in space. Examples of such coherent radiation sources inferred from observations include planetary magnetospheres (emission from auroral regions), solar corona (solar radio bursts), pulsars (pulsed emission in radio).

Although there has been no observational identification, a relativistic shock propagating in a magnetized collisionless plasma may also be a source of strong coherent radiation. A magnetized collisionless shock involves the reflection of particles coming into the shock by a sharp increase in the magnetic field at the shock front. The reflected particles gyrate around the upstream magnetic field and are on average accelerated by traveling along the direction of the upstream motional electric field. Within the shock transition layer where both the reflected and incoming populations coexist, the momentum distribution of the particles should appear as a ring-like distribution in the plane perpendicular to the local magnetic field. Such a ring-like distribution may become unstable against several different types of plasma instabilities. In the case of a highly relativistic shock, the synchrotron maser instability (SMI) [2] may be excited from a relativistic ring distribution of electrons (or positrons). The instability results from the resonance between an electromagnetic wave of extraordinary mode (X-mode) and the n-th harmonics of the cyclotron motion of the relativistic particles. One may write the resonance condition as $\omega = n\omega_{ce}$ where ω is the electromagnetic wave frequency and ω_{ce} is the relativistic cyclotron frequency of an electron. Since the frequency of an X-mode electromagnetic wave is always higher than the cyclotron frequency, higher harmonics resonance $n > 1$ is needed in general for the instability. The emission of X-mode electromagnetic waves via the SMI at magnetized relativistic shocks was demonstrated by earlier studies using fully self-consistent Particle-In-Cell (PIC) simulations in one dimension (1D) [3–5]. Being the first principles approaches for numerical modeling of collisionless plasmas, the PIC simulations require extensive computational resources and two-dimensional (2D) and three-dimensional (3D) simulation were not possible in the old days. Meanwhile, the same unstable distribution function may excite the Weibel instability (WI) [6] if the ambient magnetic field is sufficiently small. Since the mode is unstable for the wave vector perpendicular to the shock normal, multidimensional simulations are essential to reproduce this instability.

With ever-increasing computational capabilities, recent PIC simulations of collisionless shocks have been performed routinely in 2D and sometimes even in 3D

[7–9]. Some earlier multidimensional PIC simulation studies of magnetized relativistic shocks reported that the coherent precursor waves were not observed, and suggested that the coherent radiation found in 1D may be an artifact of the reduced dimensionality. However, one has to be careful in interpreting the results because the numerical resolutions used in the earlier studies might not be sufficient. We expect that the high-frequency and short-wavelength electromagnetic radiation is sensitive to the choice of the mesh size and may quickly be damped at low resolutions. Furthermore, a PIC simulation involving a relativistic bulk flow becomes unstable against a numerical Cherenkov instability. A typical strategy is to use a digital filtering technique to eliminate unphysical short-wavelength fluctuations, which would also suppress physical electromagnetic waves even if they present. These numerical issues have been an obstacle that makes it difficult to estimate the emission efficiency accurately.

Concerning astrophysical applications, such intense radiation from a relativistic shock drew attention as it may play a key role for the acceleration of cosmic rays (CRs) possibly to ultra-high energies [10–12]. The origin of CRs has been a long-standing puzzle in astrophysics, in particular at its highest energy part or Ultra-High-Energy Cosmic Rays (UHECRs). The first order Fermi acceleration, which remains as the leading mechanism for the CR acceleration, has a difficulty at relativistic magnetized shocks because of the inefficiency of particle diffusion across the magnetic field line. Therefore, an alternative model for the CR acceleration at a relativistic magnetized shock is of great interest to the astrophysics community.

Consider an intense and coherent electromagnetic wave packet propagating in an electron-proton plasma. It pushes electrons via the ponderomotive force (or a wave pressure) in the direction of the wave propagation. The protons, on the other hand, are left behind because of the much larger inertia. A charge separation will then develop, which produces a large-amplitude electrostatic field called a Wakefield. It propagates with the group velocity of the wave packet, which is very close to the speed of light. A particle trapped in an electrostatic potential may then be accelerated linearly by the propagating wave packet [13]. This mechanism, called the Wakefield acceleration, has been known as an efficient particle acceleration mechanism in the context of the laser-plasma interaction but may provide a means to accelerate CRs if indeed a relativistic shock emits intense electromagnetic radiation.

The relatively new scheme for the CR acceleration has a number of advantages over the standard first-order Fermi acceleration. Nevertheless, the model is not yet sophisticated enough especially because we do not know the intensity of radiation emitted from a relativistic shock.

In this paper, we present our recent 2D simulation results that accurately quantified the coherent wave emission efficiency at relativistic magnetized shocks in electron-positron (pair) plasmas. In particular, we investigated the dependence on the upstream magnetization parameter defined as the ratio between the Poynting flux to the particle kinetic energy flux. The results suggest that the emission efficiency in general is lower than in 1D, but nevertheless always remains finite for a parameter range we have investigated. Astrophysical implications of the obtained results are briefly discussed. More detailed discussion may be found in our recent publications [14, 15].

19.2 Simulations

19.2.1 Method and Setup

The PIC scheme is the standard numerical method for studying collisionless plasma dynamics in a fully self-consistent manner. It deals with the motions of super-particles in continuous phase space. The electromagnetic field is defined on a mesh and typically advanced in time using a finite difference scheme. The charge and current densities at the mesh points are calculated by taking velocity moments of the particle distribution function and used for updating the electromagnetic field by solving Maxwell's equations. The Lorentz forces acting onto the particles may be obtained by interpolating the electromagnetic fields defined at neighboring cells.

We used a 2D PIC simulation code [16, 17] that employs a magic CFL (Courant-Friedrichs-Lewy) number to minimize the effect of a numerical Cherenkov instability. The performance of the code concerning the suppression of the numerical instability was investigated in detail by [17]. In this study, we only present 2D simulation results with a fixed numerical resolution, but the resolution dependence was checked by performing a separate numerical convergence study in a 1D simulation box [14].

We used an injection method to generate a shock wave in the simulation box. Namely, particles with both positive and negative charges are injected from one side of the box with a relativistic bulk velocity. The injected particles are reflected at the other side of the box. The reflected and incoming particles interact with each other, which results in the particle heating and deceleration of the incoming bulk flow. Because of the symmetry, the plasma bulk flow becomes zero as a result of this heating. When an injection speed is supersonic, a shock wave forms in between the freshly injected (the upstream) plasma and the heated plasma (the downstream). The shock then propagates toward the incoming plasma to satisfy the Rankine-Hugoniot relationships. The simulation frame thus corresponds to the downstream rest frame.

We take the x direction to be parallel to the shock normal and place the injection and reflection boundaries at the upper and lower bounds in x, respectively. We performed the simulations in the x–y plane and assumed that everything is initially homogeneous in the y direction. Accordingly, the boundary condition is periodic in the y direction.

In this study, we only consider a magnetized perpendicular shock, which is defined as a shock with the upstream ambient magnetic field perpendicular to the shock normal (or parallel to the shock surface). We may thus arbitrarily choose the ambient magnetic field in the y–z plane. Because of the 2D simulation box, the shock dynamics, in general, is dependent on the choice of the upstream magnetic field direction even if all the other parameters are fixed. More specifically, if the magnetic field is in the z direction or the out-of-plane configuration, the particle gyromotion is contained in the simulation plane, while the motion parallel to the magnetic field line is not appropriately taken into account. On the other hand, if it is taken to be in the y direction or the in-plane configuration, the situation becomes opposite. It is difficult to predict the differences (if any) in between the two cases without actually performing the simulations. Therefore, we performed simulations both with the in-plane and out-of-plane configurations. In this way, we investigated possible artifacts of the 2D assumption and tried to infer the fully 3D situation as much as possible.

Our primary motivation in this study is to obtain a quantitative estimate of the efficiency of electromagnetic radiation and also understand the physic behind. Specifically, we investigated the emission efficiency at a relativistic shock in an electron-positron plasma to reduce the computational cost. We note that this choice makes it impossible to investigate the efficiency of the Wakefield acceleration itself as it requires a finite inertia difference between positive and negative charges. Nevertheless, high-frequency electromagnetic radiation is produced only by leptons. We thus think that the emission efficiency measured in this study will be a reasonable estimate even for an electron-proton plasma. This conjecture should be checked by performing simulations for an electron-proton plasma in the future.

We used a fixed injection Lorentz factor of $\gamma_1 = 40$ in all the simulations presented in this paper. The emission efficiency was then measured as a function of the magnetization parameter:

$$\sigma_e = \frac{B_1^2}{4\pi \gamma_1 N_1 m_e c^2} = \left(\frac{\omega_{ce}}{\omega_{pe}}\right)^2$$

for both the in-palne and out-of-plane configurations independently. The dependence on σ_e is important for astrophysical applications because it is common to use the magnetization parameter to characterize the properties of relativistic jets. Note that $\omega_{pe} = \sqrt{4\pi N_1 e^2/\gamma_1 m_e}$, $\omega_{ce} = eB_1/\gamma_1 m_e c$ are the electron plasma and cyclotron frequencies; B_1 and N_1 are the ambient magnetic field strength and number density in the far upstream region. The notations for the other quantities are standard. Note that we use the CGS units throughout in this paper.

A cell size of $\Delta x/c/\omega_{pe} = 1/40$, the number of particle per cell in the upstream of $N_1 \Delta x^2 = 64$ were used with the number of cells of $(N_x, N_y) = (20,000, 1680)$ for the x and y directions, respectively.

19.2.2 Shock Structures

We first discuss the simulation results obtained with the out-of-plane magnetic field configuration. The overall shock structure for $\sigma_e = 3 \times 10^{-1}$ is shown in Fig. 19.1. The upstream plasma on the right-hand side traveling toward the left is compressed at a well-developed shock at $x/c/\omega_{pe} \simeq 290$, which may be identified by a sharp density increase. One observes large-amplitude ($\delta B/B_0 \sim 1$), and short-wavelength magnetic fluctuations in the upstream region of the shock. We confirmed that these are the precursor electromagnetic waves, which were generated at the shock and propagating toward the upstream with a group velocity greater than the shock speed.

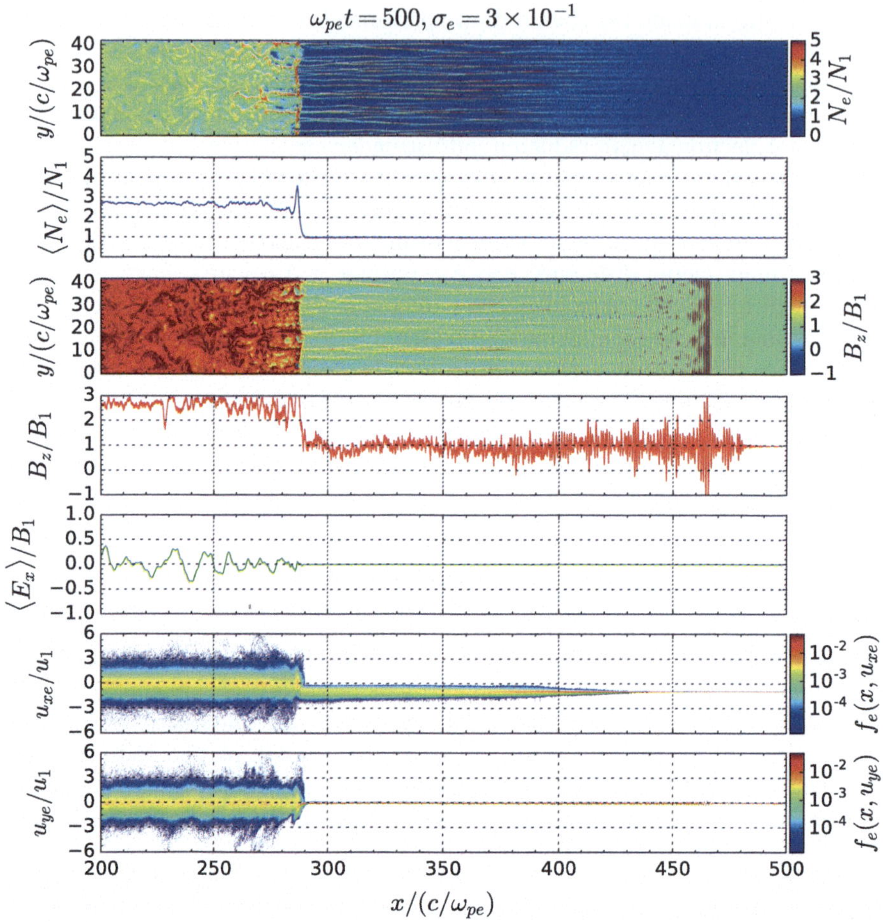

Fig. 19.1 Shock structure for $\sigma_e = 3 \times 10^{-1}$ with the out-of-plane configuration at $\omega_{pe}t = 500$ (after [14]). From top to bottom, the electron density, y-averaged electron density, magnetic field B_z, slice of magnetic field along x, y-averaged electric field E_x, phase space plots in the u_x–x and u_y–x space for electrons, respectively

Another finding in this simulation is the clear filamentary structures in density in the precursor region. We note that the filaments are generated well ahead of the shock transition region, and thus not a result of the WI (which is not active at this moderate magnetization). Therefore, the formation of the filaments should be attributed to a nonlinear interaction between the intense electromagnetic waves and the upstream plasma. Although we have not yet fully understood the formation mechanism, a nonlinear wave-wave coupling (or a parametric instability) of the large-amplitude precursors is likely to be the cause. The formation of the density filaments thus indicates that the emitted electromagnetic waves are strong and coherent.

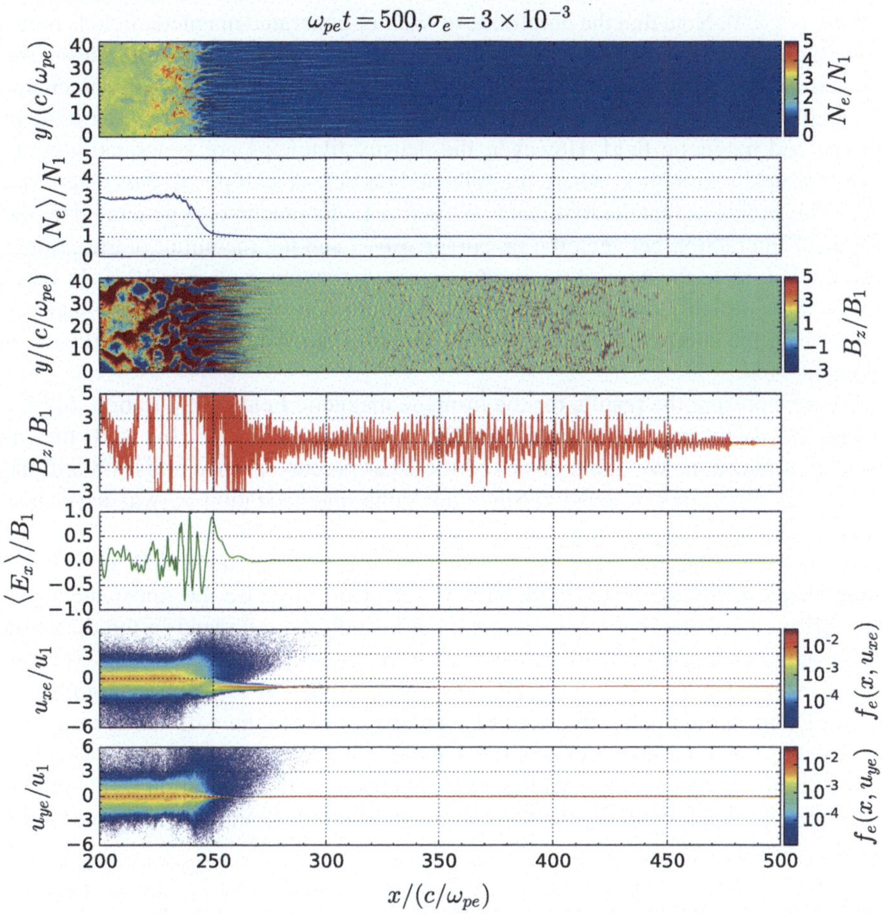

Fig. 19.2 Shock structure for $\sigma_e = 3 \times 10^{-3}$ with the out-of-plane configuration at $\omega_{pe}t = 500$ (after [14]). The format is the same as Fig. 19.1

Note that one may identify the heating of the incoming plasma already in the deep precursor $x/c/\omega_{pe} \sim 300$–350. The emission efficiency at the shock should then be modified because it is the pre-heated plasma that enters into the shock and excites the instability. We confirmed that the shock structure and the precursor wave amplitude become almost stationary at this stage. In the following, we measured the wave power well after such a quasi-steady state has been established. Therefore, the nonlinear feedback effect of the wave emission to the shock itself has already been taken into account.

In Fig. 19.2, the simulation results obtained with $\sigma_e = 3 \times 10^{-3}$, again with the out-of-plane configuration, is shown with the same format. Again, we observed the large-amplitude precursor waves and density filaments. In addition, the lower ambient magnetic field strength (or low σ_e) activates the WI in the shock transition region

$x/c/\omega_{pe} \sim 250$. Note that the amplitude of Weibel-generated magnetic field is much larger than the upstream magnetic field, but the color for the B_z-panel saturates in the shock transition region because we chose a color scale that emphasizes the precursor wave emission. It is known that the WI generates filamentary structures both in density and magnetic field. However, the density filaments are generated already in the far upstream region where the reflected particles (with positive u_x) were not seen. This indicates that the filamentary structure in the precursor is generated by the nonlinear interaction between the precursor waves and the incoming plasma rather than the WI, as in the case of the moderate magnetization $\sigma_e = 3 \times 10^{-1}$. In other words, the simulation result demonstrates the emission of intense electromagnetic waves from the shock that is largely dominated by the Weibel-generated turbulent magnetic field.

We now present the results for the in-plane magnetic field configuration. Shown in Fig. 19.3 is the result obtained with the same $\sigma_e = 3 \times 10^{-3}$, but with the in-plane configuration. We again observed the clear precursor wave emission and the filamentary structures in density. Since the same magnetization σ_e was used, it is natural that the WI is prominent in the shock transition region.

We note that the precursor electromagnetic waves were found both in y and z components of the magnetic field. This is surprising because the linear theory of the SMI predicts that the wave magnetic field should be polarized in the direction of the ambient magnetic field (waves on the X-mode dispersion brunch). Indeed, the waves were polarized only in the z direction for the out-of-plane configuration. Similarly, one would naturally expect the polarization is in the y direction in the in-plane configuration. The waves polarized in the z direction are therefore unexpected. As we discuss later, we think that the waves of unexpected polarization were emitted associated with the Weibel-generated large-amplitude magnetic fluctuations in the shock transition region.

19.2.3 Emission Efficiency

Figure 19.4 displays the compilation of our simulation results. The precursor wave power normalized to the upstream flow kinetic energy $\varepsilon = \delta B^2/8\pi\gamma_1 N_1 m_e c^2$ is shown as a function of magnetization σ_e. Since the upstream flow kinetic energy provides the free energy for the SMI and the resulting electromagnetic wave emission, one may interpret ε as the energy conversion efficiency. Note that we included both y and z components of the magnetic field fluctuations in the estimate of the wave power.

At moderate magnetization $\sigma_e \sim 10^{-1}$, the energy conversion efficiency is approximately a few % both for the in-plane and out-of-plane configurations. The efficiency was lower than the corresponding 1D results by about one order of magnitude. This reduction may be due to the feedback effect of the emission. The upstream plasma is pre-heated already far ahead of the shock, which induces the filaments in density. The pre-heating will reduce the growth rate, especially at high wavenumbers. Also,

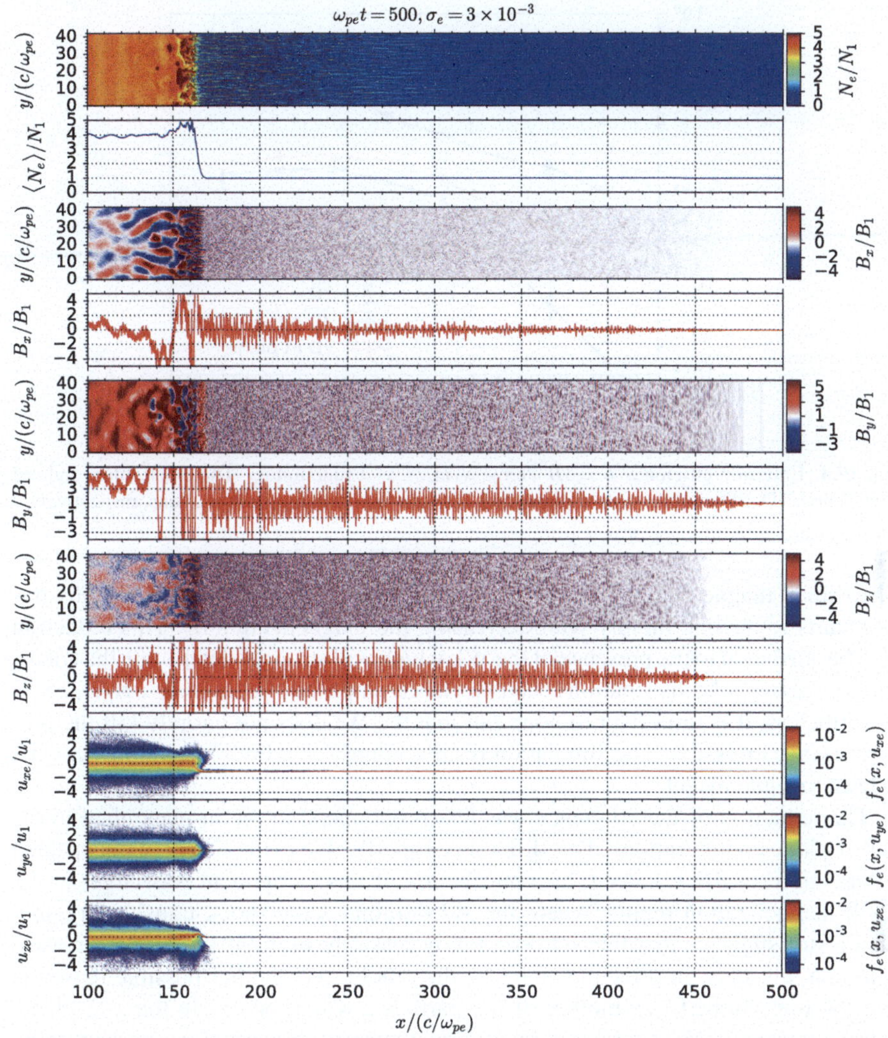

Fig. 19.3 Shock structure for $\sigma_e = 3 \times 10^{-3}$ with the in-plane configuration at $\omega_{pe}t = 500$ (after [15]). From top to bottom, the electron density, y-averaged electron density, B_x, 1D slice of B_x along x, B_y, 1D slice of B_y along x, B_z, 1D slice of B_z along x, phase space plots in the u_x–x, u_y–x, and u_z–x space for electrons, respectively

the filaments eventually interact with the shock and generate the inhomogeneity in the transverse direction at the shock. The inhomogeneity will reduce the coherence of the ring-like distribution and also the growth rate. Despite the reduction in the emission efficiency, the emitted power in 2D has yet remained strong enough to drive relativistic transverse motions of electrons [14, 15].

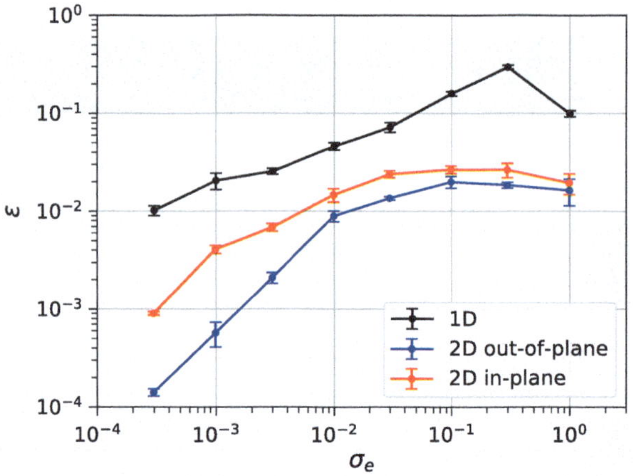

Fig. 19.4 Emission efficiency $\varepsilon = \delta B^2/8\pi\gamma_1 N_1 m_e c^2$ as a function σ_e. The black, blue, and red lines denote 1D, 2D with the out-of-plane, and 2D with the in-plane configurations, respectively

At small magnetization $\sigma_e \lesssim 10^{-2}$, the efficiency for the out-of-plane configuration starts to decline rapidly with decreasing the magnetization σ_e. This reduction may be attributed to the presence of the WI, which will strongly deteriorate the coherence of the ring-like distribution because of large-amplitude fluctuations $\delta B/B_1 \gg 1$. Nevertheless, it is important to point out that the WI does not entirely kill the precursor wave emission. The simulation results suggest that both instabilities coexist at least to some extent.

The reduction of the efficiency was less pronounced in the in-plane than in the out-of-plane configurations. This difference is partly due to the waves of unexpected polarization (or δB_z). Indeed, the contribution of this component was larger in the small magnetization regime where the WI dominates the shock transition region. This result suggests that the WI does not necessarily prohibit the precursor wave emission, but may, in principle, enhance the efficiency. Another possible reason is that the unperturbed gyromotion of the particles, which occurs in the x–z plane, is less affected by the turbulence in the this geometry. Nevertheless, as decreasing the magnetization, the large-amplitude turbulent magnetic field $\delta B_z/B_1 \gg 1$ in the shock transition region will make this effect relatively unimportant.

The discrepancy at low σ_e found in between the two magnetic field configurations clearly indicates that fully 3D simulations are needed for a more quantitative and accurate estimate. Nevertheless, the emission efficiency is maintained at a relatively high level in both cases. Therefore, we believe that the emission of intense precursor waves is indeed an intrinsic nature of a magnetized relativistic shock.

19.3 Summary and Discussion

We have performed 2D PIC simulations of magnetized relativistic shocks in pair plasmas. The results have demonstrated that the emission of intense electromagnetic radiation reported previously with 1D PIC simulations is not an artifact of the low dimensionality. At least, shocks with magnetization $\sigma_e \gtrsim 10^{-4}$ emit strong electromagnetic precursor waves.

The critical parameter for the Wakefield acceleration model is the so-called strength parameter defined as $a = eE/m_e c\omega$, where E is the wave electric field: an efficient particle acceleration has been observed in 2D PIC simulations with a simplified setup when the injected electromagnetic wave pulse satisfies $a > 1$. By assuming the scaling law $a \propto \gamma_1 \sqrt{\varepsilon}$, Iwamoto et al. [15] estimated a condition for γ_1 and σ_e that needs to be satisfied for the emitted wave to be strong enough $a > 1$. The conclusion was that the Wakefield acceleration at a highly relativistic shock favorably with a moderate magnetization remains as a possible mechanism for the acceleration of UHECRs. In this scenario, highly relativistic jets from gamma-ray bursts are the most plausible sites of the production of UHECRs.

Since both the SMI and WI grow from the same free energy source, we anticipated that the precursor wave emission would cease if the growth rate of the WI dominates over the SMI. The simulation results have clearly shown that this is not the case. The reason for this may be understood by the small-scale nature of the turbulence generated by the WI. The Lorentz force of the random turbulent magnetic field δB as seen from a relativistic particle is proportional to $\delta B \lambda / c$ where λ is the coherence length of the turbulence. The factor λ / c indicates the transit time of a particle going through the coherence length. On the other hand, the average Lorenz force due to the ambient magnetic field B_0 is proportional to B_0 / ω_{ce}. It is possible to show that the average Lorenz force is always greater than the random one for the Weibel-generated turbulence with the coherence length on the order of the electron skin depth $\lambda \sim c/\omega_{pe}$. Therefore, on average, a particle in the turbulence performs a complete gyromotion. We note that the growth rate of the SMI is on the order of the cyclotron frequency. Therefore, the coherent charge-particle bunches can develop during this gyromotion and eventually strong electromagnetic waves are emitted.

However, the presence of the WI is obviously not completely negligible. The turbulence gives random kicks in the coherent gyromotion of the particles, which gives rise to an effective temperature to the ring-like distribution in momentum. A finite temperature, in general, reduces the growth rate of the SMI in particular at short wavelengths or high harmonic numbers n. On the other hand, for the wave-particle resonance to occur with the brunch of high-frequency electromagnetic waves (so that the wave can escape upstream), the wave frequency should be larger than the plasma frequency $\omega = n\omega_{ce} \gtrsim \omega_{pe}$ (or the cut-off frequency in a strict sense). This leads to the condition $n \gtrsim \sigma_e^{-1/2}$ required for the harmonic numbers. As the magnetization σ_e decreases, the heating by the WI will become more and more prominent, and the growth rates at high n numbers will decrease. At the same time, the required harmonic numbers n will become larger and larger. We thus expect that there will be a critical threshold in σ_e, below which the precursor wave emission will diminish. A much larger computational resource is needed to prove this conjecture.

It is important to mention that once a charged-particle bunch is generated, the magnetic field amplified by the WI may help to enhance the emission simply because the radiated power is proportional to the magnetic field energy density in the synchrotron theory. Our interpretation for the higher emission efficiency of the unexpected component (δB_z) in the in-plane configuration is that the polarization is locally of X-mode type with the perturbed magnetic field oscillates in the direction of the local magnetic field vector. Because of the large-amplitude turbulence in the shock transition region, the magnetic field may be locally in the z direction. If a pre-existing charged bunch travels through a region of large B_z, a large-amplitude wave with the unexpected polarization may be emitted. The actual power of coherent radiation in the Weibel-dominated regime should be determined by the competition between the efficiency of charged-bunch generation and the magnetic field amplification. The answer to the question can only be given by conducting direct 3D simulations.

In this study, we employed an electron-positron plasma to estimate the emission efficiency. As we mentioned earlier, this makes sense because the emission will be generated only by electrons even at a relativistic electron-proton plasma shock. However, this implicitly assumes that the energy transfer between the two constituents does not occur. If the electrons are able to absorb energy from the protons, the efficiency should be rescaled by the effective kinetic energy available for the electrons. This would make the system more complex and therefore interesting. Simulation results for an electron-proton shock will be reported elsewhere in the future.

Acknowledgements This work was supported in part by JSPS KAKENHI Grant Numbers 17H02966, 17H06140, 17H02877. This work used the computational resources of Cray XC30 and computers at Center for Computational Astrophysics, National Astronomical Observatory of Japan, the K computer provided by the RIKEN Advanced Institute for Computational Science, and the HPCI system provided by Information Technology Center, Nagoya University through the HPCI System Research Project (Project ID: hp150263, hp170158, hp180071).

References

1. D.B. Melrose, Rev. Mod. Plasma Phys. **1**(1), 5 (2017). https://doi.org/10.1007/s41614-017-0007-0
2. M. Hoshino, J. Arons, Phys. Fluids B: Plasma Phys. **3**(3), 818 (1991). https://doi.org/10.1063/1.859877
3. A. Langdon, J. Arons, C. Max, Phys. Rev. Lett. **61**(7), 779 (1988). https://doi.org/10.1103/PhysRevLett.61.779
4. M. Hoshino, J. Arons, Y.A. Gallant, A.B. Langdon, Astrophys. J. **390**, 454 (1992). https://doi.org/10.1086/171296
5. Y.A. Gallant, M. Hoshino, A.B. Langdon, J. Arons, C.E. Max, Astrophys. J. **391**, 73 (1992). https://doi.org/10.1086/171326
6. E.S. Weibel, Phys. Rev. Lett. **2**(3), 83 (1959). https://doi.org/10.1103/PhysRevLett.2.83
7. L. Sironi, A. Spitkovsky, Astrophys. J. **726**(2), 75 (2011). https://doi.org/10.1088/0004-637X/726/2/75

8. L. Sironi, A. Spitkovsky, J. Arons, Astrophys. J. **771**(1), 54 (2013). https://doi.org/10.1088/0004-637X/771/1/54
9. Y. Matsumoto, T. Amano, T.N. Kato, M. Hoshino, Phys. Rev. Lett. **119**(10), 105101 (2017). https://doi.org/10.1103/PhysRevLett.119.105101
10. P. Chen, T. Tajima, Y. Takahashi, Phys. Rev. Lett. **89**(16), 161101 (2002). https://doi.org/10.1103/PhysRevLett.89.161101
11. Y. Lyubarsky, Astrophys. J. **652**(2), 1297 (2006). https://doi.org/10.1086/508606
12. M. Hoshino, Astrophys. J. **672**(2), 940 (2008). https://doi.org/10.1086/523665
13. T. Tajima, J.M. Dawson, Phys. Rev. Lett. **43**(4), 267 (1979). https://doi.org/10.1103/PhysRevLett.43.267
14. M. Iwamoto, T. Amano, M. Hoshino, Y. Matsumoto, Astrophys. J. **840**(1), 52 (2017). https://doi.org/10.3847/1538-4357/aa6d6f
15. M. Iwamoto, T. Amano, M. Hoshino, Y. Matsumoto, Astrophys. J. **858**(2), 93 (2018). https://doi.org/10.3847/1538-4357/aaba7a
16. Y. Matsumoto, T. Amano, T.N. Kato, M. Hoshino, Science (New York, N.Y.) **347**(6225), 974 (2015). https://doi.org/10.1126/science.1260168
17. N. Ikeya, Y. Matsumoto, Publ. Astron. Soc. Japan **67**(4), 64 (2015). https://doi.org/10.1093/pasj/psv052

Part V
Nanostructures and Functional Materials

Chapter 20
Laser-Induced Deposition of Metal and Hybrid Metal-Carbon Nanostructures

Alina Manshina

Abstract This chapter considers the problem of laser-matter interaction in the sense of laser-initiated chemical processes and subsequent formation of solid nanostructured materials. The main attention is focused to the effects originated as a result of laser impact onto heterogenic systems such as 'solid-liquid' interfaces, and in particular for the case when the chemical activity of the heterogenic system as a whole is mainly determined by the laser-induced effects in liquid phase. That is why all the issues connected to the laser ablation phenomena are out of the chapter scope. Heterogenic systems (substrate-solution interfaces) are extremely curious targets for the laser irradiation because of a much wider diversity of chemical processes as compared with homogeneous systems (solutions or solids). Another peculiarity of the selected system is the possibility of the chemical activity tuning by selecting either photochemical or thermochemical mode of laser exposure. This laser-induced approach proved to be efficient for one-step formation of the metal and hybrid metal-carbon nanostructures on the surface of the various kinds of substrates. The peculiarities of the laser-induced processes for the case of different solutions (electrolyte solutions or solutions of organometallic complexes) and different regimes of laser irradiation (thermal or photo-induced) are considered; as well as the possibility of the directed control of the composition of the deposited nanostructures is presented. It is important to note that the essential advantage of the laser-induced processes under discussion is the spatial control of the deposited structures that are formed in the laser-affected area of the substrate. The last circumstance together with well controlled composition and morphology ensure diverse functional properties of the laser-deposited nanostructures.

20.1 Introduction

Successful development of inorganic and organic chemistry methods has disclosed for the modern science such novel objects as nanostructures and nanomaterials that

A. Manshina (✉)
Institute of Chemistry, St. Petersburg State University, 26 Universitetskii prospect, 198504 Petrodvorets, St. Petersburg, Russia
e-mail: a.manshina@spbu.ru

K. Yamanouchi et al. (eds.), *Progress in Photon Science*, Springer Series in Chemical Physics 119, https://doi.org/10.1007/978-3-030-05974-3_20

exhibit unique and well-controllable physical/chemical and functional characteristics. A no less important result has been the revealing of fundamental chemical regularities, mechanisms, kinetics and thermodynamics of processes enabling controllable synthesis of nanoscale or nanostructured solid-state compounds. At present, nanostructures and nanomaterials form a vast class of objects systematized with respect to various characteristics: chemical composition, morphology, dimensionality, structure, etc. Analysis of the modern science development trends in the field of synthetic approaches to obtaining such objects and their application areas demonstrates an interest to optimize and simplify techniques for their fabrication, on the one hand, and to create multifunctional nanomaterials more complicated in composition, morphology and structure, on the other hand [15, 26, 28]. Especial attention is given to obtaining and studying metallic (including bi- and multi-metallic) nanostructures possessing the plasmon properties, carbon nanomaterials of various dimensionalities (zero-dimension 0D, one-dimensional 1D, and also 2D and 3D structures), and multiphase (hybrid) nanomaterials.

The interest to multimetallic nanostructures is first of all caused by their absolutely novel properties arising due to spatial confinement and increase in the ratio between the numbers of surface and bulk atoms which takes place in going to the nanoscale state; moreover, if two or more metals are combined in one and the same structure, there arises an additional possibility to precisely control the electron subsystem (and, hence, the entire set of physical/chemical and functional characteristics) by selecting the types of combined metals, their ratio, and morphology of the system as a whole.

The materials obtained by combining two or more phases are regarded as hybrid nanomaterials; at that the component sizes and total sizes of the hybrid structures remain within the nanoscale range. The hybrid nanomaterials possess a wide spectrum of important properties that may be gained by controllably varying such parameters as their composition, structure and morphology. In addition, combining of different components in the same structure and their mutual influence enables obtaining materials that exhibit not only additive properties of the components but also new ones caused by synergetic effects. Since the possibility of fabricating graphene and other synthetic allotropic modifications of carbon has been demonstrated, an interest has drastically risen to carbon nanostructures and hybrid materials based on them, e.g., metallocarboedrens M[8]C[12], pipodes M@C@HT, graphene and carbon nanotubes with metal nanoparticles, metal/carbon nanoparticles of the core/shell structure, etc. [1, 2, 5, 31, 35]. The conventional procedure for synthesizing such multicomponent materials complicated both in structure and morphology is typically based on multistage processes needing either preliminary obtaining of the composite nanocomponents and development of a technique for their subsequent integration, or stage-by-stage synthesis in which the necessary components are being successively formed on the initial matrix surface [14, 16, 32, 33]. As a rule, such multistage approaches enable obtaining of nanostructures of only one type, while the synthesis techniques themselves are rather time- and labor-consuming. Development of novel approaches to synthesizing hybrid substances of complicated composition, structure and morphology with preset characteristics might provide not only obtaining the target product for practical purposes but also a progress in fundamental understanding of new chemical processes. One of promising approaches suggesting diverse methods for initiating chemical reactions and, hence, the possibility of controlling the nanoscale structures fabrication, is the use of laser radiation for thermochemical, photochemical or spectral-selective excitation of chemical systems.

Studies in this area become at present more and more topical because of unique properties of structures and materials obtainable in this way and also of the opportunities ensured by directed modification of their properties due to varying in a wide range the parameters of both the laser radiation and irradiated medium. Let us note that one of the most important factors taken into account in assessing the perspectives of solving the up-to-date problems based on the laser-induced chemical processes considered in this paper is the controllability of characteristics of the synthesized solid substances and materials.

Thus, all the above emphasizes the urgency of developing new techniques for synthesizing nanomaterials of complicated composition and morphology.

20.2 Laser-Induced Metal Deposition from Electrolyte Solutions

The necessity of solving technological problems of creating various-purpose metal coatings on fashioned surfaces has predetermined successful development of the technique of liquid-phase chemical deposition of metals. Reactions of the metal chemical deposition are oxidation and reduction resulting in metal formation. Potential ability to chemical reduction is characteristic of a large number of metals. However, thermodynamically possible reactions may be kinetically deferred so that their rates diminish to zero. As it appeared later, laser radiation is able to eliminate kinetical troubles, especially at the initial stages of the process. One more factor that has caused the interest to laser-induced liquid-phase metal deposition is the possibility of localizing the chemical deposition process and creating metal coatings just in the area exposed to the laser irradiation.

Development of this approach resulted in successful experiments for localized laser-induced deposition of various metals onto substrates from electrolyte solutions [8, 10, 13, 34]. This technique was entitled as "laser-induced liquid-phase metal deposition" (LCLD). Later general regularities were revealed which determine compositions of the solutions to be used for laser-induced metal deposition. Compositions of solutions used for laser-induced deposition are generally similar to those for traditional chemical metallization; however, differ from the latter in concentrations. Their mandatory components are the metal source (organic or inorganic metal component), chelating additive, pH-stabilizer (buffer), acid/alkali and reducing agent dissolved in distilled water or organic solvent. For example the traditional composition of the electrolyte solution is [23]: 0.1M $CuSO_4$, 0.2M KNa-tartarate (Rochelle salt) ($KNaC_4H_4O_6 \times 4H_2O$), 0.125M NaOH, and 6M HCOH (formaldehyde).

The reaction is believed to run spontaneously if the difference between the redox potentials is sufficiently large. However, in depositing local metal layers with a laser, it is desirable to use reducing and oxidizing agents with a small potential difference since large potential differences frequently cause fast, sometimes almost instant, formation of the reduction product in the entire liquid phase bulk. To prevent fast reactions, ligands or chelates should be added to the solutions. They form sufficiently strong complexes with ions of the metal under reduction and decrease the potential difference due to shifting the redox-potential of the "metal ion/metal" pair towards negative values. The ligands play also another role: they prevent formation of metal hydroxides in the alkaline medium, which enlarges the area of metal deposition limited by the electrode potential and acid-base properties of the solution. Moreover, this made it thermodynamically possible to use as reducing agents substances whose potential is shifted towards positive values. The method of liquid-phase laser deposition may be used to locally deposit different metals like Ni, Co, Fe, Cu, Ag, Au, Pd, Rh, Ru, St, Sn, Pb, In, etc.

Prior to starting the laser-induced deposition, the electrolyte solution shall be brought into contact with the substrate. After that, the laser beam shall be focused on the interface through the liquid layer or the substrate (see Fig. 20.1).

The radiation absorption increases the liquid-substrate interface temperature, which gives rise to various non-equilibrium states in the heated area. They, in their turn, initiate a number of extra processes:

- migration of reagents from the liquid phase bulk into the laser-irradiated area due to the liquid concentration and temperature gradients;
- changes in the substrate properties in the heated area;
- reactions (reagent decomposition, reduction or oxidation) at the substrate/liquid interface;
- deposition of the solution components and removal of the reaction products from the irradiated area.

In the process of the laser-induced liquid-phase metal deposition, the laser radiation may be regarded as a heat source (radiation absorption with subsequent transformation into thermal energy) and/or as a photon source (release of electrons, free radicals and the like for reducing metal ions). Typically, the thermal (or pyrolytic) reaction needs low photon energy ($h\nu < 2$ eV), while realization of the photochemical

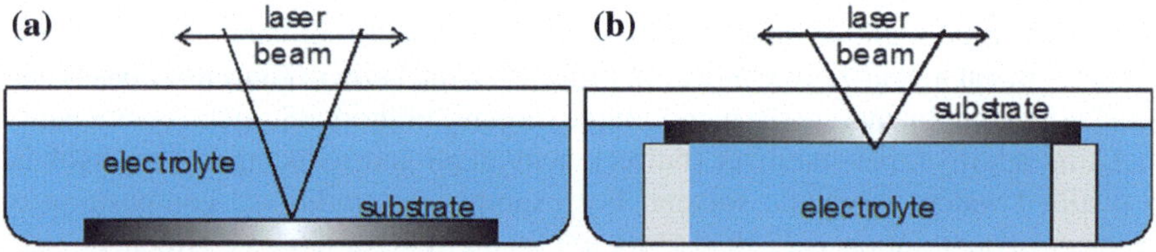

Fig. 20.1 The laser-induced deposition scheme for nontransparent (**a**) and transparent (**b**) for laser radiation substrate

(or photolythic) process requires higher energies (2–5 eV). Therewith, the increase in the laser power leads to the increase in the rate of the substrate metallization in both reaction modes. However, while increasing the injected energy one should keep in mind possible variations in the rates of other reactions and arising of extra effects (changes in the particle diffusion rate in the solution and on the substrate surface, formation of bubbles, liquid convection, melting, surface solidification or ablation, etc.).

Note that, regardless of a considerable progress achieved in the field of laser-induced liquid-phase metal deposition, namely, demonstration of the possibility of depositing various metals on various substrates (including dielectrics, semiconductors and polymers), physical and chemical characteristics of the deposited metallic structures still remain lower than those of metallic structures fabricated by alternative methods (e.g., optical lithography). Here we mean first of all the morphology and electrical resistance of the metallic structures, i.e., characteristics determining the LCLD process functional characteristics and possibility of its practical utilization in fabricating, for instance, microchips and electrical contacts. Figure 20.2a presents the image of a copper structure deposited on the surface of a polyamide substrate under the argon laser irradiation.

To obtain a continuous current-conducting structure, deposition was performed in the mode of multiple scanning (namely, a mode with a larger number of scans) [13]. As it is shown in the picture, the deposited structure is characterized by an extremely nonuniform morphology.

Studying of the dependence of the deposited structure resistance on the rate of laser-beam scanning in a wide power range (50–130 mW) showed that the deposited structure resistance is always extremely high and varies from 200 to 2000 Ω for copper structures 1 mm long (Fig. 20.2b). The estimate of the specific resistance of deposited copper structures appeared to be about 10^2 Ω mm^2/m which is higher than that of bulk copper by 4 orders of magnitude (0.0175 Ω mm^2/m). The main reason for such a high specific resistance of the deposited structures is probably formation

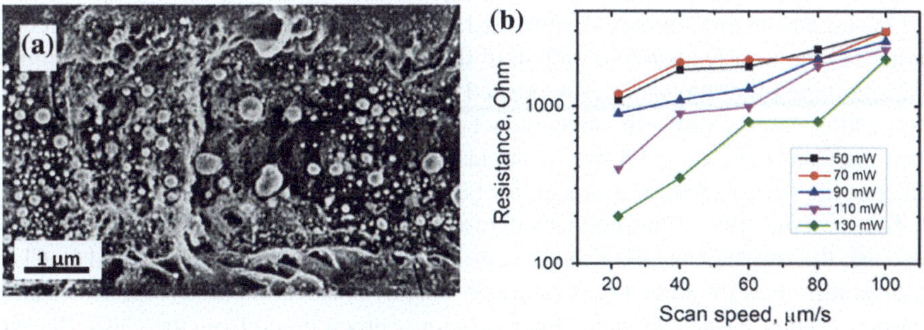

Fig. 20.2 **a** SEM-image of a copper structure deposited by using the argon laser radiation on the polyamide substrate surface in the multiple scanning mode (number of scans: 8); **b** resistance of copper structures deposited by 8 scans versus the scanning rate in the range of the laser radiation power (50–130 mW)

of a surface oxide and, more important, extremely nonuniform morphology and high granularity of the deposit. Note that laser radiation of high power (130 mW) and low scanning rate produces structures with the lowest resistance (about 200 Ω). The laser-induced deposition at high powers and low scanning rates results in local melting and more uniform morphology of the deposited metal.

A similar effect of forming structures characterized by high electrical resistance and nonuniform morphology is observed also in the case of palladium laser-induced deposition [9, 11, 12]. Using the multiple-scanning mode at a lower scanning rate and higher laser radiation power it is possible to decrease the electrical resistance of deposited Pd structures by 1 or 2 orders of magnitude. However, side effects associated with the substrate destruction significantly reduce the deposited metal adhesion.

So, the most serious drawback is the above-mentioned high nonuniformity of the deposited metal structure leading to high specific resistance of the structures fabricated (10–100 times higher than the bulk metal specific resistance). In addition, the up-to-date technique for creating continuous conducting structures needs the multiple scanning mode (up to 25 sequential scans) which is technologically low-efficient.

However, the morphology and properties of the resulting metal deposits are found to be considerably dependent on the applied parameters (the laser power, the scan velocity, the number of scans) [22–24]. It has been demonstrated that the temperature of the electrolyte solution greatly influences the deposition process. Namely, by increasing the temperature of the solution one can reduce the threshold laser power required to initiate the metal deposition, and, as a result to decrease the negative effect of laser heating [24].

The deposition process was also found to be significantly affected by a salt precursor ($CuSO_4$- and $CuCl_2$-based electrolyte solutions) as shown in [21]. Figure 20.3 shows morphology of the structures deposited from the $CuSO_4$- and $CuCl_2$-based electrolyte solutions at different laser power (continuous wave (CW) argon laser was used as a source). It can be easily seen that the morphology of the deposits depends strongly on the laser power, and on the salt precursor.

On comparing the samples obtained at 100 and 150 mW one can conclude that the higher laser power leads to the growth of the deposit dimensions for both solutions. It is significant that the samples deposited from the $CuCl_2$-based solution are characterized by a more uniform and compact structure in comparison with the deposits from the $CuSO_4$-based solution. At the laser power above 400 mW the destructive effect of the laser radiation was seen for both kinds of the solutions.

To reveal the effect of the copper salt precursor on the conductive properties of the deposits, the resistances (R) of the structures obtained from the $CuSO_4$- and $CuCl_2$-based solutions at the same temperature 55 °C, have been compared (Fig. 20.4). The comparison has shown that the copper structures precipitated from the $CuCl_2$-based electrolyte solution have a resistance which is 4–5 times lower than the resistance of the structures deposited from the $CuSO_4$-based solution [21].

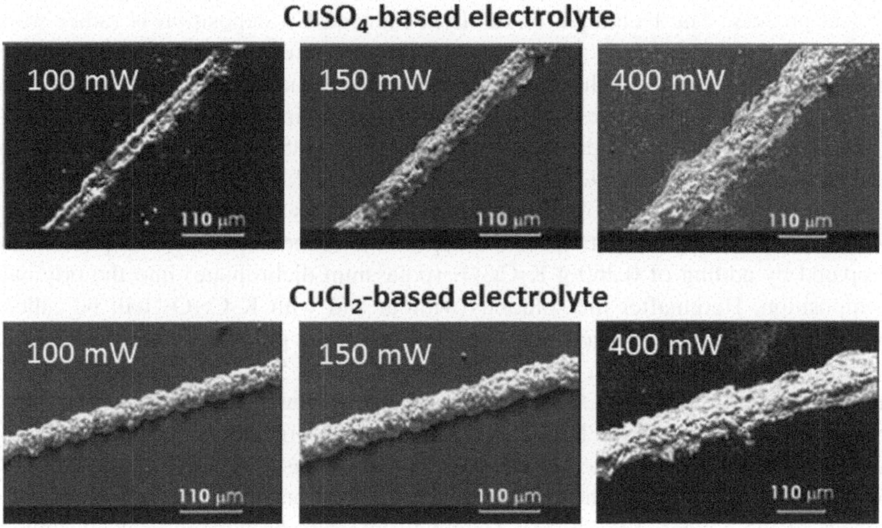

Fig. 20.3 SEM photos of the structures deposited from the $CuSO_4$- and $CuCl_2$-based electrolyte solutions at the temperature of 308 K, and at different laser powers

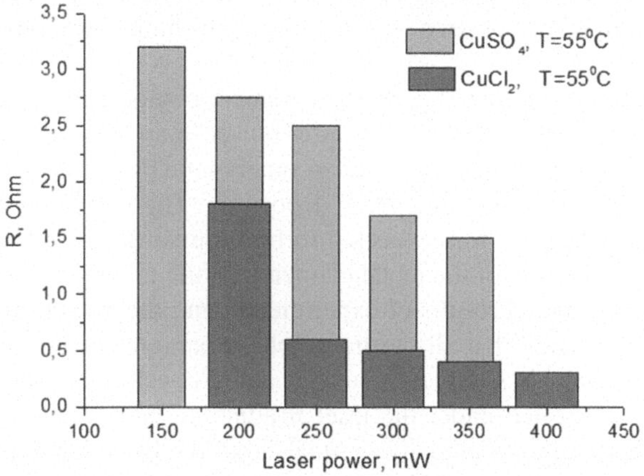

Fig. 20.4 The dependence of the resistance (*R*) on the laser power for the structures deposited from the $CuSO_4$- and $CuCl_2$-based electrolyte solution at the same solution temperature

20.3 Laser-Induced Deposition of Hetero-Metallic Structures from Liquid Phase

The intensive experiments in the LCLD also revealed possibility of the laser-induced precipitation of the heterometallic structures from the liquid solution. The metal pair copper (Cu) and chromium (Cr) was chosen for the co-deposition with the

LCLD process. The technique of the laser-induced Cu deposition is rather well developed; furthermore, it is known that there is good interaction between Cu and Cr in the case of the chemical chromium plating of copper and copper-based alloys. The chemical composition of the electrolyte was the following (content of the chemical components in 500 ml of distilled water): 0.260 g $CuSO_4$, 1.320 g KNa tartrate (Rochelle salt) ($KNaC_4H_4O_6 \cdot 4H_2O$), 0.386 g NaOH, 2.2 g HCOH (formaldehyde) and 2.5 g CH_3OH (methanol). All the used chemicals were of the analytical grade purity. The electrolyte solution for the deposition of the heterometallic phase was prepared by adding of 0.260 g $K_2Cr_2O_7$ (potassium dichromate) into the original composition. Hereinafter the solutions without and with $K_2Cr_2O_7$ will be called original and modified, respectively. The CW argon laser generated in the multiwave regime (454.6, 457.9, 465.8, 476.5, 488.0, 496.5, 501.7, 514.5 nm) was used for the metal precipitation by the LCLD method. The laser power was equal to 300 mW. Figure 20.5a, b shows the SEM photos of the structures precipitated from the original and modified solutions at the same deposition conditions [20]. EDX analysis of the precipitated structures has shown that in the case of the original solution the deposits consist of the copper (Fig. 20.5a). The small intensity broadband peak illustrates the presence of the trace amounts of the impurities in the deposit. In the case of the structures deposited from the modified electrolyte solution along with the copper the chromium was found (Fig. 20.5b). The intensity of the chromium peaks, which are two times higher than the copper peaks, testifies to the high chromium content in the resulted deposits.

In such a way the optimization of the chemical composition of the electrolyte solution and the deposition parameters (laser power, scan velocity, number of laser scans) as well as search and investigation of other metal pairs for LCLD will open a way of controllable heterometallic phase deposition. The LCLD method of heterometallic structures formation is expected to be promising for modern technology oriented on the production of the metal elements on the materials' surfaces.

Based on the presented results of experimental and theoretical investigations, a conclusion may be made that all the liquid phase parameters and deposition conditions considered in this study extremely unambiguously affect the properties of metallic structures formed under the laser irradiation. Note that there are no unified metal-deposition modes equally optimal in all the cases. In depositing metals by the LCLD technique under different experimental conditions (chemical composition and concentration characteristics of the electrolyte solution, laser irradiation parameters), it is necessary to analyze the negative and positive aspects of any given mode or parameter and optimize the deposition conditions taking into account certain requirements for the desired result.

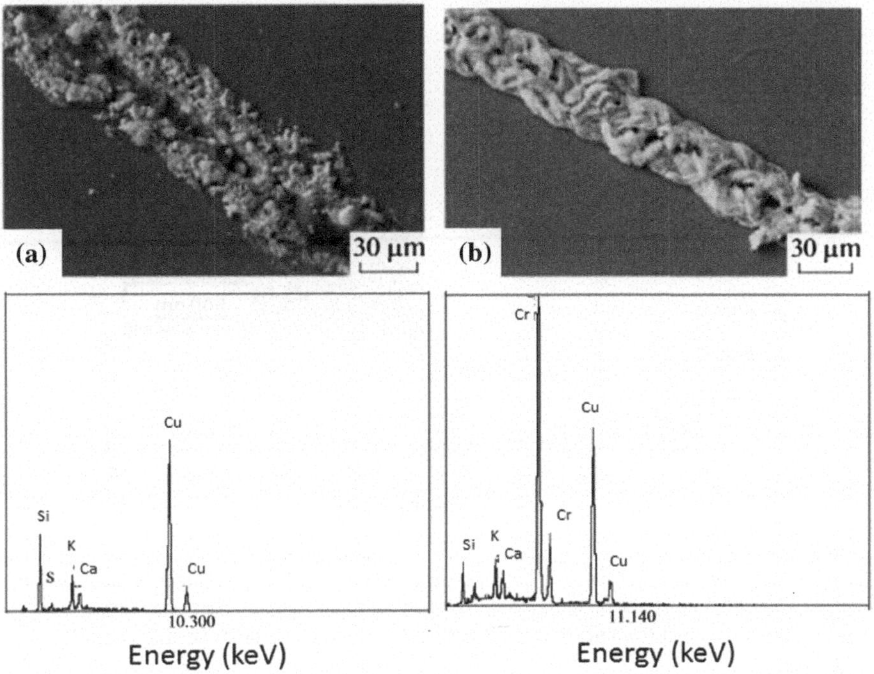

Fig. 20.5 The SEM photos of the structures precipitated from the original and modified solutions at the same deposition conditions: **a** the structure deposited from the original solution, **b** the structure deposited from the modified solution. EDX analysis of the structures deposited from the original (**a**) and modified (**b**) solutions

20.4 Laser-Induced Deposition of Hybrid Metal-Carbon Nanostructures from Solutions of Organometallic Complexes

20.4.1 Thermally-Induced Deposition from Solutions of Organometallic Complexes

As it was demonstrated in the previous paragraph, the technology of the homogeneous heterometallic phase deposition can be based on the laser-induced reduction of metal ions independently existing in liquid phase. However, in this case the reduction processes of individual ions may result in the formation of separated phases of individual metals or may provide the heterometallic phase of uncontrolled chemical composition in terms of metal elements ratio. A promising alternative approach is the use of the solutions based on the heterometallic precursors with the given ratio of the different metals. The precursors of this sort may ensure the required chemical composition and homogeneity of the deposited heterometallic phase. In this respect

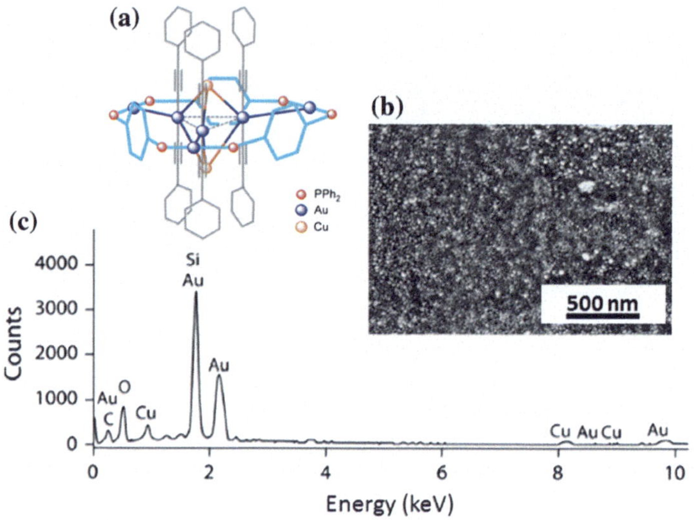

Fig. 20.6 a Schematic structure of supramolecular Au–Cu complexes $[Au_3(Ph_2PC_6H_4PPh_2)_3(Au(C_2Ph)_2)_3Cu_2](PF_6)_2$ **b** SEM photo of the structure deposited from the dimethylformamide solution of at laser power 300 mW and scan velocity 0.01 mm/s. EDX analysis of the structure deposited from the dimethylformamide solution of supramolecular complex

the heterometallic transition metal polynuclear complexes containing cluster core with the ions or the neutral metal atoms bound by the direct metal–metal bonding are of particular interest. The synthesis and the chemistry of these compounds are studied rather well and described in a few fundamental reviews and books [6, 29]. These studies resulted in the synthesis and characterization of a new class of the supramolecular heterometallic complexes, which demonstrated unprecedented structural patterns and variety of laser-induced phenomena. The main structural element of these compounds consists of the central bimetallic cluster core, stabilized by the alkynyl ligands, and the external gold(I)-phosphine belt anchored to the central part by the Au–Au bonds [18]. An example of such sort supramolecular complex with Au–Cu core is presented in Fig. 20.6a. The bimetallic cluster of this supramolecular complex is formed by six Au and two Cu atoms connected by direct metal-metal bonds.

Using of the supramolecular complexes as a precursor allows simplifying the composition of the liquid phase for the LCLD, and preparing solutions of organometallic complexes in appropriate solvent [18]. The morphology of the structures deposited from solution of the complex (Fig. 20.6a) is rather uniform and can be characterized as the close packed homogeneously distributed nanoparticles (Fig. 20.6b) with the average dimension 20–30 nm. The deposition process was realized with the CW argon laser in a wide range of laser power (5–500 mW) that provides localized heating of the laser-affected area. It is worth noting that the variation of the deposition parameters (laser power, scan velocity, number of laser scans) does not lead to the essential variation of the morphology.

The energy dispersive X-ray (EDX) analysis showed that the deposits consist of both Au and Cu, the signals of which are clearly marked in Fig. 20.6c. The spectrum also displays the signals of the substrate material (soda-lime glass) together with the low intensity broadband peaks corresponding to the trace amounts of impurities originated from the substrate.

The mechanism of the heterometallic phase formation in this case most probably consists in the laser-initiated thermal decomposition (pyrolysis) of the supramolecular complex adsorbed at the laser-induced active sites on the surface followed by the reduction of the bimetallic core by alkynyl anionic ligands that resulted in local precipitation of homogeneous heterometallic macro-pattern.

20.4.2 Photo-Induced Deposition from Solutions of Organometallic Complexes

Another important feature of the laser-induced precipitation from the solutions of organometallic precursors is possibility of the deposition process under the low intensity laser radiation that favorably differs from the electrolyte solutions deposition at high laser intensities. The photochemical instability of the supramolecular complex allows for its decomposition under mild illumination conditions thus avoiding side effects of thermal phenomena (delocalization of the deposition process, thermal decomposition of the deposits, etc.). It was found that the LCLD process can be realized also at He–Cd laser irradiation (continuous wave, $\lambda = 325$ nm, $P = 0.1\ \mathrm{W/cm^2}$). The choice of He-Cd laser was determined by the spectral position of an intense absorption band of the used supramolecular complexes [19].

We found that such approach favors formation of hybrid nanostructures with precise composition, while the characteristics of laser irradiation define their structure and morphology [27]. As the liquid phase the solution of organometallic precursor $[Au_{12}Ag_{12}(C_2Ph)_{18}Br_3(PPh_2(C_6H_4)_3PPh_2)_3](PF_6)_3$ was used for the LCLD process [27]. Figure 20.7a presents the SEM image of the deposited nanostructures that were found to be nanoparticles with the average diameter about 20–30 nm. Figure 20.7b shows the result of energy dispersive X-ray spectroscopy (EDX) analysis, which confirms that the deposited NPs consist of carbon, silver, and gold. The components ratio was found to be about 90/5/5 at.% that coincides very well to the composition of the central core in the precursor supramolecular complexes.

The mechanism of hybrid Au–Ag–C NPs formation was found to be photoinduced decomposition of the precursor molecule followed by transformation of the cluster core (red-ox intramolecular reaction and aggregation of heterometallic species) to give the hybrid metal/carbon nanostructures with multi-yolk-shell architecture. In this case the composition of the complex cluster core predefines the composition of

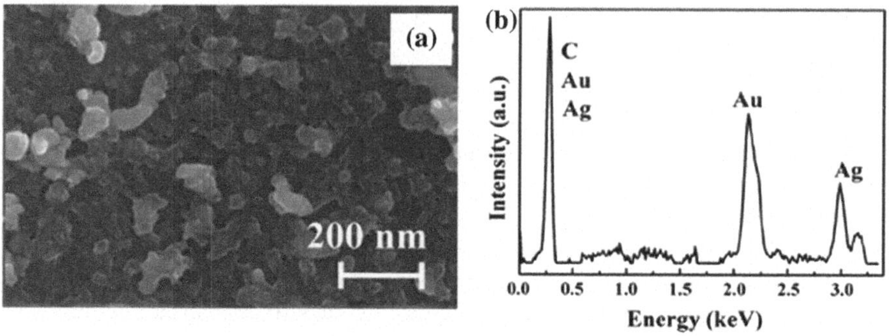

Fig. 20.7 **a** Typical SEM image of an Au–Ag–C nanostructure. **b** EDX spectrum of the deposited nanoparticles

the reduced heterometallic species, the carbon-rich organic ligand provides building blocks for carbonaceous matrix [19].

Further experiments demonstrated possibility LCLD process from the solutions of different metalorganic complexes that opens the way of deposition nanostructures with highly controlled chemical composition [17, 30]. Additional fine tuning of properties of carbon-encapsulated bimetallic nanoparticles prepared with the LCLD method and the directed control of their composition, morphology and functionality (the Surface-Enhanced Raman Scattering – SERS) was found to be performed by the annealing procedure under normal conditions from 100 to 400 °C [25].

20.5 Functional Properties of Hybrid Metal-Carbon Nanostructures

The deposited hybrid metal-carbon nanostructures revealed extremely promising functionality that is determined by their plasmon resonance characteristics of metal inclusions as well as by biocompatibility provided by carbon phase. The most striking example of plasmon NPs' properties is SERS. To study the availability of the Au–Ag–C NPs for the SERS analysis we used human serum albumin (HSA) as a test substance, which belongs to the class of proteins with poor detectability [19]. Spatially resolved Raman spectra for 0.1 μl drop of HSA water solution (concentration 10^{-6} M) onto substrate with Au–Ag–C NPs are presented in Fig. 20.8. Line scan across agglomerate of HSA molecules demonstrates prospectivity of Au–Ag–C nanoparticles for the SERS label-free detection.

Moreover the combination of metal and carbon phases is promising for synergetic effect due to the analyte adsorption by carbonaceous matrix and simultaneous Raman signal enhancement by the metal nanoinclusions. As a test experiment, the substrate with hybrid Au–Ag–C NPs was immersed into ethanol solution of organic chromophore rhodamine 6 g (10^{-6} M) for various periods of time. Then, the substrate

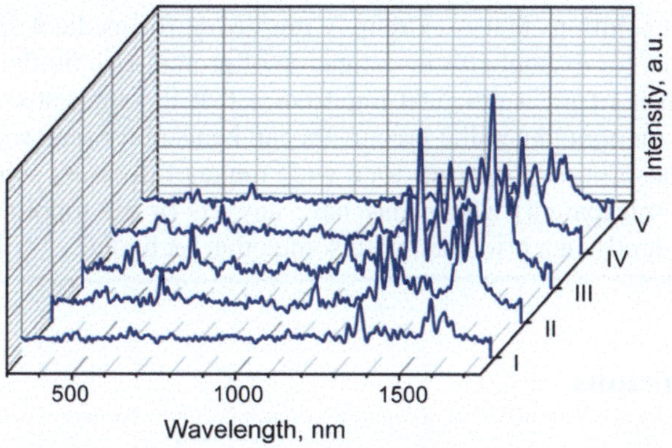

Fig. 20.8 Raman spectra measured from a line scan (I–V points with 0.25 μm distance) across agglomerate of HSA molecules on the substrate with Au–Ag–C NPs

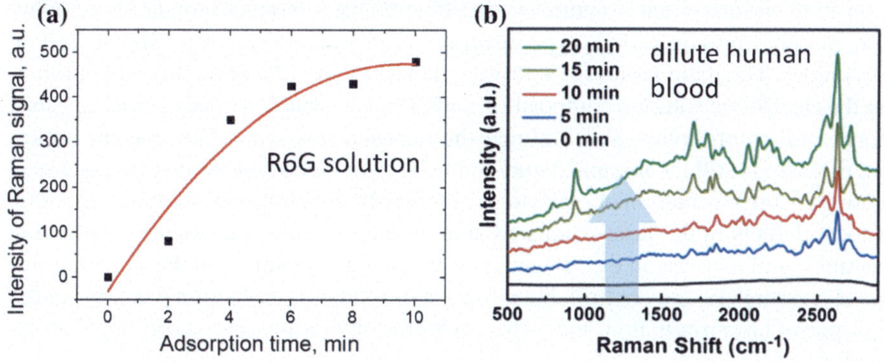

Fig. 20.9 **a** Dependence of the rhodamine 6 g Raman signal (610 cm^{-1}) on the adsorption time, **b** Raman spectra of dilute human blood (contains blood plasma as well as red blood cells) in deionized water (2 gl^{-1} blood, 10^{-7} M) measured at various immersion times (0–20 min), arrow shows immersion time increase

was taken out from the solution, and the Raman spectrum from surface of deposited NPs was recorded. One can see that immersion time increase leads to increase and then saturation of the Raman signal of rhodamine 6 g peak at 610 cm^{-1} (Fig. 20.9a). The adsorption properties of the carbonaceous shell were also confirmed for the dilute human blood (2 g of human blood in a liter of deionized water) for different time frames from 0 to 20 min (cf. Fig. 20.9b) [3].

The increase in the Raman intensity with longer immersion times indicates an effective adsorption of blood on the carbon shell. Moreover the multiplex detection capacity for the mixture of the human blood solution (20 gl^{-1}) and anthracene solutions (1.78×10^{-3} gl^{-1}) was observed [3]. All the presented results demonstrate availability of the hybrid metal-carbon nanostructures for the detection and identification of ultra-small fractions low concentrated analytes including complex multicomponent solutions that is extremely promising for medical application and clinical diagnosis, for example for investigation of protein-rich biofluids and cancer cell detection. It is of great practical importance that the suggested laser-induced approach based on organometallic precursors can be used for targeted imparting of plasmonic and adsorption properties for a wide range of different substrates (transparent and non-transparent) which may have any 2D or 3D topological structure created independently in a prior step with some other technology [4, 7, 30].

20.6 Conclusions

Numerous experiments on laser irradiation of the substrate/solution interface opened a new field of research that has great fundamental interest and promising practical importance. These studies have resulted in technological recommendations for depositing various metal structures with different characteristics (conductivity, adhesion, morphology) on various-type substrates (e.g., ceramics, porous silicon, Teflon, polyimide). The main issues in depositing metals by the LCLD method are optimizing the electrolyte solution composition and the laser radiation dose in order to gain a more regular morphology and minimize the electrical resistance. However the important problem is still the thermal nature of the laser-induced deposition process. Local heating ensures formation of catalytic centers on the substrate surface due to creation of local defects in the laser-irradiated area, acceleration of the oxidation-reduction reactions, and increase in the efficiency of the metal deposition on the substrate surface. Nevertheless, the controllable initiation of oxidation-reduction reactions under high-power laser irradiation, and, hence, at high temperatures, needs additional investigation.

In this sense the variation of the LCLD process and using of organometallic complexes solutions as a liquid phase seems more promising, as allows realization of the deposition process in the "photo-initiated mode" without thermal decomposition of the solution components and precipitated structures. In this case the composition of the organometallic precursor predefines the composition of the precipitated nanostructures, whereas the deposition parameters dictate morphology of the NPs formed. Altogether it allows synthesis of hybrid multiphase nanomaterials (in particular combination of carbonaceous matrix with metal or bi-metal nanoinclusions) with controlled composition and morphology. The other ways of preparation of analogous metal-carbon structures suggested earlier is a sophisticated multistage organometallic chemistry, which appears to be even more complicated in the synthesis of hybrid multi-metallic nanoscale materials. Thus the laser-induced transformation of supramolecular precursors resulting in formation of hybrid metal-carbon nanoparticles is promising strategy for preparation of materials of this type with a desired composition, morphology and architecture. Another important peculiarity of the hybrid metal-carbon nanostructures is wider application areas compared to monophasic materials (metallic or carbon nanostructures) such as SERS in combination with adsorption properties, promising for detection of ultra-low-volumes and concentrations of bio-agents and hazardous compounds.

Acknowledgements This work was financially supported by Russian Foundation for Basic Research (RFBR 17-03-01284) and St. Petersburg state university grant for equipment #12.40.1342.2017. Experimental investigations were carried out in "Center for Optical and Laser materials research", Centre for Geo-Environmental Research and Modelling (GEOMODEL), Centre for Molecular and Cell Technologies and "Interdisciplinary Resource Center for Nanotechnology" (St. Petersburg State University).

References

1. M. Ashino, D. Obergfell, M. Haluska et al., Atomically resolved mechanical response of individual metallofullerene molecules confined inside carbon nanotubes. Nat. Nanotechnol. **3**, 337–341 (2008). https://doi.org/10.1038/nnano.2008.126
2. M. Ashino, D. Obergfell, M. Haluska et al., Atomic-resolution three-dimensional force and damping maps of carbon nanotube peapods. Nanotechnology **20**, 264001 (2009). https://doi.org/10.1088/0957-4484/20/26/264001
3. M.Y. Bashouti, A. Manshina, A. Povolotckaia et al., Direct laser writing of μ-chips based on hybrid C-Au-Ag nanoparticles for express analysis of hazardous and biological substances. Lab Chip 1742–1747 (2015). https://doi.org/10.1039/c4lc01376j
4. M.Y. Bashouti, A.V. Povolotckaia, A.V. Povolotskiy et al., Spatially-controlled laser-induced decoration of 2D and 3D substrates with plasmonic nanoparticles. RSC Adv. **6**, 75681–75685 (2016). https://doi.org/10.1039/C6RA16585K
5. C.C. Berkdemir, A.W. Castleman Jr., J.O. Sofo, A.W. Castleman, Metal-substituted Ti8C12 metallocarbohedrynes: toward less reactive clusters as building blocks of cluster-assembled materials. Phys. Chem. Chem. Phys. **14**, 9642–9653 (2012). https://doi.org/10.1039/c2cp40509a
6. P. Braunstein, Interplay between bridging groups and metal-metal bonds in heterometallic clusters. Mater. Chem. Phys. **29**, 33–63 (1991). https://doi.org/10.1016/0254-0584(91)90004-E
7. I.E. Kolesnikov, T.Y. Ivanova, D.A. Ivanov et al., In-situ laser-induced synthesis of associated YVO_4:Eu^{3+}@SiO_2@Au-Ag/C nanohybrids with enhanced luminescence. J. Solid State Chem. **258**, 835–840 (2018). https://doi.org/10.1016/j.jssc.2017.12.021
8. K. Kordas, K. Bali, S. Leppävuori et al., Laser direct writing of copper on polyimide surfaces from solution. Appl. Surf. Sci. **154**, 399–404 (2000). https://doi.org/10.1016/S0169-4332(99)00455-9
9. K. Kordás, K. Bali, S. Leppävuori et al., Laser direct writing of palladium on polyimide surfaces from solution. Appl. Surf. Sci. **152**, 149–155 (1999). https://doi.org/10.1016/S0169-4332(99)00326-8
10. K. Kordas, J. Bekesi, R. Vajtai et al., Laser-assisted metal deposition from liquid-phase precursors on polymers. Appl. Surf. Sci. **172**, 178–189 (2001). https://doi.org/10.1016/S0169-4332(00)00852-7
11. K. Kordás, S. Leppävuori, A. Uusimäki et al., Palladium thin film deposition on polyimide by CW Ar^+ laser radiation for electroless copper plating. Thin Solid Films **384**, 185–188 (2001). https://doi.org/10.1016/s0040-6090(00)01829-0
12. K. Kordas, L. Nanai, K. Bali et al., Palladium thin film deposition from liquid precursors on polymers by projected excimer beams. Appl. Surf. Sci. **168**, 66–70 (2000). https://doi.org/10.1016/S0169-4332(00)00592-4

13. K. Kordás, L. Nanai, G. Galbacs et al., Reaction dynamics of CW Ar^+ laser induced copper direct writing from liquid electrolyte on polyimide substrates. Appl. Surf. Sci. **158**, 127–133 (2000). https://doi.org/10.1016/S0169-4332(99)00602-9
14. H. Kumar, Y.K. Mishra, S. Mohapatra et al., Compositional analysis of atom beam co-sputtered metal-silica nanocomposites by Rutherford backscattering spectrometry. Nucl. Instrum. Methods Phys. Res. Sect. B Beam Interact. Mater. Atoms **266**, 1511–1516 (2008). https://doi.org/10.1016/j.nimb.2008.01.027
15. X. Li, J. Zhu, B. Wei, Hybrid nanostructures of metal/two-dimensional nanomaterials for plasmon-enhanced applications. Chem. Soc. Rev. (2016) https://doi.org/10.1039/c6cs00195e
16. L.M. Liz-Marzán, M. Giersig, P. Mulvaney, Synthesis of nanosized gold-silica core-shell particles. Langmuir **12**, 4329–4335 (1996). https://doi.org/10.1021/la9601871
17. A. Manshina, *Laser-Inspired Chemical Transformations*, in Springer Series in Chemical Physics (2017), pp. 243–251
18. A.A. Manshina, A.V. Povolotskiy, A.V. Povolotskaya et al., Laser-induced heterometallic phase deposition from solutions of supramolecular complexes. Surf. Coatings Technol. **206**, 3454–3458 (2012). https://doi.org/10.1016/j.surfcoat.2012.02.010
19. A.A. Manshina, E.V. Grachova, A.V. Povolotskiy et al., Laser-induced transformation of supramolecular complexes: approach to controlled formation of hybrid multi-yolk-shell Au-Ag@a-C: H nanostructures. Sci. Rep. **5**, 12027 (2015). https://doi.org/10.1038/srep12027
20. A.A. Manshina, T. Ivanova, A. Povolotskiy, Laser-induced deposition of hetero-metallic structures from liquid phase. Laser Phys. **20**, 1532–1536 (2010). https://doi.org/10.1134/S1054660X10110162
21. A.A. Manshina, A.V. Povolotskiy, T.Y. Ivanova et al., Effect of salt precursor on laser-assisted copper deposition. Appl. Phys. A **89**, 755–759 (2007). https://doi.org/10.1007/s00339-007-4164-9
22. A.A. Manshina, A.V. Povolotskiy, T.Y. Ivanova et al., Laser-induced copper deposition on the surface of an oxide glass from an electrolyte solution. Glas. Phys. Chem. **33**, 209–213 (2007). https://doi.org/10.1134/s1087659607030030
23. A. Manshina, A. Povolotskiy, T. Ivanova et al., Laser-assisted metal deposition from $CuSO_4$-based electrolyte solution. Laser Phys. Lett. **4**, 163–167 (2007). https://doi.org/10.1002/lapl.200610090
24. A. Manshina, A. Povolotskiy, T. Ivanova et al., $CuCl_2$-based liquid electrolyte precursor for laser-induced metal deposition. Laser Phys. Lett. **4**, 242–246 (2007). https://doi.org/10.1002/lapl.200610093
25. A. Manshina, A. Povolotskiy, A. Povolotckaia et al., Annealing effect: controlled modification of the structure, composition and plasmon resonance of hybrid Au-Ag/C nanostructures. Appl. Surf. Sci. **353**, 11–16 (2015). https://doi.org/10.1016/j.apsusc.2015.06.048
26. A.V. Povolotskaya, A.V. Povolotskiy, A.A. Manshina, Hybrid nanostructures: synthesis, morphology and functional properties. Russ. Chem. Rev. **84**, 579–600 (2015). https://doi.org/10.1070/RCR4487
27. A. Povolotskiy, A. Povolotckaia, Y. Petrov et al., Laser-induced synthesis of metallic silver-gold nanoparticles encapsulated in carbon nanospheres for surface-enhanced Raman spectroscopy and toxins detection. Appl. Phys. Lett. **103**, 113102 (2013). https://doi.org/10.1063/1.4820841
28. C. Sanchez, P. Belleville, M. Popall, L. Nicole, Applications of advanced hybrid organic-inorganic nanomaterials: from laboratory to market. Chem. Soc. Rev. **40**, 696–753 (2011). https://doi.org/10.1039/c0cs00136h
29. E. Sappa, A. Tiripicchio, P. Braunstein, Selective metal-ligand interactions in hetero-metallic transition metal clusters. Coord. Chem. Rev. **65**, 219–284 (1985). https://doi.org/10.1016/0010-8545(85)85026-8
30. S. Schlicht, A. Kireev, A. Vasileva et al., A model electrode of well-defined geometry prepared by direct laser-induced decoration of nanoporous templates with Au-Ag@C nanoparticles. Nanotechnology **28**, 65405 (2017). https://doi.org/10.1088/1361-6528/aa536a
31. S.E. Shin, H.J. Choi, J.Y. Hwang, D.H. Bae, Strengthening behavior of carbon/metal nanocomposites. Sci. Rep. **5**, 16114 (2015). https://doi.org/10.1038/srep16114

32. S. Tang, S. Vongehr, X. Meng, Carbon spheres with controllable silver nanoparticle doping. J. Phys. Chem. C **114**, 977–982 (2010). https://doi.org/10.1021/jp9102492
33. W. Wang, S.A. Asher, Photochemical incorporation of silver quantum dots in monodisperse silica colloids for photonic crystal applications. J. Am. Chem. Soc. **123**, 12528–12535 (2001). https://doi.org/10.1021/ja011262j
34. M. Wehner, F. Legewie, B. Theisen, E. Beyer, Direct writing of gold and copper lines from solutions. Appl. Surf. Sci. **106**, 406–411 (1996). https://doi.org/10.1016/S0169-4332(96)00388-1
35. C. Wu, J. Maier, Y. Yu, Generalizable synthesis of metal-sulfides/carbon hybrids with multiscale, hierarchically ordered structures as advanced electrodes for lithium storage. Adv. Mater. **28**, 174–180 (2016). https://doi.org/10.1002/adma.201503969

Chapter 21
Crystal Lattice Defects as Natural Light Emitting Nanostructures in Semiconductors

Oleg Vyvenko and Anton Bondarenko

Abstract The review summarizes previous and very recent data on the luminescent properties of natural low-dimensional nanostructures in tetrahedrally coordinated semiconductors which are two particular types of extended crystal lattice defects: stacking faults (SF) and dislocations. Experimental data obtained in diamond-like, zink-blend and wurtzite lattice structures revealed intrinsic luminescence bands with the specific for given material properties for both types of the defects. The data are discussed in the framework of suggested theoretical models and non-solved issues of the interpretation are underlined. An unexpected intrinsic luminescence of quantum barriers formed by SF in sphalerite crystals caused by its dipole moment due to spontaneous polarization in wurtzite phase SF of the width even as narrow as dislocation dissociation width as well as the presence of quantum well formed by SF in wurtzite type crystals are suggested to be included for the revision of the previous interpretation of the dislocation-related luminescence mechanisms. The possibility of usage of the extended defects as active elements of light-emitting devices is considered.

21.1 Introduction

Low-dimensional structures such as quantum wells (QW), quantum wires (QWR), and quantum dots (QD) are the basis of modern semiconductor photonics. Of particular interest are zero-dimensional (0D) and one-dimensional (1D) systems due to singularities in their N (E) state density, which leads to an enhancement of probability of carrier recombination and, consequently, of the efficiency of light-emitting devices (LED) and a lowering of the lasing threshold. In addition, one-dimensional systems are also interesting from the point of view of the phenomenon of quasi-one-dimensional conductivity, which can find application in the development of future

O. Vyvenko (✉) · A. Bondarenko
Institute of Physics, St. Petersburg State University, Uljanovskaja 1, 198504 St. Petersburg, Petrodvoretz, Russia
e-mail: vyvenko@nano.spbu.ru

A. Bondarenko
e-mail: anton.bondarenko@spbu.ru

K. Yamanouchi et al. (eds.), *Progress in Photon Science*, Springer Series in Chemical Physics 119, https://doi.org/10.1007/978-3-030-05974-3_21

generations of high-performance field-effect transistors. Today, low-dimensional structures are manufactured mainly using epitaxial growth methods, which make it possible to obtain very efficient light-emitting semiconductor devices based on QWs and QDs [1, 2]. The technology of growth of such structures is based on layer-by-layer growth and self-assembly for QDs has reached the industrial level. The fabrication of quasi-1D structures needs to include additional technological steps and is far more complicated. Several attempts to develop the methods of the growth of quasi-1D structures were reported previously but no one brought their desirable quality. The etching of quantum wires through a lithographic pattern from grown QW suffers from the heterogeneity of the chemical composition over the QW area, which may not be noticeable when working with a 2D system, but it critically affects the properties of QWR. The epitaxial overgrowing of pre-prepared grooves, using the segregation of atoms of some elements with higher mobility, such as In and Al, at sites of local minima of potential energy [3] is actively developing but a rather low density of the structures were obtained so far. The T-shaped QWRs obtained by overgrowth on the cleavage of a structure with a QW with another QW [4] have demonstrated good optical properties in laboratory studies but this approach is very inefficient economically for industrial use. Finally, the direct epitaxial growth of nanowires (NW) [5] enables to grow their high density array with well controlled chemical composition but a variation of NW diameters commonly occurs and there is a lack of technology to reliably manipulate grown NWs for the device manufacturing.

In this review, an alternative approach to creating low-dimensional structures is considered, based on the use of intrinsic extended defects in crystalline semiconductors: dislocations and stacking faults (SF). The main advantage of this approach is the absence uncontrolled temporal and spatial variation of chemical composition which are unavoidable by the epitaxial growth of heterostructures. The dislocation line inside the crystal breaks the translational symmetry of the crystal near its core forming a particular atomic configuration. In the most tetrahedrally coordinated semiconductors the core consists of a composition of packing defect ribbon, staking fault (SF), and two bounded partial dislocations (PD) that has been experimentally established by high resolution transmission electron microscopy (HR-TEM) [6] and that coincides well with predicted by the density functional theory (DFT) methods [7]. Translational symmetry and the long-acting attracting potential of the field of mechanical stresses are the reason for the formation of 1D bands of allowed energy states, which turns a dislocation into a 1D QWR natural for any crystal [8]. The properties of dislocations in semiconductors, such as the increased mobility of carriers [9] and dislocation luminescence [10, 11], which are most demanded from the point of view of applications, are due precisely to the presence of these 1D bands. Besides, the dislocations can form nodes at the points of intersection of dislocations that can be considered as 0D quantum dots.

21.2 One- and Two-Dimensional Defects in Semiconductors with Face Centered Cubic (FCC) and Hexagonal Closest Packed (HCP) Lattices

21.2.1 Structure and Energetics

Most semiconductors with cubic crystal system have either diamond cubic structure for elemental semiconductors (group IV semiconductors) or zinc-blende (ZB), or sphalerite, crystal structure for compound semiconductors (II–VI, III–V semiconductors) [12]. Both diamond and ZB crystal structures follow face-centered cubic (fcc) Bravais lattice with two atoms in its basis set (Fig. 21.1 a, b) at (0, 0, 0) and (¼, ¼, ¼). In diamond structure basis set atoms are of the same kind whereas in ZB they differ, which adds ionic component to covalent bonds. Each atom in diamond and ZB structures is four-fold coordinated by its neighbors forming a tetrahedra, which orientation makes two atoms in basis set distinguishable and opposite directions along<111>not equivalent, denying inversion symmetry operation.

Dislocations are defined as the line bounded the region of non-complete shear displacement of one part of the crystal with respect to another. Main characteristics of a dislocation are its direction and the direction and the value of the displacement called as Burgers vector. Allowed Burgers vectors for perfect dislocation in diamond

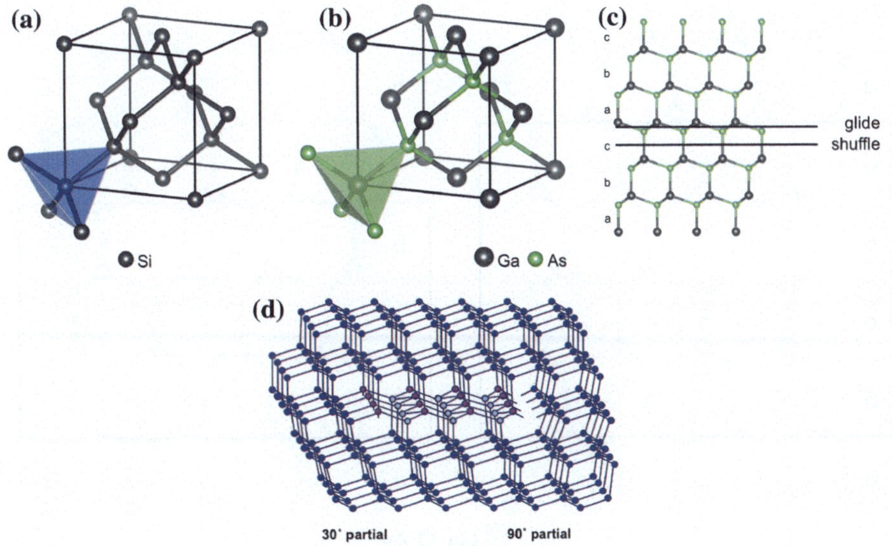

Fig. 21.1 Ball and stick model of diamond-like structure (**a**) and of zinc blende crystal structure (**b**). Tetrahedra show four-fold symmetry of bonds. Zinc blende stacking (**c**), atomic structure of the core of dissociated 60° dislocation in diamond [13] consisting of SF ribbon and bounded partials (**d**)

and ZB structures are ½<110> the same as for fcc lattice according to Frank's energy criterion [14]. For dislocations with Burgers vector ½<110> there are three possible slip planes: (100), (110), and (111). However, only the (111) slip plane was observed experimentally in fcc lattice. In (111) slip plane only two perfect dislocations are possible. Screw dislocation, and 60°-dislocation [15].

From the other hand, fcc lattice can be viewed as a stack of close-packed atomic layers in (111) planes. Atoms of consequent layers occupy one of three different sets of positions labeled a, b, and c. The perfect fcc lattice has stacking sequence of either abcabc or cbacba. The presence of additional layer in diamond and ZB stacking allows for two possible sets of slip planes. One is the glide set between close-spaced layers, the other is the shuffle set between the couples of the close-spaced ones. Transmission electron microscopy (TEM) experiments using weak beam technique [16] showed, that in diamond-like structures dislocations belong mostly to glide set and almost always are dissociated into partial dislocations bounding stacking fault (SF) ribbon (Fig. 21.1d) [17, 18].

SF is 2D defect, which arises when perfect lattice abcabc stacking is disturbed by either removal (intrinsic SF) or insertion (extrinsic SF) of a layer. In fcc lattice one may construct one intrinsic SF abcbca and one extrinsic SF abcbab (Fig. 21.2) [19]. If SF is surrounded by a perfect crystal from all sides it must be bound by 30° and 90° partial dislocations (left and right in Fig. 21.1d) [15, 17–19].

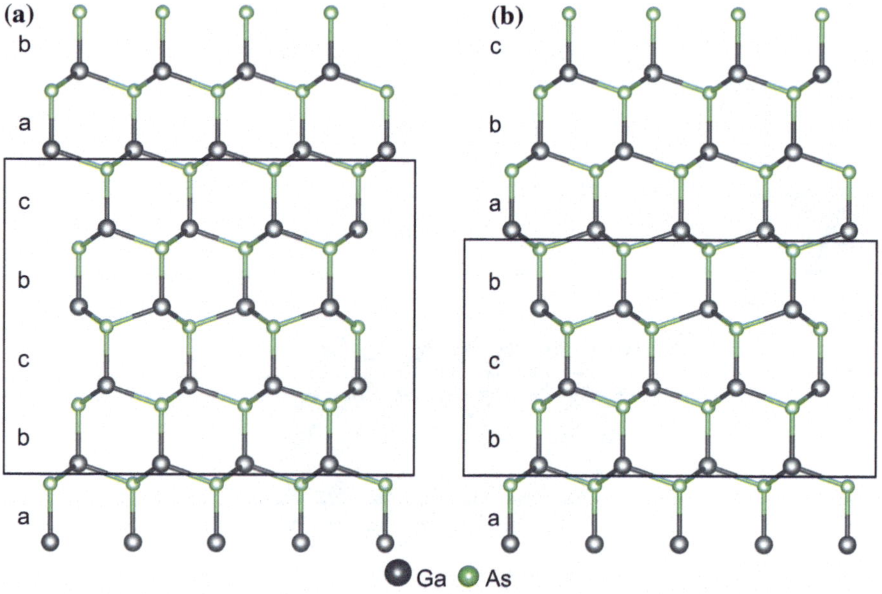

Fig. 21.2 Ball and stick model of GaAs zinc blende structure with intrinsic (**a**) and extrinsic (**b**) stacking faults. Projection along [110], vertical axis is [111]. Rectangle frames outline hexagonal stacking patterns bcbc

SF in diamond or ZB crystal structure can be considered as an inclusion of very thin hexagonal wurtzite phase bcbc-stacking layer (Fig. 21.2).

The wurtzite crystal structure (WZ) is a hexagonal type lattice containing atoms of two different elements. Two parameters a and c are only needed to completely describe WZ-type lattice, where a is side of the base of vertical prism of the primitive unite cell and c is the height of the prism. Atoms of each type form two hexagonal close-packed (HCP) sublattices positioned directly one upon another and separated in ideal WZ structure along the [0001] c-axis by 3/8c. As well as in ZB crystal structure each atom in WZ lattice is tetrahedrally coordinated, however, unlike in cubic ZB lattice the bonds do not have to be of the same length to keep the lattice symmetry. Actually, some binary semiconductors (almost all III-nitrides, except BN) always crystallize in WZ structure because of high difference in size of the atoms, which leads to formation of short bonds with large ionicity between atoms of close spaced layers in basal planes and longer bonds along vertical c-axis of the primitive unit cell [20, 21]. This makes c-axis a singular polar axis and is responsible for piezoelectricity known for many semiconductors with WZ crystal structure. Large ionic component also plays crucial role in performance of devices based on WZ-structure semiconductors due to emergence of large internal electric fields caused by spontaneous polarization [22].

In perfect WZ crystal structure stacking sequence along the c-axis has abab pattern. As it was first analyzed by Frank and Nicholas in [19] HCP lattice may have three types of stacking faults shown in Fig. 21.3. These are two intrinsic stacking faults I_1 and I_2, and extrinsic stacking fault E, which can be viewed as the inclusions of ZB-layers with abcabc stacking pattern. Each SF independently on its type must be bound by partial dislocations in (0001) plane unless it is coming out to the surface of the crystal.

In WZ-type lattice the shortest translation vector is primitive unit cell vector a 1/3 [−12−10], which allows four perfect dislocations in the (0001) plane [23]. These are 0° screw (a-screw) dislocation along [−12−10], 90° edge dislocation along [−1010], 60° dislocation along [−2110], and 30° dislocation along [−1100]. Each of these dislocations can dissociate into partials by the reactions described in [24]. Besides there is perfect, non-dissociated, prismatic 90° edge dislocation along [0001].

21.2.2 Dislocations as Quantum Conducting Wires

The dislocation electronic bound states were studied both experimentally and theoretically because they determine electrical activity of dislocations. Density functional theory (DFT) calculations [25, 26] predict that the cores of dislocations are heavily reconstructed, which lead to absence of any dangling bonds shown in Fig. 21.1d. This explains why DFT calculations do not predict any dislocation-related deep levels in the band gap for many materials studied, but only shallow bound states near to the band edges. These shallow states originate from long-range elastic strain fields, self-consistently accounted for by DFT methods. However, DFT methods are

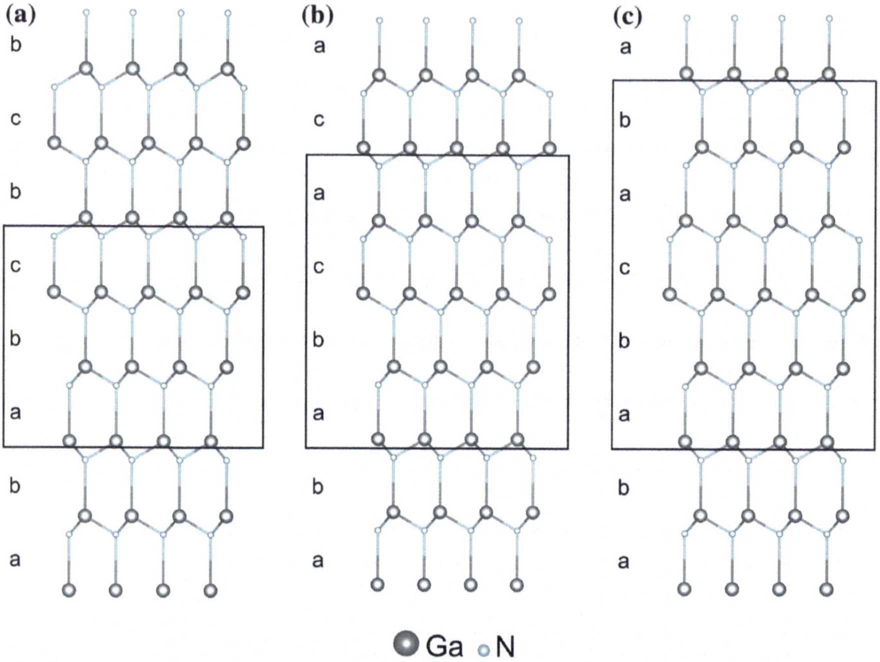

Fig. 21.3 Possible types of stacking faults in wurtzite crystal structure. Intrinsic stacking faults I_1 (**a**) and I_2 (**b**), and extrinsic stacking fault E. Projection along [10–10] vertical axis along wurtzite c-axis [0001]

only precise in calculation of total energy of the supercell and cannot provide exact energy positions for shallow bound levels in the band gap. To overcome this limitation some efforts were undertaken to calculate the shallow bounds states of dislocation deformation potential using the envelope function technique in the effective mass approximation [8, 27, 28]. Using this approach it was shown [8] that for many diamond and ZB semiconductors the deformation potential is responsible for formation of 1D bands for both electrons and holes with binding energy of about 50–80 meV.

1D dislocation bands must be responsible for enhanced conductivity of dislocations observed in many experiments [9, 29–32]. In CdS 1D-like enhanced conductivity of screw dislocations was detected by direct current measurement [29]. In Ge 1D conductivity of 60° dislocations was well established by both DC and AC measurements [30, 31]. In Si enhanced conductivity of dislocations was observed only after the hydrogenation procedure [30], which is known to passivate deep levels [33, 34]. However, in GaAs no evidence of dislocation conductivity was found for both screw and 60° dislocations [31]. Enhanced conductivity of threading dislocations in GaN was observed by means of scanning spread resistance measurement [32].

The development of direct wafer bonding technique [35] allowed for the experimental investigation of charge carrier transport mechanisms of individual dislocation in Si by preparation of tunnel field effect transistor with defined number and type of dislocations [9]. It was demonstrated that individual dislocations form additional tunneling paths result in formation of only one or a few bands in the band gap.

21.2.3 Stacking Faults as Quantum Wells

Electronic states of a SF can originate from bounding partial dislocations or from the fact, that stacking fault is actually a thin layer of another crystalline phase of the same material. Theoretical investigations [36] have shown that in Si stacking fault gives level with energy of about Ev + 0.1 eV, which depend on SF width. Later calculations [37] showed that this energy level does not depend on the structure of bounding dislocation but on the overlap of the wave functions of SF states, which increase when partials are close together. The energy of SF states in Si was predicted in the range Ev + 0.075 eV for infinite SF width and Ev + 0.21 eV for SF ribbon width of 2.3 nm.

In compound semiconductors ZB and WZ phases possess different band gaps. In all known cases band gap of WZ exceeds that of ZB as it is summarized in Table 21.1.

As the result SF in WZ phase being an insertion of ZB one forms QW while SF in ZB forms a barrier at least for one type of charged carriers.

Exact alignment of band structure is unknown in many cases. A good example is WZ GaN where luminescence measurement on SF-rich samples revealed new bands which allowed for development of theoretical model of intrinsic SF in GaN as a compressed by dislocation deformation potential cubic phase inclusion [39]. The model predicts a quantum well for electrons $\Delta Ec = 0.122$ eV and a barrier for holes $\Delta Ev = 0.062$ eV for I_2-type SF. Later, this model was enhanced in [22, 40] by accounting for the electric field of spontaneous polarization of WZ GaN, which allowed from one hand to explain the peculiarities of SF-luminescence and from the other hand to estimate spontaneous polarization in GaN (for more details see Sect. 21.6).

Table 21.1 Difference of the band gap energy in eV between WZ and ZB crystal structure for common binary semiconductors (see [38] and references therein)

	WZ	ZB	ΔE (WZ − ZB)
GaN	3.4	3.2	0.2
CdS	2.5	2.37	0.13
ZnS	3.77	3.68	0.09
SiC-2H-6H	2.9–3.3	2.3	0.6–1.0
ZnSe	2.8	2.7	0.1
GaAs	1.46	1.42	0.035–0.055

21.3 Dislocation-Related Luminescence in Ge and Si

Historically, dislocation related luminescence (DRL) was found firstly in Ge [41] in 1957 and nearly 20 years later in Si [10]. Afterwards, a lot of the papers were published until the very recent time. More detailed reviews about the state-of-the-art status of experimental data and of the mechanism understanding can be found elsewhere [42–46].

The most results on DRL in Si were obtained on the samples with dislocations introduced by plastic deformation though similar data were reported for misfit dislocations at heterointerfaces [47], and more recently, on dislocation networks produced by silicon wafer bonding [48, 49].

It was discovered that the DRL spectra (4a) consisted of a set of narrow lines in both materials. In silicon four main lines denoted as D1–D4 in the ranges 0.8–1.0 eV were divided in two groups D1–D2 and D3–D4 with well distinguished properties [51]. D3–D4 lines could be observed only at very low temperature while D1–D2 remained rather intensive even at room temperature. Cathodoluminescence (CL) with a high spatial resolution revealed that D3–D4 stemmed from dislocation lines whereas D1 in the areas between them [47]. Besides, a strong enhancement of D1 DRL due to light contamination with transition metals was established [47, 52]. The thermal stability of D1 DRL gave the hope to use it for silicon based light emitter compatible with CMOS technology and numerous investigations were performed (for recent review see [45]) and examples of D1 DRL based device fabrication were described [33, 53]. Nevertheless, an exact knowledge about the origin and mechanism of D1 DRL is still absent.

Finally, it was established that that D3 line is the phonon replica of D4 and that the latter exhibited a fine structure that can appear by varying the deformation conditions, and disappear by the annealing at sufficiently high temperature [50, 54]. It is significant that the spectral positions of the observed D4 DRL fine structure components practically never changed (Fig. 21.4b). The individual lines were numbered with integers [50] and it was postulated that radiative transitions of pairs of point centers participate in the processes, and the distance between the centers in such pairs was proposed to be defined by the SF ribbon width in the core of dissociated dislocations being the discrete parameter influencing the spectral frequency of the radiation.

The thorough investigations of the dislocation photoluminescence in Ge crystals in [44, 55] led to a more detailed picture of the origin of the spectral series (see Fig. 21.5). The importance of the dissociation of dislocations into partial dislocations (or partials) separated by stacking fault with a set of discrete widths was revealed similarly to the DRL in silicon. Actual direct experimental evidence of this was obtained in [55] by applying a specially oriented additional load, which caused either an increase or a decrease in the dissociation of the dislocations producing mostly short-wavelength or long-wavelength regions of the dislocation photoluminescence spectrum, respectively.

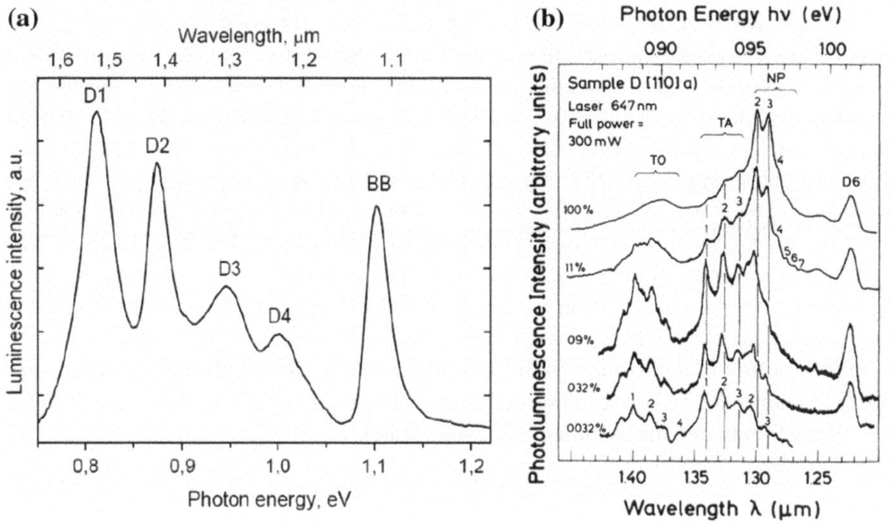

Fig. 21.4 **a** Four main lines of dislocation-related luminescence in Si [10], and **b** excitation dependence of a DRL spectrum as typical of all of the two-stage deformed Si samples. Reproduced with permission from [50]. Copyright (1986) by the American Physical Society

Fig. 21.5 Photoluminescence spectrum of Ge sample with the dislocation density $ND = 3 \times 10^5\ \mathrm{cm}^{-2}$: (1) experimental points, (2) total approximating curve, (3) Gaussian lines at energies $En \leq 0.55$ eV, (4) Gaussian lines at energies $Em > 0.55$ eV, and (5) the $d8$ line, from [44]

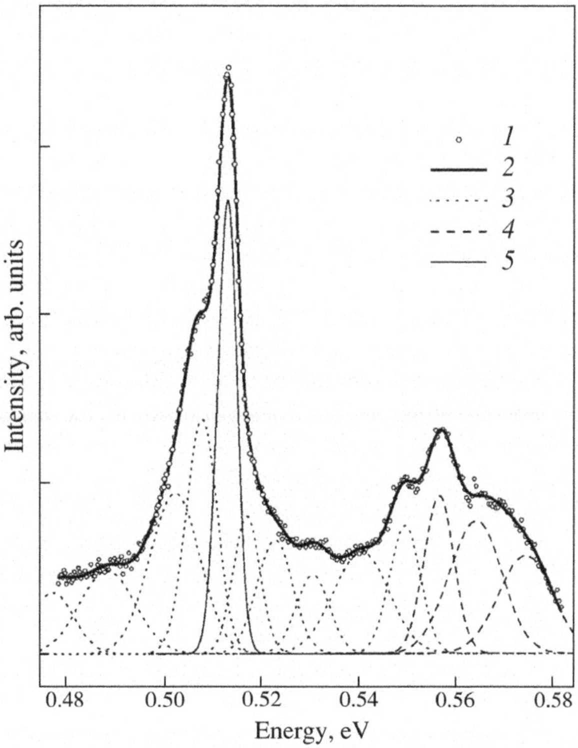

Though different empirical expression [50, 55] to describe the position of the spectral components and different shapes of the dislocation potential were considered [56, 57] a consensus about the DRL mechanism was established as the recombination of excitons bound to the shallow deformation related electronic states 90° partial dislocation. The localization energy of exciton E_l defined as the difference between the band gap energy and DRL exciton band was found of the order 100 meV for both Ge and Si in accordance with the sum of the shallow hole and electron states of dislocation [8, 58] but the exciton binding energy was as small as few meV that was ascribed to lowering of effective mass due to 1D character of the electronic system. The impact of the deformation potential of closely spaced 30° partial dislocation was considered as a small perturbation that varies the energy of the mentioned states being a function of the distance between the partials (or SF width) that can take only discrete values. SF itself was assumed neither has own electronic states, nor exhibit any impact on dislocation-related electronic states.

21.4 Dislocation Related Luminescence in ZB Semiconductors

DRL was also found in many monocrystalline ZB lattice II-VI compound semiconductors. The most detailed information about that was obtained for ZnSe [59]. First report about that was published by Dean [60] who found an unusually narrow line labelled as Y in a defect-rich material. In the following it was established that during the growth of ZnSe layers on (001) GaAs substrates by molecular beam epitaxy (MBE) Se(g)-type misfit dislocations first nucleate in pseudomorphic films of high quality above some critical thickness. The individual segments of the Se(g) misfit dislocation appears to be the only local emission centers responsible for Y-line luminescence.

In the insert of Fig. 21.6a the panchromatic CL map of ZnSe/GaAs (100) for a 165 nm epitaxial layer is displayed showing bright contrasts at Se(g) misfit dislocations. The CL contrasts lines are exactly in the [1–10] direction, which corresponds to the Se(g) subsystem misfit defect configuration. The CL emission is found to be dominantly along the [1–10] Se(g) misfit dislocations. The segments of the misfit segments are therefore the active radiative recombination centers. The panchromatic CL signal of misfit segments was found to exceed the matrix CL emission intensity by almost one order of magnitude.

The low energy spectrum displayed in Fig. 21.6a corresponds to Y luminescence line and its first LO phonon replica originating from the segments of Se(g) dislocation.

In the Y luminescence spectrum in Fig. 21.6a besides the main peak at 2.612 eV a number of weaker additional peaks can be resolved at both high and low energy sides. The fine structure of the Y line appears to be of a serial nature as it is shown in Fig. 21.6b [61]. The Y luminescence peak series has been explained by a model based on the assumption of 60° Se(g) dislocation segments to be dissociated into

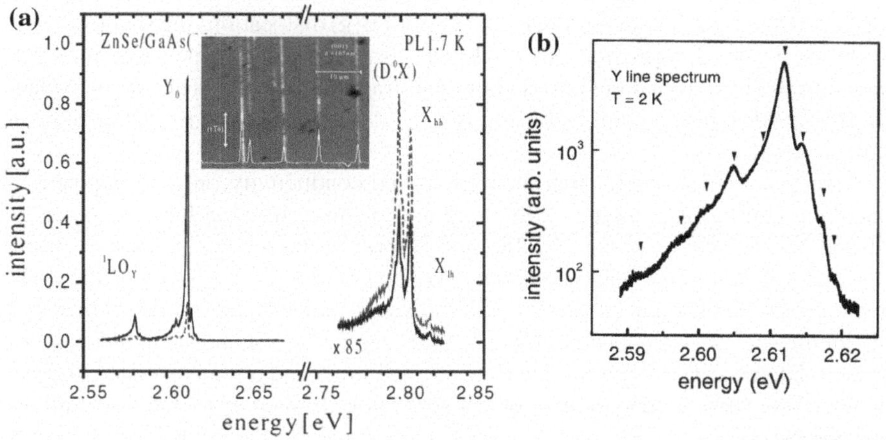

Fig. 21.6 **a** ZnSe/GaAs (001) polarization-dependent PL spectrum. In the insert the panchromatic SEM-CL map (10 kV, 10 K) exhibit bright CL contrasts at Se(g)-type misfit dislocations. The defects are radiative recombination centers as evidenced by the CL intensity profile. Reproduced with permission from [59]. Copyright (2000) IOP Publishing. **b** The fine structure of the Y-line spectrum appears to be of a serial nature. Reproduced with permission from [61]. Copyright (1999) Elsevier

30° and 90° Shockley partials. Nucleation of dissociated Se(g) misfit dislocations has also been verified by TEM examinations of ZnSe/GaAs (001) heterostructures [62]. The CL maps presented in the insert of Fig. 21.6a taken with spectral resolution revealed that the spectral position of Y line varied along the dislocation lines. That was explained to be due to variations of the separation between the partials in the frame of the model described above for DRL in Si and Ge and served an additional confirmation of the validity of this model.

The high energy PL spectrum shown in Fig. 21.6a is dominated by the recombination of the free and donor-bound exciton. The emission from the layer matrix and defect-induced luminescence (Y0) show the polarization dependence: Y luminescence is dominant along the dislocation lines, whereas the free exciton emission is dominant along perpendicular to them directions. Optical polarization parallel to the dislocation line direction is expected by assuming a one dimensional carrier potential [63]. This explained a low value Hyang-Rhys factor ($S = 0.1$) which was previously established to be fingerprint of low-dim system [64]. The high value derived for the defect bound exciton energy of 27 meV [63] confirmed this assumption as well.

It was also found that dissociated screw dislocation segments did not exhibit any luminescence. Thus, the 30° partial dislocations have been excluded as radiative recombination centers. This gave an experimental evidence that the 90° partial dislocation is the most favorable candidate for bound exciton states responsible for Y luminescence.

DRL due to Te(g) but not due to Cd(g) dislocations was found in CdTe [65]. Thus, the dislocations in the corresponding A(g)-subsystems show defect-induced non-radiative recombination properties. For both B(g) dislocations the shallow levels produced by the deformation potential were assumed to be responsible for DRL optical transitions. By taking into account the strain field only, it is unable to explain the obvious different recombination activity of A(g)- and B(g)-type dislocations in II/VI materials.

ZnSe is a semiconductor that exhibits n-type conductivity only. In opposite to that ZnTe can be only p-type. Two sharp and intense emission bands denoted as Y1 and Y2 with the maxima of 2.185 and 2.150 eV and a weak phonon coupling were investigated in the heteroepitaxial ZnTe layers in [66]. The study of the ZnTe layers grown on different substrates with thicknesses of 0.5–3.2 μm has shown that this luminescence is produced in the interface region that contains a high density of structural defects. TEM [67] revealed an array of Lomer and 60° misfit dislocations at the interface. Also, a small number of stacking faults, limited by partial dislocations, was observed.

The temperature dependence of these band spectral positions revealed the excitonic character of the recombination. Besides, a fine structure of Y1 and Yz bands was observed. In samples with a high intensity of Y1 and Y2 bands it is possible to distinguish LO-phonon replicas, with the values of the Huang-Rhys factor ($S<0.01$ for Y1 bands and $S = 0.2$ for Y2 bands) that are similar values obtained for DRL in other materials reviewed above.

The luminescence intensity of the Y bands decreases rapidly when the sample warms up and above 80 K these bands disappeared completely. Between 2 and 30 K, the estimated activation energy is about 8 meV, while above 30 K it is about 120 meV. The first activation energy is very close to the binding energy of free exciton in ZnTe (~12 meV), The second one is similar to the localization energy for the hole at acceptor-like centers (~70–100 meV). The first process was ascribed to thermal dissociation of the bound exciton, which liberates the electron, and the second one is due to the activation of the hole bound to a dislocation [66].

Though no direct evidence of dislocation origin of Y band was presented the similarity of the properties of Y band in ZnTe allows us to ascribe this band origin to the dissociated misfit dislocations.

To summarize the properties of DRL band in sphalerite type semiconductors one has to note that the Huang-Rhys factor S is rather small reflecting low-dimensional character of the bound exciton. On the other side the localization energy E_l of DRL band is rather big for all materials (for CdTe ($S = 0.2$, $E = 130$ meV) [65], for ZnSe ($S = 0.2$, $E = 220$ meV) [59] and for ZnTe ($S = <0.1$, $E = 210$ meV for Y1 and $S = 0.2$, $E = 250$ meV for Y2) [66]) and seems to exceed the sum of the energies of shallow electron and hole states due to dislocation deformation potential calculated theoretically [8] by the value that is larger than exciton binding energy being of the order not more than 30 meV.

One of the reason might be that in the above reviewed papers the role of SFs was considered as a separator between the partial dislocations only without taking into account their particular properties in compound semiconductors. That is why we include in this review two recent studies dealt with the SFs in ZnSe and GaAs.

21.5 One Dimensional Defect Formed by Intersection of the (111)-Stacking Fault and Quantum Well in ZnSe

Recently, the results of study of intersections of SF with QW was proposed to consider as the natural 1D system [68]. The samples were grown by molecular beam epitaxy (MBE) on a GaAs:Si (001) substrate covered by an epitaxial GaAs buffer layer of a 0.2 μm thickness. The heterostructures comprised a ZnSe QW with a thickness either 20 or 10 nm, embedded in $Zn_{0.84}Mg_{0.16}S_{0.12}Se_{0.88}$ barriers with thicknesses of 100 nm, and a 2-nm-thick ZnSe cap layer. This design excluded the interpretation of the narrow lines as related to SFs in a bulk material like in [69], because in this case they would emerge near band edge of the barrier material significantly higher in energy than the emission of the ZnSe QW. Besides, the composition of the solid alloy in the barriers was chosen to realize the pseudomorphical growth when the in-plane lattice parameters of GaAs and ZnMgSSe are equal excluding, thus, the formation of misfit dislocations described in the previous Section.

TEM investigations revealed long-extended SFs, with the length of about 10 μm intersecting the sample surface along the direction [1–10] and the short SF pairs which were perpendicular to the extended ones (see Fig. 21.7e). In this way, dislocation-free SF ribbons with the width of 20 or 10 nm and with the length of 100 and 10 μm served as the subject of the micro-photoluminescence investigations carried out at a temperature of 10 K.

New PL peaks were found at the positions of SFs whose were not observed far away from them and were ascribed to the excitons bound to the intersection of the QW and SFs. The lines were red shifted with respect to the free exciton peak by (22–28) meV for the wide QW and by (12–16) meV for the narrower one. The μ-PL spectra for the short SF and QW width of 20 nm are shown in Fig. 21.8 for two polarizations of the excitation. They possess the linewidth <0.5 meV, which suggests the pronounced exciton localization. The polarization degree of narrow lines obtained for the excitation power 10 μW is shown in Fig. 21.8 and exceeds 30%. For long SFs the maximum polarization degree is about 25% and the line was broader that.

As it was pointed above, the SF can be viewed as a two-monolayer-thick inset of a WZ crystal phase embedded in a zinc blende (ZB) structure (see Fig. 21.2). Since the band gap of bulk WZ ZnSe is larger than that of bulk ZB ZnSe, it seems surprising that the exciton can be bound to SFs. However, the WZ phase of ZnSe, being pyroelectric, should possess piezoelectric and spontaneous polarizations, similarly to III-nitride QW structures, which provide an internal electric field along the [111]-type direction. This leads to the offsets eU of the bands at the SF plane, as shown in Fig. 21.7b. Such a band profile was assumed to result in the formation of bound excitons with the electron and hole localized at opposite sides of SF.

DFT calculations presented in [68] in agreement with other recent results [70] predicted that the band gap in the WZ phase is about 100 meV larger than in the ZB phase. The estimated internal electric field in the WZ region was $F = 5 \times 10^5$ V/cm that is comparable to that in GaN [71] giving the value of $eU = 45$ meV and

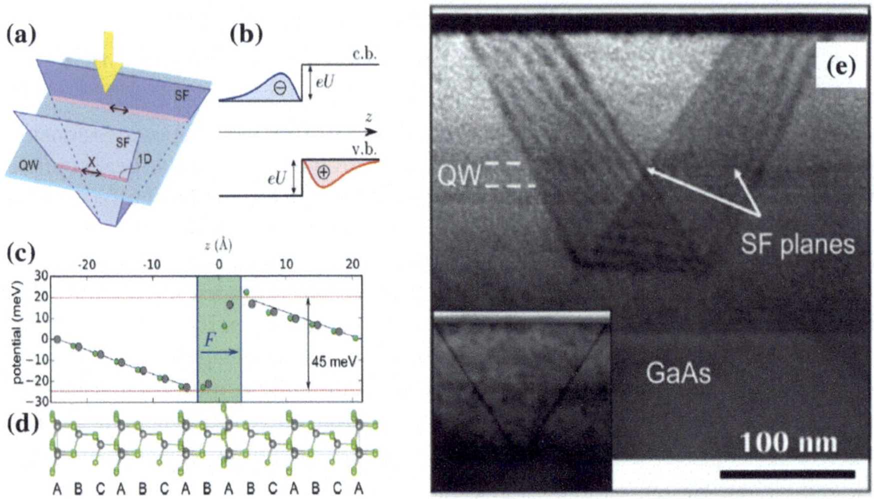

Fig. 21.7 **a** Sketch of SFs planes intersecting a QW. The incident light (yellow arrow) creates bound excitons (X) in this region, which are polarized predominantly along the 1D intersections. **b** Simplified band diagram of a SF in bulk ZnSe and the probability densities of electron and hole in a bound exciton. The band offsets, equal for the conduction and valence bands, are produced by an intrinsic electric field ***F*** within the SF region. **c** The electrostatic potential extracted from the DFT calculation and **d** elementary cell used in this calculation. **e** TEM image of the SFs originating in lower barrier. Reproduced with permission from [68]. Copyright (2018) John Wiley and Sons

Fig. 21.8 Polarization resolved μ-PL spectra after subtraction of a background related to the main excitonic peak. Reproduced with permission from [68]. Copyright (2018) John Wiley and Sons

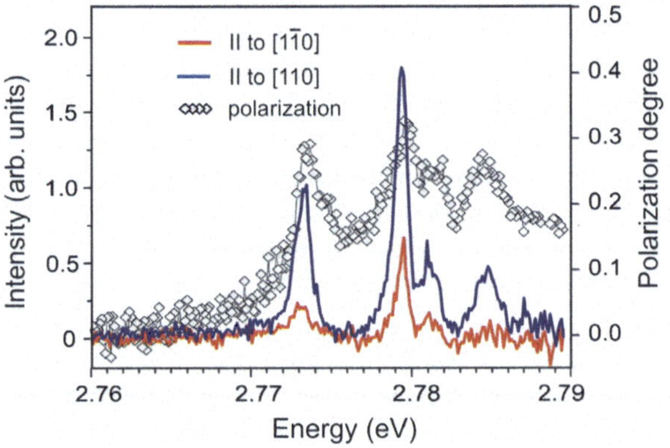

calculated exciton binding energy was very close the experimental value for SF of 20 nm width.

The width of the dislocation-free SF ribbon in [68] was 10–20 nm (as defined by the QW width) that is similar to the its width in the core of the dissociated dislocation. The direction of the polarization of the samples with the 1D intersections of the QW and SFs was also the same as for DRL in the ZnSe samples with misfit dislocations. Though the exact band diagram in the vicinity of SF remained unknown the existence

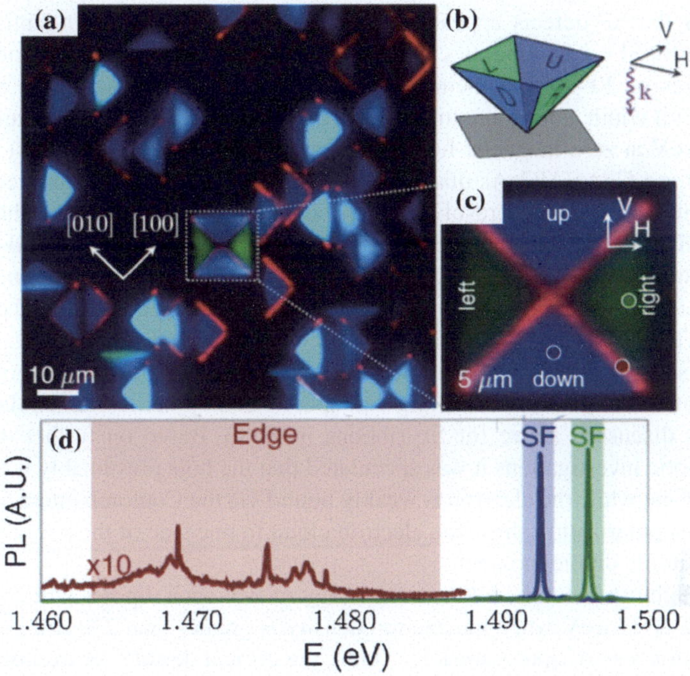

Fig. 21.9 **a** Confocal scan of SF structures. The image is formed by coloring emission in different wavelength bands as red, blue, or green, as depicted in (**d**). Excitation at 1.53 eV, 100 μW, and 1.9 K excite and collect H polarization [see **b**]. **b** Diagram of SF pyramid. The up, down, left, and right SFs are labeled, along with the H and V polarizations. **c** Detail of SF pyramid structure. **d** Low-power PL spectra at colored dots in **c**. Polarizations: blue, excite and collect H; green and red, excite and collect V. Broadband luminescence is observed from the SF edges (red). Reproduced with permission from [67]. Copyright (2016) by the American Physical Society

of additional attraction between the electrons and holes due dipole character of SF charge was demonstrated.

The importance of the spontaneous polarization was recently also demonstrated in GaAs [67] where giant value of the permanent dipole moment of two-dimensional excitons bound to a single SF was found.

SF structures were grown from the substrate epilayer interface during epitaxial growth. They were formed in a 10 μm GaAs layer MBE grown on 100 nm AlAs separated with a 5 nm/5 nm AlAs/GaAs superlattice with a semi-insulating (100) GaAs substrate. Two types of SFs were observed in the layer: a SF pyramid with four possible {111} surfaces and a SF pair defect. They nucleate near the substrate-epilayer interface during epitaxial growth.

Figure 21.9a represents confocal image of SF structures by coloring emission in different wavelength bands as red, blue, or green, as depicted in Fig. 21.9d taken at 1.9 K. Different color of two adjacent faces of a pyramid is due to different spectral line position in two perpendicular light polarization shown in Fig. 21.9b.

Most of the SF defects appear as single triangles, which were identified as a pair of nearby SFs. The binding energy of excitons to a pair of SFs depends varied in the ranges of 20–30 meV being defined by the distance between the SFs. The narrowest full width at half-maximum (FWHM) of the SF PL line in the sample was about 80 μeV at zero magnetic field, that is twice less than the narrowest reported linewidth for a GaAs/AlGaAs quantum well of 130 μeV [72]. This unprecedented homogeneity allowed one to resolve the SF–bound-exciton fine structure that helped to define an absolute value of the exciton dipole moment that was retrieved from the results of the investigations of magneto-Stark effect. The giant dipole moment value was estimated as an elemental charge times 10 nm whereas in conventional excitonic systems, typical electron-hole separations are on the order of several nm.

The existence of such dipole moment of a SF bound exciton was interpreted as a consequence of symmetry breaking and spontaneous polarization similar to that like it was discussed above for SF ribbons in ZnSe. Based on further details of magneto-optic investigations it was speculated that the hole presumably is localized in the SF plane while the electron is weakly bound via the Coulomb interaction. The spontaneous polarization shifts the electron cloud to one side of the SF, resulting in a giant excitonic dipole moment.

As a rough estimate, the interaction energy of two such dipoles will exceed the SF FWHM of 77 μeV when the exciton density is greater than 230 μm^{-2}. Using a wave-function size of approximately 10 nm, the critical density for exciton overlap in the 2D potential is 10,000 μm^{-2}. Therefore, the SF-bound exciton system could show sizable dipole-dipole interactions and may demonstrate coherent phenomena at reasonable exciton densities.

The dislocations bounded SFs which are expected to be of Lomer type in this case are seen in the PL-map in brown. Their DRL spectra denoted as "Edge" in Fig. 21.9e were red shifted with respect to free exciton by 40–60 meV. One can distinguish DRL broad bands with numerous narrow lines. The particular components of the fine spectral were not discussed in [67] but one may assume that the broad bands are due to the variation of distance between dislocations in one SF pyramid like it was discussed for DRL from dissociated dislocations in other material in the previous Sections.

Summarizing the results of this Section one may conclude that SFs in sphalerite type semiconductors possess a property that are unexpected for wider band gap inclusions, being rather quantum barrier than QW but, nevertheless, attractive to exciton making them as a perfect perspective object for further study of many electron system.

This property might be one of the reason why DRL in ZnSe and CdTe could be observed for B(g) but not A(g) dislocations since the SF polarity is the same for both types dislocations but shallow states of deformation potential of the edge dislocations are situated at opposite sides of SF. However, it is early to make definite conclusion and additional, in the first row theoretical, investigation are needed.

21.6 Dislocation-Related Luminescence in Wurtzite Type Semiconductors

21.6.1 Non-polar Lattice SiC

In wurtzite type semiconductors SFs are inclusions with a smaller band gap than the matrix (see Table 21.1). Though the exact band alignment is not known in many cases it is expected that SFs form quantum wells at least for one type of carrier. Dissociation width of dislocations of wurtzite type compounds with partly ionic bonding is usually rather small but for silicon carbide with pure covalent bonding the energy SF is so low that SF width reaches macroscopic sizes that enables to obtain the information about optical properties of SF and bounding partial dislocations by means spectrally resolved optical microscopy.

First of all, one has to mention that despite of a very high hardness of SiC the degradation in the active region was found by forward operating of 4H-SiC PiN diodes. The degradation was caused by spontaneous formation of planar defects identified as basal plane SF bounded by Shockley partial dislocations. Such phenomenon is known a recombination enhanced dislocation glide (REDG) is believed to be responsible for the lateral expansion of SFs. According to the conventional model [73] a part of the electron-hole recombination energy is redirected into nonradiative sites along the dislocation line to aid formation and migration of kinks, thus dramatically reducing the activation barrier for glide. Apart from this strong REDG effect, theoretical models interpret an SF in 4H-SiC as a 2D QW for the conduction band electrons [74]. Since entrapment of electrons in the QWs leads to reduction of the mean electronic energy, a faulted *n*-type crystal is supposed to be more stable than a perfect one. The concept of quasi-Fermi level was suggested [75] to explain the reversible expansion and contraction of SF regions in a degraded 4H-SiC PiN diode, or a virginal 4H-SiC crystal illuminated by above-bandgap laser irradiation. The expansion of SFs is the result of SF energy becoming negative when the quasi-Fermi level for electrons E_{nF} is raised above the SF energy level, thus lowering the total energy of the crystal.

As it was included in the first part of this review there are three types of SFs in wurtzite lattice, single layer (I1 or 1SF), double layer (I2 or 2SF) and E type. Among them only I2 possess glide, However, exclusively 1SF 1SFs expand in presence of the *e-h* plasma, whereas 2SF occur in highly *n*-doped material only This issue is beyond of the scope of this review and will not be discussed.

Figure 21.10 a, b represents spectrally selective electroluminescence (EL) imaging in the degraded 4H-SiC diodes taken at 510 nm (2.5 eV) and at 700 nm (1.8 eV). One can see that the first image represents planar-like defects while the second one consists of the lines only. The principal radiative channels are evident from the differential PL spectra plotted in Fig. 21.10c. The stacking faults demonstrate characteristic peaks associated with E_C–0.3 eV and E_C–0.6 eV split-off levels, which are consistent with first principles calculations for 1SF and 2SF related quantum wells [74]. It is interesting to note that partial dislocations bordering 1SF and 2SF type faults exhibit

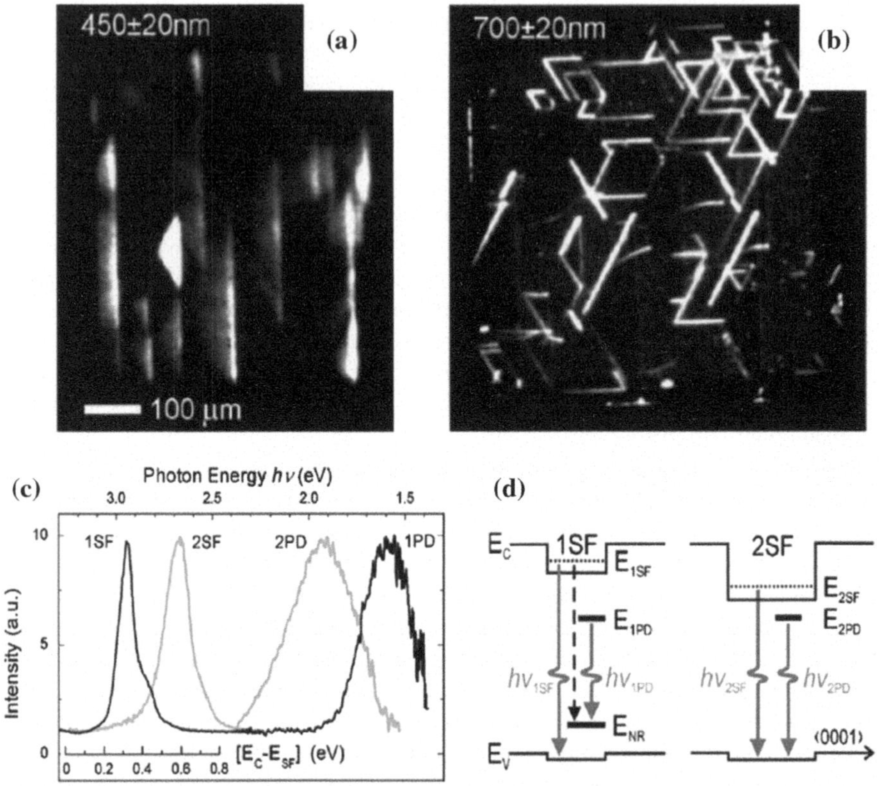

Fig. 21.10 **a**, **b** Spectrally selective plan-view EL imaging in a degraded 4H-SiC p-i-n diode. Defects of various types can be revealed by spectral filtering of EL emission: **a** only the stacking faults can be observed in a wavelength range of 450–620 nm; **b** the bounding partial dislocations and the bright spots of threading dislocations are revealed in the 700–620 nm range; **c** Spectral signatures of normalized PL at 300 K_ of single- _1SF_ and double-layer _2SF_ stacking faults and corresponding bounding partial dislocations, labeled 1PD and 2PD, respectively; **d** Schematic of the bandgap arrangement of 4H-SiC containing 2D quantum-wells formed by 1SF and 2SF in [0001] direction and enclosed by PDs lying along [11–20]. Nonradiative process aiding REDG is marked by a broken arrow; solid arrows indicate radiative transitions. Reproduced with permission from [76]. Copyright (2006) by the American Physical Society

clearly dissimilar spectral signatures corresponding to optical transitions of about 1.6 eV and 1.9 eV, respectively. It was established that luminescence is more intensive for moving Si(*g*) partials edges but a weak emission from the immobile C(*g*) edges also takes place [76]. The luminescence spectra obtained separately from the gliding and the stationary partials appear fairly similar, the latter exhibiting slightly broader high-energy shoulder.

Based on results of luminescence and deep level spectroscopy investigations a collective energy level diagram of recombination activity at 1SF and 2SF type faults was proposed [76] shown in Fig. 21.10d, which accounts for concomitant radiative and nonradiative processes at mobile Shockley partials as well as for radiative centers along stationary PDs bounding the in-grown SFs. The small valence-band discontinuity and phonon interaction do not notably change the model and are ignored for simplicity. As can be seen in the diagram, the characteristic PL of the PDs bounding 1SF and 2SF type faults then correspond to, respectively, optical transitions from the radiative center directly to the valence band ($h\nu_{2PD} = E_R \sim 1.9$ eV) and from E_R to the hole-trap E_T introduced by the Si(*g*) partial ($h\nu_{1PD} = E_R - E_T \sim 1.6$ eV). In contrast to electrically active and therefore highly mobile Si(*g*) partials, no midgap levels are introduced by the partials bounding the in-grown 2SFs, resulting both in blue shifted optical signature (towards 1.9 eV) 1234 and absence of REDG effect.

The origin of DRL of the partials in SiC in not established firmly so far. Recent investigation of DRL polarization dependence revealed some unexpected results [77]. In this work the monochromatic PL images were acquired through band-pass filters passing light of 1.64 eV for 30° -Si(g) and 6° PDs, light of 1.34 eV for 30° C(g) PDs, and light of 2.95 eV for SSFs.

It was found that the polarization of the PL from the for 30° -Si(g) and 6° PDs was distinctly polarized perpendicularly to the dislocation lines. This opposite to the DRL polarization in all other semiconductors reviewed above where it was parallel to dislocation lines. The only model that predicted non-parallel dislocation polarization was reported in [78]. It considers the anisotropic effective mass of holes and concluded that the polarization must be parallel to their Burgers vector, in disagreement with the polarization direction in the present results. It was suggested in [77] that the luminescence originates from some dislocation centers of intrinsic nature binding carriers with an anisotropic wave function.

The PL from for 30° C(g) PDs was not polarized and not uniform along the dislocation line but was interrupted at places on the dislocation lines. One might explain these two facts by assuming either that the luminescence originates from impurities inhomogenously distributed along dislocation lines which are centers of radiative or non-radiative recombination.

21.6.2 Dislocation-Related Luminescence in Wurtzite Polar Lattice Compounds

DRL in polar wurtzite compounds was found and investigated in CdS and CdSe [79–86] as well as in GaN [87–93]. DRL in these materials arose from the freshly introduced basal screw dislocations (frequently called as a-screw dislocations) as opposite to cubic lattice semiconductors where 60° dislocations were the sources of DRL. Beside the listed materials, recently luminescence related to some extended defect was reported for ZnO [94] which characterized by a low localization energy of about 40 meV, narrow line and extremely low Huang-Rhys factor indicated about strong confinement of electron-hole pairs. However, the type of the responsible dis-

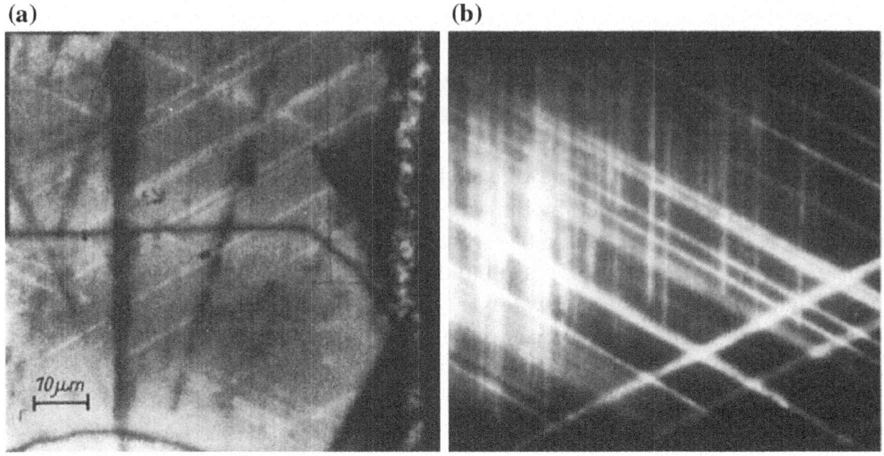

Fig. 21.11 **a** SEM-CL micrograph (20 kV, 80 K) [79] and **b** optical micrograph from (0001) surface of deformed CdS sample. The bright stripe shaped contrasts indicating a local increase of integral luminescence intensity stretched out in {1–210} directions. T = 77 K. Reproduced with permission from [83]. Copyright (1991) John Wiley and Sons

location, if any, remained unknown and the results of this study will not be discussed here.

The most extended information was obtained for CdS. The first reports about DRL in CdS was published in [79, 95] before DRL in silicon was discovered. The luminous dislocations could be viewed both by SEM CL and in optical microscope as they presented in Fig. 21.11a and 21.11b respectively.

The directions of the straight lumibous lines were {1-210} that unambiguously indentify them as a-screw dislocations. The luminous line were observed in plastically deformed crystal by uniaxial compression [80, 83, 84, 86], by identation [82, 83, 96] and by scratching [82] at room temperature. It was established that a-screw dislocations exhibited the same luminescent properties in both basal and prismatic glide systems and DRL could be observed below 150 K. Moreover, the motion of dislocation was observed even at liquid helium temperature [83, 84]. TEM study [97] revealed that the dislocations in basal glide system of all types are dissociated into partials with the SF ribbons of several nm and can glide in the dissociated state.

DRL spectrum taken from the area with multiple dislocations revealed a band of spectral lines shifted with respect to band gap by 100–120 meV. DRL fine structure consists of two doublet components which, in turn, split in two other under application of uniaxial elastic deformation as it is seen in Fig. 21.12. Detailed investigation of the evolution of the spectral positions of the doublet components allowed one to define the symmetry of the radiation center as C_s. It was concluded that the single screw dislocation has two kinds of such centers twofold degenerated in {1-210} plane which are situated immediately in the core of the dislocations. So, each of the splitting line

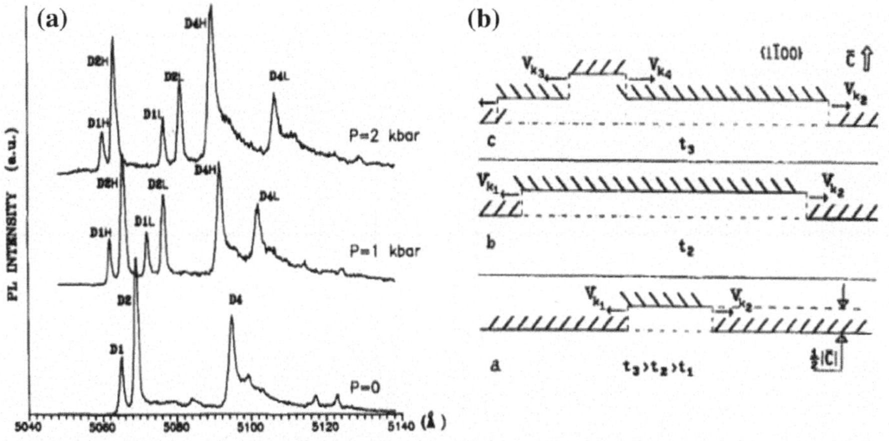

Fig. 21.12 **a** Photoluminescence spectra at T = 6 K showing the dislocation luminescence in CdS sample taken at different uniaxial pressure under an angle of 600 to the screw dislocations. **b** Schematic diagram of the space- and time-nonstationarity of the dislocation domains caused by the distribution alteration of one-dimensionally ordered Cs, centers in the screw dislocation core. Reproduced with permission from [85]. Copyright (1993) AIP publishing

components corresponds strictly to a certain type of screw dislocation—right or left with respect to their expansion direction.

In addition as it was noted in [84, 85] the moving screw dislocations observed in polarized light, with the electric-field vector E of the light forming an angle of 45–65° with the C axis, appeared to have a domain structure. As a result, screw dislocations that are continuous in unpolarized light appear in polarized light as separate scintillating segments or domains whose number can both increase and decrease with time in a discontinuous manner. When the vector E deviates from the direction indicated, the nonstationary nature of the DE becomes less noticeable. It appears that the domain structure depends on the temperature, the loading, and heavily on the optical excitation level. Also, it has been found that when a dislocation domain is quenched in one polarization it rises in the other, and vice versa. Thus, the DRL polarization directions in CdS was non-parallel to the dislocation lines and, so, different from DRL in cubic lattice materials.

The model of twofold oriented radiation center in the dislocation core proposed in [85] is schematically shown in Fig. 21.12b. According to this model, the nonstationary domain structure of screw dislocations is directly related to the reorientation of one-dimensionally ordered C, centers. Therefore, the glide of a screw dislocation segment along the C axis because of the formation and subsequent expansion of double kinks will be inevitably followed by the dislocation center reorientation. This model explained qualitatively the increase in domain scintillation frequency with increasing deformation temperature and applied stress. The strong effect of the optical excitation level on the nonstationary character of the domain structure may well be due to the stimulating action of laser illumination on the nucleation of

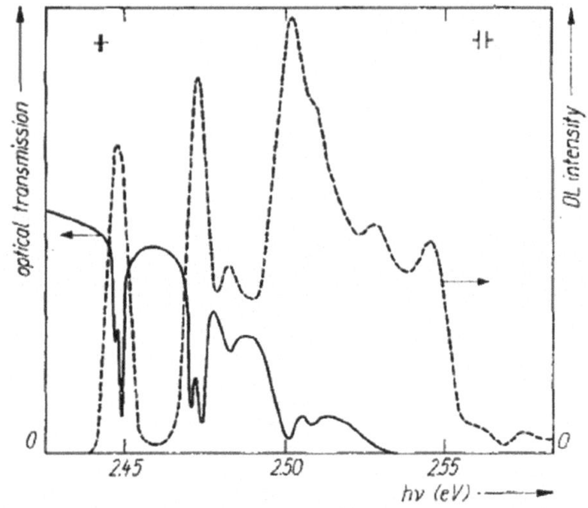

Fig. 21.13 Optical transmission spectrum (full line) and excitation spectrum of dislocation luminescence (hv = 2.45 eV, dashed line) at T = 4.2 K in plastically deformed CdS. Reproduced with permission from [80]. Copyright (1988) John Wiley and Sons

kinks from nonradiative processes on the dislocations similar to conventional REDG model.

One should underline that the model implicitly assumes that the screw dislocation has a perfect core structure as it glides in [0001] whereas according to the mentioned TEM data most of the observed moving dislocations are dissociated though in some cases of coalesce of partials into perfect core were reported as well [97]. The motion of the screw dislocation in C direction was also reported in [82, 98] where the dislocations in the near-surface region introduced by indentation or scratching of the basal surface disappeared completely after sample exposure at room temperature.

The mechanism for DRL of screw dislocation in CdS was discussed in [44] based on the coincidence of their energetic peak positions in absorption and excitation and luminescence spectra (Fig. 21.13). These facts together with a small FWHM of the optical dislocation bands lead to the conclusion about more probable the model of excitons bound to the metastable point defect complexes generated in plastically deformed samples.

Finally, screw dislocations were shown to accompany with an electrical level in the forbidden gap with binding energy of about 100 meV [98]. The level exhibits metastable behavior having an enthalpy for electron emission of 0.4 eV and a capture barrier of about 0.3 eV. One may not exclude that such center plays a role in DRL due the closeness of its binding energy to the DRL localization energy but no evidence for that is available up to now.

GaN crystals are grown mostly heteroexpitaxially for LED and HEMT production. This results in a high, more than 10^6 cm^{-2}, density of grown-in dislocations. That is why a few attempts to establish correlation between recombination properties of GaN and the presence of the dislocations have been made.

There are several reports of the investigations on a macroscale, mostly with PL, of the samples with various dislocation density. Shreter et al. [99] reports on correlation between the intensity of the luminescence bands around 3.4 eV at 11 K in MOCVD GaN on SiC with the density of threading dislocations. Two other luminescence bands Y_7 (3.2 eV) and Y_4 (3.35 eV) found in thin MBE grown GaN films and were related to edge-type threading dislocations in [100]. However, they were not registered in other samples with the same threading edge-type dislocations density [101]. This might be explained with an extrinsic nature of Y_4 and Y_7 responsible luminescence centers, i.e. with either some native point defect or impurity atoms segregated on edge-type threading dislocation [100]. The structure of dislocations introduced by bulk plastic deformation at 950 °C of HVPE GaN and their PL spectra were investigated by Yonenaga et al. [102]. It was reported that the dislocation glide destroyed yellow luminescence band with appearance of the new 1.79, 1.92, and 2.40 eV bands. Based on the TEM results and on the impact of the post-annealing on PL spectra the authors ascribed observed luminescent properties evolution to segregation of V_{Ga}-O_N complexes at prismatic edge dislocations. However, due to an unavoidably high initially dislocation density in as-grown materials the conclusions about the origin of dislocation-related luminescence in GaN made from the data on macro scale are arguable.

There are only limited number of the published works with direct observations of luminous dislocations.

CL study in TEM reveals a band at 3.29 eV of grown-in partial dislocations terminated basal SF [103]. Albrecht et al. [88] investigated the foils prepared from indented at 370 °C of (000–1) GaN using the same technique as in [103] and found an another broad band at 2.9 eV due to a-type 60° dislocations in basal plane while a-screw dislocations were reported to recombine non-radiatively. The 2.9 eV band was observed with a background of another intense broad luminescence band that hindered to give clear interpretation of its origin.

Recently, iron doped semi-insulating (SI) GaN with freshly introduced perfect a-screw dislocations was found to exhibit intense narrow luminescence line with the peak energy of about 3.35 eV at low temperatures [87]. The main DRL peak accompanied with the phonon replica (Fig. 21.14a).

Temperature and power dependences of dislocation-related luminescence (DRL) spectrum allowed to ascribe DRL to exciton bound at the dislocations. The intensity of the DRL decreased with the temperature with an activation energy of 16 meV and the peak could not be registered above 100 K. Huang-Rhys factor derived from the relation of integrated intensity of the zero-phonon peak and its first phonon replica gave extremely small value of $S = 0.017$. TEM study revealed perfect character of a-dislocation cores. Polarization-dependent measurements showed DRL polarization along the dislocation line direction. The intensity ratio of DRL peak measured for polarization along and perpendicular to the dislocation line was 8:1.

The obtained DRL localization energy of 140 meV is surprisingly large for perfect core dislocations owing only shear strain around it. The matter is that Γ point s-like conduction band minimum must be not sensitive to such kind of strains, and only quantum confined levels for the holes can be formed by the dislocations which are not deeper than 0.1 eV according simple single band model [8, 105]. In more sophisticated calculations Albrecht et al. [87] and later Belabbas et al. [104] demonstrated

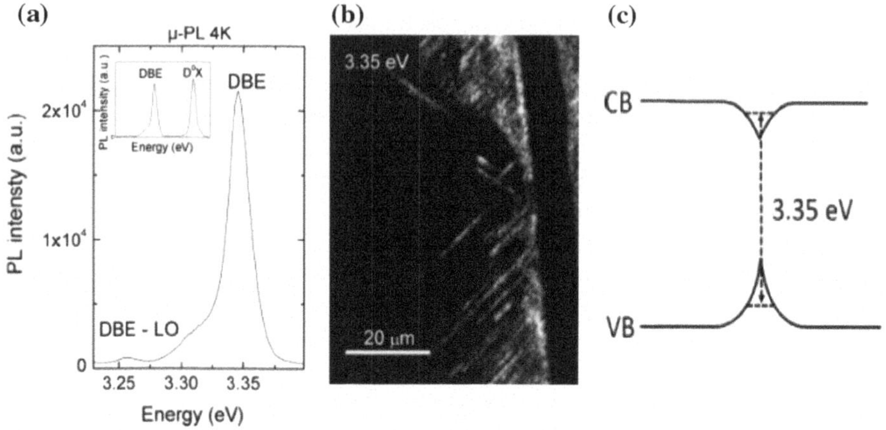

Fig. 21.14 **a** PL spectrum of a single a-type screw dislocation from a micro-photoluminescence map in semi-insulating GaN (0001) taken at 4 K. The inset shows the donor-bound exciton (D0X) and the dislocation-bound exciton in comparison. **b** Monochromatic cathodoluminescence map taken at an energy of 3.346 eV. Straight bright lines correspond to a-type screw dislocations aligned along <1–120>. Reproduced with permission from [87]. Copyright (2014) by the American Physical Society. **c** Band diagram at screw dislocation with DRL optical transition

the existence of the conduction band bending at screw dislocations in GaN arose as a result of the addition of the higher conduction bands into consideration. However, the energy of the electron quantum confined level was not reported, though the knowledge of its value is crucial to explain the observed exciton localization energy which is a sum electron and hole level energies like it is depicted in the diagram of DRL optical transition in Fig. 21.14c.

An intense luminescence band with a lower than in [87] energy of 3.15–3.18 eV at 70 K related to a-screw dislocations was observed independently by two other groups [89, 106]. The results of the investigation [89, 106] were obtained on dislocations introduced by nano- or micro-indentation or scratching of specially undoped low-ohmic GaN. Besides, it was found in [92] that DRL band 3.15–3.18 eV is frequently accompanied with the band of 3.33 eV at 70 K the position of which is rather close to DRL in semi-insulating GaN [87] but as it was shown in [92] has different origin.

Figure 21.15a represents panchromatic CL map near the scratch on c-plane reproduced from [92]. Small dashed line squares with numbers 1, 2, 3 mark the positions of the electron beam where the CL-spectra of Fig. 21.15b were acquired in spot mode. The spectrum in the point 1 (solid line in Fig. 21.15b) corresponds to well-known spectrum of as-grown GaN crystals with the most intense line of free exciton (FE) at 3.46–3.47 eV, and two other lines at 3.27 and 3.4 eV investigated in more details in [101].

Careful examination of the map recognizes two kinds of bright CL-contrasts clearly visible on the dark background. The first one is the straight stripes stretched-out in {12–10} directions being as long as up to 50 μm making a-screw dislocations

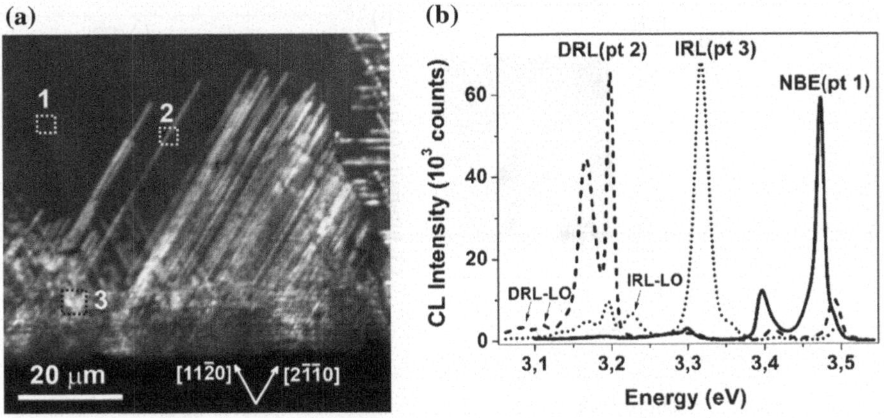

Fig. 21.15 **a** CL panchromatic image near the scratch (black bottom stripe) on c-plane of GaN; **b** CL spectra taken at the positions of electron beam marked in Fig. 21.15a as 1, 2, 3. T = 70 K. Reproduced with permission from [92]. Copyright John Wiley and Sons (2017)

as the most possible candidate for their origin. On the spectrum of the bright stripes (position 2 in Fig. 21.15a), shown as dashed line in Fig. 21.15b, an additional doublet of intense narrow lines at 3.1–3.2 eV is clearly visible. The spectral features are evidently a property of a-screw dislocations, i.e. DRL.

There are also two satellites red-shifted by 90 meV to the main doublet, the shift value that concise well with LO-phonon energy in GaN. A rough estimation of the intensity ratio of the phonon replicas to the ones of the main components from the data in Fig. 21.15b gives a rather low the Huang-Rhys factor $S < 0.1$. Full width at half maximum (FWHM) of main DRL components was reported to be close to that for FE (about 10–16 meV at 70 K). The temperature and strain variation of spectral position of DRL and FE coincided also very well [107]. It was also pointed out that the integrated intensity of DRL lines was significantly higher than FE's intensity far away from the scratch under the same excitation conditions [92]. All listed features gave arguments to interpret DRL in that spectral region as exciton bound at a-screw dislocations [93] whose spectral position in low-ohmic n-GaN differs from that for semi-insulating GaN reported in [87] due to different a-screw dislocation core structure in the materials of different conductivity level. CL spectra acquired from the regions near the indentation prick were identical for indentation of either basal or prismatic planes. This fact allowed one to conclude [93] that DRL spectrum belong only to straight a-screw dislocations gliding in both prismatic and basal plane, while prismatic edge dislocations as well as all non-straight, curvilinear dislocations in basal planes act as non-radiative recombination centers.

The second bright contrast type in Fig. 21.15a is bright spots that are mostly concentrated not so far away from the scratch. Dotted line in Fig. 21.15b is the spectrum taken in the electron beam position 3 of a bright spot in highly dislocated region.

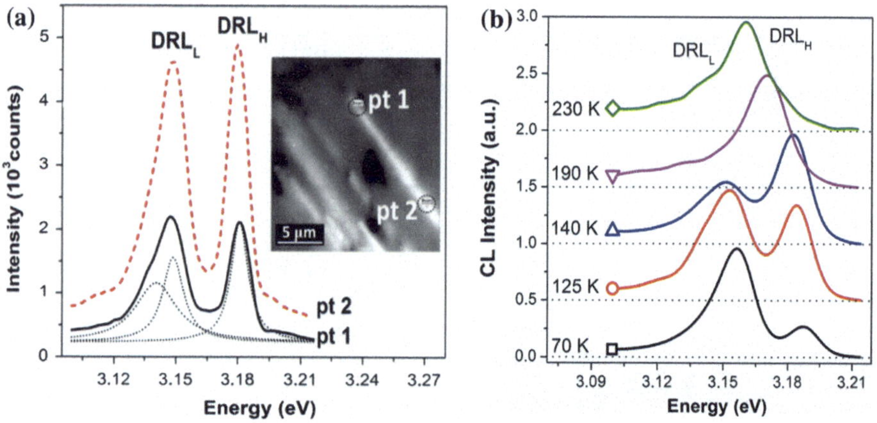

Fig. 21.16 **a** DRL spectra in two points of an alone a-screw dislocation marked as pt 1 and pt 2 in CL-map in the insert. Black dot lines are results of fitting DRL spectrum in pt 1 by Lorentz functions. T = 70 K, $E_b = 8$ keV, $I_b = 500$ pA; **b** Temperature variation of DRL spectral doublet shape. Reproduced with permission from [93]. Copyright AIP publishing (2018)

The new spectral feature at this point is a strong luminescence band IRL (intersection-related luminescence) with the peak energy of 3.3 eV while the intensity of DRL is reduced drastically indicating that the IRL band stems on the cost of DRL band. The appearance of IRL band was found in [92] also at the intersections of the screw dislocations intentionally created by two closely spaced indentations. To explain this fact it was proposed that the dislocation nodes formed by the reactions between intersecting screw dislocations are responsible for the IRL.

It was reported [93] that presence of the DRL doublet structure did not depend on dislocation density and was the property of individual a-screw dislocation. Figure 21.16a demonstrates two DRL spectra acquired in two points pt 1 and pt 2 of a single dislocation line shown on CL map in the insert of Fig. 21.16a as well as the approximation of the doublet components with Lorenz function (dotted lines). The high-energy component DRL_H seems to be a single line while the low energy one DRL_L possesses an unresolved fine structure.

The doublet components exhibit also the different behavior of their intensity and shape upon temperature changes as it is demonstrated in Fig. 21.16b. The DRL_H line does not change significantly its intensity and shape up to 190 K. The full width at half maximum (FWHM) of DRL_H in the range 70–190 K changes from about 16 to 25 meV coinciding well with the absolute change of FWHM of FE line. The intensity DRL_L line at temperatures below 120 K exceeds well that of DRL_H but their ratio rapidly decrease with the temperature. Besides, DRL_L quickly broadens with temperature increase and above 170 K the DRL_L line could be recognized only as a little hump on the low energy tail of DRL_H line (Fig. 21.16b). The doublet structure disappears completely at the room temperature turning into a single DRL band with FWHM of about 90 meV broadened from low energy side.

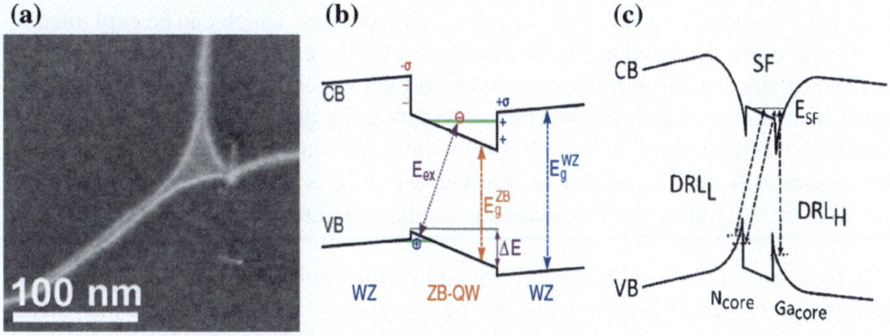

Fig. 21.17 **a** STEM image of a dissociated screw dislocation forming an extended node (Reproduced with permission from [93] Copyright AIP publishing (2018)); **b** Schematic band profile of a stacking fault band alignment in GaN. The polarization induced sheet charges σ leading to the QCSE and the resulting change in the transition energy Eex by ΔE are indicated; **c** Sketch of band diagram of the dissociated screw dislocation showing DRL transitions between energy level of SF ribbon and two 30° partial dislocations including both deformation and Coulomb potentials of partials

The investigations of the evolution of DRL spectrum and dislocation structure after high temperature treatment [91] revealed significant influence of the annealing on dislocation arrangement, while main DRL spectral features measured at the room and lower temperatures remained unchanged. The straight luminous dislocations increased their length from about 40 to more than 120 μm simultaneously with dislocation density decrease close to the scratch after short treatments at 500–750 K.

The core structure of luminous freshly introduced a-screw dislocations was investigated by TEM and STEM [93]. It was found that the dislocation cores were dissociated into two partials bounding stacking fault ribbon. The SF ribbon width was about 5 nm, which varied a little for different dislocations and also along the dislocation line. TEM investigations revealed also the formation of extended dislocation nodes. In Fig. 21.17a STEM HAADF image of dissociated screw dislocations with a typical extended node is reproduced from [93]. One can note that the typical diameter of the extended node of about 15–20 nm exceeds well the width of the dissociation of straight dislocation segments.

The fact that a-screw dislocations are dissociated allows to explain why straight dislocations introduced by scratching of basal surface were stable upon high temperature annealing. The reason for that must be impossibility of the motion of the dislocation dissociated in the basal (0001) plane perpendicular to that plane.

SF ribbon in the core of dissociated dislocation can be considered as an inclusion of GaN sphalerite phase, which has the band gap of 3.27 eV, and therefore forms a quantum well (QW) inside the wurtzite matrix [39, 108] with the conduction band offset of 0.27 eV. Luminescent line of double-layered SF I2, located in the core of extended a-screw dislocation, is red-shifted towards FE energy by about 0.15 eV. Its energy position was reported to vary in the range 3.32–3.36 eV (see recent review [40]). The spectral position of I2 SF line coincides precisely with the position of observed IRL-band of a-screw dislocations intersections, which can be explained by large area of the extended nodes that exceeds well the exciton Bohr radius [101].

The red shift of SF luminescence with respect expected from theoretically predicted in a simple model [39] was proposed [40] to be due to the spontaneous polarization of wurtzite phase that is absent in sphalerite one that changes the shape of QW potential from rectangular to triangle one as it is depicted in the diagram of Fig. 21.17b. This idea was transferred to explain the DRL properties in low-ohmic GaN.

In the cores of the two 30° partials, bounding the double layered I2 SF ribbon, are predominantly either Ga or N atoms. Being charged Ga^+ and N^- cores of the partials induce an additional electric field [104]. This additional electric field superpose with the electric field of spontaneous polarization [40], which leads to the shift of the energy levels of the Ga^+ and N^--terminated partials with respect to each other. In Fig. 21.17c the band diagram in the scope of a model, which takes into account electric field of partials, spontaneous polarization, and deformation potential, is plotted. For Ga^+/N^--terminated partials the deformation and the Coulomb potentials are both attractive for electrons in the conduction band (CB) and for holes in the valence band (VB) respectively. Together with the QW of SF, these attractive potentials form a deeper quantum confined level for electrons in CB (shown in Fig. 21.17c as E_{SF}) that results in net negative total charge of the dislocation and in the appearance of a saddle point of the sum potential [109]. Because Coulomb potential of Ga^+ and N^- is repulsive for the holes and electrons respectively, the total potential for charge carriers remain attractive only in a close vicinity of the corresponding dislocation core.

The exciton Bohr radius in the bulk GaN crystal is about 3 nm [101], which is comparable with the width of the SF ribbon. In this way, the overlap of electron and hole wave functions is sufficient to form a quasi-1D electronic system along the SF ribbon and two kinds of well distinguished optical transitions in the core of the dissociated dislocations are depicted in the diagram Fig. 21.17c. They are the direct transition between the states of the same partial (DRL_H) and the indirect transition between electron state of Ga^+ and hole state of N^- partials (DRL_L).

A strong localization of closely spaced electron and hole wave functions for the direct exciton explains the thermal stability of DRL_H line caused by a high value of exciton binding energy of 1D system [110].

The indirect exciton possesses a weaker overlapping of electron and hole wave functions, and accordingly, a lower binding energy. On the other side, the density of the holes attracted by negatively charged N-partial line exceeds that at positively charged Ga-partial. As the results, DRL_L transition probability can become larger than one of DRL_H at sufficient low temperatures but becomes less with temperature increase. Previously such kind of behavior (quantum confined Stark effect) was observed for QW structures in external electric field as well as for SFs in GaN in spontaneous polarization induced electric field [110].

21.7 Summary

In this report we summarized the data about dislocation-related and stacking fault-related luminescence in semiconductors with diamond-like, zink-blend and wurtzite lattice structure. Both types of defect were found to possess intrinsic luminescence bands with the specific for given material properties. Despite of extensive investigation of these phenomena an unambiguous understanding of all details of the radiative recombination processes is still lacking.

Important point that previously has not been considered so far is the existence intrinsic luminescence of quantum barriers formed by SFs in sphalerite crystals caused by its dipole moment due to spontaneous polarization in thin wurtzite layer that was shown to play an important role even when SF-ribbon is as narrow as dislocation dissociation width. This fact might help to understand existing issues and stimulate further experimental and theoretical investigations, especially the last ones.

Among the reviewed dislocation-related luminescent bands the only two of them has been reported to survive at room temperature that makes them as candidate for the usage in applications. These are D1 luminescence in Si (0.8 eV) and DRL in GaN (3.2 eV).

The origin of D1 is still under the debate. It is not known whether D1 is an intrinsic or extrinsic property of dislocations in Si that makes impossible to estimate a value of theoretically reachable efficiency of LEDs.

DRL intensity in GaN was shown to exceed well that of band-band radiation and an intrinsic character is undoubtful. Exact theoretical investigations and the development of the technology to prepare electroluminescent devices based on DRL GaN are needed to make final conclusions about its applicability in practice.

References

1. S. Ruvimov, P. Werner, K. Scheerschmidt, U. Gösele, J. Heydenreich, U. Richter, N. N. Ledentsov, M. Grundmann, D. Bimberg, V.M. Ustinov, A.Y. Egorov, P.S. Kop'ev, Z.I. Alferov, Phys. Rev. B **51**, 14766 (1995)
2. M.V. Rakhlin, K.G. Belyaev, G.V. Klimko, I.S. Mukhin, D.A. Kirilenko, T.V. Shubina, S.V. Ivanov, A.A. Toropov, Sci. Rep. **8** (2018)
3. Q. Zhu, J.D. Ganière, Z.B. He, K.F. Karlsson, M. Byszewski, E. Pelucchi, A. Rudra, E. Kapon, Phys. Rev. B **82** (2010)
4. T. Someya, H. Akiyama, H. Sakaki, Phys. Rev. Lett. **76**, 2965 (1996)
5. V.G. Dubrovskii, G.E. Cirlin, I.P. Soshnikov, A.A. Tonkikh, N.V. Sibirev, Y.B. Samsonenko, V.M. Ustinov, Phys. Rev. B **71** (2005)
6. P.B. Hirsch, J. Phys. Colloq. **40**, C6 (1979)
7. M. Heggie, R. Jones, Philos. Mag. Part B **48**, 365 (1983)
8. J.-L. Farvacque, P. François, Phys. Status Solidi B **223**, 635 (2001)
9. M. Reiche, M. Kittler, H. Übensee, M. Krause, E. Pippel, Jpn. J. Appl. Phys. **53**, 04EC03 (2014)
10. N.A. Drozdov, A.A. Patrin, V.D. Tkachev, JETP Lett. **23** (1976)

11. O.S. Medvedev, O.F. Vyvenko, A.S. Bondarenko, Semiconductors **49**, 1181 (2015)
12. P. Yu, M. Cardona, *Fundamentals of Semiconductors: Physics and Materials Properties*, 4th edn. (Springer, Berlin Heidelberg, 2010)
13. I. Yonenaga, Eng. Fract. Mech. **147**, 468 (2015)
14. J.P. Hirth, J. Lothe, *Theory of Dislocations* (Krieger Pub. Co., 1982)
15. J. Hornstra, J. Phys. Chem. Solids **5**, 129 (1958)
16. D.J.H. Cockayne, I.L.F. Ray, M.J. Whelan, Philos. Mag. J. Theor. Exp. Appl. Phys. **20**, 1265 (1969)
17. I.L.F. Ray, D.J.H. Cockayne, Proc. R. Soc. Lond. A **325**, 543 (1971)
18. A.M. Gómez, P.B. Hirsch, Philos. Mag. J. Theor. Exp. Appl. Phys. **36**, 169 (1977)
19. F.C. Frank, J.F. Nicholas, Lond. Edinb. Dublin Philos. Mag. J. Sci. **44**, 1213 (1953)
20. F. Bechstedt, U. Grossner, J. Furthmüller, Phys. Rev. B **62**, 8003 (2000)
21. A. García, M.L. Cohen, Phys. Rev. B **47**, 4215 (1993)
22. J. Lähnemann, Phys. Rev. B **86** (2012)
23. Y.A. Osipiyan, I.S. Smirnova, Phys. Status Solidi B **30**, 19 (1968)
24. Y.A. Osipyan, I.S. Smirnova, J. Phys. Chem. Solids **32**, 1521 (1971)
25. R. Jones, A. Umerski, P. Sitch, M.I. Heggie, S. Öberg, Phys. Status Solidi A **138**, 369 (1993)
26. S.P. Beckman, X. Xu, P. Specht, E.R. Weber, C. Kisielowski, D.C. Chrzan, J. Phys. Condens. Matter **14**, 12673 (2002)
27. R. Landauer, Phys. Rev. **94**, 1386 (1954)
28. S. Winter, Phys. Status Solidi B **79**, 637 (1977)
29. C. Elbaum, Phys. Rev. Lett. **32**, 376 (1974)
30. V.V. Kveder, R. Labusch, Y.A. Ossipyan, Phys. Status Solidi A **92**, 293 (n.d.)
31. R. Labusch, J. Phys. III **7**, 1411 (1997)
32. I. Yonenaga, Y. Ohno, T. Yao, K. Edagawa, J. Cryst. Growth **403**, 72 (2014)
33. V. Kveder, M. Badylevich, E. Steinman, A. Izotov, M. Seibt, W. Schroter, Appl. Phys. Lett. **84**, 2106 (2004)
34. A. Loshachenko, A. Bondarenko, O. Vyvenko, O. Kononchuk, Phys. Status Solidi C **10**, 36 (2013)
35. U.M. Gösele, H. Stenzel, M. Reiche, T. Martini, H. Steinkirchner, Q.-Y. Tong, Solid State Phenom. **47–48**, 33 (1996)
36. S. Marklund, Phys. Status Solidi B **108**, 97 (1981)
37. N. Lehto, Phys. Rev. B **55**, 15601 (1997)
38. C.-Y. Yeh, S.-H. Wei, A. Zunger, Phys. Rev. B **50**, 2715 (1994)
39. Y.T. Rebane, Y.G. Shreter, M. Albrecht, Phys. Status Solidi A **164**, 141 (1997)
40. J. Lähnemann, U. Jahn, O. Brandt, T. Flissikowski, P. Dogan, H.T. Grahn, J. Phys. Appl. Phys. **47**, 423001 (2014)
41. R. Newman, Phys. Rev. **105**, 1715 (1957)
42. S. Shevchenko, A.N. Tereshchenko, Solid State Phenom. **131–133**, 583 (2007)
43. D.B. Holt, B.G. Yacobi, *Extended Defects in Semiconductors* (Cambridge University Press, 2007)
44. S.A. Shevchenko, A.N. Tereshchenko, Phys. Solid State **49**, 28 (2007)
45. V.V. Kveder, M. Kittler, Mater. Sci. Forum **590**, 29 (2008)
46. M. Reiche, M. Kittler, Crystals **6**, 74 (2016)
47. V. Higgs, E.C. Lightowlers, S. Tajbakhsh, P.J. Wright, Appl. Phys. Lett. **61**, 1087 (1992)
48. T. Sekiguchi, S. Ito, A. Kanai, Mater. Sci. Eng. B **91–92**, 244 (2002)
49. T. Mchedlidze, O. Kononchuk, T. Arguirov, M. Trushin, M. Reiche, M. Kittler, Solid State Phenom. **156–158**, 567 (2009)
50. R. Sauer, C. Kisielowski-Kemmerich, H. Alexander, Phys. Rev. Lett. **57**, 1472 (1986)
51. Y.S. Lelikov, Y.T. Rebane, S. Ruvimov, A.A. Sitnikova, D.V. Tarhin, Y.G. Shreter, Phys. Status Solidi B **172**, 53 (1992)
52. V. Higgs, M. Goulding, A. Brinklow, P. Kightley, Appl. Phys. Lett. **60**, 1369 (1992)
53. X. Yu, W. Seifert, O.F. Vyvenko, M. Kittler, T. Wilhelm, M. Reiche, Appl. Phys. Lett. **93**, 2 (2008)

54. R. Sauer, J. Weber, J. Stolz, E.R. Weber, 13 (n.d.)
55. A.N. Izotov, A.I. Kolyubakin, S.A. Shevchenko, E.A. Steinman, Phys. Status Solidi A **130**, 193 (1992)
56. V.Y. Kravchenko, JETF **80**, 9 (1995)
57. Y.S. Lelikov, Y.T. Rebane, Y.G. Shreter, *Structure of Property Dislocations Semiconductors* (eds. by S.G. Roberts, Holt, P.R. Wilshaw, Bristol, 1989), p. 119
58. J.L. Farvacque, P. Franc, Phys. B Condens. Matter **274**, 995 (1999)
59. U. Hilpert, J. Schreiber, L. Worschech, L. Höring, M. Ramsteiner, W. Ossau, G. Landwehr, J. Phys.: Condens. Matter **12**, 10169 (2000)
60. P.J. Dean, A.D. Pitt, M.S. Skolnick, P.J. Wright, B. Cockayne, J. Cryst. Growth **59**, 301 (1982)
61. L. Worschech, W. Ossau, A. Waag, G. Landwehr, U. Hilpert, J. Schreiber, Y.T. Rebane, Y.G. Shreter, Phys. B **273–4**, 895 (1999)
62. A. Rosenauer, T. Reisinger, F. Franzen, G. Schütz, B. Hahn, K. Wolf, J. Zweck, W. Gebhardt, J. Appl. Phys. **79**, 4124 (1996)
63. Y.G. Shreter, Y.T. Rebane, O.V. Klyavin, P.S. Aplin, C.J. Axon, W.T. Young, J.W. Steeds, J. Cryst. Growth **159**, 883 (1996)
64. D.J. Mowbray, O.P. Kowalski, M.S. Skolnick, M. Hopkinson, J.P.R. David, Superlattices Microstruct. **15**, 313 (1994)
65. J. Schreiber, L. Höring, H. Uniewski, S. Hildebrandt, H.S. Leipner, Phys. Status Solidi A **171**, 89 (1999)
66. A. Naumov, K. Wolf, T. Reisinger, H. Stanzl, W. Gebhardt, J. Appl. Phys. **73**, 2581 (1993)
67. T. Karin, X. Linpeng, M.M. Glazov, M.V. Durnev, E.L. Ivchenko, S. Harvey, A.K. Rai, A. Ludwig, A.D. Wieck, K.-M.C. Fu, Phys. Rev. B **94** (2016)
68. D.S. Smirnov, K.G. Belyaev, D.A. Kirilenko, M.O. Nestoklon, M.V. Rakhlin, A.A. Toropov, I.V. Sedova, S.V. Sorokin, S.V. Ivanov, B. Gil, T.V. Shubina, Phys. Status Solidi RRL—Rapid Res. Lett. **12**, 1700410 (2018)
69. P. Corfdir, P. Lefebvre, J. Appl. Phys. **112**, 053512 (2012)
70. F. Boutaiba, A. Belabbes, M. Ferhat, F. Bechstedt, Phys. Rev. B **89**, 245308 (2014)
71. J. Lähnemann, O. Brandt, U. Jahn, C. Pfüller, C. Roder, P. Dogan, F. Grosse, A. Belabbes, F. Bechstedt, A. Trampert, L. Geelhaar, Phys. Rev. B **86**, 081302 (2012)
72. S.V. Poltavtsev, Y.P. Efimov, Y.K. Dolgikh, S.A. Eliseev, V.V. Petrov, V.V. Ovsyankin, Solid State Commun. **199**, 47 (2014)
73. K. Maeda, S. Takeuchi, Jpn. J. Appl. Phys. **20**, L165 (1981)
74. H. Iwata, U. Lindefelt, S. Öberg, P.R. Briddon, Phys. Rev. B **65**, 033203 (2001)
75. A. Galeckas, J. Linnros, P. Pirouz, Phys. Rev. Lett. **96**, 1 (2006)
76. A. Galeckas, A. Hallén, S. Majdi, J. Linnros, P. Pirouz, Phys. Rev. B **74**, 233203 (2006)
77. H. Rii, T. Hidekazu, T. Michio, M.I. Kohei, M. Koji, Appl. Phys. Express **6**, 011301 (2013)
78. M.A. Razumova, V.N. Khotyaintsev, Phys. Status Solidi B **174**, 165 (n.d.)
79. O. Brümmer, J. Schreiber, Ann. Phys. **483**, 105 (1972)
80. N.I. Tarbaev, J. Schreiber, G.A. Shepelskii, Phys. Status Solidi A **110**, 97 (1988)
81. A. Hoffmann, J. Christen, J. Gutowski, Adv. Mater. Opt. Electron. **1**, 25 (1992)
82. O. Vyvenko, A. Zozime, Mater. Sci. Eng., B **24**, 105 (1994)
83. V.D. Negrii, Y.A. Osipyan, N.V. Lomak, Phys. Status Solidi A **126**, 49 (1991)
84. V.D. Negrii, J. Cryst. Growth **117**, 672 (1992)
85. V.D. Negrii, J. Appl. Phys. **74**, 7008 (1993)
86. N.I. Tarbaev, G.A. Shepel'skii, Semiconductors **32**, 580 (1998)
87. M. Albrecht, L. Lymperakis, J. Neugebauer, Phys. Rev. B **90**, 241201 (2014)
88. M. Albrecht, H.P. Strunk, J.L. Weyher, I. Grzegory, S. Porowski, T. Wosinski, J. Appl. Phys. **92**, 2000 (2002)
89. O.S. Medvedev, O.F. Vyvenko, A.S. Bondarenko, Semiconductors **49**, 1181 (2015)
90. O.S. Medvedev, O.F. Vyvenko, A.S. Bondarenko, V.Y. Mikhailovskii, V.E. Ubyivovk, P. Peretzki, M. Seibt, A.I.P. Conf, Proc. **1748**, 020011 (2016)
91. O. Medvedev, O. Vyvenko, A. Bondarenko, Phys. Status Solidi C **1700111**, 1 (2017)
92. O. Medvedev, O. Vyvenko, Phys. Status Solidi RRL—Rapid Res. Lett. **1700297** (2017)

93. O. Medvedev, O. Vyvenko, E. Ubyivovk, S. Shapenkov, A. Bondarenko, P. Saring, M. Seibt, J. Appl. Phys. **123**, 161427 (2018)
94. M.R. Wagner, G. Callsen, J.S. Reparaz, J.-H. Schulze, R. Kirste, M. Cobet, I.A. Ostapenko, S. Rodt, C. Nenstiel, M. Kaiser, A. Hoffmann, A.V. Rodina, M.R. Phillips, S. Lautenschläger, S. Eisermann, B.K. Meyer, Phys. Rev. B **84**, (2011)
95. O. Brümmer, J. Schreiber, Krist. Tech. **9**, 817 (1974)
96. O.F. Vyvenko, A.A. Istratov, Phys. Solid State **36**, 1794 (1994)
97. D.J.H. Cockayne, A. Hons, J.C.H. Spence, Philos. Mag. A **42**, 773 (1980)
98. A.A. Istratov, O.F. Vyvenko, J. Appl. Phys. **80**, 4400 (1996)
99. Y.G. Shreter, Y.T. Rebane, T.J. Davis, J. Barnard, Darbyshire, M. Steeds, J.W. Perry, Camb. J. Online AU **449**, (1996)
100. M.A. Reshchikov, D. Huang, L. He, H. Morkoç, J. Jasinski, Z. Liliental-Weber, S.S. Park, K.Y. Lee, Phys. B Condens. Matter **367**, 35 (2005)
101. M.A. Reshchikov, H. Morkoç, J. Appl. Phys. **97**, 061301 (2005)
102. I. Yonenaga, H. Making, S. Itoh, T. Goto, T. Yao, J. Electron. Mater. **35**, 717 (2006)
103. R. Liu, A. Bell, F.A. Ponce, C.Q. Chen, J.W. Yang, M.A. Khan, Appl. Phys. Lett. **86**, 021908 (2005)
104. I. Belabbas, J. Chen, M.I. Heggie, C.D. Latham, M.J. Rayson, P.R. Briddon, G. Nouet, Model. Simul. Mater. Sci. Eng. **24**, 075001 (2016)
105. Y.T. Rebane, Phys. Rev. B **48**, 14963 (1993)
106. J. Huang, K. Xu, Y.M. Fan, M.T. Niu, X.H. Zeng, J.F. Wang, H. Yang, Nanoscale Res. Lett. **7**, 150 (2012)
107. O.S. Medvedev, O.F. Vyvenko, A.S. Bondarenko, J. Phys. Conf. Ser. **690**, 012008 (2016)
108. J.R.L. Fernandez, F. Cerdeira, E.A. Meneses, J.A.N.T. Soares, O.C. Noriega, J.R. Leite, D.J. As, U. Köhler, D.G.P. Salazar, D. Schikora, K. Lischka, Microelectron. J. **35**, 73 (2004)
109. M. Trushin, O.F. Vyvenko, Solid State Phenom. **205–206**, 299 (2013)
110. J. Lähnemann, O. Brandt, C. Pfüller, T. Flissikowski, U. Jahn, E. Luna, M. Hanke, M. Knelangen, A. Trampert, H.T. Grahn, Phys. Rev. B **84**, 155303 (2011)

Chapter 22
Light-Induced Processes in Porphyrin-Fullerene Systems

Alexander S. Konev

Abstract Porphyrin-fullerene dyads are representative class of donor-acceptor molecular systems capable of undergoing photoinduced charge separation, the phenomenon that is the key process in solar energy conversion systems either in organic solar cells or photoredox catalysis. The charge-separated state generated in covalently linked dyad is a highly polarized electronically excited state. Formation of this state typically proceeds as a result of the relaxation of a locally excited state of higher energy, which is populated upon the initial excitation of the dyad. The review of computational and spectroscopic studies of the excited states in porphyrin-fullerene covalently linked dyads is given in the present chapter.

22.1 Introduction

Donor-acceptor dyads capable of undergoing photoinduced charge separation are useful in both understanding of photoredox processes occurring in the natural photosynthesis and in the design of catalytic systems for artificial photon initiated redox reactions mimicking photosynthesis [1, 2]. Donor-acceptor photoreactive combinations include porphyrins [3], mesytylene [4], carotenoids [5], polythiophens [6] and phtalocyanines [7] as donors and fullerenes C_{60} [3], C_{70} [8], carbon nanotubes [9], acridinium salts [4] and porphyrins [10] as acceptors. Probably the most studied are porphyrin-fullerene dyads and more complicated molecular ensembles bearing porphyrin-fullerene system as the key fragment.

The photoinduced charge separation in porphyrin-fullerene dyads might be used for solar energy harvesting in photoinduced charge-separation type solar cells [3, 11–14]. In particular, Yamada [14] have recently reported superiority of a porphyrin-fullerene covalent dyad over porphyrin-fullerene blend in p-i-n organic photovoltaic

A. S. Konev (✉)
Institute of Chemistry, St. Petersburg State University, Universitetskii pr. 26 Peterhof, St. Petersburg 198504, Russian Federation
e-mail: a.konev@spbu.ru

K. Yamanouchi et al. (eds.), *Progress in Photon Science*, Springer Series in Chemical Physics 119, https://doi.org/10.1007/978-3-030-05974-3_22

cells. Recent DFT-based finding that Zn-porphyrin-fullerene dyad should show high rate of an electron transfer to graphene support [15] inspires further studies in this direction.

The ability of porphyrin-fullerene dyads to undergo photoinduced charge separation was well documented in a number of reviews which summarize the experimental achievements of the groups of Sakata [12], Guldi [16], D'Souza [3], Imahori [11, 17], Lemmetyinen [18], Nakamura [19] and Ito [20]. To achieve better understanding of the charge separation phenomenon in porphyrin-fullerene dyads a large number of computational research was reported, the latest review given by Agnihotri [21].

In this chapter, quantum chemical and spectroscopic description of the issues related to the charge separation process in covalently linked porphyrin-fullerene dyads are combined to obtain picture, which reflects both theoretical and experimental data and to provide the reader with the introductory bibliographic guide on the topic.

22.2 Electronic Structure of Porphyrin-Fullerene Dyads

The electronic structure of the ground state of porphyrin-fullerene dyads can be ideally described as consisting of fullerene-localized, porphyrin-localized and mixed molecular orbitals, with HOMO localized on porphyrin and LUMO—on fullerene fragment and charge separated state corresponding to $HOMO^1LUMO^1$ electronic configuration. A representative example of the idealized non-interacting case can be the electronic structure of dyad **1c** (Fig. 22.1) with large separation of the chromophores [22]. Actual degree of mixing of the atomic orbitals belonging to porphyrin or fullerene fragment upon formation of molecular orbitals depends on the geometry of the dyad.

According to the computational study of Zandler and D'Souza [23] at B3LYP/3-21G(*) level of theory, close face-to-face arrangement of the porphyrin and fullerene moieties, where the fullerene spheroid is situated above the center of the porphyrin plane, results in some mixing of the atomic orbitals of the donor and acceptor and endows the charge-transfer character to the ground state in Zn- or Mg-metallated

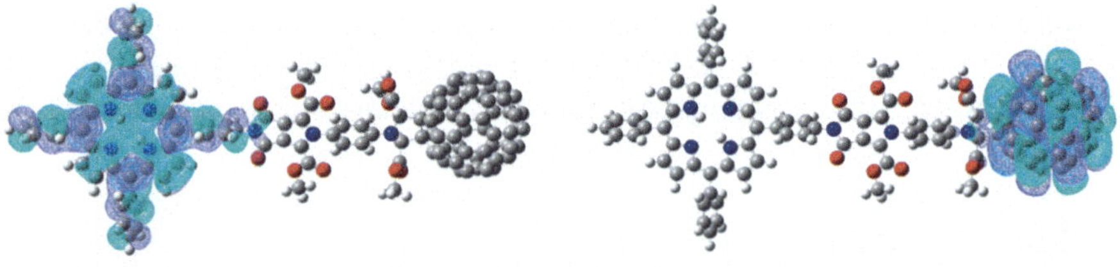

Fig. 22.1 HOMO (left) and LUMO (right) orbitals of dyad **1c** (B3LYP, 6-31G(d,p))

Packman-type porphyrin-fullerene dyads **Zn-2** and **Mg-2** (Fig. 22.2). However, no such interaction was reported for the corresponding free-base dyad **2** (Fig. 22.2). The authors explain this observation by the increased separation between the chromophores in the case of free-base dyad [23].

This is in line with the studies of Lemmetyinen et al. who demonstrated the dependence of the degree of charge transfer on the interchromophore distance. Interchromophore interaction in the ground state was reported for Packman-type porphyrin-fullerene covalent dyad **3a**, where chromophores were connected by a rigid benzanilide-based linker. The interchromophore distance and the charge transfers were found to be 6.3–7.9 Å and 0.09–0.15e respectively depending on the functional used [24]. Shorter interchromophore distance of 5.8 Å (SVWN functional), achieved in a specific conformer of a cyclophane-type dyad **4**, increases the charge-transfer character of the ground state to ca. 0.23e (SVWN functional) [25]. The conformer with larger separation of the chromophores (6.1 Å, SVWN functional) demonstrated lesser degree of charge transfer (0.12e, SVWN functional) [25]. In accord with the observed increase of charge transfer upon decrease of the interchromophore distance is the increased delocalization of HOMO and LUMO over both porphyrin and fullerene systems [25]. In-plane positioning of the porphyrin fragment with respect to the fullerene core (dyad **5**) tends to decrease the degree of charge transfer in the ground state (0.08e, 6.89 Å interchromophore distance, B3LYP functional) [26].

In this connection, a crucial effect of van der Waals corrections on the predicted molecular geometry and, hence, electronic structure, reported in the recent work of Hukka et al. [27] should be mentioned. According to the authors, application of either Grimme [28] or Elstner [29] dispersion correction schemes affects strongly the optimal geometry of porphyrin-fullerene dyad **3b** resulting in delocalization of the HOMO orbital localized on porphyrin moiety for non-corrected structure over the adjacent linker [27].

Experimentally, the interchromophore interaction in the ground state was demonstrated by observing red shift and broadening of the Soret band as compared to the fullerene-free porphyrin [30–34]. In these systems, a new broad absorption feature called charge-transfer (CT) band appears in the near IR, which was initially assigned to the direct population of the charge-separated state from the ground state [35] but was shown later to include formation of exciplex state which has a variable degree of charge transfer [31]. The definition of range for this absorption feature varies from 700–750 nm [30] to 650–750 nm [36] for free-base dyad, from 650–800 nm [31, 35] to 600–1000 nm [32] for Zn-porphyrin-fullerene dyad and from 650–900 to 750–1000 nm for Mg-porphyrin-fullerene dyad [33]. The charge transfer character of the exciplex is evidenced from the red shift of the CT band in the series non-metallated (**4b**), Zn-metallated (**Zn-4b**) and Mg-metallated (**Mg-4b**) dyad, which repeats the order of oxidation potential decrease [36]. In line with this is the observed red shift of the CT band of Zn-metallated dyad **Zn-4a** upon ligation with Cl^- or Br^- ions due to the stabilization of the polarized exciplex state [37].

Formation of the exciplex state is also confirmed by observation of new emission bands at 730 nm for non-metallated dyad **4b** and at 810 nm [31] (730 nm [38]) for

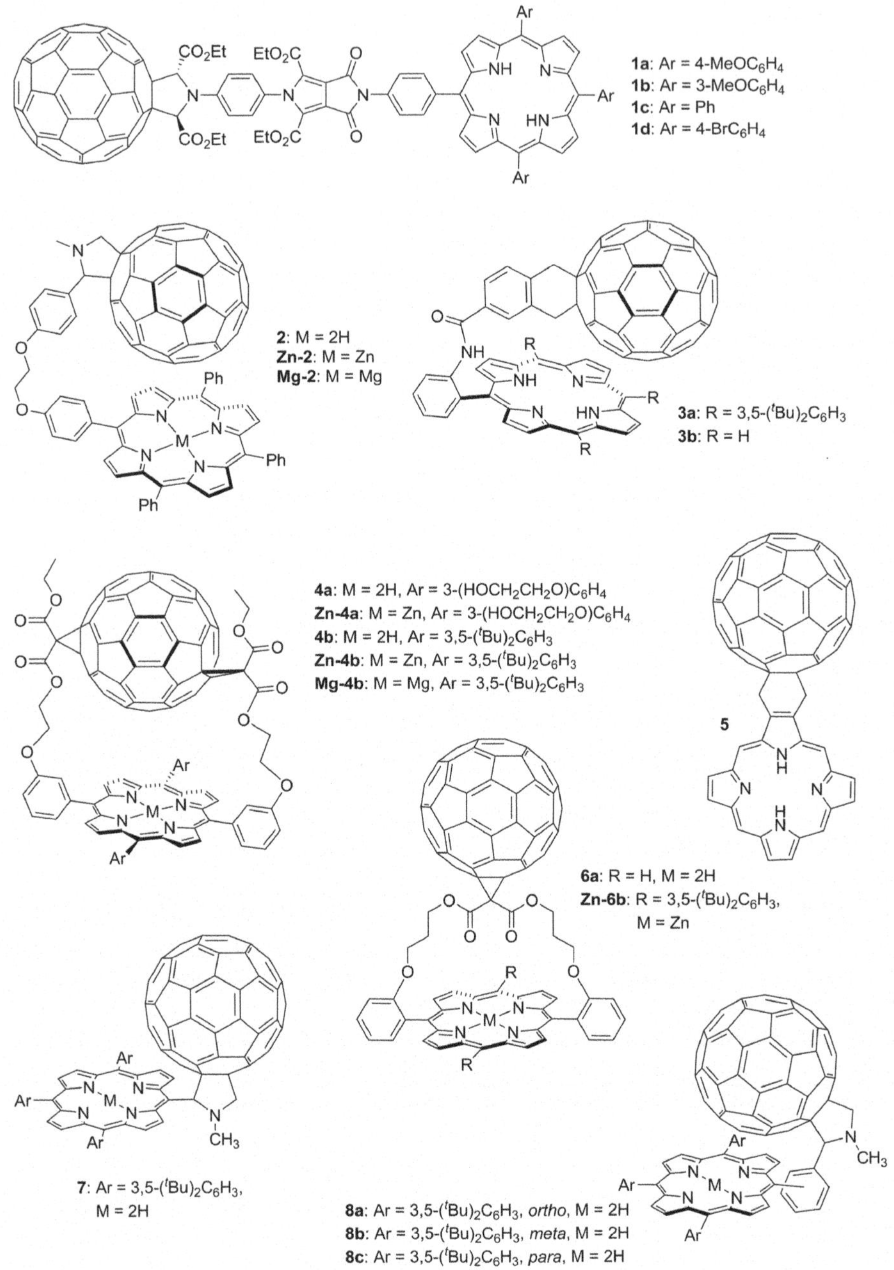

Fig. 22.2 Structural formulae of dyads **1–8**

metallated dyad **Zn-4b**. The assignment of this band is in line with the observed strong decrease of the emission intensity, shift of the emission maximum to the red and broadening of the spectral band upon increase of the solvent polarity, characteristic of exciplex-type emission [30, 31].

In accord with the computational results, electronic interaction of the chromophores decreases upon increase in the separation of fullerene and porphyrin moieties. For example, porphyrin-fullerene dyads **6** with parachute topology and large interchromophore distance, show no appreciable electronic interaction of the chromophores in the ground state [39]. The disappearance of the interchromophore interaction was traced in the series of porphyrin-fullerene dyads where porphyrin was attached either directly to fulleropyrrolidine core (**7**, 8.9 Å center-to-center distance) or via ortho- (**8a**, 9.7 Å), meta- (**8b**, 10.9 Å) or para-substituted (**8c**, 12.6 Å) phenyl ring. The red-shift of the Soret band was observed only for dyads with close arrangement of the chromophores (dyads **7** and **8a**) [38].

22.3 Electronically Excited States of Porphyrin-Fullerene Dyads

The TDDFT studies on electronically excited states and the transitions between them for porphyrin-fullerene dyads were reported in a number of works for various dyads.

Qualitative analysis of the computational results shows that the electronically excited states can be classified to (i) states corresponding to local excitation of either porphyrin or fullerene chromophore, which are populated when the electron is excited from an orbital localized on the corresponding chromophore to the orbital localized on the same chromophore, and to (ii) charge transfer states, which are populated when the electron transition occurs between orbitals localized on different chromophores (Fig. 22.3). For example, excitation of an electron from HOMO to LUMO+4 orbital in dyad **1c** can be considered as local excitation of the porphyrin fragment (Fig. 22.4). Similar to the description of the electronic structure, this classification is an ideal case and pure separation between the above types of electronically excited states is not always observed.

For example, Theodorakopoulos et al. report HOMO–LUMO excitation (pure case of charge transfer) to contribute only 68% to the lowest charge-separated state (1.67 eV, B3LYP) in dyad **9** (Fig. 22.5) [40]. An example of the pure case of charge separated state can be found in the work of Krasnov et al. who found the lowest electronically excited state (1.87 eV, oscillator strength 0.02) in dyad **5** to originate solely from HOMO to LUMO excitation [26]. These results agree well with the conclusion of Cramariuc et al. [24] that the contribution of HOMO–LUMO one-electron transition to the charge-separated state depends on the interchromophore distance: the shorter the distance, the larger is the interaction of the chromophores and, hence, the larger is "contamination" of the charge-separated state by local one-electron transitions. An example of well separated local excitations and charge separated states

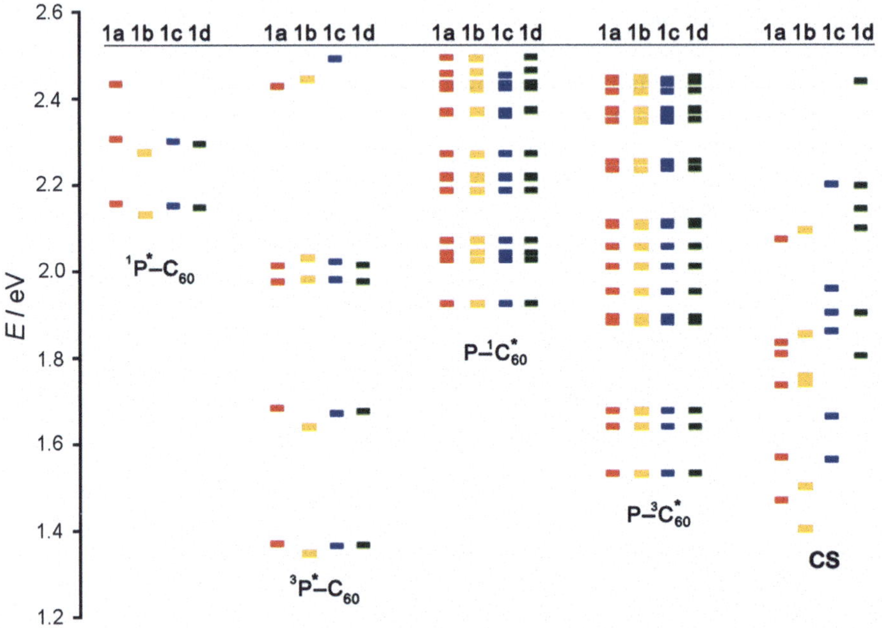

Fig. 22.3 Classification of electronically excited states in dyads **1a–d**. Data taken from [22]

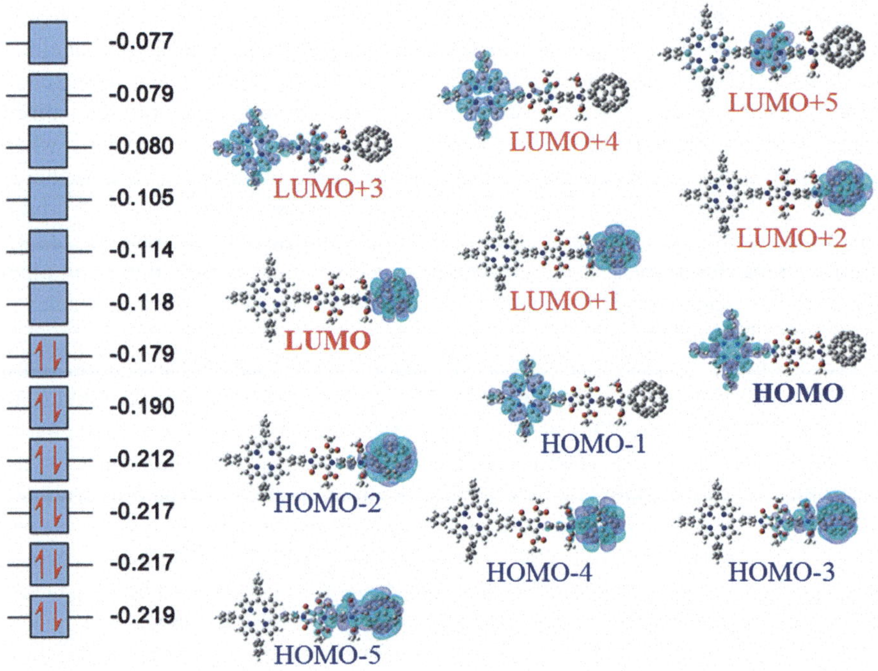

Fig. 22.4 Illustration on local excitation and charge-transfer excitation. Data taken from [22]

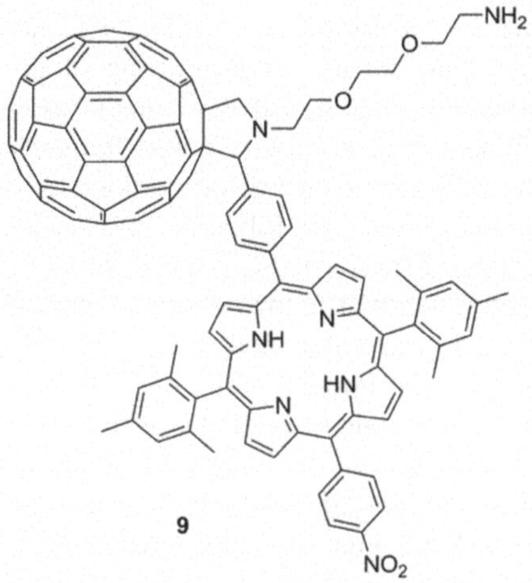

9

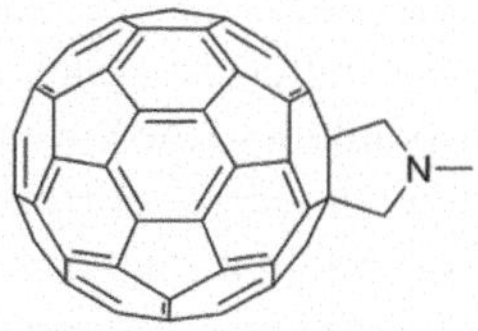

10

Fig. 22.5 Structural formulae of dyad **9** and pyrrolofullerene **10**

can be found in a porphyrin-fullerene dyads **1a–d** with large interchromophore distance [22]. The substituents in the porphyrin ring show no effect on the energy of fullerene-locally excited states (which is expected) and only a little effect on the energy of porphyrin-locally excited states, but affect dramatically the energy of the charge separated state (Fig. 22.3) [22]. Triplet excited states for both porphyrin and fullerene local excitations have lower energies than the corresponding singlet states, while for charge-separated states no difference was observed for singlet and triplet states [22].

Quantitative prediction of the energy of the charge-separated state for porphyrin-fullerene dyads is a rather complicated task due to the large size of the explored systems. An assessment of the performance of TDDFT and wavefunction based methods is given in the works of Cramariuc et al. for dyads **3** [24] and **4a** [25]. The best match with the experimental value was observed for B3LYP and PBE0 (1.6 eV vs. 1.7–1.8 eV) functionals, other DFT functionals (SVWN, PBE) underestimating the ground state to charge separated state transition energy by ca. 0.7 eV [24, 25] and wavefunction based methods (CCS, CIS(D), CC2) overestimating it by 0.9–1.3 eV [25].

Quantitative performance of TD DFT methods for excited singlet porphyrin states is somewhat better: for dyads **3** [24] and **4a** [25] the calculated excitation energy for the first porphyrin excited singlet state ranged from 1.8 to 2.0 eV (SVWN and PBE functionals) versus the experimental value of 1.89 [24]–1.91 eV [25]. However, for another set of dyads, **1**, TD DFT method showed larger discrepancy with the experimental value: 2.15 eV (B3LYP functional) versus 1.92 eV (experiment) [22]. An interesting observation is a low dependence of the predicted porphyrin excitation energy on the interchromophore distance [24, 25].

Direct assessment of the quantitative performance of TD DFT methods for excited singlet fullerene states is complicated by superimposing of the spectral bands corresponding to the lowest fullerene-based transitions with the spectral bands of much larger intensity corresponding to the porphyrin-based transitions. However, such assessment can be obtained by comparing the computational results obtained for porphyrin-fullerene dyads, where the fullerene core is attached to the molecule as pyrrolofullerene structural fragment, with the results obtained for pyrrolofullerenes not containing porphyrin part. The lowest singlet excited state for *N*-methylpyrrolofullerene (**10**, Fig. 22.5) was calculated to be 1.89 eV (B3LYP) [40], which is 0.11 eV higher than the experimental value of 1.78 eV [41]. The energy of the lowest fullerene singlet excited state in dyads **1** was found to comprise 1.93 eV (B3LYP) [22], just 0.04 eV higher than the energy of the corresponding state in *N*-methylpyrrolofullerene. It is possible to conclude that the attachment of the porphyrin fragment and the linker in dyads **1** has negligible effect on the fullerene localized transitions, which justifies comparison of the energy of the locally excited singlet fullerene state with the experimental values for *N*-methylpyrrolofullerene. So, overestimation of 0.15 eV can be used as the assessment of the performance of TD DFT method (B3LYP) for computing the energy of the excited singlet fullerene state.

22.4 Experimental Observation of Electronically Excited States in Porphyrin-Fullerene Dyads

Population of the electronically excited states in porphyrin-fullerene dyads and the dynamics of their relaxation have been studied in numerous works summarized in the reviews cited in the introduction to the chapter. In this section, the spectral features characteristic for identifying various states are of central interest.

In the studies on photodynamics of the excited states in porphyrin-fullerene dyads, the dyad is excited either in Q band (532, 565 or 590 nm excitation wavelengths were reported) or Soret band (387 [34, 42], 400–403, 425 or 430 nm excitation wavelengths were reported) range (typical Zn-porphyrin spectrum is given on Fig. 22.6). Longer excitation wavelengths excite porphyrin chromophore more selectively. At 387 nm, fullerene chromophore has significant light absorption (ε ca. 7500 dm^3 mol^{-1} cm^{-1} for pyrrolofullerene **10**) [43]. Both porphyrin and fullerene chromophores absorb at 400–403 nm in molar ratio estimated as 4:1 [44]. Pumping at 590 nm allows to achieve the ratio of excitation of porphyrin and fullerene chromophores of 15:1, as it was reported for dyad **3a** [44].

The excitation with 400–430 nm lasers populates porphyrin-excited singlet state corresponding to the Soret band, while excitation by 532 or 590 nm lasers populates lower-lying porphyrin excited singlet state corresponding to Q-bands. The porphyrin excited state corresponding to the Soret band (B-state) is known to undergo fast internal conversion to the vibrationally excited level of the Q state in porphyrins

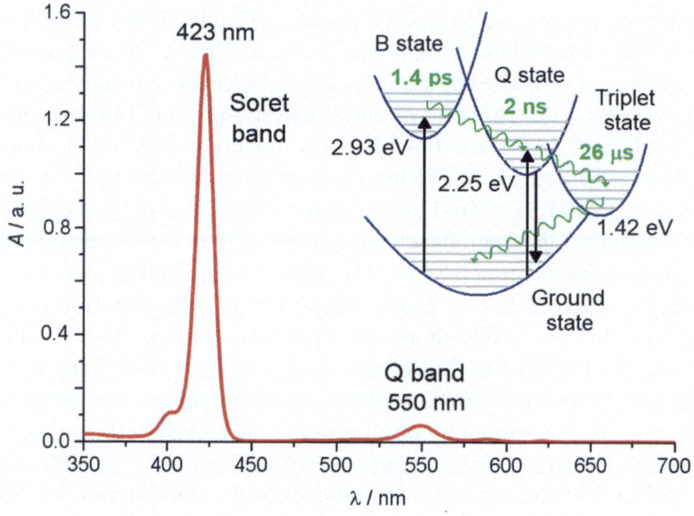

Fig. 22.6 UV-Vis absorption spectrum of porphyrin **Zn-11a** (Fig. 22.7)

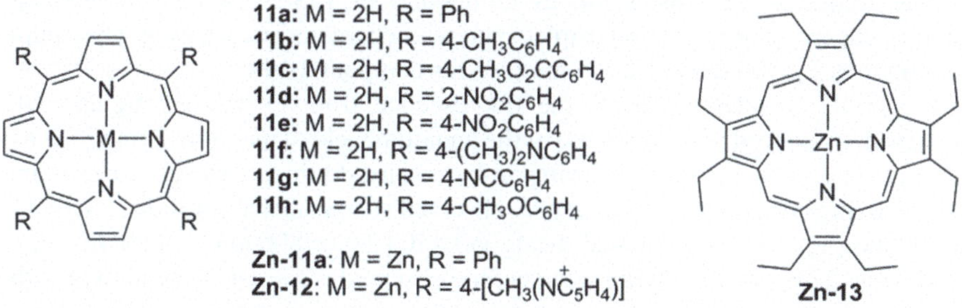

Fig. 22.7 Structural formulae of porphyrins **11a–h** and **Zn-11a**

(Fig. 22.6). For example, the intrinsic lifetime of the Soret band in benzene solutions comprises 68 fs for *meso*-tetraphenylporphyrin (**11a**, Fig. 22.7) [45] and 1.4 ps for Zn *meso*-tetraphenylporphyrinate (**Zn-11a**) [46]. For porphyrin **11a**, the internal conversion of B-state to Q-state was shown to proceed via a two-channel path: either as a direct conversion of B-state to Q_x-state or via intermediate population of Q_y state [47]. Taking into account rather weak interchromophore interactions in porphyrin-fullerene dyads, similar ultrafast processes are expected to occur in porphyrin-fullerene dyads upon excitation at the Soret band (425/430 nm), making Q state the common intermediate for excitation with both 532 and 425/430 nm lasers.

The Q state in porphyrins undergoes either non-radiative or radiative relaxation to the ground state. The radiative relaxation is seen as fluorescence around 655–690 nm (average wavelength of fluorescence protons, e.g.: 685 nm for porphyrin **11a**) for free-base porphyrins and around 595–635 nm (average wavelength of fluorescence protons, e.g.: 635 nm for porphyrin **Zn-11a**) for Zn porphyrinates [48]. The emission signal represents two well-resolved peaks, 650 nm/720 nm for porphyrin **11a** (benzene) and 600 nm/650 nm for **Zn-11a** (benzene) being representative example [48]. The Q-state fluorescence lifetime, τ_F, in porphyrins is typically several ns. For example, τ_F comprises 6–16 ns for *meso*-tetraarylporphyrins **11a–h** without heavy atoms [49] and ca. 2 ns for **Zn-11a** [46]. The quantum yield of the fluorescence from Q state of porphyrins is not so high, comprising ca. 0.03–0.15 for porphyrins **11a–h** [49] and 0.02–0.03 for **Zn-11a** [46]. Intersystem crossing to the triplet state of porphyrins constitutes the main deactivation path of the Q-state, the quantum yield of triplet state formation being 0.82 for **11a** and 0.88 for **Zn-11a** [50]. Low temperature (77 K) phosphorescence of **11a** was reported as weak emission (ϕ_P $4 \cdot 10^{-5}$) centered around 865 nm, while more intensive emission (ϕ_P 0.012) with maxima at 780 nm 875 nm for **Zn-11a**, the lifetimes of the triplet state being 6 and 26 ms respectively [50]. Population of the triplet states in porphyrins can be monitored in transient absorption experiment by observing the intensive (ε 10^4 dm^3 mol^{-1} cm^{-1} order of magnitude) triplet-triplet absorption with maximum at ca. 430 and 780 nm for free-base **11a** [51] and ca. 470, 745 and 845 nm for **Zn-11a** [49, 51, 52].

The first excited singlet state of pyrrolofullerenes has a characteristic transient absorption maximum around 900 nm [43] (886 nm for compound **10**) [53] and has the lifetime of ca. 1 ns (1.3 ns for compound **10**) [53]. This state can undergo intersystem crossing populating triplet state with characteristic transient absorption maximum at ca. 700 nm [43] (705 nm [53] for compound **10**).

In porphyrin-fullerene dyads the fluorescence from Q band is significantly quenched, evidenced as the decrease of the quantum yield of the short-living (1–2 ns) $^1P^*$-C_{60} state, analog of the Q-state in porphyrins. In addition to intersystem crossing to long-living (several μs)$^3P^*$-C_{60} state, observed as transient absorption at 470 [44] or 850 nm [32] for Zn-metallated dyads and 450 [22] or 800 nm [32] for free-base dyads (Fig. 22.8), the fluorescence quenching observed is caused by creation of additional relaxation paths upon attaching of the fullerene moiety to the porphyrin. These paths may include either energy transfer, when the energy surplus of $^1P^*$-C_{60} state promotes population of the P-$^1C_{60}{}^*$ state, or charge transfer, when $^1P^*$-C_{60} state is relaxed to highly polarized charge-separated state. The ratio of these paths depends on the molecular structure of the dyad and on the solvent used in the experiment. In general, polar solvents such as benzonitrile promote formation of the charge-separated state [3, 17, 42, 44], while non-polar solvents like toluene tend to direct the relaxation towards energy transfer path [42].

The energy transfer is usually monitored by observing the emission at 700–750 nm [17, 32, 42, 44], typical of pyrrolofullerenes [54, 55] and benzofullerenes [56] (λ_{max}^{em} 715 nm), which results from the radiative relaxation of P-$^1C_{60}{}^*$ state to the ground state in pyrrolofullerene- or benzofullerene-based porphyrin-fullerene dyads. The use of emission data for reference pyrrolofullerenes or benzofullerenes for the assignment of the emission at 700–750 nm accords well with the computational studies reporting rather low mutual influence of porphyrin and fullerene electronic systems in dyads. Taking into account non-zero absorption of the fullerene chromophore on the excitation wavelength, the energy transfer should be confirmed by analyzing

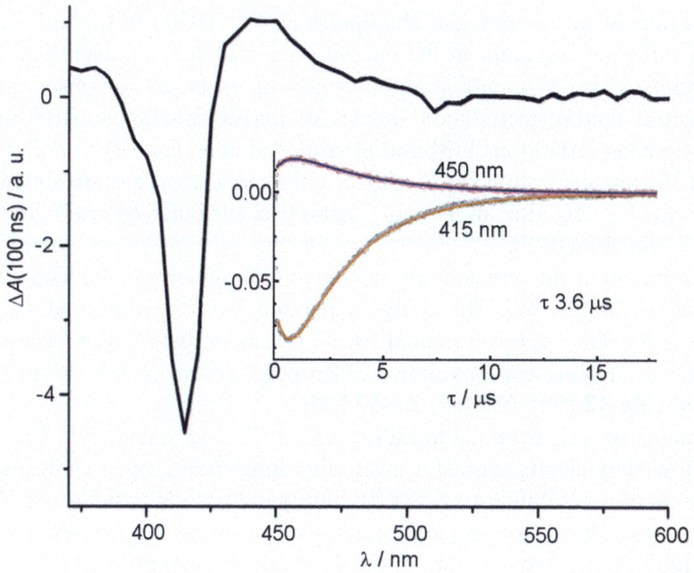

Fig. 22.8 Transient absorption at 450 nm in dyad **1c** due to the formation of $^3P^*$-C_{60} state. The spectrum recorded in deoxygenated $(CH_2Cl)_2$, λ_{ex} 532 nm, time delay 100 ns. The inset: 450 and 415 ΔA kinetic profiles. Data taken from [22]

excitation spectra for 700 nm emission which should follow the porphyrin rather than pyrrolofullerene absorption pattern [42].

Similar to the case of $^1P^*$-C_{60} state, population of the P-$^1C_{60}{}^*$ state after energy transfer often leads to formation of P-$^3C_{60}{}^*$ state via intersystem crossing, which can be monitored by observing fullerene triplet-triplet absorption with maximum at ca. 700 nm in transient absorption experiments on a ns timescale [42].

The charge transfer process in dyads with relatively strong interaction of porphyrin and fullerene is monitored by observing the NIR absorption at ca. 700–800 nm and emission at ca. 750–850 nm [31, 35, 36]. However, in such systems formation of the charge-separated state is accompanied by exciplex formation, which complicates the analysis. In dyads with little or zero interaction of the chromophores in the ground state, monitoring of the charge separated state is performed by observing transient absorption at ca. 900–1050 nm range (transient absorption signals at 900 [31, 34], 920 [44], 1000 [3, 32, 38], 1020 [3, 42] and 1040 nm [39] wavelengths were reported as diagnostic features by different authors) attributed to the local excitation of the fullerene fragment in the charge-separated state and transient absorption at ca. 600–650 nm (transient absorption signals at 600 [32], 620 [32], 630 [34], 640 [3, 38] or 650 nm [31, 39, 44] were reported as diagnostic features by different authors), attributed to the local excitation of the porphyrin fragment in the charge-separated state. The latter marker is usually well observed in Zn-metallated systems but is hard to observe in free-base dyads.

The assignment of the transient absorption at 900–1050 nm to the local excitation of the fullerene fragment in the charge-separated state, conventionally called the absorption of the C_{60}-radical-anion fragment, is based on observation of a similar spectral feature in reduced species of pyrrolofullerenes (1010 nm) [43], methanofullerenes (1040 nm) [43] and pristine fullerene C_{60} (917, 995, 1064 nm) [57]. The 1064 nm absorption in C_{60} radical-anion appears as a sharp signal of large intensity (ε ca. $1.2 \cdot 10^5$ dm^3 mol^{-1} cm^{-1}) and was attributed to symmetry-allowed (0–0) t_{1u}–t_{1g} transition [57].

The assignment of the transient absorption at 600–650 nm to the local excitation of the porphyrin fragment in the charge-separated state, conventionally called the absorption of the Zn-porphyrin radical-cation fragment, is based on observation of a similar spectral feature upon oxidation of Zn-porphyrinates (600–700 nm, **Zn-11a** [58], 700 nm, **Zn-12** [59], 650 nm, **Zn-13** [60]).

Intersystem crossing from the initially formed ^{1}P*-C_{60} state to ^{3}P*-C_{60} state can be also followed by electron transfer generating long-living triplet charge-separated state, which is of special interest for solar energy conversion systems [3].

The lifetimes of the charge-separated states vary from ps to μs timescale, depending on the multiplicity, the structure of the dyad and the solvent used. Close face-to-face alignment of the chromophores results in short-living charge-separated states (ps to 3.5 ns [61, 62]), while large interchromophore distance and face-to-edge alignment tend to favor formation of long-lived charge-separated states [22].

22.5 Conclusion

Computational studies of porphyrin-fullerene covalent dyads predict for them electronic structure which favors charge separation process. In many dyads, a good separation of the chromophores is observed in the ground state, with two sets of orbitals localized each one on a specific chromophore. In particular, HOMO is localized on the porphyrin and LUMO—on the fullerene fragment, which creates the prerequisite for the existence of a highly polarized electronically excited state with $HOMO^1LUMO^1$ configuration. Time-dependent DFT methods predict that localization of the electronic systems of the chromophores in porphyrin-fullerene dyads results in formation of such electronically excited state called charge separated state along with the locally excited states.

The qualitative picture of the electronic states obtained by DFT calculations is supported fairly well with the experimental data: formation of each one of five classes of the electronically excited states was observed by means of either steady-state or transient spectroscopy. The spectral features characteristic of the ground state, charge-separated state, ^{1}P*-C_{60}, ^{3}P*-C_{60}, P-$^1C_{60}$* and P-$^3C_{60}$* states are collected together on Fig. 22.9. Horizontal bars represent spectral ranges where spectroscopic features, either absorption or emission bands, attributed to the ground or a specific electronically excited state have been reported in the literature. The lowest locally excited singlet states, either ^{1}P*-C_{60} or P-$^1C_{60}$* have typically the lifetimes of 1–2 ns,

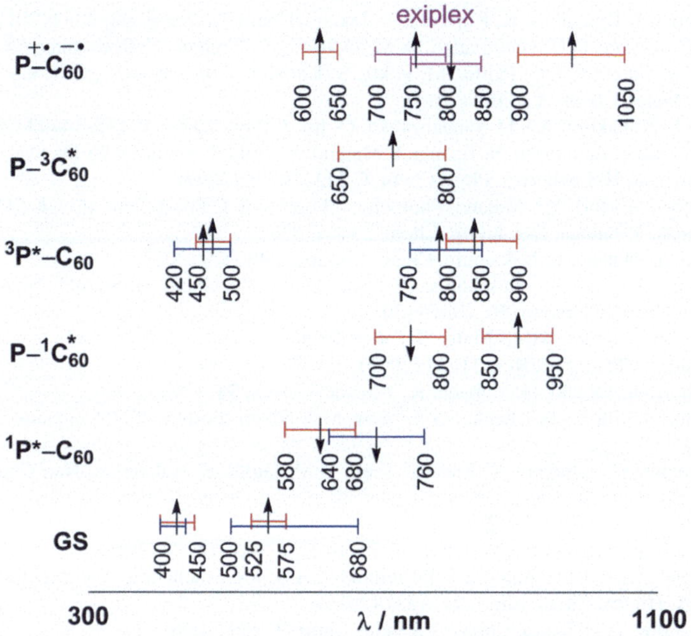

Fig. 22.9 Schematic representation of characteristic spectral features for different types of electronically excited states in porphyrin-fullerene dyads. Arrow up stands for absorption, arrow down stands for emission. Blue bars show spectral ranges characteristic for free-base dyads, in red—for Zn-metallated dyads or common for free base and metallated dyads. Magenta bars show spectral features of exciplex

while locally excited triplet states have the lifetimes up to several μs. The lifetime of the charge-separated state depends strongly on the structure of the compound, the multiplicity of the state and the solvent used in the experiment, varying from ps to μs timescale.

Acknowledgements The STEPS program funded by JSPS Inter-University Exchange Program and St. Petersburg State University—JTI joint program (Grant No. 12.54.1266.2016) are gratefully acknowledged.

References

1. G.N. Lim, C.O. Obondi, F. D'Souza, Angew. Chem. Int. Ed. **55**, 11517 (2016)
2. N. Martín, Chem. Commun. 2093 (2006)
3. F.D. Souza, O. Ito, Chem. Soc. Rev. **41**, 86 (2012)
4. S. Fukuzumi, K. Ohkubo, T. Suenobu, Acc. Chem. Res. **47**, 1455 (2014)
5. E.N. Durantini, A. Moore, T.A. Moore, D. Gust, Molecules **5**, 529 (2000)
6. M. Fujitsuka, K. Matsumoto, O. Ito, T. Yamashiro, Y. Aso, T. Otsubo, Res. Chem. Intermed. **27**, 73 (2001)

7. H. Imahori, T. Umeyama, K. Kurotobi, Y. Takano, Chem. Commun. **48**, 4032 (2012)
8. M. Vasilopoulou, D.G. Georgiadou, A.M. Douvas, A. Soultati, V. Constantoudis, D. Davazoglou, S. Gardelis, L.C. Palilis, M. Fakis, S. Kennou, T. Lazarides, A.G. Coutsolelos, P.J. Argitis, Mater. Chem. A **2**, 182 (2014)
9. T. Arai, S. Nobukuni, A.S.D. Sandanayaka, O. Ito, J. Phys. Chem. C **113**, 14493 (2009)
10. T. Umeyama, T. Takamatsu, N. Tezuka, Y. Matano, Y. Araki, T. Wada, O. Yoshikawa, T. Sagawa, S. Yoshikawa, H. Imahori, J. Phys. Chem. C **113**, 10798 (2009)
11. H. Imahori, Y. Mori, Y.J. Matano, Photochem. Photobiol. C Photochem. Rev. **4**, 51 (2003)
12. H. Imahori, Y. Sakata, Eur. J. Org. Chem. 2445 (1999)
13. T. Ichiki, Y. Matsuo, E. Nakamura, Chem. Commun. **49**, 279 (2013)
14. Y. Tamura, H. Saeki, J. Hashizume, Y. Okazaki, D. Kuzuhara, M. Suzuki, N. Aratani, H. Yamada, Chem. Commun. **50**, 10379 (2014)
15. N. Kafle, A. Buldum, AIMS Mater. Sci. **4**, 505 (2017)
16. D.M. Guldi, Chem. Soc. Rev. **31**, 22 (2002)
17. H. Imahori, K. Tamaki, H. Yamada, K. Yamada, Carbon **38**, 1599 (2000)
18. H. Lemmetyinen, N. Tkachenko, A. Efimov, M.J. Niemi, Porphyr. Phthalocyanines **13**, 1090 (2009)
19. J. Sukegawa, C. Schubert, X. Zhu, H. Tsuji, D.M. Guldi, E. Nakamura, Nat. Chem. **6**, 899 (2014)
20. O. Ito, Chem. Rec. **17**, 326 (2017)
21. N.J. Agnihotri, Photochem. Photobiol. C Photochem. Rev. **18**, 18 (2014)
22. A.S. Konev, A.F. Khlebnikov, P.I. Prolubnikov, A.S. Mereshchenko, A.V. Povolotskiy, O.V. Levin, A. Hirsch, Chem. Eur. J. **21**, 1237 (2015)
23. M.E. Zandler, F. D'Souza, Comptes Rendus Chim. **9**, 960 (2006)
24. O. Cramariuc, T.I. Hukka, T.T. Rantala, H. Lemmetyinen, J. Phys. Chem. A **110**, 12470 (2006)
25. O. Cramariuc, T.I. Hukka, T.T. Rantala, H. Lemmetyinen, J. Comput. Chem. **30**, 1194 (2009)
26. P.O. Krasnov, Y.M. Milytina, N.S. Eliseeva, Internet Electron. J. Mol. Des. **9**, 20 (2010)
27. T. Karilainen, O. Cramariuc, M. Kuisma, K. Tappura, T.I. Hukka, J. Comput. Chem. **36**, 612 (2015)
28. S. Grimme, J. Comput. Chem. **27**, 1787 (2006)
29. M. Elstner, P. Hobza, T. Frauenheim, S. Suhai, E. Kaxiras, J. Chem. Phys. **114**, 5149 (2001)
30. V. Vehmanen, N.V. Tkachenko, H. Imahori, S. Fukuzumi, H. Lemmetyinen, Spectrochim. Acta A **57**, 2229 (2001)
31. V. Chukharev, N.V. Tkachenko, A. Efimov, D.M. Guldi, A. Hirsch, M. Scheloske, H. Lemmetyinen, J. Phys. Chem. B **108**, 16377 (2004)
32. F. D'Souza, S. Gadde, M.E. Zandler, A. Klykov, M.E. El-Khouly, M. Fujitsuka, O. Ito, J. Phys. Chem. A **106**, 12393 (2002)
33. M.E. El-Khouly, Y. Araki, O. Ito, S. Gadde, A.L. McCarty, P.A. Karr, M.E. Zandler, F. D'Souza, Phys. Chem. Chem. Phys. **7**, 3163 (2005)
34. E. Krokos, C. Schubert, F. Spänig, M. Ruppert, A. Hirsch, D.M. Guldi, Chem. Asian J. **7**, 1451 (2012)
35. H. Imahori, N.V. Tkachenko, V. Vehmanen, K. Tamaki, H. Lemmetyinen, Y. Sakata, S. Fukuzumi, J. Phys. Chem. A **105**, 1750 (2001)
36. V. Chukharev, N.V. Tkachenko, A. Efimov, H. Lemmetyinen, Chem. Phys. Lett. **411**, 501 (2005)
37. A.H. Al-Subi, M. Niemi, J. Ranta, N.V. Tkachenko, H. Lemmetyinen, Chem. Phys. Lett. **531**, 164 (2012)
38. N.V. Tkachenko, H. Lemmetyinen, J. Sonoda, K. Ohkubo, T. Sato, H. Imahori, S. Fukuzumi, J. Phys. Chem. A **107**, 8834 (2003)
39. M.A. Fazio, A. Durandin, N.V. Tkachenko, M. Niemi, H. Lemmetyinen, D.I. Schuster, Chem. Eur. J. **15**, 7698 (2009)
40. I.D. Petsalakis, N. Tagmatarchis, G. Theodorakopoulos, J. Phys. Chem. C **111**, 14139 (2007)
41. M. Maggini, G. Scorrano, M. Prato, J. Am. Chem. Soc. **115**, 9798 (1993)

42. S. Vail, D.I. Schuster, D.M. Guldi, M. Isosomppi, N. Tkachenko, H. Lemmetyinen, A. Palkar, L. Echegoyen, X. Chen, J.Z.H. Zhang, J. Phys. Chem. B **110**, 14155 (2006)
43. D.M. Guldi, M. Prato, Acc. Chem. Res. **33**, 695–703 (2000)
44. H. Imahori, K. Hagiwara, M. Aoki, T. Akiyama, S. Taniguchi, T. Okada, M. Shirakawa, Y. Sakata, J. Am. Chem. Soc. **118**, 11771 (1996)
45. K.Y. Yeon, D. Jeong, S.K. Kim, Chem. Commun. **46**, 5572 (2010)
46. A. Lukaszewicz, J. Karolczak, D. Kowalska, A. Maciejewski, M. Ziolek, R.P. Steer, Chem. Phys. **331**, 359 (2007)
47. S.Y. Kim, T. Joo, J. Phys. Chem. Lett. **6**, 2993 (2015)
48. P.G. Seybold, M. Gouterman, J. Mol. Spectrosc. **31**, 1 (1969)
49. A.J. Harriman, Chem. Soc. Faraday Trans. **2**(77), 1695 (1981)
50. A.J. Harriman, Chem. Soc. Faraday Trans. **2**(76), 1978 (1980)
51. L. Pekkarinen, H. Linschitz, J. Am. Chem. Soc. **82**, 2407 (1960)
52. M. Gouterman, J. Mol. Spectrosc. **6**, 138 (1961)
53. D.M. Guldi, K.D. Asmus, J. Phys. Chem. A **101**, 1472 (1997)
54. F. D'Souza, M.E. Zandler, P.M. Smith, G.R. Deviprasad, A. Klykov, M. Fujitsuka, O. Ito, J. Phys. Chem. A **106**, 649 (2002)
55. D.M. Guldi, M. Maggini, G. Scorrano, M. Prato, J. Am. Chem. Soc. **119**, 974 (1997)
56. J.L. Anderson, Y.Z. An, Y. Rubin, C.S. Foote, J. Am. Chem. Soc. **116**, 9763 (1994)
57. M.A. Greaney, S.M. Gorun, J. Phys. Chem. **95**, 7142 (1991)
58. J. Fajer, D.C. Borg, A. Forman, D. Dolphin, R.H. Felton, J. Am. Chem. Soc. **92**, 3451 (1970)
59. P. Neta, A. Harriman, J. Chem. Soc. Farad. Trans. **81**(2), 123 (1985)
60. J.H. Fuhrhop, D. Mauzerall, J. Am. Chem. Soc. **91**, 4174 (1969)
61. D.I. Schuster, Carbon **38**, 1607 (2000)
62. D.I. Schuster, P. Cheng, P.D. Jarowski, D.M. Guldi, C. Luo, L. Echegoyen, S. Pyo, A.R. Holzwarth, S.E. Braslavsky, R.M. Williams, G. Klihm, J. Am. Chem. Soc. **126**, 7257 (2004)

Chapter 23
Optical and Magnetic Functionalities on Molecule-Based Magnetic Materials

Koji Nakabayashi, Shin-ichi Ohkoshi and Szymon Chorazy

Abstract In this chapter, the optical and magnetic properties of multifunctional cyanido-bridged metal assemblies are introduced. Cyanido-bridged metal assemblies draw much attention due to their magnetic properties and functionalities. As for magnetic properties, they have an advantage to show long-range magnetic ordering due to strong magnetic couplings between magnetic metal ions via cyanide. In addition, the cyanido-bridged metal assemblies can acquire structural diversity and various electronic states by combination of metal ions and ligand, resulted in their functionalities. For examples, introduction of transition metal ions showing charge transfer and spin crossover could allow switching of spin states by external stimuli, and metal assemblies containing lanthanide ions are expected to show luminescence and slow magnetic relaxation. Herein, some cyanido-bridged metal assemblies with unique characters of photoinduced magnetization, luminescence, and slow magnetic relaxation are presented.

23.1 Introduction

Studies on multifunctional materials where various magnetic, electric, and optical properties coexist are of interest in chemistry research [1, 2]. To design multifunctional materials, building blocks, which determine the chemical and physical properties, should be selected appropriately. From the viewpoint of designability, molecule-based materials such as metal complexes have received much attention because different combinations of metal cations and ligands provide various

K. Nakabayashi · S. Ohkoshi (✉)
Department of Chemistry, School of Science, The University of Tokyo, 7-3-1 Hongo, Bunkyo-ku, Tokyo 113-0033, Japan
e-mail: ohkoshi@chem.s.u-tokyo.ac.jp

K. Nakabayashi
e-mail: knakabayashi@chem.s.u-tokyo.ac.jp

S. Chorazy
Faculty of Chemistry, Jagiellonian University, Ingardena 3, 30-060 Kraków, Poland
e-mail: chorazy@chemia.uj.edu.pl

K. Yamanouchi et al. (eds.), *Progress in Photon Science*, Springer Series in Chemical Physics 119, https://doi.org/10.1007/978-3-030-05974-3_23

structures and desirable functionalities [3–7]. Among such materials, molecule-based magnetic materials are promising candidates for multifunctional materials. In the last two decades, multifunctional molecule-based magnetic materials, which combine magnetic phenomena of long-range magnetic ordering, slow magnetic relaxation, or spin transition with another functionality, have been reported [8–12]. Two-dimensional (2D) and three-dimensional (3D) coordination compounds with strong magnetic coupling between spin centers can exhibit long-range magnetic ordering, whereas non-centrosymmetric compounds are expected to show magnetic circular dichroism (MCD), magneto-chiral dichroism (MChD), and nonlinear magneto-optical effects. In an oxalate-bridged Mn^{II}–Cr^{III} ferromagnetic network with chiral counterions, an enhanced MChD effect has been observed due to the long-range magnetic ordering below the Curie temperature (T_C) of 7 K [13]. An enhanced MCD effect has been realized in a cyanido-bridged Mn^{II}(L)–Cr^{III} ferrimagnet with $T_C = 38$ K (L = (*S*)-1,2-diaminopropane) [14] and a cyanido-bridged Mn^{II}(L′)–Nb^{IV} with $T_C = 23.5$ K ferrimagnet (L′ = (*S*/*R*)-α-methyl-2-pyridinemethanol) [15]. A few reports have introduced chirality in low-dimensional systems displaying slow magnetic relaxation [16, 17], and the MCD effect has been demonstrated for an Mn_6–[(*R*)-sao] cluster ((*R*)-sao = chiral salicylaldoxime derivative) [18]. As for other magneto-optical effects, second harmonic generation (SHG), which is a nonlinear optical effect, is considerably enhanced by the onset of long-range magnetic ordering. This effect, called magnetization-induced second harmonic generation (MSHG), has been reported in Co–Cr Prussian blue analogs [19, 20], a cyanido-bridged Mn–Nb ferrimagnet [21, 22], a cyanido-bridged Mn–Mo ferrimagnet [23], and a Mn–Cr oxalate ferromagnet [24].

Some molecule-based magnets display ferroelectricity and proton conduction. For example, ferroelectricity and ferromagnetism coexist in rubidium manganese hexacyanoferrate [25]. Prussian blue analogs of $Co[Cr(CN)_6]_{2/3}\cdot zH_2O$ and $V[Cr(CN)_6]_{2/3}\cdot zH_2O$ show a high proton conduction exceeding 10^{-3} S cm^{-1} [26]. Moreover, $V[Cr(CN)_6]_{2/3}$ with a high magnetic phase transition temperature of 313 K exhibits an interference effect between magnetic ordering and proton conduction, which is likely caused by the distortion of the 3D hydrogen bonding network by magnetostriction below the T_C.

Photoresponsive magnets with photomagnetism such as photoinduced magnetization are some of the most interesting materials [27–31]. A cyanido-bridged Co–W metal assembly, $[\{Co^{II}(4\text{-methylpyridine})(\text{pyrimidine})\}_2\{Co^{II}(H_2O)_2\}\{W^V(CN)_8\}_2]\cdot 4H_2O$, exhibits photoinduced ferromagnetic ordering with a high T_C of 48 K and a large magnetic coercive field (H_c) of 27,000 Oe [32]. These T_C and H_c values are the highest in photoinduced magnetization systems. Furthermore, in a chiral cyanido-bridged Fe–Nb assembly, (±)–$Fe_2[Nb(CN)_8](4\text{-bromopyridine})_8\cdot 2H_2O$, spin-crossover-induced SHG, light-reversible spin-crossover, long-range magnetic ordering, and photoswitching of MSHG have been observed [33].

Luminescent magnets are an attractive target in molecule-based materials. Molecule-based magnets often have a lighter color in the visible light region compared to that of conventional magnets composed of metal oxides and other inorganic compounds. Light color materials are preferable for efficient luminescence in the visible light region. To date, some molecule-based magnetic compounds have been reported [34, 35].

23.2 Photomagnets and First-Principles Calculations

Numerous studies on optically switchable materials have been conducted toward applications as optical memory devices. Examples reported to date include metal oxides, photochromic compounds, chalcogenides, spin-crossover complexes, and cyanido-bridged metal assemblies. Cyanido-bridged Co–W metal assemblies are known to exhibit photoinduced ferromagnetism with a high T_C and H_c [30–32]. This phenomenon is attributed to an optical charge-transfer-induced spin transition (CTIST) from the Co^{III} (low spin, $S = 0$)–W^{IV}($S = 0$) phase to the Co^{II} (high spin, $S = 3/2$)–W^{V}($S = 1/2$) phase. However, the electronic structure of the optical transition and the mechanism for photomagnetic effects remain unclear.

A recent study revealed the electronic structure of a 2D cyanido-bridged Co–W metal assembly, ($H_5O_2^+$)[Co(4-bromopyridine)$_2${W(CN)$_8$}] [36]. Figure 23.1 shows the crystal structure of the cyanido-bridged Co–W metal assembly. The Co–N distances at the axial positions coordinated by 4-bromopyridine are 1.96–1.97 Å longer than those at equatorial positions coordinated by cyanide (1.88–1.91 Å). The average distance of 1.92 Å indicates trivalent Co ions. As a specific character, this Co–W metal assembly possesses $H_5O_2^+$ ions. Thus, the assembly is exposed to an acidic circumstance of $H_5O_2^+$ ions even in the solid state. The presence of IR peaks due to O–H$^+$–O groups at 930, 1090, 1310, 1400, and 1720 cm^{-1} confirms the existence of $H_5O_2^+$ ions. The blue crystal of the Co–W metal assembly shows an optical absorption band around 1.8 eV (700 nm), which is derived from Co^{III}. The magnetic susceptibility measurement demonstrates that the value is almost zero, revealing the electronic state of Co^{III}_{ls} ($S = 0$)–W^{IV} ($S = 0$) is between 2 and 390 K. This is a first example where the Co^{III}_{ls} ($S = 0$)–W^{IV} ($S = 0$) state appears over a wide temperature range between 2 and 390 K. This is because the $[W(CN)_8]^{4-}$ ion is more stable under acidic conditions.

First principles calculations reveal the optical transitions in ($H_5O_2^+$)[Co(4-bromopyridine)$_2${W(CN)$_8$}]. The periodic structure calculation was conducted based on the crystal structure determined by single-crystal X-ray diffraction analysis. Figure 23.2 shows the band structure and the density of states. The top of the valence band just below the Fermi energy (E_F) is mainly composed of W^{IV} and N states. The bottom of the conduction band just above E_F consists of Co^{III} and N states. The calculated optical absorption spectrum reproduces the observed spectrum well (Fig. 23.2b). As shown in Fig. 23.2c, the lowest-energy transition is from the valence band mainly containing the d_{z2} orbitals of W^{IV} and the p orbitals of N to the conduction band mainly derived from the d_{z2} orbitals of Co^{III} and the sp orbitals of N. Bader analysis provides charge density distributions of Co 0%, W 63%, N 36%, and C 0% for the valence band and Co 63%, W 1%, N 29%, and C 5% for the conduction band, indicating that the charge transfer from W^{IV} to Co^{III} is the lowest energy transition.

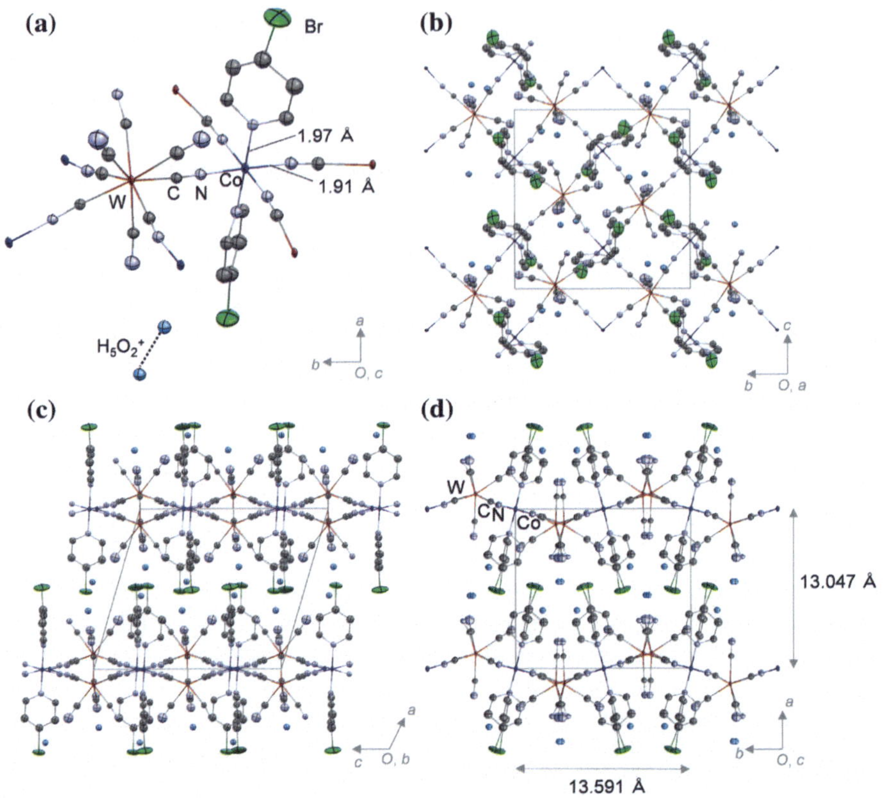

Fig. 23.1 Crystal structure of the two-dimensional cyanido-bridged Co–W metal assembly, $(H_5O_2^+)[Co(4\text{-bromopyridine})_2\{W(CN)_8\}]$. **a** Coordination geometry around Co and W. Views along **b** the a-axis, **c** the b-axis, and **d** the c-axis. Red, blue, light purple, gray, light blue, and green ellipsoids indicate W, Co, N, C, O, and Br, respectively. Hydrogen atoms are omitted for clarity. Reproduced from [36] with permission from the Royal Society of Chemistry

In addition, the charge density of the N atoms is considerably large, suggesting it contributes to the transition. Considering the orbital angular momentum and the parity of the p and sp orbitals, the transition from p to sp is allowed. Consequently, the charge transition process is accelerated through the orbitals of the bridging cyanides.

The photoresponsivity of the Co–W metal assembly was studied. Irradiating the assembly with a 785-nm CW laser light at 240 mW cm^{-2} and 4 K decreases the optical absorption around 1.8 eV (700 nm) and causes an absorption around 2.3 eV (540 nm), which is assigned to a charge transfer band from Co^{II} to W^{V} (Fig. 23.3). Thus, the photoirradiation induces a charge transfer phase transition from Co^{III}–W^{IV} to Co^{II}–W^{IV} phases with a drastic color change from blue to red. The phase transition provides a drastic change of magnetic properties. Additionally, irradiation with 785-nm light (220 mW m^{-2}) at 3 K induces spontaneous magnetization with T_C of 27 K, as shown in the magnetization (M) versus temperature (T) plots (Fig. 23.4).

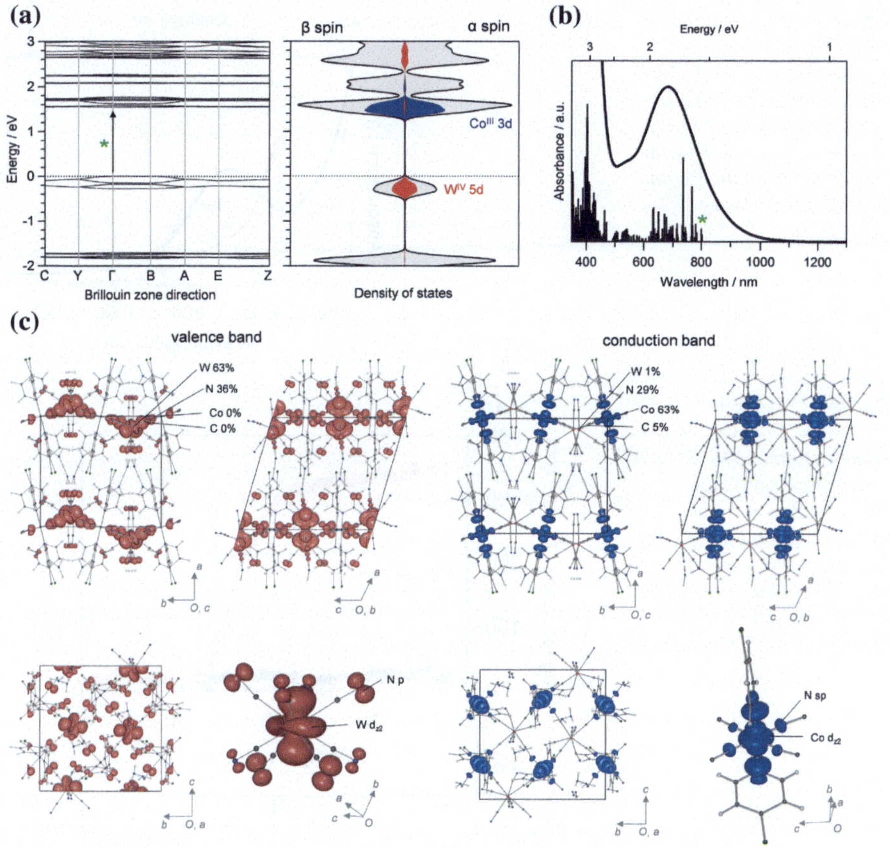

Fig. 23.2 First-principles calculations for $(H_5O_2^+)[Co(4\text{-bromopyridine})_2\{W(CN)_8\}]$. **a** Band structure (left) and density of states (DOS) near the Fermi energy level (right). Red, blue, and gray areas show the partial DOS for W^{IV} 5d, Co^{III} 3d, and the total DOS, respectively. **b** Calculated optical absorption spectrum. Black bars represent the calculated absorption strengths, while the line with an asterisk denotes the lowest-energy transition. **c** Charge density maps of the valence (left) and the conduction (right) bands. Reproduced from [36] with permission from the Royal Society of Chemistry

The M versus external magnetic field (H) plot after photoirradiation shows a coercive field of 2000 Oe at 2 K and a magnetization value of 3.0 μ_B at 50 kOe. The observed magnetization value almost agrees with the expected saturation magnetization value assuming ferromagnetic coupling between Co^{II} ($S = 1/2$, $g = 13/3$) and W^V ($S = 1/2$, $g = 2$). The photoinduced phase relaxes to the initial phase upon heating. These photothermal changes are reversible.

The powder XRD measurements before and after photoirradiation and their Rietveld analyses reveal the crystal structure of the photoinduced phase. The lattice constants increase by 3% in the 2D layer corresponding to the *bc*-plane compared to

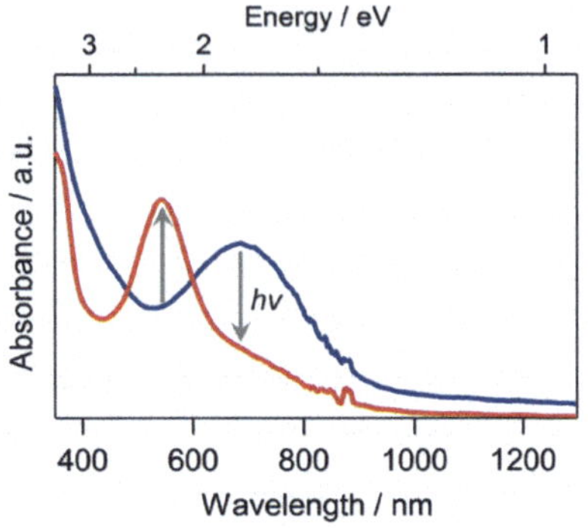

Fig. 23.3 UV-vis spectra of $(H_5O_2^+)[Co(4\text{-bromopyridine})_2\{W(CN)_8\}]$ before (blue line) and after photoirradiation (red line). Reproduced from [36] with permission from the Royal Society of Chemistry

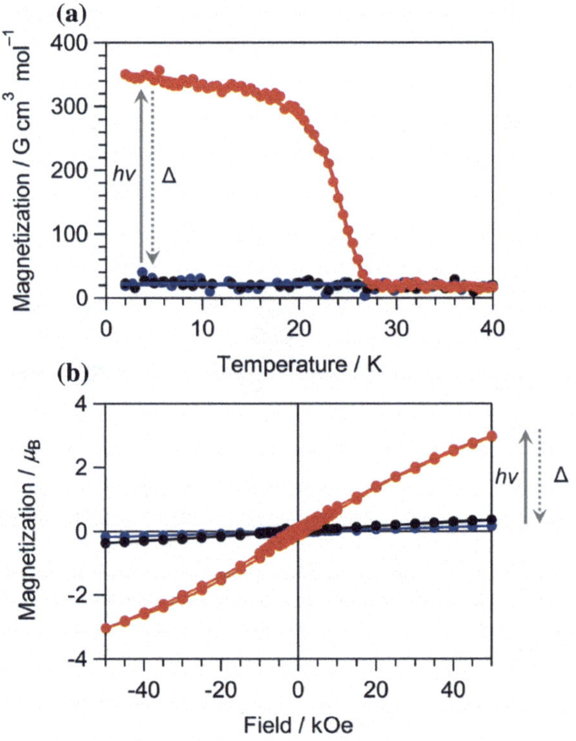

Fig. 23.4 Magnetic properties of $(H_5O_2^+)[Co(4\text{-bromopyridine})_2\{W(CN)_8\}]$ before (blue line), after photoirradiation (red line), and after heating up to 80 K (black). **a** *M-T* plots, **b** *M-H* plots. Reproduced from [36] with permission from the Royal Society of Chemistry

those of the original phase. The Co–N average distance of 2.05 Å, which is longer than that of the original phase (1.92 Å), indicates that photoirradiation changes the valence state from Co^{III} to Co^{II}. By increasing the temperature up to 60 K, the crystal structure of the photoinduced phase returns to that of the original phase. Among cyanido-bridged Co–W photomagnets, this is the first example where the crystal structure of the photoinduced phase is determined.

In another embodiment of photomagnetism for cyanido-bridged metal assemblies, a Cu–Mo metal assembly of $\{[Cu^{II}(H_2O)]_2[Mo^{IV}(CN)_8]\}\cdot 2H_2O$ was studied [37]. This compound is a known photomagnet, but until recently its crystal and electronic structures, which are necessary to discuss the optical transitions and the mechanism of photomagnetic effects, were unclear. Improving the crystallinity of the Cu–Mo metal assembly by using a gel method allows single-crystal XRD analysis, revealing the 3D cyanido-bridged Cu–Mo framework (Fig. 23.5). Based on this crystal structure, first-principles calculations of the Cu–Mo metal assembly were performed.

In the calculations, a primitive unit cell with a $\{Cu_4Mo_2C_{16}N_{16}H_{16}O_8\}$ composition composed of two stoichiometric $\{Cu_2MoC_8H_8N_8O_4\}$ molecular fragments, was used. The Cu^{II} 3p/3d electrons, the Mo^{IV} 4p/4d electrons, the O 2p electrons, the H 1s electrons, and the CN– 2p electrons were considered, resulting in a total of 300 occupied bands. Half of the bands are occupied by α spins (1st to 150th valence bands: v1α–v150α), while the other half are occupied by β spins (1st to 150th valence

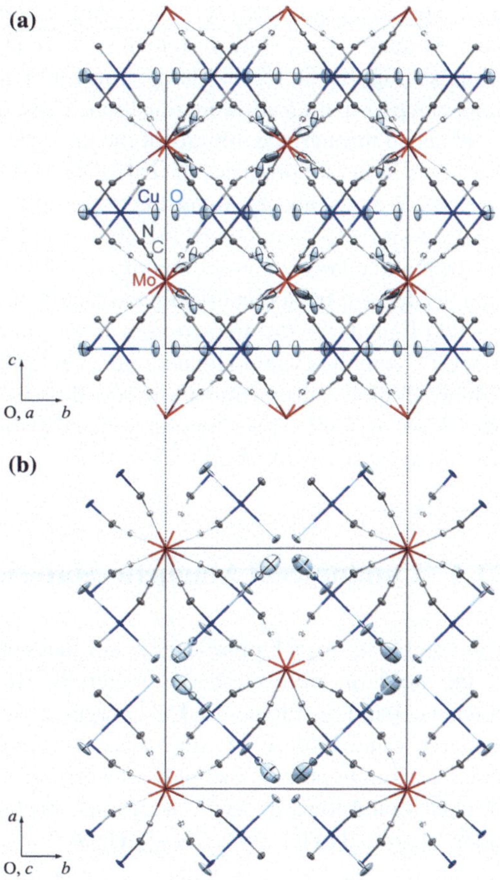

Fig. 23.5 Crystal structure of the 3D cyanido-bridged Cu–Mo metal assembly, $\{[Cu^{II}(H_2O)]_2[Mo^{IV}(CN)_8]\}\cdot 2H_2O$, along **a** the *a* axis and **b** the *c* axis. Red, blue, dark gray, gray, and light blue ellipsoids indicate Mo, Cu, N, C, and O, respectively. Hydrogen atoms are omitted for clarity. Reproduced from [37] with permission from John Wiley and Sons

bands: v1β–v150β). Figure 23.6 shows the calculated total and partial DOS in the range from –4 to 4 eV, where the occupied bands below E_F as well as the unoccupied bands in an area just above E_F (the conduction bands for or spins from c1α/β to c12α/β) are presented. The occupied bands just below the E_F level (v1α/β and v2α/β) mainly consist of the Mo^{IV} states, whereas the unoccupied bands just above the E_F level (c1α/β and c2α/β) are composed of the Cu^{II} states. The contributions from the nitrogen, carbon, and oxygen atoms exist in the valence bands below −1 eV and the conduction bands above +1.2 eV. Figure 23.7 shows the details about the band structure near the E_F level. The optical transition of the band gap is a direct transition with a band gap of 1.2 eV exactly at the point corresponding to the transition between the occupied v1α band and the unoccupied c1α band.

Figure 23.7b depicts the charge density maps for the v1α and c1α bands, indicating that the Mo^{IV} d_{z2} orbital and the N p_z orbital mainly contribute to the valence band of v1α. The conduction band of c1α is formed mostly by the Cu^{II} d_{x2-y2} orbital with an admixture of the N sp_x orbitals. Bader analysis gives charge density distributions with Cu 1.52%, Mo 63.96%, C 0.00%, and N 34.44% for the top of the valence band and Cu 68.14%, Mo 5.48%, C 8.76%, and N 17.60% for the bottom of the conduction band. Similar to the Co–W system mentioned above, the contribution of the nitrogen orbitals is significant for the charge transfer in the Cu–Mo system because the transition from p to sp is an allowed transition. Thus, the charge transfer from Mo^{IV} to Cu^{II} is enhanced through the orbitals of the nitrogen atoms in the bridging cyanides. The electronic structure produces the calculated UV-vis absorption spectrum (Fig. 23.8). The visible-range absorption is due to the optical transitions from the valence bands of v1α/β and v2α/β composed mainly of the Mo^{IV} orbitals to the conduction bands of c1α/β and c2α/β composed mainly of the Cu^{II} orbitals. The absorption in the UV range is derived from metal-to-ligand charge transfer (MLCT) from the Mo^{IV} to the cyanido ligand, the d-d transition in Mo^{IV}, and the ligand-to-metal charge transfer (LMCT) from the cyanido ligand to the Cu^{II}. These results demonstrate that visible-light irradiation causes a transition from the v1α to the c1α/β, that is, a charge transfer from Mo^{IV} to Cu^{II}. This supports a mechanism of photoinduced charge transfer in Cu–Mo systems, as reported previously.

23.3 Luminescent Magnetic Materials

The combination of luminescence and magnetic properties is an interesting topic in the field of molecule-based materials. To construct switchable luminescent magnets, building blocks of Ln^{III}, organic ligands, and cyanide metallates were selected. Lanthanide (3+) can act as luminophore and spin source. Organic ligands are luminophores and photosensitizers that induce energy transfer from an organic ligand towards Ln^{III}. A cyanido-bridged Tb^{III}(Box)–W^V metal assembly, $\{[Tb^{III}(Box)_2(dmf)_2]–[W^V(CN)_8]\}\cdot H_2O$ (Box = 2,2′-Bis(2-oxazoline), dmf = *N*,*N*′-dimethylformamide), was prepared by mixing solutions of the building blocks [38]. Figure 23.9 shows the crystal structure of the compound, revealing a 2D cyanido-

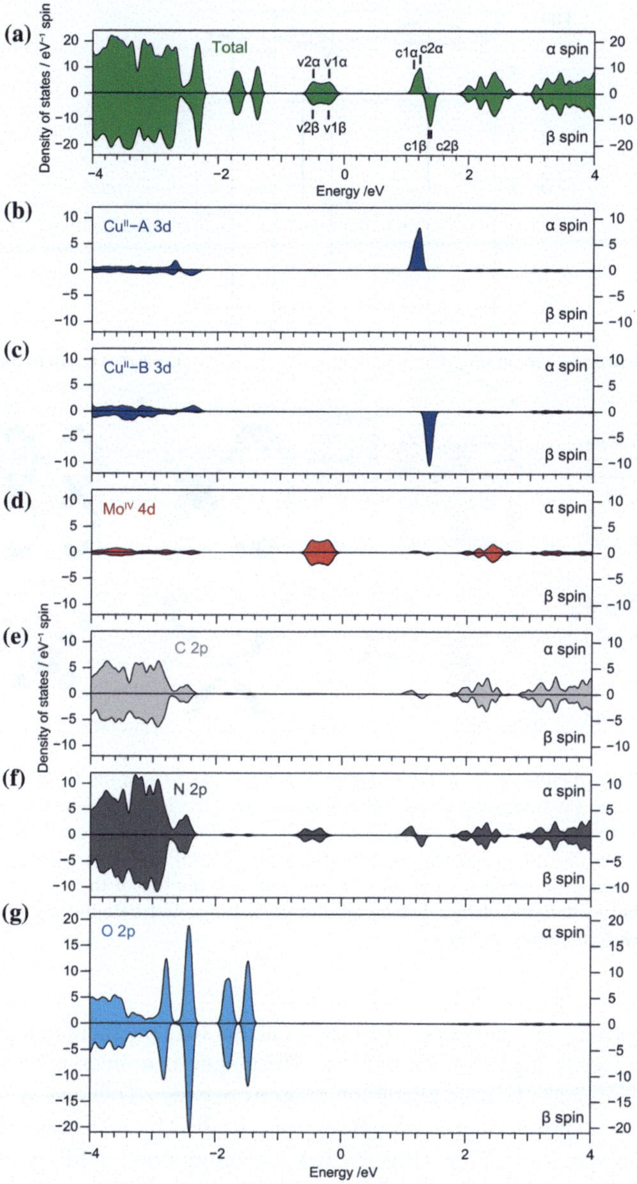

Fig. 23.6 Total and partial DOS of $\{[Cu^{II}(H_2O)]_2[Mo^{IV}(CN)_8]\}\cdot 2H_2O$. Cu^{II}–A and Cu^{II}–B represent two possible positions of Cu^{II} in the crystal structure. v1α/β and v2α/β show the first and second-highest valence bands for α or β spin, respectively. c1α/β and c2α/β indicate the first and second-highest conduction bands for α or β spin, respectively. Reproduced from [37] with permission from John Wiley and Sons

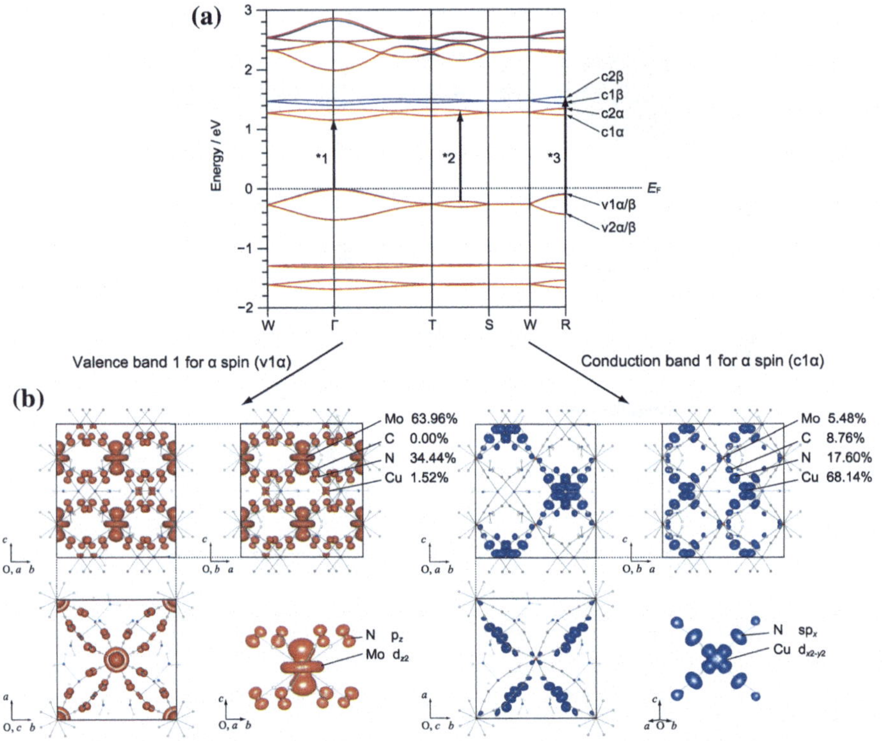

Fig. 23.7 **a** Band structure of $\{[Cu^{II}(H_2O)]_2[Mo^{IV}(CN)_8]\}\cdot 2H_2O$ near the E_F level. Red and blue lines represent α and β spin, respectively. Black arrows of *1, *2, and *3 indicate three selected transitions from the valence to the conductions bands at various *k* points, corresponding to three component lines in the calculated UV/Vis spectrum (Fig. 23.9b). **b** Charge-density maps for the bottom of the conduction band (c1α at the Γ point, top) and the top of the valence band (v1α at the Γ point, bottom) with the charge density around the metal centers. Reproduced from [37] with permission from John Wiley and Sons

bridged Tb^{III}(Box)–W^V network. The direct-current (dc) magnetism reveals a long-range magnetic ordering below T_C of 2.4 K. The magnetization vs. external magnetic field at 1.8 K suggests ferromagnetic interactions between Tb^{III} and W^V.

The solid-state luminescence properties were studied by measuring the emission and excitation spectra at 77 K (Fig. 23.10). The compound exhibits two different luminescent colours depending on the wavelength of the UV excitation. Irradiating with 260-nm excitation light provides an emission spectrum with strong sharp peaks at 490, 545, 585, and 622 nm and weaker peaks at 647, 667, and 678 nm, resulting in green luminescence. This emission is attributed to intra f^8 $^5D_4 \rightarrow {}^7F_{0\text{-}6}$ transitions in Tb^{III}. Irradiating with 340-nm light induces a dominant broad band at 650 nm and a weak broad band at 435 nm that are accompanied by the weak peaks of Tb^{III} center luminescence. The dominant broad emission produces red luminescence.

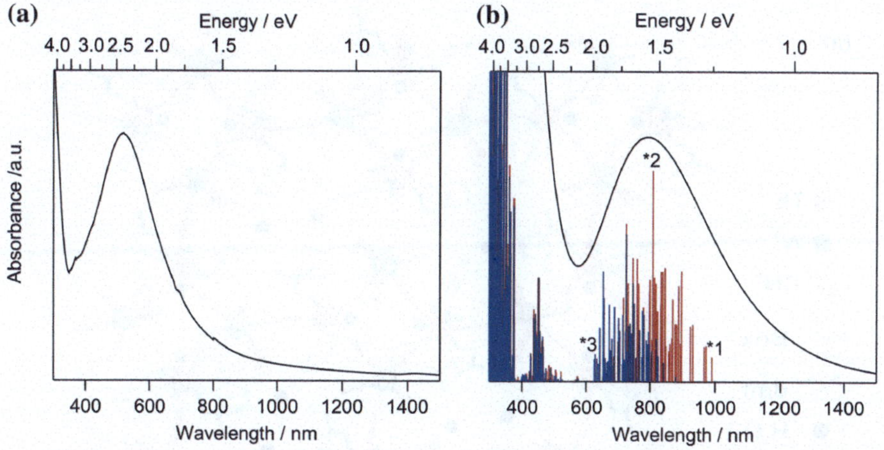

Fig. 23.8 a Experimental and **b** calculated UV/Vis spectrum of $\{[Cu^{II}(H_2O)]_2[Mo^{IV}(CN)_8]\}\cdot 2H_2O$. Red and blue lines indicate the calculated absorption strengths of the α and β spin bands, respectively. Reproduced from [37] with permission from John Wiley and Sons

The emission at 650 nm is assigned to $^3(\pi - \pi^*)T_1 \rightarrow S_0$ phosphorescence of the Box ligand. Since Box alone does not exhibit phosphorescence but instead shows the blue fluorescence of the $^1(\pi - \pi^*)S_1 \rightarrow S_0$ transition, the phosphorescence is due to the heavy atom effect of Tb^{III} accelerating the $S_1 \rightarrow T_1$ intersystem crossing followed by the $T_1 \rightarrow S_0$ transition. Thus, irradiating this compound with UV light below 300 nm only populates the excited states of Tb^{III} because Box does not possess such high energy levels. As a result, a green emission is observed. On the contrary, UV light above 300 nm provides both excited states of Tb^{III} and Box. Energy transfer efficiently occurs to the T_1 state of Box, leading to a red emission.

A cyanido-bridged Dy^{III}–Co^{III} metal assembly, $\{[Dy^{III}$(3-hydroxypyridine)$_2(H_2O)_4][Co^{III}(CN)_6]\cdot H_2O\}$, was prepared [39], and single-crystal XRD analysis revealed a zig-zag chain structure (Fig. 23.11). The alternate-current (ac) susceptibility measurements show a SMM behavior of the intrachain Dy complexes, and analyses using the generalized Debye model give a thermal energy barrier of ΔE of 266(12) cm^{-1} with a relaxation time τ_0 of $3.2(2) \times 10^{-11}$ s. This is one of the largest energy barriers among emissive SMMs. Below 10 K, the $\ln\tau$ versus T^{-1} plot becomes temperature-independent due to the significant quantum tunneling of magnetization (QTM) effect (Fig. 23.12). The QTM is partially suppressed by applying external magnetic fields of 1 and 2 kOe. Due to the high ΔE, magnetization hysteresis loops are observed below 6 K for a field sweep rate of 10 Oe/s. The QTM effect dominating at low magnetic fields and temperatures may lead to the characteristic butterfly shape of the M-H loops.

The cyanido-bridged Dy^{III}–Co^{III} metal assembly shows a white light emission when excited by 312 nm light (Fig. 23.13). The emission spectrum with 481, 576,

(a)

(b)

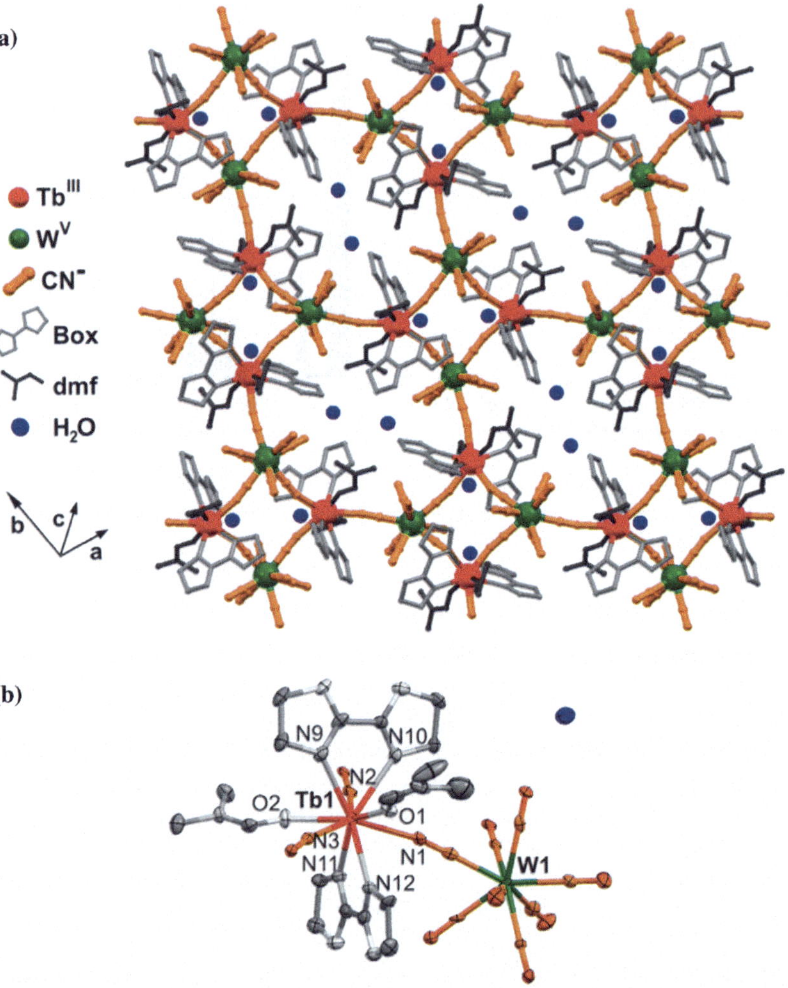

Fig. 23.9 Crystal structure of the two-dimensional cyanido-bridged Tb–W metal assembly, $\{[Tb^{III}(Box)_2(dmf)_2]–[W^{V}(CN)_8]\}\cdot H_2O$. **a** 2D structure and **b** the coordination geometry around Tb^{III} and W^{V}. Hydrogen atoms are omitted for clarity. Adapted with permission from [38]. Copyright 2014 American Chemical Society

and 660 nm is assigned to $^4F_{9/2} \rightarrow {}^6H_{15/2,13/2,11/2}$ transitions on Dy^{III}. In the excitation spectrum, the broad band at 312 nm, which is assigned to 3-hydroxypyridine and $[Co^{III}(CN)_6]^{3-}$, is predominant, indicating an energy transfer from the organic ligand and Co^{III} to Dy^{III}. At a low temperature of 3.5 K, a high-resolution emission spectrum was rather collected, and it can be deconvoluted into eight components, which are considered to be the crystal field splitting of $^6H_{15/2}$ multiples. The energy difference between the two lowest doublets is 211(31) cm^{-1}. This energy corresponds to the

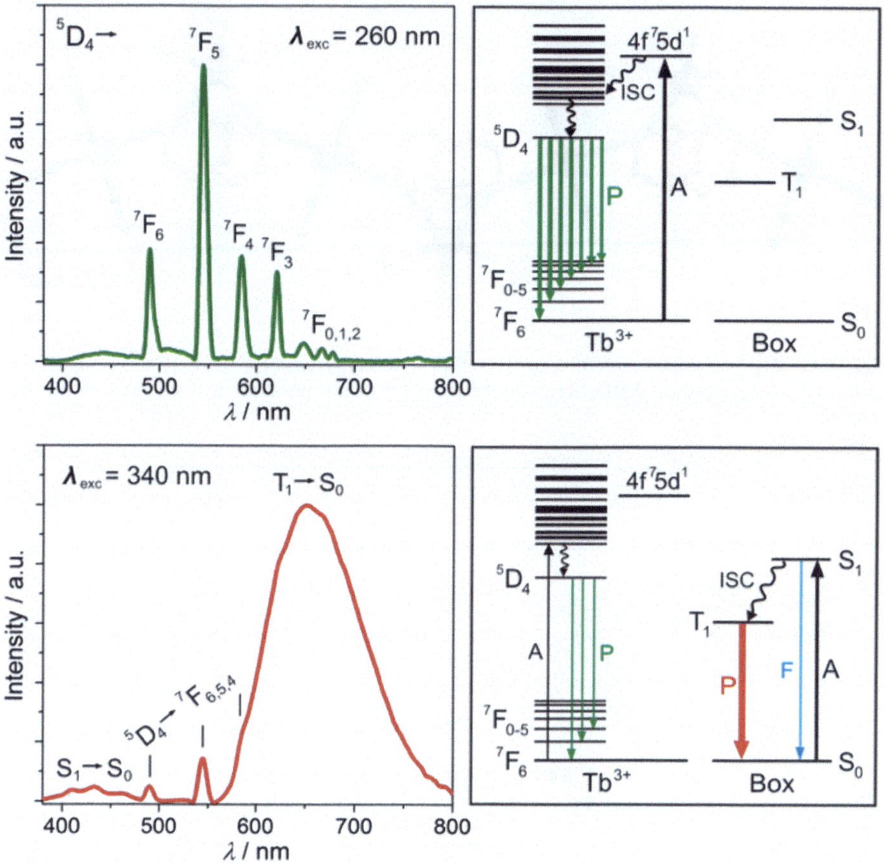

Fig. 23.10 Emission spectra of $\{[Tb^{III}(Box)_2(dmf)_2]–[W^V(CN)_8]\}{\cdot}H_2O$ at $T = 77$ K excited by 260 nm (upper left) and 340 nm (bottom left) light. Right schematic illustration represents the energy level diagrams with electronic transitions. A = absorption, P = phosphorescence, F = fluorescence, ISC = intersystem crossing. Reprinted with permission from [38].
Copyright 2014 American Chemical Society

zero-field energy barrier, which is close to the value of 266(12) cm^{-1} estimated from the magnetic analysis mentioned above. The cyanido-bridged Dy^{III}–Co^{III} metal assembly is a unique multifunctional material exhibiting a white light emission and SMM behavior.

This chapter presents recent examples of multifunctional cyanido-bridged metal assemblies exhibiting photomagnetism or luminescence. Although the aforementioned Co^{II}–W^V and Cu^{II}–Mo^{IV} metal assemblies have well known photomagnetism, the electronic structures were not clear. Recently, first principles calculations have revealed their electronic structures. This knowledge will allow researchers to design new photomagnets with higher functionalities. The combination of luminescence and long-range magnetic ordering (or single-molecule magnetic behavior) is a sophis-

Fig. 23.11 Zig-zag chain structure of $\{[Dy^{III}(3\text{-hydroxypyridine})_2(H_2O)_4][Co^{III}(CN)_6]\cdot H_2O\}$ revealed by single-crystal XRD. Reproduced from [39] with permission from John Wiley and Sons

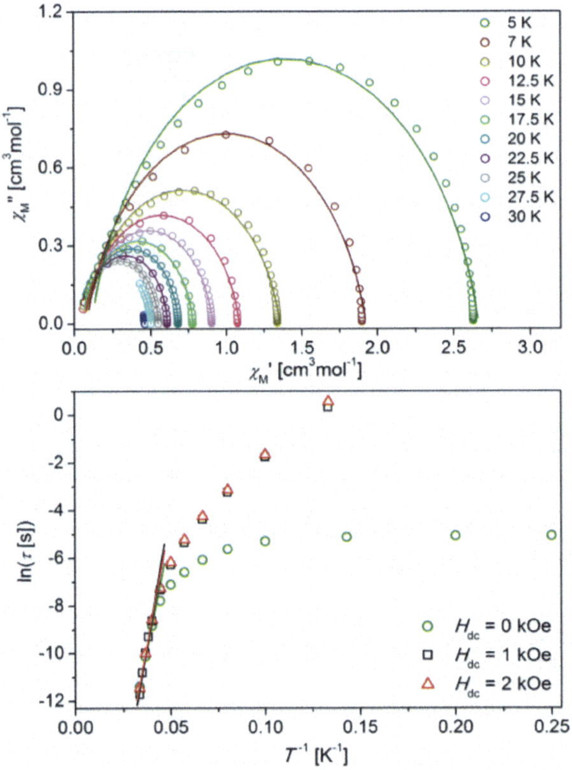

Fig. 23.12 Alternate-current (ac) magnetism of $\{[Dy^{III}(3\text{-hydroxypyridine})_2(H_2O)_4][Co^{III}(CN)_6]\cdot H_2O\}$: the $\chi_{M''} - \chi_{M'}$ plots at indicated T (upper) and the relaxation time (τ) as $\ln(\tau)$ versus T^{-1} plots in various H_{dc} of 0, 1, and 2 kOe. Solid lines in the upper figure are fitted using the generalized Debye model, whereas the lines in the bottom figure are fitted to the Arrhenius law. Reproduced from [39] with permission from John Wiley and Sons

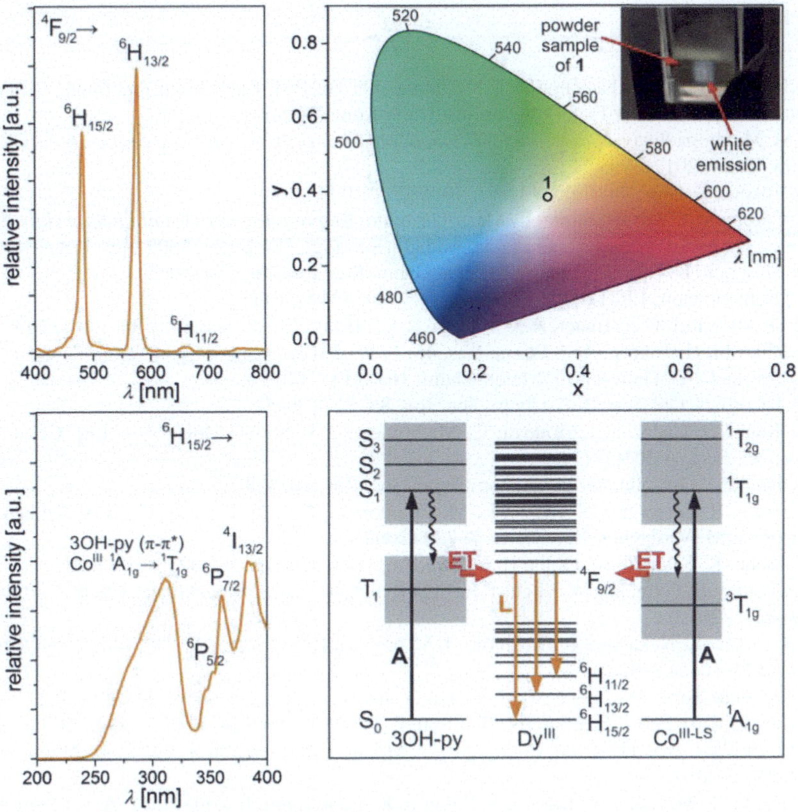

Fig. 23.13 Solid state emission (upper left, λ_{exc} = 312 nm) and excitation (bottom left, λ_{em} = 576 nm) spectra of {[Dy^{III}(3-hydroxypyridine)$_2$(H_2O)$_4$][Co^{III}(CN)$_6$]·H_2O}, 1, at T = 300 K. Emission color is presented on the CIE 1931 chromaticity diagram (upper right). Inset shows the photo of the white emission from the powder sample. Related energy level diagram is presented (bottom right; A = absorption, ET = energy transfer, L = lanthanide luminescence). Reproduced from [39] with permission from John Wiley and Sons

ticated functionality for molecule-based materials and should inspire new ideas in related research areas.

References

1. A.K.T. Lau, J. Lu, V.K. Varadan, F.K. Chang, J.P. Tu, P.M. Lam, *Multi-functional Materials and Structures* (Trans Tech Publications, Hong Kong, 2008)
2. S.M. Mukhopadhyay, *Nanoscale Multifunctional Materials: Science and Applications* (Wiley, New Jersey, 2012)
3. H. Tokoro, S. Ohkoshi, Dalton Trans. **40**, 6825 (2011)
4. E. Coronado, C. Gimnez-Saiz, C. Marti-Gastaldo, *Engineering of Crystalline Materials Properties*, ed. by J.J. Novoa, D. Braga, L. Addadi (Springer, Dordrecht, 2008), pp. 173
5. D. Maspoch, D. Ruiz-Molina, J. Veciana, Chem. Soc. Rev. **36**, 770 (2007)
6. P. Dechambenoit, J.R. Long, Chem. Soc. Rev. **40**, 3249 (2011)
7. M.D. Allendorf, C.A. Bauer, R.K. Bhakta, R.J.T. Houk, Chem. Soc. Rev. **38**, 1330 (2009)
8. S. Ohkoshi, H. Tokoro, Acc. Chem. Res. **45**, 1749 (2012)
9. E. Coronado, D. Gatteschi, J. Mater. Chem. **16**, 2513 (2006)
10. H. Tokoro, S. Ohkoshi, Bull. Chem. Soc. Jpn. **88**, 227 (2015)
11. O. Kahn, O. Cador, J. Larionova, C. Mathoniere, J.-P. Sutter, Mol. Cryst. Liq. Cryst. Sci. Technol. Sect. A **305**(1) (1997)
12. C. Train, M. Gruselle, M. Verdaguer, Chem. Soc. Rev. **40**, 3297 (2011)
13. C. Train, R. Gheorghe, V. Krstic, L.-M. Chamoreau, N.S. Ovanesyan, G.L.J.A. Rikken, M. Gruselle, M. Verdaguer, Nat. Mater. **7**, 729 (2008)
14. K. Inoue, K. Kikuchi, M. Ohba, H. Okawa, Angew. Chem. Int. Ed. **42**, 4810 (2003)
15. S. Chorazy, R. Podgajny, W. Nitek, T. Fic, E. Gçrlich, M. Rams, B. Sieklucka, Chem. Com-mun. **49**, 6731 (2013)
16. S. Chorazy, K. Nakabayashi, K. Imoto, J. Mlynarski, B. Sieklucka, S. Ohkoshi, J. Am. Chem. Soc. **134**, 16151 (2012)
17. J. Ferrando-Soria, D. Cangussu, M. Eslava, Y. Journaux, R. Lescouézec, M. Julve, F. Lloret, J. Pasan, C. Ruiz-Perez, E. Lhotel, C. Paulsen, E. Pardo, Chem. Eur. J. **17**, 12482 (2011)
18. J.M. Bradley, A.J. Thomson, R. Inglis, C.J. Milios, E.K. Brechin, S. Piligkos, Dalton Trans. **39**, 9904 (2010)
19. T. Nuida, T. Matsuda, H. Tokoro, S. Sakurai, K. Hashimoto, S. Ohkoshi, J. Am. Chem. Soc. **127**, 11604 (2005)
20. S. Ohkoshi, K. Arai, Y. Sato, K. Hashimoto, Nat. Mater. **3**, 857 (2004)
21. Y. Tsunobuchi, W. Kosaka, T. Nuida, S. Ohkoshi, Cryst. Eng. Comm. **11**, 2051 (2009)
22. D. Pinkowicz, R. Podgajny, W. Nitek, M. Rams, A.M. Majcher, T. Nuida, S. Ohkoshi, B. Sieklucka, Chem. Mater. **23**, 21 (2011)
23. M. Komine, K. Imoto, Y. Miyamoto, K. Nakabayashi, S. Ohkoshi, Eur. J. Inorg. Chem., 1367 (2018)
24. C. Train, T. Nuida, R. Gheorghe, M. Gruselle, S. Ohkoshi, J. Am. Chem. Soc. **131**, 16838 (2009)
25. E. Pardo, C. Train, H. Liu, L.M. Chamoreau, B. Dhkil, K. Boubekeur, F. Lloret, K. Nakatani, H. Tokoro, S. Ohkoshi, M. Verdaguer, Angew. Chem. Int. Ed. **51**, 8356 (2012)
26. S. Ohkoshi, H. Tokoro, T. Matsuda, H. Takahashi, H. Irie, K. Hashimoto, Angew. Chem. Int. Ed. **46**, 3238 (2007)
27. S. Ohkoshi, K. Imoto, Y. Tsunobuchi, S. Takano, H. Tokoro, Nat. Chem. **3**, 564 (2011)
28. E.S. Koumousi, I.-R. Jeon, Q. Gao, P. Dechambenoit, D.N. Woodruff, P. Merzeau, L. Buisson, X. Jia, D. Li, F. Volatron, C. Mathonière, R. Clérac, J. Am. Chem. Soc. **136**, 15461 (2014)
29. O.N. Risset, P.A. Quintero, T.V. Brinzari, M.J. Andrus, M.W. Lufaso, M.W. Meisel, D.R. Talham, J. Am. Chem. Soc. **136**, 15660 (2014)
30. S. Ohkoshi, Y. Hamada, T. Matsuda, Y. Tsunobuchi, H. Tokoro. Chem. Mater. **20**, 3048 (2008)
31. S. Ohkoshi, S. Ikeda, T. Hozumi, T. Kashiwagi, K. Hashimoto, J. Am. Chem. Soc. **128**, 5320 (2006)
32. N. Ozaki, H. Tokoro, Y. Hamada, A. Namai, T. Matsuda, S. Kaneko, S. Ohkoshi, Adv. Funct. Mater. **22**, 2089 (2012)

33. S. Ohkoshi, S. Takano, K. Imoto, M. Yoshikiyo, A. Namai, H. Tokoro, Nat. Photonics **8**, 65 (2014)
34. J.-M. Rueff, J.-F. Nierengarten, P. Gillot, A. Demessence, O. Cregut, M. Drillon, P. Ra-bu, Chem. Mater. **16**, 2933 (2004)
35. E. Chelebaeva, J. Larionova, Y. Guari, R.A.S. Ferreira, L.D. Carlos, F.A. Almeida Paz, A. Trifonov, C. Guerin, Inorg. Chem. **48**, 5983 (2009)
36. Y. Miyamoto, T. Nasu, N. Ozaki, Y. Umeta, H. Tokoro, K. Nakabayashi, S. Ohkoshi, Dalton Trans. **45**, 19249 (2016)
37. Y. Umeta, S. Chorazy, K. Nakabayashi, S. Ohkoshi, Eur. J. Inorg. Chem., 1980 (2016)
38. S. Chorazy, K. Nakabayashi, S. Ohkoshi, B. Sieklucka, Chem. Mater. **26**, 4072 (2014)
39. S. Chorazy, M. Rams, K. Nakabayashi, B. Sieklucka, S. Ohkoshi, Chem. Eur. J. **22**, 7371 (2016)

Chapter 24
Tailoring Spins and Orbitals in Spin–Orbitronic Interfaces Probed by X-Ray Magnetic Circular Dichroism

Jun Okabayashi

Abstract This review summarizes recent X-ray magnetic circular dichroism (XMCD) studies for interfacial perpendicular magnetic anisotropy. The nature of chemical bonding and electron occupancies bring novel properties at the interfaces. The control of orbital magnetic moments could enable to reveal and utilize novel promising phenomena beyond spintronics. Detailed investigations of anisotropic orbital magnetic moments at the interfaces in Fe|MgO and Co|Pd cases are discussed with brief introduction of theoretical background of XMCD spectra and brief history of this technique.

24.1 Introduction

Magneto-optics has been developed to investigate electronic and magnetic states. Linearly polarized light passes through transient magnets leading to an output of circularly polarized lights, referred to as Faraday effect. In case of a reflection geometry, similar effects are observed as Kerr effect. These effects originate from the difference in absorption coefficients between right- and left-circularly polarized lights, leading to magnetic circular dichroism (MCD). For the incident beam, the ultraviolet region has been utilized. Since the development of synchrotron radiation, in particular, circularly polarized beams in the soft- and hard-X-ray regions, magneto-optical techniques with MCD using circularly polarized X-rays have been extensively developed, referred to as XMCD techniques.

We summarize the history of XMCD. Erskine and Stern theoretically predicted the Ni-M-edge XMCD through the off-diagonal parts of the conductivity tensor σ_{xy}, which is 10^{-1} of the diagonal part σ_{xx} [1]. In 1987, Schütz et al. experimentally observed the XMCD of the Fe K-edge [2]. Using soft X-rays, Chen et al. succeeded in the measurements of XMCD of Ni L-edges [3]. Theoretical investigations of XMCD have also been proceeded with the experimental developments. Selection rules for

J. Okabayashi (✉)
Research Center for Spectrochemistry, The University of Tokyo, Hongo 7-3-1, Bunkyo, Tokyo 113-0033, Japan
e-mail: jun@chem.s.u-tokyo.ac.jp

K. Yamanouchi et al. (eds.), *Progress in Photon Science*, Springer Series in Chemical Physics 119, https://doi.org/10.1007/978-3-030-05974-3_24

optical absorption permit a difference of quantum numbers of $\Delta l = \pm 1$, which corresponds to an excitation from p to d states in transition-metal (TM) compounds. For 3d TMs, the L-edge absorption energies are located at the soft-X-ray regions. Magneto-optical sum rules, proposed by Carra and Thole, enabled quantitative estimations of spin and orbital magnetic moments [4, 5]. In 1995, C. T. Chen et al. reported conventional methods for the estimations of spin and orbital magnetic moments from sum rules using Fe and Co layers [6]. XMCD and its spectral analysis using sum rules became one of the standard techniques to deduce element-specific magnetic properties. In 1998, Koide et al. investigated the angular dependence of XMCD in order to reveal the magnetic anisotropy [7]. As the investigations of magnetic anisotropy were limited to phenomenological studies using magnetization measurements, studies from the point of view of electronic structure were required. The angular dependence in XMCD provides information about the anisotropy in spins and orbitals; it becomes a powerful technique to investigate the element-specific magnetic properties, particularly in layered interfacial structures [8]. After 2000, XMCD techniques have been continuously extended and combined with other techniques, which led to the development of XMCD microscopy [9] and time-resolved measurements [10].

A schematic of the XMCD beam-line is shown in Fig. 24.1. Soft X-rays in the range of 50–1,200 eV can be utilized for X-ray absorption spectroscopy (XAS), XMCD, and photoemission spectroscopy; the setup is constructed at the BL-7A, Photon Factory, High-Energy Accelerator Research Organization (KEK) in Tsukuba, Japan [11]. For XMCD, a 1.2-T electromagnet has been installed at the end station. As the beam comes from the bending magnets, circularly polarized lights for XMCD are obtained using the edges from the beam center, which is adjusted by a mirror in the beam-line. The beam and magnetic-field directions are fixed to be parallel, and the sample surface normal direction is rotated for the angular-dependent XMCD. Total-electron-yield (TEY) modes, which detect the drain currents excited from the sample at 3 nm from the sample surface, are utilized for XMCD measurements. In addition to TEY measurements, fluorescent yield modes are also employed, which probe to a depth of approximately 100 nm from the sample surface.

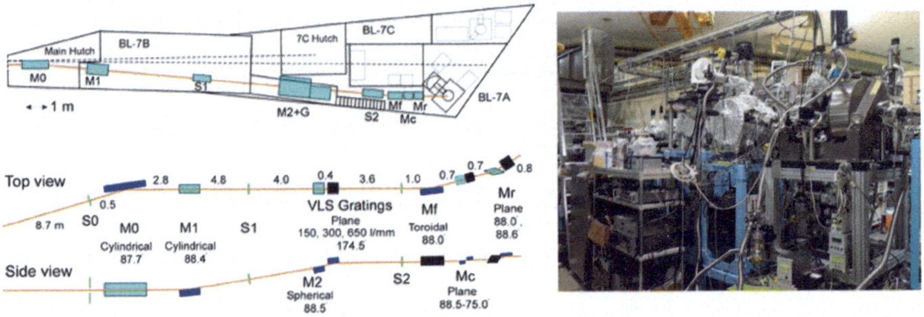

Fig. 24.1 Illustration of a beam-line and photograph of the end station

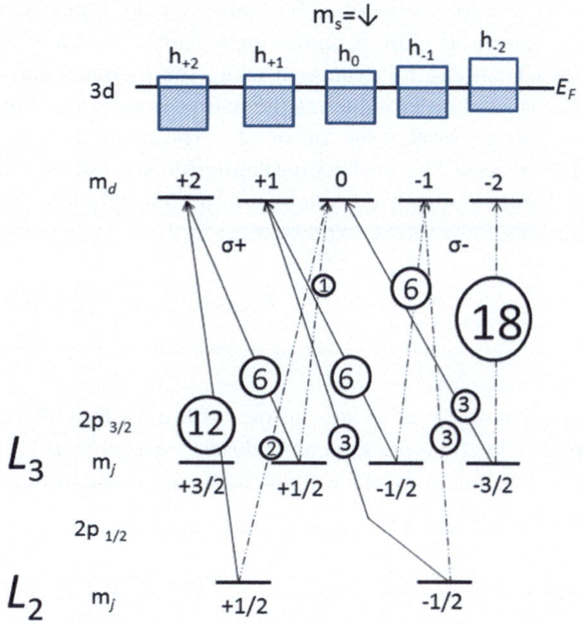

Fig. 24.2 Energy diagram considering the spin–orbit coupling in $3d$ states. The numbers in circles are transition probabilities from $2p$ to $3d$ states dependent on the quantum numbers

As XMCD analysis methods, magneto-optical sum rules for spin and orbital magnetic moments were established. These methods are summarized here. In particular, we focus on the magnetism in $3d$ TMs. As shown in Fig. 24.2, $3d$ states are split to up and down states by exchange interaction. Considering the spin–orbit interaction, the optical transition probabilities from $2p$ to $3d$ states are proportional to the Clebsch–Gordan coefficients. Un-occupancies of d states can be expressed as hole numbers h_{+2}, h_{+1}, h_0, h_{-1}, and h_{-2} for the five states. Estimated values of spin and orbital angular momenta can be defined as:

$$\langle L \rangle = -(2h_{+2} + h_{+1} + 0h_0 - h_{-1} - 2h_{-2})\hbar$$
$$\langle S \rangle = -(h_{+2} + h_{+1} + h_0 + h_{-1} + h_{-2})\hbar/2.$$

Using the coefficients in Fig. 24.2, the excitations by circularly polarized lights are estimated as:

$$L_3 : \Delta I_{L3} \propto 6h_{+2} + 6h_{+1} + 3h_0 - 6h_{-1} - 18h_{-2}$$
$$L_2 : \Delta I_{L2} \propto 12h_{+2} + 3h_{+1} - 2h_{-1} - 3h_{-2}.$$

Therefore, a total summation leads to:

$$\Delta I_{L3} + \Delta I_{L2} \propto 9(2h_{+2} + h_{+1} - h_{-1} - 2h_{-2}) \propto 9\langle L \rangle.$$

If the intensities of ΔI_{L3} and ΔI_{L2} are equal with different signs, the orbital angular momentum becomes quenched. $\langle S \rangle$ cannot be explicitly deduced from the XMCD; it can be expressed using the higher orders in the multipole expansion. The second order is the magnetic dipole term m_{T}, which exhibits the effect of spin anisotropy through the tensor Q_{ij}. The relation $m_{\mathrm{T}} = \sum_{\beta} Q_{\alpha\beta} S_{\alpha}$ is satisfied [12]. The m_{T} term does not become negligible in symmetry-broken surfaces and interfaces. In highly symmetric crystals, m_{T} can be negligible. Based on these considerations, the sum rules can be expressed as:

$$m_{\mathrm{orb}} = -\frac{4}{3}\left[\frac{\Delta I_{L3}+\Delta I_{L2}}{I_{L3}+I_{L2}}\right] n_h$$
$$m_{\mathrm{spin}} - 7m_{\mathrm{T}} = -2\left[\frac{\Delta I_{L3}-2\Delta I_{L2}}{I_{L3}+I_{L2}}\right] n_h$$

in a unit of μ_{B}. In an anisotropic system, such as interfaces, the in-plane and out-of-plane directions are separately described as $m_{\mathrm{orb}}^{\perp}$, $m_{\mathrm{orb}}^{\parallel}$, $m_{\mathrm{T}}^{\perp}$, and $m_{\mathrm{T}}^{\parallel}$. Considering the relationship: $m_{\mathrm{T}}^{x} + m_{\mathrm{T}}^{y} + m_{\mathrm{T}}^{z} = 0$, the m_{T} in the in-plane and out-of-plane directions are related by:

$$m_{\mathrm{T}}^{\perp} + 2m_{\mathrm{T}}^{\parallel} = 0$$

Further, the angular dependences of m_{orb} and m_{T} are expressed as:

$$m_{\mathrm{orb}}^{\theta} = m_{\mathrm{orb}}^{\perp}\cos^2\theta + m_{\mathrm{orb}}^{\parallel}\sin^2\theta$$
$$m_{\mathrm{T}}^{\theta} = m_{\mathrm{T}}^{\perp}\cos^2\theta + m_{\mathrm{T}}^{\parallel}\sin^2\theta$$

Using these five equations, the five values of m_{spin}, $m_{\mathrm{orb}}^{\perp}$, $m_{\mathrm{orb}}^{\parallel}$, $m_{\mathrm{T}}^{\perp}$, and $m_{\mathrm{T}}^{\parallel}$ are deduced. In particular, in the special case of magic angle obtained by: $3\cos^2\theta - 1 = 0$, $m_{\mathrm{T}}^{\theta} = -m_{\mathrm{T}}^{\parallel}(3\cos^2\theta - 1) = 0$ can be obtained in the geometry of $\theta = 54.7°$.

Using the above considerations, the element-specific spin and orbital magnetic moments are estimated quantitatively from the XMCD line-shape analyses. In this Chapter, possible origins of novel properties deduced from XMCD analyses are introduced for some systems. Interfaces induce novel properties, not present in the bulk form. The control of the interface atomic structure paves the way for novel artificial material designs.

24.2 Importance of Spintronics and Spin-Orbitronics

In this section, I briefly explain the importance of spintronics, utilized in applications such as magneto-resistive random access memory (MRAM). The research field of spin-orbitronics has attracted recent interests regarding the control of the spin–orbit interaction at thin-film interfaces, which paves the way for a new research field beyond spintronics. In order to fabricate functional high-performance devices, advanced material growth and understanding of the interfacial properties are required.

In the research field of spintronics, atomically controlled multilayered structures have been investigated, providing giant magnetoresistance (GMR) effects, discovered in Fe/Cr multilayers [13]. Prof. Fert and Prof. Grünberg were awarded with a Nobel prize in 2007 owing to the importance of the GMR effects, which accelerated the MRAM technology. Multilayers combining different types of elements exhibit novel properties beyond those in the bulk form. In particular, memory devices using magnetic states are employed in data storage. In order to improve the memory-device performances, the spin switching from the in-plane directions in the films is shifted to out-of-plane switching owing to the high densities in data storage technology. For this purpose, the perpendicular magnetic anisotropy (PMA) becomes important in the development of spintronic devices. However, the emergence of PMA in the bulk form requires non-symmetric distorted crystal structures such as spinel-type structures including oxides or alloys with Pt. Spins in 3*d* TMs and large spin–orbit coupling constants in 4*d* or 5*d* TMs combine at the film interfaces, providing PMA properties including out-of-plane magnetism of spins in 3*d* TMs and induced magnetism in heavy-metal elements. Extensive efforts have been devoted to fabricate atomically flat interfaces exhibiting PMA. Further, studies without using heavy-metal elements are important to develop low-cost memory devices. Recent studies have been focused on the control of spin–orbit interaction at interfaces, which brings novel properties for magnetization switching; e.g., spin–orbit torque could be employed for a low-power operation in the MRAM technology. Therefore, precise measurements of interfacial electronic and magnetic properties are strongly desired.

Spin-orbitronics is a novel research field, emerging after the advancements in electronics and spintronics. Figure 24.3 shows a comparison between electronics, spintronics, and orbitronics. Electronics involves the control of electrons, with devices such as transistors. The conjugate field is the electric field. In spintronics, developed by an analogy of electronics using electron spins, spins are controlled. The magnetic field is considered as the conjugate field. On the other hand, orbitronics involves a control of orbital states in the same manner [14]. Orbital states are controlled by the electron occupancies of 3*d* states, as shown in Fig. 24.2, which can be detected by XMCD. However, unlike electrons or spins, the orbital moments cannot be easily understood. Modulation of orbital occupancies is achieved by controlling lattice distortion, crystal field, or external electric field. Therefore, symmetry-broken interfaces with different types of elements are a promising platform to develop orbitronic systems.

Several phenomena have been revealed using element-specific characterizations, as illustrated in Fig. 24.4. "Induced magnetism" in non-magnetic elements can be detected by XMCD. Orbital magnetism can also be detected by XMCD; the orbital moment anisotropy is crucial at film interfaces [15]. Exchange coupling phenomena are clearly detected; ferromagnetic and antiferromagnetic types of coupling can be analyzed [16]. Orbital ordering through spin–orbit coupling is also related to the orbital magnetic moments, and can be detected by XMCD [17]. In addition, the orbital ordering is related to lattice distortions. Both structural and electronic properties have

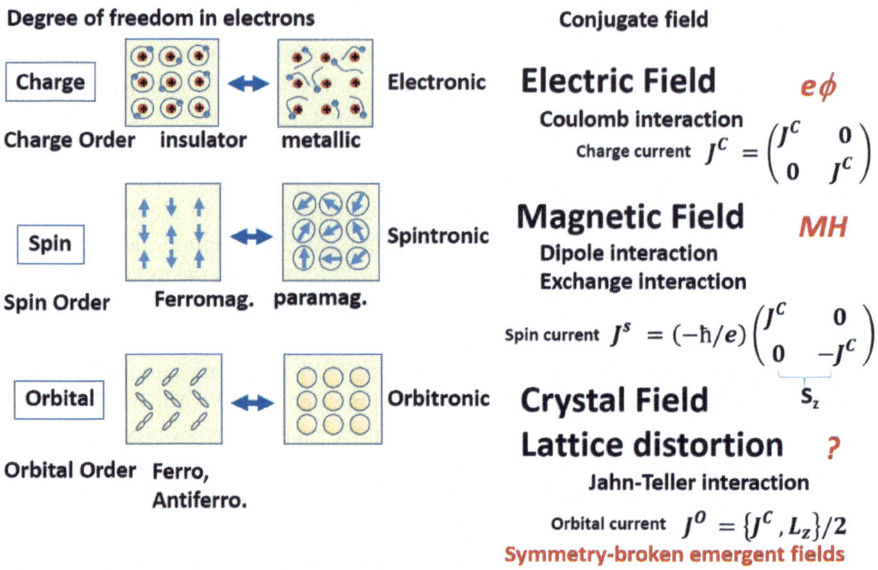

Fig. 24.3 Comparison of electronics, spintronics, and orbitronics

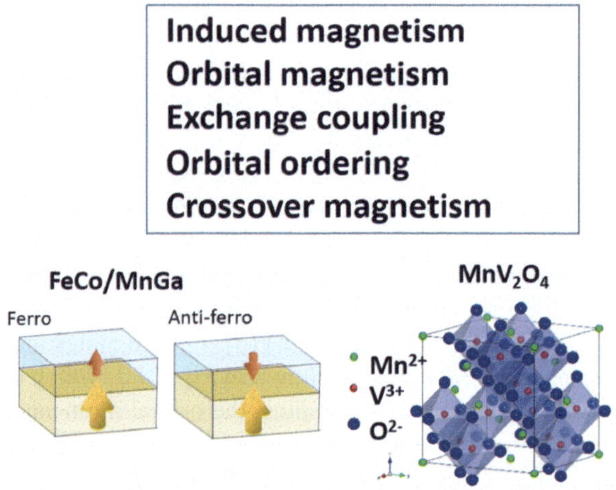

Fig. 24.4 Investigation of novel spin-orbitronic phenomena using XMCD [16, 17]

to be investigated explicitly. In addition, for spinel-type or perovskite-type transition-metal (TM) oxides, the control of crystal fields has been demonstrated.

Considering the above potentials, XMCD-related investigations are discussed in the following sections: (1) PMA at Fe/MgO interface and (2) possible origin of the PMA in Co/Pd multilayers.

24.3 Interfacial PMA in Fe/MgO Probed by XMCD[1]

Since the discovery of large tunnel magnetoresistance in single-crystal MgO-based magnetic tunnel junctions, the interfaces between MgO and magnetic layers have been investigated thoroughly [18]. Recently, large PMA energies of 0.21 MJ/m^3 have been developed by utilizing the interfaces between MgO and CoFeB transition metal alloys that are comparable to the perpendicular magnetic anisotropy (PMA) in Co/Pt multilayers without using heavy-metal elements [19]. This finding has initiated the development of high-performance MRAM using the interface PMA. For a fundamental understanding of the induction of PMA at the interface between a ferromagnetic layer and a MgO barrier layer, the electronic and magnetic structure of the interface between an ultrathin Fe layer and MgO must be clarified explicitly. Density functional theory (DFT) calculations revealed that a larger PMA appears at the interface between Fe and MgO than that of Co/MgO interfaces because of the difference in 3*d* orbital occupancies [20]. It has also been reported that the oxygen stoichiometry at the interface between Fe and MgO strongly depends on the PMA energies; namely, the over- and under-oxidization at the interface reduces the PMA. Experimentally, Koo et al. found that the PMA energies depend on the interface conditions and are controllable by the post-annealing process at the ultrathin 0.7-nm-thick Fe/MgO interface [21]. The maximum interfacial PMA is reported as 2.0 mJ/m^2. The values of PMA are of the same order of magnitude as those estimated by theoretical calculations. In order to investigate the PMA energy (K_{eff} [unit in J/m^3]), it is necessary to evaluate microscopically the orbital magnetic moments along parallel and perpendicular directions to the surface. Spin-orbit coupling is an essential factor concerning the origin of the interface PMA.

Until now, the origin of the large PMA in Fe/MgO has not yet been established because of the necessity of a large saturation magnetic field along the hard magnetization axis. Angular-dependent XMCD enables us to deduce the anisotropic orbital magnetic moments and investigate the large PMA at the Fe/MgO interface. In particular, Fe/MgO systems with different annealing conditions, which give rise to different PMA values, can provide the interpretation of the relationship between anisotropic orbital magnetic moments and PMA energies. In this section, we discuss the anisotropic interface orbital magnetic moments of ultrathin Fe facing a MgO layer by using angular-dependent XMCD and the PMA energies deduced from the orbital magnetic moments.

Samples were grown by using an ultra-high vacuum electron-beam evaporation on MgO (001) substrates. The sample structures are shown in Fig. 24.5. After cleaning the MgO (001) substrate at 1000 °C, a 5-nm-thick MgO layer was deposited on the substrate at 450 °C and a 30-nm-thick Cr buffer layer at 150 °C. Subsequently, the annealing process was performed at 800 °C in order to prepare the flat surface. A 0.7-nm-thick Fe layer, which corresponds to 5 monolayers (MLs), was deposited on the Cr buffer layer at 150 °C and a MgO layer was also grown on the Fe layer at 150 °C.

[1] This section is partly reproduced from J. Okabayashi, J. W. Koo, H. Sukegawa, S. Mitani, Y. Takagi, and T. Yokoyama, Appl. Phys. Lett. **105**, 122408 (2014), with the Permission of AIP Publishing.

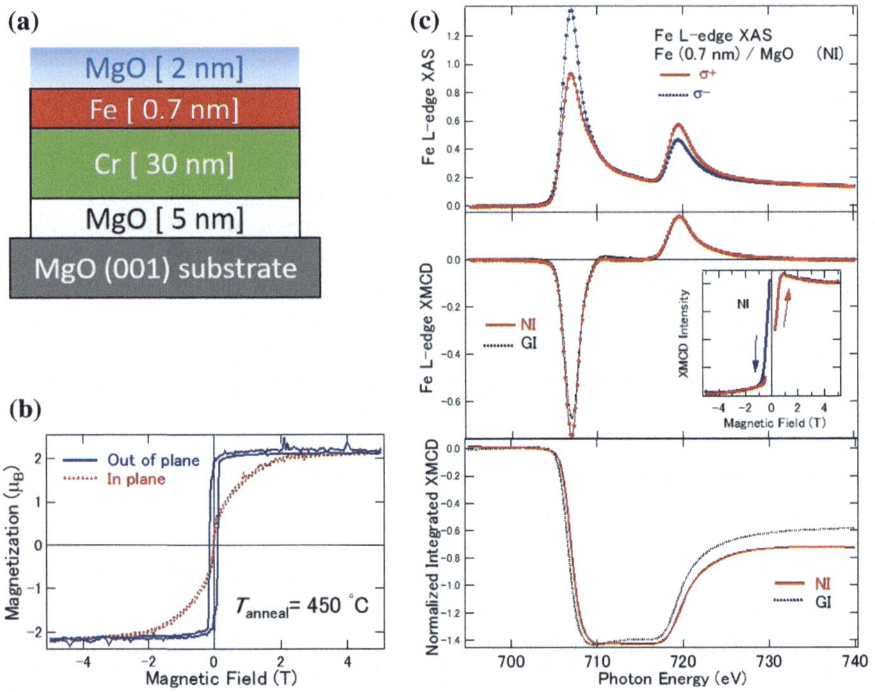

Fig. 24.5 **a** Sample structure with film thickness. **b** Magnetization curves measured by superconductive quantum interference devices. **c** XAS, XMCD, and integrals of XMCD in NI and GI geometries [24]

After the deposition of the MgO capping layer, two kinds of samples prepared with the above conditions were post-annealed separately at 450 °C in order to enhance the PMA. Details of the surface and interface conditions and the fabrication procedures are reported in [21].

For XMCD measurements for Fe *L*-edges, magnetic fields (H_{ext}) of ±5 T were applied using a superconducting magnet along the incident polarized soft X-rays in order to sufficiently saturate the magnetization along the direction of the magnetically hard axis. The total electron yield mode was adopted by detecting the drain currents from the samples. We changed the magnetic field directions in order to obtain right- and left-hand-side polarized X-rays while fixing the polarization direction of the incident X-ray. Angular-dependent XMCD was performed by rotating the angle between the incident beam and the direction of the sample's surface normal from the surface normal to 60°; these geometries are defined as normal incidence (NI) and grazing incidence (GI), respectively. In the case of the NI configuration, where both the photon helicity and the magnetic field directions are normal to the surface, the X-ray absorption processes involve the normal direction components of the orbital angular momentum ($m_{orb}^{\perp}$). The GI configuration mainly allows the detection of only the in-plane orbital angular momentum components ($m_{orb}^{\parallel}$).

Figure 24.5 shows the XAS of a 0.7-nm-thick Fe/MgO interface after the annealing at 450 °C. The XMCD taken at the NI and GI geometries and the integrals of the Fe $L_{2,3}$ absorption edges XMCD spectra are also shown. Distinct metallic peaks are evident in the XAS of the Fe $L_{2,3}$-edges, which indicates that no atomically mixed layer formation with oxygen atoms occurred at the interface, even after the 450 °C annealing process. The XMCD spectra in the NI and GI setups display a distinct difference in the intensity between the L_3-edges while the L_2-edges show almost similar intensity. The measured XMCD signal for the Fe L_3-edge from the NI geometry was larger than that from the GI geometry, which suggests that the large orbital magnetic moments are induced when the H_{ext} is perpendicular to the film plane. The magneto-optical sum rule indicates that the integrated areas of both negative L_3 and positive L_2 peaks are proportional to the orbital magnetic moments. The residuals of the integrals of both L_3 and L_2-edges in the XMCD spectra are larger in the NI configuration than in the GI one, indicating that the large orbital magnetic moments remain in the NI setup. Bottom panel of Fig. 24.5 shows the integrated XMCD signals of the Fe L-edges for both NI and GI setups. A difference can be clearly observed in the residuals of the integrals for both L_3 and L_2 peaks. These integrated XMCD spectra indicate that the large orbital magnetic moments are enhanced in the NI geometry compared with those in GI one. This is reasonable for a Fe/MgO interface with a PMA related to the orbital magnetic moments.

Using the magneto-optical sum rules for the estimation of orbital and spin magnetic moments, we list the results with the NI and GI geometries in Table 24.1. The effective spin magnetic moments (m_s^{eff}) values were determined only from the GI geometry because the magic angle geometry of 57.3° from the surface normal can theoretically neglect the magnetic dipole terms. For the application of the sum rules, we assumed the hole numbers of the Fe 3d states to be 3.4 as a standard value of Fe bulk.[25] The results listed in Table 24.1 show that the orbital magnetic moments with $m_{orb}^{\perp}$ of 0.30 and $m_{orb}^{//}$ of 0.21 μ_B were calculated, where Δm_{orb} is defined as $\Delta m_{orb} = m_{orb}^{\perp} - m_{orb}^{//}$. Considering the Bruno relationship: $K \simeq (\xi/4)\alpha\Delta m_{orb}$, where ξ is the spin-orbit coupling constant, and α is the band-structure parameter, the PMA energies are proportional to Δm_{orb}. We obtained $K = 130$ μeV/atom, which corresponds to a PMA value of 1.48 mJ/m^2, assuming a Fe lattice constant of 0.287 nm with a body-centered-cubic structure facing the MgO at the interface.

Here, we discuss the origin of the PMA at the interface of Fe/MgO. Since the excitation processes in XMCD are regarded to be the atomic excitations from the core to unoccupied states, one can estimate the magnetic anisotropy energy per atom through the Bruno's relation by using the element-specific orbital magnetic moments. Since the diamagnetic and shape-anisotropic components of the magnetic anisotropy energy do not depend on the anisotropic orbital magnetic moments, the interface PMA values in units of J/m^2 were directly estimated. The contribution of K_{eff} should be calculated by using the diamagnetic components and the ultrathin Fe

Table 24.1 Spin and orbital magnetic moments estimated from the XMCD sum rules for Fe at the Fe/MgO interface. The in-plane ($m_{orb}^{\parallel}$) and out-of-plane ($m_{orb}^{\perp}$) components are listed in units of μ_B. The interfacial PMA amplitudes (K_i) obtained from XMCD in units of mJ/m^2 are also shown with those from SQUID. [24]

	$m_{orb}^{\perp}$ [μ_B]	$m_{orb}^{\parallel}$ [μ_B]
m_{spin} [μ_B]	–	2.08
m_{orb} [μ_B]	0.30	0.21
K_i^{XMCD}	1.48 mJ/m^2 (132 μeV/Fe)	
K_i^{VSM}	2.01 mJ/m^2	

layer thickness, which is the same process as the estimation of PMA values using DFT calculations.

Next, we discuss the PMA values obtained by XMCD and the comparison with other experimental and theoretical works reported in the literature. The DFT calculations determine a magneto-crystalline anisotropy energy of 0.2 meV/atom in a free-standing Fe with 1 ML in thickness, a value of 0.9 meV/atom at the Fe (1 ML)/MgO interface, and 1.5 meV/atom at the MgO/Fe (1ML)/MgO sandwiched structures [21]. These results suggest that the PMA of the Fe/MgO interface is enhanced with a double-facing interface. Considering that the obtained PMA value of 0.13 meV/(Fe atom) through the XMCD measurement for Fe (5 MLs)/MgO structure, the result is somewhat smaller than the estimated value from the DFT calculation (0.2 meV/Fe atom). It is noted that the value for $m_{orb}^{//}$ can be underestimated, because the angle for the X-ray incident in the NI configuration was not exactly parallel to the film plane i.e. 57.3°. Therefore, the PMA values can be estimated to be small relative to those from DFT calculations. Nevertheless, the estimated value of PMA through angular dependent XMCD for Fe (5 MLs)/MgO (001) structure, without an applied external electric field, was turned out to be comparable to that from the literature on the modulation of PMA by an electric field [22].

Considering the above results, the origin of PMA at the Fe/MgO interface can be described by the anisotropic orbital magnetic moments induced by the spin-orbit interaction at the interface. The anisotropic orbital magnetic moments, resulting from the modulation of the occupancies of the Fe 3*d* states at the interface, were obtained through the sum-rule analysis of the XMCD spectra. Figure 24.6 displays the schematic energy diagram of the Fe 3*d* states at the Fe/MgO interface. First, we discuss from the view point of molecular orbitals theory. The chemical bonding between Fe $3d_{z^2}$ and O $2p_z$ orbitals stabilizes the bonding states as shown in Fig. 24.6. Filling in these orbitals are not completely occupied, Fermi level locates near between these two atomic states. Second, interfacial electronic structures are discussed by ligand-field theory. The Fe 3*d* levels split into e_g and t_{2g} states due to the crystal field. At the surface or the interface accompanied with the symmetry breaking, the degenerated e_g states of the $d_{x^2-y^2}$ and d_{z^2} orbitals split further because of the surface field and the hybridization with the O $2p_z$ orbital. The spin-orbit interaction induces further splitting depending on the direction of the $\boldsymbol{H}_{ext}$ and the magnetic quantum

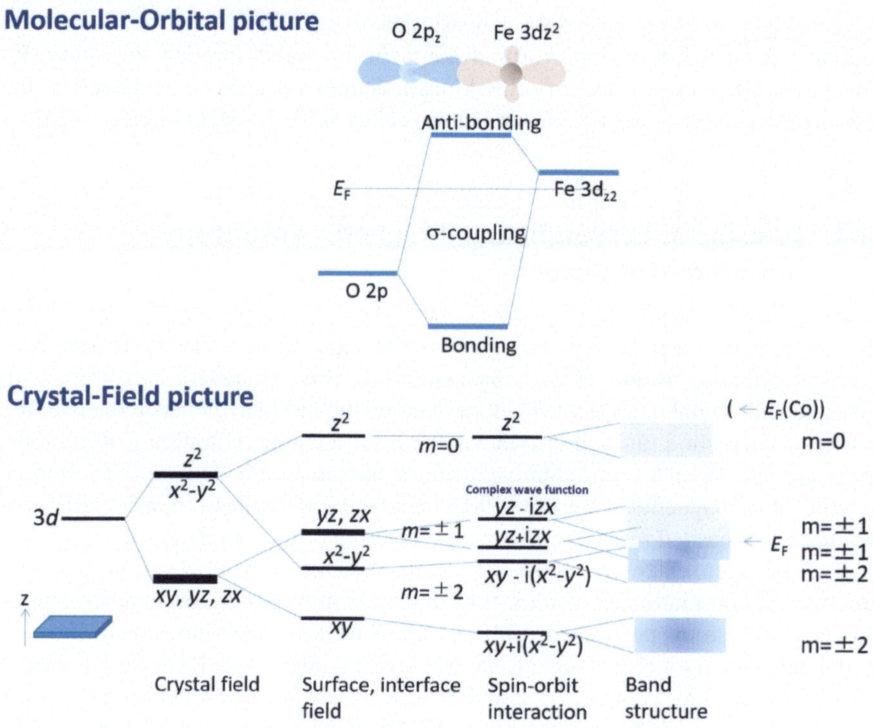

Fig. 24.6 Schematic diagrams in molecular-orbital scheme and ligand-field scheme for Fe $3d$ states

number m $(0, \pm 1, \pm 2)$. The d_{yz} and d_{zx} orbitals corresponding to $m = 1$ consist of the complex orbitals denoted as the d_{yz+izx} and d_{yz-izx} states. The anisotropy of charge occupancy between these complex states results in the anisotropic orbital magnetic moments. As was also indicated in the DFT calculations, the Fe $3d_z^2$ states are pushed up above the Fermi level through the hybridization with the O $2p_z$ orbital and the charge occupancies are modulated [20]. In addition, only when the $\boldsymbol{H}_{\text{ext}}$ perpendicular to the film plane, the modulation in the electron occupancies results in the enhanced $m_{\text{orb}}^{\perp}$. Therefore, due to the enhancement of $m_{\text{orb}}^{\perp}$ resulted from the spin orbit interaction and the hybridization between Fe $3d_{z^2}$ and O $2p_z$ orbitals, the large PMA can be expected for the Fe/MgO interface. It is well coincided with our XMCD results for the NI configuration, which showed enhanced $m_{\text{orb}}^{\perp}$. Note that the case of Co/MgO, Fermi level shifts to upper because of large charge number in $3d$ states, the gradation of orbital occupancies become small, resulting in the decrease of orbital moment enhancement.

In summary of this section, we have studied the interface PMA in ultrathin Fe/MgO (001) using angular-dependent XMCD. We found that the anisotropic orbital magnetic moments determined from the analysis of XMCD contribute to the large PMA energy, whose values depend on the annealing temperature. The large PMA energies deduced from the magnetization measurements are almost consistent with those estimated from the anisotropic orbital magnetic moments through the spin-orbit interaction. The enhancement of orbital magnetic moments can be explained by the hybridization between the Fe $3d_{z^2}$ and O $2p_z$ states at the Fe/MgO interface [23].

24.4 Interfacial Perpendicular Magnetic Anisotropy in Co/Pd Multilayers[2]

The interplay between 3*d* transition metals (TMs) and 4*d* or 5*d* TMs has been considered to offer opportunity of studying an interface-driven magnetic anisotropy with both spin and orbital degrees of freedom; namely the interplay between the two, the spin-orbit interaction through interfacial chemical bonding. Ultrathin Co/Pd multilayers are one of such representative artificial nanomaterials that exhibit *interface* perpendicular magnetic anisotropy (PMA), and the development of artificially synthesised PMA has led researchers to the expectation of ultra-high density recording media. Since then, extensive efforts have been made for studying electronic and spin structures of the interfaces of ultra-thin magnetic multilayers and nanostructures. Studies on Co atoms performed using XMCD have suggested the enhancement of orbital magnetic moments at the interfacial Co that is adjacent to Pd or Pt. It has been reported that the PMA emerges due to the cooperative effects between spin moments in 3*d* TMs and large spin-orbit interactions ξLS in the non-magnetic 4*d* or 5*d* TMs, where ξ is the spin-orbit coupling constants, L is orbital angular momentum, and S is the spin angular momentum. The Co/Pd interfaces and multilayers have also been employed to demonstrate the photo-induced precession of magnetisation [25, 26], the creation of skyrmions using the interfacial Dzyaloshinskii-Moriya interaction [27], and magnetisation reversal using the spin-orbit torque phenomena [28]. Despite the abovementioned careful studies and interesting trials with Co/Pd interfaces, the interfacial PMA, in particular the mechanism of *anisotropic* orbital magnetic moments, has not been fully understood for both Co and Pd sites. Bruno and van der Laan theoretically proposed an orbital moment anisotropy in 3*d* TMs within the second-order perturbation of the spin-orbit interaction (the weak coupling) for more than half-occupied electrons [29, 30]. However, in the case of strong spin-orbit coupling in 4*d* or 5*d* TMs, the validity of this perturbative formula has been debated. In order to study the mechanisms of PMA in Co/Pd multilayers, the contributions of orbital magnetic moments in each element should be explicitly considered.

However, it is challenging to study the anisotropy of the orbital magnetic moments of both Co and Pd elements using one specific experiment, due to the challenges in detection of the induced magnetic moments, of Pd in particular. In this study, we aim to overcome the above experimental challenges using the x-ray photon energy region that is common for both Co and Pd. In other words, a single experiment is

[2]This section is partly reproduced from J. Okabayashi, Y. Miura, and H. Munekata, Scientific Reports **8**, 8303 (2018), in accordance with the Creative Commons Attribution (CC BY) license.

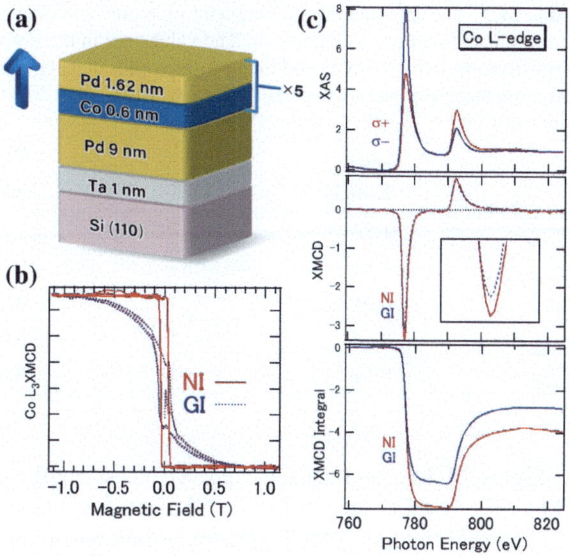

Fig. 24.7 **a** Stacked structures and **b** magnetic field dependence of the Co L_3-edge XMCD for a photon energy of 778.0 eV, in the NI and GI configurations. **c** L-edge XAS, XMCD, and XMCD integrals of the Co/Pd multilayers that exhibit PMA [31]

performed using the Pd $M_{2,3}$-edge absorption at ~530 eV, and Co $L_{2,3}$-edge absorption at ~770 eV taken by electron yield mode, instead of using different energy ranges for the Pd L-edges within 3.2 keV taken by fluorescence yield mode. The optical transitions from $2p$ to $3d$ states in $3d$ TMs, and those from $3p$ to $4d$ states in $4d$ TMs have the same transition probabilities. Therefore, the same transition processes could occur, even though the cross-section for the M-edges is smaller than that for L-edges. Furthermore, we adopt the angle-dependent XMCD that is a powerful experimental technique for studying anisotropic orbital magnetic moments. Taking into account that the X-ray absorption spectrum (XAS) attributed to the Pd $M_{2,3}$-edges ($3p$ to $4d$ transition) overlaps with that of the O K-edge absorption region, the surface oxide components have to be carefully removed, or the sample has to be prepared *in situ* in a detection chamber, in order to detect the Pd M-edge absorption signals. In this section, we have chosen the Ar-ion sputtering methods for surface cleaning. Experiments thus implemented lead us to the finding that the orbital magnetic moments in Co are anisotropic, whereas those in Pd are isotropic.

Ultrathin $[Co/Pd]_5$ multilayered structure grown by sputtering methods [25] was studied: Co (0.69 nm)/Pd (1.62 nm) that have PMA, which corresponds to four monolayers (MLs) of Co and eight MLs thickness of Pd. Cross-sectional transmission electron microscopy images have shown ⟨111⟩-oriented layered metallurgical structures with a good interface abruptness between Co and Pd [25]. Element-specific magnetisation hysteresis curves in normal incidence (NI) and oblique or grazing incidence (GI) geometries, taken at magic angle of 54.7° ($\cos^2\theta = 1/3$) from the sample surface normal, for Co are shown in Fig. 24.7. The photon energies are fixed at 778 eV for Co L_3-edge.

Table 24.2 The spin and orbital magnetic moments and magnetic dipole terms for perpendicular and in-plane directions in Sample A. The values are in the units of μ_B and are compared with the estimations from the XMCD and the first-principles density-functional-theory (DFT) calculations. Experimental error bars are estimated about 10% in Co and 20% in Pd for the applications of XMCD sum rules

	Co		Pd	
	XMCD	DFT	XMCD	DFT
$m_{spin}^{\perp}$	1.82	1.87	0.25	0.31
$m_{spin}^{\parallel}$	1.81	1.87	0.24	0.31
$7m_T$	0.01	–	0.01	–
$m_{orb}^{\perp}$	0.14	0.128	0.02	0.032
$m_{orb}^{\parallel}$	0.11	0.096	0.02	0.033

Figure 24.7 shows the Co L-edge XAS and angular-dependent XMCD that exhibit PMA. The XMCD spectra for the NI and GI geometries, and the integrals of the Co $L_{2,3}$ absorption edges XMCD spectra, are shown in Fig. 24.7b and 24.7c, respectively. There are clear differences in XAS spectra, depending on the relative helicities of the incident beam. As the XAS spectra obtained from the NI and GI configurations are identical, only the XAS spectra in the NI configuration are shown. In Fig. 24.7b, the XMCD spectra for the NI and GI configurations show a distinct difference in intensity between the L_3-edges, while the L_2-edges have almost similar intensity profiles. The measured XMCD signal for the Co L_3-edge in the NI geometry was larger than that in the GI geometry when the effective field is perpendicular to the film plane. The magneto-optical sum rule indicates that the integrated areas of both negative L_3 and positive L_2 peaks are proportional to the orbital magnetic moments. The residuals of the integrals of both L_3 and L_2-edges in the XMCD spectra are larger in the NI configuration, compared to the GI configuration, which indicates that large orbital magnetic moments are observed for the NI configuration as shown in Fig. 24.7c. Assuming the Co hole number is 2.49 [6], we deduce that the orbital moments of the perpendicular and in-plane components for Sample A, $m_{orb}^{\perp}$ and $m_{orb}^{\parallel}$, are 0.14 μB and 0.10 μB, respectively. Note that the setup of $m_{orb}^{\parallel}$ cannot detect the perfect in-plane contribution and almost half of $m_{orb}^{\perp}$ and $m_{orb}^{\parallel}$ are mixed which is proven by the MH curves in Co L_3-edge. Further, the spin magnetic moment (m_s) and magnetic dipole moment (m_T) of Co are 1.2 μ_B and 0.01 μ_B, respectively, with the uncertainties of ±10%. These values are listed in Table 24.2 by comparing with those estimated from the DFT calculations.

Figures 24.8 show the XMCD signals of Sample A for the Pd M-edges, after the removal of surface contamination. The signals that emerge due to the O K-edge absorption are removed by the Ar ion sputtering, hence clear XMCD signals are observed. These signals are induced by the proximity effects with the Co layers. We note that the intensity scale of the Pd M-edge is two-orders of magnitude smaller than that of the Co L-edge, due to the difference in photo-ionisation cross-section. The Pd $M_{2,3}$-edge XAS line shapes exhibit satellite structures that appear at 542 and 567 eV,

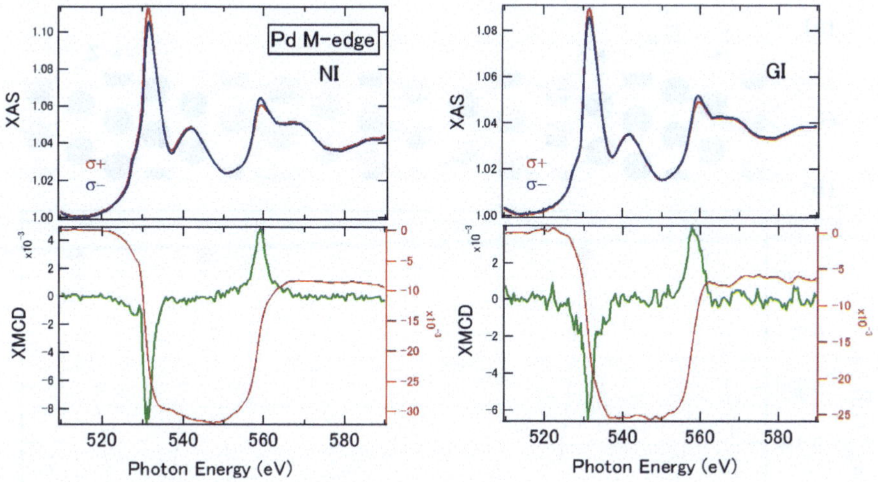

Fig. 24.8 *M*-edge XAS and XMCD of the Co/Pd multilayers that exhibit PMA. XAS at the NI configuration in left panel and the GI configuration at right panel. XMCDs acquired for the NI and GI configurations, respectively. The integrals of the XMCD are also plotted in the figures [31]

with higher photon energies of the main absorption peaks. They are obtained from the transitions from $3p$ to $5s$ states, and do not contribute to the XMCD signals. The Pd *M*-edge XMCD line shapes are almost identical in NI and GI setups, which suggests isotropic orbital moments in Pd, within the detection limits. Conventional magneto-optical sum rule analysis, often employed for $3d$ TMs (from $2p$ to $3d$ transition), can be applicable for the $3p$ to $4d$ transition in Pd *M*-edge XMCD with the same transition probabilities as that of the $2p$ to $3d$ transition. Assuming the hole number of the Pd $4d$ states is 1.10, for the NI configuration, the estimates of the spin and orbital magnetic moments are 0.25 μ_B and 0.02 μ_B, respectively, with the error bars of $\pm 20\%$ because of the estimations of XAS spectral integrals include the ambiguities. In contrast to Co, the Pd XMCD line shapes remain almost unaffected in the angular dependence (within the detection limits). This indicates that the isotropic finite orbital moments in Pd do not directly contribute to the PMA.

We consider the (111) interfaces, as illustrated in Fig. 24.9a. The chemical bonding along the z-direction is staggered between Co and Pd, which causes the non-perfect σ bonding at the interface. Therefore, the isotropic distribution of orbital moments in Pd, and the anisotropic orbital moments in the Co layer at the interface promotes the PMA at the Co/Pd interface. On the other hand, it has been reported that the Pd (001) orientation cannot stabilise the PMA at the interface. Furthermore, three monolayers are necessary for the periodic stacking in (111) orientation, as shown in Fig. 24.9a. Therefore, (111) orientation stacks are necessary for the PMA, due to the tuning of the interfacial hybridisation strength.

In order to analyse the origin of the PMA, we computed the orbital moment anisotropy at each atomic site. Spin and orbital magnetic moments in Co and Pd

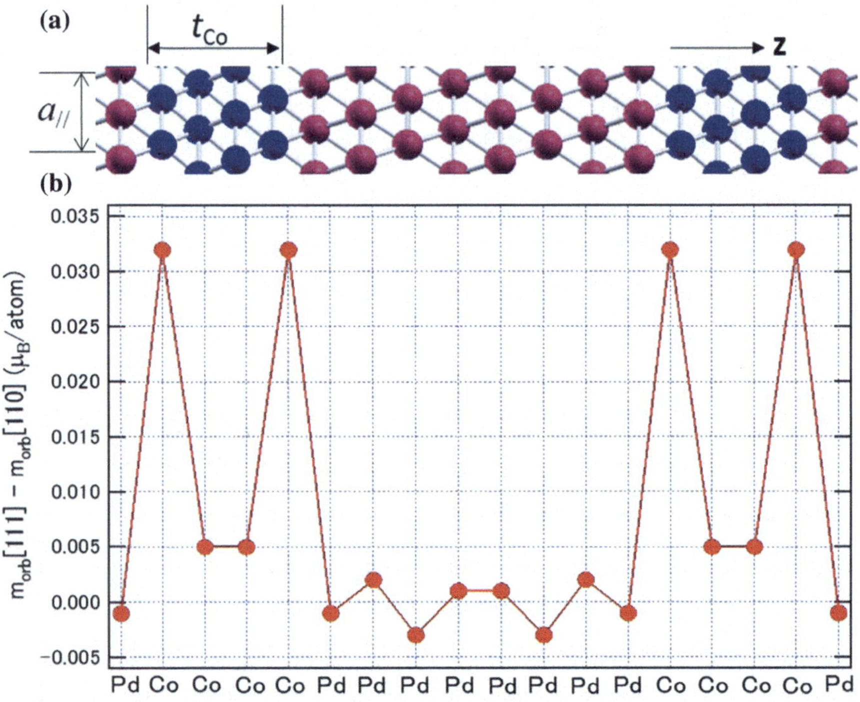

Fig. 24.9 **a** A schematic of the Pd(8ML)/Co(4ML)(111) multilayer. **b** Layer-resolved orbital moment anisotropy of the Pd(8ML)/Co(4ML), using $a_{\|} = 0.391$ nm, in DFT calculations [31]

sites are listed in Table 24.2. As shown in Fig. 24.9b, the orbital moment anisotropy of the Co atoms at the Co(4 ML)/Pd(8 ML) interfacial layer is enhanced and has a value of 0.033 μ_B. We emphasise that the PMA of the Co monolayer that is next to the Pd monolayer can be augmented primarily by the contribution of the interfacial layer. The PMA in a Co monolayer decreases as the distance between this monolayer and the Co/Pd interface increases, which is expected due to the bulk-like Co-Co bonding. On the other hand, we estimate that the orbital moment anisotropy induced in Pd is very small compared with that in Co, even in the Pd layer that is next to Co (Fig. 24.9b). The induced magnetic moments in the Pd monolayer emerge due to the *d*-orbital hybridisation between neighboring Co 3*d* and Pd 4*d* states at the interface, which qualitatively coincides with the independence of orbital magnetic moment anisotropy in Pd sites deduced from the XMCD.

Figure 24.10a, b represent the contributions of the crystalline magnetic anisotropy on the anisotropy energy at each atomic site. Four types of spin transition processes occur between the occupied and unoccupied states within the second-order perturbation of the spin-orbit interaction. The "up-down" process implies a virtual excitation from an occupied up-spin state to an unoccupied down-spin state in the second-order perturbation. For Co sites, the transition between down-down spin states is dominant, as shown in Fig. 24.10a. This suggests the conservation of spin states in the transition, which can be explained using the Bruno model assuming a large spin splitting. The Co sites exhibit positive energies in total, which confirms the orbital-moment-driven PMA. In contrast, for Pd, the spin-flipped transitions between up-down and down-up states become dominant due to the small band splitting, hence both spin-preserved and spin-flipped processes occur near the Fermi level. Element- and orbital-resolved density of states (DOS) for the Co 3*d* and Pd 4*d* states are shown in Fig. 24.10c and 24.10d, respectively. The DOS of both Co $3d(xy, x^2-y^2)$ and $3d(yz, zx)$ orbitals at the interface contribute to the PMA because the $3d(xy, x^2-y^2)$ and $3d(yz, zx)$ orbitals become dominant at the Fermi level in the minority-spin state. These states provide large matrix elements of L_z, $<3d(x^2-y^2)\downarrow|L_z|3d(xy)\downarrow>$ and $<3d(yz)\downarrow|L_z|3d(zx)\downarrow>$, for the second order perturbation of the spin-orbit interaction, leading to an enhancement of the perpendicular components of the orbital magnetic moments. The DOS of the Co 3*d* states is clearly split, with a spin magnetic moment of 1.89 μ_B. We used the lattice constant of $a = 0.391$ nm for the equilibrium condition. The DOS of Pd 4*d* states is also split, due to the proximity with the Co layers, as shown in Fig. 24.10d. For the Pd 4*d* states, we estimated that the induced spin magnetic moment is 0.311 μ_B. The Pd 4*d* states exhibit a small splitting at the interfaces, whereas the DOS at the Fermi level is large, which is considered as a Stoner-type ferromagnetism. The matrix elements of $<4d(yz)\downarrow|L_z|4d(x^2-y^2)\uparrow>$ are dominant in Pd sites, which favours to in-plane anisotropy. The spin-conserved transition in Co and spin-flipped transitions in Pd are illustrated in Fig. 24.10e. The enhancement of orbital moments of Co is explained by the spin conserved transition derived from the band structures. The spin-orbit coupling in heavy-metal elements causes a quadrupole-like formation by the spin-flip transitions, resulting in the magnetic dipole term (m_T), however it does not contribute to the anisotropy of the orbital moments. These results explain both the angular dependence of the Co *L*-edge and the Pd *M*-edge XMCD spectral line shapes.

Considering the results of XMCD and DFT calculations, we discuss the quadrupole-like contribution of the interfacial Pd layer. The spin sum rule includes not only the spin moment m_s, but also the magnetic dipole term m_T, and reveals the effective spin magnetic moment $m_s^{eff} = m_s + 7m_T$. Here, the m_T can be separated from the angular dependence of XMCD, as the GI configuration cancels out the m_T term for the magic angle geometry of 54.7° ($\cos^2\theta = 1/3$) with respect to the surface normal. Our Pd XMCD results indicate that the Pd orbital moments induced at the interface are isotropic. Note that m_T is an order of magnitude smaller than the orbital moments, i.e., 0.01 μ_B or less, comparable with the detectable limits. The element-specific PMA energy *K* that includes the m_T term beyond the Bruno model which is derived from only orbital moment anisotropy Δm_{orb} can be expressed theoretically as:

$$K \sim \frac{1}{4\mu_B}\xi\Delta m_{orb} - \frac{21}{2\mu_B}\frac{\xi^2}{\Delta_{ex}}\Delta m_T \tag{24.1}$$

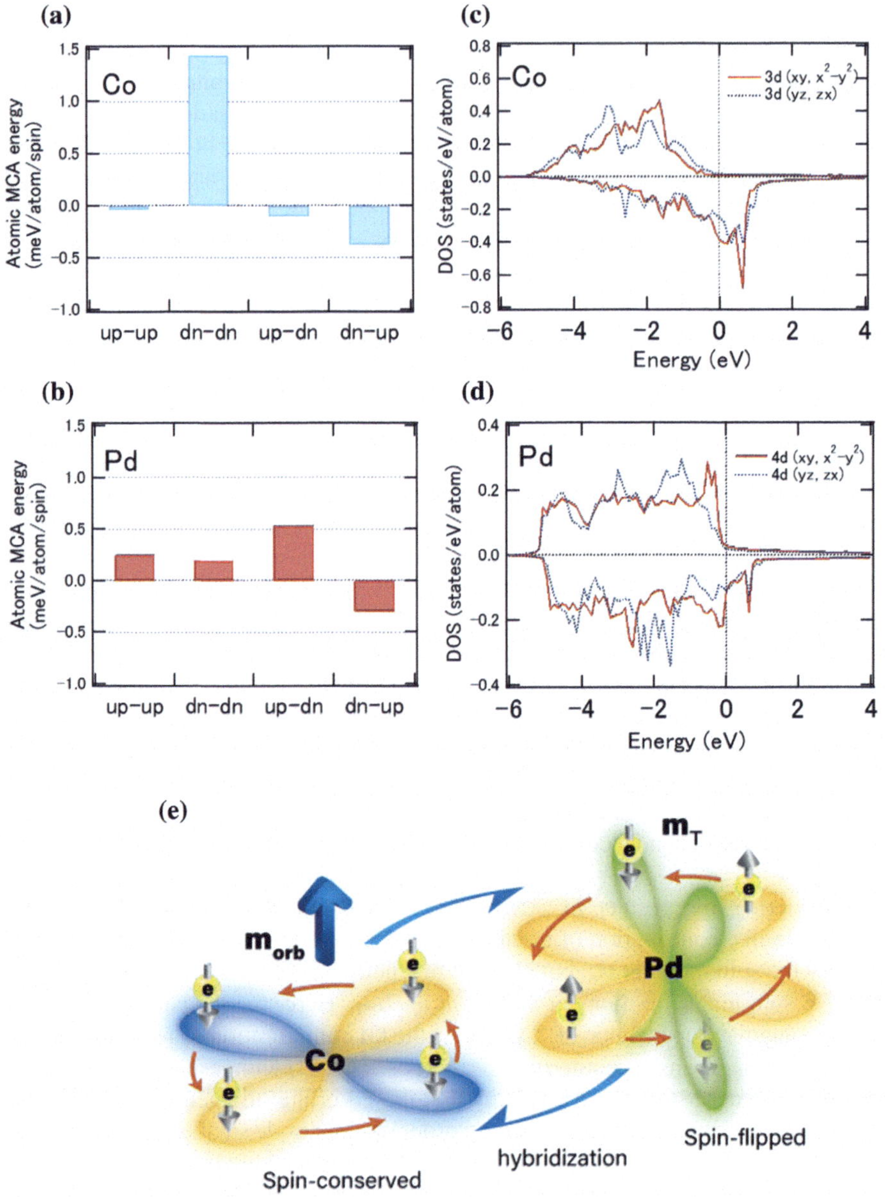

Fig. 24.10 Bar graph of the second-order perturbative contribution of the spin-orbit interaction to the MCA energy at the interfacial atomic sites of **a** Co and **b** Pd for the Pd(8ML)/Co(4ML), using $a_{\parallel} = 0.391$ nm. Spin-resolved local density of states (LDOS) of the $d(xy, x^2 - y^2)$ and $d(yz, zx)$ states for the interfacial **c** Co and **d** Pd sites, for the Pd(8ML)/Co(4ML), using $a_{\parallel} = 0.391$ nm. **e** A schematic of the electron hopping in Co and Pd at the interface. $3d(xy, x^2 - y^2)$ orbitals in Co and both Pd $4d(xy, x^2 - y^2)$ and $4d(yz, zx)$ orbitals are illustrated [31]

where Δ_{ex} is the exchange splitting between spin-up and spin-down bands. Δm_T satisfies the relation of $\Delta m_T = m_T^{\perp} - m_T^{\parallel}$ with $m_T^{\perp} = -2m_T^{\parallel}$. For Co, we estimate the first and second terms in (24.1) contributing 10^{-4} and 10^{-5} eV, respectively, in Co, assuming $\Delta_{ex\,Co} = 3$ eV, and $\xi_{Co} = 70$ meV, and the orbital moment anisotropy becomes dominant even in the finite m_T value. For Pd, the orbital anisotropy that corresponds to the first term becomes almost zero and the second term becomes dominant in the order of 10^{-4} eV because of small $\Delta_{ex\,Pd} = 200$ meV and large $\xi_{Pd} = 110$ meV, even if the m_T in Pd is as small as 0.01 μ_B. The relatively large spin-orbit coupling constant and small Pd exchange splitting contribute to the appearance of PMA by means of the second term in (24.1) with quadrupole-like interactions. The second term in Pd is comparable or smaller than the orbital moment anisotropy in Co. Therefore, the first term in (24.1) of orbital moment anisotropy in Co and the second term related to m_T in Pd contribute dominantly to E_{MCA} through the interfacial proximity effects. Furthermore, the contribution to the first term of (24.1) in Co is underestimated because Fig. 24.9b suggests the enhancement of Δm_{orb} at the interfacial layer. XMCD detects the signals from all layers, which suppresses the contribution from the interfacial layers. However, the enhancement of the orbital moment anisotropy can be concluded qualitatively.

Finally, we discuss the hybridization and spin-orbit coupling at the Co/Pd interfaces. Our findings indicate that the orbital moments in Pd are isotropic. This suggests that the magnetic dipole transitions can be essential, due to the mixing of spin-up and spin-down states in the Pd 4*d* states, even though the m_T values are much smaller than m_{orb}. As another physical origin of PMA, the facts that the radii of the Pd 4*d* orbitals are larger than those of Co 3*d* are also important, which results in the decrease of the anisotropy of the orbital moments and the increase of lattice strains. These effects couple at the interfaces, which enhances the PMA in Co/Pd multilayers. For relatively thick Co layers, the shape anisotropy in Co governs and suppress the effects of PMA at the interface, resulting in the in-plane anisotropy. The origin of PMA at the Co/Pd interface can be explained by the interfacial Co orbital moment anisotropy and spin-flipped processes at the Pd sites through the strong hybridization at Co/Pd interface, as illustrated by the magnetic dipoles in Fig. 24.10e.

In summary of this section, we have investigated the origin of PMA at Co/Pd interfaces using the angle–dependent XMCD and DFT calculations. The Co 3*d* orbital states are anisotropic, while the Pd 4*d* orbital states are isotropic. In contrast to the large spin splitting in Co 3*d* states, the induced spin splitting in Pd 4*d* states at the interface exhibits a combination of up and down spin transitions that accompany the quadrupole-like states in Pd. In other words, the anisotropy of the orbital moments in Co is enhanced at the interface through the proximity with Pd and orbital moment anisotropy is not induced in Pd even in the facing layer on Co.

24.5 Theory of Interfacial Perpendicular Magnetic Anisotropy

Considering the above mentioned cases of Fe/MgO and Co/Pd interfaces, the microscopic origin of interfacial PMA can be discussed using theoretical formulations. In the simple case, the anisotropy of orbital magnetic moments is proportional to PMA assuming the large exchange splitting between majority and minority bands. When the exchange splitting is small and mixing of up and down spin states, the spin-flipped contribution has to be formulated. Generally, the magnetocrystalline energy E_{MCA} can be the sum of the two contributions,

$$E_{\mathrm{MCA}} \sim E^{\downarrow\downarrow}(\sigma) + E^{\uparrow\downarrow}(\sigma)$$

where $\downarrow\downarrow$ and $\uparrow\downarrow$ represent the spin transition of spin-conserved and spin-flipped cases, respectively. σ represents the directions of spins (x, y, and z). Spin-orbit interaction H_{SO} can be written as

$$H_{\mathrm{SO}} = \xi\{1/2(l_+s_- + s_+l_-) + l_zs_z\}$$

by using ladder operators. The term of $l_+s_- + s_+l_-$ corresponds to the spin-flipped process. The l_zs_z term corresponds to the orbital moment anisotropy in spin-conserved state. Within the second-order perturbation of spin-orbit interaction, assuming the lattices elongated to z axis, E_{MCA} can be deduced by the comparison between out-of-plane and in-plane directions and formed as follows [32]:

$$E_{\mathrm{MCA}} \sim -\frac{\xi}{4}(l_z - l_x) + \xi^2 \frac{|\langle\downarrow|l_z|\uparrow\rangle|^2 - |\langle\downarrow|l_x|\uparrow\rangle|^2}{\Delta_{\mathrm{ex}}} \tag{24.2}$$

In this formulation, the cases of $l_z - l_x$ contributions become positive coefficients. Since $l_z^2 - l_x^2$ can be written to be $-1/2(l^2 - 3l_z^2)$, where $l^2 = l_x^2 + l_y^2 + l_z^2$, the second term is modified.

$$|E_{\mathrm{MCA}}| \sim \frac{\xi}{4}\left|\Delta l_\uparrow - \Delta l_\downarrow\right| - \xi^2 \frac{|\langle\downarrow|l^2 - 3l_z^2|\uparrow\rangle|}{2\Delta_{\mathrm{ex}}} \tag{24.3}$$

In this notation, positive amplitude of E_{MCA} stabilizes the PMA. Therefore, defining the quadrupole-like tensor formulation $\mathrm{T} = l^2 - 3l_z^2$, the contribution from the second term become negative sign in the case of positive sign in the first term. Further, the matrix elements of second term in (24.3) are the connection between up and down states, suggesting that the mixing of majority and minority bands at the Fermi level is essential to enhance this term. While the first term is expressed as the anisotropy of orbital magnetic moments, the second term implies the changes of orbital shapes.

In the case of negative T, the second term also favors the PMA. As the definition of $\mathrm{T}_i = \sum_j \mathrm{Q}_{ij} S_j$, $\mathrm{T}<0$ represents the spin quadrupole-like feature elongated to z-axis with $3z^2-r^2$-like prolate shape (cigar shaped along z-axis). On the other hand, T>0 is defined as x^2-y^2-like oblate type feature (donut shaped in xy plane). Because of the matrix elements of l_x in the spin-flipped transitions, magnetic quantum number has to change between ± 1 and ± 2 and between ± 1 and 0. Former transition exhibits the $3z^2-r^2$-like prolate shape and later transition relates to x^2-y^2-like oblate shape considering the shapes of wave functions. Beyond the orbital moment anisotropy, the second term is also the candidate of PMA. In order to enhance the second term, ξ must be large and Δ_{ex} be small. As discussed in Co/Pd interface, the m_T term in Pd 4d states contributes to PMA though large ξ and small Δ_{ex}. In the case of Fe/MgO case, ξ of Fe atom is as small as 50 meV and large exchange splitting, resulting in the negligible second term contribution. Therefore, the cases which second term becomes dominant are limited in the cases using heavy-metal elements which cooperate with 3d TMs through the orbital hybridization at the interfaces.

As for the comparison with the first-principles calculation and XMCD, it should be noted about the difference in the second term estimations. Above mentioned theoretical formulation of spin-flipped term can be estimated by the calculation as $E(\uparrow\downarrow)$ + $E(\downarrow\uparrow)$. However, in XMCD spin sum rule, the m_T term—(21/2) $(\xi^2/\Delta_{ex})\Delta m_T$ written in the previous section includes not only spin-flipped term but also spin-conserved states in principles in the notation of $\{-2E(\downarrow\downarrow) - 2E(\uparrow\uparrow) + E(\uparrow\downarrow) + E(\downarrow\uparrow)\}$. As defined by van der Laan [30], the correction of spin-conserved term must be included when estimating the PMA energy.

24.6 Summary and Outlook in XMCD Spectroscopy

Recent XMCD studies were reviewed to clarify the PMA. The nature of chemical bonding and electron occupancies bring novel properties at interfaces. The control of orbital magnetic moments could enable to reveal and utilize novel promising phenomena beyond spintronics. As a next step, manipulation of orbital moments is required. For this purpose, the introduction of lattice strain by ferroelectric materials is a promising approach owing to the electric field control of strain in ferroelectric materials. Junctions between ferromagnets and ferroelectric materials introduce lattice strain into ferromagnets, and control the orbital moments reversibly by electric field. Another approach of analysis involving XMCD is to use microscopy with a high spatial resolution. In addition, not only element-specific but also spatial mapping of spins and orbitals are desired for interface spin-orbitronic analyses including topological properties.

Finally, for tailoring PMA at the interfaces, the enhancement of second term in (24.2) and (24.3) is essential. One of the interesting targets is $Mn_{3-x}Ga$. The Mn $3d$ orbital magnetic moments are almost quenched, resulting in no orbital moment anisotropy. Spin-flipped transitions become a key factor using even in small ξ in the distorted lattices. Therefore, the material designs of spin-orbitronics link to the modulation of lattice parameter, which opens up novel research fields in spin-orbitronics.

Acknowledgements The author acknowledges to the collaboration with Dr. S. Mitani and Dr. H. Sukegawa in NIMS for the parts of section 3 and Dr. Y. Miura in NIMS and Prof. H. Munekata in Tokyo Institute of Technology for the parts of section 4. These works are partly supported by KAKENHI.

References

1. J.L. Erskine, E.A. Stern, Phys. Rev. B **12**, 5016 (1975)
2. G. Schütz, W. Wagner, W. Wilhelm, P. Kienle, R. Zeller, R. Frahm, G. Materlik, Phys. Rev. Lett. **58**, 737 (1987)
3. C.T. Chen, F. Sette, Y. Ma, S. Modesti, Phys. Rev. B **42**, 7262 (1990)
4. B.T. Thole, P. Carra, F. Sette, G. van der Laan, Phys. Rev. Lett. **68**, 1943 (1992)
5. B.T. Paolo Carra, M. Thole, M. Altarelli, X. Wang, Phys. Rev. Lett. **70**, 694 (1993)
6. C.T. Chen, Y.U. Idzerda, H.-J. Lin, N.V. Smith, G. Meigs, E. Chaban, G.H. Ho, E. Pellegrin, F. Sette, Phys. Rev. Lett. **75**, 152 (1995)
7. T. Koide, H. Miyauchi, J. Okamoto, T. Shidara, A. Fujimori, H. Fukutani, K. Amemiya, H. Takeshita, S. Yuasa, T. Katayama, Y. Suzuki, Phys. Rev. Lett. **87**, 257201 (2001)
8. J. Stöhr, J. Mag. Mag. Mater. **200**, 470 (1999)
9. J. Stöhr, H.C. Siegmann, Magetism, (Springer)
10. A. Kirilyuk, A.V. Kimel, T. Rasing, Rev. Mod. Phys. **82**, 2731 (2010)
11. K. Amemiya, H. Kondoh, T. Yokoyama, T. Ohta, J. Electron Spectrosc. Relat. Phenom. **124**, 151 (2002)
12. J. Stöhr, H. Konig, Phys. Rev. Lett. **75**, 3748 (1995)
13. G. Binasch, P. Grünberg, F. Saurenbach, W. Zinn, Phys. Rev. B **39**, 4828 (1989)
14. Y. Tokura, N. Nagaosa, Science **288**, 562 (2000)
15. J. Okabayashi, H. Sukegawa, Z. Wen, K. Inomata, S. Mitani, Appl. Phys. Lett. **103**, 102402 (2013)
16. J. Okabayashi, K.Z. Suzuki, S. Mizukami, J. Mag. Mag. Mater. **460**, 418 (2018)
17. J. Okabayashi, S. Miyasaka, K. Hemmi, K. Tanaka, S. Tajima, H. Wadati, A. Tanaka, Y. Takagi, T. Yokoyama, J. Phys. Soc. Jpn. **84**, 104703 (2015)
18. S. Yuasa, T. Nagahama, A. Fukushima, Y. Suzuki, K. Ando, Nat. Mater. **3**, 868 (2004)
19. S. Ikeda, K. Miura, H. Yamamoto, K. Mizunuma, H.D. Gan, M. Endo, S. Kanai, J. Hayakawa, F. Matsukura, H. Ohno, Nat. Mater. **9**, 721 (2010)
20. A. Hallal, H.X. Yang, B. Dieny, M. Chshiev, Phys. Rev. B **88**, 184423 (2013)
21. J.W. Koo, S. Mitani, T.T. Sasaki, H. Sukegawa, Z.C. Wen, T. Ohkubo, T. Niizeki, K. Inomata, K. Hono, Appl. Phys. Lett. **103**, 192401 (2013)
22. R. Shimabukuro, K. Nakamura, T. Akiyama, T. Ito, Physica E **42**, 1014 (2010)
23. K. Nakamura, T. Akiyama, T. Ito, M. Weinert, A.J. Freeman, Phys. Rev. B **81**, 220409R (2010)
24. J. Okabayashi, J.W. Koo, H. Sukegawa, S. Mitani, Y. Takagi, T. Yokoyama, Appl. Phys. Lett. **105**, 122408 (2014)
25. K. Yamamoto, T. Matsuda, K. Nishibayashi, Y. Kitamoto, H. Munekata, IEEE Trans. Magnet. **249**, 3155 (2013)
26. C. Boeglin, E. Beaurepaire, V. Halte, V. Lopez-Flores, C. Stamm, N. Pontius, H.A. Dürr, J.-Y. Bigot, Nature **465**, 458 (2010)
27. S.D. Pollard, J.A. Garlow, J. Yu, Z. Wang, Y. Zhu, H. Yang, Nature Commun. **8**, 14761 (2017)
28. M. Jamali, K. Narayanapillai, X. Qiu, L.M. Loong, A. Manchon, H. Yang, Phys. Rev. Lett. **111**, 246602 (2013)
29. P. Bruno, Phys. Rev. B **39**, 865 (1989)
30. G. van der Laan, J. Phys.: Condens. Matter **10**, 3239 (1998)
31. J. Okabayashi, Y. Miura, H. Munekata, Sci. Rep. **8**, 8303 (2018)
32. D. Wang, R. Wu, A.J. Freeman, Phys. Rev. B **47**(14), 932 (1993)

Chapter 25
Interaction of Laser Radiation with Explosives, Applications and Perspectives

Yuriy Tverjanovich, Andrey Tverjanovich, Anatoliy Averyanov, Maksim Panov, Mikhail Ilyshin and Mikhail Balmakov

Abstract This chapter provides a brief overview of the main directions in research and application of the interaction of laser radiation with explosives. Historically the first application of such interaction based on thermal initiation of explosives is briefly characterized. The main methods of remote detection of explosives using laser radiation are listed. Particular attention is paid to the areas of research that have been recently formed such as spectral selective resonance interaction of laser radiation with explosives and explosives modified by nano-additives. It was noted that depending on the choice of the optical absorption band of the explosives, its excitation can lead either to the effective activation of an explosive or to its decomposition, which is not accompanied by a significant thermal effect. The latter case can be used for remote detection of the explosives and, partly, for passivation of their surface. Finally, it was demonstrated that the absorbing and refractive light nano-additives are able to reduce the threshold intensity of initiation of explosives by laser radiation, while keeping the resistance of explosives to impact or thermal effects that provides the safety conditions of working with them.

Y. Tverjanovich (✉) · A. Tverjanovich · A. Averyanov · M. Panov · M. Balmakov
Institute of Chemistry, St. Petersburg State University, Universitetskiy ave. 28, St. Petersburg, Russia
e-mail: y.tveryanovich@spbu.ru

A. Tverjanovich
e-mail: andr.tver@yahoo.com

M. Panov
e-mail: m.s.panov@spbu.ru

M. Balmakov
e-mail: balmak1@yahoo.com

M. Ilyshin
St. Petersburg State Institute of Technology, Moskovsky ave. 26, St. Petersburg, Russia
e-mail: explaser1945@yandex.ru

K. Yamanouchi et al. (eds.), *Progress in Photon Science*, Springer Series in Chemical Physics 119, https://doi.org/10.1007/978-3-030-05974-3_25

25.1 Introduction

The first experiments focused on a study of interaction between energy-saturated materials and laser radiation have been performed simultaneously with development of laser technology in the second half of the twentieth century. In most cases, the infrared He–Ne lasers available at that time were applied in this regard. It was assumed that the direct transfer of thermal energy to the substance is provided in the infrared region and, as a result, this is the easiest way to ensure rapid combustion of a substance to be initiated, and subsequent ignition of brisant explosives. The further development of works in this area lies in implementation of high-power lasers for the direct initiation of brisant explosives in order to exclude the necessity to use highly sensitive initiating explosives. However, the use of high-power lasers is difficult to implement and expensive. It was considered that in their absence the probability of accidental initiation of ammo decreases. The number of works on laser initiation of energy-saturated materials has increased several times over the past ten years, which proves their relevance. The recent success in the understanding of the theory of the initiation process under the action of laser radiation and in the technical development of laser technologies also provide the conditions for further progress of the aforementioned approach. Laser initiation opens wide opportunities for application in those areas where modern safety standards are implemented and restrictions to traditional methods of initiation are presented: space technology and rocket science, mining and blasting with high explosion hazard, oil production, and others. Furthermore, laser initiation makes it possible to create complex explosive systems insensitive to electromagnetic interference and temperature fluctuations and easy to combine into multi-level networks, which makes its application more safe and technological. An additional factor is that the laser initiation process consumes significantly less energy than the thermal type of initiation, in turn, the cost of such initiators is lower than usual ones and their size and weight are also significantly less. For example, the initiation of many pyrotechnic compounds requires low radiation power (mJ), which is accessible for many commercial lasers. Another striking advantage of such initiators is their reusability. At the present, the main approaches describing the process of laser initiation of explosives can be divided into two separate directions: thermal and shock. In the first case, the beam falls on the target causing a self-sustaining combustion process, which then proceeds into explosive combustion or detonation. In the second case, the detonation occurs due to the energy of shock initiation caused by particles flying at high speed, which are formed upon laser radiation.

One more actual and perspective way of using lasers is the methods of detecting various substances. There are a lot of different laser methods for detection of energetic materials. But traditional detection methods are somewhat limited because the broad spectral features of many explosive vapors make phase sensitive detection methods difficult. One of the common problems of a large part of these methods is the following. Large molecules, which include the majority of explosives, usually have weak and poorly resolved optical transitions. This circumstance complicates their detection by spectroscopic methods. At present, remote sensing of energy-saturated substances using resonant laser-explosives interactions is a promising direction of solving this problem.

Another modern directions in the field of interaction of laser radiation with explosives are using resonant laser-explosive interaction and nanoparticle additives. Introduction to the explosive composition some chemically inactive nanoparticles can reduce the threshold power of initiation by laser radiation. The low chemical activity of these nanoparticles allows to achieve this goal without reducing the stability and safety of the explosives.

In the sections below, an overview of the research in this direction will be reviewed.

25.1.1 Laser Thermal Initiation of Explosives

In many modern studies, it was observed that ignition or initiation of materials by laser radiation result in thermal inhomogeneity of the process, which is manifested by formation of hot spots [1]. It is assumed that during irradiation a local center of about 0.1 μm is formed, which has enough energy to ignite an initiator. For that reason, the temperature of hot spot should be approximately 700 °C or more, and its lifetime should be not less than 10 μs. There is a possibility to form a number of such hot points; however, formation of a "critical" hot spot induces detonation or ignition. The process of thermal initiation consists of a complex of different physical and chemical stages. For example, the initiation process of hexogen (RDX—Royal Demolition Explosive, Hexogen) can be described as follows [2–4]. In the beginning, the surface of the sample is heated by a laser beam (at atmospheric pressure in argon with CO_2 laser heat flux from 35 to 600 W/cm^2) and the temperature profile of the material in the solid phase is formed. Then, when the material reaches melting temperature (T_m), a quasi-equilibrium two-phase zone (mushy zone) is formed, in which both the solid and liquid phases are present. Further, the molten liquid phase is formed, which moves deep into the material and transfers thermal energy. This process accompanied by decomposition and formations of gaseous phase.

In the next stage, a torch is formed on the surface of the material followed by the intense evaporation of gaseous products from the surface of the material and subsequent sharp increase in pressure. Finally, if the heat flux is strong enough to provide a self-sustaining exothermic reaction, the ignition process occurs (Fig. 25.1).

As a rule, the mathematical modeling methods are used to study the influence of various factors of laser radiation on explosives. Abdulazeem et al. [5] perform theoretical calculations for lead azide using the following mathematical equation:

$$\rho c \frac{\partial T}{\partial t} = k \frac{\partial^2 T}{\partial x^2} + \rho q A e^{-E/RT} + \alpha I(t) e^{-\alpha x},$$

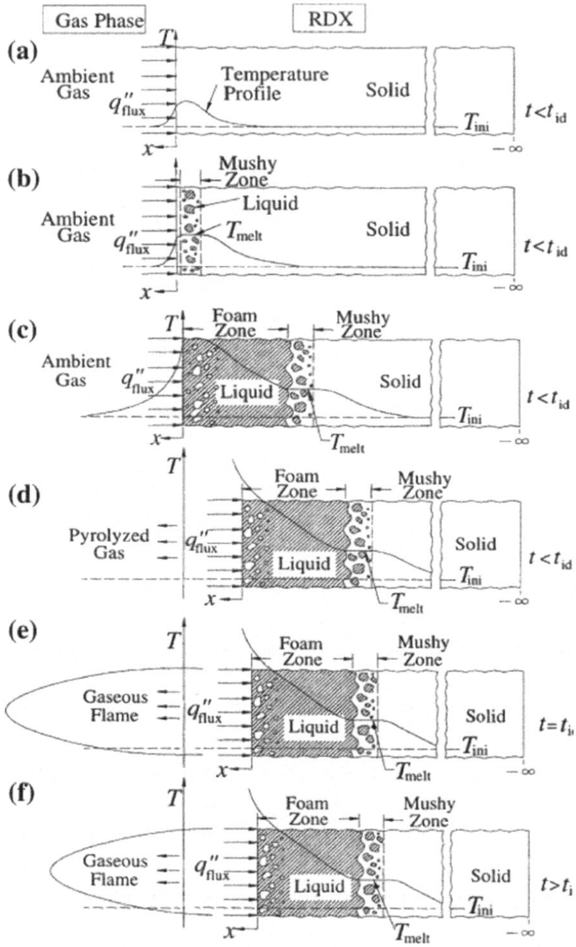

Fig. 25.1 Processes involved in laser-induced ignition of RDX [2]

where, $k\frac{\partial^2 T}{\partial x^2}$—the value of heat transfer taking into account coefficient of thermal conductivity, $\rho q A e^{-E/RT}$—the amount of heat generated during the chemical reaction, $\alpha I(t)e^{-\alpha x}$—the amount of heat produced by laser radiation.

As a result, the dependencies of the surface temperature and the initiation time on the laser radiation density at different values of the heat transfer coefficient were determined (Fig. 25.2a, b). In general, the performed calculations are consistent with the experimental data. In turn, some inconsistency between theory and experiment can be explained by insufficient consideration of an influence of the local zones of the thermal inhomogeneity observed during laser-induced initiation (Fig. 25.2c, d).

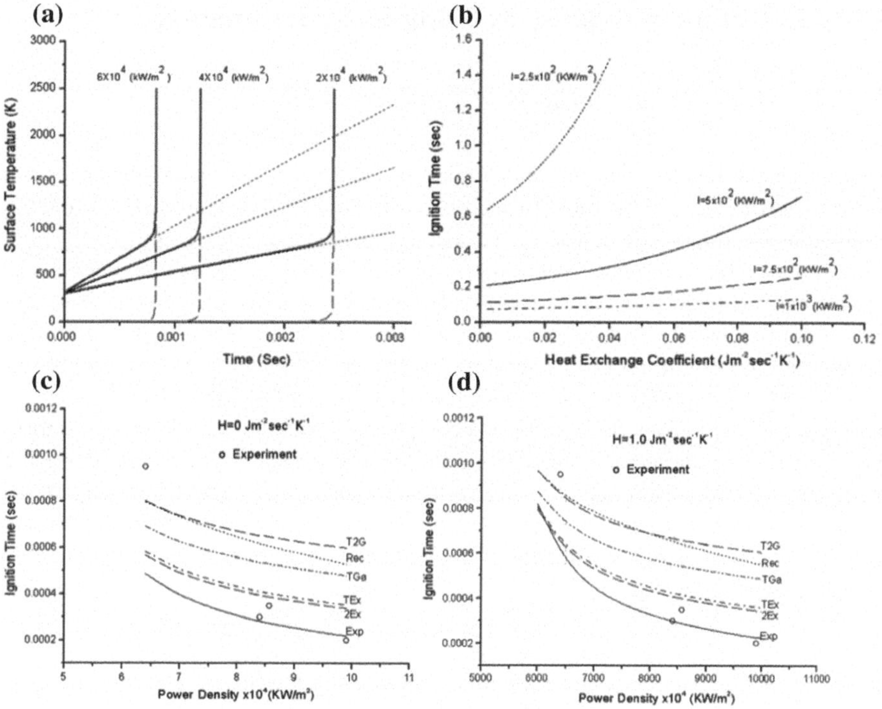

Fig. 25.2 The calculated and obtained experimentally dependencies of the surface temperature and the initiation time on the laser radiation density at different values of the heat transfer coefficient [5]

25.1.2 Laser Shock Initiation of Explosives

The well-known alternative way to initiate explosives is to use the energy of plasma produced during laser ablation of a metal. The shock wave generated by the flying particles directed on the initiating explosive creates an inertial impact that transforms into a detonation front. Typically, a metal coating is made of a thin metal film placed on the transparent "window" or tip of the optical fiber. The laser pulse (Nd:YAG laser, laser energy 67–375.7 mJ) vaporizes the metal and forms a plasma. In the implementation of this mechanism of initiation of hexogen (RDX), octogen (HMX), trinitrotoluene (TNT), and hexanitrostilbene (HNS) the following processes occur: melting and evaporation of a metal, plasma formation, shock wave formation, and, finally, detonation of an explosive. In this case, the main materials used as target are aluminum and copper. For fabrication of multilayer films based on carbon, magnesium, germanium, titan, titan oxide, aluminum oxide, and zinc sulfide are also used [6].

25.2 Detection of Explosives Using Laser Irradiation

25.2.1 Main Laser Methods of Explosives Detection

Currently, methods for the remote detection of the energy-saturated substances is a perspective direction in the modern security strategies [7]. There are two approaches applicable for remote detection of explosives. It is standoff detection (fully non-contact method) and remote detection, in which an analytical equipment contacts the sample explosive, whereas an operator is located at a safe distance. Standoff identification of explosives is more promising and demanded, however, more difficult to implement and is limited to low saturated vapor pressures upon normal conditions (e.g., for TNT—9 ppb (~$1.7 \cdot 10^{-3}$ Pa), for RDX—6 ppt (~$4 \cdot 10^{-6}$ Pa)). Moreover, new types of explosives have complicated spectral characteristics and, in the most cases, are poorly studied and the appropriate spectral databases are absent. However, many substances of this class have high adhesion and contain a relatively large trace amount of molecules on the surface per unit area, which facilitates their determination. Modern trends in development of methods and equipment for remote detection are presented below [8].

25.2.1.1 Laser-Induced Breakdown Spectroscopy (LIBS) or Laser Spark Emission Spectroscopy

This method determines the elemental composition of the target substance based on registration of the characteristic lines of emission spectrum of laser plasma [8–12]. For that purpose, the pulsed lasers operating in the UV (area of electronic transitions of most explosives) or IR (area of vibrational-rotational transitions) parts of the spectrum are typically used. For example, the analysis of such explosives as RDX, HMX, TNT, pentaerythritol tetranitrate (PETN), C4, A5, M43, LX-14, and JA2 was successfully performed. Thus, this method has a high sensitivity (nano- and picograms); it is relatively simple and applicable for determination of the most of explosives.

25.2.1.2 Raman Spectroscopy (Spectroscopy of Raman Scattering)

This approach is based on detection of the characteristic Raman spectra of the individual compounds upon laser irradiation [13]. The main disadvantage of this method is insufficient intensity of the scattered radiation of the compounds of interest, which causes the implementation of highly sensitive detectors. In addition, the systems of scanning and accumulation (multiplication) of the signal are required. Moreover, the method is sensitive to external light illumination and luminescence of the object of research and other compounds localized in the scanning zone, which complicates the detection process [14, 15]. In order to increase the intensity of the Raman signal it is necessary to use the radiation sources in the range of the electronic transitions of molecules (Resonance Raman scattering (RRS)) that allows us to achieve intensity (cross-section of Raman scattering) of thousand times higher than at wavelengths of the visible region for some explosives. For example, in [16] it was shown that the cross-section of Raman scattering at 229 nm for the studied explosives exceeds its value in the visible region of the spectrum by about three orders of magnitude. Gares et al. [17] suggested to use a UV-Raman method, in which the detection occurs right after when molecules of explosive are promoted to the excited state by the UV light. This type of experiments was conducted for TNT, RDX, $NaNO_3$, and NH_4NO_3. It is considered that the optimal distance for remote detection of various explosive compounds using Raman spectroscopy is about 500–1000 m [18].

25.2.1.3 Coherent Anti-Stokes Raman Scattering (CARS) Spectroscopy

This method is based on the phasing molecular vibrations in the field of resonant bi-harmonic pump and subsequent coherent scattering of the probe wavelength [13, 19]. This makes CARS popular for standoff detection of explosive remainders on the surfaces of objects. At the same time, the use of the narrow-band lasers allows achieving high spectral resolution of the Raman bands. Furthermore, CARS has a number of advantages over the method previously discussed, but at the same time, there are difficulties, which complicate its implementation in portable devices designed for remote detection of trace remainders of explosives.

25.2.1.4 Laser-Induced Fluorescence of Products PF-LIF (PD-LIF)

This technique deals with determination of explosives based on the products of their decomposition. In this regard, the analysis is performed by means of detection of the characteristic set of the fragmentation products on the basis of detection of their fluorescence [20–25]. The molecules of the most explosives have weak transitions, whereas the transitions corresponding to the products of their fragmentation are intense and well-studied, therefore, they can be easily identified. The detection process can be divided into two steps: rapid heating and dissociation of the molecules of explosives caused by laser irradiation are followed by the release of a large number of volatile products. As a rule, the spectral characteristics of such decomposition products have simple structure. In the next step, the detection of fluorescence signal of primary decomposition products of explosives as well as their assignment are conducted. This approach has great perspective for further development and with a high degree of probability allows to identify explosives or their mixtures. Here are main advantages of this method: the possibility to use a single source for fragmentation and excitation, high selectivity of detection of functional groups, low influence of optical interference, and relatively intense fluorescence signal.

25.2.1.5 IR Spectroscopy of the Products of Laser Photofragmentation of Explosives (MIR-PF)

This method is based on principles, which are similar to the previous approach differing only in the fact that after irradiation (usually at a wavelength of about 1.5 μm) the registration of gaseous products is carried out using IR spectroscopy. The signal in the form of reflected (scattered) radiation is recorded using IR camera [26–30].

25.3 An Investigation of Laser-Explosive Interaction from New Perspectives

25.3.1 Optimization of Laser Initiation of Explosive—Resonant Laser Explosive Interaction

At the very beginning, research in the field of interaction of laser radiation with photosensitive explosives was largely slowed down by a limited number of high-tech, reliable, powerful, and commercial available lasers. In these conditions, on the one hand, it was technically difficult to find a laser suitable for excitation of the specific absorption bands of an explosive, however, on the other hand, it was useless since the sufficiently powerful laser can initiate almost any explosive. As a result, due to thermal initiation the laser beam interacts with the surface of an explosive by heating it up to a critical temperature [31]. The development of laser technologies and the creation of a wide range of lasers with various characteristics operating at different wavelengths, including those that correspond to the of optical absorption of explosives, provide the conditions for the study of the resonance interaction of laser radiation with explosives.

Thus, it is possible to distinguish two new directions in research of light-sensitive explosives: the resonant interaction with laser radiation and influence of nanoparticles on the mechanism of interaction of explosives with laser radiation.

Here are some aspects of the resonant interaction of explosives with laser radiation. The mechanism of the photo-induced decomposition of HNS (2,2′, 4,4′, 6,6′-hexanitrostillbene) and the resulting decomposition products were studied using UV spectroscopy, electron paramagnetic resonance (EPR), and XPS (X-ray photoemission spectroscopy) [32]. The HNS particles dissolved in acetonitrile were irradiated with a monochromatic source of UV radiation (365 nm) at a temperature of 24 °C. Then, the irradiated HNS was studied using optical absorption spectroscopy within the range of 190–900 nm, as well as X-ray photoemission spectroscopy (XPS) after 2 and 8 h of irradiation. After irradiation, new peaks on the HNS absorption spectrum centered at 330–350 nm appear, the intensity of which increases in proportion to the irradiation time. These peaks are associated with elimination of NO_2-group from the benzene ring of the HNS molecule (Fig. 25.3).

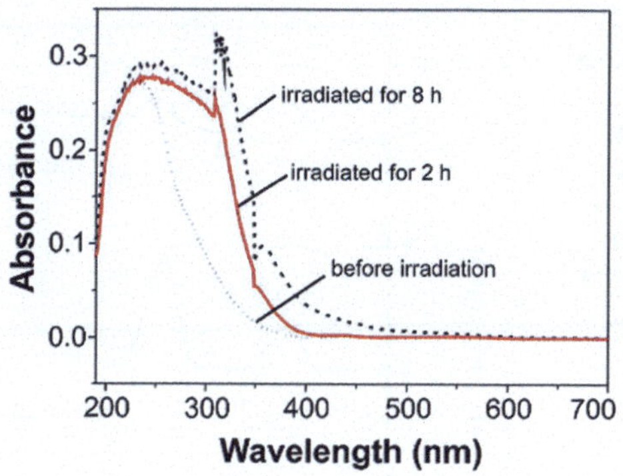

Fig. 25.3 UV-Vis spectra of HNS before and after UV irradiation [32]

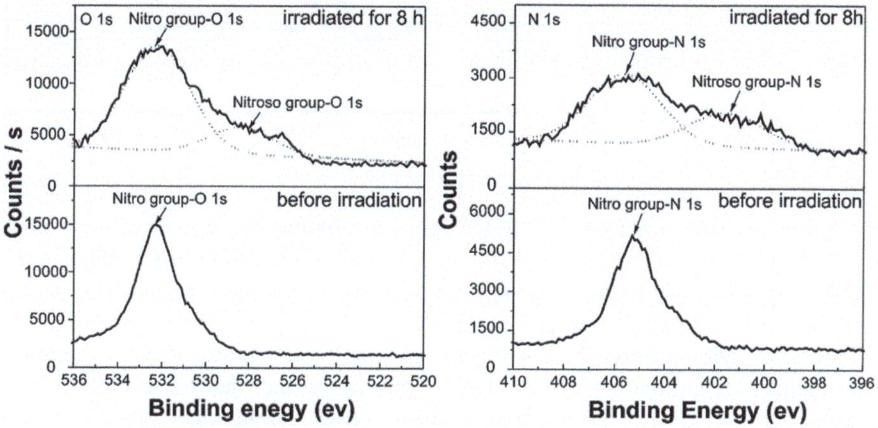

Fig. 25.4 XPS O 1s and N 1s spectra of HNS before and after irradiation, the solid line is the XPS curve and the short dotted line is the fitting curve [32]

The data presented in Fig. 25.3 are confirmed by X-ray photoemission spectroscopy (XPS), electron paramagnetic resonance (EPR), and liquid chromatography-mass spectrometry (LC-MS). The XPS results demonstrate a change in the intensity of peaks N 1s (401 eV) and O 1s (528 eV) when comparing spectra before and after irradiation of HNS. This indicates a breaking of a C–NO_2 bond and the elimination of oxygen from the nitro group (Fig. 25.4). The EPR analysis reveals the presence of free •NO_2 radicals, which are formed during the photocleavage of the C–NO_2 bond in the benzene ring. Thus, the decomposition reaction proceeds mainly through the breaking of a C–NO_2 bond and the elimination of oxygen atoms from the NO_2 groups.

Kakar et al. recorded the absorption spectra (Fig. 25.5) of TATB (triaminotrinitrobenzene or 2,4,6-triamino-1,3,5-trinitrobenzene) before and after irradiation of different intensity (5 min, ~10^{10} photons/s; 5 min, ~10^{14} photons/s), using X-ray

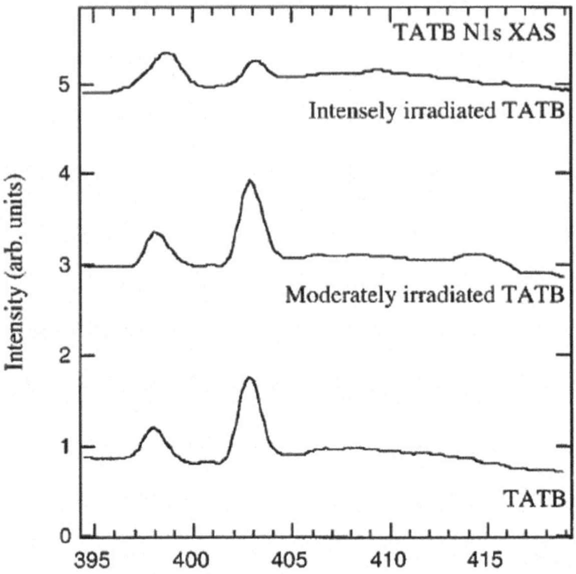

Fig. 25.5 N 1s absorption spectra of TATB recorded at 403 eV: bottom-a fresh TATB film, middle-TATB film exposed to moderate dose of radiation for 5 min, top-TATB film exposed to intense dose of radiation for 5 min [33, 34]

absorption spectroscopy (XAS) [33, 34]. After irradiation, the signals of the transitions N 1s (C–NO_2) → π* (NO), C 1s (C–NO_2) → π* (NO), and O 1s (C–NO_2) → π* (NO) are significantly reduced, which indicates the elimination of the NO_2 group from the benzene ring of the TATB molecule.

Particular interest deserves a study of resonance interaction of laser radiation with cobalt (III) aminates [35–42]. This is due to the fact that these high-energy materials are sensitive to laser radiation, have a high detonation speed (~7 km/s) in comparison with the industrial initiating explosives (e.g., detonation speed for lead azide is 5.5 km/s) [43], have a short area of transition between burning and detonation, and their sensitivity is comparable to the brisant explosives [44]. Thus, all mentioned above make these high-energy materials safe in terms of initiation [45].

Here are the following high-energy materials used in the current study: (5-Nitrotetrazolato-N_2) Pentaammin-Cobalt (III) Perchlorate (NCP); bis-[cis-(5-Nitrotetrazolato–N_2)]}tetraaminecobalt (III) perchlorate (BNCP); (1,5-diaminotetrazole–N_2) pentaamminecobalt (III) perchlorate (DPCP); (5-trinitrometiltetrazole–N_2) pentaamminecobalt (III) perchlorate (TPCP); Aquapentaammincobalt (III) perchlorate (APCP). APCP is used as a precursor for the synthesis of the aforementioned compounds. NCP has successfully passed industrial tests at Geofizika science-and-production company as the main component of the explosion converter in the blasting-perforation equipment applied for drilling of deep oil and natural-gas wells. BNCP is used for space-rocket complexes as one of the most effective explosives.

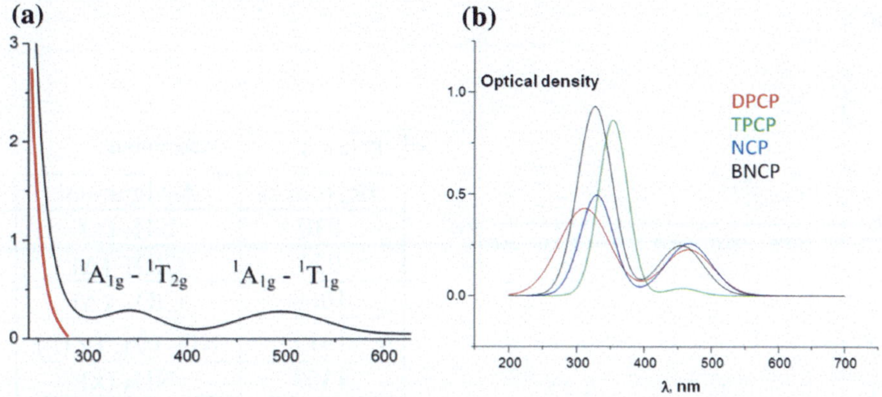

Fig. 25.6 **a** The absorption spectrum of APCP consisting of the charge transfer ligand-to-metal absorption band (highlighted in red) and two d–d absorption bands, **b** The absorption bands corresponding to d–d transitions of other studied cobalt (III) complexes

The 0.022 M aqueous solutions of these complexes were prepared and placed in cuvettes with a volume of 0.2 cm^2 to determine the absorption band frequencies. For example, the absorption spectra of APCP in UV and visible regions [35, 41] consist of the charge transfer metal-to-ligand absorption band and two absorption bands corresponding to the d–d transitions of cobalt (Fig. 25.6a). The d–d absorption bands of other energy-saturated complexes are shown in Fig. 25.6b. Figure 25.6 demonstrates that the third harmonic of the YAG-Nd laser ($\lambda = 355$ nm, pulse duration is 20 ns, pulse frequency 12 kHz, and laser power of 1.2 W) and the semiconductor laser operating at 470 nm (laser power of 0.2 W) can be used for excitation of these bands. In addition, the IR spectra of DPCP, TPCP, BNCP, and NCP were measured (Fig. 25.7) in order to identify the decomposition products [35, 46, 47].

The spectra of all studied complexes are characterized by the following main modes (Fig. 25.7d). One of them are absorption bands associated with vibrations of ClO_4^- ion. These vibrations are characterized by two very intense bands centered at 1070 and 620 cm^{-1}, as well as a weak absorption band centered at 930 cm^{-1} [48]. These absorption bands were observed for all studied complexes. The absorption bands corresponding to vibrations of the intrasphere ligand NH_3 lie in the region of 3200 and 3300 cm^{-1} [49].

Raman spectra of DPCP, TPCP, NCP, and BNCP were recorded for the analogical purpose [36, 38]. Figure 25.8 presents the Raman spectra of NCP and its decomposition producs.

According to Fig. 25.8, there are three intense Raman signals of ClO_4^- centered at 470, 630, and 940 cm^{-1} [50]. These Raman signals were observed for all studied complexes. There are two Raman signals corresponding to vibrations of the intrasphere ligand NH_3 appeared in the region of 1320 cm^{-1} [50–52].

The comparison of the spectra of initial complexes and liquid products of their photolytic decomposition reveals weakening of the intensities of the bands associated

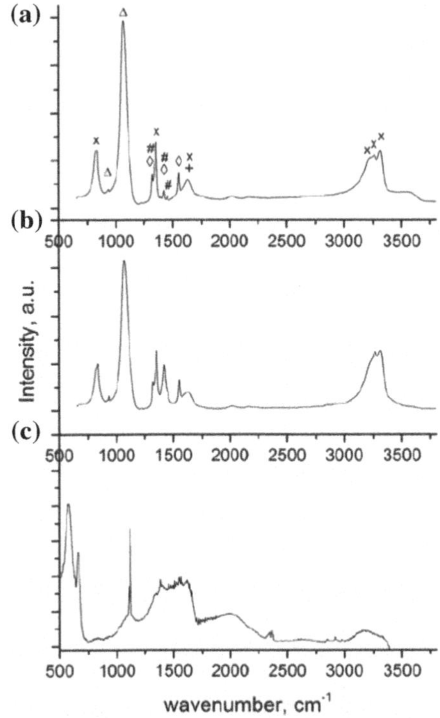

(d) IR (cm⁻¹) Assignment

IR, (cm^{-1})	Assignment
830	NH_3 (x)
934	ClO_4^- (Δ)
1068	ClO_4^- (Δ)
1317	NO_2 (◊), Tz (#)
1353	NH_3 (x)
1421	NO_2 (◊), Tz (#)
1451	Tz (#)
1549	NO_2 (◊)
1629	H_2O (+), NH_3 (x)
3221	NH_3 (x)
3261	NH_3 (x)
3312	NH_3 (x)

Fig. 25.7 **a** IR spectrum of NCP; IR spectra of liquid, **b** and solid, **c** products of photolytic decomposition of NCP upon irradiation at 355 nm, **d** the results of assignment of the obtained IR absorption bands

to intrasphere NH_3 ligand and Co–N, preservation of the intensity of the signals assigned to the outersphere ClO_4 ligand, and the increase in the intensities of the bands corresponding to nitrogen oxides.

In addition to the products remaining in a solution, a solid precipitate is formed because of photolytic decomposition of the complexes. The more detailed study of this precipitate showed that it is composed by cobalt and oxygen forming a mixture of cobalt oxides and hydroxides. Moreover, the increase in the intensity of the light flux leads to decrease of amount of hydroxides in the mixture.

As can be seen from Fig. 25.9, the described above picture is similar in both cases: at 355-nm excitation of the cobalt $^1A_{1g} \rightarrow {}^1T_{2g}$ transition by the third harmonic of the YAG-Nd laser and at 470 nm excitation of the cobalt $^1A_{1g} \rightarrow {}^1T_{1g}$ transition by a semiconductor laser.

The thermal decomposition of cobalt (III) complexes was also studied using mass spectrometry (ionizing radiation 70 eV) [40]. The results of these studies for APCP and NCP demonstrate that the transition from low (150 °C) to high (250 °C) temperatures results in the decrease of the amount of ammonium in the decomposition products, the increase of the average oxidation degree of nitrogen and carbon, and the

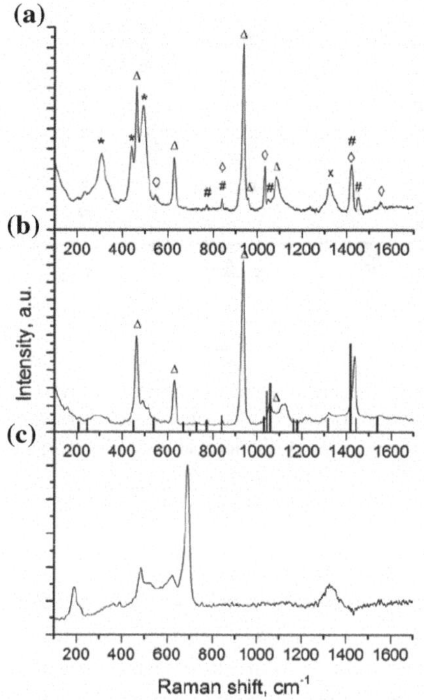

(d)

Raman shift (cm^{-1}) Assignment

Raman shift, (cm^{-1})	Assignment
307	Co-N (*)
440	Co-N (*)
463	ClO_4^- (Δ)
494	Co-N (*)
544	NO_2 (◊)
625	ClO_4^- (Δ)
774	Tz (#)
839	Tz (#), NO_2 (◊)
937	ClO_4^- (Δ)
959	ClO_4^- (Δ)
1034	NO_2 (◊)
1053	Tz (#)
1085	ClO_4^- (Δ)
1322	NH_3 (x)
1419	NO_2 (◊), Tz (#)
1453	Tz (#)
1550	NO_2 (◊)

Fig. 25.8 **a** Raman spectrum of NCP; Raman spectra of liquid, **b** and solid, **c** products of photolytic decomposition of NCP upon irradiation at 355 nm, **d** the results of assignment of the obtained Raman bands

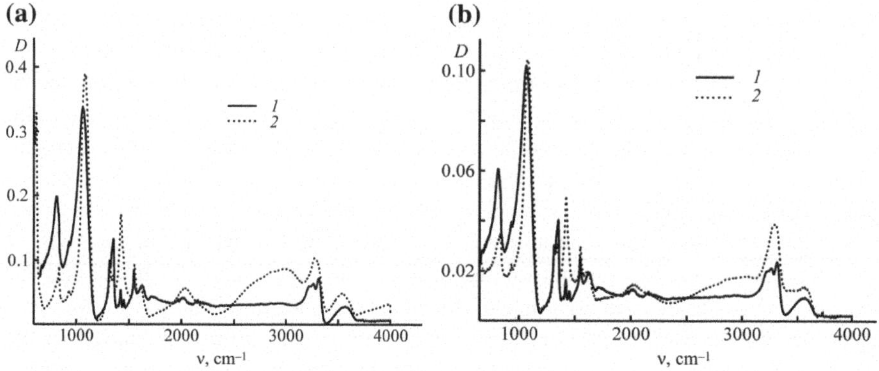

Fig. 25.9 IR spectra of NCP recorded before (1) and after (2) excitation at 470 (**a**) and 355 (**b**) nm

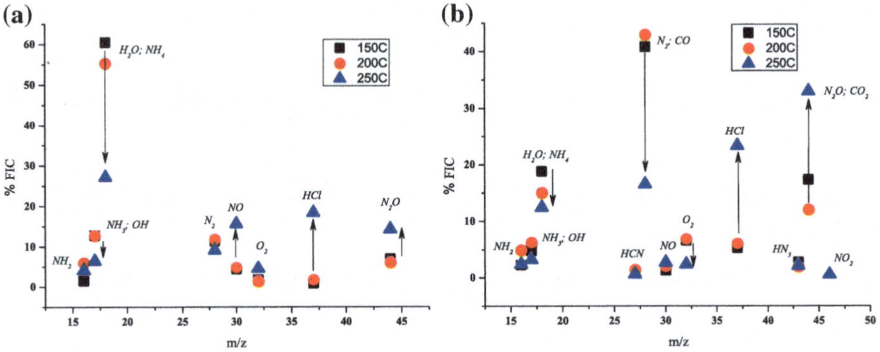

Fig. 25.10 The dependence of the proportion of ions in the whole mass spectrum on the ratio of their mass-to-charge for the products of pyrolysis of APCP (**a**) and NCP (**b**) at three temperatures. The arrows indicate the marked changes induced by the increase of the temperature of pyrolysis. The signals of HCl ions with different isotopes of chlorine are combined in one

increase of the amount of HCl (Fig. 25.10). The latter is the result of the decomposition of the perchlorate ion indicating a change in the mechanism of oxidation. Thus, the main oxidizing agents during the pyrolysis of the studied complexes at temperatures up to 200 °C are the central cobalt ion and the nitro group of 5-nitrothiazole (in the case of NCP). On the other hand, the outersphere perchlorate ion becomes the main oxidizing agent at 250 °C.

Golubev et al. performed the density functional theory (DFT) quantum-chemical calculations of the primary decomposition (with minimal energy of excitation) of NCP and BNCP using Gaussian 09 [53, 54]. The following possible primary decomposition products were calculated: ammonia, (NH_3), perchloric acid ($HClO_4$), 5-nitrotetrazolato (N_4HCNO_2), and the nitro group (NO_2). It was observed that the main primary mechanism of decomposition of both complexes is the elimination of the ammonia molecule with concurrent rearrangement of the remaining molecular fragment. At the same time, the temperature effect of this decomposition mechanism is minimal. The authors claim that the calculations are consistent with the results of mass spectrometry [55].

According to the discussion above, it can be concluded that the mechanism of low-temperature decomposition (not higher than 200 °C) corresponds to the decomposition of the complex via breaking of the weakest bonds and is similar to the photolytic decomposition of cobalt (III) amminates upon the 350 nm excitation of the d–d absorption bands, and, in opposite to high-temperature decomposition regime, is not accompanied by a significant thermal effect.

Thus, the resonance photolysis of the studied complexes conserves the perchlorate ions in the decomposition products, which can be easily detected by Raman spectroscopy. The photolysis itself is a low-temperature process that does not lead to detonation, and its solid products (cobalt oxides and hydroxides) passivate the surface of explosives [56].

25.3.1.1 Absorbing Additives

The development of nanotechnology provides new opportunities for modification of explosives. For example, the light-absorbing nanoparticles (soot, metal nanoparticles, graphene, nanotubes) are added to the explosives in order to concentrate absorbed light energy in a smaller volume [57]. Another direction is the introduction of the light-scattering nanoparticles, which extend the optical path of light in the surface layer of an explosive leading to a similar result.

It was demonstrated [58] that the addition of carbon to RDX reduces the energy of initiation compared to a pure RDX. The initiation was carried out by Nd:YAG diode pumped laser (laser power of 2.6 W and pulse duration of 200 ms) operated at 808 nm. This laser was used to study the interaction of laser radiation with HMX, RDX, PENT, and TNT, as well as their modifications with carbon (graphite, soot). Adding 1–3% carbon reduces the energy of initiation of RDX by 10–15%, probably, as noted by the authors, due to reducing the losses caused the reflection of the laser beam, which are typically about 20%.

Kalenskii et al. studied the dependence of the absorption coefficient of laser radiation (Nd:YAG, 1064 nm) of lead azide on the size of embedded lead nanoparticles [59]. It was observed that the maximum absorption coefficient (1.18) corresponds to the diameter of lead nanoparticles equals to 74 nm. The authors used the "hot spot" model in order to explain the obtained results.

The effect of the modification of the explosive compositions under the resonant laser radiation have not been previously studied. A study of the modification effect on the threshold of initiation was conducted on pressed polycrystalline samples of NCP with additions of different amounts of graphene, fullerenes (light-absorbing nanoparticles), and detonation nanodiamonds (nanoparticles with high refractive index).

At 350 nm irradiation, corresponding to d–d transitions of cobalt ion, upon reaching the threshold power the darkening of the sample surface was observed (photolytic decomposition of NCP and the formation of cobalt oxides), whereas, no explosive initiation was noticed. It was also noted that the increase in the concentration of graphene and fullerenes nanoparticles increased the threshold radiation power. This can be explained by the fact that these nanoparticles take up some part of energy and convert it into the thermal energy (Fig. 25.11b). Thus, the resulting laser power was not enough on the one hand for the photolytic initiation on the other hand, for thermal initiation, since even the radiation power (70–160 mW/mm^2) was an order of magnitude less than the required value (Fig. 25.11a).

IR spectroscopy revealed a sufficiently intense absorption band centered at 1554 nm corresponding to the first harmonic of the valence vibrations of N–H bonds. Thus, in order to observe the thermal initiation of NCP, the tunable femtosecond laser operated at 1554 nm (pulse duration of 100 fs and pulse frequency of 76 MHz) was used for the resonant excitation of this absorption band. An increase in the concentration of graphene up to 3% (in the weight percent) led to a decrease in the initiation threshold by an order of magnitude (Fig. 25.11a). This can be explained by the increase in the absorption coefficient o NCP. The further increase in the concen-

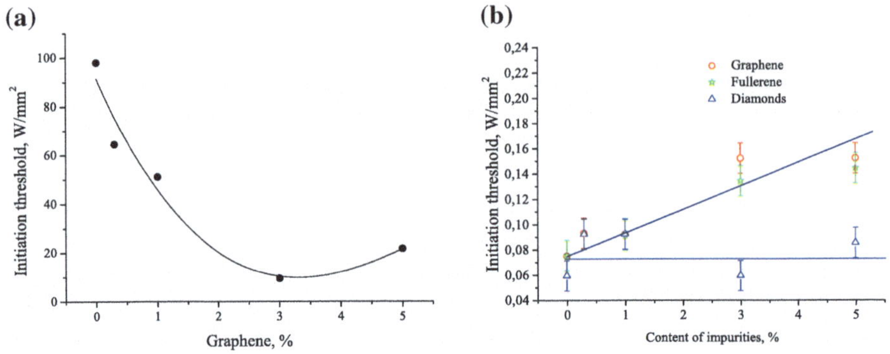

Fig. 25.11 The influence of the additives of nano-disperse materials (in wt%) on the threshold of initiation of NCP by, **a** femtosecond 1554 nm laser and, **b** Nd:YAG laser operated at 355 nm

tration of graphene led to the opposite effect, apparently, due to an increase in the outflow of a heat from the NCP layer that absorbs the bulk of the incident radiation.

25.3.1.2 Scattering Additives

The effect of additives of diffusers of light (e.g. nanodiamonds) on the explosives are poorly understood in comparison with additives based on highly absorbing nanoparticles. The use of nanodiamonds as an additive for the laser initiation of the composites based on the complex perchlorates was studied by Ilyushin et al. [57]. In this work, the composites with additives of concentrations varied between 0.5 and 5 mass% were investigated. Introduction of nanodiamonds provides sensitization of the composites with respect to laser radiation (Nd:YAG laser). The energy dependence of the initiation threshold on the additive concentration represents a curve with a minimum of about 3 mass%. The authors explain this observation by the increase of the internal optical path length of the pump radiation in the sample due to the scattering of laser radiation by ultrafine diamonds. We have studied the influence of concentration of nanodiamonds as lenses of radiation on the change in the initiation threshold at resonant excitation corresponding to the d–d of transition of cobalt ion. The obtained results showed that the modification of NCP perchlorate by detonation nanodiamonds did not affect the magnitude of the threshold power (Fig. 25.11b). Apparently, this is due to the fact that the increase in the length of the optical path as a result of the scattering effect of such additives is critical only for materials with a low optical absorption coefficient.

25.4 Conclusions

Modern directions in the field of the explosive treatment by laser light lie in an investigation of the interactions between the resonant laser radiation and the absorbing and scattering light nanoparticles used as additives to explosives, and affecting the initiation threshold. The implementation of this approach to cobalt (III) nitrotetrazolato amminates allowed us to obtain the following results. For these compounds, three groups of resonance absorption bands have been established: the first group—the overtones of valence vibrations at wavelengths of 1.5 and more than 2 μm; the second group—two absorption bands corresponding to d–d transitions of cobalt ion located in the region of 300–500 nm, and the third group—the charge transfer of ligand-to-metal absorption band located in the short-wavelength region. Photolytic decomposition of the aforementioned complexes was carried out using laser radiation. It was found that irradiation of cobalt (III) amminates at 355 and 470 nm, corresponding to d–d transitions of cobalt ion, leads to "soft" photolytic decomposition providing the conditions for passivation of the surface of the explosives. In this scenario, perchlorate anion as one of the decomposition products is produced, which can be easily detected and identified. Addition to NCP of the highly absorbing carbon nanoparticles allows to significantly reduce the initiation threshold at the wavelengths corresponding to the valence vibrations.

Acknowledgements This work was supported by the Russian Foundation for Basic Research, project no. 16-29-01056-ofi_m. Measurements were partly made at the resource center of St. Petersburg State University "Optical and Laser Methods for Analysis of Substances".

References

1. M.J. Gifford, W.G. Proud, J.E. Field, Development of a method for qualification of hot-spots. Thermochim. Acta **384**, 285–290 (2002)
2. Y.-C. Liau et al., Laser-induced ignition of RDX monopropellant. Combust. Flame **126**, 1680–1698 (2001)
3. J.M. McAfee, The deflagration to detonation transition, in *Shock Wave Science and Technology Reference Library. 5: Non-Shock Initiation of Explosives*, ed. by B.W. Asay (Springer, Berlin, 2010), pp. 483–535
4. M.D. Furnish, N.N. Thadhani, Y. Horie, Am. Inst. Phys. (Melville, NY) 878–881 (2002)
5. M.S. Abdulazeem et al., Int. J. Therm. Sci. **50**, 2117–2121 (2011)
6. S. Ruiqi, W. Lizhi, Z. Wei, Z. Haonan, Laser ablation of energetic materials, in *Laser Ablation—From Fundamentals to Applications*. http://dx.doi.org/10.5772/intechopen.71892
7. L.A. Skvortsov, Laser methods for detection of the explosives traces on the surfaces of distant objects. Quantum Electron. **42**(1) (2012)
8. L.A. Skvortsov, E.M. Maksimov, Quantum Electron. **40**(7), 565 (2010)
9. D. Kremers, L. Radziemsky, *Laser-Indused Breakdown Spectroscopy* (Technosphere, Moscow, 2009)
10. A. Popov, T. Labutin, N. Zorov, Mosc. Univ. Chem. Bull. **50**(6), 453 (2009)
11. F. De Lucia, A. Samuels, R. Harmon, R. Walters, K. McNesby, A. LaPointe, R. Winkel, A. Miziolek, IEEE Sens. J. **5**, 681 (2005)

12. J. Gottfried, F.D. Lucia, C.J. Munson, A. Miziolek, Anal. Bioanal. Chem. **395**, 283 (2009)
13. V. Demtreder, *Laser Spectroscopy. Basic Principles and Experimental Technique* (Science, Moscow, 1985)
14. S. Sharma, P. Lucey, M. Ghosh, H. Hubble, K. Horton, Spectrochim. Acta A **59**, 2391 (2003)
15. J. Carter, J. Scaffidi, S. Burnett, B. Vasser, S. Sharma, S. Angel, Spectrochim. Acta A **61**, 2288 (2005)
16. D. Tuschel, A. Mikholin, B. Lemoff, S. Asher, Appl. Spectrosc. **64**(4), 425 (2010)
17. K.L. Gares et al., Review of explosive detection methods and the emergence of standoff deep UV resonance Raman. J. Raman Spectrosc. **47**, 124–141 (2016)
18. B.H. Hokra et al., Single-shot stand-off chemical identification of powers using random Raman lasing. PNAS **111**(34), 12320–12324 (2014)
19. S.A. Ahmanov, N.I. Koroteev, UFN **123**, 405 (1977)
20. T. Arusi-Parpar, D. Heflinger, R. Lavi, Appl. Opt. **40**, 6677 (2001)
21. T. Arusi-Parpar, R. Lavi, Remote detection of explosives by enhanced pulsed laser photodissotiation/laser-induced fluorescence method, in *Paper Presented at the NA TO Advanced Research Workshop on Stand-off Detection of Suicide Bombers and Mobile Subjects Pfinztal* (2006), pp. 13–14
22. C. Wynn, S. Palmacci, R. Kunz, M. Rothshild, Lincoln Lab. J. **17**(2), 27 (2008)
23. C. Wynn, R. Palmacci, K. Kunz, K. Clow, M. Rothshild, Proc. SPIE Int. Soc Opt. Eng. **6954**, 695407 (2008)
24. C. Wynn, R. Palmacci, K. Kunz, K. Clow, M. Rothshild, Appl. Opt. **37**(31), 5767 (2008)
25. J. White, F. Akin, H. Oser, R. Crosley, Appl. Opt. **50**(1), 74 (2011)
26. C. Bauer, P. Geiser, J. Burgmeier, J. Holl, W. Schade, Appl. Phys. B Lasers Opt. **85**, 251 (2006)
27. C. Bauer, J. Burgmeier, C. Bohling, W. Schade, J.C. Holl, in *Proceedings of the NATO Advanced Research Workshop on Stand-off Detection of Suicide-Bombers and Mobile Subjects* (Springer, The Netherlands, 2006), p. 27
28. U. Willer, M. Saraji, A. Khorsandi, P. Geisher, W. Schade, Opt. Lasers Eng. **44**, 699 (2006)
29. C. Bauer, A. Sharma, U. Willer, J. Burgmeier, B. Braunschweig, W. Schade, S. Blaser, L. Hvozdara, A. Mffler, G. Holl, Appl. Phys. B **92**(3), 327 (2008)
30. C. Bauer, U. Willer, W. Schade, Opt. Eng. **49**, 111126 (2010)
31. D. Edward, A. Krechetov, A. Mitrofanov, D. Nurmukhametov, M. Kuklja, J. Phys. Chem. C **115**, 6893–6901 (2011)
32. Y. Sun, X. Tao, Y. Shu, F. Zhong, UV-induced photodecomposition of 2,2′, 4,4′, 6,6′-hexanitrostillbene (HNS), Mater. Sci.-Pol. **31**(3), 306–311 (2013), http://www.materialsscience.pwr.wroc.pl/. https://doi.org/10.2478/s13536-013-0105-9
33. S. Kakar et al., Phys. Rev. B. **62**, 15666 (2000)
34. J.W. McDonald et al., J. Energ. Mater. **19**, 101 (2001)
35. A.S. Tverjanovich, A.O. Averyanov, M.A. Ilyushin, YuS Tverjanovich, A.V. Smirnov, Effect of laser radiation on tetrazolate ammine cobalt III complexes. Bull. SpbSIT (TU) **26**(52), 3–7 (2014)
36. A.S. Tverjanovich, A.O. Averyanov, M.A. Ilyushin, YuS Tverjanovich, A.V. Smirnov, The Raman spectra of nitrotetrazolo(lato) ammine cobalt III perchlorates. Bull. SpbSIT (TU) **27**(53), 8–10 (2014)
37. A.S. Tverjanovich et al., Universum. **12**(19) (2015)
38. Int. J. Energ. Mater. Chem. Propuls. **15**(2), 113–122 (2016)
39. G.O. Abdrashitov, A.O. Aver'yanov, M.D. Bal'makov, M.A. Ilyushin, A.S. Tverjanovich, Yu.S. Tver'yanovich, Decomposition of Pentaammineaquacobalt (III) Perchlorate under laser radiation action. Russ. J. Gen. Chem. **87**(7), 1451–1455 (2017)
40. M.A. Ilyushina, Yu.S. Tverjanovich, A.S. Tverjanovich, A.O. Aver'yanov, A.V. Smirnov, I.V. Shugalei, On the mechanism of Cobalt(III) aminates pyrolysis. Russ. J. Gen. Chem. **87**(11), 2600–2604 (2017)
41. A.S. Tverjanovicha, A.O. Aver'yanov, M.A. Ilyshin, Yu.S. Tverjanovich, A.V. Smirnov, Decomposition of Cobalt(III) Nitrotetrazolato Amminates under the action of laser light. Russ. J. Gen. Chem. **88**(2), 226–231 (2017)

42. M.A. Ilyushin, A.V. Smirnov, V.N. Andreev, I.V. Tselinskii, I.V. Shugalei, O.M Nesterova. Russ. J. Gen. Chem. **85**(13), 1620 (2015)
43. A.V. Smirnov, M.A. Ilyushin, I.V. Tselinskii, Synthesis of Cobalt(III) Ammine complexes as explosives for safe taking charges. Russ. J. Appl. Chem. **77**(5), 794–796 (2004)
44. M.A. Ilyushin, A.M. Sudarikov, I.V. Tselinskii, *Metallic Complexes in High-Energy Materials* (LGU im A. S. Pushkina Publ., St. Petersburg, 2010), p. 188
45. M.A. Ilyushin, I.V. Tselinskii, A.A. Kotomin, *High Power Substances for Arsenal of Initiation* (SPbGTI (TU) Publ., St. Petersburg, 2013), p. 176
46. Int. J. Energ. Mater. Chem. Propuls. **15**(2), 113–122 (2016)
47. Eng. J. Gun. Than. **88**(2) (2017)
48. JTh Kloprogge, D. Wharton, L. Hickey et al., Infrared and Raman study of interlayer anions CO_3^{2-}, NO_3^-, SO_4^{2-} and ClO_4^- in Mg/Al-hydrotalcite. Am. Miner. **87**(5–6), 623–629 (2002)
49. E. Ingier-Stocka, M. Maciejewski, Thermal decomposition of $[Co(NH_3)_6]_2(C_2O_4)_3 \cdot 4H_2O$: I. Identification of the solid products. Thermochim. Acta. **354**, 45–57 (2000)
50. E. Mikulia, A. Migdal-Mikulia, N.S. Gorskaa Wrobelb, J. Sciesinskic, E. Sciesinskac, Phase transition and molecular motions in $[Co(MH_3)_6](ClO_4)_3$ studied by differential scanning calorimetry and infrared spectroscopy. J. Mol. Struct. 651–653 (2003)
51. Sigma-Aldrich, Catalog of Raman spectra, Hexamminecobalt (III) chloride (2012)
52. H.A. Block, Vibrational study of the hexamminecobalt (III) ion. Trans. Faraday Soc. **55**, 867–875 (1959)
53. V.K. Golubev, M.A. Ilyushin, The primary mechanism of decomposition of nitrotetrazolium of cobalt(III)//Doha. **87**(2), 312 (2017)
54. V.K. Golubev, M.A. Ilyushin, Primary decomposition mechanism of Cobalt (III) Nitrotetrazolatoammine complexes. Russ. J. Gen. Chem. **87**, 286 (2017)
55. A.S. Tverjanovich, A.O. Aver'yanov, M.A. Ilyushin, YuS Tverjanovich, A.V. Smirnov, Russ. J. Appl. Chem. **88**(2), 226 (2015)
56. A.S. Tverjanovich, e.a. Patent RF 2636525 (2016)
57. M.A. Ilyushin et al., Effect of additives of ultra fine carbon particles on the laser initiation threshold of a polymer is a photosensitive explosive composition. Chem. Fiz **24**(10), 49–56 (2005)
58. M. Harkoma, Confinement in the diode laser ignition of energetic materials, Thesis for the degree of Doctor of Technology to be presented with due permission for public examination and criticism in Sahkotalo Building, Auditorium S1, at Tampere University of Technology, 2010
59. A.V. Kalenskii et al., Paradox of small particles in the pulsed laser initiation of explosive decomposition. Combust. Explos. Shock. Waves **52**(2), 234–240 (2016)

Index

K. Yamanouchi et al. (eds.), *Progress in Photon Science*, Springer Series in Chemical Physics 119, https://doi.org/10.1007/978-3-030-05974-3

G

H

I

K

L

M

The manufacturer's authorised representative in the EU is Springer Nature Customer Service Centre GmbH, Europaplatz 3, 69115 Heidelberg, Germany. If you have any concerns regarding our products, please contact ProductSafety@springernature.com

Printed and bound by CPI Group (UK) Ltd, Croydon, CR0 4YY
30/09/2026
02225443-0020